Microsensors

Edited by

Richard S. Muller

Professor of Electrical Engineering and Computer Sciences
University of California, Berkeley
Co-director: Berkeley Sensor & Actuator Center

Roger T. Howe

Associate Professor of Electrical Engineering and Computer Sciences
University of California, Berkeley
Associate Director: Berkeley Sensor & Actuator Center

Stephen D. Senturia

Professor of Electrical Engineering
Massachusetts Institute of Technology

Rosemary L. Smith

Assistant Professor of Electrical Engineering and Computer Sciences
University of California, Davis

Richard M. White

Professor of Electrical Engineering and Computer Sciences
University of California, Berkeley
Co-director: Berkeley Sensor & Actuator Center

A volume in the IEEE PRESS Selected Reprint Series,
prepared under the sponsorship of the IEEE Electron Devices Society.

 IEEE PRESS

The Institute of Electrical and Electronics Engineers, Inc., New York

IEEE Order Number: PC0257-6

Library of Congress Cataloging-in-Publication Data

Microsensors / edited by Richard S. Muller ... [et al.].
    p.   cm.
''Published under the sponsorship of the IEEE Electron Devices Society.''
Includes bibliographical references.
ISBN 0-87942-245-9
  1. Miniature electronic equipment—Design and construction.
  2. Detectors—Design and construction. 3. Microelectronics.
  I. Muller, Richard S. II. IEEE Electron Devices Society.
TK7870.M4575   1990
681.2—dc20   90-4745
                                                                  CIP

Contents

Preface

I think no virtue goes with size.

—the Titmouse, EMERSON

THE microsensor is differentiated from previous sensors by its small size and by the techniques used in its manufacture. The microsensors we consider here produce an electrical output that corresponds to some physical, chemical, or biological quantity, such as pressure, heat, magnetic field, a particular vapor, or a biological molecule. Microsensors are often formed by processes that have been perfected to fabricate integrated circuits *and*, in many cases, by additional specialized fabrication steps; an example is the use of a sacrificial layer that is ultimately etched away to leave a free-standing micrometer-dimensioned beam, bridge, or membrane.

Some microsensors incorporate integrated circuits on a single die with the sensor elements. Although this complicates the fabrication process and can limit the operating-temperature range for the sensor, it often leads to superior performance at an acceptable cost. These *integrated* microsensors may provide a more linear output than that of the sensor itself, or an output having a digital format that can readily be handled by associated data-logging or display systems. In some cases, on-chip circuitry for impedance transformation permits the use of capacitive, piezoelectric, or other readout elements that would not function if they were remotely located from the sensor.

Microsensors seem sure to find countless applications in transportation, consumer products, process control, environmental monitoring, and health care. At this writing (Fall 1988), microsensors for measuring pressure are widely available commercially, and the production of silicon microaccelerometers is increasing rapidly. Many other types of microsensors have been described and are progressing from research and development towards manufacture and sale as a result of worldwide activity. The seventy-one papers reprinted here include contributions from Europe, North America, and Asia, reflecting the wide geographical distribution of microsensor work.

The small size of microsensors is a convenience, a contribution to their potentially low cost, and a source of new conceptual challenges. The figure included here,[1] comparing roughly the scale of microsensors and other quantities, suggests that with microsensors one enters a domain of new phenomena. Films as thin as one-hundred nanometers, having features with transverse dimensions ranging from a few micrometers up to a few millimeters, are commonly used. Spacings between microsensor parts may be one micrometer or less, a distance comparable with mean-free-pathlengths for collisions of ambient gas molecules and with thermal diffusion

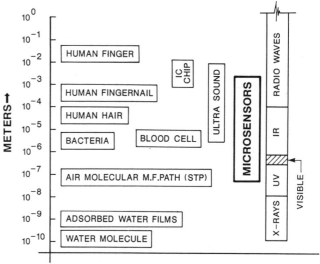

COMPARATIVE SCALE OF MICROSENSORS

lengths. Hence, gas flow and heat flow can behave very differently in microsensors than in macroscopic structures. Further, the mechanical properties of thin-film microsensor materials depend upon their fabrication process and generally differ from those of the bulk materials that are customarily used for mechanical or electrical components. Several papers reprinted in this volume consider these issues.

The challenges of microsensors are inherently multidisciplinary, providing subjects for research within the disciplines of engineering, materials science, electronics, and chemistry. A prime purpose of this collection of papers is to explore these challenges and to consider ways in which they have been met in the past and may be addressed in the future. After an initial overview of microsensor uses and opportunities (Part 1), we consider microsensor fabrication techniques in Part 2; the importance of fabrication in determining what can be designed and built is reflected by the fact that this is the longest section in the volume. The papers in Part 3 focus on microsensor transduction principles; only a few articles concerned with optical transduction are included in order to keep the length of this volume reasonable. No papers on microactuators or on the marriage of scanning tunneling microscopy with microsensors are included because these topics are too new. Circuits associated with sensors for readout and signal conditioning are the subjects of Part 4. Applications are introduced in many of the papers in these early sections since most microsensor projects culminate in the construction and testing of actual devices. However, the papers in Part 5 focus particularly on microsensors for specific uses.

We have used several criteria when selecting papers for reprinting. Had the paper been referred to frequently in the sensor literature? Had we, the Editors, found the paper to be seminal, memorable, or a unique source of information? Would students and workers new to the field learn from the paper? What about the paper's length, availability in libraries,

---

[1] After K. S. Udell *et al.*, "Microsensors for heat transfer and fluid flow measurements," in *Proc. First World Conference on Experimental Heat Transfer, Fluid Mechanics, and Thermodynamics*, Dubrovnik, Yugoslavia, September 4–9, 1988.

its timeliness, and its timelessness? Selecting papers was difficult, and many papers cited in the bibliography (Part 6) would have been reprinted in a lengthier volume. The cutoff for selections was the date that the list of contents was sent to the IEEE Press. No papers published after November 1988 are included.

Of course, we used computers extensively while compiling this volume and the bibliography that is Part 6. As an aid to readers interested in an electronic data base of citations to the microsensor literature, the authors have prepared a much more extensive set of references that is obtainable at a low cost on either Apple Macintosh or IBM PC diskettes by contacting the following source.

Microsensor Reference Diskette
Engineering, University of California Extension
2223 Fulton St.
Berkeley, CA 94720

Phone: (415) 642-4151 FAX: (415) 643-8683

The diskette contains (in a pure ASCII file) citations to all papers reprinted here, the entire contents of *Sensors and Actuators* since its inception, the papers presented at the major sensor conferences since 1982, and other selected references. As of January, 1989, approximately 2000 references were entered onto the diskette. Diskette users will, of course, be able to increment this data base.

Part 1
Review Papers

A Sensor Classification Scheme

RICHARD M. WHITE, FELLOW, IEEE

Abstract—We discuss a flexible and comprehensive categorizing scheme that is useful for describing and comparing sensors.

IN virtually every field of application we find sensors that transform real-world data into (usually) electrical form. Today many groups around the world are investigating advanced sensors capable of responding to a wide variety of measurands. In an attempt to facilitate comparing sensors and obtaining a comprehensive overview of them, we present here a scheme for categorizing sensors.

Sensor classification schemes range from the very simple to the complex. Extremes are the often-seen division into just three categories (physical, chemical, and biological) and the finely subdivided hierarchical categories used by abstracting journals. The scheme to be described here is flexible, intermediate in complexity, and suitable for use by individuals working with computer-based storage and retrieval systems. It is derived from a Hitachi Research Laboratory communication.

Tables I–VI, containing possible sensor characteristics, appear in order of degree of importance for the typical user. If we take for illustration a *surface acoustic-wave oscillator accelerometer*, these entries might be as follows: the *measurand*—acceleration; *technological aspects*—sensitivity in frequency shift per g of acceleration, short- and long-term stability in hertz per unit time, etc.; *detection means*—mechanical; *sensor conversion phenomena*—elastoelectric; *sensor materials*—the key material is likely an inorganic insulator; and *fields of application*—many, including automotive and other means of transportation; marine, military, and space; and scientific measurement.

Table I lists alphabetically most measurands for which sensors may be needed under the headings: acoustic, biological, chemical, electric, magnetic, mechanical, optical, radiation (particle), and thermal. A convention adopted to limit the number of Table I entries is that any entry may represent not only the measurand itself but also its temporal or spatial distribution. Thus, the entry "Amplitude" under the heading "Optical" could apply to a device that measures the intensity of steady infrared radiation at a point, a fast photodiode detecting time-varying optical flux, or a camera for visible light imaging.

Manuscript received August 29, 1986; revised October 27, 1986.

The author is with the Department of Electrical Engineering and Computer Sciences and the Electronics Research Laboratory, University of California, Berkeley, CA 94720.

IEEE Log Number 8612468.

With a particular measurand, one is primarily interested in sensor characteristics such as sensitivity, selectivity, and speed of response. These are termed "technological aspects" and listed in Table II. Table III lists the detection means used in the sensor.

Tables IV and V are of interest primarily to technologists involved in sensor design and fabrication. Entries in Table IV are intended to indicate the *primary* phenomena used to convert the measurand into a form suitable for producing the sensor output. The entries under "Physical" are derived from the interactions among physical variables diagrammed in Fig. 1. This is a modification and simplification of the diagrams used by Nye [1] and Mason [2] to show binary relations among the common physical variables.

Most sensors contain a variety of materials (for example, almost all contain some metal). The entries in Table V should be understood to refer to the materials *chiefly* responsible for sensor operation. Finally, an alphabetical list of fields of application comprises Table VI.

USES FOR THE CLASSIFICATION SCHEME

A useful scheme should facilitate comparing sensors, communicating with other workers about sensors, and keeping track of sensor progress and availability. Categorizing might help one think about new sensing principles that could be explored, and Table II might serve as a checklist to consult when considering commercial sensors.

All the entries in the tables have been given unique alphanumeric identifiers to facilitate use with computerized file systems such as electronic spreadsheets and databases used for storing information about sensors. The identifiers can be used as well in the keyword field of the lesser-known bibliographic utility *refer,* a part of the Unix operating system package, that enables a user to create and easily retrieve entries from a personalized database of citations to journal articles, books, and reports.

SENSOR EXAMPLES

We consider several examples to illustrate how terms in the tables can be used to characterize sensors.

Diaphragm Pressure Sensor: Differential pressure distorts a thin silicon diaphragm in which the deflection is inferred from the change of the values of resistors diffused into the diaphragm. The measurand is pressure, A6.5; the primary detection means is mechanical, C5; the sensor conversion phenomenon (piezoresistance) is elastoelec-

Reprinted from *IEEE Trans. Ultrason. Ferroelec. Freq. Contr.*, vol. UFFC-34, no. 2, pp. 124–126, March 1987.

TABLE I

A. MEASURANDS

A1. Acoustic
- A1.1 Wave amplitude, phase, polarization, spectrum
- A1.2 Wave velocity
- A1.3 Other (specify)

A2. Biological
- A2.1 Biomass (identities, concentrations, states)
- A2.2 Other (specify)

A3. Chemical
- A3.1 Components (identities, concentrations, states)
- A3.2 Other (specify)

A4. Electric
- A4.1 Charge, current
- A4.2 Potential, potential difference
- A4.3 Electric field (amplitude, phase, polarization, spectrum)
- A4.4 Conductivity
- A4.5 Permittivity
- A4.6 Other (specify)

A5. Magnetic
- A5.1 Magnetic field (amplitude, phase, polarization, spectrum)
- A5.2 Magnetic flux
- A5.3 Permeability
- A5.4 Other (specify)

A6. Mechanical
- A6.1 Position (linear, angular)
- A6.2 Velocity
- A6.3 Acceleration
- A6.4 Force
- A6.5 Stress, pressure
- A6.6 Strain
- A6.7 Mass, density
- A6.8 Moment, torque
- A6.9 Speed of flow, rate of mass transport
- A6.10 Shape, roughness, orientation
- A6.11 Stiffness, compliance
- A6.12 Viscosity
- A6.13 Crystallinity, structural integrity
- A6.14 Other (specify)

A7. Optical
- A7.1 Wave amplitude, phase, polarization, spectrum
- A7.2 Wave velocity
- A7.3 Other (specify)

A8. Radiation
- A8.1 Type
- A8.2 Energy
- A8.3 Intensity
- A8.4 Other (specify)

A9. Thermal
- A9.1 Temperature
- A9.2 Flux
- A9.3 Specific heat
- A9.4 Thermal conductivity
- A9.5 Other (specify)

A10. Other (specify)

TABLE II

B. TECHNOLOGICAL ASPECTS OF SENSORS
- B1 Sensitivity
- B2 Measurand range
- B3 Stability (short-term, long-term)
- B4 Resolution
- B5 Selectivity
- B6 Speed of response
- B7 Ambient conditions allowed
- B8 Overload characteristics
- B9 Operating life
- B10 Output format
- B11 Cost, size, weight
- B12 Other (specify)

TABLE III

C. DETECTION MEANS USED IN SENSORS
- C1 Biological
- C2 Chemical
- C3 Electric, Magnetic, or Electromagnetic Wave
- C4 Heat, Temperature
- C5 Mechanical Displacement or Wave
- C6 Radioactivity, Radiation
- C7 Other (specify)

TABLE IV

D. SENSOR CONVERSION PHENOMENA

D1. Biological
- D1.1 Biochemical transformation
- D1.2 Physical transformation
- D1.3 Effect on test organism
- D1.4 Spectroscopy
- D1.5 Other (specify)

D2. Chemical
- D2.1 Chemical transformation
- D2.2 Physical transformation
- D2.3 Electrochemical process
- D2.4 Spectroscopy
- D2.5 Other (specify)

D3. Physical
- D3.1 Thermoelectric
- D3.2 Photoelectric
- D3.3 Photomagnetic
- D3.4 Magnetoelectric
- D3.5 Elastomagnetic
- D3.6 Thermoelastic
- D3.7 Elastoelectric
- D3.8 Thermomagnetic
- D3.9 Thermooptic
- D3.10 Photoelastic
- D3.11 Other (specify)

TABLE V

E. SENSOR MATERIALS
- E1 Inorganic
- E2 Organic
- E3 Conductor
- E4 Insulator
- E5 Semiconductor
- E6 Liquid, gas or plasma
- E7 Biological substance
- E8 Other (specify)

TABLE VI

F. FIELDS OF APPLICATION
- F1 Agriculture
- F2 Automotive
- F3 Civil engineering, construction
- F4 Distribution, commerce, finance
- F5 Domestic appliances
- F6 Energy, power
- F7 Environment, meteorology, security
- F8 Health, medicine
- F9 Information, telecommunications
- F10 Manufacturing
- F11 Marine
- F12 Military
- F13 Scientific measurement
- F14 Space
- F15 Transportation (excluding automotive)
- F16 Other (specify)

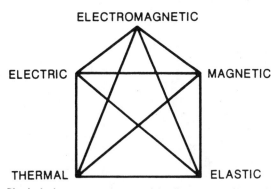

Fig. 1. Physical phenomena represented by lines connecting nodes that represent physical fields.

tric, D3.7; and the key sensor material is an inorganic semiconductor, E1 and E5.

SAW Vapor Sensor: A polymethylmethacrylate (PMMA) polymer coating in the propagation path of a surface acoustic wave delay-line oscillator absorbs vapor, causing mass loading and hence a change of wave velocity and oscillator frequency. The measurand is chemical concentration, A3.1; the primary detection means is mechanical, C5; the sensor conversion phenomenon is physical transformation (a vapor becomes an absorbed constituent), D2.2; and the key sensor material is the organic polymer, E2. If, for greater selectivity, the polymer were altered so that it reacted chemically with only one type of vapor, chemical transformation, D2.1, would be the pri-

mary conversion phenomenon. If the polymer were replaced with an immobilized antibody to detect a particular antigen, biochemical transformation would be involved, D1.1.

Fiber Optic Magnetic Field Probe: A magnetostrictive nickel film deposited on an optical fiber in an interferometer is distorted by an external magnetic field, causing a photoelectrically detected change of light level at the interferometer. The primary detection means is mechanical, C5, and secondarily electromagnetic waves are involved, C3. The fundamental conversion phenomenon is elastomagnetic, D3.5, involving primarily a metallic film, E3, and an insulating fiber, E4. Since fiber optic sensors constitute an important identifiable class, one might key all such sensors similarly, for example by specifying under "Other" in "Detection Means" a category "C7.1 fiber-optic."

ACKNOWLEDGMENT

Research supported by the Berkeley Integrated Sensor Center, an NSF/Industry/University Cooperative Research Center. Discussions with Leslie Field and Ted Zellers are also gratefully acknowledged.

REFERENCES

[1] J. F. Nye, *Physical Properties of Crystals.* Oxford: Oxford Univ., 1957.
[2] W. P. Mason, *Crystal Physics of Interaction Processes.* New York: Academic. 1966, see Figs. 1.1 and 1.2.

SMART SENSORS; WHEN AND WHERE?*

S. MIDDELHOEK and A. C. HOOGERWERF

Department of Electrical Engineering, Delft University of Technology, 2600 GA Delft (The Netherlands)

(Received July 30, 1985; accepted September 13, 1985)

Abstract

Silicon as a sensor material is very promising, because it permits the integration of the sensing element and the signal-processing circuit on one chip, the so-called 'smart sensor'. An overview of physical transduction effects in silicon, presently available integrated sensors, desirable on-chip signal processing functions and academic and industrial results in the field of smart sensors will be presented.

1. Introduction and definitions

The further penetration of micro-electronic components into traditionally non-electronic products is being thwarted by the non-availability of sensors having a performance/price ratio comparable to that of micro-electronic components.

Many technologies for the fabrication of sensors have been proposed. Silicon as a sensor material is very promising, because it shows many large effects, makes batch fabrication possible and, moreover, permits the integration of the sensing element and the signal-processing circuit on one chip, leading to so-called 'smart sensors'. In the sensor literature no agreement exists with respect to terminology. In this paper the term 'smart sensor' is used for a device in which one or more sensing elements and some signal-conditioning circuits are integrated on one and the same silicon (or gallium-arsenide) chip. To distinguish these devices from devices in which the sensing element and the electronics, *e.g.*, a microcomputer, are mounted in close proximity, preferably in the same housing, one could adopt the terms 'integrated smart sensors' and 'hybrid smart sensors'.

In order to design a smart sensor successfully for a certain measurand it is necessary that: (1) silicon shows a suitable physical signal-conversion effect; (2) an integrated sensor based on this effect can be designed and fabricated; (3) suitable signal-processing circuits can be found; and (4) integration of the sensing element and the circuits on one chip using standard silicon planar technology is feasible. In this paper an overview of these requirements will be given. Moreover, current academic and industrial research in the field of smart sensors will be briefly reviewed. The paper concludes with an evaluation of the current situation and an attempt to predict when and where smart sensors will be properly put to use.

2. Signal-conversion effects in silicon

In a sensor a measurand such as light intensity, temperature or displacement is converted into an electronic signal. The fact that signal conversion in a sensor is always based on energy conversion makes it practical to distinguish five non-electrical signal domains: radiant, mechanical, thermal, magnetic and chemical [1].

TABLE 1

Signal domains with examples

Radiant signals	light intensity, wavelength, polarization, phase
Mechanical signals	force, pressure, vacuum, flow, tilt, thickness
Thermal signals	temperature, temperature gradient, heat, entropy
Magnetic signals	field intensity, flux density, permeability
Chemical signals	concentration, toxicity, pH, reduction potential

In order to obtain the value of a measurand, the measurand is very often first converted into a form that can more easily be detected by the sensor in question. For instance, a displacement can be converted into a light intensity change, which can be easily detected by a photodiode. A flow of air is converted into a temperature difference, which can be detected by an integrated thermocouple. When silicon is suitable for the fabrication of sensors, the minimum requirement is that it shows sufficiently large physical effects [2]. In Table 2 the most important effects are given.

TABLE 2

Physical effects for sensors in silicon

Radiant signals	photovoltaic effect, photoelectric effect, photoconductivity, photomagneto-electric effect
Mechanical signals	piezoresistivity, lateral photoelectric effect, lateral photovoltaic effect
Thermal signals	Seebeck effect, temperature dependence of conductivity and junctions, Nernst effect
Magnetic signals	Hall effect, magnetoresistance, Suhl effect
Chemical signals	ion-sensitive field effect

*Paper presented at the Third International Conference on Solid-State Sensors and Actuators (Transducers '85), Philadelphia, PA, U.S.A., June 11 - 14, 1985.

It appears that silicon shows sufficient effects. It is a pity, though, that silicon is not piezoelectric. Most effects such as the Hall effect, the Seebeck effect, the piezoresistance, etc. are rather large. One problem with silicon is that its sensitivities to, say, strain, light and magnetic field show a rather large cross sensitivity to temperature.

When silicon does not display the proper effect, it is also possible to deposit layers of materials with the desired sensitivity on top of a silicon substrate. Thus it is possible to make a magnetic-field sensor based on a thin Ni–Fe layer or a pressure sensor based on a ZnO layer deposited on top of a silicon substrate.

3. Silicon sensors

Many integrated silicon sensors employing the above effects have been produced. In this section the most important devices will be briefly presented. The sensors are grouped according to the already-mentioned five signal domains.

Radiant signal domain

Silicon can be used to construct sensors for a wide spectrum of electromagnetic radiation, from gamma rays to infrared. Photoconductors, photodiodes and phototransistors can easily be fabricated with planar silicon technology. Line and image sensors based on metal–insulator–silicon structures have been produced, and a low-cost solid state camera based on charge-coupled devices (CCD) is now being introduced into the market. Silicon appears to be a very suitable material for this field [4].

More recently, silicon has also been used to detect nuclear radiation. It appears that with silicon, detectors with a very high energy and spatial resolution can be obtained [5].

Mechanical signal domain

Because silicon is not piezoelectric only the piezoresistance effect can be used. This effect is rather large, because the average mobility of electrons and holes in silicon is strongly affected by the application of strain.

Based on this effect numerous force and pressure sensors have been designed and, with adequate specifications and acceptable prices, are now commercially available [6].

By combining a piezoresistor, diffused in a cantilevered beam or a piezoelectric layer, with a silicon mass serving as an inertial reference, it is possible to make miniature accelerometers.

By making effective use of the lateral photoelectric effect it is possible to make very accurate position and angular sensors. With a specially designed photodiode it is possible to obtain x- and y-signals exactly proportional to the coordinates of the light spot, where a light beam is incident on the diode. When the diode is given a circular shape, an angular sensor results [7].

Silicon has also been shown to be rather suitable for the measurement of air or gas velocities. When a slightly heated silicon structure, on which two temperature measuring devices are integrated, is brought into an air flow a temperature difference proportional to the square root of the flow velocity is measured [8].

Thermal signal domain

All electron devices in silicon show temperature dependences which are usually considered to be very undesirable. Therefore, the literature on this topic is extensive. The best-documented effect is the temperature dependence of the forward characteristic of a diode. This effect can also be used in a transistor structure. The most favourable structure is based on the use of two bipolar transistors with a constant ratio of the emitter-current densities. This can be achieved by operating two identical transistors at two different collector currents or two transistors with unequal emitter areas and equal collector currents. In both cases the difference in the base–emitter voltages is proportional to the absolute temperature [9]. Another temperature-sensing device can be obtained by integrating thermocouples consisting of evaporated aluminium films and diffused p-type and n-type layers. The Seebeck effect in silicon is very large, so that one thermocouple can already yield a 1 mV/K temperature difference [10].

Magnetic signal domain

Silicon is non-magnetic, and, therefore, the charge carriers cannot be affected in a direct way by changing the magnetization. Fortunately, silicon lends itself rather well to the construction of Hall plates and transistor structures that are sensitive to magnetic fields. In these devices the rather small mobility of the charge carriers in silicon plays a minor role, so that usable output signals can be obtained. In magnetoresistors the change of the resistance is proportional to the square of the mobility, so that silicon, with its rather low electron and hole mobility, is not very suitable here. Dual-collector or drain structures have been shown to be suitable as magnetic-field sensors. With four-collector structures it is even possible to construct sensors capable of simultaneously measuring the magnitude and the direction of a magnetic-field vector [11]. Because the magnetoresistance effect is rather large in magnetic layers, sensors have been constructed in which a thin magnetic Ni–Fe film is deposited on top of a silicon chip that also contains electronic circuits [12].

Chemical signal domain

The demand for better process control for biomedical, automotive and environmental applications has encouraged many laboratories to undertake silicon chemical sensor research. The ion-sensitive field-effect transistor (ISFET) is the best-known result of this effort [13]. When an ISFET with properly chosen ion-selective gate insulators is immersed in an electrolyte, the change of the drain current is a measure of the concentration of the ions or the pH.

In other MOSFET structures palladium is used as a gate material. Because palladium films can adsorb hydrogen, which in turn can diffuse to the palladium–oxide interface, the threshold voltage is a measure of the hydrogen concentration [14]. When suitable polymer layers are applied, FET structures appear to be sensitive to many more gases such as carbon monoxide and methane. Structures that can operate as accurate humidity sensors can also be designed in silicon.

Chemical sensors have to be immersed in the often rather corrosive media they have to measure. Therefore, the encapsulation of chemical sensor is a very difficult and as yet unsolved problem.

Summarizing, it can be said that during the last two decades academic and industrial research has generated a rather large collection of useful silicon sensors, many of which are already commercially available. This assortment of devices is strong enough to justify our attempting to see whether it is feasible to combine sensing elements and electronic circuits on one silicon chip.

The question that then remains to be answered is which signal-conditioning circuits are appropriate and whether integration is desirable.

4. Desirable on-chip signal processing

The silicon sensors outlined above display a wide variation in output signals (voltages, currents, resistances and capacitances), output formats (analog or digital) and characteristics (noise, drift, offset, cross sensitivity, etc.) and require ample signal conditioning (amplification, voltage-to-current or analog-to-digital conversion, etc.) Up to now the required signal processing has often been performed with specially designed electronic circuits on boards or chips. These circuits are connected to the sensors by wires or are sometimes encapsulated together with one or more sensors in the same housing [15]. When the substrate material for both the sensor and the electronic circuits is silicon, the question of whether the integration of the sensor and circuits on one chip is possible and profitable is justified.

Indeed, in the past decade this approach has been followed; full integration might bring the following advantages:

(a) Better signal-to-noise ratio

The electrical output signal of most sensors is rather weak and, therefore, if this signal is transmitted through long wires a lot of noise might be added. On-the-spot amplification can considerably improve the signal-to-noise ratio.

(b) Improvement of characteristics

(1) Non-linearity; most sensors show some non-linearity. Using on-chip feedback systems or look-up tables might improve linearity.

(2) Cross-sensitivity; most sensors show an undesirable sensitivity to strain and temperature. This can be counteracted by incorporating relevant sensing elements and circuits in the same chip.

(3) Offset; if low-cost mass-produced sensors are needed, offset is an inconvenient parameter, as the abatement of offset requires expensive trimming procedures. Moreover, offset tends to drift. At present it seems that an offset reduction method called the 'sensitivity variation method' eases the problem and can moreover be implemented on the same chip [16].

(4) Parameter drift; component values tend to be a function of time. This can lead to a change in sensitivity, offset, linearity, etc. Based on on-chip accurate current or voltage sources, well-defined values of the measurand and can be generated at the sensor location, allowing periodic and automatic calibration.

(5) Frequency response; by proper feedback methods the frequency response of a sensor can be markedly improved.

Clearly on-chip integration of these circuits offers important advantages.

(c) Signal conditioning and formatting

(1) Analog-to-digital conversion; nearly all sensors generate analog signals. However, at present, the system in which the sensor is used will most often be based on digital techniques, e.g., a microprocessor. Therefore, on-chip analog-to-digital conversion is very desirable. Moreover, the noise immunity of digital signals is very good. The output signal of a sensor can be encoded in several forms, e.g., serial, parallel, frequency, phase and pulse rate.

(2) Impedance matching; most sensors have high internal impedances, which makes them very susceptible to noise. Therefore, on-chip impedance transformation is advantageous.

(3) Output formatting; for the applications engineer the large number of different sensor principles and output formats is very confusing. When electronic circuits can be added to the sensing elements on the same chip, it might be possible in the future to manufacture a whole family of sensor devices for a variety of measurands, all with the same output format. Then measurement systems can easily be composed of sensors and processors.

(4) Conditioning; on-chip electronic circuits can perform all kinds of desirable functions. Many sensors give much more data than required. Superfluous information could be dumped on-chip before it floods the processor. Circuits can also be used to carry out averaging of the sensor signal, thus making it possible to measure signals much smaller than the average noise signal. It might also be possible to subsequently adjust the sensitivity or the output format of a sensor by applying voltages or other signals to it.

(d) Speculative applications

In the future the integration of sensors and circuits on one chip might allow very sophisticated measurement and control systems to be designed. Once the more simple ideas with respect to smart sensors are generally accepted and put into practice, one can begin thinking of integrating complete systems on one chip. Conceivably, one might generate a microprocessor family with integrated sensors for different recurring parameters. It might also be possible to make a multiple sensor structure containing many sensing elements. Then the desired sensing function could be activated by electrical means. In a chemical sensor many sensing elements, each sensitive to a different gas, could be integrated. Careful processing of all the responses might give a reliable measure of gas composition. Should the fabrication of a high-quality sensor prove too difficult, it might be possible to design a structure consisting of a multitude of sensing elements, where the definite output signal is obtained by some averaging or voting process.

It is conceivably also possible to design failure-proof sensors or sensor structures that have to work in corrosive ambients, where for each subsequent measurement a new sensor can be activated. It is clear that very interesting sensor concepts are possible on paper, but that there is still a long way to go before such sensors will be realized.

5. Smart sensors

Though some very convincing advantages are attached to smart sensors, only a few such devices are at present commercially available. Because the few industries active in this field are rather reluctant to describe their results in the scientific literature, it is not easy to glean information about these devices.

Radiant signal domain

The commercially most interesting device for the future is the CCD camera. Two-dimensional CCD cameras with on-chip readout shift registers and signal amplification already exist. CCD line scanners are also available [17]. Photodiodes with on-chip amplification, voltage stabilization and temperature compensation are on the market. By also including a Schmitt trigger on the chip, a light discriminator can be obtained. Photodiode arrays are extensively used in cameras for exposure control and autofocusing, where analog as well as digital signal processing is applied. Because telecommunication along glass fibres is very attractive, a lot of effort in the field of smart sensors is taking place. However, these smart sensors are not based on silicon, but on III-V compounds [18].

Mechanical signal domain

Silicon pressure sensors have been available for years, but did not achieve important commercial successes until recently, when the automotive industry become interested in such devices. At present, numerous devices based on the piezoresistance effect in silicon are available with on-chip amplification and temperature compensation. There is also a pressure sensor based on pressure-sensitive silicon with a frequency output [19]. Silicon sensors based on pressure-sensitive capacitor structures have been reported, in which some signal processing is performed on-chip [20]. Based on non-standard (ZnO) processing steps, accelerometers with some on-chip signal processing have also been reported [21].

Thermal signal domain

The temperature behaviour of silicon electron devices has received a great deal of study. Therefore several smart temperature sensors are commercially available. These devices may contain an operational amplifier, but devices with a frequency output are also known. In a limited range these sensors show a linear output as a function of temperature [9].

Magnetic signal domain

Silicon Hall platelets are rather popular, but show offset and are sensitive to strain and temperature. Devices on the market include those with on-chip amplification, offset and temperature compensation. When a Schmitt trigger is added, a device with two output levels is obtained. When combined with a small permanent magnet a useful contactless switch results. These are being applied by the millions in computer keyboards [22].

Because the magnetoresistance effect in magnetic layers is much larger than that of silicon, some magnetic-field sensors are based on a silicon substrate containing the signal-processing circuits and a magnetic-field sensitive part obtained by depositing a magnetic layer on the oxide-covered silicon substrate [23].

Chemical signal domain

There is a great demand here for devices that can sense chemical parameters, usually some concentration of a material A in a material B, where the aggregation state can vary.

In a smart chemical sensor the sensing element must be exposed to the chemical measurand, whereas the electronic circuit must be carefully protected. In view of these severe limitations one need not be surprised that so few devices are now available. In one project an ISFET is combined with a MOSFET to protect it against electrical breakdown, and to permit some temperature and drift compensation. In another smart sensor, ten ISFETS with on-chip multiplexing and amplification are used. Circuits could be added to average the outputs of the ten sensors or to vote for the most probable output [24].

6. Conclusions — smart sensors when and where?

Silicon's many physical effects can be put to good use in making sensors. Based on these effects, a myriad of sensors for nearly all physical and chemical measurands has been designed and usually experimentally proved to function. Integrating electronic circuits on the sensor chip makes it possible to choose the most favourable interface between sensor and signal processing. In view of these facts one would expect that a large number of smart sensors designed along these lines would already be available. A careful literature search indeed reveals a large variety of smart sensors. Yet in view of the immense efforts in this field over the last decades, one might expect an even larger number. This would seem to call for a re-evaluation of the smart-sensor field. Smart sensors are made with the same technology as integrated circuits. The IC industry is and has to be paranoid with respect to yield and production numbers. Therefore, in general this industry will only get involved in smart sensors if a very large market can be envisaged and the production of the smart sensor does not require non-standard processing steps. The consumer and automotive markets are proof of this. Only in periods when the IC market shows apparently unavoidable backslides might the IC industry suddenly and temporarily become interested in non-standard sensors. The smaller markets will only develop when the more specialized companies in the measurement and control field need the smart-sensor concept to stay in business. In view of the growing competition in this field, it looks as if that moment has now come. The government and the industries in Japan have understood the potential of the sensor field very well and are apparently determined to conquer a very important share of this and the associated measurement and control systems market.

It is to be expected that, in general, two groups of industries will pursue the smart-sensor concept. The first group, the measurement and control systems industry, with in-house silicon processing capabilities, is in a good position to reap the benfits of smart sensors. In this industry the broad knowledge necessary to be successful in this field might already be available.

The second group, the smaller instrumentation companies, will run into more difficulties in entering this market. However, in view of the expected strong growth of the custom-IC market, it is possible that small instrumentation companies cooperating with the so-called custom houses might also gain access to the smart sensor market.

In conclusion, silicon is a very suitable material for smart sensors. A lot of intensive research efforts will still be required to reap the benefits of the smart-sensor concept, but based on experience with the already-existing devices, it is to be expected that in the coming decade a large number of successful smart sensors will emerge.

References

1 S. Middelhoek and D. J. W. Noorlag, Three-dimensional representation of input and output transducers, Sensors and Actuators, 2 (1981) 29 - 41.
2 S. Middelhoek and D. J. W. Noorlag, Signal conversion in solid-state transducers, Sensors and Actuators, 2 (1982) 211 - 228.
3 S. Middelhoek and D. J. W. Noorlag, Silicon micro-transducers, in B. E. Jones (ed.), Instrument Science and Technology, Vol. 2, Adam Hilger Ltd, Bristol, 1983, pp. 33 - 44.
4 V. Hartel, Optoelectronics, McGraw-Hill, New York, 1978.
5 W. R. Th. ten Kate and C. L. M. van der Klauw, Experimental results on an integrated strip detector with CCD-readout, Nucl. Instrum. Methods, A228 (1) (1985) 105 - 109.
6 M. Poppinger, Silicon diaphragm pressure sensors, Proc. ESSDERC'85, Aachen, 1985, to be published.
7 D. J. W. Noorlag, Quantitative analysis of effects causing nonlinear position response in position sensitive photodetectors, IEEE Trans. Eelectron Devices, ED-29 (1982) 158 - 161.
8 J. H. Huijsing, J. P. Schuddemat and W. Verhoef, Monolithic integrated direction-sensitive flow sensor, IEEE Trans. Electron Devices, ED-29 (1982) 133 - 136.
9 G. C. M. Meijer, A low power easy-to-calibrate temperature transducer, IEEE J. Solid-State Circuits, SC-17 (1982) 609 - 613.
10 A. W. van Herwaarden, The Seebeck effect in silicon ICs, Sensors and Actuators, 6 (1984) 245 - 254.
11 V. Zieren and S. Middelhoek, Magnetic field vector sensor based on a two-collector transistor structure, Sensors and Actuators, 2 (1982) 251 - 261.
12 T. Usuki, S. Sugiyama, M. Takeuchi, T. Takeuchi and I. Igarashi, Integrated magnetic sensor, Proc. 2nd Japanese Sensor Symposium 1982, pp. 215 - 217.
13 A. A. Saaman and P. Bergveld, A classification of chemically sensitive semiconductor devices, Sensors and Actuators, 7 (1985) 75 - 87.
14 I. Lundstrom, Hydrogen sensitive MOS structures, Part 1: Principles and applications, Sensors and Actuators, 1 (1981) 403 - 426.
15 J. E. Brignell and A. P. Dorey, Sensors for microprocessor-based applications, J. Phys. E: Sci. Instrum., 16 (1983) 952 - 958.
16 S. Kordic, V. Zieren and S. Middelhoek, A novel method for reducing the offset of magnetic-field sensors, Sensors and Actuators, 4 (1983) 55 - 61.
17 R. Melen and D. Buss (eds.), CCD: Technology and Applications, IEEE Press, New York, 1977.
18 M. Ito, K. Nakai and T. Sakurai, Monolithic integration of a metal-semiconductor-metal photodiode and a GaAs preamplifier, IEEE Electron Devices Lett., EDL-5 (1984) 531 - 532.
19 S. Sugiyama, T. Suzuki, M. Takigawa and I. Igarashi, Miniature piezoresistive strain and pressure sensors with on-chip circuitry, Proc. 3rd Japanese Sensor Symposium 1983, pp. 209 - 213.
20 A. Hanneborg, T. E. Hansen, P. A. Ohlckers, E. Carlson, B. Dahl and O. Holwech, A new integrated capacitive pressure sensor with frequency modulated output, Proc. Transducers '85, Philadelphia, 1985, pp. 186 - 188.
21 P. L. Chen, R. S. Muller and A. P. Andrews, Integrated silicon PI-FET accelerometer with proof mass, Sensors and Actuators, 5 (1984) 119 - 126.
22 R. Allan, Sensors in silicon, High Technol., 4 (1985) 43 - 50.
23 M. Hirata and S. Suzuki, Integrated magnetic sensor, Proc. 1st Japanese Sensor Symposium, 1981, pp. 305 - 310.
24 D. Yu and Y. H. Xu, CMOS multiplex circuit for multiple pH sensors, EDC Lab. Memo No. 290 (1982).

(Invited)   Approaches to Intelligent Sensors

Hiro Yamasaki

Faculty of Engineering The University of Tokyo
7-3-1, Hongo, Bunkyo-ku, Tokyo 113, Japan

Sensing systems incorporated with dedicated signal processing functions are called intelligent sensors or smart sensors. The objectives of the intelligent sensors are to enhance design flexibility and realize new sensing functions, and additional objectives are to reduce loads on central processing units and signal transmission lines by distributed information processing.

Technical approaches to the intelligent sensors can be divided into three different categories: (1) New functional materials. (2) New functional structure. (3) Integration with computers.

Typical examples of the approaches are described. These are ranging from single chip sensing devices integrated with microprocessors to big sensor arrays utilizing synthetic aperture techniques, and from two dimensional functional materials to human sensory systems.

§1.   The Objectives of Intelligent Sensors

Various physical phenomena and effects can be applied for signal conversion in sensor devices. However some of their applications are limited due to inherent nonlinearity or influencial errors due to change in other quantities.

An accurate and flexible compensation of numerical signal processing can eliminate some of the constraint in present sensor techniques. Thus freedom of sensor design can be substantially enlarged. In other words, new sensor devices can be developed by the use of new phenomenna which can not be utilized without numerical compensation or linearization. This is an advantage of intelligent sensors which can be realized at the early stage.[1]

Physical quantities which are clearly defined by mathematical models at a specialized point can be measured accurately by the use of certain sensor device. However, identification of multi-dimensional state or image are difficult task for sensing devices at present. Diagnosis of facilities or fault detection of systems are urgent social needs for sensing techniques, but very few sensors can meet the requirement.

Effective means to overcome above mentioned weak points of the sensor techniques are to develop sensing systems combining multiple sensors, arrays of sensors with computer aided measurement techniques. Intelligent sensors are one of powerful implementation of sensing systems with distributed functions of information processing.

As the result of expanded design flexibility of device and the system approach, new functions are realized. This is the most important objectives of intelligent sensors. The second objective is to reduce loads on central processing unit and signal transmission lines by distributed information processing.

The first objective is especially important in the area of fault detection systems, remote sensings and multi dimensional measurement. The second one is important in applications relating to big systems such as space satellites.

A general image of intelligent sensors seems to be one chip device in which sensor and microcomputer are integrated, but there are other different technical approaches to realize the objectives. The technical approaches can be divided into three categories as follows: (1) New functional materials. (2) New functional structure. (3) Integration with computers.

§2.   Approach Through New Functional Materials

For obtaining informations of two or three dimensional objects, for example tactile senses, sheet of conductive rubber and PVDF are useful as functional material for tactile sensors, a special simple algorithm is developed for recognition of object's shape and calculation of the center of force.[2]

Enzymes and microbes are very powerful to realize high selectivity for a specified substance, and they can minimize signal processing to reject the effects of coexisted components.

Shape memory metals can make sensors unificated with actuators. In addition to the unificated structure, their memory function may be useful for unique applications.

§3. Approach Through Functional Mechanical Structure

If signal processing function is implemented in the mechanical structure or the form of sensors, processing of the signal is simplified and the rapid response can be expected.

Let us discuss a spatial filter for example, an object having a two dimensional optical pattern $f(x, y)$ is projected through a spatial filter $g(\xi, \eta)$, and its output is focused and converted into electrical signal $e(t)$ is given by eq.(1).

$$e(t) = \int\!\!\int_D g(\xi, \eta)[f(x, y, t)*h(\xi, \eta)]\mathrm{d}\xi\,\mathrm{d}\eta \quad (1)$$

where $h(\xi, \eta)$: point spread function, $*$: convolution.

as seen in eq. (1), two dimensional convolution is carried out in optical configuration. Typical application of the spatial filter technique is noncontact velocity measurement with simple hardware.

We can tell three dimensional directions of sound sources with two ears. We can also identify the direction of sources even in the median plane.

The identification seems to be made based on the direction dependency of pinnae responses. Obtained impulse responses are shown in Fig. 1, in which signals are picked up by small electret microphone inserted in the external ear canal, and spark of electric discharge is used as the sound source. Differences can be easily observed.[3]

Usually at least three sensors are necessary for identification of three dimensional localization. So pinnae are supposed to act as a kind of signal processing hardware with inherent special shapes. We are studying on this mechanism utilizing synthesized sounds which are made by convolutions of impulse responses and natural sound and noise.[4]

Not only human ear systems, sensory systems of man and animals are good examples of intelligent sensing system with functional structure.

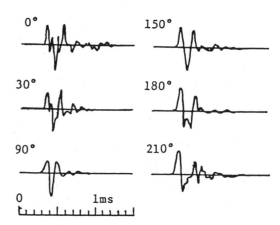

Fig. 1. Pinna impulse responses in the median plane.

The most important feature of such intelligent sensing systems is integration of multiple functions; sensing and signal processing, sensing and actuating, signal processing and signal transmission.

Our fingers are typical example of integration of sensors and actuators, our eyes are too. Signal processing for noise rejection such as lateral inhibition are carried out in the course of signal transmission in nervous system.

§4. Approach Through Integration with Computer Part 1

The most popular image of intelligent sensor is an integrated device combinig sensor with micro computer, however, such sensor is not realized yet.

A development stages of such intelligent sensors is illustrated in Fig. 2.[5] Four separate fuctional blocks: sensor, signal conditioner, A/D converter and microprocessor are gradually coupled on single chip then turn into a direct coupling of sensor and microprocessor.

In my opinion we are in the second stage at present. Let us discuss some examples in the stage.

Several results are reported about research on single chip pressure sensor device on which

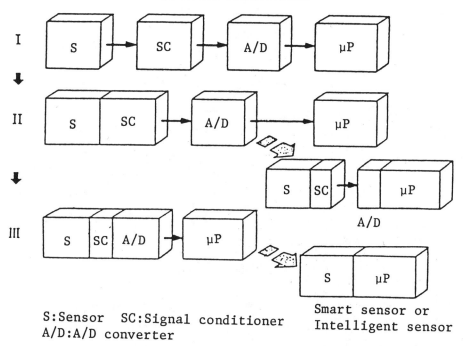

S:Sensor SC:Signal conditioner
A/D:A/D converter

Smart sensor or
Intelligent sensor

Fig. 2. Development trends in integration with microprocessors.

analog circuit for simple signal processing are integrated.[6,7] The circuits are amplifier, temperature compensation circuit, oscillator etc. (Fig.3) No results on single silicon chip sensor device integrated with microprocessor is reported so far. A problem of insulation between circuits and sensors on a common silicon substrate is reported. The problem may limit the maximum range of signals.[8]

Usually sensor device should be exposed to severe environmental conditions, but microprocessor device is relatively sensitive to ambient condition and electromagnetic noise induction. Reliability of processor in severe atmosphere may be another problem. However, in the case of noncontact measurement or image sensor, early realization will be possible.

Fig. 4 shows a recent example of intelligent differential pressure transmitter for process instrumentation.

A single chip device including a silicon diaphragm differential pressure sensor, an absolute pressure sensor and a temperature sensor is used. The output signals from these

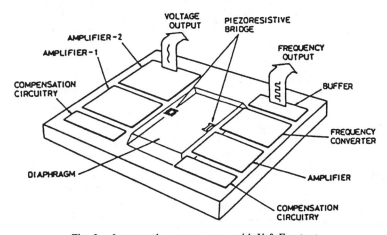

Fig. 3. Integrated pressure sensor with V & F output.

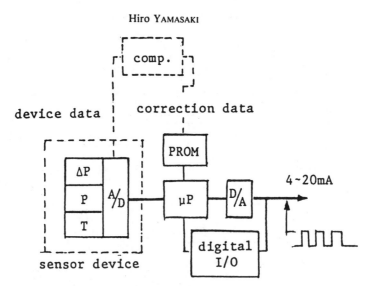

Fig. 4. Digital smart Δp transmitter.

three sensors are applied to a microprocessor via A/D converter both on separate chip. The processor calculates the output and at the same time compensates effects of absolute pressure and temperature numerically. The data for compensation of each sensor device is measured in the manufacturing process and loaded in ROM of the processor respectively. Thus the compensation can be precise, the accuracy of 0.1% is obtained.

The transmitter has a pulse communication ability through two wire analog signal line and digital communication interface. Remote span and zero setting, remote diagnosis and other functions can be performed by the use of digital communication means.

The range of analog output signal is the IEC standard of 4–20 mA DC, so total circuits including microprocessor should work within 4 mA, the problem is overcome by CMOS circuit approach.[9]

Sensors having frequency output are advantageous in interfacing with microprocessors. Frequency output density sensor and pressure sensor which are compensated and linearized by dedicated processors are reported.[10]

A new approach to chemical intelligent sensors is proposed. Six thick film gas sensors which have different sensitivity for various gases are combined, and the sensitivity pattern is recognized by microcomputer. Materials and sensed gases are shown in Table 1. Fig. 5 shows several examples of sensor conductivity patterns for organic and inorganic gases.

Typical patterns are memorized and identified by dedicated microprocessor utilizing similarity analysis of patterns. Maximum sensitivity of 1 ppm is reported.[11,12]

§5. Approach Through Integration with Computer Part 2

A more advanced function of coupled system of sensor and computer is observed in synthetic aperture sensing system.

The ultimate resolution of an optical system is determined by the ratio of its aperture size to the wave-length. As the wave-length of electromagnetic wave and that of ultra sonic wave is much longer than that of visible light, a big aperture is necessary. A radio telescope, for example, its aperture, the diameter of parabola antenna is required to be several tens of kilometers for similar resolution of optical one. Such big antenna can neither be built nor be driven, even if the design itself is possible.

Table I. Materials & sensed gases of thick film gas sensors.

Sensor	Material	Sensible gas
S_1	ZnO	Organic gases
S_2	ZnO (Pt doped)	Same above (low Alcohol sensitivity)
S_3	WO_3	H_2, CO, C_2H_5OH
S_4	WO_3 (Pt doped)	Same above (low Alcohol sensitivity)
S_5	SnO_2	Reducing gases
S_6	SnO_2 (Pd doped)	Same above (high Methane sensitivity)

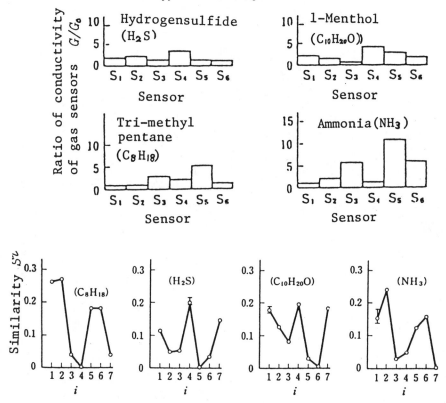

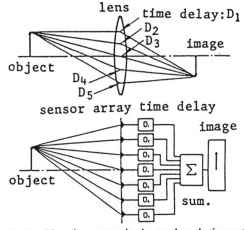

Fig. 5. Sensor sensitivity patterns for various gases.

The problem can be solved by the use of the synthetic aperture techniques. Array of small aperture sensors coupled with computers can be the substitution for array size aperture, and realize higher resolution.

Fig. 6 shows the basic principle of synthetic aperture system corresponding to a dioptric system using lens. Each light from the object passes different path and with different time delay due to lens and then focus into the image. Each path seems to have different length, but all of them have the same propagation length.

In the synthetic aperture system, the outputs of each sensors which different time lag is added to, and they are summed. The sum makes constructed image. Configuration of two system seems different but their functions are same.

The direction of light axis and the focus of the synthetic aperture system can be driven by adjusting time delays in computer, so very rapid scanning is possible. Even simultaneous focusing into more than one point is realizable if parallel signal processing is available. Such

focusing function is never realized in physical systems.

This approach improved the resolution of radar and radio telescope substantially. Now we are working on the development of a high resolution sonar and a measuring system of velocity vector distribution in the sea water by the use of the synthetic aperture technique.

Fig. 6. Dioptric system using lens and synthetic aperture techniques

15

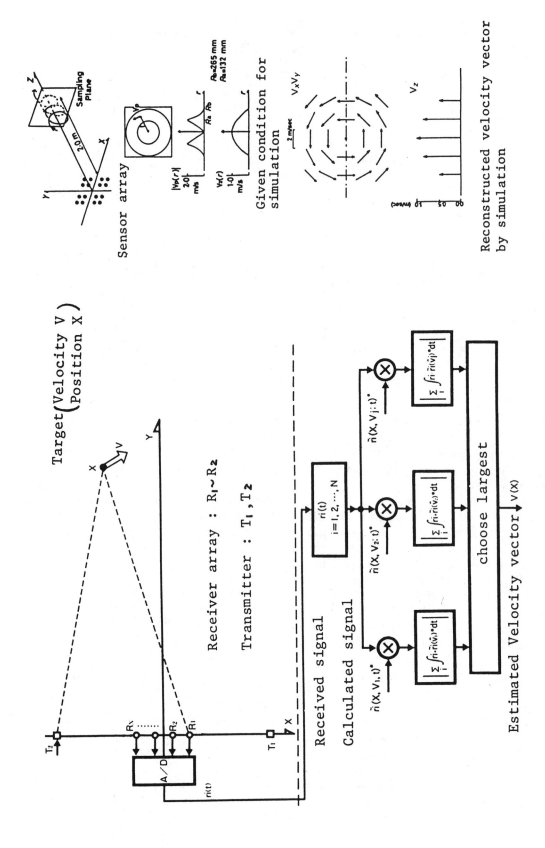

Fig. 7. Measuring system of velocity vector distribution utilizing synthetic aperture & numerical reconstruction of sound field. (left) Concept of the system, (right) A result of oomputer simulation.

The concept of our system and an example of computer simulation results are shown in Fig. 7.[13]

This technique seems to be a very powerful approach for higher spatial resolution with small size point sensors which have poor resolution, thus it is an effective counter measure to overcome the weak point of present sensor techniques described previously.

§6. Approach Through Integration with Computer Part 3

Another possible sensor function which is realized by integration of intelligence is self repair for fault tolerant system.

A heavy redundant system, for instance, 2 out of 3 system is in practical use with simple logic. The function of self diagnosis with smarter algorithm can make a sensing system less redundant but fault tolerant. One such approach is proposed.

In this system, "Analog parity" is used for fault detection, localization and correction of error due to the fault. The analog parity is calculated from the linear combination of multiple sensor outputs.

The outputs of various sensors and the analog parities are processed together in a computer, the fault detection and error correction are performed.[14]

This approach can be considered an extension of error correction code in digital technique to analog signal.

§7. Conclusion —Future Images of Intelligent Sensors—

Fig. 8 shows a concept of an intelligent area image sensing system integrated on single chip device. Research of this device is planned in the R & D Project of Basic Technology for Future Industries which is promoted by MITI. A future image of intelligent sensors can be seen in the Figure.[15]

The device consists of multi-layer structure, each layer performs different function based on physical properties of layer materials. A number of light sensing devices are arrayed on the top layer, signal transmission devices are built in the second layer, memories are in the third, computing devices are in the fourth, and power supplies are in the bottom layer.

An image processing such as feature extraction and edge enhancement can be performed in the three-dimensional multifunctional structure. This is just like the retina of our eyes. As previously described in this paper, the important feature of sensing systems of man and animals is such integration of multi functions and distributed signal processing.

It can be said that a future image or target of our intelligent sensors are the sensing systems of man and advanced animals, in which the three approaches are combined together.

It is important to note that the approach to the future image is not single but three

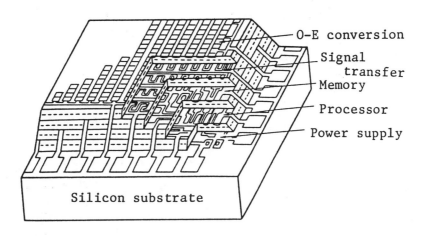

Fig. 8. Concept of integrated image sensing system with multi-layer structure.

different appoaches should be equally considered.

In this paper, these approaches are discussed with typical examples for purpose mentioned.

References

1) W. G. Wolber: Smart sensors, ACTA IMEKO 1982, 3 (1982).

2) M. Ishikawa & M. Shimojo: A Tactile sensor using pressure sensitive conductive rubber, Preprint 2nd Sensor Symposium 95 (1982).

3) Y. Hiranaka & H. Yamasaki: Envelope representations of pinna impulse responses relating to 3-dimensional localization of sound sources, J. Acoustical Society of America, 73(1), 291 (1983).

4) Y. Hiranaka et al: Envelope representation of pinna responses & localization trials by computer simulated sound, Note H-83-31, WG on auditory sense, Acoustical Society of Japan (1983).

5) Mackintosh International, Sensors, Vol. 2 4 (1981).

6) S. Sugiyama et al: Integrated piezo resistive pressure sensor with both voltage & frequency output, Abstracts Solid state Transducer 83 115, (1983).

7) K. Yamada et al: A piezoresistive integrated pressure sensor, Abstacts Solid-state Transducer 83 113 (1983).

8) W. H. Ko et al: Capacitive pressure transducers with integrated circuits, Sensors & Actuators Vol. 4 403 (1983).

9) Digital smart transmitter DSTJ 3000, J.SICE 22(12) 1054 (1983).

10) I. Ohno: Frequency dependent sensors & their advantages, J.SICE 23(3) 327 (1984).

11) M. Kaneyasu et al: Odor detection by integrated thick film gas sensor, Preprint of 22th SICE Meeting, 271 (1983).

12) NE Report, Nikkei Electronics Sept. 12 (1983), 78.

13) Y. Tamura & H. Yamasaki: Non-contact measurement of velocity vector distribution by numerical reconstruction of reflected sound field, Trans. SICE, 18(1) 44 (1982).

14) S. Fujitani: A fault torelant measurement system, SICE 8th Systems Symposium, 427 (1982).

15) S. Kataoka: Sensors & human society, J. IEEJ 102(5) 345 (1982).

STRATEGIES FOR SENSOR RESEARCH

Richard S. Muller

Director: Berkeley Integrated Sensor Center
An NSF/Industry/University Cooperative Research Center
Department of Electrical Engineering and Computer Sciences
and the Electronics Research Laboratory
University of California, Berkeley, CA 94720

ABSTRACT

Sensor research presents a strong multi-disciplinary challenge. The over-riding goal of research in this area, considering the unprecedented advances in information processing already achieved by integrated circuits, is to provide a more effective integration of systems and sensors. Strategies that can help to reach this goal need: (a) to recognize the scope of the talents required to do an effective job, (b) to focus on the interfaces between areas of expertise, and (c) to forge links enabling effective communication between the disciplines. The field of integrated sensors can profitably borrow from proven experience with integrated circuits and computing systems by identifying and developing versatile technologies.

Research results have established the use of thin films of ZnO on silicon ICs as a successful example of a versatile sensor technology. This technology and some of its applications are discussed with a specific focus on its use to build an IC-processed microphone. The microphone consists of a miniature diaphragm pressure transducer fabricated by combining micromachining procedures (to produce a thin silicon-nitride diaphragm) with ZnO thin-film processing. The diaphragm, 2 μm in thickness, is the thinnest yet reported for a piezoelectric readout structure of relatively large area (3 x 3 mm^2).

INTRODUCTION

In the sense intended by the title for this paper, a strategy is "skillful management to attain a goal." The *goal*, of course, is to make major advances in interfacing electronics with nonelectrical signals-- in coupling the unprecedented advances of integrated circuits with the real-world variables on which they must operate. An important element of a successful strategy is to have an organization that enables one to address problems effectively. The approach that we have taken within the university framework is to form a Center for Research that is interactive with industrial goals and needs. Features of the Center (the Berkeley Integrated Sensor Center or "BISC") are highlighted in Fig. 1.

BERKELEY INTEGRATED SENSOR CENTER

An NSF/Industry/University Cooperative Research Center

*Department of Electrical Engineering and Computer Sciences
and the Electronics Research Laboratory, University of California, Berkeley*

- Provides Focus on Expanding Multi-Disciplinary Field
- Admirably Suited to University Research
- Symbiotic Interface with Industrial Partners

Center Program funded by National Science Foundation, Sept. 1, 1986.

Figure 1. Some features of the Berkeley Integrated Sensor Center.

A next element of a successful strategy is to "pick a winner" if at all possible. With this in mind, we have focused on the silicon IC microfabrication process. A proven, powerful, and versatile technology, the IC process embodies advantages that can be employed in new ways to contribute heavily to success in the sensor field. Some of the features of the microfabrication process that are especially valuable for sensor design are: (1) unprecedented dimensional control, (2) availability of a variety of materials for sensor fabrication from: polycrystalline silicon layers, epitaxial silicon layers, oxide, nitride, resist layers, and from metals such as Al, Au, W and NiCr, (3) a vast library of proven circuits, (4) highly developed computer aids, and the very important option of integrating sensors and circuits.

Among the sensors that we have already produced using processes based on IC fabrication techniques are the following:
- Microbeam accelerometer
- Polysilicon tangential force sensor
- Polysilicon vapor-flow sensors
- Vapor sensors using acoustic waves
- Heat-of-reaction sensor
- Vapor sensor using vibrating polysilicon beams
- Miniaturized microphone
- Integrated carrier-domain magnetometer

Reprinted with permission from *Transducers '87, Rec. of the 4th Int. Conf. on Solid-State Sensors and Actuators*, 1987, pp. 107–111.
Copyright © 1987 by the Institute of Electrical Engineers of Japan.

- Proximity and tactile array sensors
- Pyroelectric airflow sensors
- Integrated infrared sensor

Because of its inherent interdisciplinary nature, a successful strategy for sensor research must involve experts from several fields. In order to be effective in meeting the interdisciplinary challenge of sensor research, we have found it imperative to:

- recognize the multi-disciplinary breadth,
- form bridges between the disciplines,
- focus on the interfaces.

The first two of these goals are self explanatory; the third represents the results of experience. Within a university the goals enumerated above can best be met if graduate students from differing departments are brought together in a single facility to work side-by-side on mutually useful research areas. We have found the synergism between students in electrical engineering, mechanical engineering, chemical engineering, and materials science to hold the greatest importance. Professors automatically become interactive once their students are working together. Thus, for example, a specialist in mechanisms in the Mechanical Engineering Department is directing research on the mechanical properties of polysilicon that provides a necessary foundation for research in electrical engineering on sensor applications.

Another strategy for successful research on silicon-based sensors is to identify and develop versatile technologies that can be exploited to produce classes of novel devices. We call such technologies *generic* because of their wide applicabilities. A *generic* technology of considerable importance is that used to make micromechanical elements of polysilicon. Another is the compatible production of active piezoelectric and pyroelectric thin films with integrated circuits. The candidate film for this *generic* technology in our research center is zinc oxide.

GENERIC TECHNOLOGY: ZnO THIN FILMS

We have used thin films of zinc oxide in conjunction with silicon planar processing techniques to form a number of integrated sensors. Among those fabricated and demonstrated in our laboratory are:

(1)   cantilever-beam accelerometer [1],

(2)   SAW convolver [2],

(3)   chemical vapor sensor [3],

(4)   anemometer [4],

(5)   infrared detector array [5],

(6)   chemical reaction sensor [6],

(7)   tactile sensor array [7],

(8)   infrared charge-coupled device imager [8],and

(9)   miniature diaphragm microphone [9].

The compatibility of the ZnO-on-Si process with con-

ventional IC fabrication was demonstrated in a multisensor chip [8]. The operation of all sensors was based on the piezoelectric or pyroelectric effects in the zinc-oxide films.

Film Deposition: Although a number of techniques that have been used for ZnO thin-film deposition, planer-magnetron sputtering appears to give the best piezoelectric and pyroelectric characteristics [10]. Using planer magnetron sputtering, highly oriented ZnO films have been deposited on SiO_2, polycrystalline silicon, and bare silicon substrates. X-ray diffraction measurements indicate preferential c- axis orientation with a single diffraction peak at 33.8° as shown in Fig. 2.

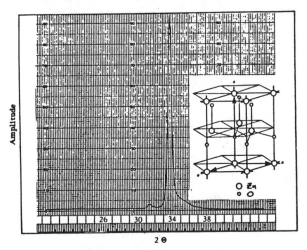

Fig. 2. Typical x-ray diffraction spectrum showing the preferential c- axis orientation in ZnO thin films prepared by planer magnetron sputtering.

We have found the best thin-film crystallinity at a sputtering power of 200 W with a 10 mTorr ambient gas mixture consisting of an equal mix of oxygen and argon. The distance between the substrate and the target in our system is 4 cm and the substrate temperature is maintained at 230°C during deposition.

Material Characterization: Pyroelectric and piezoelectric properties have been measured by electrical techniques and correlated with the x-ray diffraction results. The measured pyroelectric coefficient at T = 300 K is $p^\sigma = 1.4 \times 10^{-9} Ccm^{-2}K^{-1}$ and the most important piezoelectric coefficient d_{33} is $14.4 \times 10^{-12} CN^{-1}$, in good agreement with the range of reported values in crystalline ZnO [11,12]. The temperature dependence of these coefficients is shown in Figs. 3 and 4. For typical sensor applications in 1 μm-thick films at T = 300 K, signal levels are roughly 5.2 mV/gm (piezoelectric effect) and 150 mV/K (pyroelectric effect).

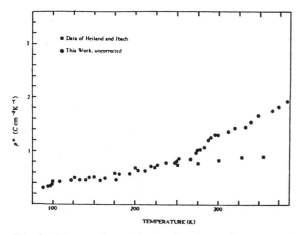

Fig. 3. Measurement of pyroelectric coefficient versus temperature in a 1.0–μm-thick ZnO film.

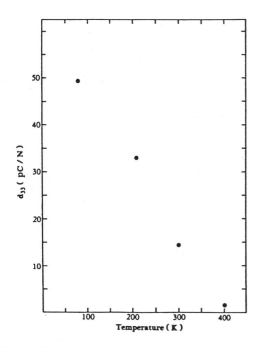

Fig. 4. Temperature dependence of the piezoelectric coefficient d_{33}.

Sensors using ZnO thin films as piezoelectrics show a near dc response because typical film thicknesses (1μm) are much thinner than a Debye length (120μm). Static charge decay times longer than 32 days have been measured in these films when they have been encapsulated by 0.4 μm-thick layers of SiO_2 [13]. Resistivities of $3 \times 10^7 \Omega$-cm and relative permittivities of $\varepsilon_r = 10.3$ characterize these thin films.

The performance of sensors made with these films and on-chip NMOS circuits demonstrates the versatility of ZnO thin-film technology for integrated-sensing applications based on either the piezoelectric or pyroelectric effect. A recent example of research making use of ZnO films, together with micromechanics, is an IC-processed microphone [9].

IC PROCESSED ZnO-FILM MICROPHONE

The most widely applied microprocessed silicon sensors are diaphragm pressure transducers, particularly for applications at near atmospheric pressures. We have concentrated in our laboratory on sensors for very low pressures and have succeeded through the application of micromechanics and thin-film technologies in producing a device that is sensitive to signals in the μbar range, a sensitivity level associated with audio microphones. One of these sensors has operated over the course of weeks in an audio link to a radio receiver producing acceptable sound quality when driving an amplifier and small speaker. The microphone picks up a low-level audio signal from an output loudspeaker driven by the radio receiver. The sensor has been constructed on a silicon wafer using a silicon-nitride diaphragm and a thin piezoelectrically active ZnO film as the mechanical-to-electrical transducer.

Previous diaphragm sensors with piezoelectric readouts have been reported by Royer, Holmen, Wurm and Aadland [14] and by Smits, Tilmans and Lammerink [15]. These other sensors used diaphragms made of single-crystal silicon. We have found that control of the thickness and of the latent stress in the diaphragm is better when it is made of silicon nitride than we have achieved using silicon. The silicon-nitride layer was deposited essentially stress-free using techniques similar to those described by Sekimoto and co-workers [16]. The diaphragms are mechanically robust and planer when deposited at higher deposition temperatures and higher gas ratios (between the reactant gas SiH_2Cl_2 and the nitrogen source NH_3) than are usual in conventional IC applications of nitride.

Figure 5 (a) shows a top view and Figure 5 (b) shows the cross section of the micromachined sensor.

TOP VIEW LAYOUT

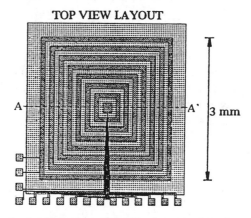

Figure 5. (a) Top view of the diaphragm microphone. The 8 annular patterns delineate the Al and polysilicon electrodes which are connected to 16 different bonding pads. ZnO covers the entire diaphragm area of 3×3 mm²

CROSS SECTION OF THE SENSOR THROUGH A-A'

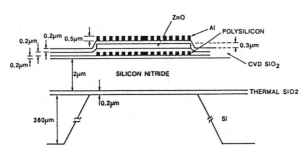

Figure 5. (b) Cross section of the sensor through A-A'.

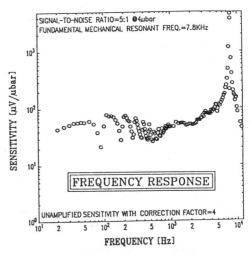

Figure 6. Frequency response of the microphone. The values plotted represent the unamplified response at the piezoelectric electrode.

The piezoelectric ZnO film is patterned to cover the 3-mm square diaphragm that is formed of Si_3N_4. The ZnO is encapsulated with CVD SiO_2 (0.2μm thick deposited at 450°C). Both the aluminum top electrodes and the polysilicon bottom electrodes are segmented with annular patterns. We used the segmented annular patterns to provide flexibility for experimental evaluation of the sensor.

Five masks were used to fabricate the microphone. The main steps are described in the following. Silicon nitride is deposited at 835°C over thermally grown oxide using low-pressure CVD with dichlorosilane to ammonia at a gas ratio of 5 to 1. This deposition condition produces almost stress-free silicon-nitride films. Anisotropic etchant (EPW) is used to form the silicon-nitride diaphragm by etching the wafer from the backside. The Si_3N_4 diaphragm has a thickness of 2 μm. The polysilicon electrodes are then formed on the diaphragm over CVD SiO_2. Next, another layer of CVD SiO_2 is deposited, and over this a 0.3μm layer of ZnO is sputter-deposited with *c*-axis oriented perpendicularly to the plane of the diaphragm. Aluminum is sputter-deposited after CVD SiO_2 is laid down to encapsulate the ZnO and contact windows have been opened. Patterning and sintering of Al complete the fabrication process. An advantage of this processing sequence is that front-to-backside alignment and protection from EPW etchant problems are avoided because the diaphragm is formed early in the process. If the sensor were produced with on-chip circuits, however, it would probably be better to form the diaphragm in the last stages of the process.

Results: Plotted on Fig. 6 are values of the unamplified microphone sensitivity (μV per μbar versus frequency.

The variation of sensitivity from 20 Hz to 4 kHz is approximately 9dB with the typical sensitivity being 50 μV/μbar. A fundamental mechanical resonance in the diaphragm occurs at 7.8kHz with $Q > 20$ as predicted from the theory of plate vibration. The signal-to-noise ratio at 4μbar is 5:1. Figure 7 shows the response of the sensor to a step change in air pressure.

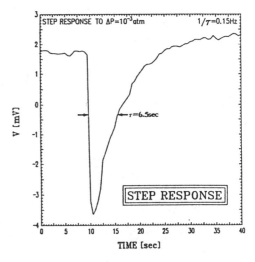

Figure 7. Response of the sensor to a step change of air pressure.

As can be seen in Fig. 7, the decay time is 6.5 seconds which means that the sensor can be operated down to 0.15 Hz. We expected to get this low-frequency response because the ZnO layer is encapsulated in CVD SiO_2.

The successful fabrication and test of this IC-processed microphone provides a powerful demonstration of the utility of the generic ZnO technology. The microphone is the first sensor to employ ZnO on a silicon-nitride diaphragm and the first to employ a diaphragm as thin as $2\mu m$.

CONCLUSIONS

The experience gained in the development of integrated circuits can be put to effective use for advances in integrated sensors. Technology is the central issue; mastering it and extending it is a multi-disciplinary challenge. The development of versatile, generic technologies provides an avenue for progress in the field. A successful example of such a generic technology is the compatible production of oriented ZnO piezoelectric and pyroelectric films on silicon ICs.

Acknowledgements: Research described has been carried out by graduate student E.S. Kim and by recent graduates of the Berkeley program, notably Drs. Dennis Polla and Roger Howe. We also acknowledge the highly valued assistance of Ms. Katalin Voros and the staff of the Berkeley Microfabrication Laboratory and comments by R. Hijab. Prof. R.M. White collaborates on all aspects of the Berkeley sensors program.

REFERENCES

(1)   P.-L. Chen, R. S. Muller, R. D. Jolly, G. L. Halac, R. M. White, A. P. Andrews, T. C. Lim, and M. E. Motamedi, "Integrated Silicon Microbeam PIFET Accelerometer," *IEEE Trans. Electr. Dev. ED-29*, 27-32 (1983).

(2)   A. E. Comer and R. S. Muller, "A New ZnO on Silicon Convolver Structure," *IEEE Electr. Dev. Lett. EDL-3*, 118-120 (1982).

(3)   C. T. Chuang and R. M. White, "Sensors utilizing thin-membrane SAW oscillators," *IEEE Ultrasonics Symp.*, Chicago, IL, (1981).

(4)   D. L. Polla, R. S. Muller, and R. M. White, "Monolithic Zinc Oxide on Silicon Pyroelectric Anemometer," *IEEE Int. Electr. Dev. Mtg.*, 639-642 Washington, DC, (1983).

(5)   D. L. Polla, R. S. Muller, and R. M. White, Fully-Integrated ZnO on Silicon Infrared Detector Array," *IEEE Int. Electr. Dev. Mtg.*, 382-384, San Francisco, CA, (1984).

(6)   D. L. Polla, R. M. White, and R. S. Muller, "Integrated Chemical-Reaction Sensor," *Third Int. Conf. on Solid-State Sensors and Actuators,* 33-36, Philadelphia, PA, (1985).

(7)   D. L. Polla, W. T. Chang, R. S. Muller, and R. M. White, "Integrated Zinc Oxide-on-Silicon Tactile Sensor Array," *IEEE Int. Electr. Dev. Mtg.*, 133-136, Washington, DC, (1985).

(8)   D. L. Polla, R. M. White, and R. S. Muller, "Integrated Multi-Sensor Chip," *IEEE Electr. Dev. Lett.*, EDL-7, 254-256, (1986).

(9)   E.S. Kim and R.S. Muller, "IC-Processed Piezoelectric Microphone," *submitted for publication, March, 1987.*

(10) T. Shiosaki, T. Yamamoto, A. Kawabata, R.S. Muller, and R.M. White, "Fabrication and Characterization of ZnO Piezoelectric Films for Sensor Devices," *IEEE Int. Electr. Dev. Mtg,* 151-154, Washington DC, Dec 3-5, 1979.

(11) G. Heiland and H. Ibach, "Pyroelectricity in Zinc Oxide," *Solid State Commun. 4, 353* (1966).

(12) Landolt-Bornstein Tables Vol. 15, edited by O. Madelung, H. Schulz, and H. Weiss, Springer-Verlag, Berlin, (1980).

(13) P.L. Chen, R.S. Muller, R.M. White, and R. Jolly, "Thin-Film ZnO-MOS Transducer with Virtually DC Response," *Proc. IEEE Ultrasonics Symp.*, Boston, MA, (1980).

(14) M. Royer, J.O. Holmen, M.A. Wurm, and O.S. Aadland, "ZnO on Si Integrated Acoustic Sensor," *Sensors and Actuators, 4, 357-363*, (1983).

(15) J.G. Smits, H.A.C. Tilmans, and T.S.J. Lammerink, "Pressure Dependence of Resonant Diaphragm Pressure Sensor," *Transducers '85, IEEE-ECS Int. Conf. on Solid-State Sensors and Actuators,* 93-96, Philadelphia, PA, (1985).

(16) M. Sekimoto, H. Yoshihara, and T. Ohkubo, "Silicon Nitride Single-Layer X-Ray Mask," *J. Vac. Science and Tech., 21,* 1017-1021, (1982).

(Invited)

Integrated Sensors:
Key to Future VLSI Systems

K. D. WISE

Solid-State Electronics Laboratory
Department of Electrical Engineering and
Computer Science, University of Michigan
Ann Arbor, Michigan 48109

Integrated solid-state sensors are rapidly emerging as an important key to future progress in the microelectronics industry. Coupling electronics to the non-electronic world is probably the most challenging task associated with many segments of future VLSI, and sensor research is currently playing an increasingly vital role in improving the performance of existing VLSI systems, opening new market areas, automating the fabrication of VLSI circuits, and exploring areas which may suggest future breakthroughs for microelectronics. Recent progress in integrated sensors in highlighted in this paper by considering research on ultrasensitivie solid-state pressure sensors, thin-dielectric structures for thermal sensing, and a microprobe with on-chip signal processing for use in studies of information processing in neural systems. The microprobe combined precision micromachining, custom integrated interface electronics, and new packaging techniques in ways that are also applicable to a much broader array of emerging solid-state sensors.

§1. Introduction

The rapid and sustained progress made in the area of microelectronics over the past three decades has had a revolutionary impact on many aspects of society. Driven by continuing advances in solid-state process technology, the number of transistors which can be successfully integrated on a single chip of silicon has been increasing at the rate of two orders of magnitude per decade, with corresponding decreases in functional cost. Microcomputers are now pervasive in most instrumentation and control applications, and very-large-scale-integrated (VLSI) systems on a chip are sparking fundamental changes in the implementation of other complex systems, including many associated with defense, consumer products, and telecommunications.

The continuing progress in microelectronics has been possible, in part, because of the continuous opening of new markets for electronic products. The increased market volumes have helped to absorb the increasingly high cost of fabrication equipment and VLSI chip design; however, many of the future markets for VLSI systems are dependent on not only low-cost control, but on interfacing that control to a non-electronic host. Data acquisition and sensing (and also actuation) are already more difficult problems than control/signal-processing for many systems. Sensors are already more expensive than many microcomputers and may be more limiting in terms of system performance. In a very real sense, sensors have not kept pace with the rapid developments in the control/signal-processing area and now represent the weak link in many systems. Integrated sensors are thus the focus for increasingly intense efforts worldwide, and substantial progress is being made. As these efforts continue, sensor development can be expected to exert substantial leverage on the broader microelectronic scence in at least four ways: 1) improved sensors are required to obtain improved performance in many existing product areas, since they are currently the limiting elements; 2) sensors hold the key to the successful entry of VLSI into a variety of new market areas, including those in health care, consumer products, and automated manufacturing; 3) sensors are a vital part of successfully automating the VLSI chip fabrication process itself and thus are important in VLSI manufacturing productivity; and 4) sensors hold the key to better understanding of biological/cellular systems, which may contribute to long-term fundamental improvements in the electronic art.

This paper will illustrate recent progress in

the world of integrated sensors[1] and their roles in the evolution of VLSI itself. Such devices consist of four distinct and important components: custom thin films for transduction, selectivity, or structural support[2]; silicon microstructures; precision interface electronics; and microcomputer-based signal processing techniques. Visible imaging arrays[3] currently represent the most highly developed examples of integrated sensors. These devices have evolved rapidly because of clear market applications, relatively standard packaging, and compatibility with conventional IC processes. However, even here, custom thin films have had to be developed for color selectivity and other functions[4,5], and the understanding of optical effects in the dielectric and semiconducting films over the active device areas has been the subject of considerable work. Area imagers exceeding 1.2 M pixels per chip and compatible with high-definition television have recently been reported[6] and probably represent the largest chips produced by the semiconductor industry.

Since the sensor area is very broad, this paper will concentrate on three areas of high current interest. First, progress in solid-state pressure sensors for application in process automation and robotics is discussed. Second, recently-developed thin-dielectric structures for thermal-based sensing of gas flow and infrared radiation are explored. Finally, work on an implantable multielectrode recording array for use in studies of information processing in neural structures is described. This last chip involves silicon micromachining, thin-films, and custom interface electronics in ways which are typical of many evolving sensors.

§2. Silicon Capacitive Pressure Sensors

One of the earliest of the silicon sensors was the piezoresistive pressure sensor, which is probably the most developed of the non-imaging devices now commercially available. Growing from important beginnings in the 1950s[7] and early 60s,[8,9] this device is relatively well understood in both full-bridge[10,11] and shear-based[12] designs. Fully integrated sensors having on-chip circuitry for signal amplification and other conditioning have been reported by several manufacturers.[13,14] These devices are linear in their pressure response but require

compensation for offset and slope variations, both with pressure and with temperature.

Capacitive pressure sensors[15-17] are more nonlinear in their pressure response than their piezoresistive counterparts, but they are also significantly more sensitive to pressure and less sensitive to temperature. Uncompensated temperature coefficients are of the order of 50 ppm/°C (equivalent to about 3 Pa/°C). The basic structures are now the subject of substantial development efforts as shown in Fig. 1. Bulk-diaphragm structures realize a pressure-variable capacitance between the silicon diaphragm (formed by selective anisotropic etching) and a selectively-metallized glass plate. The glass is electrostatically-bonded to the silicon and is thermally-matched to it. For diaphragms between 0.5 μm and 20 μm in thickness, the diaphragm is conveniently formed using a boron etch-stop. Since the zero-pressure capacitance values are very small (typically 1–5 pF), on-chip readout electronics is required. A hermetic seal between the device and the outer world and satisfac-

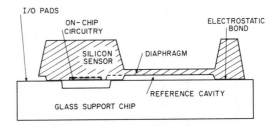

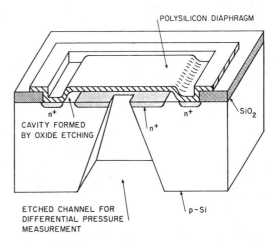

Fig. 1. Silicon Capacitive Pressure Sensing Structures. (Above) Bulk-diaphragm structure with on-chip circuitry; (Below) Deposited-diaphragm structure using an undercut (sacrificial) oxide layer.

tory lead transfers between the transducer and the onchip electronics are the major challenges. For deposited-diaphragm structures, the diaphragm material (usually polysilicon or silicon nitride) is deposited over a sacrificial oxide layer which is subsequently removed to form a plate which is moveable in response to differential pressure. Stress in the diaphragm material limits the plate size which can be formed and dominates the variability in offset and sensitivity. These structures are still evolving but have been the basis for several interesting devices.[18-20]

Figure 2 shows the substantial range of pressure (and applications) for capacitive silicon pressure sensors. With the current interest in evolving these structures, it is important to understand the scaling behavior of these devices both to very high and very low pressures. There is substantial interest in devices capable of operation at pressures of >10^6 Pa in order to couple electronic control to hydraulic systems. There is also considerable interest in sensors designed for low-cost pressure and flow measurements, particularly in low-pressure processing ambients.

Until recently, there had been relatively few efforts aimed at understanding the scaling behavior of silicon pressure sensors. Fundamental limits for ultrasensitive devices are set by Brownian noise[21] and by circuit noise sources.[22] Brownian noise is equivalent to about 10^{-4} Pa, while circuit noise corresponds to about 10^{-2} Pa. Thus, using deposited or bulk diaphragm structures, it should be possi-

ble to define pressure sensors capable of resolving pressures of about 0.01 Pa (0.1 μmHg). Such silicon devices are more sensitive than any yet reported and are promising for applications in both pressure and mass flow control. For their realization, however, the intrinsic stress in the thin diaphragm must be accurately controlled. Boron-doped silicon diaphragms have sufficient internal tensile stress that their pressure sensitivity is substantially reduced when compared with lightly-doped diaphragms. Methods for accurate stress control in silicon films have not yet been reported, and dielectrics such as silicon nitrode may offer an attractive alternative.[23]

In some applications, arrays of pressure sensors are required. One such application of current interest is that of tactile imaging arrays for use in robotics. Figure 3 shows the structure and electrical schematic of a silicon tactile imager[24] based on a bulk-diaphragm capacitive pressure cell. This structure utilizes a dielectric layer on the silicon diaphragm to provide electrical isolation between horizontal row lines recessed in the silicon and vertical column lines selectively metallized on the glass. Thus, the overall structure forms a miniature capacitive keyboard in which all dimensions can be controlled to whthin one micron or better. The pressure/force sensitivity of the structure can be scaled over several orders of magnitude by varying the thickness of the silicon diaphragm. The overlying pad acts as a simple force transmitter, so that hysteresis/compression effects there do not significantly affect imager performance. For an 8 × 8-element array having a spatial resolution of 2 mm, the uncompensated accuracy and resolution are equivalent to over 8 bits for a temperature span of 50°C. Such arrays address the high-performance end of tactile imaging requirement and may also be useful for a variety of other applications where multi-point measurements of pressure/force are required.

§3. Thin-Film Structures for Thermal Sensing

Thermal effects have been used for many years as the basis for non-integrated transducers. Devices in this area have included sensors for flow, gas composition, and in-

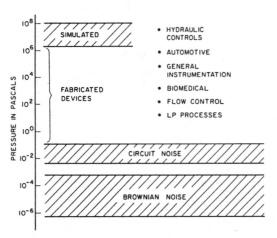

Fig. 2.   Pressure Range for Silicon Capacitive Pressure Sensors along with Application Areas.

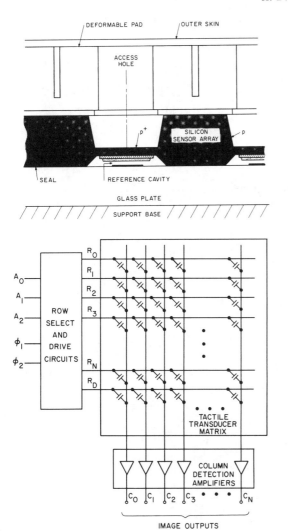

Fig. 3. Structure and Electrical Schematic of a Silicon Tactile Imager Based on a Capacitive Cell. The electronics can be off-chip, simplifying the fabrication of the device.

frared radiation. Recently, substantial interest has been created by the use of thin-film structures on silicon[25] to provide thermal isolation between on-chip transducers (which sometimes much run at elevated temperatures) and integrated electronics for signal readout.

Figure 4 shows the cross-section of a thermopile-based infrared imager developed recently.[26] A thin dielectric window is formed using multiple layers of silicon dioxide and silicon nitride, balanced so as to produce a composite film in modest tension. As a final step in the process, this window is formed by anisotropically removing the silicon over a por-

tion of the chip. The window supports a series-connected array of thermocouples (a thermopile) whose hot junctions are on the window and whose cold junctions are over the chip rim. Incident radiation over a broad spectral band ($<0.3\,\mu m$ to $>30\,\mu m$) warms the window, and the thermopile converts the temperature change into an output voltage.

Figure 5 shows a top view of a 32-element thermal imager developed using this structure. The imager contains on-chip multiplexing for signal access as well as circuitry for the independent measurement of the Seebeck coefficient and the ambient chip temperature. The device employs two staggered 16 element linear arrays to achieve an effective vertical spatial resolution of $300\,\mu m$. The time constant is less than 10 mSec with a responsivity of 60–100 V/W, a window size of 0.4 mm × 0.8 mm × 1.3 μm, and a dynamic range spanning nearly seven orders of magnitude. Such windows can be realized with high yield, operate over an ambinent temperature range of several hundred degrees, and are promising for a variety of thermographic and spectroscopic applications in industrial process control. The thin-dielectric technology is also applicable to a wide variety of other important sensing structures, including integrated gas flowmeters[27-29] and pressure sensors.[30]

§4. A Multichannel Microprobe for Neural Recording

The combination of silicon micromachining, thin sensing films, and custom readout electronics is well illustrated in a sensor recently developed for the simultaneous recording of spatially separated activity from multiple neurons in the nervous system. Such probes are required for exploring the circuit/system-level organization of the nervous system and for the development of closed-loop neural prostheses. Understanding the techniques used in neural signal processing and pattern recognition is particularly important since it may spark new architectures for future computers realized using VLSI technology. At the cellular level, improved sensing structures may also lead to the development of biological and molecular devices to extend electronics beyond the current solid-state.

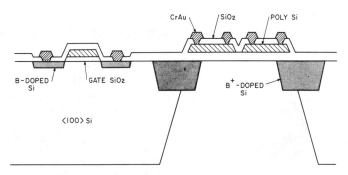

Fig. 4. Cross-Section of a Silicon Thermal Imager. A thin dielectric window thermally isolates the hot junctions of a thermopile detector from the chip rim, which remains at ambient temperature and contains readout electronics.

Fig. 5. Top View of a Portion of a 32-Element Thermal Imager. Each window measures 0.4 mm × 0.8 mm × 1.3 μm and contains 40 series-connected thermacouples. The total chip size is 5.5 mm × 11 mm.

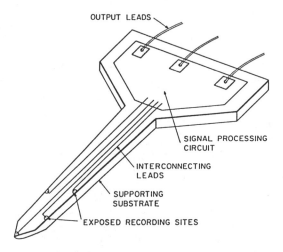

Fig. 6. A Multichannel Multiplexed Recording Array for Studies of Information Processing in Neural Structures.

Figure 6 shows the probe structure.[31,32] A micromachined silicon substrate supports an array of thin-film conductors which are insulated above and below by deposited films of silicon nitride and silicon dioxide. Inlayed gold recording sites are formed over the ends of the conductors by openings in the upper dielectrics and are connected to signal processing electronics at the rear of the structure.

Fabrication of the probe is based on the use of a deep-diffused boron layer[32] to define the probe substrate. After the completion of wafer processing, an ethylene diamine-pyrocatechol etch is used to dissolve the wafer. None of the probe materials are attacked so that following dissolution, only the probes remain. The process has a yield exceeding 80 percent. Passive probes (without electronics) require only four masks and single-sided processing on wafers of normal thickness.

Since the recording sites are very small (typically 50 μm²), the recording channel impedances are high (about 10 Mohms at 1 KHz). For a probe capable of long-term implantation in the cortex, on-chip signal processing is required to amplify and buffer the low-amplitude (250 μV) neural signals and multiplex them to the outside world. This approach reduces the number of output leads and the tethering they represent and also eases the encapsulation requirements on the output leads. On-chip circuit compatibility is achieved by restricting the deep boron diffusion used to define the probe shank to only the perimeter of the rear part of the substrate, leaving an area of lightly-doped silicon in which to fabricate the circuitry. Although the boron diffusion requires 12–15 hours at 1175°C, the circuit area can be completely masked using silicon nitride or silicon dioxide. Following circuit processing, the silicon lifetime, inferred from junction leakage current, has been measured at about 0.2 μSec; MOS threshold values are also undisturbed by the deep diffusion.

Figure 7 shows a block diagram of the on-chip electronics for use on a ten-channel probe.[33] Neural signals are amplified and presented to an analog multiplexer, which is driven by a two-phase dynamic shift register and on-chip clock. The resulting analog time-multiplexed (sampled) signal is fed off-chip via a broadband output buffer. The per-channel voltage gain is 100 with a bandwidth of 100 Hz–6 KHz. A framing pulse is inserted as an eleventh channel and is stripped off externally to regenerate the sample clock, allowing demultiplexing of the recorded signals. Since the input DC voltage drift from the recording site is many times the amplitude of the neural signals, the DC level at the input to the preamplifiers is stabilized by using the leakage current of the input protection devices near zero bias to polarize the recording surface. This technique reduces a 50 mV variation in the recording site potential to less than 0.1 mV at the preamplifier input.

The probe can be latched into a self-test mode by momentarily pulsing the power supply from 5 V to 8 V. A 50 mV 1 KHz test signal is capacitively coupled to each of the electrodes, where it is amplified and transmitted to the outside world as a measure of the electrode impedance. This impedance is a reasonable measure of the recording ability of an electrode in tissue. Thus, test capability is realized for both electrode impedance and the on-chip electronics.

Inorganic films of silicon nitride and silicon dioxide have been adequate for the encapsulation of the recording sites, while the polyimide PI-2555 has been satisfactory over the circuitry. The output leads present the greatest packaging challenge since they flex as the cortex moves with respect to the skull. Parylene-C is at present the most promising candidate for the output leads, and the high output drive capability of the on-chip electronics definitely

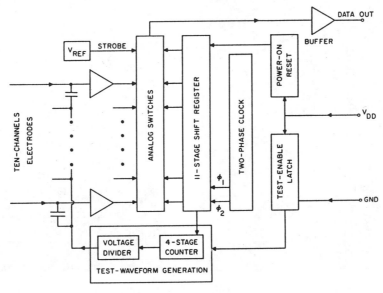

Fig. 7. Block Diagram of the Implantable Electronics for use on a Ten-Channel Recording Array.

helps to reduce the packaging requirements in this area.

Figure 8 shows two views of a ten-channel probe with on-chip MOS circuitry. The chip dissipates 5 mW from a single 5 V supply, requires only three leads (power, ground, and data), and has an active circuit area of 1.3 mm^2 in 6 μm E/D NMOS technology. The probe can be realized with high yield and is 15 μm thick and 4.7 mm long overall. The shank tapers from a width of about 160 μm at its base to less than 20 μm at the tip. Considerably smaller shanks are possible using reduced feature sizes for the conductors.

§5. Future Sensing Approaches

Integrated sensors offer the prospect of considerable leverage on the larger microelectronics industry and are an important key in extending it into new areas. Improvements in sensing structures, process technology, circuit techniques, and system organization in this evolution will all be important. In the area of custom thin films for transduction, substantial progress can be expected over the next few years, particularly through the use of techniques such as molecular-beam epitaxy to tailor the films for certain parameter sensitivities.[34] These devices will likely employ silicon or gallium arsenide substrates together with mixed electrical and perhaps optical signal processing to realize high-performance selective sensing arrays for a wide variety of applications.

Acknowledgements

Many individuals contributed substantially to the work reported here. In particular, the author would like to thank K. J. Chun for his work on the tactile imager, H. L. Chau for research on pressure sensor scaling limits, I. H. Choi for the development of the thermal imager, and K. Najafi for his work on the multichannel microprobe. H. Masuda and M. Aoki of the Hitachi Central Research Laboratories and T. Mochizuki of the Toshiba Semiconductor Device Engineering Laboratory also contributed to the microprobe development. The financial support provided by the US Air Force Office of Scientific Research, the Semiconductor Research Corporation, and the National In-

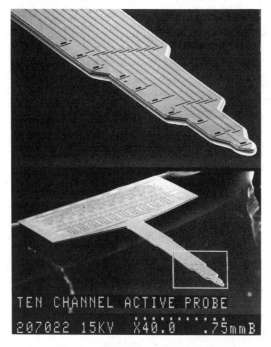

Fig. 8. SEM Photomicrographs of a Ten-Channel Implantable Microprobe. (a) View of the probe tip (above) and the entire chip. The recording sites are spaced 100 μm apart. (b) Close-up of the rear section of the prove, containing the readout circuitry.

stitutes of Health for these efforts is also gratefully acknowledged.

References

1) K. D. Wise: Integrated Silicon Sensors: Interfacing Electronics to a Non-Electronic World, *Sensors and Actuators* 3, pp. 229–237, July 1982.

2) K. D. Wise: The Role of Thin Fims in Integrated Solid-State Sensors, *Journ. Vac. Sci. Tech.* May 1986.

3) T. J. Tredwell: High-Density Solid-State Image Sensors, *Digest IEEE Int. Conf. on Solid-State Sensors and Actuators* pp. 424–429, June 1985.

4) N. Koike, T. Nakano, A. Sasano, Y. Taniguchi, I. Takemoto and T. Fujita: Heat- and Light-Resistance Characteristics of an MOS Imaging Device with Monolithically Integrated Color Filters, *IEEE Trans. Electron Devices* 32, pp. 1475–1479, August 1985.

5) Y. Ishihara and K. Tanigaki: A High Photosensitivity IL-CCD Image Sensor with Monolithic Resin Lens Array, *Digest IEEE Int. Electron Dev. Meeting* pp. 497–500, December 1983.

6) I. Akiyama, T. Tanaka, E. Oda, T. Kamata, K. Masubuchi, K. Arai and Y. Ishihara: A 1280 × 980 Pixel CCD Image Sensor, *Digest IEEE Int. Solid-State Circ. Conf.* pp. 96–97, February 1986.

7) C. S. Smith: Piezoresistance Effect in Germanium and Silicon, *Phys. Rev.* 94, pp. 42–49, April 1954.

8) O. N. Tufte, P. W. Chapman and D. Long: Silicon Diffused-Element Piezoresistive Diaphragms, *J. Appl. Phys.* 33, pp. 3322–3327, 1962.

9) O. N. Tufte and E. L. Stelzer: Piezoresistive Properties of Silicon Diffused Layers, *J. Appl. Phys.* 34, pp. 313–318, 1963.

10) Y. Kanda: A Graphical Representation of the Piezoresistance Coefficients in Silicon, *IEEE Trans. electron Devices* pp. 64–70, January 1982. See also, Y. Kanda: Optimum Design Considerations for Silicon Pressure Sensors using a Four-Terminal Gauge, *Sensors and Actuators* pp. 199–206, October 1983.

11) Samaun, K. D. Wise and J. B. Angell: An IC Piezoresistive Pressure Sensor for Biomedical instrumentation, *IEEE Trans. Biomed. Engr.* 20, pp. 101–109, March 1973.

12) J. E. Gragg, W. C. McCulley, W. B. Newton and C. E. Derrington: Compensation and Calibration of a Monolithic Four-Terminal Silicon Pressure Transducer, *Diqest IEEE Solid-State Sensor Conf.* pp. 21–27, June 1984.

13) K. Yamada, M. Nishihara, R. Kanzawa and R. Kobayashi: A Piezoresistive Integrated Pressure Sensor, *Sensors and Actuators* pp. 63–70, September 1983.

14) S. Sugiyama, M. Takigawa and I. Igarashi: Integrated Piezoresistive Pressure Sensor with Both Voltage and Frequency Output, *Sensors and Actuators* pp. 113–120, September 1983.

15) Y. S. Lee and K. D. Wise: A Batch-Fabricated Silicon Capacitive Pressure Transducer with Low Temperature Sensitivity, *IEEE Trans. Electron Devices* 29, pp. 42–48, January 1982.

16) W. H. Ko, M. H. Bao and Y. D. Hong: A High-Sensitivity Integrated-Circuit Capacitive Pressure Transducer, *IEEE Trans. Electron Devices* 29, pp. 48–56, January 1982.

17) S. K. Clark and K. D. Wise: Pressure Sensitivity in Anisotropically-Etched Thin-Diaphragm Pressure Sensors, *IEEE Trans. Electron Devices* 26, pp. 1887–1896, December 1979.

18) R. S. Hijab and R. S. Muller: Micromechanical Thin-Film Cavity Structures for Low-Pressure and Acoustic Transducer Applications, *Digest IEEE Int. Conf. on Solid-State Sensors and Actuators* pp. 178–181, June 1985.

19) R. Howe and R. S. Muller: Integrated Resonant-Microbridge Vapor Sensor, *Digest Int. Electron Devices Meeting* pp. 213–216, December 1984.

20) Y. C. Tai, R. S. Muller and R. T. Howe: Polysilicon Bridges for Anemometer Applications, *Digest IEEE Int. Conf. on Solid-State Sensors and Actuators* pp. 354–357, June 1985.

21) H. L. Chau and K. D. Wise: Noise Due to Brownian Motion in Ultrasensitive Solid-State Pressure Sensors, Submitted to *IEEE Trans. Electron Devices*, in Review.

22) H. L. Chau and K. D. Wise: Scaling Limits in Batch-Fabricated Silicon Pressure Sensors, *Digest IEEE Int. Conf. on Solid-State Sensors and Actuators* pp. 174–177, June 1985.

23) H. Guckel and D. W. Burns: A Technology for Integrated Transducers, *Digest IEEE Int. Conf. on Solid-State Sensors* pp. 90–92, June 1985.

24) K. J. Chun and K. D. Wise: A High-Performance Tactile Imager Based on a Capacitive Cell, *IEEE Trans. Electron Devices* 32, pp. 1196–1201, July 1985.

25) G. R. Lahiji and K. D. Wise: A Batch-Fabricated Silicon Thermopile Infrared Detector, *IEEE Trans. Electron Devices* 29, pp. 14–22, January 1982.

26) I. H. Choi and K. D. Wise: A Silicon Thermopile-Based Infrared Sensing Array for Use in Automated Manufacturing, *IEEE Trans. Electron Devices* 33, pp. 72–79, January 1986.

27) R. G. Johnson, R. E. Higashi, P. J. Bohrer and R. W. Gehman: Design and Packaging of a Highly-Sensitive Microtransducer for Air Flow and Differential Pressure Sensing Applications, *digest IEEE Int. Conf. on Solid-State Sensors and Actuators* pp. 358–360, June 1985.

28) K. Petersen, J. Brown and W. Renken: High-Precision High-Performance Mass-Flow Sensor with Integrated Laminar Flow Channels, *Digest IEEE Int. Conf. on Solid-State Sensrs and Actuators* pp. 361–363, June 1985.

29) O. Tabata: Fast-Response Silicon Flow Sensor with an On-chip Fluid Temperature Sensing Element, *IEEE Trans. Electron Devices* 33, pp. 361–365, March 1986.

30) A. W. van Herwaarden, P. M. Sarro and H. C. Meijer: Integrated Vacuum Sensor, *Sensors and Actuators* pp. 187–196, November 1985.

31) S. L. BeMent, K. D. Wise, D. J. Anderson, K. Najafi and K. L. Drake: Solid-State Electrodes for Multichannel Multiplexed Intracortical Neuronal Recording, *IEEE Trans. Biomed. Engr.* **33**, pp. 230–241, February 1986.

32) K. Najafi, K. D. Wise and T. Mochizuki: A High-Yield IC-Compatible Multichannel Recording Array, *IEEE Trans. Electron Devices* **32**, pp. 1206–1211, July 1985.

33) K. Najafi and K. D. Wise: An Implantable Multielectrode Array with On-Chip Signal Processing, *Digest 1986 IEEE Int. Solid-State Circ. conf.* pp. 98–99, February 1986.

34) K. Takahashi: Tailor-Made Sensing Materials by MBE/-MOCVD Technology, *Digest IEEE Int. Conf. on Solid-State Sensors and Actuators* pp. 274–277, June 1985.

FRONTIERS IN SOLID STATE BIOMEDICAL TRANSDUCERS

Wen H. Ko

Electrical Engineering & Applied Physics Dept., and
Electronics Design Center
Case Western Reserve University
Cleveland, Ohio 44106

ABSTRACT

The explosive advances recently made in microelectronics, microcomputers and instrumentation techniques have created a vast opportunity and demand for biomedical measurement and control systems. The lack of reliable biomedical transducers is one of the bottlenecks in realizing the potential for better biomedical research as well as health care. The need for transducers in closed loop prostheses, artificial organs, diagnosis, care of the ill in both hospitals and at home, is discussed. A list of measurements for various physical, chemical and psychological parameters and the transducers needed to make these measurements is given.

Solid state electronics and technology important to transducer research is outlined with examples of sensors and actuators being developed at Case Western Reserve University. Future directions and concepts for new sensors and actuators are discussed.

INTRODUCTION

The explosive advances in VLSI, microcomputers, signal analysis and instrumentation techniques during the last decade has created a tremendous opportunity and demand for biomedical measurement and control systems. The lack of suitable transducers that meet biomedical requirements is the most serious bottleneck in realizing the potential for better biomedical research as well as health care. Biomedical transducers not only should have the sensitivity, selectivity and time stability of industrial devices, but often require minimization of size, weight and power consumption. They also must be packaged so that they are biocompatible to body tissues while the device is protected from corrosive body fluids. A great deal of work has been devoted to the development of input sensors and output actuators for use in industrial control, robotics and automation, consumer appliances, transportation, safety monitoring, measurement instruments, bioscience and health care, especially those fabricated with solid state electronic technology. [1] This article attempts to review the developing biomedical instrumentation and control systems as well as the transducers they are waiting for, to summarize the state of the art solid state transducers that may meet biomedical application needs, and to discuss, subjectively, the future directions in biomedical transducer research. A brief outline of the presentation is given below.

FRONTIERS OF BIOMEDICAL INSTRUMENTATION AND THE NEED FOR TRANSDUCERS

VLSI components and microprocessors make possible many new reliable, high performance, biomedical instrumentation systems to carry out sophisticated diagnostic and control functions. The general trends are: (1) Intelligent systems having logic, computing and analysis functions built-in in addition to data acquisition. (2) Complex systems in which many related parameters are measured and correlated to suggest and project the patient's status. (3) Continuous on-line monitoring of the critically ill. (4) Noninvasive or less invasive monitoring. (5) Patient care outside of the hospital. (6) Closed loop control of prostheses and therapeutic systems.

Many interesting biomedical systems are being studied that will have a strong impact on our society. Most of them need new or better transducers to make them practical for the general patient population.[2] A summary list of closed loop control biomedical systems is given in Table I. By electrical stimulation of nerve or muscle, many disabled patients can regain some of the lost body functions. It has been demonstrated that "functional stimulation" can be used to: (1) recover the use of a paralyzed hand, (2) allow patients in wheelchairs to stand up and walk within a controlled environment, (3) control epilepsy, and (4) control heart rate according to the body's needs. It also shows great potential in sensory aids for the deaf and blind and control of urinary incontinence. When proper sensors are available, it is probable that an artificial pancreas for diabetes and an implant system to control hypertension can be realized.

Using computer controlled walking [3] as an example, 32 or more electrodes are implanted into various leg muscles that activate the walking action. They have to be controlled to contract and generate coordinated forces at the proper time during a walking cycle. In order to control walking with a dedicated computer one needs the sensors for: (1) angle positions of the ankle, knee, hip (2) upper body posture, (3) pressure distribution of both feet in contact with the

Reprinted from *Rec. of the IEEE Int. Electron Devices Meeting*, 1985, pp. 112–115.

ground, (4) position or acceleration of limb sections, as well as output sensors to let the patient know some of this information to indicate how the paralyzed leg is doing in the walking cycle. For long-term use of the system, these sensors and part of the systems will have to be implanted. Therefore, the packaging, the communication between sensors and central control unit have to be developed.

Another example is the on-line monitoring of blood gas in critically ill patients. pH, PO_2, PCO_2 and oxygen saturation sensors that can be assembled on the catheter tip to monitor blood gas continuously at the bedside would be a very valuable diagnostic instrument. The ion-sensitive field effect transistor [4,5] and other thin film or thick film blood gas sensors [6,7] are being developed. Both the time stability and packaging of these sensors remain bottlenecks which need to be resolved.

Many instruments for diagnosis and biomedical research that require physical and chemical sensors are being developed. The on-line measurement of enzymes, protein and parameters at the cellular level are important as the basis of treatment as well as furthering the knowledge of life sciences.

TRANSDUCER TECHNOLOGY AND MICROMACHINING

In order to design and develop new transducers, new technology, in addition to solid state electronics and integrated circuits, will be needed. The field of micromachining, where silicon or other materials can be machined or fabricated in micrometer dimensions, has been growing for the last decade. [8,9] This technology is particularly important for transducers and is being developed by transducer research groups. The major micromachining functions are: [1]

Precision Etching and Etch Stop
Wet etching, using chemical and electrochemical techniques, has been developed for silicon, metal and other materials. Dry etching techniques using plasma, ion beam and spark erosion have been reported and can be used to machine three-dimensional structures from a substrate, either semiconductor or others. Silicon diaphragms thinner than 1 micron with an area greater than 1 $(mm)^2$, micro-cavities, beams, bridges in micron scales, etc. have been fabricated in laboratories.

Bonding
Three-dimensional structures for transducers can also be fabricated by bonding different layers together. For example, silicon or GaAs can be electrostatically bonded to pyrex glass (#7740) to form a hermetically sealed unit with a joint flatness under a micron. Other bonding techniques use sputtered glass, both Corning 7740 and low melting temperature glasses, as a sealing layer to bond silicon to silicon substrates or silicon to metal. Metal compounds have also been used as a brazing material to seal silicon to metal, ceramics, etc.

Selective Deposition
Many techniques have been developed that can selectively deposit layers of conductive, semiconductive and insulating materials of various properties on the substrate. These include: evaporation, sputtering, ion beam sputtering, plasma and chemical vapor deposition (CVD). Single crystal silicon, polysilicon, SiO_2, Si_3N_4, Al_2O_3 and organic films have been deposited on silicon and other substrates to form three-dimensional microstructures. Single chip multiple sensors can also be fabricated with these techniques.

Feedthrough, Holes and Packaging
In transducer design there is need for insulated electrical connections between bonded layers which require holes and micro-chambers fabricated on the substrate and particularly require a method to package those transducers for implant or indwelling applications where the devices have to be in communication with the biological system. At the same time, leads and signal processing circuits have to be protected from the corrosive and highly conductive fluids in the body. The literature reported includes the following methods [1]:
1. Diffused Al column with thermal gradient to form conductive paths across a silicon wafer.
2. Anisotropic etched back contact such that lead connection to outside is not exposed to sensor environment
3. Plasma etched holes for feedthrough
4. Laser drilled holes
5. Spark erosion holes
6. Other techniques such as ion beam milling, centrifugal etching, etc.

BIOMEDICAL TRANSDUCERS

The development of transducers for biomedical applications evolved from discrete large devices to miniaturized devices to solid state transducers and is progressing to integrated sensors [1- F] The active areas of research include transducers for: (1) implant sensing and stimulating electrodes and micro-electrode arrays for cell measurement; (2) pressure, force and flow measurements of blood, body fluids and air; (3) sound and ultrasound arrays; (4) linear and angular position, displacement and acceleration of body parts; (5) other physical parameters such as touch, vibration and volume; (6) chemical ionic concentrations in body fluids; (7) enzyme protein and molecular activities in the body; (8) gaseous composition for anesthesia and in the respiratory system; (9) measurements to determine behavioral condition; and (10) control devices to alter body function or to supplement abnormal body function such as are used in prostheses and regulatory devices used in animal research. A few selected examples will be described to illustrate the evolution and existing problems.

Electrodes
Electrodes are used to sense voltage, current and impedance and to deliver stimulating currents to nerves or muscles to activate the desired body response. The main problems to be overcome in

the development of sensing electrodes are the electrode-tissue interface stability, the corrosion and body reaction to electrodes, and the noise pick-up. Metal electrodes, especially the Ag-AgCl electrode, have been used extensively in health care. Microelectrodes with tips around 1μ diameter are used for cellular level sensing. They are traditionally made of glass tubes filled with electrolytes or metals such as tungsten. Silicon technology has been used to fabricate microelectrodes and arrays of electrodes for neurophysiological studies. Followers and amplifiers were integrated with the electrode tip to provide high input impedance and low output impedance such that the environmental noise problem can be avoided. An insulated electrode was developed that has a layer of stable oxide insulation covering the metal or other conductor surface of this electrode. The insulated electrode using Ta_2O_3 or TiO_2 Si_2N_3 surface has achieved good stability and can be used as both a sensing and stimulating device. In order to alleviate the electrode-electrolyte interface problem, a scheme of microelectrode arrays is being investigated where a micropocket is formed surrounding the metal electrode and then filled with a high conductivity electrolyte. The pocket has an opening to interface with body electrolytes and the opening serves as the effective electrode surface. Electronics can be integrated on the substrate of these microelectrode arrays to improve the signal quality. Figure 1 illustrates the evolution of the electrodes.

For stimulation, metal electrodes are used; the leads are coiled to provide flexibility and avoid fatigue. The stimulating charge per pulse per unit area of the Pt electrode has to be limited to 30μ coulomb/cm^2--pulse, to avoid surface corrosion and pH change near the electrode that might damage the tissue nearby. Insulated electrodes and pocket electrodes are being investigated as stimulating electrodes. The major problems are (1) corrosion of electrode and biocompatability, (2) the breakage of the electrode when implanted, (3) the relative motion between the electrode and the tissue. New materials and construction techniques are needed.

Pressure, Force, Flow Transducers

The basic structure of pressure, force and flow transducers is generally the same if the mechanical strain is used for the measurement. (Force = Pressure X Area. Flow = (Pressure drop/flow resistance)). Other principles such as thermal dilution, ultrasound or light Doppler effect, have been used for the measurement of flow. Figure 2 illustrates the evolution of pressure sensors -- from conventional piezoresistive or piezoelectric pressure sensors to silicon substrate capacitive sensors, to integrated sensors with analog and digital output that is compatible with computer circuits [1-F]

Position, Velocity and Acceleration Sensors

Both linear and angular position, velocity and acceleration sensors are needed for closed loop prosthetic systems. Although these sensors were developed for industrial uses [1], biomedical applications have special requirements due to the body structure and further research in this area is required. As an example, consider the knee angle measurement for computer controlled walking. Because the knee is not a simple two linkage joint, it is difficult to obtain relative angular motion on the knee or at any implant position. Two gravitational sensors were proposed to relate the orientation of the upper and lower leg to the direction of gravity, thus obtaining a differential angle. However, the frequency response (10 hertz) and the interference of linear and angular acceleration of the leg made it inoperative. However, when the knee is bent, the surface on one side of the knee joint stretches. The stretch may be an indication of the angle. A capacitive stretch and angle sensor was designed and evaluated with favorable results.

In the chemical sensor area, interesting developments are:

Blood Gas and Electrolyte Concentration Sensors

Ion Sensitive Field Effect Transistors, ISFET, use the exposed gate insulator to the electrolyte to measure ionic concentration in body electrolytes. pH, K^+, Na^+, Ca^{++}, Ma^{++}, Cl^-, F^-, etc. can be measured with various modified insulators or by adding a membrane to the gate insulator [5]. Thin and thick film sensors for PO_2, PCO_2 and other gas and ionic species in solution are also being developed using silicon substrates and photolithography. Multiple sensors and reference electrodes can be integrated on the chip with the exposed area on the order of 10 μ^2, thus alleviating many problems due to flow and protein deposition [6]. Fiber optics have been used for oxygen saturation measurement in blood. Improvement in signal analysis and sensor design should lead to the development of an on-line catheter-type monitoring device. [9]

Gas Sensors [1-D,E]

H_2 gas can be monitored with Pd or Pt gate MOSFET. The hydrogen permeates through the metal gate into the metal insulator interface producing a change in the threshold voltage of the FET. At 150°C, 10 ppm of H_2 can be detected. Other reducing and oxidizing gases can be monitored by the impedance of SnO_2, ZnO, TiO_2 and other oxide surfaces. The combustive gas monitor used in gas heated homes and various laboratories already has a sizable commercial market. Humidity can be measured by the dew point and the change of capacitance and resistance of an interdigitated electrode structure. These oxides can be deposited on silicon substrates and then modified with various catalytic materials deposited on the surface to sense various gases with improved sensitivity and selectivity.

Enzyme, Protein and Molecular Sensors [1-E,F]

By enclosing, trapping or absorbing enzymes on substrate materials on the surface of chemical sensors, the product of the enzymatic reaction can be sensed as a means of detecting the presence of enzymes or the substrate. Similar techniques can be used for body protein detection and the measurement of other molecular substances in biomedical research and health care. The possibility for on-line determination of these proteins will greatly improve health care and advance our understanding of living systems. The stability and the packaging are the major problems.

FUTURE TRENDS

The future focus will probably be in:
1. New materials and principles for transducer
2. Integrated sensors with active circuits on the chip
3. Multiple sensors or sensor array integrated with signal processing functions to obtain higher performance than the single device
4. Micropower and computer compatible
5. New design and fabrication Techniques.

Transducer development requires a multidisciplinary collaboration and team effort. Important steps have been initiated and their promising results generate many speculations. In order to realize the potentials and to meet future needs , more concentrated efforts will be required.

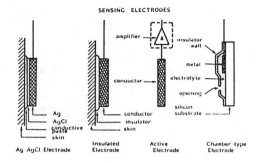

Figure 1

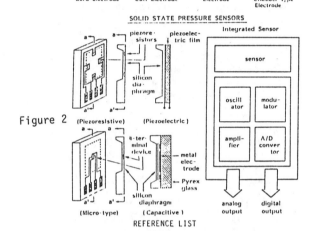

Figure 2

TABLE 1. CLOSED LOOP CONTROL MEDICAL SYSTEMS

A. THERAPEUTIC SYSTEMS

DISEASE	TREATMENT	SENSORS REQUIRED
CARDIAC ARRHYTHMIA	PHYSIOLOGIC PACEMAKERS	PRESSURE, FLOW pH, PO_2, PCO_2
PARAPLEGIA and QUADRIPLEGIA	CONTROL of LIMBS with NERVE BYPASS SYSTEM	ELECTRODES, POSITION, TOUCH, and ANGLE
SENSORY AIDS	IMPLANTABLE HEARING and VISUAL AIDS	SENSORS, MICROPHONE and STIMULATORS
RESPIRATORY APNEA	DIAPHRAGM or NERVE PACING	PRESSURE, FLOW, PCO_2 and STIMULATOR
URINARY INCONTINENCE	BLADDER CONTROL SYSTEM	PRESSURE and CONTROL VALVE
HYPERTENSION	AUTOMATIC BLOOD PRESSURE CONTROL	BLOOD PRESSURE, STIMULATOR
EPILEPSY	CEREBELLAR STIMULATION	EEG SENSORS and STIMULATORS
HYDROCEPHALUS	ICP and VENTRICULAR SIZE CONTROL	PRESSURE, VOLUME, CONTROL VALVES
DIABETES	AUTOMATIC INSULIN INFUSION PUMP	GLUCOSE SENSOR and CONTROLLED PUMP
PSYCHOLOGICAL SYSTEMS	DRUG RELEASE SYSTEMS and ELECTRICAL STIMULATION	VARIOUS SENSORS

B. ARTIFICIAL ORGANS

ORGANS	HARDWARE SYSTEM	SENSORS REQUIRED
ARTIFICIAL HEART	BLOOD PUMP and CONTROL	PRESSURE, FLOW PO_2 and PCO_2
ARTIFICIAL PANCREAS	INSULIN PERFUSION SYSTEM	FLOW, PRESSURE and GLUCOSE SENSOR
ARTIFICIAL KIDNEY	DIALYSIS MACHINE	URIC ACID, FLOW and PRESSURE
ARTIFICIAL LARYNX	BODY CONTROLLED OSCILLATORS	EMG SENSORS and ACTUATORS

C. DRUG DELIVERY SYSTEMS

DISEASE	TREATMENT	SENSORS REQUIRED
PAIN, CANCER AND OTHERS	DRUG PUMP	DRUG LEVEL SENSOR and ACTUATORS

REFERENCE LIST

(1) A. Sensors and Actuators, (ed.S.Middelhoek) 1981-
    B. IEEE Trans.Ed-26,12;1979 and ED-29,1;1982
    C. Proc.Workshops on Biomedical Sensors,Case Institute of Tech. CWR University,Cleveland,Ohio 44106
    D. Proc. 1st and 2nd Sensor Symposium,IEE Soc. of Japan,1981,82.
    E. Proc. Int. Meeting on Chem. Sensors,Japan,1983(ed.S.Seiyama)
    F. Tech.Digest 1985 Int.Solid State Sensors and Actuators Conf. Philadelphia,USA.(EDC ,CWR University,Cleveland OH 44106)
(2) W.Ko,"Solid State Physical Transducers for Biomedical Research," IEEE Trans.BME-32, Dec. 1985
(3) E.B. Marsolais,et.al."Functional Walking in Paralyzed Patients by means of Electrical Stimulation," Clin. Ortho. and Rel.Res. 175:pp30-36,1983.
(4) P. Bergveld,"Development,Operation and Application of the Ion Sensitive Field Effect Transistor as a Tool for Electrophysiology," IEEE Trans.BME-19:342,1972
(5) P.Cheung,W.Ko,C.Fung,A.Wong,"Theory,Fabrication,Testing and Clinical Response ofISFET," In Theory ,Design and Biomedical Applicationsof Solid State Chemical Sensors(ed.P.Cheung et al.) Boca Raton, Florida,CRC Press,1978,pp 91-118.
(6) V.Karagounis,L.Lun and C.C.Liu,"A Thick Film Multiple Component Cathode Three-electrode Oxygen Sensor," IEEE Trans.BME-32,1985
(7) O.Prohaska,"New Developments in Miniaturized Electrochemical Sensors," Tech. Digest, 1985 Int.Conf. on S.S.Sensors and Actuators, June 1985,pp 402-405.
(8) K. Petersen,"Siliczon as a Mechanical Material,"Proc.IEEE 70: 420,May ,1979.
(9) Workshop on Micromachining and Micropackageng of Transducers" Nov.1984,EDC,CWR University. Proc. to be published by Elsevier Scientific Publishing Co. Nov. 1985.

ACKNOWLEDGEMENTS
This work is partially supported by NIH grants RR-80057,RR-02024 and NS-19174

Part 2
Fabrication Technology and Material Properties

NEW, promising microsensor structures have been created by employing materials and microfabrication technologies developed for the integrated-circuits industry. The reprints collected in this section provide a sampling of these plus some materials and methods specific to microsensors. Because it is impossible to present a comprehensive review of all processes and materials, we have selected here articles of a more fundamental nature with special applicability to microsensors. There are five subsections; materials, silicon micromachining, surface micromachining, integration with electronic-fabrication processes, and packaging.

Silicon and silicon compounds are commonly employed in the fabrication of microsensors. Most of the electronic properties and fabrication technologies of these materials are known because of their extensive use in integrated circuits. Their mechanical, optical, and chemical properties have been under more recent study as a result of applications to microsensors. The first paper on materials is a review of silicon, especially its unique mechanical and chemical properties, by Petersen. In the second paper, Senturia underscores the need for basic materials properties in the design of sensors. These papers are followed by two papers on polysilicon thin films. Obermeier, Kopystynski, and Nießl review the properties and applications of this electronic, and more recently, mechanical material. Guckel, Burns, Rutigliano, Showers, and Uglow demonstrate how polysilicon microstructure relates to its physical properties. Zinc oxide, a piezo- and pyroelectric thin film used for microsensors, is reviewed by Muller. The last article in this subsection introduces porous silicon—a new form of this material to be added to the arsenal of physical and chemical modifications that have been used in microsensor structures.

A second group of articles is a representative sampling of the broad range of etchant systems and techniques used presently in micromachining. Micromachining refers to a sequence of deposition and etching processes that produce three-dimensional microstructures in, or on the surface of, an otherwise planar substrate. Examples include cavities, grooves, and via holes, and movable parts such as diaphragms and cantilever beams. The method by which 3-D structures are formed *in* a silicon substrate is referred to as silicon micromachining, which often makes use of anisotropic etchants. These etchants exhibit etch-rate anisotropy with respect to crystallographic orientation. The properties of two of the most commonly used anisotropic etchants, potassium hydroxide (KOH) and ethylene–diamine–pyrocatechol (EDP), are discussed in an article by Seidel. A paper by Kloeck, Collins, de Rooij, and Smith discusses the electrochemical etch-stop, which is important for the precise control of microstructural dimensions.

In a third subsection, the processes by which 3-D structures are formed in thin films *on top of* a substrate, collectively referred to as surface micromachining, are highlighted. Papers describe techniques for releasing both fixed and movable structures, the sealing of cavities, and electroforming and molding. Both inorganic and organic micromachined structures are represented.

The integration of microsensors with microelectronics is challenged by the constraints of the planar process and the very tight control on material properties required to produce functioning electronic devices. Degraded performance or complete failure can result from the addition of high-temperature steps, the deposition of intermediary or additional thin-film layers, or the introduction of new, possibly contaminating materials to an established integrated-circuit process. Careful consideration of the interaction of process steps and materials in the design of a process flow is required for what is commonly referred to as IC compatibility. Articles in this subsection present microsensor designs and processes for the successful combination of the materials and methods described in many of the preceding articles with electronic devices and circuitry. The first article, by Takahashi and Matsuo, describes the combination of silicon micromachining and IC fabrication to produce a needle-shaped probe with at-site signal processing. Several articles follow that combine mechanical and/or chemical three-dimensional structures and planar electronic devices. A concluding paper, by Polla, Muller, and White, presents a family of microsensor structures that utilizes zinc-oxide thin films and integrated-electronic devices.

The packaging of microsensors is complicated by the simultaneous requirements of sensor interaction with the environment and the isolation of active electronics and input–output (I/O) connections. Corrosion protection for I/O is of particular concern for sensing in wet environments. The overall size of the packaged part is often a constraint for biomedical sensors. System–design considerations as they pertain to packaging are discussed in the first article of this subsection by Senturia and Smith. This is followed by three articles, each describing different packaging technologies. An application of micromachining to chemical-microsensor packaging is described by Smith and Collins. Ko, Suminto, and Yeh review bonding and sealing techniques, including anodic bonding, eutectic bonding, and low-temperature glass sealing. Petersen, Barth, Poydock, Brown, Mallon, and Bryzek present an application of the relatively new technique of silicon thermal bonding to the packaging of silicon pressure transducers.

This collection of papers was chosen to familiarize the reader with current materials and technologies for microsensor design and fabrication. The designs and methods presented are meant to be instructive, not exclusive, and to assist the reader by stimulating new ideas.

Silicon as a Mechanical Material

KURT E. PETERSEN, MEMBER, IEEE

Abstract—Single-crystal silicon is being increasingly employed in a variety of new commercial products not because of its well-established electronic properties, but rather because of its excellent mechanical properties. In addition, recent trends in the engineering literature indicate a growing interest in the use of silicon as a mechanical material with the ultimate goal of developing a broad range of inexpensive, batch-fabricated, high-performance sensors and transducers which are easily interfaced with the rapidly proliferating microprocessor. This review describes the advantages of employing silicon as a mechanical material, the relevant mechanical characteristics of silicon, and the processing techniques which are specific to micromechanical structures. Finally, the potentials of this new technology are illustrated by numerous detailed examples from the literature. It is clear that silicon will continue to be aggressively exploited in a wide variety of mechanical applications complementary to its traditional role as an electronic material. Furthermore, these multidisciplinary uses of silicon will significantly alter the way we think about all types of miniature mechanical devices and components.

I. INTRODUCTION

IN THE SAME WAY that silicon has already revolutionized the way we think about electronics, this versatile material is now in the process of altering conventional perceptions of miniature mechanical devices and components [1]. At least eight firms now manufacture and/or market silicon-based pressure transducers [2] (first manufactured commercially over 10 years ago), some with active devices or entire circuits integrated on the same silicon chip and some rated up to 10 000 psi. Texas Instruments has been marketing a thermal point head [3] in several computer terminal and plotter products in which the active printing element abrasively contacting the paper is a silicon integrated circuit chip. The crucial detector component of a high-bandwidth frequency synthesizer sold by Hewlett-Packard is a silicon chip [4] from which cantilever beams have been etched to provide thermally isolated regions for the diode detectors. High-precision alignment and coupling assemblies for fiber-optic communications systems are produced by Western Electric from anisotropically etched silicon chips simply because this is the only technique capable of the high accuracies required. Within IBM, ink jet nozzle arrays and charge plate assemblies etched into silicon wafers [5] have been demonstrated, again because of the high precision capabilities of silicon IC technology. These examples of silicon micromechanics are not laboratory curiosities. Most are well-established, commercial developments conceived within about the last 10 years.

The basis of micromechanics is that silicon, in conjunction with its conventional role as an electronic material, and taking advantage of an already advanced microfabrication technology, can also be exploited as a high-precision high-strength high-reliability mechanical material, especially applicable wherever miniaturized mechanical devices and components must be integrated or interfaced with electronics such as the examples given above.

The continuing development of silicon micromechanical applications is only one aspect of the current technical drive toward miniaturization which is being pursued over a wide front in many diverse engineering disciplines. Certainly silicon microelectronics continues to be the most obvious success in the ongoing pursuit of miniaturization. Four factors have played crucial roles in this phenomenal success story: 1) the active material, silicon, is abundant, inexpensive, and can now be produced and processed controllably to unparalleled standards of purity and perfection; 2) silicon processing itself is based on very thin deposited films which are highly amenable to miniaturization; 3) definition and reproduction of the device shapes and patterns are performed using photographic techniques which have also, historically, been capable of high precision and amenable to miniaturization; finally, and most important of all from a commercial and practical point of view, 4) silicon microelectronic circuits are batch-fabricated. The unit of production for integrated circuits—the wafer—is not *one* individual saleable item, but contains hundreds of identical chips. If this were not the case, we could certainly never afford to install microprocessors in watches or microwave ovens.

It is becoming clear that these same four factors which have been responsible for the rise of the silicon microelectronics industry can be exploited in the design and manufacture of a wide spectrum of miniature mechanical devices and components. The high purity and crystalline perfection of available silicon is expected to optimize the *mechanical* properties of devices made from silicon in the same way that *electronic* properties have been optimized to increase the performance, reliability, and reproducibility of device characteristics. Thin-film and photolithographic fabrication procedures make it possible to realize a great variety of extremely small, high-precision mechanical structures using the same processes that have been developed for electronic circuits. High-volume batch-fabrication techniques can be utilized in the manufacture of complex, miniaturized mechanical components which may not be possible by any other methods. And, finally, new concepts in hybrid device design and broad new areas of application, such as integrated sensors [6], [7] and silicon heads (for printing and data storage), are now feasible as a result of the unique and intimate integration of mechanical and electronic devices which is readily accomplished with the fabrication methods we will be discussing here.

While the applications are diverse, with significant potential impact in several areas, the broad multidisciplinary aspects of silicon micromechanics also cause problems. On the one hand, the materials, processes, and fabrication technologies are all taken from the semiconductor industry. On the other hand, the applications are primarily in the areas of mechanical en-

Manuscript received December 2, 1981; revised March 11, 1982. The submission of this paper was encouraged after the review of an advance proposal.

The author was with IBM Research Laboratory, San Jose, CA 95193. He is now with Transensory Devices, Fremont, CA 94539.

Reprinted from *Proc. IEEE*, vol. 70, no. 5, pp. 420–457, May 1982.

TABLE I

	Yield Strength (10^{10} dyne/cm^2)	Knoop Hardness (kg/mm^2)	Young's Modulus (10^{12} dyne/cm^2)	Density (gr/cm^3)	Thermal Conductivity (W/cm°C)	Thermal Expansion (10^{-6}/°C)
•Diamond	53	7000	10.35	3.5	20	1.0
•SiC	21	2480	7.0	3.2	3.5	3.3
•TiC	20	2470	4.97	4.9	3.3	6.4
•Al$_2$O$_3$	15.4	2100	5.3	4.0	0.5	5.4
•Si$_3$N$_4$	14	3486	3.85	3.1	0.19	0.8
•Iron	12.6	400	1.96	7.8	0.803	12
SiO$_2$ (fibers)	8.4	820	0.73	2.5	0.014	0.55
•Si	7.0	850	1.9	2.3	1.57	2.33
Steel (max. strength)	4.2	1500	2.1	7.9	0.97	12
W	4.0	485	4.1	19.3	1.78	4.5
Stainless Steel	2.1	660	2.0	7.9	0.329	17.3
Mo	2.1	275	3.43	10.3	1.38	5.0
Al	0.17	130	0.70	2.7	2.36	25

•Single crystal. See Refs. 8, 9, 10, 11, 141, 163, 166.

gineering and design. Although these two technical fields are now widely divergent with limited opportunities for communication and technical interaction, widespread, practical exploitation of the new micromechanics technology in the coming years will necessitate an intimate collaboration between workers in *both* mechanical *and* integrated circuit engineering disciplines. The purpose of this paper, then, is to expand the lines of communication by reviewing the area of silicon micromechanics and exposing a large spectrum of the electrical engineering community to its capabilities.

In the following section, some of the relevant mechanical aspects of silicon will be discussed and compared to other more typical mechanical engineering materials. Section III describes the major "micromachining" techniques which have been developed to form the silicon "chips" into a wide variety of mechanical structures with IC-compatible processes amenable to conventional batch-fabrication. The next four sections comprise an extensive list of both commercial and experimental devices which rely crucially on the ability to construct miniature, high-precision, high-reliability, *mechanical* structures on silicon. This list was compiled with the primary purpose of illustrating the wide range of demonstrated applications. Finally, a discussion of present and future trends will wrap things up in Section VIII. The underlying message is that silicon micromechanics is not a diverging, unrelated, or independent extension of silicon microelectronics, but rather a natural, inevitable continuation of the trend toward more complex, varied, and useful integration of devices on silicon.

II. MECHANICAL CHARACTERISTICS OF SILICON

Any consideration of mechanical devices made from silicon must certainly take into account the mechanical behavior and properties of single-crystal silicon (SCS). Table I presents a comparative list of its mechanical characteristics. Although SCS is a brittle material, yielding catastrophically (not unlike most oxide-based glasses) rather than deforming plastically (like most metals), it certainly is not as fragile as is often believed. The Young's modulus of silicon (1.9×10^{12} dyne/cm^2 or 27×10^6 psi) [8], for example, has a value approaching that of stainless steel, nickel, and well above that of quartz and most other borosilicate, soda-lime, and lead-alkali silicate glasses [9]. The Knoop hardness of silicon (850) is close to quartz, just below chromium (935), and almost twice as high as nickel (557), iron, and most common glasses (530) [10]. Silicon single crystals have a tensile yield strength (6.9×10^{10}

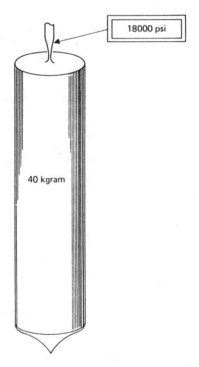

18000 psi

40 kgram

Fig. 1. Stresses encountered commonly in silicon single crystals are very high during the growth of large boules. Seed crystals, typically 0.20 cm in diameter and supporting 40-kg boules, experience stresses over 1.25×10^8 Pa or about 18 000 psi in tension.

dyne/cm^2 or 10^6 psi) which is at least 3 times higher than stainless-steel wire [8], [11]. In practice, tensile stresses *routinely* encountered in seed crystals during the growth of large SCS boules, for example, can be over 18 000 psi (40-kg boule hanging from a 2-mm-diameter seed crystal, as illustrated in Fig. 1). The primary difference is that silicon will yield by fracturing (at room temperature) while metals usually yield by deforming inelastically.

Despite this quantitative evidence, we might have trouble intuitively justifying the conclusion that silicon is a strong mechanical material when compared with everyday laboratory and manufacturing experience. Wafers do break—sometimes without apparent provocation; silicon wafers and parts of wafers may also easily chip. These occurrences are due to several factors which have contributed to the misconception that silicon is mechanically fragile. First, single-crystal silicon is normally obtained in large (5–13-cm-diameter) wafers, typically only 10–20 mils (250 to 500 μm) thick. Even stainless steel of these dimensions is very easy to deform inelastically. Silicon chips with dimensions on the order of 0.6 cm × 0.6 cm, on the other hand, are relatively rugged under normal handling conditions unless scribed. Second, as a single-crystal material, silicon has a tendency to cleave along crystallographic planes, especially if edge, surface, or bulk imperfections cause stresses to concentrate and orient along cleavage planes. Slip lines and other flaws at the edges of wafers, in fact, are usually responsible for wafer breakage. In recent years, however, the semiconductor industry has attacked this yield problem by contouring the edges of wafers and by regularly using wafer edge inspection instruments, specifically designed to detect mechanical damage on wafer edges and also to assure that edges are properly contoured to avoid the effects of stress concentration. As a result of these quality control improvements, wafer breakage has been greatly reduced and the intrinsic strength of silicon is closer to being realized in practice during wafer handling. Third, chipping is also a potential problem with brittle materials such as SCS. On whole wafers, chipping occurs for the same qualitative reasons as breaking and the solutions are identical. Individual die, however, are subject to chipping as a result of saw- or scribe-induced edge damage and defects. In extreme cases, or during rough handling, such damage can also cause breakage of or cracks in individual die. Finally, the high-temperature processing and multiple thin-film depositions commonly encountered in the fabrication of IC devices unavoidably result in internal stresses which, when coupled with edge, surface, or bulk imperfections, can cause concentrated stresses and eventual fracture along cleavage planes.

These factors make it clear that although high-quality SCS is intrinsically strong, the apparent strength of a particular mechanical component or device will depend on its crystallographic orientation and geometry, the number and size of surface, edge, and bulk imperfections, and the stresses induced and accumulated during growth, polishing, and subsequent processing. When these considerations have been properly accounted for, we can hope to obtain mechanical components with strengths exceeding that of the highest strength alloy steels.

General rules to be observed in this regard, which will be restated and emphasized in the following sections, can be formulated as follows:

1) The silicon material should have the lowest possible bulk, surface, and edge crystallographic defect density to minimize potential regions of stress concentration.

2) Components which might be subjected to severe friction, abrasion, or stress should be as small as possible to minimize the total number of crystallographic defects in the mechanical structure. Those devices which are never significantly stressed or worn could be quite large; even then, however, thin silicon wafers should be mechanically supported by some technique—such as anodic bonding to glass—to suppress the shock effects encountered in normal handling and transport.

3) All mechanical processing such as sawing, grinding, scribing, and polishing should be minimized or eliminated. These operations cause edge and surface imperfections which could result in the chipping of edges, and/or internal strains subsequently leading to breakage. Many micromechanical components should preferably be separated from the wafer, for example, by etching rather than by cutting.

4) If conventional sawing, grinding, or other mechanical operations are necessary, the affected surfaces and edges should be etched afterwards to remove the highly damaged regions.

5) Since many of the structures presented below employ anisotropic etching, it often happens that sharp edges and corners are formed. These features can also cause accumulation and concentration of stress damage in certain geometries. The structure may require a subsequent isotropic etch or other smoothing methods to round such corners.

6) Tough, hard, corrosion-resistant, thin-film coatings such as CVD SiC [12] or Si_3N_4 should be applied to prevent direct mechanical contact to the silicon itself, especially in applications involving high stress and/or abrasion.

7) Low-temperature processing techniques such as high-pressure and plasma-assisted oxide growth and CVD depositions, while developed primarily for VLSI fabrication, will be just as important in applications of silicon micromechanics. High-temperature cycling invariably results in high stresses within the wafer due to the differing thermal coefficients of expansion of the various doped and deposited layers. Low-temperature processing will alleviate these thermal mismatch stresses which otherwise might lead to breakage or chipping under severe mechanical conditions.

As suggested by 6) above, many of the structural or mechanical disadvantages of SCS can be alleviated by the deposition of passivating thin films. This aspect of micromechanics imparts a great versatility to the technology. Sputtered quartz, for example, is utilized routinely by industry to passivate IC chips against airborne impurities and mild atmospheric corrosion effects. Recent advances in the CVD deposition (high-temperature pyrolytic and low-temperature RF-enhanced) of SiC [12] have produced thin films of extreme hardness, essentially zero porosity, very high chemical corrosion resistance, and superior wear resistance. Similar films are already used, for example, to protect pump and valve parts for handling corrosive liquids. As seen in Table I, Si_3N_4, an insulator which is routinely employed in IC structures, has a hardness second only to diamond and is sometimes even employed as a high-speed, rolling-contact bearing material [13], [14]. Thin films of silicon nitride will also find important uses in silicon micromechanical applications.

On the other end of the thin-film passivation spectrum, the gas-condensation technique marketed by Union Carbide for depositing the polymer parylene has been shown to produce virtually pinhole-free, low-porosity, passivating films in a high polymer form which has exceptional point, edge, and hole coverage capability [15]. Parylene has been used, for example, to coat and passivate implantable biomedical sensors and electronic instrumentation. Other techniques have been developed for the deposition of polyimide films which are already used routinely within the semiconductor industry [16] and which also exhibit superior passivating characteristics.

One excellent example of the unique qualities of silicon in the realization of high-reliability mechanical components can be found in the analysis of mechanical fatigue in SCS structures. Since the initiation of fatigue cracks occurs almost exclusively at the *surfaces* of stressed members, the rate of fatigue depends strongly on surface preparation, morphology, and defect density. In particular, structural components with highly polished surfaces have higher fatigue strengths than those with rough surface finishes as shown in Fig. 2 [17]. Passivated surfaces of polycrystalline metal alloys (to prevent intergrain diffusion of H_2O) exhibit higher fatigue strengths than unpassivated surfaces, and, for the same reasons, high water vapor content in the atmosphere during fatigue testing will significantly decrease fatigue strength. The mechanism of fatigue, as these effects illustrate, are ultimately dependent on a surface-defect-initiation process. In polycrystalline ma-

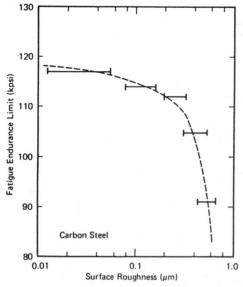

Fig. 2. Generally, mechanical qualities such as fatigue and yield strength improve dramatically with surface roughness and defect density. In the case of silicon, it is well known that the electronic and mechanical perfection of SCS surfaces has been an indispensable part of integrated circuit technology. Adapted from Van Vlack [17].

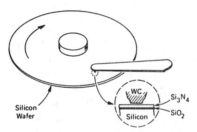

Fig. 3. A rotating MNOS disk storage device demonstrated by Iwamura et al. [21]. The tungsten–carbide probe is in direct contact with the nitride-coated silicon wafer as the wafer rotates at 3600 r/min. Signals have been recorded and played back on such a system at video rates. Wear of the WC probe was a more serious problem than wear of the silicon disk.

terials, these surface defects can be inclusions, grain boundaries, or surface irregularities which concentrate local stresses. It is clear that the high crystalline perfection of SCS together with the extreme smoothness and surface perfection attainable by chemical etching of silicon should yield mechanical structures with intrinsically high fatigue strengths [18]. Even greater strengths of brittle materials can be expected with additional surface treatments [9]. Since hydrostatic pressure has been shown to increase fatigue strengths [19], any film which places the silicon surface under compression should decrease the initiation probability of fatigue cracks. Si_3N_4 films, for example, tend to be under tension [20] and therefore impart a compressive stress on the underlying silicon surface. Such films may be employed to increase the fatigue strength of SCS mechanical components. In addition, the smoothness, uniformity, and high yield strength of these thin-film amorphous materials should enhance overall component reliability.

A new rotating disk storage technology which has recently been demonstrated by Iwamura et al. [21] not only illustrates some of the unique advantages derived from the use of silicon as a mechanical material but also indicates how well silicon, combined with wear-resistant Si_3N_4 films, can perform in demanding mechanical applications. As indicated in Fig. 3,

data storage was accomplished by an MNOS charge-storage process in which a tungsten carbide probe is placed in direct contact with a 3-in-diameter silicon wafer, rotating at 3600 r/min. The wafer is coated with 2-nm SiO_2 and 49-nm Si_3N_4, while the carbide probe serves as the top metal electrode. Positive voltage pulses applied to the metal probe as the silicon passes beneath will cause electrons to tunnel through the thin SiO_2 and become trapped in the Si_3N_4 layer. The trapped charge can be detected as a change in capacitance through the same metal probe, thereby allowing the signal to be read. Iwamura et al. wrote and read back video signals with this device over 10^6 times with little signal degradation, at data densities as high as 2×10^6 bits/cm^2. The key problems encountered during this experiment were associated with wear of the tungsten carbide probe, not of the silicon substrate or the thin nitride layer itself. Sharply pointed probes, after scraping over the Si_3N_4 surface for a short time, were worn down to a 10-μm by 10-μm area, thereby increasing the active recording surface per bit and decreasing the achievable bit density. After extended operation, the probe continued to wear while a barely resolvable 1-nm roughness was generated in the hard silicon nitride film. Potential storage densities of 10^9 bits/cm^2 were projected if appropriate recording probes were available. Contrary to initial impressions, the rapidly rotating, harshly abraided silicon disk is not a major source of problems even in such a severely demanding mechanical application.

III. MICROMECHANICAL PROCESSING TECHNIQUES

Etching

Even though new techniques—and novel applications of old techniques—are continually being developed for use in micromechanical structures, the most powerful and versatile processing tool continues to be etching. Chemical etchants for silicon are numerous. They can be isotropic or anisotropic, dopant dependent or not, and have varying degrees of selectivity to silicon, which determines the appropriate masking material(s). Table II gives a brief summary of the characteristics of a number of common wet silicon etches. We will not discuss plasma, reactive-ion, or sputter etching here, although these techniques may also have a substantial impact on future silicon micromechanical devices.

Three etchant systems are of particular interest due to their versatility: ethylene diamine, pyrocatechol, and water (EDP) [22]; KOH and water [23]; and HF, HNO_3, and acetic acid CH_3OOH (HNA) [24], [25]. EDP has three properties which make it indispensable for micromachining: 1) it is anisotropic, making it possible to realize unique geometries not otherwise feasible; 2) it is highly selective and can be masked by a variety of materials, e.g., SiO_2, Si_3N_4, Cr, and Au; 3) it is dopant dependent, exhibiting near zero etch rates on silicon which has been highly doped with boron [26], [27].

KOH and water is also orientation dependent and, in fact, exhibits much higher (110)-to-(111) etch rate ratios than EDP. For this reason, it is especially useful for groove etching on (110) wafers since the large differential etch ratio permits deep, high aspect ratio grooves with minimal undercutting of the masks. A disadvantage of KOH is that SiO_2 is etched at a rate which precludes its use as a mask in many applications. In structures requiring long etching times, Si_3N_4 is the preferred masking material for KOH.

HNA is a very complex etch system with highly variable etch rates and etching characteristics dependent on the silicon dopant concentration [28], the mix ratios of the three etch

TABLE II

Etchant (Diluent)	Typical Compositions	Temp °C	Etch Rate (μm/min)	Anisotropic (100)/(111) Etch Rate Ratio	Dopant Dependence	Masking Films (etch rate of mask)	References
HF HNO$_3$ (water, CH$_3$COOH)	10 ml 30 ml 80 ml	22	0.7-3.0	1:1	≤10^{17}cm^{-3} n or p reduces etch rate by about 150	SiO$_2$ (300Å/min)	24,25,28,30
	25 ml 50 ml 25 ml	22	40	1:1	no dependence	Si$_3$N$_4$	
	9 ml 75 ml 30 ml	22	7.0	1:1	-----	SiO$_2$ (700Å/min)	
Ethylene diamine Pyrocatechol (water)	750 ml 120 gr 100 ml	115	0.75	35:1	≥7×10^{19} cm^{-3} boron reduces etch rate by about 50	SiO$_2$ (2Å/min) Si$_3$N$_4$ (1Å/min) Au,Cr,Ag,Cu,Ta	20,26,27,35, 43,44
	750 ml 120 gr 240 ml	115	1.25	35:1			
KOH (water, isopropyl)	44 gr 100 ml	85	1.4	400:1	≥10^{20} cm^{-3} boron reduces etch rate by about 20	Si$_3$N$_4$ SiO$_2$ (14Å/min)	23,32,33,36, 37,38,42
	50 gr 100 ml	50	1.0	400:1			
H$_2$N$_4$ (water, isopropyl)	100 ml 100 ml	100	2.0	----	no dependence	SiO$_2$ Al	40,41
NaOH (water)	10 gr 100 ml	65	0.25-1.0	----	≥3×10^{20} cm^{-3} boron reduces etch rate by about 10	Si$_3$N$_4$ SiO$_2$ (7Å/min)	34

components, and even the degree of etchant agitation, as shown in Fig. 4 and Table II. Unfortunately, these mixtures can be difficult to mask, since SiO$_2$ is etched somewhat for all mix ratios. Although SiO$_2$ can be used for relatively short etching times and Si$_3$N$_4$ or Au can be used for longer times, the masking characteristics are not as desirable as EDP in micromechanical structures where very deep patterns (and therefore highly resistant masks) are required.

As described in detail by several authors, SCS etching takes place in four basic steps [30], [31]: 1) injection of holes into the semiconductor to raise the silicon to a higher oxidation state Si$^+$, 2) the attachment of hydroxyl groups OH$^-$ to the positively charged Si, 3) the reaction of the hydrated silicon with the complexing agent in the solution, and 4) the dissolution of the reacted products into the etchant solution. This process implies that any etching solution must provide a source of holes as well as hydroxyl groups, and must also contain a complexing agent whose reacted species is soluble in the etchant solution. In the HNA system, both the holes and the hydroxyl groups are effectively supplied by the strong oxidizing agent HNO$_3$, while the flourine from the HF forms the soluble species H$_2$SiF$_6$. The overall reaction is autocatalytic since the HNO$_3$ plus trace impurities of HNO$_2$ combine to form additional HNO$_2$ molecules.

$$HNO_2 + HNO_3 + H_2O \rightarrow 2HNO_2 + 2OH^- + 2h^+.$$

This reaction also generates holes needed to raise the oxidation state of the silicon as well as the additional OH$^-$ groups necessary to oxidize the silicon. In the EDP system, ethylene diamine and H$_2$O combine to generate the holes and the hydroxyl groups, while pyrocatechol forms the soluble species Si(C$_6$H$_4$O$_2$)$_3$. Mixtures of ethylene diamine and pyrocatechol

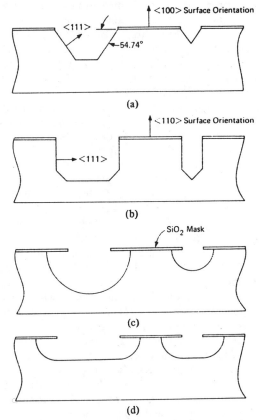

Fig. 4. A summary of wet chemically etched hole geometries which are commonly used in micromechanical devices. (a) Anisotropic etching on (100) surfaces. (b) Anisotropic etching on (110) surfaces. (c) Isotropic etching with agitation. (d) Isotropic etching without agitation. Adapted from S. Terry [29].

without water will not etch silcon. Other common silicon etchants can be analyzed in the same manner.

Since the etching process is fundamentally a charge-transfer mechanism, it is not surprising that etch rates might be dependent on dopant type and concentration. In particular, highly doped material in general might be expected to exhibit higher etch rates than lightly doped silicon simply because of the greater availability of mobile carriers. Indeed, this has been shown to occur in the HNA system (1 : 3 : 8) [28], where typical etch rates are 1–3 μm/min at p or n concentrations $>10^{18}$ cm^{-3} and essentially zero at concentrations $<10^{17}$ cm^{-3}.

Anisotropic etchants, such as EDP [26], [27] and KOH [32], on the other hand, exhibit a different preferential etching behavior which has not yet been adequately explained. Etching decreases effectively to zero in samples heavily doped with boron ($\sim10^{20}$ cm^{-3}). The atomic concentrations at these dopant levels correspond to an average separation between boron atoms of 20–25 Å, which is also near the solid solubility limit (5×10^{19} cm^{-3}) for boron *substitutionally* introduced into the silicon lattice. Silicon doped with boron is placed under tension as the smaller boron atom enters the lattice substitutionally, thereby creating a local tensile stress field. At high boron concentrations, the tensile forces became so large that it is more energetically favorable for the excess boron (above 5×10^{19} cm^{-3}) to enter interstitial sites. Presumably, the strong B–Si bond tends to bind the lattice more rigidly, increasing the energy required to remove a silicon atom high enough to stop etching altogether. Alternatively, since this etch-stop mechanism is *not* observed in the HNA system (in which the HF component can readily dissolve B$_2$O$_3$), perhaps the boron oxides and hydroxides initially generated on the silicon surface are not soluble in the KOH and EDP etchants. In this case, high enough surface concentrations of boron, converted to boron oxides and hydroxides in an intermediate chemical reaction, would passivate the surface and prevent further dissolution of the silicon. The fact that KOH is not stopped as effectively as EDP by p$^+$ regions is a further indication that this may be the case since EDP etches oxides at a much slower rate than KOH. Additional experimental work along these lines will be required to fully understand the etch-stopping behavior of boron-doped silicon.

The precise mechanisms underlying the nature of chemical anisotropic (or orientation-dependent) etches are not well understood either. The principal feature of such etching behavior in silicon is that (111) surfaces are attacked at a much slower rate than all other crystallographic planes (etch-rate ratios as high as 1000 have been reported). Since (111) silicon surfaces exhibit the highest density of atoms per square centimeter, it has been inferred that this density variation is responsible for anisotropic etching behavior. In particular, the screening action of attached H$_2$O molecules (which is more effective at high densities, i.e., on (111) surfaces) decreases the interaction of the surface with the active molecules. This screening effect has also been used to explain the slower oxidation rate of (111) silicon wafers over (100). Another factor involved in the etch-rate differential of anisotropic etches is the energy needed to remove an atom from the surface. Since (100) surface atoms each have two dangling bonds, while (111) surfaces have only one dangling bond, (111) surfaces are again expected to etch more slowly. On the other hand, the differences in bond densities and the energies required to remove surface atoms do not differ by much more than a factor of two among the various planes, so it is difficult to use

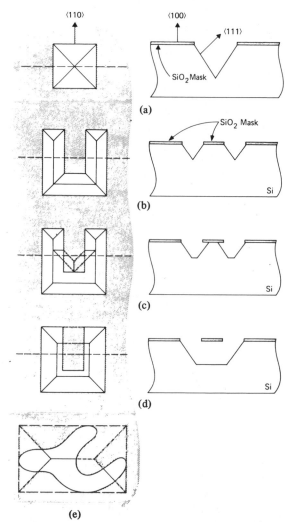

Fig. 5. (a) Typical pyramidal pit, bounded by the (111) planes, etched into (100) silicon with an anisotropic etch through a square hole in an oxide mask. (b) Type of pit which is expected from an anisotropic etch with a slow convex undercut rate. (c) The same mask pattern can result in a substantial degree of undercutting using an etchant with a fast convex undercut rate such as EDP. (d) Further etching of (c) produces a cantilever beam suspended over the pit. (e) Illustration of the general rule for anisotropic etch undercutting assuming a "sufficiently long" etching time.

these factors alone to explain etch rate differentials in the range of several hundred or more [33] which is maintained over a relatively large temperature range. This implies that some screening effects must also play a role. It seems likely that the full explanation of anisotropic etching behavior is a combination of all these factors.

Since anisotropic etching will be a particularly useful tool in the micromachining of structures described below, some detailed descriptions of the practical engineering aspects of this complex subject are deserved.

Consider a (100) oriented silicon wafer covered with SiO$_2$. A simple rectangular hole etched in the SiO$_2$ (and oriented on the surface in the (110) directions) will result in the familiar pyramidal-shaped pit shown in Fig. 5(a) when the silicon is etched with an anisotropic etchant. The pit is bounded by (111) crystallographic surfaces, which are invariably the slowest etching planes in silicon. Note that this mask pattern consists only of "concave" corners and very little undercutting of the mask will occur if it is oriented properly. Undercutting due to mask misalignment has been discussed by several workers including Kendall [33], Pugacz-Muraszkiewicz [34], and Bassous [35]. The more complicated mask geometry shown in Fig. 5(b) includes two convex corners. Convex corners, in general, will be undercut by anisotropic etches at a rate determined by the magnitude of the maximum etch rate, by the etch rate ratios for various crystallographic planes, and by the amount of local surface area being actively attacked. Since the openings in the mask can only support a certain flux of reactants, the net undercut etch rate can be reduced, for example, by using a mask with very narrow openings. On the other hand, the undercut etch rate can be increased by incorporating a vertical etch stop layer (such as a heavily boron-doped buried layer which will limit further downward etching); in this case, the reactant flux from the bottom of the etched pit is eventually reduced to near zero when the etch-stopping layer is exposed, so the total flux through the mask opening is maintained by an increased etch rate in the horizontal direction, i.e., an increased undercut rate.

In Fig. 5(b), the convex undercut etch rate is assumed to be slow, while in Fig. 5(c) it is assumed to be fast. Total etching time is also a factor, of course. Convex corners will continue to be undercut until, if the silicon is etched long enough, the pit eventually becomes pyramidal, bounded again by the slow etching (111) surfaces, with the undercut portions of the mask (a cantilever beam in this case) suspended over it, as shown in Fig. 5(d). As an obvious extension of these considerations [34], a general rule can be formulated which is shown graphically in Fig. 5(e). If the silicon is etched long enough, any arbitrarily shaped closed pattern in a suitable mask will result in a rectangular pit in the silicon, bounded by the (111) surfaces, oriented in the (110) directions, with dimensions such that the pattern is perfectly inscribed in the resulting rectangle.

As expected, different geometries are possible on other crystallographic orientations of silicon [35]–[38]. Fig. 4 illustrates several contours of etched holes observed with isotropic etchants as well as anisotropic etchants acting on various orientations of silicon. In particular, (110) oriented wafers will produce vertical etched surfaces with essentially no undercut when lines are properly aligned on the surface. Again, the (111) planes are the exposed vertical surfaces which resist the attack of the etchant. Long, deep, closely spaced grooves have been etched in (110) wafers as shown in Fig. 6(a). Even wafers not exactly oriented in the (110) direction will exhibit this effect. Fig. 6(b) shows grooves etched into a surface which is 10° off the (110) direction—the grooves are simply oriented 10° off normal [36]. Note also that the four vertical (111) planes on a (110) wafer are not oriented 90° with respect to each other, as shown in the plan view of Fig. 6(c).

Crystallographic facet definition can also be observed after etching (111) wafers, even though long times are required due to the slow etch rate of (111) surfaces. The periphery of a hole etched through a round mask, for example, is hexagonal, bounded on the bottom, obviously, by the (111) surface [39]. The six sidewall facets are defined by the other (111) surfaces; three slope inward toward the center of the hole and the other three slope outward. The six inward and outward sloping surfaces alternate as shown in Fig. 7.

Electrochemical Etching

While electrochemical etching (ECE) of silicon has been studied and basically understood for a number of years [45]–[47], practical applications of the technique have not yet been fully realized. At least part of the reason ECE is not now a popular etching procedure is due to the fact that previous

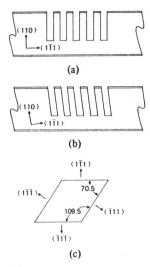

Fig. 6. Anisotropic etching of (110) wafers. (a) Closely spaced grooves on normally oriented (110) surface. (b) Closely spaced grooves on misoriented wafer. (c) These are the orientations of the (111) planes looking down on a (110) wafer.

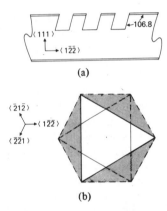

Fig. 7. Anisotropic etching of (111) silicon surfaces. (a) Wafer cross section with the steep sidewalls which would be found from grooves aligned along the (122) direction. (b) Top view of a hole etched in the (111) surface with three inward sloping and three undercut sidewalls, all (111) crystallographic planes.

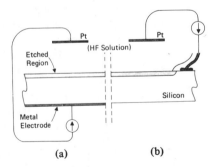

Fig. 8. Uniform electrochemical etching of wafer surfaces has been practiced in the past by making electrical contact either to the back (a) or to the front (b) of the wafer (with suitable protection for the current carrying leads). A positive voltage applied to the silicon causes an accumulation of holes at the silicon/solution interface and etching occurs. A negatively biased platinum electrode in the HF-based solution completes the circuit.

implementations of ECE offered no real advantage over the conventional, isotropic, dopant-dependent formulations discussed in the preceding section. As shown by Fig. 8(a) and (b), in typical ECE experiments electrical contact is made to the front or back of the wafer (the contacted region suitably protected from the etching solution, e.g., with wax or a special holding fixture) and the wafer is either totally immersed or is slowly lowered into the solution while a constant current flows between the positively biased silicon electrode and the negative platinum electrode. Since etching is still, principally, a matter of charge transfer, the fundamental steps are the same as discussed above. The etchants employed, however, are typically HF/H_2O solutions. Since H_2O is not as strong an oxidizing agent as HNO_3, very little silicon etching occurs (<1 Å/min) when the current flow is zero. Oxidation, then, is promoted by applying a positive voltage to the silicon which causes an accumulation of holes in the silicon at the Si/solution interface resulting in an accumulation of OH^- in the solution at the interface. Under these conditions, oxidation of the silicon surface proceeds very rapidly while the oxide is readily dissolved by the HF. Holes, which are transported to the negative platinum electrode (cathode) as H^+ ions, are released there in the form of hydrogen gas bubbles. In addition, excess hole-electron pairs can be created at the silicon surface by optical excitation, thereby increasing the etch rate.

Since the oxidation rate is controlled by current flow and optical effects, it is again clear that the etching characteristics will depend not only on dopant type and resistivity but also on the arrangement of p and n layers in the wafer interior. In particular, ECE has been employed successfully to remove heavily doped substrates (through which large currents are easily conducted) leaving behind more lightly doped epi-layer membranes (which conduct smaller currents, thereby etching more slowly) in all possible dopant configurations (p on p^+, p on n^+, n on p^+, n on n^+) [48], [49].

Localized electrochemical jet etching has been used to generate small holes or thinned regions in silicon wafers. A narrow stream of etchant is incident on one side of a wafer while a potential is applied between the wafer and the liquid stream. Extremely rapid etching occurs at the point of contact due to the thorough agitation of the solution, the continual arrival of fresh solution at the interface, and the rapid removal of reacted products.

A more useful electrochemical procedure using an anisotropic etchant has been developed by Waggener [50] for KOH and more recently by Jackson et al. [51] for EDP. Instead of relying on the electric current flowing through the solution to actively etch the silicon, a voltage bias on an n-type epitaxial layer is employed to stop the dissolution of the p-type silicon substrate at the n-type epitaxial layer. This technique has the advantage of retaining all the anisotropic etching characteristics of KOH and EDP without the need for a buried p^+ layer. Such p^+ films, while serving as simple and effective etch-stop layers, can also introduce undesirable mechanical strains in the remaining membrane which would not be present in the electrochemically stopped, uniformly doped membrane.

When ECE is performed at very low current densities, or in etchant solutions highly deficient in OH^- (such as concentrated 48-percent HF), the silicon is not fully oxidized during etching and a brownish film is formed. In early ECE work, the brownish film was etched off later in a conventional HNA slow silicon etch, or the ECE solution was modified with H_2SO_4 to minimize its formation [47]. This film has since been identified as single-crystal silicon permeated with a dense network of very fine holes or channels, from much less than 1 μm to several micrometers in diameter, preferentially oriented in the direction of current flow [52], [53]. The thickness of the layer can be anywhere from micrometers up to many mils. Porous silicon, as it is called, has a number of interesting properties. Its average density decreases with increasing applied current

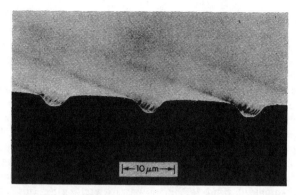

Fig. 9. SEM profile of laser-etched grooves [56]. The horizontal bar indicates 10 μm. Conditions were 100 torr Cl_2, 5.5-W multiline argon-ion laser, $f/10$ focusing, single scan at 90 μm/s. Photo courtesy of D. Ehrlich.

density to as low as 10 percent of normal silicon. Since it is so porous, gases readily diffuse into the structure so that the high-temperature oxidation, for example, of a relatively thick ($\sim$4-μm) porous silicon layer can be completed in a very short time (30 min at 1100°C) [52]. Several studies have been undertaken to determine the feasibility of using such deeply oxidized porous silicon regions as a planarizing, deep IC isolation technique [54]. The porous regions are defined by using Si_3N_4 masking films which are attacked relatively slowly by the concentrated HF ECE solution. Problems, however, encountered in the control and elimination of impurities trapped in the porous silicon "sponge-like" material, stress-related effects, and enhanced leakage currents in devices isolated by this technique have been difficult to overcome. Mechanical devices, on the other hand, may not be restricted by these disadvantages.

Besides magnifying the effective thermal oxidation rates, porous silicon can also be chemically attacked at enormously high rates. As expected, the interiors of the pores provide a very large surface area for exposure to the etchant solution. Wafers covered with 100-μm-thick porous silicon layer, for example, will actually shatter and explode when immersed in fast-etching HNA solutions.

Gradations in the porosity of the layer can be simply realized by changing the current with time. In particular, a low current density followed by a high current density will result in a high-porosity region covered with a low-porosity film. Since the porous region is still a single crystal covered with small holes (reported to be near 100 Å on the surface), it is not surprising that single-crystal epitaxial layers have been g own over porous silicon regions, as demonstrated by Unagami and Seki [55]. Once the thickness of the epi-layer corresponds to several times the diameter of the surface pores, it has been verified that the layer will be a uniform single crystal since the crystallinity of the substrate was maintained throughout, despite its permeation with fine holes.

A relatively new tool added to the growing list of micromechanical processing techniques is laser etching. Very high instantaneous etch rates have been observed when high-intensity lasers are focussed on a silicon surface in the presence of some gases. In particular, 20–30 MW/cm^2 of visible argon-ion laser radiation, scanned at rates of 90 μm/s in atmospheres of HCl and Cl_2 produced 3-μm-deep grooves [56], as shown in Fig. 9. At least part of the etching reaction occurs solely as a result of local thermal effects. It has been known for some time that silicon will be vigorously attacked by both these gases at temperatures above about 1000°C. Recent experiments in laser annealing have verified that silicon can easily be raised above the melting point at these power densities. There is still some controversy concerning the magnitude of photochemical effects, which might aid in the dissociation of the chlorine-based molecules and enhance the etch rate. In a typical reaction, for example,

$$4HCl + Si_{solid} \rightarrow 2H_2 + SiCl_4.$$

Although many applications in the area of IC fabrication have been suggested for laser etching, the fact that the laser must be scanned over the entire wafer and the etching therefore takes place "serially," net processing time per wafer will necessarily by very high in these applications. For example, a 20-W laser at a power density of 10^7 W/cm^2 etching a 1-μm layer will require over 100 h to completely scan a 4-in-diameter wafer even if etch rates of 100 μm/s are realized. Laser etching is clearly applicable only in special micromachining processing requirements such as the various contours which may be required in print-heads, recording-heads, or other miniature mechanical structures integrated with electronics on the same silicon ship. Versatile as they are, conventional, isotropic, anisotropic, electrochemical, and ion-etching processes exhibit a limited selection of etched shapes. On the other hand, the significant key advantage of laser etching is that nearly any shape or contour can be generated with laser etching in a gaseous atmosphere simply by adjusting the local exposure dose continuously over the etched region. Such a capability will be extremely useful in the realization of complex mechanical structures in silicon.

Epitaxial Processes

While the discussion up to this point has concentrated on material removal as a micromachining technique, material addition, in the form of thin film deposition or growth, metal plating, and epitaxial growth are also important structural tools. Deposited thin films have obvious applications in passivation, wear resistance, corrosion protection, fatigue strength enhancement (elaborated on in Section II), and as very thin, high-precision spacers such as those employed in hybrid surface acoustic wave amplifiers and in other thin-film devices. On the other hand, epitaxy has the important property of maintaining the highly perfect single-crystal orientation of the substrate. This means that complex vertical and/or horizontal dopant distributions (i.e., fast and slow etching regions for subsequent micromachining by etching) can be generated over many tens of micrometers without compromising the crystal structure or obviating subsequent anisotropic processes. Etch-stop layered structures are important examples and will be considered in more detail in Section VI. Fig. 10(a), however, briefly illustrates two simple configurations: hole A is a simple etch-stop hole using anisotropic etching and a p^+ boron-doped buried layer while hole B is a multilevel hole in which the epilayer and a portion of the lightly doped substrate have been anisotropically etched from the edge of the p^+ buried region. One obvious advantage of these methods is that the depth of the hole is determined solely by the thickness of the epi-layer. This thickness can be controlled very accurately and measured even before etching begins. Such depth control is crucial in many micromechanical applications we will discuss later, particularly in fiber and integrated optics.

Where the goal of IC manufacturing is to fabricate devices as small as possible (indeed, diffusions deeper than a few micrometers are very difficult and/or time-consuming), a necessary feature of most micromechanical processing techniques is the

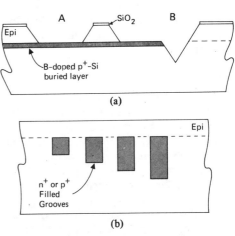

(a)

(b)

Fig. 10. (a) Since anisotropic etchants such as KOH and EDP exhibit reduced etch rates on silicon heavily doped with boron, many useful structures have been realized by growing epi over a diffused region to form a buried etch-stop layer. (b) Diagram showing how epitaxial silicon could be grown preferentially [57] in vertical-walled grooves. Doped grooves with large cross sections (>25 × 25 μm) can then be buried beneath an ordinary epi-layer.

ability to generate structures on the order of tens or even hundreds of micrometers. Both etching and epitaxial deposition possess this property. Epitaxial silicon can be grown at rates of 1 μm/min, so that layers even greater than 100 μm are readily attainable. In addition, the process parameters can be accurately controlled to allow the growth of complex three-dimensional patterns. For example, since the growth rate depends critically on temperature and gas-mixing dynamics, increased deposition rates can be observed at the *bottom* of deep, narrow, anisotropically etched grooves. In this way, Runyan *et al.* [57] (and later Smeltzer) were able to completely fill 10-μm-wide grooves (up to 100 μm deep) epitaxially with negligible silicon growth over the rest of the wafer surface. The simultaneous addition of HCl gas during the growth process is required to obtain these unusual results. Since HCl gas is an isotropic silicon etchant at these temperatures, the silicon which is epitaxially grown on the outer surface is immediately etched away in the flowing gas stream. Silicon grown in the poorly mixed atmosphere of the grooves, however, etches at a much slower rate and a net growth occurs in the groove. Heavily doped, buried regions extending over tens of micrometers are easily imagined under these circumstances as indicated in Fig. 10(b). After refilling the grooves with heavily doped silicon, the surface has been lightly etched in HCl and a lightly doped layer grown over the entire wafer. These results could not be obtained by conventional diffusion techniques. One implementation of such structures which has already been demonstrated is in the area of high-power electronic devices [58], to be discussed below in more detail. Such a process could also be used in mechanical applications to bury highly doped regions which would be selectively etched away at a later stage to form buried channels within the silicon structure.

Finally, a limited amount of work has been done on epitaxial growth through SiO$_2$ masks. Normally under these conditions, SCS will grow epitaxially on the bare, exposed crystal while polycrystalline silicon is deposited on the oxide. This mixed deposit has been used in audio-frequency distributed-filter, electronic circuits by Gerzberg and Meindl at Stanford [59]. At reduced temperatures, however, with HCl added to the H$_2$ and SiCl$_4$ in the gas stream no net deposits will occur on the SiO$_2$ while faceted, single-crystal, epitaxial pedestals will grow on the exposed regions since polysilicon is etched by the HCl at a faster rate than the SCS [60]. Such epitaxial projections may find use in future three-dimensional micromechanical structures.

Thermomigration

During 1976 and 1977, Anthony and Cline of GE laboratories performed a series of experiments on the migration of liquid eutectic Al/Si alloy droplets through SCS [61]–[67]. At sufficiently high temperatures, Al, for example, will form a molten alloy with the silicon. If the silicon slice is subjected to a temperature gradient (approximately 50°C/cm, or 2.0°C across a typical wafer) the molten alloy zone will migrate toward the hotter side of the wafer. The migration process is due to the dissolution of silicon atoms on the hot side of the molten zone, transport of the atoms across the zone, and their deposition on the cold side of the zone. As the Al/Si liquid region traverses the bulk, solid silicon in this way, some aluminum also deposits along with the silicon at the colder interface. Thermomigration hereby results in a p-doped trail extending through, for example, an n-type wafer. The thermomigration rate is typically 3 μm/min at 1100°C. At that temperature, the normal diffusion rate of Al in silicon will cause a lateral spread of the p-doped region of only 3–5 μm for a migration distance of 400 μm (the full thickness of standard silicon wafers).

Exhaustive studies by Anthony and Cline have elucidated much of the physics involved in the thermomigration process including migration rate [62], p-n junction formation [64], stability of the melt [65], effect of dislocations and defects in the silicon bulk, droplet morphology, crystallographic orientation effects, stresses induced in the wafer as a result of thermomigration [67], as well as the practical aspects of accurately generating, maintaining, and characterizing the required thermal gradient across the wafer. In addition, they demonstrated lamellar devices fabricated with this concept from arrays of vertical junction solar cells, to high-voltage diodes, to negative-resistance structures. Long migrated columns were found to have smaller diameters in (100) oriented wafers, since the droplet attains a pyramidally tapered point whose sides are parallel to the (111) planes. Migrated lines with widths from 30 to 160 μm were found to be most stable and uniform in traversing 280-μm-thick (100) wafers when the lines were aligned along the (110) directions. Larger regions tended to break up into smaller independent migrating droplets, while lines narrower than about 30 μm were not uniform due to random-walk effects from the finite bulk dislocation density in the wafer. Straight-line deviations of the migrated path, as a result of random walk, could be minimized either by extremely low ($\ll 100$/cm^2) or extremely high ($> 10^7$/cm^2) dislocation densities. On the other hand, the dislocation density in the recrystallized droplet trail is found to be essentially zero, not unexpected from the slow, even, liquid-phase epitaxy which occurs during droplet migration. Dopant density in the droplet trail corresponds approximately to the aluminum solid solubility in silicon at the migration temperature $\sim 2 \times 10^{19}$ cm^{-3}, which corresponds to $\rho = 0.005$ $\Omega \cdot$ cm. The p-type trail from a 50-μm-diameter aluminum droplet migrated through a 300-μm-thick n-type wafer would, therefore, exhibit less than 8-Ω resistance from front to back and would be electrically well-isolated from other nearby trails due to the formation of alternating p-n junctions, as shown in Fig. 11.

Nine potential sources of stress (generated in the wafer from the migrated regions) have been calculated by Anthony and

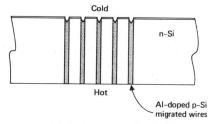

Fig. 11. In some applications of silicon micromechanics, it is important to connect the circuitry on one side of a wafer to mechanical structures on the other side. Thermomigration of Al wires, discussed extensively by Anthony and Cline [61]–[67], allows low-resistance (<8-Ω), close-spaced (<100-μm) wires to be migrated through thick (375-μm) wafers at reasonable temperatures (~1100°C) with minimal diffusion (<2 μm).

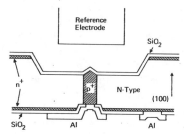

Fig. 12. Structure of the gate-controlled diode of Wen and Zemel [69]. Circuitry is on the bottom (protected) side of the wafer, while the sensor electrode is on the top. The p$^+$ feedthrough was accomplished by thermomigration of Al from the circuit side to the sensor side of the wafer. For ionic concentration measurements, an appropriate ion-sensitive membrane must be deposited over the oxide on the sensor side. Figure courtesy of C. C. Wen.

Cline. Maximum stresses intrinsic to the process (i.e., those which are present even when processing is performed properly) are estimated to be as high as 1.39×10^9 dyne/cm^2, which can be substantially reduced by a post-migration thermal anneal. Although the annealed stress will be about two orders of magnitude below the yield point of silicon at room temperature, it may increase the susceptibility of the wafer to fracture and should be minimized, especially if a large number of migrated regions are closely spaced.

One obvious utilization of thermomigration is the connection of circuitry on one side of a wafer to a mechanical function on the other side. Another application may be the dopant-dependent etching of long narrow holes through silicon. Since the work of Anthony and Cline, the thermomigration process has been used to join silicon wafers [68] and to serve as feedthroughs for solid-state ionic concentration sensors (see Fig. 12) [69]. Use of thermomigrated regions in power devices is another potential application. Even more significantly, laser-driven thermomigration has been demonstrated by Kimerling et al. [70]. Such a process may be extremely important in practical implementations of these migration techniques, especially since the standard infrared or electron-beam heating methods used to induce migration are difficult to control uniformly over an entire wafer.

Field-Assisted Thermal Bonding

The use of silicon chips in exposed, hostile, and potentially abrasive environments will often require mounting techniques substantially different from the various IC packaging methods now being utilized. First reported by Wallis and Pomerantz in 1969, field-assisted glass–metal thermal sealing [71] (sometimes called Mallory bonding after P. R. Mallory and Co., Inc., where Wallis and Pomerantz were then employed) seems to

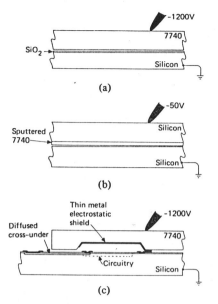

Fig. 13. Field-assisted thermal bonding can be used to hermetically bond (a) 7740 glass to silicon (bare or oxidized) or (b) silicon to silicon simply by heating the assembly to about 300°C and applying a voltage. Glass can be bonded to IC chips (c) if the circuitry is first protected by etching a shallow (~10-μm) well in the glass and depositing a grounded metal shield inside the well [76].

fulfill many of the requirements for bonding and mounting micromechanical structures. The technique is simple, low temperature, high strength, reliable, and forms hermetic seals between metals and conventional alkali-silicate glasses [72]. It is also very similar to well-known high-temperature thermal bonds where the cohesive metal–oxides, which are generated during the heating process, readily mix with the viscous glass. In the case of silicon, a glass slide is placed over a polished wafer (bare *or* thermally oxidized), the assembly is heated to about 400°C, and a high voltage (~1200 V) is applied between the silicon and the metal contact to the other side of the glass. If the sample is not too large, the metal contact may be a simple point probe located near one corner as shown in Fig. 13(a). Since the negative electrode is applied to the glass, ionic conduction causes a drift of positive ions away from the glass/Si interface into the bulk of the glass. The depletion of positive ions at the interface results in a high electric field across the air gap between the two plates. Electrostatic forces here, estimated to be higher than 350 psi, effectively clamp the pieces locally, conforming the two surfaces to obtain the strong, uniform, hermetic seal characteristic of field-assisted thermal bonding. The bonding mechanism itself has been the subject of some controversy, as discussed recently by Brownlow [73]. His convincing series of deductions, however, suggest that the commonly observed initial current peak at the onset of bonding is actually dissipated in the newly formed, narrow space-charge region in the glass at the interface. This high energy-density pulse, in the early stages of bonding, was shown to be capable of increasing the interfacial temperature by as much as 560°C, more than enough to induce the familiar, purely thermal glass/metal seal. Brownlow shows how this model correlates well with several other features observed during the bonding process.

From a device viewpoint, it is important to recognize that the relative expansion coefficients of the silicon and glass should match as closely as possible to alleviate thermal stresses after the structure has cooled. This aspect of field-assisted bonding also has the obvious advantage of yielding integrated

mechanical assemblies with very small mechanical drifts due to ambient temperature variations. Corning borosilicate glasses 7740 and 7070 have both been used successfully in this regard. In addition, Brooks *et al.* [74] have even bonded two silicon wafers by sputtering approximately 4 μm of 7740 glass over one of the wafers and sealing the two as already described, with the negative electrode contacting the coated wafer as shown in Fig. 13(b). Since the glass is so thin, however, the sealing voltage was not required to be above 50 V.

A high degree of versatility makes this bonding technique useful in a wide variety of circumstances. It is not necessary to bond to bare wafers, for example; silicon passivated with thermal oxide as thick as 0.5 μm is readily and reliably bonded at somewhat higher voltage levels. The bonding surface may even be partially interrupted with aluminized lines, as shown by Roylance and Angell [75], without sacrificing the integrity or hermeticity of the seal since the aluminum also bonds thermally to the glass. In addition, glass can be bonded to silicon wafers containing electronic circuitry using the configuration shown in Fig. 13(c) [76]. The circuitry is not affected if a well is etched in the glass and positioned over the circuit prior to bonding. A metal film deposited in the well is grounded to the silicon substrate during actual bonding and serves as an electrostatic shield protecting the circuit. Applications of all these aspects will be presented and expanded upon in the following sections.

IV. Grooves and Holes

Even simple holes and grooves etched in a silicon wafer can be designed and utilized to provide solutions in unique and varied applications. One usage of etched patterns in silicon with far-reaching implications, for example, is the generation of very high precision molds for microminiature structures. Familiar, pyramidal-shaped holes anisotropically etched in (100) silicon and more complex holes anisotropically etched in (110) silicon were used by Kiewit [77] to fabricate microtools such as scribes and chisels for ruling optical gratings. After etching the holes in silicon through an SiO_2 mask, the excess SiO_2 was removed and very thick layers of nickel–phosphorus or nickel–boron alloys were deposited by electroless plating. When the silicon was completely etched away from the thick plated metal, miniature tools or arrays of tools were accurately reproduced in the metal with geometrically well-defined points having diameters as small as 50 nm. The resulting metal tools had a hardness comparable to that of file steel.

Similar principles were employed by Wise *et al.* [78] to fabricate miniature hemispherical structures for use as thermonuclear fusion targets. In these experiments, a large two-dimensional array of hemispherical holes was etched into a silicon wafer using an HNA isotropic solution, approximately as shown in Fig. 4(c). After removing the SiO_2/Cr/Au etch mask, polymer, glass, metal, or other thin films are deposited over the wafer, thereby conforming to the etched hemispherical shapes. When two such wafers are aligned and bonded, the silicon mold can be removed (either destructively by etching or nondestructively by using a low adhesion coating between the silicon and the deposited film). The resulting molded shape is a thin-walled spherical shell made from the deposited material. Fig. 14 is the process schedule for a simple metal hemishell demonstrated by Wise *et al.*

The potential of making arrays of sharp points in silicon itself by etching was employed in a novel context by Thomas and Nathanson [79], [80]. They defined a very fine grid

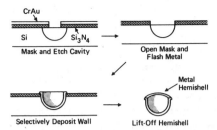

Fig. 14. Fabrication sequence for free-standing metal hemishells using an isotropic silicon-etching technique [78]. Typical dimensions of the hemishell are 350-μm diameter with a 4-μm-thick wall. Courtesy of K. D. Wise.

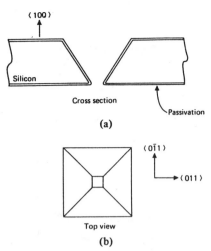

Fig. 15. (a) Cross section and (b) top view of anisotropically etched silicon ink jet nozzle in a (100) wafer developed by E. Bassous et al. [5], [43], [8].

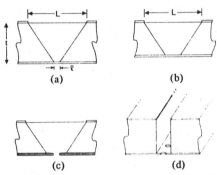

Fig. 16. A number of different methods have been developed for fabricating silicon ink jet nozzles. (a) and (b) show the errors in final nozzle size which occurs when the wafer thickness varies. (c) shows a p$^+$ membrane structure. This design yields round nozzles and also minimizes the effects of wafer thickness variations. Nozzles can be more closely spaced by using the p$^+$ membrane technique on a (110) wafer, as shown in (d) [35].

(typically 25 μm center to center) in an SiO$_2$ mask, then isotropically etched the silicon exposed in the grid lines with an HNA mixture. The isotropic etch undercuts each square segment of the oxide grid uniformly around its periphery. If the etching is quenched just after the oxide segments are completely undercut and fall from the surface, a large array of very sharply tipped silicon points is obtained. Point diameters were estimated to be about 20 nm. These silicon points, at densities up to 1.5×10^5 cm^2, were used by Thomas and Nathanson as efficient, uniform, photosensitive field emitter arrays which were imaged onto a phosphor screen closely spaced to the wafer. A more complex extension of this fabrication technique will be described below in the section on Thin Cantilever Beams.

Ink Jet Nozzles

Since anisotropic etching offers a powerful method for controlling undercutting of masks during silicon etching, these techniques are important candidates for etching high-resolution holes clear through wafers as Bassous et al. [5], [43], [81], [82] first realized and pursued extensively; see Fig. 15. Patterns etched clear through wafers have many potential applications, as will be seen below, but one of the simplest and most commercially attractive is in the area of ink jet printing technology [83], [86]. As shown in Fig. 16(a), the geometry of the pyramidal hole in (100) silicon can be adjusted to completely penetrate the wafer, the square hole on the bottom of the wafer forming the orifice for an ink jet stream. The size of the orifice (typically about 20 μm) depends on the wafer thickness t and mask dimension L according to $l = L - (2t/\tan \theta)$, where $\theta = 54.74°$ is the angle between the (100) and (111) planes. In practice, the dimension l is very difficult to control accurately because 1) wafer thickness t is not easy to control accurately and 2) small angular misalignments of a square mask will result in an effective L which is larger than the mask dimension [43], thereby enlarging l as shown in Fig. 16(b). The angular misalignment error can be eased by using a round mask (*diameter L*) which will give a square hole $L \times L$ independent of orientation, as described in Section III (and Fig. 5(e)) by the general rule of anisotropic undercutting.

Membrane structures have also been used in ink jet nozzle designs not only to eliminate the effects of wafer thickness variations, but also to permit more densely packed orifices as well as orifice shapes other than square. In one technique described by Bassous et al. [35], the wafer surface is highly doped with boron everywhere but the desired orifice locations. Next, the wafer is anisotropically etched clear through with EDP as described above, using a mask which produces an l which is 3 to 5 times larger than the actual orifice. Since EDP does not attack silicon which is highly doped with boron, a p$^+$ silicon membrane will be produced, suspended across the bot-

tom of the pit with an orifice in the center corresponding to the location previously left undoped; see Fig. 16(c). The use of a membrane can also be extended to decrease the minimum allowed orifice spacing. Center-to-center orifice spacing is limited to about 1.5 times the wafer thickness when the simple square geometries of Figs. 15, 16(a)–(c) are employed, but can be much closer using membranes. Orifice spacings in *two* dimensions can be made very small by using (110) oriented wafers and etching vertical-walled *grooves* (as described in Section III) clear through the wafer, aligned to rows of orifices on the other side fabricated by this membrane technique. The result, shown in Fig. 16(d), is a number of closely spaced rows containing arbitrarily spaced holes in a long, narrow rectangular p$^+$ membrane [35].

Deep grooves or slots etched clear through (110) silicon have been used by Kuhn et al. [87] in another important ink jet application. At a characteristic distance from the ink jet orifice, the ink stream, which is ejected under high pressure, begins to break up into well-defined droplets at rates of about 10^5 drops per second as a result of a small superimposed sinusoidal pressure disturbance. A charge can be induced on individual droplets as they separate from the stream at this point by passing the jet through a charging electrode. Once charged, the drops can be electrostatically deflected (like an electron beam) to strike the paper at the desired locations. Kuhn et al. etched

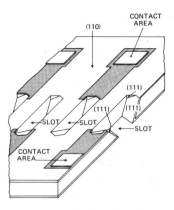

Fig. 17. Grooves anisotropically etched clear through a (110) wafer were employed as charge electrode arrays by Kuhn *et al.* [87] in an ink jet printing demonstration. A charge can be induced on individual ink droplets as they pass through the grooves by applying a voltage to the walls of the groove. Subsequently, drops are "steered" to the paper after traveling through a high electric field. Figure courtesy of L. Kuhn.

several grooves clear through (110) silicon, doped the walls of the grooves so they would be conductive, and defined contact pads connected to the doped sidewalls of the grooves, as shown in Fig. 17. By arranging for the streams to pass through these grooves right at the breakoff points, the grooves can be operated as an array of independent charge electrons. In the design of large, linear arrays of closely spaced ink jet orifices (typical spacing is less than 250 μm), where high precision miniaturized structures are required, silicon micromechanics can provide useful and viable structural alternatives, as long as the usual materials considerations (such as materials compatibility, fatigue, and corrosion) are properly taken into account.

In an effort to integrate ink jet nozzle assemblies more efficiently and completely, another experimental structure was demonstrated in which nozzle, ink cavity, and piezoelectric pressure oscillator were combined using planar processing methods [88]. Orifice channels were first etched into the surface of a (110) oriented wafer as shown in Fig. 18, using an isotropic HNA mixture. After growing another SiO$_2$ masking layer, anisotropic (EDP) etching was employed to etch the cavity region as well as a deep, vertical-walled groove (which will eventually become the nozzle exit face) clear through the wafer. The wafer must be accurately aligned to properly etch the vertical grooves according to the pattern in Fig. 19. After etching, the silicon appears as seen in Fig. 20(a). The individual chips are separated from the wafer and thick 7740 glass (also containing the supply channel) is anodically bonded to the bottom of the chips. Next, a thin 7740 glass plate (125 μm thick), serving as the pump membrane, is aligned to the edge of the nozzle exit face and anodically bonded to the other side of the silicon chip. The exit orifice, after anodic bonding, is shown in Fig. 20(b). Once the piezo-plate is epoxied to the thin glass plate, a droplet stream can be generated, exiting the orifice at the edge of the chip and parallel to the surface, as shown in Fig. 21.

This planar integrated structure was deliberately specified to conform to the prime requirement of silicon micromechanical applications—no mechanical machining or polishing and minimum handling of individual chips to keep processing and fabrication costs as low as possible. Even though the drops are ejected from the edge of the wafer in this design, the exit face is defined by crystallographic planes through anisotropic etch-

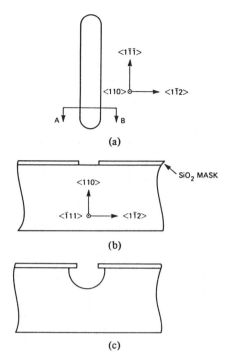

Fig. 18. Orientation and cross section of the isotropically etched nozzle for the planar ink jet assembly after etching. (a) Top view of nozzle channel. (b) Cross section *AB* before silicon etch. (c) After silicon etch. Typical channel depth is 50 μm.

ing. Any other nozzle design in which drops are to be ejected parallel to the surface would require an expensive polishing step on the edge of the chip to obtain the necessary smoothness which occurs automatically in this design as a result of inexpensive, planar, batch-processed, anisotropic etching.

Miniature Circuit Boards and Optical Benches

The packing density of silicon memory and/or circuitry chips can be greatly increased by using silicon essentially as miniature pluggable circuit boards. Two-dimensional patterns of holes have been anisotropically etched clear through two wafers, which are then bonded together such that the holes are aligned as illustrated in Fig. 22. When the resulting cavities are filled with mercury, chips with beam-lead, plated, or electromachined metal probes can be inserted into both sides of the minicircuit board. Such a packaging scheme has been under development for low-temperature Josephson-junction circuits [89]. Dense circuit packaging and nonpermanent die attachment are the primary advantages of this technique. In the case of Josephson-junction circuits, there is an additional advantage in that the entire computer—substrates for the thin-film circuits, circuit boards, and structural supports—are all made from silicon, thereby eliminating thermal mismatch problems during temperature cycling.

Perhaps the most prolific application of silicon anisotropic etching principles is miniature optical benches and integrated optics [90]–[102]. Long silicon V-grooves in (100) wafers are ideal for precise alignment of delicate, small-diameter optical fibers and permanently attaching them to silicon chips. Two cleaved fibers can be butted together this way, for example to accuracies of 1 μm or better. In addition, a fiber can be accurately aligned to some surface feature

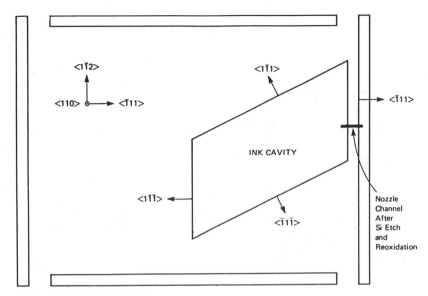

Fig. 19. Orientation of the anisotropically etched ink cavity and deep grooves. After EDP etching, all the (111) surfaces will have flat, vertical walls. Typical cavity size is about 0.5 cm.

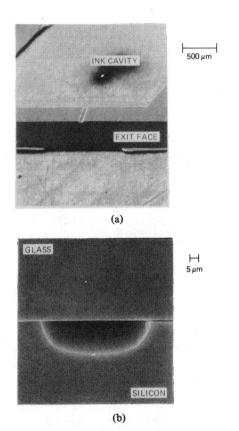

(a)

(b)

Fig. 20. (a) SEM photograph of silicon nozzle structures after the EDP etch, ready for anodic bonding. Note the nozzle channel which connects the ink cavity to the flat, vertical walls of the exit face. (b) SEM photograph of the ink jet orifice after anodic bonding; glass membrane on top, silicon on bottom.

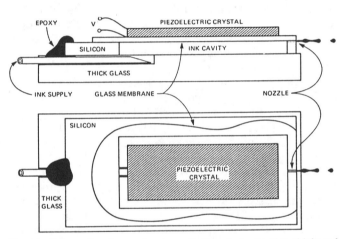

Fig. 21. Schematic of completed nozzle structure showing thick and thin glass plates anodically bonded to either side of the silicon, ink supply line, and piezoelectric ceramic epoxied to the thin glass plate. From [88].

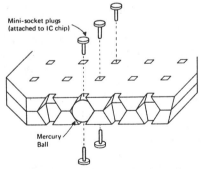

Fig. 22. Complete circuit-board assemblies are under development to optimize the packaging and interconnection of cryogenic Josephson-junction circuits and computers [89]. Miniature socket arrays are created by bonding together two silicon wafers with anisotropically etched holes and filling the cavity with mercury. Miniature plugs attached to the circuit chips themselves are inserted into both sides of the "circuit board." Silicon is used because it can be micromachined accurately, wiring can be defined lithographically, and thermal mismatch problems are alleviated.

[96], [97], [99], [101], [102]. In Fig. 23(a), a fiber output end is butted up against a photodiode, which can then be integrated with other on-chip circuitry; fiber arrays, of course, are also easily integrated with diode arrays. In Fig. 23(b), a fiber core is accurately aligned to a surface waveguiding layer,

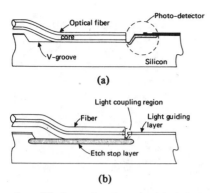

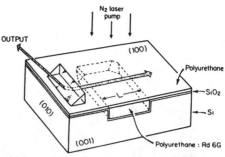

Fig. 23. Silicon is rapidly becoming the material of choice for manipulating fiber-optic components. Two examples are shown here. (a) Coupling a fiber output to a diode detector using an etched V-groove for simple and accurate fiber alignment. (b) Coupling a fiber output to a deposited thin-film optical waveguide using a buried etch-stop layer to obtain precise vertical alignment.

Fig. 25. The high-precision structures of which SCS is inherently capable have included the laser resonator shown here which was demonstrated by Hu and Kim [98]. In this case, sidewalls defined by (100) crystallographic planes have become the perfectly flat and parallel surfaces necessary for the aligned mirrors of a thin-film laser cavity. Figure courtesy of C. Hu.

filled waveguides [98]. When a shallow rectangular well, oriented parallel to the (010) and (001) directions, is etched into a (100) silicon wafer using KOH, the sidewalls of the etched well are defined by these planes and are vertical to the surface. Since the two facing walls of the cavity are ideal, identical crystallographic planes, they are perfectly parallel to each other and normal to the wafer surface. After the wafer is oxidized and spun with a polymer containing a laser dye, the two reflecting, parallel walls of the etched hole (with the dye in between) form a laser cavity. This waveguide laser was optically pumped with a pulsed nitrogen laser by Hu and Kim. Some of the radiation in the cavity itself is coupled out through leakage modes to the thin, excess layer of polymer covering the wafer surface around the laser cavity, as shown in Fig. 25. The output radiation is, of course, in the form of surface guided waves and can be coupled out by conventional integrated optics prism or grating methods.

Gas Chromatograph on a Wafer

One of the more ambitious, practical, and far-reaching applications of silicon micromechanical techniques has been the fully integrated gas chromotography system developed at Stanford by S. Terry, J. H. Jerman, and J. B. Angell [29], [103]. The general layout of the device is illustrated in Fig. 26(a). It consists of a 1.5-m-long capillary column, a gas control valve, and a detector element all fabricated on a 2-in silicon wafer using photolithography and silicon etching procedures. Isotropic etching is employed to generate a spiral groove on the wafer surface 200 μm wide, 40 μm deep, and 1.5 m long. After the wafer is anodically bonded to a glass plate, hermetically sealing the grooves from each other, the resulting 1.5-m-long capillary will be used as the gas separation column. Gas input to the column is controlled by one valve fabricated integrably on the wafer along with the column itself. The valve body is etched into the silicon wafer in three basic steps. First a circular hole is isotropically etched to form the valve cylinder. A second isotropic etch enlarges the valve cylinder while leaving a circular ridge in the bottom of the hole which will serve as the valve seating ring. Finally, holes are anisotropically etched clear through the wafer in a manner similar to ink jet nozzles such that the small orifice exists in the center of the seating ring (see Fig. 26(b)). The flexible valve sealing diaphragm, initially made from a silicon membrane, is now a thin (5–15-μm) nickel button flexed on or off by a small electrical solenoid. Both the valve body and sealing diaphragm are coated with parylene to provide conformal leak-tight sealing surfaces. The sensor, located in the output line of the column,

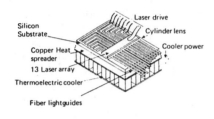

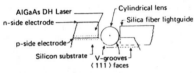

Fig. 24. The most advanced fiber-optic coupling scheme was designed and demonstrated by Crow *et al.* [100]. The output from an array of solid-state lasers was focussed into a corresponding array of optical fibers using another fiber, aligned between the laser array and the output fibers, as a cylindrical condenser lens. All the fibers are aligned by pressing them into accurately aligned V-grooves anisotropically etched into the silicon. Figure courtesy of J. Crow.

by resting the fiber on a buried etch-stop diffusion over which an epitaxial layer has been grown to an accurate thickness.

The most ambitious use of silicon as a mini-optical bench is the GaAs laser-fiber array developed by Crow *et al.* [100]. In this assembly, the light outputs from a perpendicular array of GaAs lasers, mounted on the silicon surface in Fig. 24, are coupled into an optical fiber aligned parallel to the array by one V-groove. This first fiber serves as a cylindrical lens to focus the highly divergent laser light into a perpendicular array of fibers corresponding to the laser array. The linear fiber bundle can now be maneuvered, swept, or positioned independently of the laser package. In addition, this scheme couples the laser light into the fibers very efficiently, while the silicon substrate has the important advantages of serving as an efficient heat sink for the laser array, can be processed to provide isolated electrical contacts and, potentially, on-chip driving electronics to each individual laser in the array.

In addition to fiber alignment aids, such V-grooves, when passivated with SiO_2 and filled with a spun-on polymer, have also been employed as the light-guiding structures themselves [91], [92]. A similar, highly innovative device demonstrated by Hu and Kim also made use of anisotropically etched and

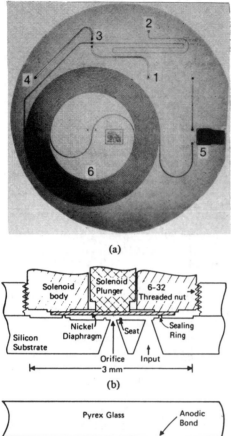

(a)

(b)

(c)

Fig. 26. The most ambitious project utilizing the mechanical properties of silicon is the Stanford gas chromatograph [29], [103]. (a) Overall view of the full silicon wafer showing 1) sample input, 2) purge input, 3) valve region, 4) exhaust of unused sample, 5) sensor region, 6) separation column. The various etched grooves are sealed by anodically bonding a glass plate over the entire wafer. A cross section of the valve assembly is drawn in (b) including the valve cavity, seating ring, and input orifice etched into the silicon as well as the thin nickel diaphragm. The thin-film thermal detector in (c) is also silicon based, consisting of a metal resistor evaporated on SiO_2, thermally isolated by etching the silicon from beneath. Figures courtesy of J. Jerman and S. Terry.

is also based on silicon processing techniques. A thin metal resistor is deposited and etched in a typical meandering configuration over a second oxidized silicon chip. Next, the silicon is anisotropically etched from the back surface of the wafer leaving an SiO_2 membrane supported over the etched hole. This hole is aligned so that the metal resistor is positioned in the center of the membrane and thus thermally isolated from the silicon substrate as shown in Fig. 26(c). The gases separated in the column are allowed to flow over the sensor before being exhausted.

Operation of the column proceeds as follows. After completely purging the system with the inert carrier gas, which flows continuously through port 2 at a pressure of about 30 psi, the valve 3 is opened and the unknown gas sample (held at a pressure higher than the purge gas) is bled into the

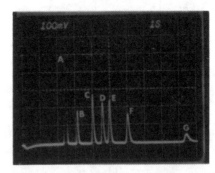

Fig. 27. Example of an output from the miniature gas chromatograph shown in Fig. 26. A) nitrogen; B) pentane; C) dichloromethane; D) chloroform; E) 111-trichloroethane; F) trichloroethylene; G) toluene. Photo courtesy of J. Jerman and S. Terry.

column through port 1 while the narrow purge supply line appears as a high impedance path to the direction of the sample flow. After introducing a sample with a volume as low as 5 nl, the valve is closed again and purge gas flushes the sample through the column 6. Since the etched capillary is filled with a gas chromatography liner, the various molecular constituents of the sample gas traverse the column at different rates and therefore exit the system sequentially. The sensor element 5 detects the variations in thermal conductivity of the gas stream by biasing the thin, deposited metal resistor at a fixed current level and monitoring its resistance. A burst of high thermal conductivity gas will remove heat from the resistor more efficiently than the low conductivity carrier gas and a small voltage pulse will be detected. A typical signal is shown in Fig. 27. Such a small chromatograph can only operate properly if the sample volume is much smaller than the volume of the column. For this reason, it is essential to fabricate the ultra-miniature valve and detector directly on the wafer with the column to minimize interfering "dead space."

A complete, portable gas chromatograph system prototype is being developed by the Stanford group which will continuously monitor the atmosphere, for example, in a manufacturing environment and identify and record 10 different gases with 10 ppm accuracy—all within the size of a pocket calculator.

Miniature Coolers

Besides the Stanford gas chromatograph, the advantageous characteristics of anodic bonding are being employed in even more demanding applications. Recognizing the proliferation of cryogenic sensing devices and circuits based on superconducting Josephson junctions, W. A. Little at Stanford has been developing a Joule–Thomson minirefrigeration system initially based on silicon anisotropic etching and anodic bonding [104]. As shown in Fig. 28, channels etched in silicon comprise the gas manifold, particulate filter, heat exchanger, Joule–Thomson expansion nozzle, and liquid collector. The channels are sealed with an anodically bonded glass plate and a hypodermic gas supply tubing is epoxied to the input and output holes. Such a refrigerator cools down the region near the liquid collector as the high-pressure gas (after passing through the narrow heat exchange lines) suddenly expands into the liquid collector cavity. Little has derived scaling laws for such Joule–Thomson minirefrigeration systems, which show that cooling capacities in the 1–100-mW range at 77 K, cool down rates on the order of seconds, and operating times of 100's of hours (with a single gas cylinder) are attainable using a total channel length of about 25 cm, 100 μm in diameter—dimensions simi-

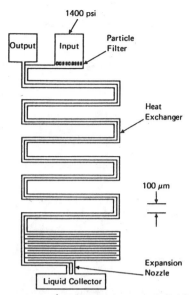

Fig. 28. Grooves etched in silicon have been proposed for the construction of miniature cryogenic refrigerators. In the Joule–Thomson system here, high pressure N_2 gas applied at the inlet expands rapidly in the collection chamber, thereby cooling the expansion region. An anodically bonded glass plate seals the etched, capillary grooves. Adapted from W. A. Little [104].

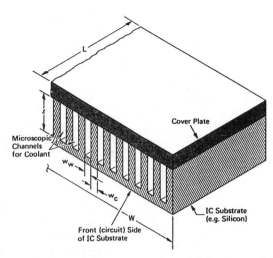

Fig. 29. Schematic view of a compact heat sink incorporated into an integrated circuit chip [105]. For a 1-cm^2 silicon IC chip, using water as the coolant, the optimum dimensions, are approximately $w_w = w_c = 57$ μm and $z = 365$ μm. The cover plate is 7740 glass anodically bonded to the silicon, and the channels are anisotropically etched into the (110) wafer with a KOH-based etchant. Thermal resistances less than 0.1 C/W were measured. Figure courtesy of D. Tuckerman.

lar to the gas chromatograph design discussed previously. These lines, however, must not only withstand the thermal shocks of repeated heating and cooling, but also survive the high internal gas pressures (as high as 1000 psi) which occur simultaneously. SCS can be designed to work well in this application because of its high strength. In addition, the glass/silicon bond is ideal not only because of its strength, but also because the nature of the bonding process presupposes an excellent match in thermal coefficients of expansion of the two materials. One disadvantage of silicon in this application is its very high thermal conductivity, even at low temperatures, which limits the attainable temperature gradient from the (ambient) inlet to the liquid collection chamber. Similar all-glass devices have already found use in compact, low-temperature IR sensors and will likely be employed in other scientific instruments from high-sensitivity magnetometers and bolometers to high-accuracy Josephson-junction voltage standards.

As the cycle times of conventional room-temperature computer mainframes and the level of integration of high-speed semiconductor bipolar logic chips continue to increase, the difficulty of extracting heat from the chips in the CPU is rapidly creating a serious packaging problem. Faster cycle times require closer packing densities for the circuit chips in order to minimize signal propagation times which are already significant in today's high-speed processors. This increased packing density is the crux of the heat dissipation problem. Maximum power dissipation capabilities for conventional multichip packaging assemblies have been estimated at 20 W/cm^2. In response to these concerns, a new microcooling technology has been developed at Stanford by Tuckerman and Pease which makes use of silicon micromachining methods [105]. As shown in Fig. 29, a (110) oriented wafer is anisotropically etched to form closely spaced, high aspect ratio grooves about $\frac{3}{4}$ of the way through the wafer. A glass plate with fluid supply holes is anodically bonded over the grooves to provide sealed fluid channels through which the coolant is pumped. Input and output manifolds are also etched into the silicon at the same time as the grooves. The circuitry to be

cooled is located on the opposite side of the wafer. Over a 1-cm^2 area, a thermal resistance of about 0.1°C/W was measured for a water flow rate of 10 cm^3/s, for a power dissipation capability of 600 W/cm^2 (at a typical temperature rise above ambient of 60°C). This figure is 30 times higher than some previously estimated upper limits.

The use of silicon in this application is not simply an extravagant exercise. Tuckerman and Pease followed a novel optimization procedure to derive all the dimensions of the structure shown in Fig. 29. For optimal cooling efficiency, the fins should be 50 μm wide with equal 50-μm spaces and the height of the fins should be about 300 μm. Fortuitously, these dimensions correspond closely to typical silicon wafer thicknesses and to typical anisotropically etched (110) structures easily realized in practice. Besides the fact that the fabrication of such miniature structures would be extremely difficult in materials other than silicon, severe thermal mismatch problems are likely to be encountered during temperature cycling if a heat-sink material other than silicon were employed here.

The microcooling technique of Tuckerman and Pease is a compact and elegant solution to the problem of heat dissipation in very dense, very-high-speed IC chips. Advantages of optimized cooling efficiency, thermal and mechanical compatibility, simplicity, and ease of fabrication make this an attractive and promising advance in IC packaging. Bipolar chips with 25 000 circuits, each operating at 10 mW per gate (250 W total) are not unreasonable projections for future CPU's, now that a practical cooling method, involving silicon micromechanics, has been demonstrated.

Applications to Electronic Devices

Various isotropic and anisotropic etching procedures have been employed many times in the fabrication of IC's and other silicon electronic devices [106], [107]. In particular, silicon etching for planarization [108], for isolation of high-voltage devices [109], [110], or for removing extraneous regions of a chip to reduce parasitics [111], [112], and in VMOS [113] (more recently UMOS [114]) transistor structures are well-

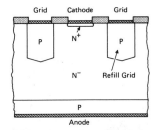

Fig. 30. The deep grid structure of a vertical-channel field-controlled thyristor [58] was accomplished by anisotropically etching deep grooves in the (110) wafer and growing p-doped silicon in the grooves by the epitaxial refill process of Runyan *et al.* and Smeltzer [57] which is shown in Fig. 10(b). Figure courtesy of B. Wessels.

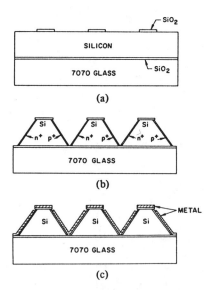

Fig. 31. Major fabrication steps for the V-groove, multijunction solar cell [116]. (a) Grow silicon dioxide layer, field assist bond oxidized wafer to glass, etch pattern windows in silicon dioxide. (b) Anisotropically etch silicon down to 7070 glass substrate, implant n^+ and p^+ regions at an angle, anneal implants. (c) Deposit metallization and alloy. Figure courtesy of T. Chappell.

known and some are used extensively in commercial products. Two areas of application in this category deserve special comment in this section, however. The first is a novel technique for producing very deep, doped regions for high-power electronic devices and is based on the epitaxial groove-filling process first demonstrated by Runyon *et al.* and Smeltzer [57] and shown schematically in Fig. 10(b). High-voltage high-power devices require deep diffusions not only to accommodate larger space-charge regions in the silicon (for increased breakdown voltages) but also to carry the larger currents for which such devices are designed. It is not unusual, for example, to schedule high-temperature diffusion cycles lasting over 100 h during some stages in the fabrication of high-power electronic devices. Furthermore, the geometries of such structures are limited because lateral diffusion rates are approximately equal to the vertical rates, i.e., diffusion in silicon is an isotropic process. By anisotropically etching grooves in (110) n-type silicon and refilling them epitaxially with p-type SCS, a process is obtained which appears effectively as an anisotropic diffusion. In this way, very deep, high aspect ratio, closely spaced diffused regions have been realized for high-speed vertical-channel power thyristors such as those demonstrated by Wessels and Baliga [58] (illustrated in Fig. 30), as well as for more complex buried-grid, field-controlled power structures [115]. Similar types of "extended" device geometries have been demonstrated by Anthony and Cline [64] using aluminum thermomigration (see Fig. 11). These micromachining techniques offer another important degree of freedom to the power device designer, which will be increasingly exploited in future generations of advanced high-power devices and IC's.

A second electronic device configuration employing the micromechanical principles discussed here is the V-groove multijunction solar cell [116]. The basic device configuration and a schematic processing schedule are shown in Fig. 31. Fabrication is accomplished by anodically bonding an SiO_2-coated silicon wafer to 7070 glass, anisotropically etching long V-grooves the full length of the wafer completely through the wafer to the glass substrate, ion-implanting p and n dopants into the alternating (111) faces by directing the ion beam at alternate angles to the surface, and finally evaporating aluminum over the entire surface at normal incidence such that the overhanging oxide mask prevents metal continuity at the top of the structure, while adjacent p and n regions at the bottom are connected in series. Solar conversion efficiencies of over 20 percent are expected from this device in concentrated sunlight conditions when the light is incident through the glass substrate. Advantages of these cells are ease of fabrication (one masking step), high voltage (~70 V/cm of cells), long effective light-absorption length (and therefore high

efficiency) because of multiple internal reflections, no light-blocking metal current collection grid on the illuminated surface, and excellent environmental protection and mounting support provided by the glass substrate. Silicon solar cells based on this technique offer dramatic improvements over present single-crystal designs and may eventually be of commercial value.

V. SILICON MEMBRANES

While the micromechanical devices and components discussed in the preceding section were fabricated exclusively by rather straightforward groove and hole etching procedures, the following applications require some additional processing technologies; in particular, dopant-dependent etching for the realization of thin silicon membranes, which have been discussed in Section III.

X-Ray and Electron-Beam Lithography Masks

An early application of very thin silicon membrane technology which is still very much in the process of development is in the area of high-precision lithography masks. Such masks were first demonstrated by Spears and Smith [117] in their early X-ray lithography work and later extended by Smith *et al.* [118]. Basically, the procedure consists of heavily doping the surface of the silicon with boron, evaporating gold over the front surface, etching the gold with standard photolithographic or electron-beam techniques to define the X-ray mask pattern, and finally etching away most of the silicon substrate from the back side of the wafer (except for some support grids) with EDP [119]. Since heavily boron-doped silicon is not as rapidly attacked by EDP (or KOH), a self-supporting membrane is obtained whose thickness is controlled by the boron diffusion depth, typically 1–5 μm. Since the boron enters the silicon lattice substitutionally and the boron atoms have a smaller radius than the silicon, this highly doped region tends to be under tension as discussed in Section III. When the substrate is etched away, then the member becomes stretched taut and appears smooth and flat with no wrinkles, cracks, or bowing. X-rays are highly attenuated by the gold layers but not by the thin silicon "substrate" [120], [121]. Several variations on this scheme have been reported. Bohlen *et al.* [122], for example, have taken the X-ray design one step further by plasma etching completely through the remaining thin p^+ silicon regions not covered by gold and using the mask structure for electron-beam proximity printing.

These same basic principles were employed as early as 1966 by Jaccodine and Schlegel [123] to fabricate thin membranes (or windows) of SiO_2 to measure Young's modulus of thermally grown SiO_2. They simply etched a hole from one side of an oxidized Si wafer to the other (using hot Cl_2 gas as the selective etchant), leaving a thin SiO_2 window suspended across the opposite side. By applying a pressure differential across this window, they succeeded in measuring its deflection and determining Young's modulus of the thermally grown SiO_2 layer. Such measurements were later expanded upon by Wilmsen *et al.* [124]. Finally, Sedgwick *et al.* [119] and then Bassous *et al.* [125] fabricated these membrane windows from silicon and Si_3N_4 for use as ultra-thin electron-beam lithography "substrates" (to eliminate photoresist line broadening due to electron backscattering exposures from the substrate) for the purpose of writing very high resolution lines and for use in generating high-transparency X-ray masks. Thin, unsupported silicon nitride windows also have the advantage, in these applications, of being in tension as deposited on the silicon wafer, in the same way that boron-doped silicon membranes are in tension. SiO_2 membranes, such as those studied by Jaccodine and Schlegel [123] and by Wilmsen *et al.* [124], on the other hand, are in compression as deposited, tend to wrinkle, bow, and distort when the silicon is etched away, and are much more likely to break.

Circuits on Membranes

The potential significance of thin SCS membranes for electronic devices has been considered many times. Anisotropic etching, together with wafer thinning, were used by Rosvold *et al.* [111] in 1968 to fabricate beam-lead mounted IC's exhibiting greatly reduced parasitic capacitances. The frequency response of these circuits was increased by a factor of three over conventional diffused isolation methods. Renewed interest in circuits on thinned SCS membranes was generated during the development of dopant-dependent electrochemical etching methods. Theunissen *et al.* [45] showed how to use ECE both for beam-lead, air-gap isolated circuits as well as for dielectrically isolated circuits. Dielectric isolation was provided by depositing a very thick poly-Si layer over the oxidized epi, etching off the SCS substrate electrochemically, then fabricating devices on the remaining epi using the poly-Si as an isolating dielectric substrate. Meek [49], in addition to extending this dielectric isolation technique, realized other unique advantages of such thin SCS membranes, both for use in crystallographic ion channeling studies, as well as large-area diode detector arrays for use in low parasitic video camera tubes.

A backside-illuminated CCD imaging device [126] developed at Texas Instruments depends fundamentally on the ability to generate high-quality, high-strength, thin membranes over large areas. Since their double level aluminum CCD technology effectively blocked out all the light incident on the top surface of the wafer, it was necessary to illuminate the detector array from the backside. In addition, backside illumination improves

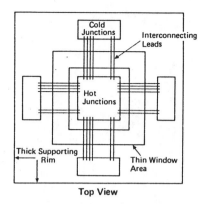

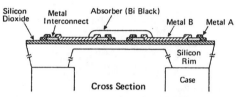

Fig. 32. Thermopile detector fabricated on a silicon membrane [127]. The hot junctions of the Au-poly-Si thermocouples are located in the central region of the membrane, while the cold junctions are located on the thick silicon rim. Efficient thermal isolation, small size, and a large number of integrated junctions result in high sensitivity and high-speed detection of infrared radiation. Figure courtesy of K. D. Wise.

spatial sensing uniformity and eliminates inference problems associated with front illumination through transparent layers. The high absorption coefficient of silicon in the visible, however, required the imager to be subsequently thinned from the backside (after circuit fabrication) to about 10 μm for efficient collection of photogenerated carriers. It was found that thin, highly uniform membranes could be realized over areas greater than 1 cm^2 with no deleterious effect on the sensitive CCD array and that these membranes exhibited exceptional strength, durability, and resistance to vibration and thermal cycling. Several such large-area CCD imaging arrays (800 × 800 pixels) will be installed in the space telescope scheduled to be launched by the Space Shuttle in 1985.

An important aspect of thin insulating membranes is that they provide excellent thermal isolation for thin-film devices deposited on the membrane. Lahiji and Wise [127] have demonstrated a high-sensitivity thermopile detector based on this principle. They fabricated up to 60 thin-film thermocouples (Bi-Sb and Au-polycrystalline Si), wired in series on a 2 mm × 2 mm × 1 μm SiO_2/p^+-Si membrane. Plan and cross-sectional views of this device are shown in Fig. 32. Hot junctions are arranged in the central membrane region while cold junctions are spaced over the thick periphery of the chip. When the membrane is coated with a thin thermal absorbing layer, sensitivities up to 30 V/W and time constants below 10 ms were observed for chopped 500 C black-body radiation incident from the etched (or bottom) surface of the wafer. Such low-mass, thermally isolated structures are likely to be commercially developed for these and related applications.

One thermally isolated silicon structure, in fact, is already commercially available. The voltage level detector of a high-bandwidth ac frequency synthesizer (Models 3336A/B/C) manufactured by Hewlett-Packard [4] is shown in Fig. 33. Two thin silicon cantilever beams with larger masses suspended in the center have been defined by anisotropic etching. The

Fig. 33. A high-bandwidth, thermal rms voltage detector [4] fabricated on silicon employs two cantilever beams with matching temperature-sensitive diodes and heat dissipation thin-film resistors on each. This device is used in the output-voltage regulation circuitry of the HP Model 3330 series of frequency synthesizers. Photo courtesy of P. O'Neil.

central masses of each beam are thermally isolated from each other and from the rest of the substrate. Fabricated on each isolated silicon island are a temperature-sensing diode and a thin-film heat-dissipation resistor. When a dc control current is applied to one resistor, the silicon island experiences a temperature rise which is detected by the corresponding diode. Meanwhile, a part of the ac output signal is applied to the resistor on the second island resulting in a similar temperature rise. By comparing the voltages of the two temperature-sensitive diodes and adjusting the ac voltage level until the temperatures of the two diodes match, accurate control of the output ac rms voltage level is obtained over a very large frequency range. This monolithic, silicon thermal converter offers the advantages of batch-fabrication, good resistor and diode parameter matching, while minimizing the effects of ambient thermal gradients. In addition, the masses of the islands are small, the resulting thermal time constants are therefore easy to control, and the single chip is simple to package.

In some applications, great advantages can be derived from electronic conduction normal to SCS membranes. In particular, Huang and van Duzer [128], [129] fabricated Schottky diodes and Josephson junctions by evaporating contacts on either side of ultrathin SCS membranes produced by p^+ doping and anisotropic etching. As thin as 400 Å, the resulting devices were characterized by exceptionally low series resistances, one-half to one-third of that normally expected from epitaxial structures. For Josephson junctions, the additional advantage of highly controllable barrier characteristics, which comes for free with silicon, could be of particular value in microwave detectors and mixers.

Large-area Schottky diodes on SCS membranes with contacts on either side have also found use as dE/dx nuclear particle detectors by Maggiore et al. [130]. Since the diodes (membranes) are extremely thin, 1–4 μm, the energy loss of particles traversing the sample is relatively small. This means that heavier ions, which typically have short stopping distances, can be more readily detected without becoming implanted in

the silicon detector itself. Consequently, higher sensitivities, less damage, and longer lifetimes are observed in these membrane detectors compared to the more conventional epitaxial detectors.

Thin, large-area, high-strength SCS membranes have a number of other applications related to their flexibility. Guckel et al. [131] used KOH and the p^+ etch-stop method to generate up to 5-cm^2 membranes as thin as 2–4 μm. They mounted the structure adjacent to an electroded glass plate and caused it to vibrate electrostatically at the mechanical resonant frequency. Since the membrane is so large (typically 0.8 × 0.8 cm), the resonant frequency is in the audio range 10–12 kHz, yet the Q is maintained at a relatively high value, 23 000 in vacuum, 200 in air.

Pressure Transducers

Certainly the earliest and most commercially successful application of silicon micromechanics is in the area of pressure transducers [132]. In the practical piezoresistive approach, thin-film resistors are diffused into a silicon wafer and the silicon is etched from the backside to form a diaphragm by the methods outlined in Section III. Although the silicon can be etched isotropically or anisotropically from the backside (stopping the etching process after a fixed time), the dimensional control and design flexibility are dramatically improved by diffusing a p^+ etch-stop layer, growing an epitaxial film, and anisotropically etching through the wafer to the p^+ layer. As Clark and Wise showed [133], the membrane thickness is accurately controlled by the epi thickness and its uniformity is much improved. The resistors are located on the diaphragm, near the edges where the strains are largest. A pressure differential across the diaphragm cause deflections which induce strains in the diaphragm thereby modulating the resistor values. Chips containing such membranes can be packaged with a reference pressure (e.g., vacuum) on one side. The first complete silicon pressure transducer catalog, distributed in August 1974 by National Semiconductor, described a broad line of transducers in which the sensor chip itself was bonded to another silicon wafer in a controlled atmosphere, as shown in Fig. 34(a), so that the reference pressure was maintained within the resulting hermetically sealed cavity. This configuration was also described in 1972 by Brooks et al. [74] who employed a modified, thin-film anodic bond (as shown in Fig. 13(b)) to seal the two silicon pieces. Silicon eutectic bonding techniques (Au, Au–Sn) and glass-frit sealing are also used frequently in these applications. The National Semiconductor transducer unit is mounted in a hybrid package containing a separate bridge detector, amplifier, and thick-film trimmable resistors. The configuration of Fig. 34(a) suffers from the fact that the pressure to be sensed is incident on the top surface of the silicon chips where the sensitive circuitry is located. Although relatively thick parylene coatings [15] cover the membrane and chip surfaces of this silicon transducer line, it is clear that a different mounting technique is required for many applications in which the unknown pressure can be applied to the less-sensitive backside.

Presently, Foxboro, National Semiconductor, and other companies frequently mount chips in a manner similar to that shown in Fig. 34(b) such that the active chip surface is now the reference side. Chips are bonded both to ceramic and to stainless-steel assemblies. Many commercial sensor units are not yet even hybrid package assemblies and signal conditioning is accomplished by external circuitry. Recently, however, the

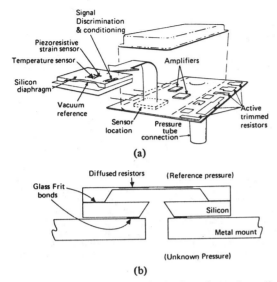

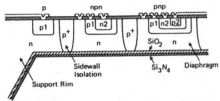

Fig. 34. Piezoresistive pressure transducers have been the earliest and most successful mechanical applications of silicon. At least eight firms now manufacture such sensors, rated for pressures as high as 10 000 psi. (a) Hybrid sensor package marketed by National Semiconductor. The resistor bridge on the silicon diaphragm is monitored by an adjacent detector/amplifier/temperature-compensation chip and trimmable thick-film resistors. Figure courtesy of National Semiconductor Corporation. A cross section of a typical mounted sensor chip is shown in (b). Chip bonding methods include eutectic bonding, anodic bonding, and glass-frit sealing.

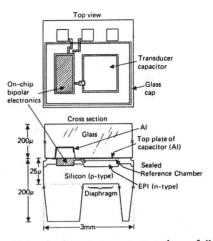

Fig. 36. One silicon diaphragm pressure transducer fully integrated with on-chip electronics is the capacitive sensor assembly demonstrated by Sander et al. [136] at Stanford. The design of this device has been directed toward implantable, biomedical applications. An etched glass plate, bonded to the silicon according to Fig. 13(c), hermetically seals the circuitry and also contains the top capacitor electrode. Figure courtesy of J. Knutti.

Fig. 35. Piezoresistive silicon pressure transducers with integrated detection and signal conditioning circuitry are now available commercially. Borky and Wise [134] have fabricated a pressure sensor (shown here in cross section) in which the bipolar circuitry is located on the deflectable diaphragm itself. Figure courtesy of K. D. Wise.

Microswitch division of Honeywell has been marketing an integrated pressure transducer chip which incorporates some of the required signal-conditioning circuitry as well as the piezoresistive sensing diaphragm itself. A further indication of future commercial developments along these lines can be seen in the fully integrated and temperature-compensated sensors demonstrated by Borky and Wise [134], and by Ko et al. [135]. A cross-sectional view of the membrane transducer fabricated by Borky and Wise, Fig. 35, shows how the signal-conditioning circuitry was incorporated on the membrane itself, thereby minimizing the chip area and providing improved electrical isolation between the bipolar transistors.

Several companies supply transducers covering a wide range of applications; vacuum, differential, absolute, and gauge as high as 10 000 psi. Specific areas of application include fluid flow, flow velocity, barometers, and acoustic sensors (up to about 5 kHz) to be used in medical applications, pneumatic process controllers, as well as automotive, marine, and aviation diagnostics. In addition, substantial experience in reliability has been obtained. One of Foxboro's models has been cycled from 0 to 10 000 psi at 40 Hz for over 5×10^9 cycles (4 years) without degradation.

Few engineering references are available in the open literature concerning the design of silicon pressure transducers. In a recent paper, however, Clark and Wise [133] developed a comprehensive stress-strain analysis of these diaphragm sensors from a finite-element approach. Dimensional tolerances, piezoresistive temperature coefficients, optimum size and placement of resistors, the effects of potential process-induced asymmetries in the structure of the membranes, and, of course, pressure sensitivities have been considered in their treatment.

The sensitivities and temperature coefficients of membrane-based, capacitively coupled (CC) sensors were also calculated by Clark and Wise and found to be substantially superior to the piezoresistive coupled (PC) sensors. For the geometry and mounting scheme, they proposed, however, (with the very thin—2-μm—capacitive electrode gap exposed to the unknown gas), it was concluded that overriding problems would be encountered in packaging and in maintaining the electrode gap free of contaminants and condensates.

Recently, however, a highly sophisticated, fully integrated capacitive pressure sensor has been designed and fabricated at Stanford by C. Sander et al. [136]. As shown in Fig. 36, the device employs many of the micromechanical techniques already discussed. A silicon membrane serves as the deflectable element; wells etched into the top 7740 glass plate are used both as the spacer region between the two electrodes of the variable capacitor and as the discharge protection region above the circuitry, the principle of which was discussed in Section III (Fig. 13(c)). Field-assisted thermal bonding seals the silicon chip to the glass plate and assures the hermeticity of the reference chamber (which is normally kept at a vacuum level). The frequency-modulated bipolar detection circuitry is designed to charge the capacitive element with a constant current source, firing a Schmitt trigger when the capacitor reaches a given voltage. Clearly the firing rate of the Schmitt trigger will be determined by the value of the capacitor—or the separation of the capacitor plates. Perhaps one of the more significant aspects of this pressure transducer design is that the fabrication procedure was carefully planned to satisfy the primary objectives and advantages of silicon micromechanics. In particular, the silicon wafer and the large glass plate are

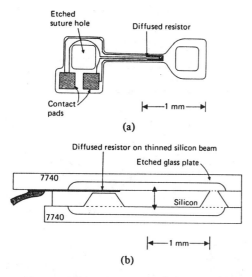

(a)

(b)

Fig. 37. Research at Stanford has extended the basic piezoresistive pressure sensor concept to complex strain sensor and accelerometer geometries for biomedical implantation applications. The strain sensor (a) contains a diffused piezoresistive element as well as etched suture loops on either end. Figure adapted from [137]. The accelerometer (b) is a hermetically sealed silicon cantilever beam accelerometer [75] sandwiched between two anodically bonded glass plates for passivation and for protection from corrosive body fluids. Figure adapted, courtesy of L. Roylance.

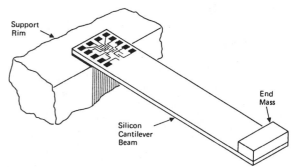

Fig. 38. The mechanical resonant frequency of a silicon cantilever beam was excited in the "Resonistor" by applying a sinusoidal current signal (at 1/2 of the resonant frequency) to a resistor on the silicon surface. These thermal fluctuations cause periodic vibrations of the beam which are detected by on-chip piezoresistive sensors. The signal from the sensor was employed in a feedback loop to detect and stabilize resonant oscillations. The function proposed for the "Resonistor" was a tuned, crystal oscillator. Adapted from Wilfinger *et al.* [138].

both processed using conventional IC techniques, both plates are anodically bonded, and only then is the entire assembly diced up into completed, fully functional transducer chips. Inexpensive batch fabrication methods, as required for practical, commercial silicon IC applications, are followed throughout.

Other Piezoresistive Devices

The principle of piezoresistance has been employed in other devices analogous to pressure transducers. J. B. Angell and co-workers at the Stanford Integrated Circuits Laboratory have advanced this technique to a high level of creativity. His group has been particularly concerned with *in vivo* biomedical applications. Fig. 37(a), for example, shows a silicon strain transducer etched from a wafer which has been successfully implanted and operated in the oviduct of a rabbit for periods exceeding a month [137]. Its dimensions are 1.7 X 0.7 mm by 35 μm thick. Two bonding pads on the left portion of the element make contact to a u-shaped resistor diffused along the narrow central bar. Two suture loops at both ends are also etched in the single-crystal transducer to facilitate attachment to internal tissue. Similar miniature strain transducers, etched from silicon, are now available commercially.

A cantilever beam, microminiature accelerometer, also intended for *in vivo* biomedial studies, is shown in Fig. 37(b). It was developed by Roylance and Angell [75] at Stanford and represents more than an order of magnitude reduction in volume and mass compared to commercially available accelerometers with equivalent sensitivity. Sutured to the heart muscle, it is light enough (<0.02 g) to allow high-accuracy high-sensitivity measurements of heart muscle accelerations with negligible transducer loading effects. It is also small enough (2 X 3 X 0.06 mm) for several to fit inside a pill which, when swallowed, would monitor the magnitude and direction of the pill's movement through the intestinal tract,

while telemetry circuitry inside the pill transmits the signals to an external receiver.

Fabrication of the silicon sensor element follows typical micromechanical processing techniques—a resistor is diffused into the surface and the cantilever beam is separated from the surrounding silicon by etching from both sides of the wafer using an anisotropic etchant. The thickness of the thinned region of the beam, in which the resistor is diffused, is controlled by first etching a narrow V-groove on the top surface of the wafer (whose depth is well defined by the width of the pattern as in the case of ink jet nozzles) and, next, a wider V-groove on the bottom of the wafer. When the etched holes on either side meet (determined by continual optical monitoring of the wafer), etching is stopped. The remaining thinned region corresponds approximately to the depth of the V-groove on the top surface. The final form is that of a very thin (15-μm) cantilever beam active sensing element with a silicon (or gold) mass attached to the free end, surrounded by a thick silicon support structure. A second diffused resistor is located on the support structure, but adjacent to the active piezoresistor for use as a static reference value and for temperature compensation. The chip is anodically bonded on both sides to two glass plates with wells etched into them. This sealed cavity protects the active element by hermetically sealing it from the external environment, provides mechanical motion limits to prevent overdeflection, yet allows the beam to deflect freely within those limits. Resonant frequencies of 500 to 2000 Hz have been observed and accelerations of less than $10^{-3}g$ have been detected. Such devices would be extremely interesting in fatigue and yield stress studies.

An early micromechanical device with a unique mode of operation was demonstrated by Wilfinger *et al.* [138], and also made use of the piezoresistive effect in silicon. As shown in Fig. 38, a rectangular silicon chip (typically 0.9 X 0.076 X 0.02 cm) was bonded by one end to a fixed holder, forming a silicon cantilever beam. Near the attached edge of the bond, a circuit was defined which contained a heat-dissipating resistor positioned such that the thermal gradients it generated caused a deflection of the beam due to thermal expansion near the (hotter) resistor, relative to the (cooler) backside of the chip. These deflections were detected by an on-chip piezoresistive bridge circuit, amplified, and fed back to the heating resistors to oscillate the beam at resonance. Since the beam

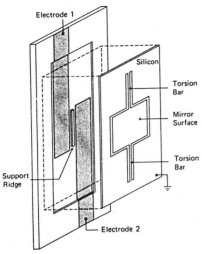

Fig. 39. Exploded view of silicon torsion mirror structures showing the etched well, support ridge, and evaporated electrodes on the glass substrate. From [139].

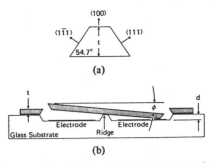

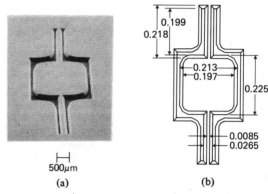

Fig. 40. (a) Cross section of the anisotropically etched torsion bar where $t = 134$ μm. (b) Cross section of the mirror element defining the deflection angle ϕ, where $d = 12.5$ μm and a voltage is applied to the electrode on the right.

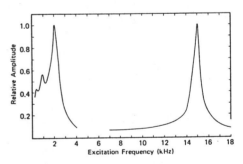

Fig. 41. (a) SEM of typical torsion mirror (tilted 60°) and (b) measured dimensions of 15-kHz mirror element (in cm). The SEM photo is a view of the mirror from the back surface where the electrostatic fields are applied.

has a very-well-defined resonant frequency and a high Q (>2000), the output from the bridge exhibits a sharp peak when the heated resistor is excited at that mechanical resonant frequency. This oscillator function has been demonstrated in the range of 1.4 to 200 kHz, and stable, high-Q oscillations were maintained in these beams continuously for over a year with no signs of fatigue.

Silicon Torsional Mirror

This section closes with the description of a device which is not actually a membrane structure, but is related to the strain-measurement mechanisms discussed above and has important implications concerning the future capabilities and potential applications of SCS micromechanical technology. The device is a high-frequency torsional scanning mirror [139] made from SCS using conventional silicon processing methods. An exploded view, shown in Fig. 39, indicates the silicon chip with the anisotropically etched mirror and torsion bar pattern, as well as the glass substrate with etched well, central support ridge, and electrodes deposited in the well. After the two pieces are clamped together, the silicon chip is electrically grounded and a high voltage is applied alternately to the two electrodes which are very closely spaced to the mirror, thereby electrostatically deflecting the mirror from one side to the other resulting in twisting motions about the silicon torsion bars. If the electrode excitation frequency corresponds to the natural mechanical torsional frequency of the mirror/torsion bar assembly, the mirror will resonate back and forth in a torsional mode. The central ridge in the etched well was found to be necessary to eliminate transverse oscillations of the mirror assembly. A cross-sectional view of the torsional bar and of the mirror deflections is shown in Fig. 40. The well-defined angular shapes in the silicon, which are also seen in the SEM (scanning-electron microscope) photograph in Fig. 41(a) (taken from the backside of the silicon chip), result, of course, from the anisotropic etchant. Fig. 41(b) gives typical device dimensions used in the results and the calculations to follow.

Reasonably accurate predictions of the torsional resonant frequency can be obtained from the equation [140]

$$f_R = \frac{1}{2\pi} \sqrt{\frac{12KEt^3}{\rho l b^4 (1 + \nu)}} \qquad (1)$$

Fig. 42. Deflection amplitude versus drive frequency for two mirrors with differing resonant frequencies.

where E is Young's modulus of silicon ($E = 1.9 \times 10^{12}$ dyne/cm^2), t is the thickness of the wafer ($t \sim 132$ μm), ρ is the density of silicon ($\rho = 2.32$ g/cm^3), ν is Poisson's ratio ($\nu = 0.09$) [141], l is the length of the torsion bar, b is the dimension of the square mirror, and K is a constant depending on the cross-sectional shape of the torsion bar ($K \sim 0.24$). For these parameters, we calculate $f_R = 16.3$ kHz, compared to the experimental value of 15 kHz for the device shown in Fig. 41. The resonant behavior of two experimental torsional mirrors is plotted in Fig. 42.

While complex damping mechanisms, including viscous air-damping and proximity effects due to closely spaced electrodes [142], dominate the deflection amplitudes near resonance, close agreement between theory and experiment can be obtained at frequencies far enough below resonance and at deflection angles small compared to the maximum deflection angle $\phi_{max} = 2d/b$, illustrated in Fig. 43. Under these restric-

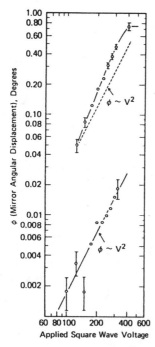

Fig. 43. Experimental deflections of torsion mirror. Resonant displacements are shown at the top, off-resonance at the bottom. Note departure from square-law dependence at resonance.

tions, it can be shown that [143]

$$\phi = \frac{\epsilon_0 V^2 l b^3 (1+\nu)}{16 K E d^2 t^4} A \qquad (2)$$

where ϵ_0 is the free-space dielectric permittivity, V is the applied voltage, d is the steady-state electrode/silicon separation, and A is an areal correction factor ($A \sim 0.8$) due to the fact that the active electrode area is somewhat less than half the area of the mirror. We can see from the lower curve in Fig. 43, that the square-law dependence on voltage is confirmed by the data and that the observed deflection amplitudes are only about 20 percent below those predicted by (2). As expected, nonlinearities in the deflection forces are also evident in Fig. 43 during operation at resonance, since the square-law dependence is not maintained.

Optically, silicon possesses an intrinsic advantage over common glass or quartz mirrors in high-frequency scanners because of its high E/ρ ratio, typically 3 times larger than quartz. Using the mirror distortion formulation of Brosens [144], $\frac{1}{3}$ smaller distortions are expected in rapidly vibrated silicon mirrors, compared to quartz mirrors of the same dimensions.

Of prime importance in the study of mechanical reliability is the calculation of maximum stress levels encountered. The maximum stress of a shaft with the trapezoidal cross section of Fig. 38(a) occurs at the midpoint of each side and is given by [145]

$$\tau_{max} \simeq \frac{16 K E}{(1+\nu)} \left(\frac{dt}{bl} \right) \qquad (3)$$

when the torsion bars are under maximum torque ($\phi = \phi_{max}$). For our geometry, this corresponds to about 2.5×10^9 dyne/cm^2 (36 000 psi), or more than an order of magnitude below the fracture stresses found in the early work of Pearson et al. [11]. Reliability, then, is predicted to be high.

This initial prediction of reliability was verified in a series of life tests in which mirrors were continuously vibrated at reso-

nance, for periods of several months. Despite being subjected to peak accelerations of over 3.5×10^6 cm/s^2 (3600 g's), dynamic stresses in the shaft of over 2.5×10^9 dyne/cm^2 (36 000 psi), 30 000 times a second for 70 days ($\sim 10^{11}$ cycles) no stress cracking or deterioration in performance was detected in the SEM for devices which had been properly etched and mounted. After a dislocation revealing etch on this same sample, an enhanced dislocation density appeared near the fixed end of only one of the torsion bars. Since this effect was observed only on one bar, it was presumed to be due to an asymmetry in the manual mounting and gluing procedure, resulting in some unwanted traverse oscillations.

These calculations and observations strongly indicate that silicon mechanical devices, such as the torsion mirror described here, can have very high fatigue strengths and exhibit high reliability. Such results are not unexpected, however, from an analysis of the mechanisms of fatigue. It is well known that, whatever the process, fatigue-induced microcracks initiate primarily at free surfaces where stresses are highest and surface imperfections might cause additional stress concentration points [19]. Since etched silicon surfaces can be extremely flat with low defect and dislocation damage to begin with, SCS structures with etched surfaces are expected, fundamentally, to possess enhanced fatigue strengths. In addition, the few microcracks which do develop at surface dislocations and defects typically grow during those portions of the stress cycle which put the surface of the material in tension. By placing the surface of the structure under constant, uniform compression, then, enhanced fatigue strengths have been observed in many materials. In the case of silicon, we have seen how thin Si_3N_4 films, while themselves being in tension, actually compress the silicon directly underneath. Such layers may be expected to enhance even further the already fundamentally high fatigue strength of SCS in this and other micromechanical applications.

A comparison of the silicon scanner to conventional, commercial electromagnetic and piezoelectric scanners is presented in Table III. The most significant advantages are ease of fabrication, low distortion, and high performance at high frequencies.

VI. Thin Cantilever Beams

Resonant Gate Transistor

Micromechanics as a silicon-based device technology was actually initiated by H. C. Nathanson et al. [147], [148] at Westinghouse Research Laboratories in 1965 when he and R. A. Wickstrom introduced the resonant gate transistor (RGT). As shown in Fig. 44, this device consists of a plated-metal cantilever beam, suspended over the channel region of an MOS transistor. Fabrication of the beam is simply accomplished by first depositing and delineating a spacer layer. Next, photoresist is applied and removed in those regions where the beam is to be plated. After plating, the photoresist is stripped and the spacer layer is etched away, leaving the plated beam suspended above the surface by a distance corresponding to the thickness of the spacer film. Typical dimensions employed by Nathanson et al. were, for example, beam length 240 μm, beam thickness 4.0 μm, beam-to-substrate separation 10 μm.

Operating as a high-Q electromechanical filter, the cantilever beam of the RGT serves as the gate electrode of a surface MOSFET. A dc voltage applied to the beam biases the transistor at a convenient operating point while the input signal electrostatically attracts the beam through the input force plate,

TABLE III

	Silicon Mirror		Electromagnetic[a]		Piezoelectric[a]	
Fabrication procedure	Batch fabrication of two lithographically processed plates		Complex mechanical assembly of many parts		Two bonded ceramic plates with separate mirror attached	
Frequency scan angle	15 kHz ±1°	50 kHz [b] ±2°	1 kHz ±30°	15 kHz ±2°	1 kHz ±5°	40 kHz ±0.1°
Power	<0.1 W dissipated in drive circuitry		≈0.5 W dissipated in assembly		<0.1 W dissipated in drive circuitry	
Relative distortion	1/3 (silicon mirror)		1 (quartz mirror)		1 (quartz mirror)	
Reliability	≈10^{12} cycles demonstrated		Very high		Very high	
Other	High voltage		High power Heavy assembly		High voltage Off-axis mirror Creep and hysteresis	

[a] See Ref. 146.
[b] Projected Performance.

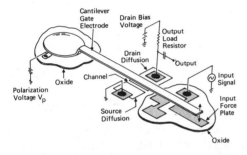

Fig. 44. The earliest micromechanical cantilever beam experiments were conceived at Westinghouse and based on the plated-metal configuration shown here. Operated as an analog filter, the input signal causes the plated beam to vibrate. Only when the signal contains a frequency component corresponding to half the beam mechanical resonant frequency are the beam motions large enough to induce an output from the underlying MOS structure [147], [148]. Figure courtesy of H. Nathanson.

thereby effectively increasing the capacitance between the beam and the channel region of the MOS transistor. This change in capacitance results in a variation of the channel potential and a consequent modulation of the current through the transistor. Devices with resonant frequencies (f_R) from 1 to 132 kHz, Q's as high as 500, and temperature coefficients of f_R as low as 90 ppm°C were described and extensively analyzed by Nathanson *et al.* They constructed high-Q filters, coupled multipole filters, and integrated oscillators based on this fabrication concept. Since the electrostatically induced motions of the beam are only appreciable at the beam resonant frequency, the net Q of the filter assembly is equivalent to the mechanical Q of the cantilever beam. Typical ac deflection amplitudes of the beams at resonance for input signals of about 1 V were ~50 nm.

Practical, commercial utilization of RGT's have never been realized for a number of reasons, some of which relate to technology problems, and some having to do with overall trends in electronics. The most serious technical difficulties discussed by Nathanson *et al.* are 1) reproducibility and predictability of resonant frequencies, 2) temperature stability, and 3) potential limitations on lifetime due to fatigue. The inherent inaccuracies suffered in this type of selective patterned plating limited reproducibility to 20–30 percent over a given wafer in the studies described here. It is not clear if

this spread can be improved to much better than 10 percent even with more stringent controls. Temperature stability was related to the temperature coefficient of Young's modulus of the plated beam material, about 240 ppm for gold (the temperature coefficient of f_R is about half this value). Although this problem could be solved, in principle, by plating low-temperature coefficient alloys, such experiments have not yet been demonstrated. Lifetime limitations due to fatigue is a more fundamental problem. Although the strain experienced by the cantilever beam is small (~10^{-5}), the stability of a polycrystalline metal film vibrated at a high frequency (e.g., 100 kHz) approaching 10^{14} times (10 years) is uncertain. Indeed, it is known, for example, that polycrystalline piezoelectric resonators will experience creep after continued operation in the 10's of kilohertz. (Single-crystal or totally amorphous materials, on the other hand, exhibit much higher strengths and resistance to fatigue.) These technological difficulties, together with trends in electronics toward digital circuits, higher frequencies of operation (>1 MHz for D/A and A/D conversion), higher accuracies, and lower voltages have conspired to limit the usefulness of devices like the RGT. The crux of the problem is that the RGT filter, while simpler and smaller than equivalent all-electronic circuits, was forced to compete on a basis which challenged well-established conventions in circuit fabrication, which did not take real advantage of its unique mechanical principles, and which pitted it against a very powerful, fast-moving, incredibly versatile all-electronic technology. For all these reasons, conceptually similar devices, which will be discussed below, can only hope to be successful if they 1) provide functions which cannot easily be duplicated by *any* conventional analog or digital circuit, 2) satisfactorily solve the inherent problems of mechanical reliability and reproducibility, and 3) are fabricated by techniques totally compatible with standard IC processing since low-cost high-yield device technologies are most likely only if well-established batch fabrication processes can be employed.

Micromechanical Light Modulator Arrays

The first condition was addressed during the early 1970's when several attempts to fabricate two-dimensional light-modulator arrays were undertaken with various degrees of short-lived success [149]–[152]. Conventional silicon circuits

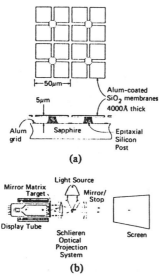

(a)

(b)

Fig. 45. Taking advantage of the excellent mechanical behavior of thermally grown SiO$_2$, Westinghouse fabricated large arrays (up to 400 000) of deflectable oxide flaps (0.35 μm thick) and demonstrated an electron-beam addressed image projection display with intrinsic image storage capability [153], [154]. (a) Top view and cross section of oxide flap geometry. The electron-beam addressing and optical projection assembly schematic is shown in (b). Figure courtesy of H. Nathanson.

certainly cannot modulate light. These membrane light modulator arrays were composed of thin ($\sim$200-nm) perforated metal sheets, suspended over support structures ($\sim$5 μm high) on various substrates. Small segments of the metal sheet behaved as independent mirror elements and could be electrostatically deflected by a scanned electron-beam or by voltages applied to underlying deposited electrodes. The preliminary device demonstrated by van Raalte [149], for example, contained 250 000 individually deflectable thin metal regions. Clearly, the light valve function had not previously been successfully implemented by any other thin-film technology.

Nevertheless, these metal-film designs did not satisfy the second and third requirements. Not only were they potentially susceptible to metal fatigue (gold has a particularly low fatigue strength), but the principles of fabrication were tedious, complex, and intrinsically difficult. In addition, portions of the fabrication process were not even remotely compatible with conventional IC techniques.

During their studies of silicon point field-emitter arrays discussed in Section IV [79], [80], Thomas and Nathanson found that large arrays of deformagraphic elements could also be built using totally conventional and compatible IC processing methods [153], [154]. In addition, while previous workers employed metal films as the deformagraphic material, they used thermally grown amorphous SiO$_2$ films with fatigue strengths expected to be substantially greater. The structure consisted of a 5-μm silicon epitaxial layer grown over sapphire and covered with a 350-nm-thick thermal oxide. First the oxide is etched into the pattern shown in Fig. 45(a), which defines the final cloverleaf shape of the deflectable oxide membranes. Next, a much thinner oxide is grown and etched away only along the grid lines separating individual light-valve elements. The crucial step involves isotropic etching of the silicon from underneath the oxide elements, undercutting the thin film until only narrow pedestals remain supporting the oxide structures. Finally, the thin oxide remaining in the four slits of each self-supported SiO$_2$ plate is etched away and the

entire array is rinsed and dried. Evaporation of a thin (30-nm) aluminum film makes the oxide beams highly reflective and deposits a metal grid through the openings in the oxide onto the sapphire substrate.

Writing an image on the Mirror-Matrix Display tube is done by raster scanning a modulated electron beam across the target. Each individual mirror element will become charged to a voltage dependent on the integrated incident electron current. Typically, the electron-beam voltage is specified to cause a large secondary electron emission coefficient. The secondaries are collected by a metal grid closely spaced to the target (100 μm) and the mirror metallizations become charged positively. When a negative voltage is then applied to the stationary metal grid on the sapphire surface, electrostatic forces mechanically deflect those elements which have become charged. The pattern of deflected and undeflected beams is illuminated from the backside through the sapphire substrate and the reflected light is projected in a Schlieren optical system such that the deflected beams appear as bright spots on a ground-glass screen. A schematic of the optical and electron-beam illumination systems is shown in Fig. 45(b). Since the metallizations on the mirror surfaces are electrically well insulated from the metal grid on the sapphire surface, charge images can be stored on the deformagraphic array for periods up to many days. At the same time, fast erasure can be accomplished by biasing the external grid negatively and flooding the array with low-energy electrons which equalize the varying potentials across the mirror surfaces.

The particular cloverleaf geometry implemented in the mirror matrix target (MMT) design exhibits very high contrast ratios since the deflected "leaves" scatter light from the optical system at 45° from the primary diffracted radiation. An opaque, cross-shaped stop placed at the focal point of the imaged light beam will only pass that portion of the light reflected from bent "leaves." Contrast ratios over 10 : 1 have been attained in this way. Over and above the optical advantages, however, the mirror matrix target satisfies all the basic requirements for a potentially practicable micromechanical application. The principles of fabrication are completely compatible with conventional silicon processing; it performs a function (light modulation and image storage) not ordinarily associated with silicon technology; and it greatly alleviates the potential fatigue problems of previous metal-film deformagraphic devices by using amorphous SiO$_2$ as the active, cantilever beam material. Unfortunately, one of the fabrication problems of the MMT was the isotropic Si etch. Since this step did not have a self-stopping feature, and the dimensions of the post are critical, small etch-rate variations over an array could have a dramatic impact on the yield of a large imaging array.

The MMT has never become a viable, commercial technology primarily because it did not offer a great advantage over conventional video monitors. Image resolution of both video and deformagraphic displays, for example, depend primarily on the number of resolvable spots of the electron-beam system itself (not the form of the target) and, therefore, should be identical for equivalent electron optics and deflection electronics. In addition, since electron-beam scan-to-scan positioning precision is typically not high, it would be very difficult to write on single individual mirrors accurately scan after scan. This means a single written picture element would actually spread, on the average, over at least four mirrors. For an optical image resolution of 10^6 PEL's, then, at least 4 $\times$ 10^6 mirrors would be required—a formidable task for high-yield photolithography. Finally, the potential savings in the deformagraphic systems due to their intrinsic storage capability is more than offset by the expensive mirror matrix target itself (equivalent in area to at least 20 high-density IC chips when yield criteria are taken into account) as well as the imaging lamp and optics, which are required *in addition to* the electron tube and its associated high-voltage circuits. Historically, high-density displays have always been exceptionally difficult and demanding technologies. While silicon micromechanics may eventually find a practical implementation for displays (especially if silicon-driving circuitry can be integrated on the same chip, matrix addressing the two-dimensional array of mirrors—all electronically), we need not be that ambitious to find important, novel, much-needed device applications amenable to silicon micromechanical techniques.

SiO₂ Cantilever Beams

The full impact of micromechanical cantilever beam techniques was not completely evident until the development of controlled anisotropic undercut etching, as described in Section III. Flexible, fatigue-resistant, amorphous, insulating cantilever beams, co-planar with the silicon surface, closely spaced to a stationary deflection electrode, and fabricated on ordinary silicon with conventional integrated circuitry on the same chip can be extremely versatile electromechanical transducers. A small, linear array of voltage-addressable optical modulators was the first demonstration of this new fabrication technique [155]. Since the method forms the basis of other devices discussed in the rest of this section, we will describe it in some detail. The silicon wafer is heavily doped with boron in those regions where cantilever beams are desired. Since this film will serve as an etch-stop layer during fabrication of the beams, as well as a deflection electrode during operation of the device, the dopant concentration at the surface must be high enough initially to effectively inhibit subsequent anisotropic etching even after the many high-temperature cycles required to complete the structure and its associated electronic circuitry. These additional high-temperature steps tend to decrease the original surface concentration. A level of 7×10^{19} cm⁻³ is usually taken to be the minimum peak value necessary to stop the etchant. Next, an epitaxial layer is grown on the wafer to a thickness corresponding approximately to the desired electrode separation. Since the buried p⁺ region is so heavily doped, the electrical quality of the epi grown over these regions may be poor and electronic devices probably should not be located directly above them. At this point, any necessary electronics can be fabricated on the epi-layer adjacent to the buried regions. After depositing (e.g., Si₃N₄) or growing (e.g., SiO₂) the insulating material to be used for the cantilever beam, a thin metal film (typically 30 to 40 nm of gold on chrome) is evaporated and delineated to form the upper electrode. Before silicon etching, care must be taken to insure that the aluminum circuit metallization is adequately passivated since the hot EDP solution will attack these films. Finally, the insulator is patterned as shown in Fig. 46(b), the exposed silicon is etched in EDP to undercut the insulator and free the cantilever beams, and the wafer is carefully rinsed and dried. Since further lithography steps are virtually impossible after the beam is freed, the anisotropic etch is always the last processing step.

As long as the patterns are correctly oriented in (110) directions, undercutting can be completely avoided in unwanted areas as described in Section III. Fig. 47 shows an optical

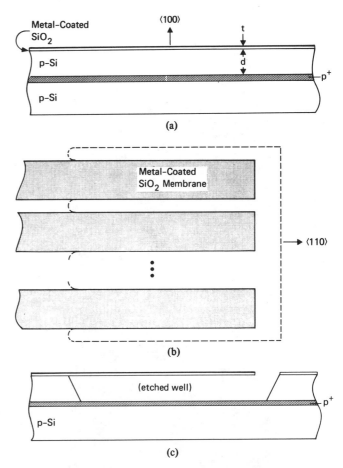

Fig. 46. (a) Cross section of the layers in the linear, cantilever-beam light modulator array. (b) The metal is etched into lines and the oxide is etched from between the lines within the dashed region (top view). (c) The silicon is anisotropically etched from under the oxide to release the beams. From [155]. Each beam can be independently deflected by applying a voltage between the top metallization and the p⁺ layer.

modulator array fabricated in this way. Since the patterns have been oriented correctly, only the beams are undercut, while the periphery of the etched rectangular hole is bounded by the (111) planes. At the same time, the buried p⁺ layer stops etching in the vertical direction. The dimensions of the beams in this first demonstration array were 100 μm long made from SiO₂ 0.5 μm thick. These are spaced 12 μm from the bottom of the well. A Cr–Au metallization, about 50 nm thick, serves as the top electrode.

Each individual beam can be deflected independently through electrostatic attraction simply by applying a voltage between the top electrode and the buried layer. For low voltages, the deflection amplitude varies approximately as the square of the voltage as indicated in Fig. 48. Both experiment and detailed calculations have shown that once the membrane tip is moved approximately a third of the way down into the etched well, the beam position becomes unstable and it will spontaneously deflect the rest of the way as a result of the rapid buildup of electrostatic forces near the tip. This effect was also analyzed by Nathanson *et al.* [147], [148] for the plated gold RGT beam. In the present geometry, the threshold voltage for spontaneous deflection can be shown to be approximately [156]

$$V_{\text{th}} = \sqrt{\frac{3Et^3d^3}{10\epsilon_0 l^4}} \qquad (4)$$

(a)

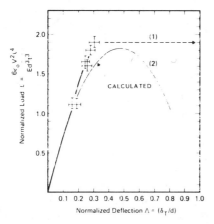

(b)

Fig. 47. SEM photographs of a completed 16-element light modulator array, ready for bonding. Note the high degree of flatness and uniformity of the beams. The SiO$_2$ beams are 100 μm long, 25 μm wide, and 0.5 μm thick.

Fig. 48. Calculated beam deflection is plotted here as the lighter line. At the threshold voltage (the peak in the calculated curve), the membrane tip will spontaneously deflect the rest of the way to the bottom of the pit, since its position is unstable along line (2). Experiment data, including the observed spontaneous threshold deflection (along line (1)), are also shown.

where l is the beam length, t is the thickness, ρ is the density, E is Young's modulus, ϵ_0 is the free-space permittivity, and d is the electrode separation. The 16-element array shown in

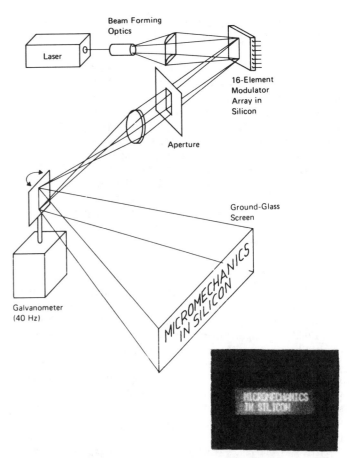

Fig. 49. Schematic of simple optical display employing the 16-beam array in Fig. 47. Each beam is operated as a single modulating element for the corresponding horizontal line in the display. The inset shows a photograph of the display projected onto the ground glass screen.

Fig. 47 was operated below threshold in the simple optical system shown in Fig. 49. A laser was focussed on all 16 mirrors simultaneously while an external aperture was adjusted such that only light reflected from bent membranes was allowed to pass. The resulting vertical line of light, now modulated at 16 points, was then scanned horizontally by a galvanometer to produce a two-dimensional display by rapidly deflecting each membrane independently in the correct sequence. An example of this display is given in the inset in Fig. 49.

Certainly, resonant frequency is an important parameter of such cantilever beams. The first mechanical resonance can be accurately calculated from [157]

$$f_R = 0.162 \frac{t}{l^2} \sqrt{\frac{E}{\rho}} K \qquad (5)$$

where K is a correction factor (close to one) depending on the density, Young's modulus, and thickness ratios of the metal layer to the insulating layer. For the array given above, $f_R = 45$ kHz. The highest resonant frequency yet observed is 1.25 MHz for the 8.3-μm-long, 95-nm-thick SiO$_2$ beams shown in Fig. 50(a).

Once the resonant frequency and the dimensions of a beam are measured, (5) can readily be used to determine Young's modulus of the insulating beam material. Fig. 50(b) shows an array of beams of various lengths fabricated specifically for such measurements. The vibrational amplitude of a reflected optical signal is plotted in Fig. 51 as a function of vibrational frequency for beams of five different lengths. Note that an

TABLE IV

	Insulator Thickness (Å)	Chromium Thickness (Å)	Assumed Insulator Density (g/cm³)	Measured Young's Modulus (10^{12} dyn/cm³)	Published Bulk Young's Modulus (10^{12} dyn/cm²)
SiO_2 (thermal-wet)	4250	150	2.2	0.57	
SiO_2 (thermal-dry)	3250	150	2.25	0.67	0.7 (fused silica)[a]
SiO_2 (sputtered)	4000	100	2.2	0.92	
Si_3N_4 (CVD)	3500	100	3.1	1.46	1.5[b]
Si_3N_4 (sputtered)	2900	100	3.1	1.3	
7059 glass (sputtered)	4200	50	2.25	0.52	0.6 (typ.)[3]
Nb_2O_5 (sputtered)	8400	50	4.47	0.85	1.6 (Nb_2O_3)[d]
α-SiC (glow discharge)	8800	50	~3.0	0.85	4.8 (single x-tal)[e]
Cr (sputtered)	---	---	7.2	1.8	2.8[f]

[a]See Refs. 164 and 165. [d]See Ref. 166.
[b]See Ref. 164. [e]See Ref. 10.
[c]See Ref. 165. [f]See Ref. 163.

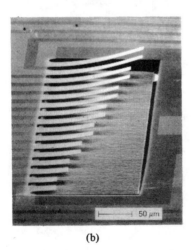

(a)

(b)

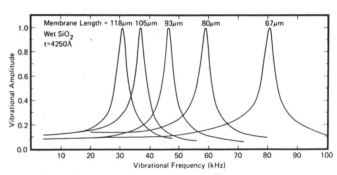

Fig. 51. Resonant behavior measurements of five SiO_2 beams with different lengths to determine Young's modulus of the thermally grown oxide layer.

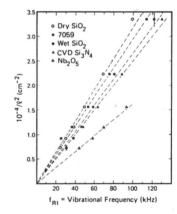

Fig. 52. Determination of Young's modulus according to (5). See Table IV for a complete listing of the films and their parameters. From [158].

Fig. 50. Various beam sizes have been fabricated from various materials for the measurements of Young's modulus. The beams in (a) are 8 μm long and less than 0.1 μm thick—they have a 1.2-MHz mechanical resonant frequency. The array in (b) is made from dry thermal SiO_2 and the beams range in length from 118 to 30 μm.

electrostatically deflected cantilever beam is a simple frequency doubler because the beam experiences an attractive force for positive as well as negative voltage swings. The resonant frequency is easily detected and can be shown, in Fig. 52, to follow the l^{-2} dependence predicted by (5). A series of measurements on a wide variety of deposited, insulating thin films was undertaken by Petersen and Guarnieri [158], the results of which are tabulated in Table IV. In the past, Young's modulus has been an extremely difficult parameter to measure in thin films due to the fragility of the samples and the concomitant problems in their handling and mounting as well as minutely deforming them and monitoring their motions [159]–[161]. This new, micromechanical technique, on the other hand, is simple, accurate, applicable to a wide range of materials and deposition methods, and useful in many thin-film microstructure studies. Chen and Muller, for example, have fabricated various composite p^+-Si/SiO_2 beams [162], as indicated in Fig. 53, studying their mechanical stability. Such measurements provide additional insight into the nature

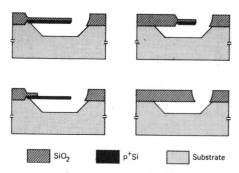

SiO₂   p⁺Si   Substrate

Fig. 53. Miniature cantilever beams of several configurations, including p⁺-Si layers, were fabricated by Jolly and Muller [162] using anisotropic etching together with the p⁺ etch-stop technique, as shown here in cross section. The warpage and stability of the beams were studied as a function of beam length (15–160 μm), width (5–40 μm), and thickness (0.2–0.8 μm). Figure courtesy of R. Muller.

of deposited films, how they differ from the corresponding bulk materials, how they depend on deposition conditions, and how multilayer films interact mechanically.

Integrated Accelerometers

While pressure transducers have been the first commercially important solid-state mechanical transducers, it is likely that accelerometers will become the next. Substantial literature already exists in this area, including the relatively large silicon cantilever beam piezoresistive device of Roylance and Angell which was described in Section V. Closely related to this structure in terms of fabrication principles is the folded cantilever beam accelerometer developed jointly by Signetics Corporation and Diax Corporation [167]. Initial work on the folded beam silicon accelerometer concentrated on piezoresistive strain sensors diffused into the silicon beam, while later studies by Chen et al. employed deposited piezoelectric ZnO sensors on identical silicon devices [168]. None of these demonstrations incorporated signal detection or conditioning circuitry on the sensor chip itself, however, as has been accomplished in pressure transducers.

Cantilever beam accelerometers, such as those mentioned above, made by etching clear through a wafer must address serious packaging problems. Special top and bottom motion-limiting plates, for example, must be included in the assembly to prevent beam damage during possible acceleration overshoots. A more problematical issue (which is also a concern in silicon pressure transducers) is the potential for unintentional, residual stresses resulting in hysteresis or drift in the detected signal. Such stresses could be developed, for example, from temperature-coefficient mismatches between the silicon and the various packaging and/or bonding materials.

Small oxide cantilever beams etched directly on the silicon surface alleviate many of the handling, mounting, and packaging problems of solid-state accelerometers since the beam itself is already rigidly attached to a thick silicon substrate. Furthermore, since the active beam element occupies just the top surface of the silicon, potential strains induced during mounting and packaging of the substrate will have little effect on the detector itself. An example of this type of accelerometer is the ZnO/SiO₂/Si composite beam structure demonstrated by Chen et al. [169]. As shown in Fig. 54, beam motions induce a voltage across the thin piezoelectric ZnO which is detected and amplified by adjacent, on-chip MOS circuitry. Careful device design and ZnO thin-film deposition techniques have yielded devices with low drift and hysteresis, potential

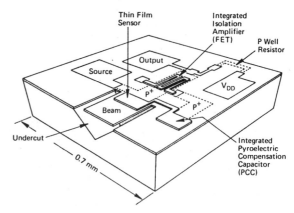

Fig. 54. Layout of the integrated, planar-processed PI-FET accelerometer demonstrated by Chen et al. [169]. As the cantilever beam vibrates, voltages induced in the thin-film, ZnO piezoelectric sensor element are detected by the on-chip MOS isolation amplifier. Figure courtesy of R. Muller.

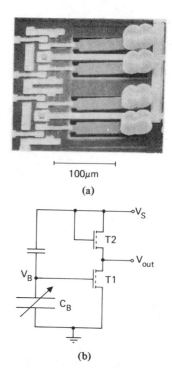

100μm

(a)

(b)

Fig. 55. The integrated accelerometer shown here in (a) consists of an SiO₂ cantilever beam sensor (loaded with a gold mass for increased sensitivity) coupled to an MOS detection circuit. The capacitance of the beam (typically 3.5 fF) is employed in a voltage divider network (b) from which small variations in the beam capacitance drive the detection transistor. From [170].

areas of concern when dealing with piezoelectric materials. By completely encapsulating the ZnO layer, for example, charge-storage times on the order of many days were observed. An advantage of the piezoelectric cantilever beam approach is that no buried p⁺ layer is required. One disadvantage is that ZnO is not yet an established commercial technology and may be difficult to adapt to standard IC fabrication procedures.

Another cantilever beam accelerometer [170] also integrated with on-chip circuitry is shown in Fig. 55. As the circuit schematic indicates, the capacitor formed by the p⁺-Si buried layer and the beam metallization are used in a capacitive voltage divider network to bias an MOS transistor in an active operating region, similar to the operating mode of the RGT. Motions

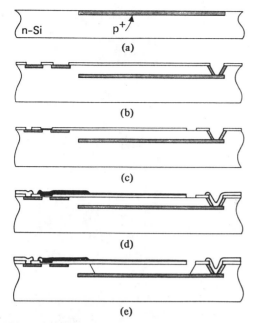

Fig. 56. Abbreviated processing schedule for the integrated acceler-ometer shown in Fig. 55. (a) Diffusion of p⁺ ground electrode and etch stop. (b) Implantation of source, drain, and the sidewalls of the anisotropically etched via. (c) Growth of gate oxide, opening of con-tact holes, and field oxide etch which defines the geometry of the cantilever beam. (d) Metallization steps including the Al circuit metal, the protective gold overcoat on the Al, and the thin Cr–Au beam metal. (e) Final step in which the silicon is anisotropically etched in a self-stopping procedure to undercut and release the metal-coated SiO₂ cantilever beam.

of the chip normal to the surface can result in movements of the cantilever beam which will modulate the source–drain voltage of the transistor by varying the voltage bias on its gate. Fabrication proceeds according to the schedule shown in Fig. 56. After diffusing the boron-doped buried layer in selected areas and growing n-type epi over the wafer, conventional Al-gate p-channel transistors are defined adjacent to the p⁺ regions. Electrical contact to the buried layer is accomplished by doping the sidewalls of a hole anisotropically etched down to the heavily doped region. This diffusion is done at the same time as the source–drain diffusion by a boron-ion-implant step. Next, Al is deposited and etched to form the metal inter-connections—even down in the hole used to contact the buried p⁺ layer.

Another very thin sputtered Cr–Au layer is deposited and etched for the electrode covering the cantilever beam. This same film is also employed as the plating base to selectively plate a protective Au layer over the already defined aluminum conductors, since the EDP etchant will attack thin Al films. Additional thick gold bumps have also been plated on the ends of the cantilever beams to increase their deflection amplitude with acceleration. Finally, the silicon is anisotropically etched in EDP to free the SiO₂ cantilever beams as described above.

The sensitivity of an SiO₂ beam of thickness t, width b, and length l with a concentrated load at the tip of mass M can be shown to be approximately

$$\Delta V/g = V_s \left(\frac{C_0}{3C_{eff}} \right) \frac{740 l^4}{Edb^2 t^3} M \quad (\text{V/g of acceleration}) \quad (6)$$

where d is the epi thickness, ϵ_0 is the free-space permittivity, and ρ and E are the density and Young's modulus of SiO₂, respectively. V_s is the circuit supply voltage and C_{eff} is the effective circuit capacitance in the voltage divider network

(~120 fF). Also

$$C_0 = \frac{\epsilon_0 lb}{d} \simeq 3.5 \text{ fF} \quad (7)$$

is the equilibrium capacitance between the beam and the buried layer. In good agreement with these calculations, sensitivities of 2.2 mV/g have been measured with loads of 0.35 μg, $V_s = -22$ V, and beam dimensions of 105 μm by 25 μm and $t = 0.5$ μm, $d = 7$ μm. Typical beam motions are about 60 nm/g of acceleration at the beam tip. Capacitance variations as small as 10 aF (corresponding to an acceleration of $0.25g$) have been detected with this sensor. Clearly, with such minute equilib-rium capacitances, these and other similar miniature solid-state accelerometers *require* on-chip integrated detection circuitry simply to maintain parasitics at a tolerable level. This trend toward higher levels of integration appears to be a continuing feature in many areas of sensor development.

Electromechanical Switches

In 1972, Frobenius *et al*. [171] reported a lithographically fabricated *threshold* accelerometer, conceptually different from the analog devices described above. Based on the plated beam technique developed for the RGT, the tip of the plated beam is allowed to make intermittent contact with another metallization on the surface during large accelerations of the chip normal to its surface. Although this was the first demon-stration of a photolithographically generated micromechanical electrical switch, the previously mentioned fatigue problem associated with plated-metal deflectable elements, in addition to its uncertain applicability, halted further development along these lines.

The insulator beam techniques used above for light modula-tors and accelerators, however, also lend themselves very well to the fabrication of small, fast, integrable, voltage-controlled electrical switching devices which are illustrated schematically in Fig. 57. Such four-terminal switches are not possible with the techniques described by Frobenius because the use of the *insulating* portion of the beam is crucial. Functional micro-mechanical switching was first realized [156] with a device like that shown schematically in Fig. 57(a). Construction of these switches follows the processing sequence shown in Fig. 58. The first few steps are identical to those employed in the optical modulator array; dope the surface with boron for the vertical etch-stop layer, grow an epitaxial spacer film (about 7 μm), deposit or grow the insulator for the cantilever beam (350-nm SiO₂), deposit a thin Cr–Au metallization (50 nm), delineate the metal lines and the insulator patterns to define the shape of the membrane, which is shown in cross section in Fig. 58(a). Plated metal crossovers and fixed electrode contact points are fabricated in a manner similar to the beams in the RGT and the threshold accelerometer of Frobenius *et al*. First a photoresist layer (PR1) is applied and patterned to provide contact holes through which gold will be plated and to define mesas over which gold will be plated to form the crossover and fixed electrode structures (Fig. 58(b)). Next, a second Cr–Au layer (0.3 μm) is evaporated over the entire wafer to serve as a plating base, and a second photoresist layer (PR2) is applied to define windows through which gold is selectively, electrochem-ically plated to a thickness of about 2 μm (Fig. 58(c)). After stripping the photoresist layers and excess plating base, the cantilever beams are released by etching the exposed silicon in EDP for about 20 min, then rinsed and dried (Fig. 58(d)).

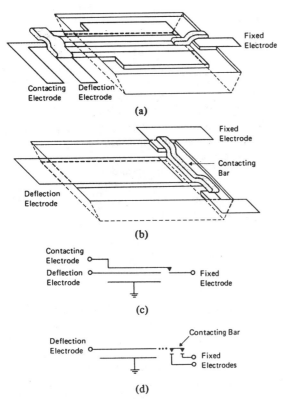

(a)

(b)

(c)

(d)

Fig. 57. Two designs of micromechanical switches. (a) The single-contact low-current design. (b) The double-contact configuration. (c) and (d) Suggested circuit representations of these devices. From [156].

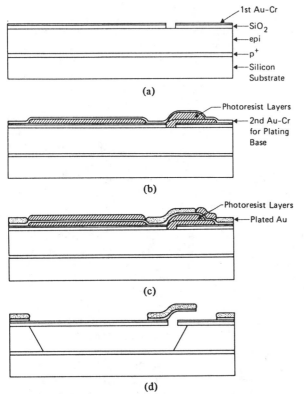

(a)

(b)

(c)

(d)

Fig. 58. Cross-sectional diagrams of a single-contact micromechanical switch at various stages during the fabrication procedure. (a) After first metal etch and oxide etch. (b) After evaporation of Au–Cr plating base. (c) After selective Au plating through photoresist holes. (d) Finished structure after photoresist stripping, removal of excess plating base, and EDP etch.

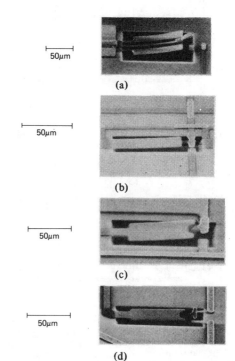

(a)

(b)

(c)

(d)

Fig. 59. Four different micromechanical switch designs. (a) Single-contact. (b) Double-contact with a contact bar as shown in Fig. 57(b). (c) and (d) Double-contact designs with two orientations of the fixed electrodes. From [172].

The SEM photographs in Fig. 59 show four different switch configurations, which can be classified as single- or double-contact designs [172]. Electrically, these devices behave as ideal, four-terminal, fully isolated, low-power, voltage-controlled switches. As a voltage is applied between the deflection electrode and the p^+ ground plane, the cantilever beam is deflected and the switch closes, connecting the contact electrode and the fixed electrode. Oscilloscope traces of pulsed switching is shown in Fig. 60. Single-contact switches are simple to operate and were the first to be demonstrated. However, their current-carrying capability is limited to less than 1 mA since all surface metallizations, which extend the length of the cantilever beam (including the current-carrying signal line leading up to the contact bar) must be thin to minimize its influence on the beam's mechanical characteristics. Much higher currents can be switched in the double-contact design since all the signal lines can be plated thick. With one extra masking step, even the fixed electrode regions under the contact bar can be plated prior to the application of the photoresist spacing layers.

While little consideration was given to the details of the contact electrode design in these early switch studies, it is clear that any practical utilization of the devices will rely critically on the ability to maximize current-carrying capability, contact force, reliability, and lifetime by optimizing the metallurgy and the geometry of the contacting electrodes. Gold, for example, is easy to electroplate and is certainly corrosion resistant but may be a poor choice as a contact metal because of its ductility and its self-welding tendency. Equally important is the design and configuration of the contacting surfaces for optimum electrical performance. Clearly, the development of micromechanical switches is in an early stage, yet the versatility of lithography and thin-film processing techniques permit a high degree of engineering design options for such micromechanical structures.

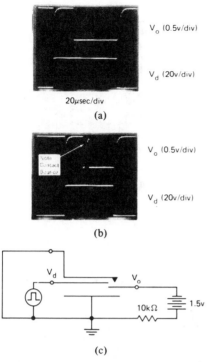

Fig. 60. (a) and (b) Oscilloscope traces of typical pulsed switching behavior for the circuit shown in (c). The deflection voltage in (a) is just above the switching threshold (~60 V), while the voltage in (b) has been increased to about 62 V. Note the delay time (~40 μs) between the application of the deflection voltage and the actual time the switch is closed. For higher deflection voltages (shown in the second oscilloscope trace), delay times are reduced and contact bounce effects are observed.

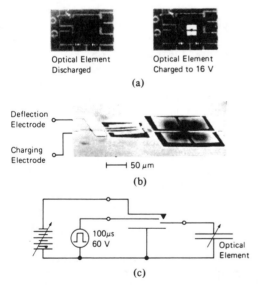

Fig. 61. A charge-storage application of micromechanical switches is illustrated by the circuit of (a) and (b). The capacitor leaves can be charged (and, therefore, deflected) by increasing the charging voltage to 16 V and pulsing the switch on for 100 μs. They will remain stored in this deflected position, shown at the right in (c), until the switch is pulsed again with the charging voltage reduced to zero, thereby releasing the leaves back to their discharged, or undeflected, positions.

Applications of these micromechanical switches range from telephone and analog signal switching arrays, to charge-storage circuits, to temperature [172] and magnetic field sensors. Fig. 61 shows an optical storage cell illustrating the high off-

state impedance of which these micromechanical switches are capable. In this device, the charge-storage capacitor is an MMT-type deflectable cloverleaf element. When the switch is activated and the cloverleaf element is charged up to 16 V, the four leaves deflect and appear bright in a dark-field illuminated microscope. Furthermore, cells can be stored for many hours in either the charged (bright) or discharged (dark) condition.

The optimum design of these switches for particular operating parameters involves tradeoffs between the three primary performance criteria of speed, current-carrying capacity, and switching voltage. While the switch resonant frequency (which is related to the switching speed) is given by

$$f_R = \frac{1}{2\pi} \sqrt{\frac{Ebt^3}{4l^3(M + 0.23m)}} \qquad (8)$$

where m is the mass of the oxide beam and M is the mass of the plated contact bar, the switching voltage can be expressed approximately as given in (4), repeated here

$$V_{th} = \sqrt{\frac{3Et^3d^3}{10\epsilon_0 l^4}}. \qquad (9)$$

From these relationships, we can understand the conflicting requirements for high performance on all aspects of device operation. Since both f_R and V_{th} depend in similar ways on the dimensions of the cantilever beam, high-speed devices imply high switching voltages. While this problem can be alleviated somewhat by reducing the electrode separation d (or epi thickness), the hold-off voltage across the contact electrodes will also be reduced. At the same time, high current capacity implies a large contact metallization M, which also limits switching speed through (8). For the geometries described here, it seems unlikely that resonant frequencies above 200–300 kHz can be realized at voltages less than 20 V in low-current (<1-mA) applications. At the other end of the spectrum, current levels above 1 A might be difficult to obtain in a single device. These micromechanical switch geometries, therefore, are not *very* high speed, nor *very* high current devices, but rather seem to fill a niche between transistors and conventional electromagnetic relays. Other micromechanical switch geometries, however, possibly related to the torsion mirror described in Section V, may be possible which have different ranges of performance.

The advantages of these switches are that they can be batch-fabricated in large arrays, they exhibit extremely high off-state to on-state impedance ratios, the off-state coupling capacitance is very small, switching and sustaining power is extremely low, switching speed is at least an order of magnitude faster than relays, and other electronic devices can easily be integrated on the same chip. Their most serious disadvantages seem to be a relatively high switching voltage (near 50 V) and relatively low current-carrying capability (probably less than 1 A). The ideal applications would be in systems requiring large arrays of medium-current switches or drivers with very low internal resistances. It is still difficult, for example, to integrate large arrays of bipolar drivers with internal resistances less than about 10 Ω.

VII. Silicon Head Technology

An area of particular importance which has already begun to feel the impact of silicon micromechanics is silicon head technology. Heads of all types have common features which are ideally solved by silicon micromechanical techniques. They

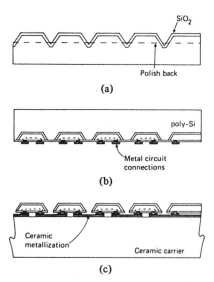

Fig. 62. Fabrication of the thermal print head found in several products manufactured by Texas Instruments is shown here. Based on earlier dielectric isolation schemes, SCS islands are embedded in a polysilicon substrate by a process of etching grooves, depositing SiO_2 followed by the thick poly-Si substrate (250 μm or more), and polishing back the single crystal just until the grooves are penetrated. Next, circuitry is fabricated in the single-crystal islands and the chip is bonded to a ceramic carrier. Finally, the polysilicon is selectively removed. Adapted from Bean and Runyan [175].

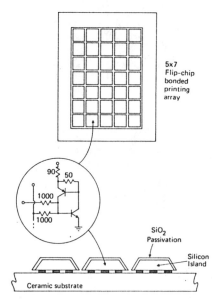

Fig. 63. A 5 × 7 array of silicon islands forms the complete print head in the T.I. Silent 700 computer terminals. Each island is about 25 μm thick and contains the bipolar power latching circuit shown, yet this thin silicon IC chip is also forcefully abraded against the paper while simultaneously thermally cycled to 260°C. Adapted from [174].

typically must be moved rapidly across some surface such as paper, magnetic media, or optical disk and so must be lightweight for high speed, accurate movement, and tracking abilities. In addition, it is always valuable, especially in multichannel heads, to have decoding and driving electronics available as close to the head as possible both to limit flexible wire attachments and to increase signal-to-noise ratios. As printing and recording technologies strive for higher and higher resolution, thin-film methods are being intensely pursued for the design and fabrication of active head elements. Micromechanics offers the potential of using thin-film methods not only for active head elements, but also for the precise micromachining of the passive structural assemblies on which the active head elements are located. Batch-fabrication, materials compatibility, simplified electronic interfaces, increased reliability, and low cost result from this strategy.

Ink jet nozzles have already been extensively discussed in Section IV and it seems probable that some variations on these designs will find their way into products in the 1980's. We will not describe these ink jet applications of silicon head technology further in this section.

Texas Instruments Thermal Print Head

Exploitation of silicon head technology has been successfully implemented in the thermal print head manufactured and sold by Texas Instruments [3]. New generations of high-speed (120 characters per second) print heads are now employed in the TI 780 series computer terminals and the 400 element Tigris plotter/printer [173], [174]. Based on the dielectric isolation process described by Bean and Runyan [175], a very abbreviated schematic of the fabrication procedure is illustrated in Fig. 62. Anisotropic etching is used to etch a grid in a silicon wafer which defines a 5 × 7 array of SCS mesas, for example. After growing an insulating film (SiO_2) over the surface, a thick (10–15-mils) layer of polysilicon is deposited, the wafer is turned upside down, and the single crystal is mechanically polished (or etched) back just until the original mesas are reached as shown in Fig. 62(b). At this stage, typical dimensions of the single-crystal mesas are 250 μm square by 25 μm thick. Conventional bipolar circuit fabrication methods are now used to define circuitry on each individual mesa as well as some peripheral circuitry in the single-crystal region surrounding the mesa array. The circuit metallization is completed by depositing solder or beam-lead connections wherever the circuit makes electrical contact to the ceramic substrate. Finally, the wafer is diced up, each individual die is flip-chip bonded to a ceramic substrate containing thick-film printed metallurgy, and the thick polysilicon layer is completely etched away. The resulting structure is shown in Fig. 63. Each mesa contains a latching circuit, a power transistor, and a power-dissipation resistor. The fabrication technique is designed to thermally isolate the mesas from each other (to inhibit thermal crosstalk) and to provide a controlled rate of heat transport to the ceramic substrate. Under normal operation, the latches on the 35 mesas are all set high or low in a pattern corresponding to the alphanumeric character to be printed, then the thermal dissipation supply voltage is pulsed and the activated mesa circuits are heated to about 260°C for 33 ms, causing dark spots to appear on the thermally sensitive paper in contact with the silicon print head.

In complete contrast to presently accepted practices for handling IC chips, in which the active silicon circuit is carefully sealed up, hidden from view, and diligently protected in sophisticated packages, it is important to realize that this thermal print head design places the silicon in direct, unprotected, and abrasive contact with the paper. Clearly, our traditional views of silicon chips as fragile components, always requiring delicate handling and careful protection, need to be modified. In addition, the micromachining techniques discussed in this review are certainly not limited to the fabrication of thermal print heads and it is likely that other electro-printing methods will also make use of the principles of silicon head technology.

VIII. Conclusion

While inexpensive microprocessors proliferate into automobiles, appliances, manufacturing equipment, instruments, and office machines, the engineering difficulties and the rapidly increasing expense of interfacing digital electronics with sensors and transducers is demanding more attention and causing lengthy development times. In addition, the general technical trends in instrumentation, communication, and input/output (I/O) devices continue to be in the direction of miniaturization for improved performance [1], [6], [7], [176] and reliability. As we have seen in this review, by employing the low-cost, batch-fabrication techniques available with silicon micromachining methods, many of these sensor and transducer I/O functions can be integrated on silicon, alongside the necessary circuitry, using common processing steps, and resulting in systems with improved performance and more straightforward implementation. Significantly, the interest of the engineering community has risen dramatically in response to these recent trends in I/O technology. Three special issues of the IEEE TRANSACTIONS ON ELECTRON DEVICES (September 1978 "Three-Dimensional Semiconductor Device Structures," December 1979 "Solid-State Sensors, Actuators, and Interface Electronics," and January 1982 "Solid-State Sensors, Actuators, and Interface Electronics") have been devoted to micromachining and integrated transducer techniques. At least three feature articles have appeared in major trade journals within the past year [177]–[179]. The first issue of a new international journal *Sensors and Actuators* devoted to research and development in the area of solid-state transducers was published in November 1980.

Why are silicon-related I/O applications enjoying such a sudden technical popularity? A large part of the answer comes from the nature of the new, microprocessor-controlled electronics market, in which the costs of the sensors, transducers, and interface electronics exceed the cost of the microprocessor itself [1], [6]. This surprising turn of events was undreamed of 7–8 years ago. Since silicon technology has been so successful in the development of sophisticated, yet inexpensive, VLSI electronic circuits, it seems an obvious extension to employ the same materials and the same fabrication principles to lower the costs of the sensors and transducers as well. The purpose of this review has been to illustrate how this extension might be accomplished for a broad range of applications.

From this applications point of view, it is clear that microprocessors and other computing systems will continue to be employed in every conceivable consumer and commercial product. Since the interface and transducer functions will largely determine the costs of such products, those who have learned to fabricate sensors, transducers, and interfaces in a cost-effective manner will be successful in the areas of computer-controlled equipment, from robots, industrial process controllers, and instruments, to toasters, automobiles, and bathroom scales, to displays, printers, and storage devices.

Given the rapidly accelerating demand for silicon-compatible sensors and transducers, and the corresponding success in demonstrating these various components, it is evident that silicon will be increasingly called upon, not only in its traditional electronic role, but also in a wide range of mechanical capacities where miniaturized, high-precision, high-reliability, and low-cost mechanical components and devices are required

in critical applications, performing functions not ordinarily associated with silicon. We are beginning to realize that silicon isn't just for circuits anymore.

Acknowledgment

This paper has benefited greatly from the many technical contributions, helpful discussions, and suggestions provided by R. Muller, K. D. Wise, E. Bassous, P. Cade, J. Knutti, A. Shartel, V. Hanchett, J. Hope, and F. Anger.

References

[1] Proceedings from the topical meeting on "The Limits to Miniaturization," for optics, electronics, and mechanics at the Swiss Federal Institute of Technology, Lausanne, Switzerland, Oct. 1980.

[2] Commercial devices have been available for some time. See, for example, the National Semiconductor catalog, *Transducers; Pressure and Temperature* for Aug. 1974.

[3] Texas Instruments Thermal Character Print Head, EPN3620 Bulletin DL-S7712505, 1977.

[4] P. O'Neill, "A monolithic thermal converter," *Hewlett-Packard J.*, p. 12, May 1980.

[5] E. Bassous, H. H. Taub, and L. Kuhn, "Ink jet printing nozzle arrays etched in silicon," *Appl. Phys. Lett.*, vol. 31, p. 135, 1977.

[6] W. G. Wolber and K. D. Wise, "Sensor development in the microcomputer age," *IEEE Trans. Electron Devices*, vol. ED-26, p. 1864, 1979.

[7] S. Middelhoek, J. B. Angell, and D.J.W. Noorlag, "Microprocessors get integrated sensors," *IEEE Spectrum*, vol. 17, p. 42, Feb. 1980.

[8] A. Kelly, *Strong Solids*, 2nd ed. (Monographs on the Physics and Chemistry of Materials). Oxford, England: Clarendon, 1973.

[9] E. B. Shand, *Glass Engineering Handbook*. New York: McGraw-Hill, 1958.

[10] *CRC Handbook of Chemistry and Physics*, R. C. Weast, Ed. Cleveland, OH: CRC Publ., 1980.

[11] G. L. Pearson, W. T. Reed, Jr., and W. L. Feldman, "Deformation and fracture of small silicon crystals," *Acta Metallurgica*, vol. 5, p. 181, 1957.

[12] K. Kuroiwa and T. Sugano, "Vapor-phase deposition of beta-silicon carbide on silicon substrates," *J. Electrochem. Soc.*, vol. 120, p. 138, 1973.
S. Nishino, Y. Hazuki, H. Matsunami, and T. Tanaka, "Chemical vapor deposition of single crystalline beta-SiC films on silicon substrate with sputtered SiC intermediate layer," *J. Electrochem. Soc.*, vol. 127, p. 2674, 1980.

[13] C. W. Dee, "Silicon nitride: Tribological applications of a Ceramic material," *Tribology*, vol. 3, p. 89, 1970.

[14] E. Rabinowicz, "Grinding damage of silicon nitride determined by abrasive wear tests," *Wear*, vol. 39, p. 101, 1976.

[15] T. E. Baker, S. L. Bagdasarian, G. L. Kix, and J. S. Judge, "Characterization of vapor-deposited paraxylylene coatings," *J. Electrochem. Soc.*, vol. 124, p. 897, 1977.

[16] L. B. Rothman, "Properties of thin polyimide films," *J. Electrochem. Soc.*, vol. 127, p. 2216, 1980.

[17] Adapted from Van Vlack, *Elements of Materials Science*. Reading, MA: Addison Wesley, 1964, p. 163.

[18] J. H. Hobstetter, "Mechanical properties of semiconductors," in *Properties of Crystalline Solids* (ASTM Special Technical Publication 283). Philadelphia, PA: ASTM, 1960, p. 40.

[19] G. Sinclair and C. Feltner, "Fatigue strength of crystalline solids," presented at the Symposium on Nature and Origin of Strength of Materials at the 63rd Annual Meeting of ASTM (ASTM Publisher, 1960, p. 129).

[20] C. M. Drum and M. J. Rand, "A low-stress insulating film on silicon by chemical vapor deposition," *J. Appl. Phys.*, vol. 39, p. 4458, 1968.

[21] S. Iwamura, Y. Nishida, and K. Hashimoto, "Rotating MNOS disk memory device," *IEEE Trans. Electron Devices*, vol. ED-28, p. 854, 1981.

[22] R. M. Finne and D. L. Klein, "A water-amine-complexing agent system for etching silicon," *J. Electrochem. Soc.*, vol. 114, p. 965, 1967.

[23] H. A. Waggener, R. C. Kragness, and A. L. Taylor, *Electronics*, vol. 40, p. 274, 1967.

[24] H. Robbins and B. Schwartz, "Chemical etching of silicon, II. The system HF, HNO_3, $HC_2H_3O_2$," *J. Electrochem. Soc.*, vol. 106, p. 505, 1959.

[25] B. Schwartz and H. Robbins, "Chemical etching of silicon, IV. Etching technology," *J. Electrochem. Soc.*, vol. 123, p. 1903, 1976.

[26] J. C. Greenwood, "Ethylene diamine-catechol-water mixture shows preferential etching of p-n junctions," *J. Electrochem. Soc.*, vol. 116, p. 1325, 1969.

[27] A. Bohg, "Ethylene diamine-pyrocatechol-water mixture shows etching anomaly in boron-doped silicon," *J. Electrochem. Soc.*, vol. 118, p. 401, 1971.

[28] H. Huraoka, T. Ohhashi, and Y. Sumitomo, "Controlled preferential etching technology," in *Semiconductor Silicon 1973*, H. R. Huff and R. R. Burgess, Eds. (The Electrochemical Society Softbound Symposium Ser., Princeton, NJ, 1973), p. 327.

[29] S. C. Terry, "A gas chromatography system fabricated on a silicon wafer using integrated circuit technology," Ph.D. dissertation, Department of Electrical Engineering, Stanford University, Stanford, CA, 1975.

[30] W. Kern, "Chemical etching of silicon, germanium, gallium arsenide, and gallium phosphide," *RCA Rev.*, vol. 29, p. 278, 1978.

[31] S. K. Ghandhi, *The Theory and Practice of Microelectronics*. New York: Wiley, 1968.

[32] J. B. Price, "Anisotropic etching of silicon with potassium hydroxide-water-isopropyl alcohol," in *Semiconductor Silicon 1973*, H. R. Huff and R. R. Burgess, Eds. (The Electrochemical Society Softbound Symposium Ser., Princeton, NJ, 1973), p. 339.

[33] D. L. Kendall, "On etching very narrow grooves in silicon," *Appl. Phys. Lett.*, vol. 26, p. 195, 1975.

[34] I. J. Pugacz-Muraszkiewicz, "Detection of discontinuities in passivating layers on silicon by NaOH anisotropic etch," *IBM J. Res. Develop.*, vol. 16, p. 523, 1972.

[35] E. Bassous, "Fabrication of novel three-dimensional microstructures by the anisotropic etching of (100) and (110) silicon," *IEEE Trans. Electron Devices*, vol. ED-25, p. 1178, 1978.

[36] K. E. Bean, "Anisotropic etching of silicon," *IEEE Trans. Electron Devices*, vol. ED-25, p. 1185, 1978.

[37] A. I. Stoller, "The etching of deep vertical-walled patterns in silicon," *RCA Rev.*, vol. 31, p. 271, 1970.

[38] D. Kendall, "Vertical etching of silicon at very high aspect ratios," *Annu. Rev. Materials Sci.*, vol. 9, p. 373, 1979.

[39] W. K. Zwicker and K. K. Kurtz, "Anisotropic etching of silicon using electrochemical displacement reactions," in *Semiconductor Silicon 1973*, H. R. Huff and R. R. Burgess, Eds. (The Electrochemical Society Softbound Symposium Ser., Princeton, NJ, 1973), p. 315.

[40] D. B. Lee, "Anisotropic etching of silicon," *J. Appl. Phys.*, vol. 40, p. 4569, 1969.

[41] M. J. Declercq, L. Gerzberg, and J. D. Meindl, "Optimization of the hydrazine-water solution for anisotropic etching of silicon in integrated circuit technology," *J. Electrochem. Soc.*, vol. 122, p. 545, 1975.

[42] D. F. Weirauch, "Correlation of the anisotropic etching of single crystal silicon spheres and wafers," *J. Appl. Phys.*, vol. 46, p. 1478, 1975.

[43] E. Bassous and E. F. Baran, "The fabrication of high precision nozzles by the anisotropic etching of (100) silicon," *J. Electrochem. Soc.*, vol. 125, p. 1321, 1978.

[44] A. Reisman, M. Berkenblit, S. A. Chan, F. B. Kaufman, and D. C. Green, "The controlled etching of silicon in catalyzed ethylene diamine-pyrocatechol-water solutions," *J. Electrochem. Soc.*, vol. 126, p. 1406, 1979.

[45] A. Uhlir, "Electrolytic shaping of germanium and silicon," *Bell Syst. Tech. J.*, vol. 36, p. 333, Mar. (1956).
D. R. Turner, "Electropolishing silicon in hydroflouric acid solutions," *J. Electrochem. Soc.*, vol. 105, p. 406, 1958.

[46] M.J.J. Theunissen, J. A. Appels, and W.H.C.G. Verkuylen, "Application of electrochemical etching of silicon to semiconductor device technology," *J. Electrochem. Soc.*, vol. 117, p. 959, 1970.
R. L. Meek, "Anodic dissolution of n⁺ silicon," *J. Electrochem. Soc.*, vol. 118, p. 437, 1971.

[47] C. D. Wen and K. P. Weller, "Preferential electrochemical etching of p⁺ silicon in an aqueous HF-H₂SO₄ electrolyte," *J. Electrochem. Soc.*, vol. 119, p. 547, 1972.

[48] H.J.A. van Dijk and J. de Jonge, "Preparation of thin silicon crystals by electrochemical thinning of epitaxially grown structures," *J. Electrochem. Soc.*, vol. 117, p. 553, 1970.

[49] R. L. Meek, "Electrochemically thinned n/n⁺ epitaxial silicon—Method and applications," *J. Electrochem. Soc.*, vol. 118, p. 1240, 1971.

[50] H. A. Waggener, "Electrochemically controlled thinning of silicon," *Bell Syst. Tech. J.*, vol. 50, p. 473, 1970.

[51] T. N. Jackson, M. A. Tischler, and K. D. Wise, "An electrochemical etch-stop for the formation of silicon microstructures," *IEEE Electron Device Lett.*, vol. EDL-2, p. 44, 1981.

[52] Y. Watanabe, Y. Arita, T. Yokoyama, and Y. Igarashi, "Formation and properties of porous silicon and its applications," *J. Electrochem. Soc.*, vol. 122, p. 1351, 1975.

[53] T. Unagami, "Formation mechanism of porous silicon layer by anodization in HF solution," *J. Electrochem. Soc.*, vol. 127, p. 476, 1980.

[54] T. C. Teng, "An investigation of the application of porous silicon layers to the dielectric isolation of integrated circuits," *J. Electrochem. Soc.*, vol. 126, p. 870, 1979.

[55] T. Unagami and M. Seki, "Structure of porous silicon layer and heat-treatment effect," *J. Electrochem. Soc.*, vol. 125, p. 1339, 1978.

[56] D. J. Ehrlich, R. M. Osgood, and T. F. Deutsch, "Laser chemical technique for rapid direct writing of surface relief in silicon," *Appl. Phys. Lett.*, vol. 38, p. 1018, 1981.

[57] W. R. Runyan, E. G. Alexander, and S. E. Craig, Jr., "Behavior of large-scale surface perturbations during silicon epitaxial growth," *J. Electrochem. Soc.*, vol. 114, p. 1154, 1967.
R. K. Smeltzer, "Epitaxial deposition of silicon in deep grooves," *J. Electrochem. Soc.*, vol. 122, p. 1666, 1975.

[58] B. W. Wessels and B. J. Baliga, "Vertical channel field-controlled thyristors with high gain and fast switching speeds," *IEEE Trans. Electron Devices*, vol. ED-25, p. 1261, 1978.

[59] L. Gerzberg and J. Meindl, "Monolithic polycrystalline silicon distributed RC devices," *IEEE Trans. Electron Devices*, vol. ED-25, p. 1375, 1978.

[60] P. Rai-Choudhury, "Chemical vapor deposited silicon and its device applications," in *Semiconductor Silicon 1973*, H. R. Huff and R. R. Burgess, Eds. (The Electrochemical Society Softbound Symposium Ser., Princeton, NJ, 1973), p. 243.

[61] H. E. Cline and T. R. Anthony, "Random walk of liquid droplets migrating in silicon," *J. Appl. Phys.*, vol. 47, p. 2316, 1976.

[62] —, "High-speed droplet migration in silicon," *J. Appl. Phys.*, vol. 47, p. 2325, 1976.

[63] —, "Thermomigration of aluminum-rich liquid wires through silicon," *J. Appl. Phys.*, vol. 47, p. 2332, 1976.

[64] T. R. Anthony and H. E. Cline, "Lamellar devices processed by thermomigration," *J. Appl. Phys.*, vol. 48, p. 3943, 1977.

[65] —, "Migration of fine molten wires in thin silicon wafers," *J. Appl. Phys.*, vol. 49, p. 2412, 1978.

[66] —, "On the thermomigration of liquid wires," *J. Appl. Phys.*, vol. 49, p. 2777, 1978.

[67] —, "Stresses generated by the thermomigration of liquid inclusions in silicon," *J. Appl. Phys.*, vol. 49, p. 5774, 1978.

[68] T. Mizrah, "Joining and recrystallization of Si using the thermomigration process," *J. Appl. Phys.*, vol. 51, p. 1207, 1980.

[69] C. C. Wen, T. C. Chen, and J. M. Zemel, "Gate-controlled diodes for ionic concentration measurement," *IEEE Trans. Electron Devices*, vol. ED-26, p. 1945, 1979.

[70] L. C. Kimerling, H. J. Leamy, and K. A. Jackson, "Photoinduced zone migration (PIZM) in semiconductors," in *Proc. Symp. on Laser and Electron Beam Processing of Electronic Materials* (The Electrochemical Society Publisher, Electronics Division), vol. 80-1, p. 242, 1980.

[71] G. Wallis and D. I. Pomerantz, "Field-assisted glass-metal sealing," *J. Appl. Phys.*, vol. 40, p. 3946, 1969.

[72] P. B. DeNee, "Low energy metal-glass bonding," *J. Appl. Phys.*, vol. 40, p. 5396, 1969.

[73] J. M. Brownlow, "Glass-related effects in field-assisted glass-metal bonding," IBM Rep. RC 7101, May 1978.

[74] A. D. Brooks and R. P. Donovan, "Low-temperature electrostatic silicon-to-silicon seals using sputtered borosilicate glass," *J. Electrochem. Soc.*, vol. 119, p. 545, 1972.

[75] L. M. Roylance and J. B. Angell, "A batch-fabricated silicon accelerometer," *IEEE Trans. Electron Devices*, vol. ED-26, p. 1911, 1979.

[76] J. H. Jerman, J. M. Pendleton, L. N. Rhodes, C. S. Sanders, S. C. Terry, and G. V. Walsh, "Anodic bonding," Stanford University Lab. Rep. for EE412, 1978.

[77] D. A. Kiewit, "Microtool fabrication by etch pit replication," *Rev. Sci. Instrum.*, vol. 44, p. 1741, 1973.

[78] K. D. Wise, M. G. Robinson, and W. J. Hillegas, "Solid-state processes to produce hemispherical components for inertial fusion targets," *J. Vac. Sci. Technol.*, vol. 18, p. 1179, 1981.
K. D. Wise, T. N. Jackson, N. A. Masnari, M. G. Robinson, D. E. Solomon, G. H. Wuttke, and W. B. Rensel, "Fabrication of hemispherical structures using semiconductor technology for use in thermonuclear fusion research," *J. Vac. Sci. Technol.*, vol. 16, p. 936, 1979.

[79] R. N. Thomas and H. C. Nathanson, "Photosensitive field emission from silicon point arrays," *Appl. Phys. Lett.*, vol. 21, p. 384, 1972.

[80] R. N. Thomas, R. A. Wickstrom, D. K. Schroder, and H. C. Nathanson, "Fabrication and some applications of large area silicon field emission arrays," *Solid-State Electron.*, vol. 17, p. 155, 1974.

[81] E. Bassous, "Nozzles formed in mono-crystalline silicon," U.S. Patent 3 921 916, 1975.

[82] E. Bassous, L. Kuhn, A. Reisman, and H. H. Taub, "Ink jet nozzle," U.S. Patent 4 007 464, 1977.

[83] R. G. Sweet, "High frequency recording with electrostatically

deflected ink jets," *Rev. Sci. Instrum.*, vol. 36, p. 131, 1965.

[84] F. J. Kamphoefner, "Ink jet printing," *IEEE Trans. Electron Devices*, vol. ED-19, p. 584, 1972.

[85] R. D. Carnahan and S. L. Hou, "Ink jet technology," *IEEE Trans. Ind. Appl.*, vol. IA-13, p. 95, 1977.

[86] Special Issue on Ink Jet Printing, *IBM J. Res. Develop.*, vol. 21, 1977.

[87] L. Kuhn, E. Bassous, and R. Lane, "Silicon charge electrode array for ink jet printing," *IEEE Trans. Electron Devices*, vol. ED-25, p. 1257, 1978.

[88] K. E. Petersen, "Fabrication of an integrated, planar silicon ink-jet structure," *IEEE Trans. Electron Devices*, vol. ED-26, p. 1918, 1979.

[89] W. Anacker, E. Bassous, F. F. Fang, R. E. Mundie, and H. N. Yu, "Fabrication of multiprobe miniature electrical connector," *IBM Tech. Discl. Bull.*, vol. 19, p. 372, 1976.
S. K. Lahiri, P. Geldermans, G. Kolb, J. Sokolwski, and M. J. Palmer, "Pluggable connectors for Josephson device packaging," *J. Electrochem. Soc. Extended Abstr.*, vol. 80-1, p. 216, 1980.

[90] L. P. Boivin, "Thin-film laser-to-fiber coupler," *Appl. Opt.*, vol. 13, p. 391, 1974.

[91] C. C. Tseng, D. Botez, and S. Wang, "Optical bends and rings fabricated by preferential etching," *Appl. Phys. Lett.*, vol. 26, p. 699, 1975.

[92] W. T. Tsang, C. C. Tseng and S. Wang, "Optical waveguides fabricated by preferential etching," *Appl. Opt.*, vol. 14, p. 1200, 1975.

[93] C. C. Tseng, W. T. Tsang, and S. Wang, "A thin-film prism as a beam separator for multimode guided waves in integrated optics," *Opt. Commun.*, vol. 13, p. 342, 1975.

[94] W. T. Tsang and S. Wang, "Preferentially etched diffraction gratings in silicon," *J. Appl. Phys.*, vol. 46, p. 2163, 1975.

[95] ——, "Thin-film beam splitter and reflector for optical guided waves," *Appl. Phys. Lett.*, vol. 27, p. 588, 1975.

[96] J. S. Harper and P. F. Heidrich, "High density multichannel optical waveguides with integrated couplers," *Wave Electron.*, vol. 2, p. 369, 1976.

[97] H. P. Hsu and A. F. Milton, "Single mode optical fiber pick-off coupler," *Appl. Opt.*, vol. 15, p. 2310, 1976.

[98] C. Hu and S. Kim, "Thin-film dye laser with etched cavity," *Appl. Phys. Lett.*, vol. 29, p. 9, 1976.

[99] H. P. Hsu and A. F. Milton, "Flip-chip approach to endfire coupling between single-mode optical fibres and channel waveguides," *Electron. Lett.*, vol. 12, p. 404, 1976.

[100] J. D. Crow, L. D. Comerford, R. A. Laff, M. J. Brady, and J. S. Harper, "GaAs laser array source package," *Opt. Lett.*, vol. 1, p. 40, 1977.

[101] H. P. Hsu and A. F. Milton, "Single-mode coupling between fibers and indiffused waveguides," *IEEE J. Quantum Electron.*, vol. QE-13, p. 224, 1977.

[102] J. T. Boyd and S. Sriram, "Optical coupling from fibers to channel waveguides formed on silicon," *Appl. Opt.*, vol. 17, p. 895, 1978.

[103] S. C. Terry, J. H. Jerman, and J. B. Angell, "A gas chromatograph air analyzer fabricated on a silicon wafer," *IEEE Trans. Electron Devices*, vol. ED-26, p. 1880, 1979.

[104] W. A. Little, "Design and construction of microminiature cyrogenic refrigerators," in *AIP Proc. of Future Trends in Superconductive Electronics* (University of Virginia, Charlottesville, 1978).

[105] D. B. Tuckerman and R.F.W. Pease, "High-performance heat sinking for VLSI," *IEEE Electron Device Lett.*, vol. EDL-2, p. 126, 1981.

[106] K. E. Bean and J. R. Lawson, "Application of silicon orientation and anisotropic effects to the control of charge spreading in devices," *IEEE J. Solid-State Circuits*, vol. SC-9, p. 111, 1974.

[107] M. J. Declerq, "A new C-MOS technology using anisotropic etching of silicon," *IEEE J. Solid-State Circuits*, vol. SC-10, p. 191, 1975.

[108] H. N. Yu, R. H. Dennard, T.H.P. Chang, C. M. Osburn, V. Dilonardo, and H. E. Luhn, "Fabrication of a miniature 8 k-bit memory chip using electron beam exposure," *J. Vac. Sci. Technol.*, vol. 12, p. 1297, 1975.

[109] C.A.T. Salama and J. G. Oakes, "Nonplanar power field-effect transistors," *IEEE Trans. Electron Devices*, vol. ED-25, p. 1222, 1978.

[110] K. P. Lisiak and J. Berger, "Optimization of nonplanar power MOS transistors," *IEEE Trans. Electron Devices*, vol. ED-25, p. 1229, 1978.

[111] W. C. Rosvold, W. H. Legat, and R. L. Holden, "Air gap isolated micro-circuits-beam-lead devices," *IEEE Trans. Electron Devices*, vol. ED-15, p. 640, 1968.

[112] L. A. D'Asaro, J. V. DiLorenzo, and H. Fukui, "Improved performance of GaAs microwave field-effect transistors with low inductance via-connections through the substrate," *IEEE Trans. Electron Devices*, vol. ED-25, p. 1218, 1978.

[113] T. J. Rodgers and J. D. Meindl, "Epitaxial V-groove bipolar integrated circuit process," *IEEE Trans. Electron Devices*, vol. ED-20, p. 226, 1973.

[114] E. S. Ammar and T. J. Rodgers, "UMOS transistors on (110) silicon," *IEEE Trans. Electron Devices*, vol. ED-27, p. 907, 1980.

[115] B. J. Baliga, "A novel buried grid device fabrication technology," *IEEE Electron Device Lett.*, vol. EDL-1, p. 250, 1980.

[116] T. I. Chappell, "The V-groove multijunction solar cell," *IEEE Trans. Electron Devices*, vol. ED-26, p. 1091, 1979.

[117] D. L. Spears and H. I. Smith, "High resolution pattern replication using soft X-rays," *Electron. Lett.*, vol. 8, p. 102, 1972.

[118] H. I. Smith, D. L. Spears, and S. E. Bernacki, "X-ray lithography: A complementary technique to electron beam lithography," *J. Vac. Sci. Technol.*, vol. 10, p. 913, 1973.

[119] T. O. Sedgwick, A. N. Broers, and B. J. Agule, "A novel method for fabrication of ultrafine metal lines by electron beams," *J. Electrochem. Soc.*, vol. 119, p. 1769, 1972.

[120] P. V. Lenzo and E. G. Spencer, "High-speed low-power X-ray lithography," *Appl. Phys. Lett.*, vol. 24, p. 289, 1974.

[121] C. J. Schmidt, P. V. Lenzo, and E. G. Spencer, "Preparation of thin windows in silicon masks for X-ray lithography," *J. Appl. Phys.*, vol. 46, p. 4080, 1975.

[122] H. Bohlen, J. Greschner, W. Kulcke, and P. Nehmiz, "Electron beam step and repeat proximity printing," in *Proc. Electrochem. Soc. Meet.* (Seattle, WA, May 1978).

[123] R. J. Jaccodine and W. A. Schlegel, "Measurement of strains at Si-SiO$_2$ interface," *J. Appl. Phys.*, vol. 37, p. 2429, 1966.

[124] C. W. Wilmsen, E. G. Thompson, and G. H. Meissner, "Buckling of thermally grown SiO$_2$ thin films," *IEEE Trans. Electron Devices*, vol. ED-19, p. 122, 1972.

[125] E. Bassous, R. Feder, E. Spiller, and J. Topalian, "High transmission X-ray masks for lithographic applications," *Solid-State Technol.*, vol. 19, p. 55, 1976.

[126] G. A. Antcliffe, L. J. Hornbeck, W. W. Chan, J. W. Walker, W. C. Rhines, and D. R. Collins, "A backside illuminated 400 × 400 charge-coupled device imager," *IEEE Trans. Electron Devices*, vol. ED-23, p. 1225, 1976.

[127] G. R. Lahiji and K. D. Wise, "A monolithic thermopile detector fabricated using integrated-circuit technology," in *Proc. Int. Electron Devices Meet.* (Washington, DC), p. 676, 1980.

[128] C. L. Huang and T. van Duzer, "Josephson tunnelling through locally thinned silicon," *Appl. Phys. Lett.*, vol. 25, p. 753, 1974.
——, "Single-crystal silicon-barrier Josephson junctions," *IEEE Trans. Magn.*, vol. MAG-11, p. 766, 1975.

[129] ——, "Schottky diodes and other devices on thin silicon membranes," *IEEE Trans. Electron Devices*, vol. ED-23, p. 579, 1976.

[130] C. J. Maggiore, P. D. Goldstone, G. R. Gruhn, N. Jarmie, S. C. Stotlar, and H. V. Dehaven, "Thin epitaxial silicon for dE/dx detectors," *IEEE Trans. Nucl. Sci.*, vol. NS-24, p. 104, 1977.

[131] H. Guckel, S. Larsen, M. G. Lagally, G. Moore, J. B. Miller, and J. D. Wiley, "Electromechanical devices utilizing thin Si diaphragms," *Appl. Phys. Lett.*, vol. 31, p. 618, 1977.

[132] O. N. Tufte, P. W. Chapman, and D. Long, "Silicon diffused-element piezoresistive diaphragms," *J. Appl. Phys.*, vol. 33, p. 3322, 1962.
A.C.M. Gieles and G.H.J. Somers, "Miniature pressure transducers with silicon diaphragm," *Philips Tech. Rev.*, vol. 33, p. 14, 1973.
Samaun, K. D. Wise, and J. B. Angell, "An IC piezoresistive pressure sensor for biomedical instrumentation," *IEEE Trans. Biomed. Eng.*, vol. BME-20, p. 101, 1973.
W. D. Frobenius, A. C. Sanderson, and H. C. Nathanson, "A microminiature solid-state capacitive blood pressure transducer with improved sensitivity," *IEEE Trans. Biomed. Eng.*, vol. BME-20, p. 312, 1973.

[133] S. K. Clark and K. D. Wise, "Pressure sensitivity in anisotropically etched thin-diaphragm pressure sensors," *IEEE Trans. Electron Devices*, vol. ED-26, p. 1887, 1979.

[134] J. M. Borky and K. D. Wise, "Integrated signal conditioning for silicon pressure sensors," *IEEE Trans. Electron Devices*, vol. ED-26, p. 1906, 1979.

[135] W. H. Ko, J. Hynecek, and S. F. Boettcher, "Development of a miniature pressure transducer for biomedical applications," *IEEE Trans. Electron Devices*, vol. ED-26, p. 1896, 1979.

[136] C. S. Sander, J. W. Knutti, and J. D. Meindl, "A monolithic capacitive pressure sensor with pulse-period output," *IEEE Trans. Electron Devices*, vol. ED-27, p. 927, 1980.

[137] H. C. Tuan, J. S. Yanacopoulos, and T. A. Nunn, "Piezoresistive force sensors for observing muscle contraction," *Stanford Univ. Electron. Res. Rev.*, p. 102, 1975.

[138] R. J. Wilfinger, P. H. Bardell, and D. S. Chhabra, "The resonistor: A frequency selective device utilizing the mechanical resonance of a silicon substrate," *IBM J. Res. Develop.*, vol. 12, p. 113, 1968.

[139] K. E. Petersen, "Silicon torsional scanning mirror," *IBM J. Res. Develop.*, vol. 24, p. 631, 1980.

[140] A. Higdon and W. D. Stiles, *Engineering Mechanics, Volume II: Dynamics*. Englewood Cliffs, NJ: Prentice-Hall, 1961, p. 555.

[141] H. F. Wolf, *Silicon Semiconductor Data*. New York: Pergamon, 1969.

[142] W. E. Newell, "Miniaturization of tuning forks," *Science*, vol. 161, p. 1320, 1968.

[143] S. P. Timeshenko and J. M. Gere, *Mechanics of Materials*. New York: Van Nostrand Reinhold, 1972, p. 516.

[144] P. J. Brosens, "Dynamic mirror distortions in optical scanning," *Appl. Opt.*, vol. 11, p. 2987, 1972.

[145] R. J. Roark and W. C. Young, *Formulas for Stress and Strain*. New York: McGraw-Hill, 1975.

[146] L. Beiser, "Laser scanning systems," in *Laser Applications Volume 2*, M. Ross. Ed. New York: Academic Press, 1974.

[147] H. C. Nathanson and R. A. Wickstrom, "A resonant-gate silicon surface transistor with high-Q bandpass properties," *Appl. Phys. Lett.*, vol. 7, p. 84, 1965.

[148] H. C. Nathanson, W. E. Newell, R. A. Wickstrom, and J. R. Davis, Jr., "The resonant gate transistor," *IEEE Trans. Electron Devices*, vol. ED-14, p. 117, 1967.

[149] J. A. van Raalte, "A new Schlieren light valve for television projection," *Appl. Opt.*, vol. 9, p. 2225, 1970.

[150] K. Preston, Jr., "A coherent optical computer system using the membrane light modulator," *IEEE Trans. Aerosp. Electron. Syst.*, vol. AES-6, p. 458, 1970.

[151] L. S. Cosentino and W. C. Stewart, "A membrane page composer," *RCA Rev.*, vol. 34, p. 45, 1973.

[152] B. J. Ross and E. T. Kozol, "Performance characteristics of the deformagraphic storage display tube (DSDT)," presented at the IEEE Intercon., Mar. 1973.

[153] J. Guldberg, H. C. Nathanson, D. L. Balthis, and A. S. Jensen, "An aluminum/SiO₂ silicon on sapphire light valve matrix for projection displays," *Appl. Phys. Lett.*, vol. 26, p. 391, 1975.

[154] R. N. Thomas, J. Guldberg, H. C. Nathanson, and P. R. Malmberg, "The mirror matrix tube: A novel light valve for projection displays," *IEEE Trans. Electron Devices*, vol. ED-22, p. 765, 1975.

[155] K. E. Petersen, "Micromechanical light modulator array fabricated on silicon," *Appl. Phys. Lett.*, vol. 31, p. 521, 1977.

[156] ——, "Dynamic micromechanics on silicon: Techniques and devices," *IEEE Trans. Electron Devices*, vol. ED-25, p. 1241, 1978.

[157] J. P. Den Hartog, *Mechanical Vibrations*. New York: McGraw-Hill, 1956.

[158] K. E. Petersen and C. R. Guarnieri, "Young's modulus measurements of thin films using micromechanics," *J. Appl. Phys.*, vol. 50, p. 6761, 1979.

[159] J. W. Beams, J. B. Freazeale, and W. L. Bart, "Mechanical strength of thin films of metals," *Phys. Rev.*, vol. 100, p. 1657, 1955.

[160] C. A. Neugebauer, "Tensile properties of thin, evaporated gold films," *J. Appl. Phys.*, vol. 31, p. 1096, 1960.

[161] J. M. Blakely, "Mechanical properties of vacuum-deposited gold," *J. Appl. Phys.*, vol. 35, p. 1756, 1964.

[162] R. D. Jolly and R. S. Muller, "Miniature cantilever beams fabricated by anisotropic etching of silicon," *J. Electrochem. Soc.*, vol. 127, p. 2750, 1980.

[163] *Metals Reference Handbook*, C. J. Smithels, Ed. London, England: Butterworths, 1976.

[164] R. E. McMillan and R. P. Misra, "Insulating materials for semiconductor surfaces," *IEEE Trans. Elec. Insulat.*, vol. EI-5, p. 10, 1970.

[165] *Corning Laboratory Glassware Catalogue*, 1978.

[166] J. F. Lynch, C. G. Ruderer, and W. H. Duckworth, *Engineering Properties of Selected Ceramic Materials*. The American Ceramic Society, OH, 1966.

[167] Air Force Tech. Rep. AFAL-TR-77-152, Air Force Avionics Lab., Wright-Patterson Air Force Base, OH, 45433.

[168] P. Chen, R. S. Muller, T. Shiosaki, and R. M. White, "Silicon cantilever beam accelerometer utilizing a PI-FET capacitive transducer," *IEEE Trans. Electron Devices*, vol. ED-26, p. 1857, 1979.

[169] P. Chen, R. Jolly, G. Halac, R. S. Muller, and R. M. White, "A planar processed PI-FET accelerometer," in *Proc. Int. Electron Devices Meet.* (Washington, DC, Dec. 1980), p. 848.

[170] K. E. Petersen, A. Shartel, and N. Raley, "Micromechanical accelerometer integrated with MOS detection circuitry," *IEEE Trans. Electron Devices*, vol. ED-29, p. 23, Jan. 1982.

[171] W. D. Frobenius, S. A. Zeitman, M. H. White, D. D. O'Sullivan, and R. G. Hamel, "Microminiature ganged threshold accelerometers compatible with integrated circuit technology," *IEEE Trans. Electron Devices*, vol. ED-19, p. 37, 1972.

[172] K. E. Petersen, "Micromechanical membrane switches on silicon," *IBM J. Res. Develop.*, vol. 23, p. 376, 1979.

[173] B. Boles, "Thermal printing applications and technology," presented at the Invitational Computer Conf., Sept. 1979.

[174] M. L. Morris, *Thermal Line Printers*, Texas Instruments product bulletin.

[175] K. E. Bean and W. R. Runyan, "Dielectric isolation: Comprehensive, current and future," *J. Electrochem. Soc.*, vol. 124, p. 5c, 1977.
K. E. Bean, "Chemical vapor deposition applications in microelectronics processing," *Thin Solid Films*, vol. 83, p. 173, 1981.

[176] H. C. Nathanson and J. Guldberg, "Topologically structured thin films in semiconductor device operation," in *Physics of Thin Films Volume 8*. New York: Academic Press, 1975.

[177] "System designers fish for microcomputer-compatible sensors and transducers," *EDN Mag.*, p. 122, Mar. 20, 1980.

[178] "Sensing the real world with low cost silicon," *Electronics*, p. 113, Nov. 6, 1980.

[179] "Ultraminiature mechanics," *Machine Des.*, p. 112, Jan. 8, 1981.

MICROFABRICATED STRUCTURES FOR THE MEASUREMENT OF

MECHANICAL PROPERTIES AND ADHESION OF THIN FILMS

Stephen D. Senturia
Microsystems Technology Laboratories
Department of Electrical Engineering and Computer Science
Massachusetts Institute of Technology
Cambridge, MA, 02139, USA

ABSTRACT

The successful design of micro-mechanical parts depends on the accuracy with which the behavior of the structures can be predicted, fabricated, and reproduced. This, in turn, depends on three critical factors: control of the mechanical properties of the materials used for fabrication, dimensional control of the structure during fabrication, and structural analysis of the primary part and its support or encapsulation. Tools for handling the latter two issues are relatively well developed. However, for many of the materials used in microfabricated structures, basic data on mechanical properties and their control through process variables is lacking. This paper reviews briefly some of the methods that have been used to develop such data, and reports in more detail on some new approaches to this problem that are particularly suited to the study of thin films having residual tensile stress. The use of these methods is illustrated with measurements on polyimide films spin-cast onto oxidized silicon substrates.

INTRODUCTION

Anyone who has attempted to design a microsensor is aware of the vast complexity of the problem. Not only is the designer required to know all about integrated-circuit and microfabrication technology, but he or she must also be knowledgeable about measurement science, and about the specific type of transducer principle being used in the microsensor design. Furthermore, because measurement devices usually cannot be critically tested until they have been packaged, the microsensor designer must also understand the constraints involved in sensor encapsulation.

Microsensors, in particular, place new and stringent requirements on the mechanical properties of the component parts. Figure 1 illllustrates the

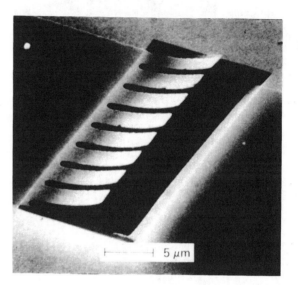

Fig. 1 Cantilevers showing curvature due to nonuniform residual stress (courtesy of Kurt Petersen, Novasensor).

problem of trying to construct a flat cantilever beam[1]. Lateral stresses, if uniform, would relax when the cantilever is etched free from its support; however, if the stresses are nonuniform (or not symmetrically balanced) through the thickness of the beam, the free-standing beam will curl, as shown in the Figure. Thus, not only must the designer know the normal mechanical properties of the materials, such as Young's modulus and Poisson's ratio, but the designer must be able to predict residual stresses. This turns out to be difficult.

The problem is not unique to mechanical sensors. Figure 2 shows a remarkable spiral peel defect in a CVD aluminum oxide film used as the chemically sensitive surface of an ion-sensitive field-effect transistor (ISFET) [2]. The origin of this defect is not rigorously known; however, we speculate that the film is under compressive stress when deposited, and chemical attack of the interface between the film

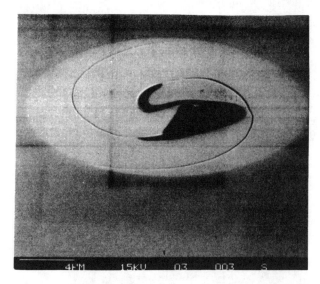

Fig. 2 Spiral fracture of CVD aluminum oxide in a chemical sensor (courtesy of Rosemary Smith, MIT).

and the substrate through a pinhole defect leads to a crack, followed by the spiraling ribbon-shaped peeling of the film as it debonds from the substrate. Smith has observed this defect pattern on many aluminum-oxide-coated devices[2]. Obviously, ISFETs with such coatings could be expected to show long-term instabilities in their behavior.

THE ANALOGY TO INTEGRATED-CIRCUIT CAD

When confronting difficult problems, Polya[3] suggests that one look to analogies for guidance. In the microsensor area, there is an excellent analogy to draw on: integrated circuit design. We are all aware of the spectacular advances in technology, and of the sophisticated methods that are now used to design integrated circuits. Indeed, CAD tools now exist for every step of the design process, from specifying the basic process with SUPREM, to understanding electrical device performance (many programs exist), to modeling circuit performance (including parasitics) with circuit simulators like SPICE. Furthermore, design checkers have been developed to verify that the final mask set actually results in the required circuit, and automatic logic checking and timing estimates can be made, all before the circuit is built. As a result, while the design cost for a new circuit may be significant, the final device can be expected to perform according to specifications.

It is instructive to ask how such a support structure for design came about. The process modeling program SUPREM required collection of an enor-

mous database of experimental information on oxidation and diffusion rates under a wide variety of circumstances, and the incorporation of that information into a digital environment suitable for solving the diffusion equation with moving boundaries. Similarly, programs for device and circuit modelling required, first, the creation of a mathematical environment for modelling, and, second, the encoding of appropriate data -- representing the cumulative experimentation of many workers over many years -- into modules of code that represented the physical behavior of the device or circuit subsection.

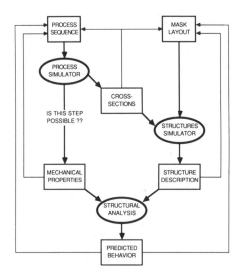

Fig. 3 Schematic CAD sequence for microsensors.

Now let us examine the case for microsensor design. Figure 3 schematically illustrates the design process and the CAD components one would like to have. A design consists of a process sequence and a mask layout. Three simulation steps are required to go from the design to predicted behavior. The PROCESS SIMULATOR, for a given process sequence, gives the residual stress, elastic modulus, Poisson's ratio, and tensile strength of each constituent material, as well as the usual device cross-section produced by SUPREM. The STRUCTURE SIMULATOR combines the cross-section output from the PROCESS SIMULATOR and mask data and fabrication tolerances (possibly expressed in terms of lithographic design rules) to produce a three-dimensional geometric description of the fabricated structure. Finally, the STRUCTURAL ANALYSIS program combines the structure description with the mechanical property data and performs a prediction of the mechanical behavior of the complete structure, including effects of the support or encapsulation, and incorporating the basic transduction principle (such as piezoresistance) into predictions of the electrical behavior of the device under various conditions.

Is this fanciful? Not necessarily. Most of these tools are already developed to some degree. Specifically, the paths leading to the structure description could be developed by combining SUPREM-like cross-sections with mask-analysis programs. Further, there are many finite element codes that can perform structural analysis, provided the model used for the structure is appropriate (in terms of selection of elements, and the handling of geometric nonlinearity and large deflections). While these finite element codes are not without problems, and must be used with great care and with continuous check against experiment, there is, nevertheless, a vast set of resources for the microsensor designer to draw on.

<u>What is missing is the ability to predict the basic mechanical property data from the process sequence!</u>

Many investigators have developed methods for measuring various mechanical properties of microelectronic materials, and several comprehensive reviews are available[4,5]. To cite a few key works, modulus and/or residual stress values have been reported for thermal oxides and CVD dielectrics[1,6-8], polysilicon [9,10], polyimides[11,12], and metals [13]. The techniques used include wafer curvature [14], cantilever studies [7,9], membrane load-deflection studies [12,15], and buckling of clamped structures due to residual stress [9,10]. However, even with the data that has been obtained from these works and from the many additional references too numerous to cite here, we do not have the ability to do reliable <u>prediction</u> of mechanical properties of microelectronic materials in the actual state in which they are deposited, or as they are modified by subsequent processing (which is so well handled by SUPREM for oxidation and diffusion).

Further, the available data are incomplete. It is true that for a material under very specific deposition conditions (possibly specific to one apparatus) one may know the state of stress and may also know the modulus. However, the tensile strength and Poisson's ratio are less well known. (Consider, for example, the number of reliability problems that are related to brittle fracture of dielectrics.) Finally, except possibly for well-characterized individual pieces of equipment,

we do not know how to predict how these properties will change if we modify the process. Until we have such insight, and can routinely draw on it during the design process, each new microsensor design will be burdened by the need to develop the relevant mechanical property data as part of the design process. This is a serious problem.

SUSPENDED MEMBRANES AND RELEASED STRUCTURES

Recent work in our laboratory has resulted in some new techniques for the study of thin-film mechanical properties [12,16]. One motivation for this work has been the hope that as our knowledge improves, <u>prediction</u> of such properties for a given process sequence may become routinely possible. The techniques are particularly well suited to in-situ measurements on films with residual tensile stress, and, hence, are very appropriate for polymer layers, such as polyimides, used as interlevel dielectrics and passivants. The remainder of this paper briefly surveys this new work.

Figure 4 shows the process sequence used for the structures. <u>Suspended membranes</u> (see Fig. 5) are prepared by first creating a silicon diaphragm by conventional anisotropic etching, then applying the coating, and, finally, removing the supporting silicon from the back with an SF_6 plasma. Sus-

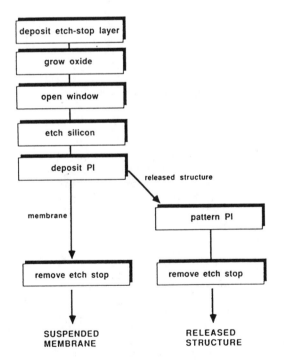

Fig. 4 Process sequences for suspended membranes and released structures.

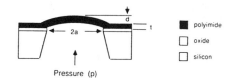

Fig. 5 Schematic cross section of the suspended membrane.

pended square polyimide membranes of thickness between 1 and 15 μm and between 1 and 25 mm on a side have been fabricated using this sequence. By pressurizing one side of the membrane, as shown in Fig. 6, and measuring the deflection, one can extract both the residual stress and Young's modulus of the membrane. The load-deflection curve for these structures is of the form

$$pa^2/dt = C_1 + C_2(d/a)^2 \qquad (1)$$

where p is the pressure, d is the center deflection, a is the square size, t is the membrane thickness, C_1 is a constant that is proportional to the residual stress, and C_2 is a constant that is proportional to Young's modulus.

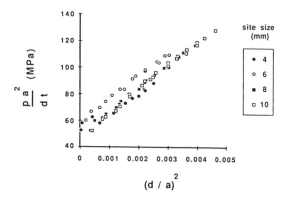

Fig. 6 Load-deflection behavior of a polyimide membrane, illustrated schematically at the top. Data for different diaphragm sizes illustrate scaling according to Equation (1).

The data in Figure 6, which were taken from a single coated wafer containing four membrane sites of different sizes, obey the scaling implied by Equation (1)[16]. The intercept yields the stress and the slope yields the modulus. For this particular sample, the stress is 17 MPa, and the modulus is 6 GPa, resulting in a residual stress-to-modulus ratio of 0.003. This stress-to-modulus ratio, also referred to as the residual strain, is what must be compared to the ultimate strain (see below)

when evaluating the potential reliability problems associated with cracking of films. We have examined a variety of polyimides, and have observed residual strains in the range from 0.0002 to 0.012, depending on the details of the polyimide chemistry. These results are being reported separately.

A released structure is made by the same process as the suspended membrane (see Fig. 4) except that the polymer film is patterned into an asymmetrical structure before removing the thin silicon support. An example is shown in Fig. 7. Once released, the wide suspended strip (width W_1) pulls on the thinner necks (total width W_2), resulting in a deflection δ from its original mask position toward the right to its final position after release. The residual tensile stress in the film is the driving force for the deformation. By varying the geometry, it is possible to create structures in which the strain in the thinner sections is very small, to others in which the ultimate strain of the film is exceeded. For structures in which the strain is small enough to be modeled with linear elastic behavior, the deflection δ can be related to the stress-to-modulus ratio (σ_O/E) as follows:

$$\sigma_O/E = \delta(W_1/L_1 + W_2/L_2)/(W_1 - W_2) \qquad (2)$$

where the geometry terms are defined in Fig. 7.

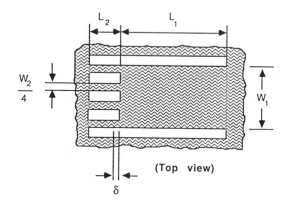

Fig. 7 Schematic top-view of axial beam released structure.

Figure 8 shows a photograph of two such released structures, one with thicker necks, the other with necks so thin that the film fractured when released. Based on the residual tensile strain of the film and the geometry of the structures that failed, it was determined that the ultimate strain of these particular films was 4.5%.

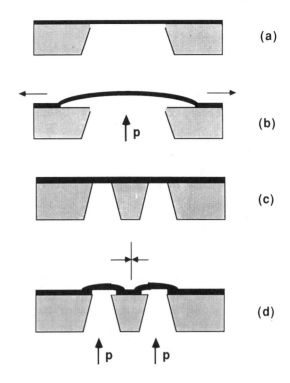

Fig. 9 Illustrating the adhesion test sites. (a) suspended membrane; (b) outward peel; (c) island structure; (d) inward peel.

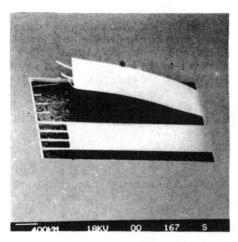

Fig. 8 Two released structures, one of which has exceeded ultimate strain of film, resulting in fracture of the necks.

ADHESION MEASUREMENTS

Adhesion of various films to one another is as important as the mechanical properties of the individual films in overall device performance and reliability. Figure 9 shows schematically how two different suspended structures are used for adhesion measurements. Fig. 9(a) shows a suspended membrane, and Fig. 9(b) shows the same membrane after it has been peeled from its substrate by an applied load. Figure 10 illustrates the P-V cycle for such an experiment, in which the original membrane is inflated, then peels, then is deflated. The P-V work in creating the new surface is given by the shaded portion of Fig. 10. It is equal to the average work of adhesion for the film-substrate interface times the area peeled during the test.

Figure 10 (c) and (d) illustrate a more versatile structure in which an island is left in the center of the suspended membrane. Peel is induced toward the center of the island. This structure is better suited for studying systems with very good adhesion. The fracture mechanics of such structures, however, must be analyzed with care in order to relate the critical pressure at which peel occurs to the strength of the adhesive bond. In preliminary measurements however, reasonably good agreement has been obtained between the P-V analysis and the analysis of the critical debond pressure.

Our work on adhesion is still in a very preliminary phase. However, microfabrication does show promise as a method of studying adhesion in-situ with a variety of very thin films.

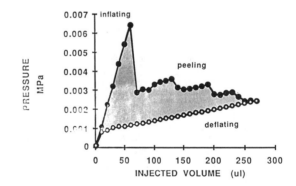

Fig. 10 Peel adhesion data for a sample of the type in Fig. 9(b). The shaded area is the work done to create the peeled surface.

CONCLUSION

The ability to predict mechanical properties is essential to sensor design, yet no CAD tool presently exists that permits such prediction from a given process sequence. Further, it is not yet clear that such a CAD tool could be created without further advances in the basic science of thin film mechanics and morphology. The creation of the CAD tool and its supporting data require quantitative, reproducible experimental methods. Many good methods exist, and we have elaborated on some new methods specifically directed toward materials with residual tensile stress. It is critical, however, that the process dependence of these mechanical properties be studied, documented, and understood in order to provide a basis for intelligent design of the next generations of microsensors.

ACKNOWLEDGEMENT

The author is indebted to Mark G. Allen, Mehran Mehregany, Martin Schmidt, Fariborz Maseeh, and Prof. Roger Howe for many useful discussions of mechanical property issues, and to Prof. Dimitri Antoniadis for discussions of the development of SUPREM and other CAD tools. Thanks also to Dr. Kurt Petersen and Prof. Rosemary Smith for providing interesting examples.

REFERENCES

1) K. E. Petersen, "Silicon as a mechanical material", Proc. IEEE, Vol. 70, 1982, pp. 420-457.

2) R. L. Smith, private communication.

3) G. Polya, How to Solve It, Princeton, Princeton University Press, 1973.

4) R. W. Hoffman, "The mechanical properties of thin condensed films", in G. Haas and R. E. Thun (eds.), Physics of Thin Films, Vol. 3, 1966, pp. 211-273.

5) D. S. Campbell, "Mechnanical properties of thin films", in L. I. Maissel and R. Glang (eds.), Handbook of Thin Films, New York, McGraw Hill, 1970, Chapter 12.

6) E. P. EerNisse, "Stress in thermal SiO_2 during growth", Appl. Phys. Lett., Vol. 35, 1979, pp. 8-10.

7) K. E. Petersen and C. R. Guarnieri, "Young's modulus measurements of thin films using micromechanics", J. Appl. Phys., Vol. 50, 1979, p. 6761.

8) A. K. Sinha, H. J. Levinson, and T. E. Smith, "Thermal stresses and cracking resistance of dielectric films (SiN, Si_3N_4, and SiO_2) on Si substrates", J. Appl. Phys., Vol. 49 (1978), pp. 2423-2416.

9) R. T. Howe and R. S. Muller, "Stress in polycrystalline and amorphous silicon thin films", J. Appl. Phys., Vol 54, 1983, pp. 4674-4675.

10) H. Guckel, T. Randazzo, and D. W. Burns, "A simple technique for the determination of mechanical strain in thin films with applications to polysilicon", J. Appl. Phys., Vol. 57, 1985, pp. 1671-1675.

11) P. Geldermans, C. Goldsmith, and F. Bedetti, "Measurement of stresses generated during curing and in cured polyimide films", in K. Mittal (ed.), Polyimides, New York, Plenum Press, 1984, pp. 695-711.

12) M. G. Allen, M. Mehregany, R. T. Howe, and S. D. Senturia, "Microfabricated structures for the in-situ measurement of residual stress, Young's modulus, and ultimate strain of thin films", Appl. Phys. Lett., in press.

13) R. W. Hoffman, "Internal stresses in thin films", in B. Schwartz and N. Schwartz (eds.), Measurement Techniques for Thin Films, New York, The Electrochemical Society, 1967, pp. 312-333.

14) R. Glang, R. A. Holmwood, and R. L. Rosenfeld, "Determination of stress in films on single crystalline silicon substrates", Rev. Sci. Instr., Vol. 36, 1965, pp. 7-10.

15) E. I. Bromley, J. N. Randall, D. C. Flanders, and R. W. Mountain, "A technique for the determination of stress in thin films", J. Vac. Sci. Technol., Vol. B1, 1983, pp. 1364-1366.

16) M. G. Allen and S. D. Senturia, "Microfabrciated test structures for adhesion measurement", Tech. Digest, Adhesion Society Meeting, Williamsburg, VA, Feb., 1987.

CHARACTERISTICS OF POLYSILICON LAYERS AND THEIR APPLICATION IN SENSORS

E. Obermeier, P. Kopystynski, R. Nießl

Fraunhofer-Institut für Festkörpertechnologie, Paul-Gerhardt-Allee 42
D-8000 Muenchen 60, Germany

Abstract: The important characteristics of boron doped LPCVD polysilicon layers with regards to sensor applications are presented. Properties such as the resistivity, temperature coefficient of the resistance, gauge factor, and long term stability are described. A pressure sensor utilizing polysilicon piezoresistors with a measurement range of 1 bar and a sensitivity of roughly 11 mV/V F.S. is discussed. Finally a polysilicon temperature sensor with on chip linearization, a sensitivity of $-3.5 \times 10^{-3} \mathrm{K}^{-1}$ and a linearity error of less than 0.5% is described.

Introduction

Polycrystalline silicon has been an important material in integrated circuit technology for many years [1]-[4]. Growing interest has been shown recently in the utilization of polysilicon as a basic material for sensors [5]-[8]. Next to the resistivity, of particular importance for the development of sensors which operate through a change in the electrical characterisitics of polysilicon are the temperature coefficient of the resistance, the strain sensitivity (gauge factor) and the long term stability of the resistors. In the following sections are described our investigations of boron doped LPCVD (low-pressure chemical vapor deposition) polysilicon structures used for the development of pressure and temperature sensors.

Characteristics of boron doped polysilicon layers

For the investigations described below polysilicon layers with a thickness of 0.5µm were deposited using LPCVD onto oxidized silicon wafers with an oxide thickness of 0.1µm. The layers were boron doped using ion implantation and then annealed in N_2 at 950°C for 30 min. Next the layers were patterned by wet chemical etching using negative photoresist as an etch mask. Metalization is accomplished through aluminum vapor deposition and another photolithographic patterning procedure. As a final step the wafers are annealed at 470°C for 20 min in an N_2 atmosphere.

Resistivity

Figure 1 shows the resitivity of boron doped polysilicon resistors as a function of doping concentration together with the resistivity of p-doped monocrystalline silicon. From this graph it can be seen that the resistivity of polysilicon layers is always higher than that of single crystal material, even when the boron concentration is very high. At low doping concentrations the resistivity climbs rapidly, so that only the impurity concentration range shown is of interest for sensor applications as described in this paper. The solid line shows the calculated values using the carrier trapping model [9].

Temperature Coefficient

The relative resistance change of boron doped LPCVD polysilicon resistors over a temperature range from -60°C to +160°C is shown in Figure 2 with the implantation dose as the varying parameter. The following points can be seen:

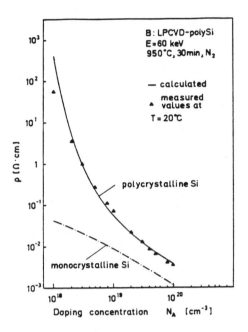

Fig. 1: Resistivity of boron doped LPCVD polysilicon as a function of doping concentration (T=20°C).

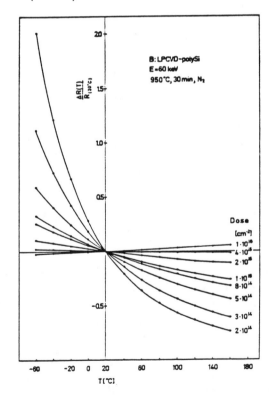

Fig. 2: Relative resistance change for boron doped LPCVD polysilicon resistors as a function of temperature.

Reprinted from *Rec. of the IEEE Solid-State Sensors Workshop*, 4 pages, 1986.

- The resistance change with temperature is not linear.
- The temperature coefficient of the resistance may be selected over a wide range, both positive and negative, through selective doping.
- The temperature dependence increases with decreasing doping concentrations.

To determine the temperature coefficient α_R of the resistors, the curves in Fig. 2 are approximated between 0° and 40° through

$$R(T) = R_{20} \exp \lfloor \alpha_R (T-T_o) \rfloor \qquad T \text{ in } °C, \ T_o = 20°C$$

where

$$\alpha_R = \frac{1}{R_{20}} \left. \frac{d\ R(T)}{dT} \right| \ T = 20°C$$

is the temperature coefficient at $T = 20°C$.

Figure 3 shows the change in the temperature coefficient α_R as a function of boron concentration N_A. From the figure it can be seen that the temperature coefficient of boron doped LPCVD polysilicon resistors can be negative, approach zero, or be positive depending on the doping concentration. This relationship is of particular importance for sensor applications. In comparison the temperature coefficient of monocrystalline p-doped silicon is shown in Fig. 3 as well [10].

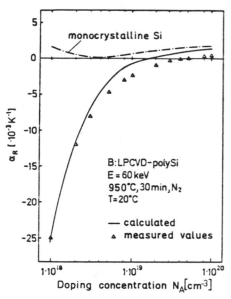

Fig. 3: Temperature coefficient of resistance for boron doped LPCVD polysilicon resistors for $T = 20°C$.

Gauge factor
For the development of sensors for pressure, force, or acceleration knowledge of the strain sensitivity of polysilicon resistors expressed through the gauge factor is of particular interest. Figure 4 shows curves of the relative resistance change of boron doped polysilicon resistors, referenced to the resistance value R_o under stress free conditions, as a function of longitudinal strain ε_1. The parameter varied is the implantation dose. It can be seen that the resistance decreases with compression and increases under tension. The resistance change with strain decreases with increasing doping concentration. This relationship is equivalent in principle to that of single crystal material [11]. The study was carried out through the utilization of a silicon cantilever beam structure with longitudinal and transverse polysilicon resistors.

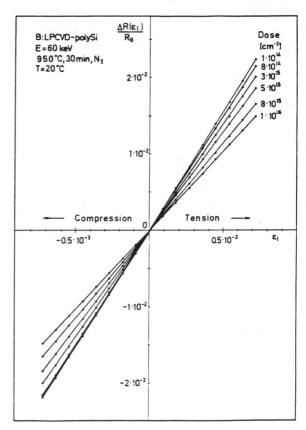

Fig. 4: Relative resistance change $(\Delta R/R_o)_1$ of boron doped polysilicon resistors as a function of the longitudinal strain ε_1.

Figure 5 shows the longitudinal and transverse gauge factors, G_1 and G_t, as a function of the doping concentration N_A. G_1 may be determined in accordance with fig. 4 through the approximation

$$G_1 = \frac{\Delta R}{R_o} \frac{1}{\varepsilon_1} \ .$$

Fig. 5: Longitudinal and transverse gauge factor as a function of the doping concentration N_A of boron doped LPCVD polysilicon resistors under tension.

The maximum value of G_1 in Fig. 5 is roughly 30 and is therefore roughly 15 times as large as metal strain gauges. In comparison with diffused piezoresistors this factor is however approximately 1/3 as large. The transverse gauge factor G_t displays an unusual characteristic which has not been explained up to this point in time. Due to this characteristic only the longitudinal gauge factor can be used for sensor applications.

In Figure 6 the longitudinal gauge factor is shown as a function of temperature. A linear reduction in G_l with increasing temperature can be observed. The temperature coefficient which can be determined from the curves shown can be expressed as

$$\alpha_l = \frac{1}{G_l} \left. \frac{dG_l}{dT} \right| \quad T = 20°C$$

and lies between $-2 \cdot 10^{-3} \ K^{-1}$ and $-1 \cdot 10^{-3} \ K^{-1}$. With increased doping the temperature coefficient decreases.

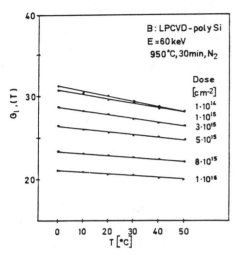

Fig. 6: Change of the longitudinal gauge factor G of boron doped LPCVD polysilicon resistors with temperature.

Longterm stability

Polysilicon resistors are capable of realizing at least as high a level of long term stability as may be expected from resistors in monocrystaline silicon since surface effects play only a secondary role in device characteristics. First results with non-passivated resistors show that at a temperature of 125°C over a time period of 1000 hours a drift of less than $5 \cdot 10^{-3}$ is obtained. It can be expected that passivation with a plasma deposited Si_3N_4 will result in a corresponding improvement in drift characteristics.

Polysilicon sensors

Pressure Sensor

Figure 7 shows the structural principle of the polysilicon pressure sensor described in this section. In this sensor the piezoresistors are no longer integrated in the single crystal silicon diaphragm, instead they are fabricated as polysilicon resistors.

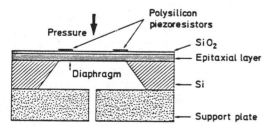

Fig. 7: Structural principle of a pressure sensor utilizing polysilicon piezoresistors.

Figure 8 shows the front and back sides of a pressure sensor chip with four polysilicon pizeoresistors which are connected to a Wheatstone Bridge. The doping of the resistors was chosen to obtain the smallest possible temperature coefficient of resistance at $N = 4.5 \cdot 10^{15}$ boron ions/cm². The resistance of the individual elements is roughly 1.8 kΩ. The diaphragm fabrication is accomplished through an anisotropic etch with a KOH/H_2O solution. A Si_3N_4 layer serves as an etch mask. The diaphragm dimensions are 1.5 x 2.5 x 0.03 mm³. The fabrication of this sensor is presented in detail in [8], and [12].

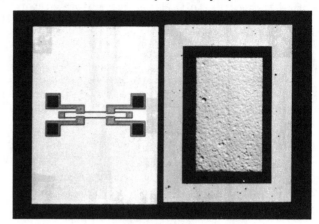

Fig. 8: Photograph of the sensor chip with polysilicon piezoresistors.

The output voltage characteristics V_o of the sensor are shown in Figure 9 as a function of pressure for the temperature range between -60°C and 200°C. At room temperature the sensor has a sensitivity of roughly 11 mV/V. The temperature coefficient of the offset voltage is 0.01% F.S./°K, of the sensitivity -0.08% F.S./°K and the linearity error is ± 0.35%. The temperature range given above is not meant to imply an operating temperature limit of the sensor as measurements up to 300°C have shown.

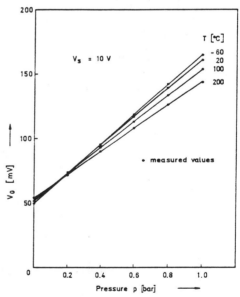

Fig. 9: Output voltage characteristics V_o of the pressure sensor as a function of pressure for temperatures between -60°C and +200°C.

Temperature Sensor

Over a limited temperature range when linearity requirements are minimal a boron doped polysilicon resistor can be directly utilized as a temperature sensor. When however a reduced linearity error over a wider temperature range is required, the resistors temperature characteristics must be linearized. A reduced linearity error may be realized through a three point linearization method in which the actual characteristic curve of the sensor corresponds with a desired line at three equidistant points [13]. The possibility of simultaneously realizing polysilicon resistors with comparatively large and very small temperature coefficients allows the placement of measurement and linearization resistors on the same chip. In this manner no external elements are is needed for the linearization of the sensor characteristics.

Figure 10 shows a polysilicon temperature sensor with on chip linearization. The resistors have varying doping concentrations and are adjusted to their specified values by laser trimming. The chip size is 1.5 x 1.5 mm². In Figure 11 the characteristic curves of this sensor for temperatures between -40°C and +140°C are displayed. The sensitivity of the sensor is -3.5 x 10⁻³ /°K. At room temperature the resistors have a value of 1 kΩ.

Fig. 10: Polysilicon temperature sensor with on-chip parallel linearization.

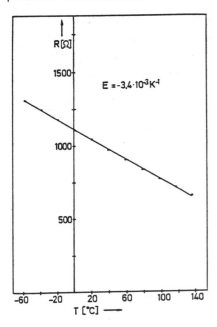

Fig. 11: Characteristic curve of the polysilicon temperature sensor shown in Figure 10.

Conclusion

From the investigations discribed here the following preferential characteristics for utilizing polysilicon layers as a sensor material have been determined:

- The layers may be deposited on electrically insulating substrates.
- The processing is compatible with semiconductor fabrication technology.
- Layers are easily structured.

- Junction isolation is not necessary thus allowing sensor operation over a larger temperature range.
- By varying the doping concentration the temperature coefficient of the resistor may be selected over a wide range, positive, negative, or zero.
- Polysilicon layers exhibit a comparatively large and isotropic piezoresistance effect.
- The layers are highly reproducible and exhibit homogenous characteristics.
- Through laser trimming the resistance values can be selectively adjusted.

From these advantages the many sided application possibilities of polysilicon layers in measurement technology applications can be clearly seen. The dielectric isolation of the layers in connection with an increased operating temperature range, the possibilities for temperature compensation and the trimming capabilities are the most important advantages of polysilicon as opposed to monocrystaline silicon.

Acknowledgement

We would like to thank I. Schulz for her valuable assistance.

References

[1] R.S. Rosler, "Low Pressure CVD Production Processes for Poly, Nitride, and Oxide," Solid State Technol., vol. 20, pp. 63-70, Apr. 1977.

[2] W. Kern, G.L. Schnable, "Low-pressure Chemical Vapor Deposition for Very Large-scale Integration Processing - A Review," IEEE Trans. Electron Devices, vol. ED-26, No. 4, pp. 647-657, Apr. 1979.

[3] F. Faggin, T. Klein, "Silicon Gate Technology," Solid-State Electr., vol. 13, pp. 1125-1144, 1970.

[4] F. Masuoka, K. Ochii, M. Masuda, K. Kobayashi, and T. Kondo, "A New High Density Full CMOS SRAM Cell Using Polysilicon Interconnetion Structure," in IEDM Technical Digest, Washington, D.C., pp. 280-283, 1985.

[5] R.T. Howe, R.S. Muller, "Polycrystalline Silicon Micromechanical Beams," J. Electrochem. Soc., vol. 130, No. 6, pp. 1420-1423, June 1983.

[6] W. Benecke, L. Csepregi, A. Heuberger, K. Kühl, H. Seidel, "A Frequency-selective, Piezoresistive Silicon Vibration Sensor," in Proc. Transducers '85, Philadelphia, pp. 105-108, 1985.

[7] H. Guckel, D.W. Burns, "A Technology for Integrated Transducers," in Proc. Transducers '85, Philadelphia, pp. 90-92, 1985.

[8] E. Obermeier, " Polysilicon Layers Lead to a New Generation of Pressure Sensors," in Proc. Transducers '85, Philadelphia, pp. 527-536, 1985.

[9] N.C.C. Lu, L. Gerzberg, and J.D. Meindl, "A Quantitative Model of The Effect of Grain Size On The Resistivity of Polycrystalline Silicon Resistors," IEEE Electron Device Lett., Vol. EDL-1, no. 3, pp. 38-41, 1980.

[10] W.M. Bullis, F.H. Brewer, C.D. Kolstad and L.J. Swartzendruber, "Temperature Coefficient of Resistivity of Silicon and Germanium Near Room Temperature", Solid-State Electr., vol. 11, pp. 639-646, 1968.

[11] O.N. Tufte, E.L. Stelzer, " Piezoresistive Properties of Silicon Diffused Layers," J. Appl. Phys., vol. 34, pp. 313-318, 1963.

[12] E. Obermeier, F.v. Kienlin, "Silizium-Drucksensor für hohe Betriebstemperaturen," in Proc. Sensor '85, Karlsruhe, pp.4.3-4.3.12, 1985.

[13] A. Burke, "Linearizing Thermistors with a Single Resistor," Electronics, pp. 151-154, June 1981.

FINE GRAINED POLYSILICON AND ITS APPLICATION TO
PLANAR PRESSURE TRANSDUCERS

H. Guckel, D. W. Burns, C. R. Rutigliano, D. K. Showers and J.Uglow
Wisconsin Center for Applied Microelectronics
University of Wisconsin
Madison, WI 53706, USA

Introduction

Electro-mechanical sensors are fundamentally three-dimensional structures. Their microminiaturization depends on fabrication techniques which either adapt two-dimensional IC-processing procedures to three-dimensional situations or involves new processing tools which recognize this aspect of sensors. Any research program which is to produce engineering results would first deal with the adaptation issue and at a later stage move to more speculative, specialized processing for advanced sensors. This did indeed happen at Wisconsin in a preconceived way. In phase I modified planar construction techniques are used and tested via the sensor which currently has the largest market share: the pressure transducer. Phase II is much more three-dimensional and involves plasma deposited photoresist as well as extensive use of reactive ion etching. Work under phase I is leaving the research area and is addressing manufacturing as well as economic issues. Phase II has produced its first research device.

Modified Planar Processing

Planar processing is well developed and, in a sufficiently planar situation, can produce minimum feature sizes of 0.5 micron. The most direct method for the incorporation of three-dimensional features involves the use of thin films as sensor construction material and also as sacrificial layers which are removed by lateral etching. Thus, a pill box may be constructed using a silicon substrate, decorating it locally with a sacrificial layer, covering with a second layer which forms the upper half of the pill box and by removing the sacrificial layer via lateral etching.[1] This procedure will fit well into planar processing if the pill box minimizes deviation from planarity and if the films are process compatible. The use of high temperature film materals is particularly attractive. However, it is also true that a pill box with predetermined mechanical features is only possible if the mechanical properties of the sensor construction films are known and can be reproduced repeatedly. The material selection, the measurement of mechanical constants and optimized processing for mechanical properties, become major issues in this type of research.

The two dominant film materials at Wisconsin are polysilicon and silicon nitride. Polysilicon when LPCVD deposited from silane has many forms. For depositions near 580°C a very fine grained silicon film is obtained whereas deposition near 640°C produces rather coarse grained or large crystallite films. Both films produce as deposited compressive strains above 0.2%. However, carefully designed anneal procedures can reduce the compressive strain significantly, and, in the case of fine grained polysilicon, reductions to less than 0.014% after annealing for three hours at 1150°C without increase in grain size have been achieved. Silicon nitride is an excellent insulator. LPCVD deposited nitride from dichlorosilane and ammonia produces a film which has a strong tensile field at stoichiometric compositions and becomes compressive for silicon rich films. It is therefore possible to produce films which are essentially strain free by choosing the deposition conditions properly. These conditions, or the local strain field for various films, have been measured experimentally.[2,3] The tools for this were developed at Wisconsin and involve three basic structures: the cantilever, the doubly supported beam and the doubly supported ring with a weak cross member.[4] They are combined in a single mask, STRAIN6, which is shown in

Reprinted with permission from *Transducers '87, Rec. of the 4th Int. Conf. on Solid-State Sensors and Actuators*, 1987, pp. 277–282.

figure 1. In this diagnostic mask the cantilevers are used to determine strain field uniformity in the direction of growth. Buckling of beams is used to measure the strain field for compressive and tensile configurations. For compressive fields doubly supported beams will buckle when the strain e exceeds

$$e \geq (\pi^2/3)(h/l)^2 \qquad (1)$$

where h is the beam thickness and l its length. Tensile fields are measured by using a doubly supported ring. The cross-member buckles under conditions which allow strain field determination.

Figure 1. Residual strain fields in deposited films are measured using a single masking step. This pattern features doubly supported beams for compressive strain fields, ring and beam structures for tensile strain fields, and cantilevers for strain field uniformity.

Tools of this type are necessary for process optimization. Strain is indeed the most important quantity which has to be measured. Tensile strength is a close second. Young's Modulus and Poison's ratio can be estimated and present a minor problem if they remain constant for a given process procedure.

The selection of polysilicon rather than silicon nitride for the pill box is based in part on thermal expansion considerations: an all-silicon cavity, and, in part, on the tensile strength of the films: silicon is very superior to silicon nitride.

The fabrication of the latest version of the Wisconsin pressure transducer involves seven masks. The first three pertain to the pill box, and the remaining four deal with piezoresistive readouts. The starting material is silicon, either (111) or (100) orientation, which is oxidized to 400Å and covered with 400Å of strain compensated LPCVD silicon nitride. Mask #1 removes the nitride over the intended cavity and is followed by an oxidation - oxide strip procedure which is in turn followed by a second oxidation which results in an oxide-filled cavity of 7500Å thickness. The second mask deals with with an essentially planar wafer and is used to remove the nitride and oxide in the anchor regions for the transducer plate. Proper anchor and etch channel configurations were determined experimentally and are typically as shown in figure 2. Deposition of polysilicon, typically with design dimensions of two micron, is followed by high temperature annealing to set the strain field. Measurements indicate less than 80 defects per 3" wafer with a haze of roughly 200Å and, of course, a very small grain size.

Mask #3 removes the polysilicon over the etch channels. These are shown in figure 3. The channel configuration: very narrow and interior to the main polysilicon structure, is a concession to photoresist integrity during subsequent processing. The removal of the sacrificial oxide layer occurs after mask #3. Improvements in the silicon texture and gap reductions to 7500Å produced major difficulties with surface tension during the final stages of the cavity etch. They had to be solved by carefully designed rinse cycles after etch completions. Experimental evidence exists which indicates that the repeatability of fine grained poysilicon for the present application is very much better than that of coarse grained polysilicon.

The design of the plate and the gap dimensions for a particular pressure range has received detailed attention. A plate theory which includes built-in strain effects has been developed via analytic techniques and has given confidence for finite element results for the pill box. This tool together with the independent measurement of longitudinal as well as transverse gage factors for piezoresistors is an absolute necessity if resistor placement on the pill box is to be accomplished in a rational and optimized manner.

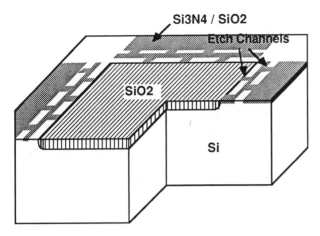

Figure 2. Cross sectional view of the WCAM Planar Processed Pressure Transducer shows location of etch channels and interior oxide. Isoplanar processing results in the oxide upper surface being coplanar with the wafer surface.

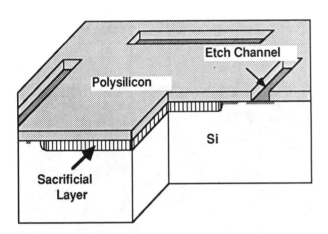

Figure 3. A layer of fine grain polysilicon is deposited using LPCVD techniques. The sacrificial oxide is removed to produce the free-standing polysilicon plate.

Resistor construction on the pill box via diffusion is a possibility. This approach was tried and abandoned in favor of deposited silicon resistors. The dielectric isolation layer over the polysilicon is achieved by a 400Å oxidation which is followed by an LPCVD deposition of 4400Å of piezoresistive polysilicon. This deposition is also the primary seal for the cavity etch channels and, as explained elsewhere, produces a good vacuum via reactive sealing. A blanket implant of 5×10^{15} cm^{-2} at 80 KeV follows. Mask #4 provides additional doping for metal contact regions and is also used to lower the resistance value of the sensing structures in unwanted strained regions. Mask #5 defines all polysilicon structures and is used in combination with a reactive ion etching procedure. A second nitride layer of 450Å is used to passivate the entire

structure. Mask #6 deals with contact opening and mask #7 defines the desired metal pattern. Figure 4 diagrams the completed device.

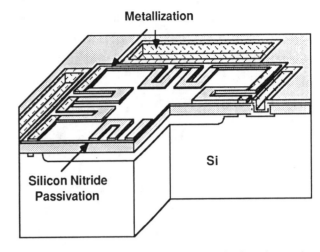

Figure 4. This view of the completed device shows the deformable plate element above an evacuated cavity. Optimal placement of deposited polysilicon piezoresistors is based on maximum pressure sensitivity with high tolerance to alignment error.

The resistor design for the device deserves special attention. The sheet resistance is near 130Ω per square with a measured TCR of 140 ppm/°C. The longitudinal gage factor is 22 and the transverse gage factor is -6 according to cantilever measurements. The possibility of using p and n resistors for a fully active bridge was investigated and postponed because of differential etch rates during RIE procedures. Instead attention was paid to resistor positioning which produces good sensitivity and allows for misalignment during mask exposure. This becomes an important issue when one considers that diaphragm sizes are typically near 100 x 100 micron and that pill boxes must be furnished with resistors of 5±.1 micron linewidths with absolute placements that must be controlled accurately if identical gages are to result. These arguments led to the decision to use a single resistor per cavity. The resistor, as indicated in figure 4, has four series connected legs. Each leg has four active meander sections. The resistor is turned at the neutral positions on the plate. The resistor sections which are parallel to the plate edge are also doped p$^+$ to eliminate them from strain detection considerations and improve the pressure sensitivity. This type of a layout is optimized towards misalignment errors as figure 5 indicates. A second research effort concerned itself with the electrical nature of the device. Nonlinearity with current and contact resistance variations had to be eliminated. This has been accomplished and excellent resistors of 19kΩ are currently in use. Thermal matching is accomplished by using an active gage and connecting it in series with an inactive gage; one in which the sacrificial layer has not been removed. This procedure produces gages which typically change by 1.6% over span and exhibit linearity to 0.1%.

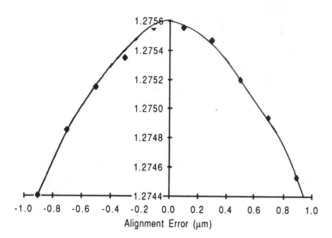

Pressure Sensitivity (% Full Scale)

Alignment Error (μm)

Figure 5. Piezoresistors designed with the turn-arounds at neutral strain points on the plate result in minimal variation of pressure transducer sensitivity to small misalignments. These calculations are for a single resistor deposited on a 116 x 116 x 2 micron plate with a compressive residual strain field of .1%, and a pressure range of 15 psia. Further reduction in the variation with resistor misalignment occurs when two or four resistors are series connected.

An evaluation of the device has to deal with mechanical performance, electronic behavior and manufacturability. Mechanically the device is very small. Thus for a vacuum transducer a 3" wafer will contain some 10,000 devices versus 400 or so for standard designs. However, all mechanical advantages are most likely overshadowed by the intrinsic overpressure stop which the device has.[4] Overpressure of 500 psi on one atmosphere designs have produced no breakage or hysterisis. Thermally the device is essentially an all silicon structure which eliminates differential expansion effects. Electronically the device is quite interesting. Thus 19kΩ resistors show standard deviation of 60 Ω. More important is the observation that the impedance level of the device and the TCR are quite designable. This implies that the data extraction circuitry is not restricted to instrumentation operational amplifiers but that time constant measurements and lock-in amplifier techniques

become feasible and add to the expected system performance.

The issue of manufacturability is an integral component of any sensor assessment. The elimination of front to back alignment as well as KOH etching is in itself good. Tighter dimensional tolerances via thin film technology is an added benefit. However, alignment and etching errors can have very bad effects and are of course more troublesome for this small device than for existing larger structures.

The ability to produce a device which works as well or better than a larger device is in itself useful. However, if size itself is not an important issue, array structures become feasible and offer many interesting possibilities. Thus, if a single resistor of value R involves a standard deviation s, an n x n array of resistors, if properly connected, will produce the resistance R with a standard

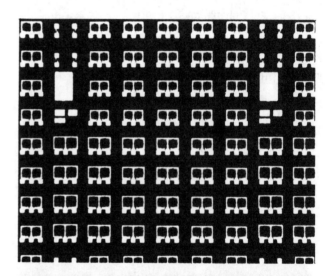

Figure 6. This micrograph (top) shows a die containing 44 pressure transducers from 5 to 25 psia, blocked in groups of four. Also visible are cantilevers and diagnostic structures for monitoring mechanical and electronic characteristics of the devices. The bonding pads are 100μm x 100μm. Die size is .180 x .200 mils. A closer view taken with an interferometer reveals the deflections of the active devices (bottom). The plate sizes are 176μm and 166μm. The serpentine resistor configuration is visible on the inactive devices.

deviation of s/n. This device is now under investigation with a 12 X 12 array of one atmosphere devices. The extension of this concept to fault tolerant systems requires different interconnections. It is also very feasible. Finally there is the issue of an extended range transducer. Multiple transducers obviously extend the dynamic range of the composite over that of a single device for a given resolution. This observation has been used in figure 6 where transducers with touch down pressures between 5 psia and 25 psia in 2 psi increments are shown. Figure 7 continues these ideas and shows the size variations which are needed to design extended range arrays. This graph is based on square plate devices. However, it should be noted that this construction technique can easily deal with rectangular and circular devices, or any planar shape diaphragms.

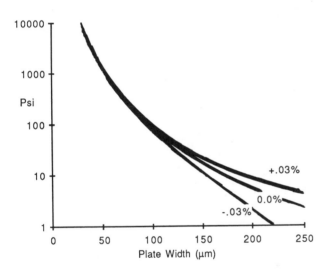

Figure 7. Design points for extended range devices show that pressure ranges can span nearly four orders of magnitude for a single film thickness. The curves are based on a thickness of 2.0μm, a gap of 7500Å, Young's modulus of 22.2 Mpsi, and Poison's ratio of 0.28. The effect of the built-in strain field is most apparent when plate dimensions approach critical dimensions, i.e. near buckling.

Three-Dimensional Processing

The pressure transducer work is for all practical purposes an extension of typical two-dimensional processing. This approach can and does produce devices. However, it is also restrictive. A more three-dimensional processing technique is highly desireable and is major goal of research efforts at Wisconsin. This work, phase II, is

based on the premise that photoresist applications via standard spinning are too restrictive. Thus the alternative: plasma polymerized photoresist is receiving detailed attention.

A second set of arguments involves the use of RIE etching. Very good results in micromachining of silicon are obtained if the material to be etched is less than, say, 50 micron in thickness. Both observations are combined in experiments which involve front to back aligned substrates which are locally thinned by KOH etching. The unetched front face is covered with chromium which forms an excellent etch mask against RIE etching in fluorine

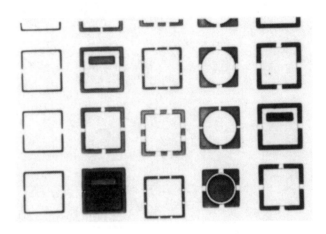

Figure 8. Recent advancements in truly 3-dimensional processing have resulted in supported silicon plates of various geometries. In the upper photo, the square plates are 1000μm x 1000μm, while the circular plates are 850μm in diameter. In the lower photo, a 1000μm x 1000μm x 50μm thick single crystal silicon plate is suspended by 6000Å thick aluminum beams on the periphery.

atmospheres. The chromium is covered with plasma deposited PMMA and exposed via X-ray lithography or a good projection printing system. Removal of the uncovered chromium in a CCl_4 plasma prepares the wafer for RIE etching in NF_3 or CF_4. Proper etching conditions produce vertical wall etches through the silicon to aluminum supports. Figure 8 shows recent results for this fabrication procedure which appears to have the promise of producing major progress in new electromechanical device realization.

References

1) H. Guckel, D. W. Burns, "A Technology for Integrated Transducers", Transducers '85, Philadelphia, 1985, pp 90-92.

2) M. Sekimoto, H. Yoshiara, T. Ohkubo, "Silicon Nitride Single Layer X-Ray Mask", Journal of Vacuum Science Technology, Vol. 21, No. 4, pp 1017-1021, Nov/Dec 1982.

3) H. Guckel, D. K. Showers, D. W. Burns, C. R. Rutigliano, and C. G. Nesler, "Deposition Techniques and Properties of Strain Compensated LPCVD Silicon Nitride Films", IEEE Sensor Workshop, Hilton Head, S. C., 1986.

4) H. Guckel, T. Randazzo, D. W. Burns, "A Simple Technique for the Determination of Mechanical Strain in Thin Films with Application to Polysilicon", J. Appl. Phys., 57(5), March, 1985, pp.1671-1675.

5) S. Sugiyama, T. Suzuki, K. Kawahata, K. Shimaoko, M. Takigawa, I. Igarashi, "Micro-Diaphragm Pressure Sensor", IEDM '86, Los Angeles, 1986, pp 184-187.

Sensors and Actuators, 4 (1983)

HEAT AND STRAIN-SENSITIVE THIN-FILM TRANSDUCERS*

RICHARD S. MULLER

Department of Electrical Engineering and Computer Sciences and the Electronic Research Laboratory, University of California, Berkeley, CA 94720 (U.S.A.)
Fulbright Research Professor, 1982 - 83, TU München and Fraunhofer Inst. für Festkörpertechnologie, Munich (F.R.G.)

Abstract

Thin-film technologies for heat and strain sensing show considerable promise for direct integration with silicon integrated circuits. When combined with on-chip processing of silicon micromechanical structures, wholly new design options for sensing systems can be considered. Miniature bolometers, thermocouple arrays and pyroelectric heat sensors are suitable thin-film heat sensors. Metallic and polycrystalline silicon piezoresistors, electrets and piezoelectric films are candidates for IC-compatible strain sensors. Features of the technology for r.f.-sputtered ZnO films on silicon are described.

1. Introduction

Thin-film technologies have been a part of planar silicon processing since its inception; the oxidized silicon wafer provides an excellent substrate for many deposition procedures. Compatibility with microcircuit production, layout geometry and passivation of the circuits are, however, non-trivial problems to be considered if thin-film sensing is to be *on-chip* integrated with associated electronics. A development that has made this integration more attractive is the achievement of micromechanical structures in silicon [1]. Of particular interest is the reliable production of thin silicon diaphragms using anisotropic etchants.

In this paper several promising thin-film applications to heat and strain sensing are considered in terms of integration with silicon microcircuitry. Instead of an exhaustive overview, the discussion is composed of selected capsule reports, mainly from the author's experience, on promising sensor

and material approaches. A final section lists several procedures used successfully to prepare ZnO films on microcircuit substrates. This description, based on work in the author's laboratory, draws attention to a thin-film material that is both piezoelectric and pyroelectric and of interest for sensor applications over a frequency range from d.c. to hundreds of MHz.

2. Heat-sensitive thin films

Heat-sensitive thin films are useful for the remote sensing of temperature through detection of the black-body radiation by warm objects. Our skin, for example, has thermal sensors that are capable of remotely sensing this radiation if the emitting body is above about 100 °C. For objects below about 500 °C, the radiation is in the infrared (IR) spectral region with characteristic wavelengths measured in micrometers. Applications for detectors sensitive to IR radiation range widely including, for example, intrusion 'people' detectors (a typical human body radiates more than 100 W), IR imaging, gas analysers, fire alarms and the sensing of temperatures in objects with low thermal conductivities or small thermal capacities.

Heat-sensitive radiation detectors can be divided into two categories, the *quantum detectors*, in which a threshold corresponding to a limiting energy for absorbed photons exists and what may be called *energy detectors*, in which a relative continuum of photon energies is detected. The quantum detectors are characterized by high response peaks and a complete lack of output at long wavelengths. Typically, cooling is needed in order to operate quantum detectors for wavelengths longer than about 1 μm, and they are not easily built from thin-film materials.

The energy detectors include thermocouples, pyroelectrics and bolometers, and are prime candidates for thin-film fabrication. These devices have a response (sensitivity *versus* frequency) that is typically nearly flat over a broad range of wavelengths extending to beyond 10 μm (characteristic of the black-body emission peak at 300 K). Non-cooled energy detectors have been built that are sensitive to temperature changes for targets as low as −100 °C [2]. The sensitivity of the energy detectors is, however, typically orders-of-magnitude lower than the maximum sensitivity of quantum detectors such as junction photocells. Closely coupled, low-noise amplification is essential for their practical application; hence their direct integration with ICs is highly desirable.

The most important parameters when comparing thermal sensors are: the *responsivity* (output voltage per incident power), the *response time* and the *minimum detectable input power* [3]. The responsivity is primarily a property of the material used and the minimum detectable power depends on the system noise. The response time is determined primarily by the thermal mass of the detector element and its support. Thin-film fabrication on small support members is very advantageous in obtaining short response times.

*Based on an Invited Paper presented at Solid-State Transducers 83, Delft, The Netherlands, May 31 - June 3, 1983.

Optimization of these parameters is greatly enhanced through their direct integration with silicon microcircuits. The utilization of microcircuit fabrication techniques to build thin-film heat sensors also affords advantages such as high-definition IR imaging, extreme miniaturization and the series connections of a device array to maximize the available signal.

Bolometers

Bolometers are energy detectors based upon a change in the resistance of materials (called bolometer elements) that are exposed to a radiation flux. Bolometer elements have been made from both metals and semiconductors. In metals, the resistance change is essentially due to variations in the carrier mobility, which typically decreases with temperature. Greater sensitivity can be obtained in high-resistivity semiconductor bolometer elements in which the free-carrier density is an exponential function of temperature, but thin-film fabrication of semiconductors for bolometers is a difficult problem.

Construction of thin-film bolometers using microcircuit techniques provides a straightforward means for their miniaturization. For example, small photo-patterned gold meander-line resistors (50×50 μm^2) were shown to provide excellent sensitivity and low noise in a bolometer deposited on a thin Al_2O_3 membrane obtained by etching away the bottom Al from an oxidized substrate. The resultant bolometer, which has a response time of 5 ms [4], could be similarly constructed on a silicon/silicon-dioxide diaphragm.

Thermocouples

Thermocouples, like bolometers, offer a flat response over the visible-to-IR spectrum and sensitivities extending to the far infrared. Utilizing microcircuit processes, thermocouples can easily be reduced to very small dimensions. This miniaturization makes possible the connection of many elements in series (thermopile) to enhance responsivity. Design of an integrated thermopile has been demonstrated using an anisotropically etched diaphragm on a silicon substrate. Thin-film Bi-Sb or Au-polysilicon thermocouples are formed using 10 μm minimum feature sizes [3]. Sixty couples make up the thermopile in which the reference junctions are placed on the thick periphery and the exposed junctions, covered by a bismuth-oxide absorbing layer, are fabricated on the thinned diaphragm.

Pyroelectrics

A third type of energy detector is based on the pyroelectric effect, in which a built-in crystalline polarization changes with temperature. Many piezoelectrics, such as ZnO, are also pyroelectric [5] and hence a technology for such films can be applied to both strain- and temperature-sensing elements. The obverse side of this coin is that an unwanted sensitivity has to be considered in the design of a given element. Temperature changes can be detected in a pyroelectric thin film either through an electrometer measurement of the voltage across the detector or through detection of the short-circuit currents that flow through it as the internal polarization varies. Single-crystal, thin (5 to 10 μm) films of $NaNO_2$ have recently been produced and shown to be suitable for pyroelectric applications [6]. Extrapolating experimental results of fabrication on mica sheets, the $NaNO_2$ films appear to be suitable for integrated fabrication on Si.

Another pyroelectric film that has also been applied as a piezoelectric is the polymer polyvinylidenefluoride PVF_2. Films of this material form the basis of a 'people sensor' that has been demonstrated to be effective at 10 m [7]. An integrated array of acoustic transducer sensors has been built using PVF_2 [8].

3. Strain-sensitive thin films

Piezoresistors

Resistance changes in stressed conductors have been the basis for wire strain gauges for many years. The strain sensitivity results from mobility variations in the deformed lattice. This effect, called piezoresistance, is also one of the earliest strain-sensing mechanisms employed in single-crystal silicon. Thin-membrane Si pressure sensors based on piezoresistance in a Wheatstone-bridge sensor were first described in 1962 [9], and are today an important type of commercial sensor.

Piezoresistance in thin-film polycrystalline Si deposited on SiO_2 is at present being studied for sensor applications [10]. The outstanding advantage of polycrystalline Si is the ease with which it is incorporated into microcircuit technologies (indeed, it is an essential part of most MOS processes). Some other polycrystalline Si properties that are interesting for sensor design are the high sheet resistance and the variation in sheet resistance (four orders of magnitude is readily obtained) under the control of the designer. Another variable is the temperature coefficient of resistance which can be varied strongly and made either positive or negative. The test of utility for polycrystalline silicon as a sensor material is the control of these variables that can be achieved.

Electrets

Electrets are materials into which metastable charge configurations are introduced by specific processing steps. Strain transducers using electrets are similar to condenser microphones, relying on a capacitive coupling that varies with strain for an electrical output. In a condenser microphone, the voltage across a capacitor biased in series with a resistor connected to a voltage source is caused to vary as the capacitance varies. The electret-film transducer, in contrast, does not require a bias voltage because the internal electret charge is detected by the capacitor.

Electrets have been produced in several ways on oxidized silicon wafers [11] and sensitivities to acoustic signals exceeding those of comparable ZnO transducers have been measured. One fabrication technique, which could conceivably be carried out as a part of IC processing, is the anodization of Al in phosphoric acid to produce Al_2O_3. Anodization produces electret films because a space charge of ions is left as the oxide films grow. When examined in a TEM, the anodic films are revealed to be cellular in nature, with central cavities open at the upper surface [12]. If these cells are filled with polar, high resistivity fluids such as methanol or pure water, they adsorb dipoles which cause a four-fold increase in the transducer output voltage. Providing an overlying sealant and upper electrode for the anodized electret films poses difficulties and typically reduces the transducer efficiency below that obtained by detection of the signal across an air gap. Transducers with sensitivities of the order of ZnO films have been fabricated by using a layer of spun-on silicone rubber covered with evaporated Al/Cu [11]. An unsolved problem with anodized electret films is an observed long-term decay in the transducer output voltage. The best films produced have an extrapolated decay to zero in the order of 18 years. Another type of electret film is produced by electron-beam irradiation of polymer layers. Irradiation of a 50 μm-thick film of Teflon® with a 20 keV electron beam led to surface-charge densities of 8×10^{-9} C cm^{-2}, capable of producing an output voltage of the order of 10 mV per nm of deflection [12].

Piezoelectrics

In a number of piezoelectric crystals there is a strong growth-rate anisotropy which can be exploited to produce highly-oriented thin films. Typically the films consist of columnar crystallites having their bases on the substrate plane. This structure is evident in the SEM photograph of a sputtered ZnO film shown in Fig. 1. Although the films are not single crystals, the uniform orientation along the vertical axes of the crystallites forming them results in a summation of the individual polarizations so that the effective piezoelectric coupling coefficients in carefully grown films typically exceed 85% of the single-crystal value. These high coupling coefficients have been achieved in films of CdS and CdSe deposited through evaporation from a quartz crucible [13]. High frequency sputtering has also been employed for both of these materials as well as for another II - VI compound, ZnO.

Sputtering is a proven means of obtaining high-quality piezoelectric ZnO [14]. An oxygen overpressure is found to improve the piezoelectric characteristics of the films. The electro-mechanical coupling coefficient of ZnO is roughly twice that of CdS (0.29 versus 0.15) and the material typically forms a hard adherent film on a variety of substrates, making it expecially interesting for transducer applications. Various properties of ZnO as well as of other piezoelectrics are given in Table 1. The properties of thin-film ZnO layers have been investigated thoroughly because of their application to surface acoustic wave (SAW) systems [15]. An example of the process complexity achieved to date is the fabrication of a new type of SAW convolver in which a very wide channel (1000 μm) integrated DMOSFET is used to generate the non-linearity required for a convolver output [16].

Surface acoustic wave structures have also been applied for vapour and pressure sensing. An interesting exploitation of micromechanics for a SAW sensing system is the use of a thinned diaphragm to couple a travelling wave from one surface of a silicon wafer (on which the system electronics are built) to an active sensing area on the other side [17]. Experiments on SAW propagation with a similar embodiment in which the diaphragm is thinner than a SAW wavelength (45 μm as compared to 76 μm) show a striking improvement in transmission loss figures as well as the potential for significantly higher frequency operation of the system than for an unetched Si wafer [18].

TABLE 1

Properties of some piezoelectrics

Material	Piezo. coeff. K_t	$\epsilon^S_{33}/\epsilon_0$	C^D_{33} N m^{-2}	Transmit/Receive parameters		Pyroel. coeff. C K^{-1} m^{-2}
				T.P. Å V^{-1}	R.P. per thick.	
ZnO	0.29	10.2	22.8×10^{-10}	0.066	0.067	6.8×10^{-6}
PVF$_2$	0.19	12	0.8×10^{-10}	0.138	0.129	2.6×10^{-5}
CdS	0.15	9.5	9.6×10^{-10}	0.045	0.053	4.1×10^{-6}
Quartz	0.09	4.5	8.8×10^{-10}	0.020	0.050	Not pyroel.

$$\text{T.P.} = \frac{K_t}{1 - K_t^2} \left(\frac{\epsilon^S_{33}}{C^D_{33}} \right)^{1/2} \qquad \text{R.P.} = \frac{K_t t}{(C^D_{33} \epsilon^S_{33})^{1/2}}$$

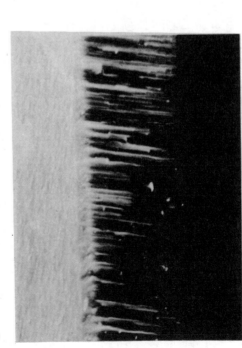

Fig. 1. Scanning electron microscope view of the fractured edge of a 10 μm thick ZnO film sputtered on an oxidized Si substrate. The columnar structure reveals the C-axis oriented crystallites.

Another compatible integrated sensor and circuit is an accelerometer, in which the strain at the support end of an etched micromechanical Si cantilever is detected by a ZnO overlay. The signals are delivered along a fully insulated line to the gate of an integrated depletion-mode MOSFET [19]. Since there is no direct bias connection, the ZnO drives an essentially pure capacitive load. The ZnO film is also highly resistive and its thickness is only a fraction of its Debye length. These two features prohibit charge relaxation and lead to essentially d.c. operation of the accelerometer [20]. This is a highly desirable capability that cannot be achieved in an unencapsulated piezoid stress sensor [21].

These applications of ZnO on Si demonstrate that integration of the various technologies typically provides new dimensions that are simply unavailable if the system is made by hybrid interconnection.

4. Technology for ZnO films on silicon

The compatible production of piezoelectric thin-film ZnO with a planar silicon circuit was reported in 1972 [22]. Since that time, applications of compatible ZnO film and IC processing have led to a number of improvements in technology. Some features of the procedures currently being used in the author's laboratory are described in the following.

A modified r.f. sputtering unit (Materials Research Corporation) is used with the target mounted below the substrate (sputter-up configuration). The system has a heated substrate and a permanent magnet is mounted below the target. The magnet has an annular north pole surrounding an axial south pole, causing magnetic flux to form arching patterns over the target as seen in Fig. 2. The function of the magnet in this planar magnetron sputtering system is to cause charged particles (mainly electrons) emitted from the bombarded source material to be returned to the target [14, 15]. This has several benefits: it reduces bombardment damage at the substrate and it

returns energy that would otherwise be carried away to the substrate making temperature control and the achievement of good growth conditions there difficult. The energy accompanying the charged particles returning to the substrate also leads to higher deposition rates. Table 2 lists typical sputtering conditions for ZnO film deposition.

TABLE 2

Sputtering conditions: ZnO films on Si

Target	Hot pressed ZnO - 99.99% pure
Target size	12.7 cm diameter
Target/substrate distance	4 cm
Substrate temperature	250 °C
Pressure	10 μm Hg
Atmosphere	80% Ar, 20% O_2
r.f. power	200 W
Deposition rate	~2 μm/hr

It is not advisable to use the sputtering system for any material other than ZnO in order to maintain control over film properties. The piezoelectric activity of the films is strongly affected by system cleanliness. The major steps carried out for each deposition sequence are the following: (1) All components in the chamber are thoroughly cleaned and baked out. (2) The chamber is baked with IR lamps for several hours and pumped to about 2×10^{-7} Torr before sputtering. (3) An initial sputter deposition onto a shutter is continued for about 30 minutes before films are laid down. (4) Extensive cold trapping within the chamber is carried out.

Two etch procedures have proved satisfactory for pattern delineation in ZnO. An etch consisting of $CH_3COOH:H_3PO_4:H_2O$ in a volume ratio 1:1:10 has an etch rate of 1.5 μm/minute. Increased percentages of H_3PO_4 strongly increase the etch rate. An etch formed from 6 g of NH_4Cl in a bath of NH_4OH and H_2O (4 ml:30 ml) has an etch rate of 0.5 μm/minute. A solution of $KOH:K_3Fe(CN)_6:H_2O$ (1 g:10 g:100 ml) has proved satisfactory for removing Al without attacking ZnO.

5. Conclusions

Integration of heat and strain-sensing thin films with Si ICs is possible with several technologies. The combination of sensing thin film and planar IC processing techniques typically offers more than the sum of each separate technology. Silicon micromechanics provides a further dimension lending promise for the development of new innovative sensor systems. The hindsight from integrated circuits shows clearly that integration of components must be thought of as far more than an improved production method; it is a development that casts a new light on the whole system design.

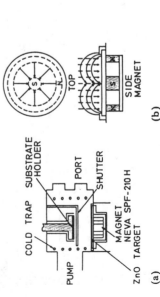

Fig. 2. Features of a planar magnetron sputtering system. (a) A cross-sectional view of the sputtering chamber; (b) schematic indication of the magnetic flux (solid lines) and the r.f. electric field energizing the sputtering ions (dotted lines).

Acknowledgements

Contributions from K. Wise, J. Bernstein, C. T. Chuang, D. Polla, P. Kleinschmidt, E. Obermeier and R. M. White are greatly appreciated. The author thanks the German Fulbright Commission and Prof. I. Ruge of the Technical University, Munich for support and hospitality during 1982 - 83. Portions of this research were supported by the National Science Foundation under grant ECS-81-20562 and by the University of California Microelectronics Program.

References

1 K. E. Petersen, Dynamic micromechanics on silicon: techniques and devices, *IEEE Trans. Electron Devices, ED-25* (1978) 1241.

2 W. Heimann and U. Mester, Noncontact determination of temperatures by measuring the infrared radiation emitted from the surface of a target, *Int. Phys. Conf. Series, No. 26* (1975) 219.

3 G. R. Lahiji and K. D. Wise, A monolithic thermopile detector fabricated using IC technology, *IEEE Trans. Electron Devices, ED-29* (1982) 14 - 22.

4 Batelle Institute e.V. Frankfurt, Germany, reported in *Elektronik*, (11) (1979) 11.

5 G. Heiland and H. Ibach, Pyroelectricity of ZnO, *Solid-State Communications, 4* (1966) 353 - 356.

6 H. Vogt, H. P. Zepf, P. Würfel, and W. Ruppel, NaNO₂ layers as pyroelectric sensors for heat radiation, *Sensoren Technologie und Anwendung, NTG-Fachberichte,* Vol. 79, Bad Nauheim, Germany, March 9 - 11, 1982, 135 - 140.

7 P. Kleinschmidt, Piezoceramic sensors, see ref. 6, pp. 189 - 195 (in German).

8 R. G. Swartz and J. Plummer, Integrated silicon PVF₂ acoustic transducer arrays, *IEEE Trans. Electron Devices, ED-26* (1979) 1921 - 1931.

9 O. N. Tufte, P. W. Chapman and D. Long, Silicon diffused-element piezoresistance diaphragms, *J. Appl. Phys., 33* (1962) 3222.

10 (a) E. Obermeier and H. Reichl, Polycrystalline Si as a material for sensors, see ref. 6, pp. 49 - 55 (in German).
(b) R. Weissel, Polysilicon as a new material for temperature and pressure sensors, see ref. 6, pp. 56 - 61 (in German).

11 J. J. Bernstein and R. M. White, Thin-film electret and condenser ultrasonic transducers, *Proc. IEEE Ultrasonics Symposium, San Diego, California, October 27 - 29, 1982.*

12 J. J. Bernstein, *Doctoral Dissertation,* Dept of EECS, University of California, Berkeley, December 1982.

13 J. Conragan and R. S. Muller, Piezoelectric field-effect strain transducers, *IEEE Proc. Solid-State, Catalog No. 70C25, SENSOR* (1970) 52 - 55.

14 T. Shiosaki, T. Yammamoto, A. Kawataba, R. S. Muller and R. M. White, Fabrication and characterization of ZnO piezoelectric films for sensor devices, *Proc. Int. Electr. Devices Mtg, Washington, D.C., 1979,* pp. 151 - 154.

15 B. T. Kuri-Yakub and J. G. Smits, Reactive magnetron sputtering of ZnO, *J. Appl. Phys., 52* (1981) 4772.

16 A. E. Comer and R. S. Muller, A new ZnO on Si convolver structure, *IEEE Electron Devices Lett., EDL-3* (1982) 118 - 120.

17 C. T. Chuang and R. M. White, Sensors utilizing thin membrane SAW oscillators, *Proc. IEEE Ultrasonics Symposium, Chicago, IL, October 14 - 16, 1981.*

18 C. T. Chuang and R. M. White, Coupling of interdigital transducer to plate modes in a slotted acoustically thin membrane, *IEEE Electron Devices Lett., EDL-4,* (1983) 35 - 38.

19 P. L. Chen, R. S. Muller, R. D. Jolly, G. L. Halac, R. M. White, A. P. Andrews, T. C. Lim and M. E. Motamedi, Integrated silicon microbeam PI-FET accelerometer, *IEEE Trans. Electron Devices, ED-29* (1982) 27 - 33.

20 P. L. Chen, R. S. Muller, R. M. White and R. Jolly, Thin-film ZnO MOS transducer with virtually d.c. response, *Tech. Digest IEEE Ultrasonics Symposium, Boston, MA, Nov. 1980.*

21 K. Klaassen, Pyroelectric side-effects in piezoelectric accelerometers, *Proc. SENSOR 1982,* Vol. 1, Essen, Germany, January 12 - 14, 1982, pp. 70 - 86.

22 E. W. Greeneich and R. S. Muller, Acoustic wave detection via a piezoelectric field-effect transducer, *Appl. Phys. Lett., 20* (1972) 156.

FORMATION AND OXIDATION OF POROUS SILICON FOR SILICON ON INSULATOR TECHNOLOGIES

G. Bomchil, R. Herino+ and K. Barla

Centre National d'Etudes des Télécommunications, B.P. 98, 38243 Meylan, France
+Université Scientifique et Médicale de Grenoble, 38000 Grenoble, France

Résumé

Les principales propriétés du silicium poreux, matériau obtenu par attaque anodique du silicium monocristallin sont discutées. Des résultats sur la cinétique de formation, la structure cristallographique, la distribution de la taille des pores, le comportement thermique et les mécanismes d'oxydation sont présentés. Les applications du matériau à la technologie SOI avec ses avantages et limitations actuelles sont décrites.

Abstract

Main properties of porous silicon, a material obtained by anodic attack of monocrystalline silicon are discussed.

A general review is presented of results about the formation mechanism, crystallographic structure, pore size distribution, thermal behaviour and oxidation mechanism. Applications of the material to the SOI technology with its present advantages and limitations are discussed.

Introduction

Over the past several years, various techniques have been developed to form thin layers of single crystal on silicon dioxide for applications in the VLSI technology. One of the techniques is based in the use of oxidised porous silicon to insulate high quality silicon. In its principle, this technique presents advantages over some other SOI techniques where device quality silicon is obtained after a recristallisation process. /1/

Despite this potential advantage, the use of oxidised porous silicon still presents problems which must be overcome to establish a reliable VLSI technology. Although some of the problems are processing dependent, others result from a still incomplete knowledge of the properties of the material.

In this review, we summarize the known properties of porous silicon layers, and we discuss some of the aspects where still work remains to be done for a better understanding of the material. The main results obtained by its use in SOI technologies are described in a last part.

I. Formation of porous silicon

I.1 Experimental

Porous silicon is obtained by making silicon the anode of an electrochemical cell where the electrolyte is a concentrated hydrofluoric acid solution with concentrations ranging from 10 to 50 %. The cathode is generally a platinum grid. The electrochemical cell configuration to obtain homogeneous porous layers must be such that current lines are well delimitated between the silicon surface and the platinum grid. The back contact to the silicon wafer is critical specially for rather high resistive substrates where the back side of the wafer must be doped to insure a good electrical contact either using metallic plating or electrolyte contact.

Although it occurs in an anodic process, porous silicon formation is associated with the evolution of hydrogen bubbles which stick to the outer surface of the wafer and also inside the pores. Various mechanical means have been used to eliminate the bubbles, but the most efficient method to obtain an homogeneous material is to use ethanoic-HF solutions which decrease the interfacial tension and favorise an easy removal of the bubbles /2/.

I.2 Kinetics and pore formation

The forming rate of porous silicon has been extensively studied /3-9/. At constant current density, the thickness of the porous layer increases linearly with the electrolysis time for layers up to about 15 microns. For thicker layers, the linear dependence changes towards a $t^{1/2}$ function that is attributed to concentration gradients in the pores /3,4/.

The variations of the forming rate with the electrolyte concentration or with the current density are related to the porosity of the material and the dissolution valence of silicon.

In the region where porous silicon is formed, the dissolution valence was found to vary between 2 to 3 when current density is increased from low to high values, the effect being more marked at low electrolyte concentration /7/. For the case of porous silicon formed in n+ substrates, the value of the dissolution valence varies within a wider range /8/.

The porosity of the layers prepared in p type substrates increases with increasing the forming current density and decreases when increasing the electrolyte concentration. Most reported works except one /6/ indicate that the porous layer is homogeneous in depth.

The porosity increases when the doping level is decreased from 10^{19} to 10^{16} cm^{-3}, and remains quite constant from 10^{16} to 10^{15} cm^{-3}. The forming rate of the porous layer follows quite well the variation of the porosity, the highest values being obtained with the lowest porosities.

In the case of n layers, the variation of porosity with forming current density and doping level is more complex /4/. In addition, the material changes its porosity for increasing layer thickness, the effect being more noticeable at low HF concentration /8/.

Although the differences between n and p substrates and the dependence with resistivity, electrolyte concentration and current densities have been well studied, they are still not related to a comprehensive formation mechanism of porous silicon. However, several different formation mechanisms have been proposed. One kind of approach is mainly based on the electrochemical and coupled chemical reactions that occur in silicon dissolution.

Silicon dissolution is assumed to occur according to the following electrochemical reaction /10,11/ :

$$Si + 2\ HF + \lambda h+ \longrightarrow SiF_2 + H+ +(2-\lambda)e- \quad (1)$$

where h+ represents the hole valence band charge transfer and e- the electron conduction band charge transfer. The physical picture behind (1) is that the primary reaction, perhaps the initial breaking of the covalent bonds, requires holes, but that electronic exchange with the conduction band also migth occur. With p substrates, holes are already available while for n silicon, holes can be supplied in highly doped material by tunneling through a thin surface barrier /12,13/. According with this reaction, the effective dissolution valence should be 2. In practice, values between 2 and 3 are found. This can be explained by assuming a partial dissolution of silicon in the tetravalent state /7/. Another explanation is possible assuming the redeposition of silicon on the pore walls as an amorphous silicon or a silicic acid film resulting from the chemical reactions of the unstable SiF$_2$ /14,7/. Mechanisms of pore formation have been proposed based on the existence of an insulating film of amorphous Si or SiF$_6$H$_2$ on the pore walls wich blocks the electrochemical reaction /6,7/.

A second kind of approach is based on surface photovoltage measurements which show that at equilibrium, the silicon surface in contact with HF solutions is in depletion /15/. According to /16/, the anodic attack starts at inhomogeneites at the atomic level in the interface. A preferential current flow down the electrolyte occurs because the semiconductor surface between pores is depleted and therefore highly resistive compared to the electrolyte and the bulk silicon at the bottom of the pores. In a similar approach /17/, it is assumed that the current density can be approximated by an expression of the type $J_0 \exp (1/r)$ where r is the local curvature radius. Using this expression and the fact that, due to depletion, silicon islands can be electrically isolated once their thickness is less than twice the thickness of the depletion layer, 2D model calculations clearly results in structures showing pores with the random distribution of the actual porous material. Although interesting from a qualitative view point, it appears that still more work is required to establish a quantitative model of porous silicon formation mechanism. We believe that any future model should be based on the nature of the energy band diagram at the semiconductor electrolyte interface, but that the influence of the Helmholtz layer must be also included /16/. Finally, the chemical aspect of the complex reactions occurring during silicon dissolution should be also considered.

II. Porous layer microstructure

Parameters seldom mentioned in the description of the material are the pore size and the pore size distribution, although they determinate altogether with the porosity the lateral thickness of the silicon walls. All the physical properties like crystalline quality, optical reponse, thermal behaviour are quite dependent on the porosity and the size of the pores.

Gas adsorption technique as used in the study of powders is an appropriate tool to evaluate the pore size and surface area of the material /19/. Main results obtained using this technique show that pore sizes are generally very small with radii between 10 to 100 Å for different preparation conditions. As shown in Figure 1, for different samples pore size distribution is quite sharp. In the case of p+ porous layers, results indicate that similar distributions are obtained for different preparation conditions (electrolyte concentration and current density) provided that the porosity is the same. On the contrary, for the case of p layers (1 ohm×cm) shown in Figure 2, a very fine structure (radius below 20 Å) is obtained and even for the same porosity, p+ and p samples display a different structure. This results confirms that the formation mechanism is very dependent on the electronic properties of the silicon substrate.

At sigth of these results, it appears that a systematic use of the adsorption technique to study the microstructure of porous silicon could provide a new insight on porous silicon and its formation mechanism.

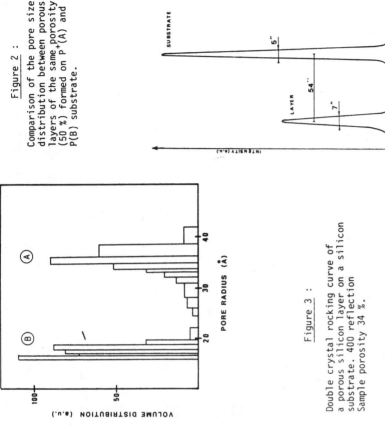

Figure 2 :
Comparison of the pore size distribution between porous layers of the same porosity (50 %) formed on P+(A) and P(B) substrate.

Figure 3 :
Double crystal rocking curve of a porous silicon layer on a silicon substrate. 400 reflection Sample porosity 34 %.

III. Structural and optical properties

A detailed study of the porous silicon structure using different x Ray techniques has shown that the diffracting properties of porous silicon prepared in highly doped p substrates are equivalent to those of a single crystal. The main results of this study /2,20/ can be summarised as follows :
1) porous silicon is a single crystal over its whole thickness even for thick films
11) the presence of a porous layer induces an overall convex elastic bending of the substrate, the porous layer being under compression.
111) the lattice parameter of porous silicon is sligthy bigger than that of silicon $\Delta a/a \simeq 10-5$, the value increasing with porosity and mean pore radius of the material. This difference in lattice parameter is at the origin of the substrate curvature.

When studying a porous layer in its substrate by double crystal x Ray diffraction, two distinct peaks occur. A typical pattern is shown in Figure 3 where two peaks are seen, the silicon peak only slightly larger. The shift between the two peaks depends on the considered Bragg reflection and can be analysed in terms of difference in lattice spacing and misorientation of porous silicon lattice planes relative to the same lattice planes of the silicon substrate. It is suggested that the lattice expansion is a consequence of the

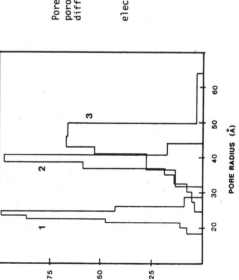

Figure 1 :
Pore size distribution for p+ porous layers formed at different current densities :
1 - 10 mA/cm²
2 - 80 mA/cm²
3 - 240 mA/cm²
electrolyte : 25 % HF, 25 % H_2O 50 % C_2H_5OH

small size of the porous silicon crystallites, the material being considered as a coherent assembly of small crystallites. The small size of the crystallites have been confirmed by results of neutron small angle scattering where the expected scattered intensity at small angles due to the small size of the silicon particles is quite strong /21/.

In the case of porous silicon prepared from highly doped n+ substrates, it was found by electron diffraction a strong dependence of the monocrystalline quality of the porous silicon surface with the density of the material. A good single crystal surface was obtained for porous silicon densities greater than 1.9 g/cm^3 /22/.

The detailed structure of porous silicon layers formed in p type wafers in a large range of resistivities from 0.01 to 25 ohmxcm have been studied using several optical techniques /23/. For the case of porous silicon layers formed in highly doped p substrates, all the single crystal characteristics of the starting wafer, the main constituents being for all densities crystalline silicon and voids. Optical densities obtained directly from the measurement of the refraction index of the material are in good agreement with densities obtained by a gravimetric method.

For the higher resistivity material, a wide scatter of results was observed with the optical density being generally lower than the gravimetric values. An important result of this optical study is the discovery that as anodised porous silicon layers prepared in resistive substrates are partially oxidised and densities measured by the gravimetric method underestimate the real porosity of the material. The optical properties of porous silicon layers were also studied in the infrared range (2-25μm) and were found to differ considerably from values obtained for the original substrate. In particular the results show that in porous silicon there is a considerable reduction of the mean carrier density. The experimental results fit a matrix model of a medium formed by crystalline silicon and voids with a pore shape parameter /24/.

IV Thermal behaviour and oxidation

High temperature annealing of porous silicon leads to a coarsening of the structure wich reduces its surface to volume ratio and modifies the pore distribution. A demonstration of this effect is shown in Figure 4. In this case the samples were annealed in vacuum at different temperatures between 450°C, and 900°C. Even at temperatures as low as 450°C the structure starts to change and at higher temperatures the whole structure collapses and large cavities surrounded by

thick silicon walls develops /25/. Such a structural evolution can be avoided by growing a thin silicon dioxide layer on the pore walls, before any further thermal treatment. An appropriate mean to obtain this result is to perform a preoxidation at 300°C in dry oxygen of the as prepared porous samples. As shown in Figure 5, pore size distributions obtained from gas adsorption experiments are practically identical for the as prepared samples and samples annealed at 800°C after the preoxidation step /26/. It is suggested that when a silicon dioxide layer is present on the pore walls, although the surface area does not change surface diffusion of silicon atoms is hindered and porous structure remains unchanged. This technique has been succesfully used to prepare samples for silicon epitaxial growth where relative high temperatures are required to clean up the surface.

When as prepared porous silicon samples are directly introduced in an oxidation furnace, a thin dioxide layer which blocks pore restructuration might be formed. The result is quite dependent on the details of the introduction procedure. To obtain reproducible results, it is better to start any oxidation procedure by a low temperature preoxidation step.

Because of the easy access through the pores inside the bulk of the porous layer of the oxidizing species, the oxidation of porous silicon is claimed to be quite easy and requiring rather low temperatures and short oxidation times. A typical porous layer with a density half of that of the silicon density with pore diameters of about 10 nm presents silicon walls of an average thickness of about 10 nm that in theory can be oxidised in a few minutes at 900°C. In fact, the obtention of a good oxide, equivalent to thermal silicon dioxide without introducing severe strains is not a straightforward step. The porous microstructures can be very different according to the substrate doping, the electrolyte concentration and the current density used in the formation of porous layers and oxidation procedure has to be adapted to the different structures.

When porous silicon film is oxidised, there are cases where step height occurs as compared with its surface before oxidation. A detailed study of the step height dependence on the oxidation time for different porous silicon densities and annealing atmosphere has provided the first ideas to understand the mechanism of porous silicon oxidation /27/. The main results of these works can be summarised as follows.

1) At the very beginning of oxidation, there is a volume expansion due to the oxidation of the silicon in the porous film
ll) In films of low porosity, the volume expansion reaches a limit value without further decrease with the oxidation time.
lll) In films of relative high porosity, the initial volume expansion is followed by an abrupt decrease corresponding to the shrinkage of the film, which occurs only when the oxidation atmosphere is wet oxygen. This phenomenon is attributed to the densification of the formed porous oxide.

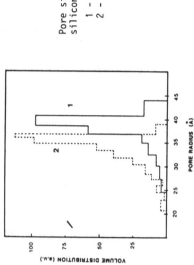

Figure 5 :

Pore size distribution of porous silicon

1 - as prepared
2 - after preoxidation at 300°C and annealing in vacuum at 800°C

VOLUME DISTRIBUTION (a.u.)

PORE RADIUS (Å)

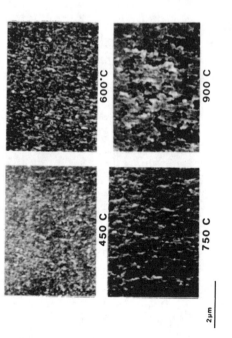

Figure 4 :

SEM pictures of porous silicon annealed under vacuum at different temperatures.

450 C   600°C

750 C   900 C

2μm

SILICON ON INSULATOR TECHNOLOGIES

Several approaches have been proposed to prepare full isolated silicon layers by oxidised porous silicon. Figure 7 shows the main processing steps of the FIPOS method (Full Insulation by Oxidised Porous Silicon) /30,31/. Porous silicon is formed only in the p type regions while the n type regions remain unchanged. Proton implantation instead of conventionnal dopants is used to create the n islands because the donnor levels created by proton implantation dissapear during the high temperature oxidation step. In this way, n channel devices can be easily prepared in the p type island. Implanted oxygen display the same behaviour /32/. However, FIPOS technology presents some processing problems for application in a VLSI process. In order to completely isolate a silicon island of width w, it is necessary to prepare porous silicon layers of a thickness $T = t + W.0.5$ where t is the thickness of the silicon island determined by the implantation profile. After full oxidation of the porous silicon layer, there always exists a thermal stress due to the difference in thermal expansion coefficient between the oxide and the silicon substrate which induces a convex warpage of the wafer. The stress value is a function of the oxidation temperature and the oxide thickness. As maximum wafer warpage that can be tolerated in the photolythography steps is about 100 μm for 4 inches wafer, there is a restriction in the thickness of the porous layer to about 8 microns. Therefore the maximum width of the silicon island is limited to 16 μm and in practice silicon island used up to date are 10 μm wide.

In addition to thermal stress other sources of stress must be considered. If the porosity of the porous silicon is less than the exact value required to account for the volume changes during oxidation, further stress at the level of each island will develop. If the porosity is greater, complete densification will lead to a good oxide, but step heigth migth appear between islands if the porosity is not the same between and under the islands. Both problems can be minimised if the densification of the oxide during the oxidation step is incomplete, but the quality of the oxide is affected and the etching rates in buffered HF solutions will considerably increase.

Further evidences to support this last point were obtained by a detailed study of the minimum time required to obtain a "good" oxide from different porous samples. It was found that this minimum oxidation time is well described by an Arrehnius type law with an activation energy of about 3ev as it is shown in Figure 6. This high value of the activation energy is much greater than that corresponding to a standard oxidation as described by the Deal-Grove model. In fact, it is quite similar to the value reported for viscous flow of silica. The transformation of porous silicon layer into thermal SiO_2 proceeds from two steps : the chemical oxidation (very fast process in the studied temperature range) followed by oxide densification through a viscous flow mechanism wich appears to be the determining step of the overall oxidation process /28,29/.

Densification times will be quite dependant on the initial structure of the porous layers. Comparison of the behaviour of P and P+ samples illustrates well this dependance. The slopes of the curves in Figure 6 are identical, indicating that the same flowing mechanism occurs, but times necessary to obtain "good" oxides differ from one order of magnitude. The shorter times obtained for P substrates are due to the much thinner structure of porous P silicon layers which leads after chemical oxidation to a porous oxide with microscopic voids and silicon walls of a much smaller size than with P+ substrates.

According to the two steps model, the time necessary to obtain a good oxide will be dependant on the conditions of viscous flow of silica and therefore on the annealing atmosphere, as it is known that activation energy for silica viscosity strongly depends on the hydroxil and oxygen content. Experimental results support this point as complete densification of porous layers at 1090°C take 4 min in wet O_2, 4 h in dry O_2 and 14 dry N_2.

If densification occurs simultaneously with chemical oxidation as it occurs in wet oxygen oxidation at temperatures higher than 1000°C, the access of the oxidant to the pores might be blocked by the densification of the outer region of the layer. It appears that for complete oxidation of thick layers or large buried layers, it is preferable to perform first a rather long chemical oxidation step at relative low temperatures.

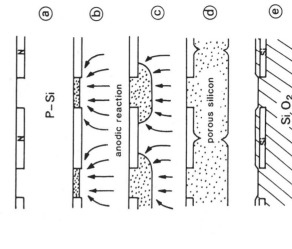

a) P-Si

b) N   N

c) anodic reaction

d) porous silicon

e) Si Si Si
   Si_2O_2
   P-Si

Figure 7 :

FIPOS processing technology :

a) starting structure
b) c) formation of porous silicon in the p-type region
d) complete insulation of islands by porous silicon
e) final structure after full oxidation of porous silicon.

Figure 6 :

Variations of the minimum oxidation time necessary to obtain "good" oxides with temperature :

A) P$^+$ porous sample
B) P porous sample

Another way to minimise the problem is obtained by ionic implantation of boron at the access window between islands wich creates a highly doped p+ surface layer. Porous silicon formed in this layer has a porosity lower than the porosity in the rest of the layer because of the differences in doping level introduced by the boron implantation. In the subsequent oxidation, the dense surface layer of porous silicon give a dense oxide with low etch rate which protects the rest of the oxide layer /31/. High pressure oxidation seems to be also a satisfactory technique to obtain good insulating layers and minimise the problems of wafer warpage in the FIPOS technique /33/.

Another interesting phenomenon seldom mentioned in the literature is described in Figure 8 a-b which show a SEM micrograph of a cross section of a FIPOS structure. As it is clearly seen, a silicon wedge in the center traversing the porous layer from the top to the bottom is present. As this wedge structure is not porous during the oxidation step, unacceptable island deformation might occur. For structures where silicon islands are covered by an insulating film during the anodic attack, the wedge is only limited to the regions near the top and the bottom of the layer. In this case, oxidation of the porous layer affects little the silicon island.

The crystalline quality of the isolated silicon layers fabricated by the FIPOS technology have been examined by TEM observations. It has been shown that dislocations are generated during the oxidation step. However dislocation density is quite dependent on the starting porous structure and the oxidation procedure, low values of dislocation densities-two orders of magnitude lower than in epitaxial layers made by SOS techniques have been reported /34/.

Using FIPOS technology, a high speed CMOS 16 K bit static RAM has been developed /35/. With the help of the high pressure oxidation procedure, a high speed 1 micron CMOS working at 60 psec/stage at V_{dd} = 5V 1 GHZ and transistor characteristics equivalent to those in bulk CMOS has been realised /36/.

Other approaches have been proposed to increase the width of the isolated islands with a minimum porous layer thickness in order to minimise stress and wafer warpage.

In the structure of Figure 9, due to different reaction rates in the ligthly doped n region, porous silicon is formed exclusively in the thin highly doped n layer. After oxidation, complete dielectric isolation of silicon islands up to 45 µm width are obtained /37/. Similar results have been obtained by using an equivalent structure formed by a thin p+ layer over a p substrate /38/. These process have potential advantages over the conventional FIPOS approach because relative large silicon islands can be isolated by thin oxidised porous layers.

Another technique for full dielectric isolation can be realised by growing an epitaxial silicon film over the porous silicon layer. This is possible because despite its porosity, porous silicon films prepared in highly doped substrate keep excellent monocrystalline properties. Using conventional chemical vapor deposition /39,40/, high temperatures (1100°C) are necessary to obtain good epitaxial films and during this high temperature process, the bulk of the porous silicon structure drastically changes and the originally fine porous structure is changed to a thick single crystal material with large voids /25,26,41/ This effect decreases the surface area of the material and oxidation of porous silicon proceeds as slowly as that of the original substrate. Instead of conventional CVD, plasma assisted CVD appears as a more appropiate technique as good epitaxial films of silicon over porous silicon were obtained at relative low temperature (770°C) and full oxidation of the underneath porous layer was therefore possible /41/. Attempts to avoid high temperature processing of the bulk of porous structures were also made by using laser annealing to recrystallise only a thin layer at the porous silicon surface. Although epitaxially reconstructed films free of grain boundaries can be obtained, they present a great number of defects /43/.

Using molecular beam epitaxy, silicon films were grown on <111> porous silicon surface at 770°C without high temperature preheating of the substrate during the cleaning procedure /44/. Although the top surface of the epitaxial films display a very smooth surface morphology, there are defects at the interface between the epitaxial layer and the porous layer. Cleaning and growth sequences must be carefully adjusted to achieve good epitaxial layers. The introduction of a neon sputter cleaning instead of Argon before epitaxy greatly reduces the defect density /45,46/.

Good results have been obtained by MBE using as starting material porous silicon layers preoxidised in dry oxigen at 300°C. Under these conditions, relative high temperatures annealing cycles are possible without affecting the porous structure and excellent epitaxial films can be obtained on the surface of porous silicon /47/. Figure 10 shows a SEM micrograph of the interface between porous silicon and the epitaxial layer. The excellent homogeneity of the epitaxial silicon as well as a sharp interface can be noticed.

All this epitaxial approach look quite promising because it results in silicon on insulator structures with a thin silicon dioxide insulating layer. Moreover the island dimensions are only determined by the maximum width of buried porous silicon that can be converted into a good thermal oxide. At present, there is no report about the possible limiting value of the island width.

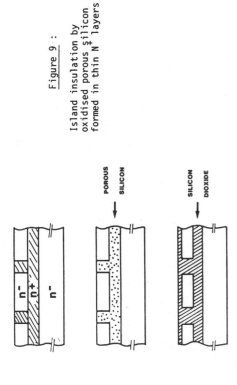

Figure 9 :

Island insulation by oxidised porous silicon formed in thin N+ layers

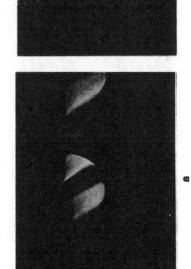

Figure 8 : SEM micrograph of cross sections in the FIPOS structure after oxidation and partial etching a) without b) with insulation of the N island during anodisation.

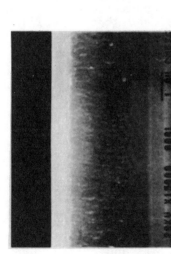

Figure 10 :

SEM micrograph of a cross-section of an MBE epitaxial silicon layer over a preoxidised porous layer.

Conclusion :

In conclusion, we can say that the various approaches for using porous silicon in a SOI technology indicate that this material is a good candidate for such applications, although all problems are not solved up to now. The physical properties of porous silicon have been quite well studied, but there are still several points which are worth to be studied in more details, as for exemple electrical and thermal conductivity or crystallographic and surface properties, which all involve different aspects of fundamental physics. Furthermore, there is still much work to be done to understand the formation mechanism of this material.

REFERENCES

1. AUVERT G., BENSAHEL D., BOMCHIL G., COLINGE J.P.
   Echo des Recherches 114(1983) 71
2. BARLA K., BOMCHIL G., HERINO R., PFISTER J.C., BARUCHEL J.
   J.Cryst. Growth 68(1984) 721
3. WATANABE Y., ARITA Y., YOKOYAMA T., IGARASHI Y.
   J. Electrochem. Soc. 122(1975) 1351
4. ARITA Y., SUNOHARA Y.
   J. Electrochem. Soc. 124(1977) 285
5. ARITA Y.J. Cryst. Growth 45(1978) 383
6. LABUNOV Y., BARANOV I., BONDARENKO V.
   Thin Solid Films 64(1979) 479
7. UNAGAMI T. J. Electrochem. Soc. 127(1980) 476
8. PARKHUTIK V.P., GLINENKO L.K., LABUNOV V.A.
   Surface Technol. 20(1983) 265
9. HERINO R., BOMCHIL G., BARLA., PFISTER J.C.
   Electrochem Soc. Meet., Detroit (1982) Ext. Abst. 223
10. UHLIR A. Bell Sys Tech. Jour. 35(1956) 333
11. TURNER D.R. J.Electrochem.Soc. 105(1958) 402
12. MEEK R.L. J.Electrochem.Soc. 118(1971) 437
13. ARITA Y. Rev.Elect.Comm.Lab. 27(1979) 41
14. MEMMING R., SCHWANDT G. Surface Science 4(1966) 109
15. MEMMING R.,SCHWANDT G. Surface Science 5(1966) 97
16. BEALE M.I.J., CHEN N.G., UREN M.J., CULLIS A.G., BENJAMIN J.D.
    Appl. Phys. Lett. 46(1985) 86
17. PFISTER J.C. Personal comm. (1985)
18. TENHUNEN H., KRUSIUS J.P.
    Electrochem. Soc.Meet.Cincinatti (1984) Ext abst 68
19. BOMCHIL G., HERINO R., R. BARLA K., PFISTER J.C.
    J. Electrochem.Soc. 130(1983) 1611
    see also BOMCHIL G., HERINO R., GINOUX J.L.
    J.Electrochem.Soc. 131(1984) 727
20. BARLA K., HERINO R., BOMCHIL G., PFISTER J.C., FREUND A.
    J.Cryst. Growth 68(1984) 727
21. BOMCHIL G., GOELTZ G.
    Small angle scattering of porous silicon. Experimental report instrument D 11
    exp 6-12 .Inst.Laue Langevin Dec.1984
22. LABUNOV V.A., BONDARENKO V.P., GLINENKO L., BASMANOV I.N.
    Akad.SSSR. Mikroelektronika 12(1984) 11
23. PICKERING C., BEALE M.I.J., ROBBINS D.J., PEARSON P.J., GREEF R.
    J.Phys.C solid state 17(1984) 6535
24. BILENKO D., ABANSHIN N., GALISHNIKOVA YU., MARKELOVA G., MYSRNKO I., KKHASNA E
    Sov.Phys.Semic. 17(1984) 1336
25. UNAGAMI T., SEKI M. J.Electrochem.Soc. 125(1980) 1339
26. HERINO R., PERIO A., BARLA K., BOMCHIL G.
    Materials Lett. 2(1984) 519
27. ARITA Y., KURANARI K., SUNOHARA Y.
    Jap.J-Appl.Phys. 15(1976) 1655
28. YON J.J. Oxydation du silicium poreux.
    DEA Microelectronique CNET(1983)
29. BARLA K., YON J.J., HERINO R., BOMCHIL G.
    Intern.Conf.Insulating Films on Semicon. TOULOUSE(1985)
30. IMAI K. Solid State Elect. 24(1981) 159
31. IMAI K., UNNO H. IEEE Trans. Elec. Dev. ED 31(1984) 297
32. CHI J.Y, HOLMSTROM R.P.
    Appl.Phys.Lett. 40(1982) 420
33. OTTOI F., ANZAI K., KITABAYASHI H., UCHINO K., MIZOKAMI Y.
    Electrochem.Soc.Meeting (1984) Ext Abs. 508
34. IMAI K., UNNO H., TAKAOKA H
    J.Cryst.Growth 63(1983) 547
35. MANO T., BABA T., SWADA H., IMAI K.
    VLSI Symp. OISO JAPAN Sept 1982
36. ANZAI K., OTOI F., OHNISHI M., KITABAYASHI H.
    IEDM (1984) 796
37. HOLMSTROM R.P., CHI J.Y.
    Appl.Phys.Lett. 42(1983) 386
38. NESBIT L.A., IEDM (1984) 800
39. ITOH T., TAKAI H., HORIUCHI M.
    Electrochem.Soc.Meet.Michis,(1981) Ext-Abs. 304
40. LABUNOV V.A., GLINENKO L.K., BONDARENKO V.P.,
    5th Int. Conf. Surf. Madrid (1983) pag 149
41. BAUMGART H., FRYE R.C., TRIMBLE L.E., LEAMY H.J. CELLER G.K.
    (1982) "Laser and e-beam interac.with solids" North Holl. pag 609
42. TAKAI H., ITOH T.,J.Electron.Mat. 12(1983) 973
43. BAUMGART H., PHILLIPP F., CELLER G.K.
    Microscopy of semicon.materials OXFORD(1983) pag 223
44. KONAKA S., TABE M., SAKAI T.
    Appl.Phys.Lett. 41(1982) 86
45. HARDEMAN R.W., BEALE M,I.J., GASSON D.B., KEEN J.M., PICKERING C., ROBINS D.J.
    Surface Science (1984)
46. BEALE R.L.J., CHEW N.G., GASSON D.B., HARDEMAN R.W., ROBBINS D.J.
    J.Vac.Sc.Tech. B3 2(1985) 732
47. ARNAUD D'AVITAYA F., CAMPIDELLI Y., BARLA K.
    Intern.Symp. on silicon MBE.Electrochem.Soc. Meet. Toronto(1985)

Section 2.2: Crystalline Semiconductor Micromachining

THE MECHANISM OF ANISOTROPIC SILICON ETCHING

AND ITS RELEVANCE FOR MICROMACHINING

H. Seidel

Messerschmitt-Bölkow-Blohm GmbH
P.O. Box 801109
D-8000 Munich 80, W. Germany

ABSTRACT

Several alkaline silicon etchants based on ethylenediamine, KOH, NaOH, and LiOH were characterized experimentally with respect to their dependence on the crystal orientation and on the boron dopant concentration of silicon. Based on these results a model for the electrochemical etch mechanism is proposed explaining the quasi-etch stop behavior and the effects of anisotropy. Finally, suggestions are given as to what etchant composition serves best for a given application.

INTRODUCTION

Anisotropic etching is a key technology for the micromachining of miniature three-dimensional structures in silicon. It is used for the fabrication of a large variety of sensors, such as for pressure, acceleration, vibration, or flow. Two etching systems are of practical interest, one based on ethylenediamine and water, with additives like pyrocatechol and pyrazine, the other consisting of purely inorganic alkaline solutions like KOH, NaOH, or LiOH. Since the effects of the different aniones were found to be minor when solutions of equal molarity are used, the discussion of the inorganic solutions will be limited to KOH.

So far, several papers have been published providing experimental data and models for describing specific aspects of the etching mechanism (1-14). In this work, an attempt is undertaken for giving a consistent model for both etching systems explaining the phenomena of anisotropy and the boron induced quasi etch stop.

EXPERIMENTAL

In this investigation KOH solutions in a concentration range from 10 - 50 %, and an ethylenediamine-based solution (EDP) type S with a composition of 1 l ethylenediamine, 133 ml water, 160 g pyrocatechol, and 6 g pyrazine were used (3). All experiments were carried out in a double-walled thermostated glass vessel with a reflux column and a nitrogen purge.

For measuring the anisotropic behavior, oxidized silicon wafers of p or n type, 1 - 10 Ω cm, with a surface orientation of (100) or (110) were patterned lithographically with a star-shaped mask, exposing blank stripes of silicon with an angular separation of one degree (9). After etching, the depth of the resulting grooves was determined by a Taly-step. The lateral underetching was measured by a linewidth measurement system (Leitz Latimet). The crystal orientation of the sidewalls of the etch grooves was determined by means of laser reflection experiments.

For determination of the boron dopant concentration of the etch rate epitaxial layers with a boron concentration ranging from $1 * 10^{19}$ cm^{-3} to $1.5 * 10^{20}$ cm^{-3} were deposited on (100) and (110) oriented wafers. The same lithography mask as described above was used on a CVD siliconnitride deposited on these wafers and the resulting depth of the grooves was determined after etching.

The etch rates of SiO_2 and Si_3N_4 were determined by ellipsometric measurements.

RESULTS

For all etchants, (111) crystal planes exposed the slowest etch rates being approximately two orders of magnitude smaller than for other principal crystal directions. For an EDP solution type S the temperature dependence of the vertical (100) and (110) etch rates, as well as the lateral (111) etch rate as determined on (100) wafers is shown in Fig. 1. Whereas etched (100) and (111) surfaces were found to be quite smooth, (110) surfaces developed grooves bounded by (331) crystal planes.

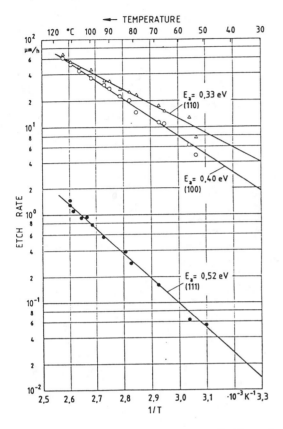

Fig. 1: Arrhenius diagram of the etch rates of the main crystal directions when using an EDP solution type S.

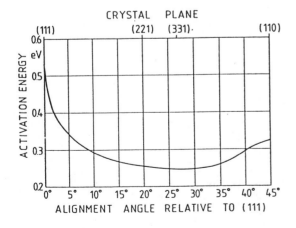

Fig. 2: Activation energy of the lateral etch rates on (100) silicon as a function of the alignment angle. Principle crystal planes corresponding to a certain angle are indicated.

For ethylenediamine-based solutions the laser reflection experiments on laterally underetched (100) and (110) wafers revealed that the etch bordering crystal planes are characterized by the Miller indices (hkk) with h = k. The maximum etch rate is obtained for the (331) crystal orientation (8, 9). The activation energies of the etch rates of different crystal planes were found to be correlated with the etch rates themselves (s. Fig. 2). A maximum of 0.52 eV is obtained for the slowest etching (111) planes, whereas the minimum of 0.25 eV corresponds to (331) with the maximal etch rate.

For KOH solutions, a maximum for the etch rate was observed for a concentration of 15 %. For higher concentrations the etch rate decreases with the fourth power of the water concentration (s. Fig. 3). KOH solutions with a concentration below 20 % were found to be of limited applicability, due to the formation of rough surfaces and a tendency for the development of insoluble white residues on the silicon surfaces. The activation energies for the etch rates of (100) and (110) crystal planes were found to be 0.60 eV. Of special interest is the development of vertical (100) planes on (100) wafers when using KOH with a concentration higher than 30 %.

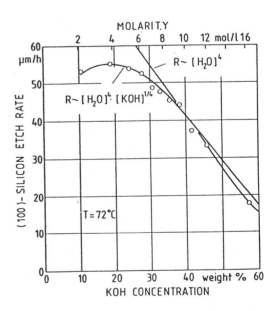

Fig. 3: (100) silicon etch rate as a function of KOH concentration.

Whereas siliconnitride proved to be a perfect masking material, not at all being attacked in alkaline solutions, the etch rate of silicondioxide shows a strong dependence on the composition of the etchant chosen. For EDP solutions this etch rate is about three orders of magnitude lower than for KOH (s. Fig. 4). The ratio of the (100) silicon etch rate to the SiO_2 etch rate as a function of temperature for EDP type S and for several KOH solutions is shown in Fig. 5.

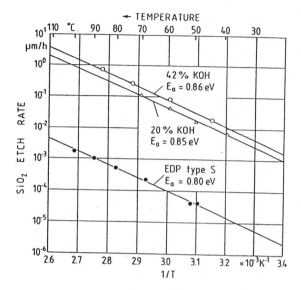

Fig. 4: Arrhenius diagram of the SiO_2 etch rate for EDP type S and KOH.

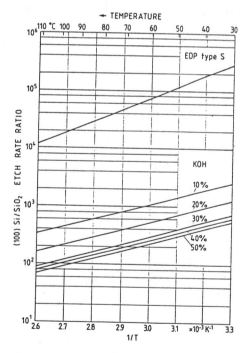

Fig. 5: Ratio of the (100) Si/SiO_2 etch rate as a function of temperature.

For boron concentrations above a critical value C_O of approximately $3 * 10^{19}$ cm^{-3} a drastic reduction of the etch rate can be observed on both (100) and (110) oriented samples. Fig. 6 shows this effect for an EDP solution type S, where the absolute etch

rate was normalized by the etch rate for lowly doped silicon. The drop is inversely proportional to the fourth power of the boron concentration. Based on experimental data, the following semi-empirical formula for the dependence of the etch rate R on the boron concentration C_B is proposed:

$$R = \frac{R_i}{(1 + (C_B/C_O)^a)^{4/a}} \qquad (1)$$

where R_i is the etch rate for lowly doped silicon. The parameter a determines the abruptness of the transition around the critical boron concentration C_O. For EDP solutions an excellent agreement with the experiment is achieved by setting a equal to 4 (s. Fig. 6).

The critical etch rate exhibits a slight temperature dependence with an activation energy of 0.025 eV (s. Fig. 6). This effect was found to be universal for all anisotropic etchants investigated (13). The Arrhenius diagram in Fig. 7 shows the activation energies of the etch rates for various boron concentrations when using an EDP solution type S. For all boron concentrations above the critical value the activation energies are approximately 0.1 eV larger than in the case of lowly doped silicon. This difference is equal to four times the activation energy of the critical boron concentration which can be deduced from the fourth power law in Eq. 1.

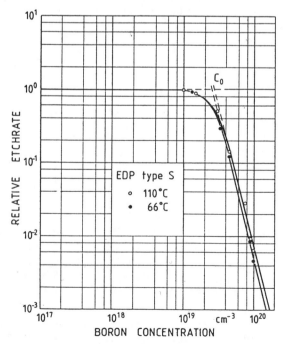

Fig. 6: Relative silicon etch rate as a function of boron dopant concentration for EDP type S.

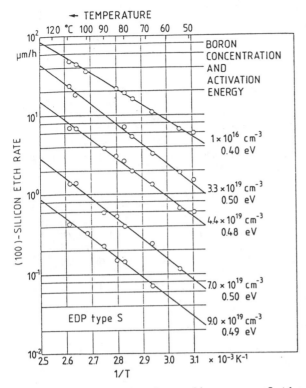

Fig. 7: Arrhenius diagram of the (100)silicon etch rate for various boron dopant concentrations.

When using KOH solutions, the critical boron concentration increases to somewhat higher values (s. Fig. 8). The transition region becomes smoother for higher KOH concentrations. This behavior can be described by Eq. 1 by setting the parameter a equal to 1.

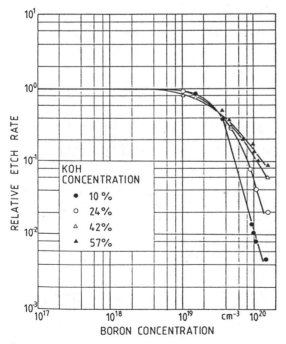

Fig. 8: (100) silicon etch rate as a function of the boron dopant concentration for various KOH solutions.

DISCUSSION

Based on the alkalinity of the solution, the development of hydrogen bubbles during the etching process, the inverse fourth power law for the etch rate in case of highly boron doped silicon and the fourth order dependence of the etch rate on the water concentration in case of high KOH concentrations, the following mechanism is proposed.

The atoms on the silicon surface react with hydroxyle ions with a simultaneous injection of four electrons into the conduction band of silicon:

$$Si + 2\ OH^- \rightarrow Si(OH)_2^{++} + 4\ e^- \quad (2)$$

This primary oxidation step is followed by a reduction reaction leading to the development of hydrogen gas:

$$4\ H_2O + 4\ e^- \rightarrow 4\ OH^- + 2\ H_2 \quad (3)$$

The positively charged complex of the primary oxidation step (Eq. 2) reacts further to a soluble silicon complex:

$$Si(OH)_2^{++} + 4\ OH^- \rightarrow$$
$$SiO_2(OH)_2^{--} + 2\ H_2O \quad (4)$$

This ion has been identified as a reaction product by Raman spectroscopy when etching with KOH (7). The overall reaction can be summarized as follows:

$$Si + 2\ OH^- + 2\ H_2O \rightarrow$$
$$SiO_2(OH)_2^{--} + 2\ H_2 \quad (5)$$

For dopant concentrations below $1 * 10^{19}\ cm^{-3}$, an extended space charge layer at the silicon surface provides a sufficiently long lifetime of the injected electrons to be able to react with the water molecules at the semiconductor/-electrolyte interface. At a critical boron concentration of $3 * 10^{19}\ cm^{-3}$ silicon degenerates (15) and starts to behave like a metal due to the dropping of the Fermi level into the valence band. As a consequence the space charge layer reduces essentially to a surface charge of a few angstrom thickness. Thus, the injected electrons immediately recombine with the holes which are present in a very large concentration making the reduction step in Eq. 3 to the rate limiting reaction. The remaining concentration of free electrons becomes inversely proportional to the concentration of holes and, thus, to the boron dopant concentration. By assuming that four electrons are involved in the electrochemical reaction, the inverse fourth power law can be explained. The temperature dependence of the critical etch rate C_0 is presumably correlated to the temperature dependence of the semiconductor-metal transition, explaining the universality of the observed activation energy of 0.025 eV for all etchants investigated.

Since EDP solutions have a much lower pH of approximately 12.5 as compared to at least 14 for KOH it can be assumed that in the former case the reaction is oxidation limited due to the relatively low concentration of hydroxyle ions. In the case of KOH for concentrations above 20 %, the reaction becomes reduction limited due to the relatively low water concentration. Since a high boron concentration also limits the reduction reaction, the smoother transition region towards a decreasing etch rate for high KOH concentrations can be explained.

The large anisotropic effects with differences in etch rates on the order of 100 : 1 are due to relatively small differences in activation energies resulting from different binding forces of the surface atoms of different crystal planes. The surface atoms of (111) planes are bonded most strongly, having only one dangling bond whereas (100) and (110) have two.

CONCLUSION

For practical applications several recommendations can be deduced from the results given above and from the proposed mechanism.

In applications where the quasi etch stop of highly boron doped silicon is taken advantage of, EDP solutions are clearly preferable. When KOH solutions are preferred for other reasons, the concentration should be chosen as low as possible where 20 % seems to be a practical lower limit because of increasing surface roughness and the likelihood of residue formation. A slight increase of the etch rate ratio between lowly and highly doped silicon can be achieved by lowering the etch temperature.

Due to differences in activation energy, the anisotropic effects generally increase when choosing a lower etch temperature. Since the overall etch rates also decrease considerably, a reasonable compromise has to be found. For KOH solutions a ratio of 400:1 between (110) and (111) has been reported (4). However, even a minute misalignment of 0.5° would lead to a

decrease of this ratio to less than 100. This is the order of magnitude that we found to be practically achievable. The same is true for EDP (s. Fig. 1).

A special effect when using KOH solutions with a concentration higher than 30 % is the formation of vertical (100) crystal planes on (100) wafers. This can be used for etching very high needles, as well as for corner undercutting compensation.

An important consideration lies in the applicability of passivation layers. Si_3N_4 has been found to be a perfect masking material for both systems. SiO_2 is an excellent passivation when using ethylenediamine based etchants. When SiO_2 is to be used as a masking material with a KOH solution, both temperature and concentration of the solution should be chosen as low as possible.

An important consideration is the safe handling of the etchants. Ethylenediamine is suspected to cause cancer and shoud therefore be handled with the greatest care possible. In this respect, KOH is clearly much less dangerous.

Summarizing, the use of ethylenediamine based solutions can strongly be recommended when the necessary safety precautions have been taken. It is superior with respect to SiO_2 passivation layers and the applicability of a boron etch stop. For applications, where anisotropy is the only requirement and the question of passivation can be handled, KOH can also be recommended.

REFERENCES

1) R. M. Finne and D. L. Klein, "A Water-Amine-Complexing Agent System for Etching Silicon", J. Electrochem. Soc., 114, 965 (1967).

2) J. B. Price, "Anisotropic Etching of Silicon with KOH-H_2O-Isopropyl Alcohol", in Semiconductor Silicon 1973, H. R. Huff and R. R. Burgess eds. (The Electrochemical Society Softbound Symposium Ser., Princeton, NJ, 1973), p. 339.

3) A. Reisman, M. Berkenblit, S. A. Chan, F. B. Kaufman, and D. C. Green, "The Controlled Etching of Silicon in Catalyzed Ethylenediamine-Pyrocatechol-Water Solutions", J. Electrochem. Soc., 126, 1406 (1979).

4) D. L. Kendall, "Vertical Etching of Silicon at Very High Aspect Ratios", Ann. Rev. Mater. Sci., R. A. Huggins ed., 9, 373 (1979).

5) H. Seidel and L. Csepregi, "Studies on the Anisotropy and Selectivity of Etchants Used for the Fabrication of Stressfree Structures", Abstract 123, p. 194, The Electrochemical Society Extended Abstracts, Montreal, Canada, May 9 - 14, 1982.

6) E. D. Palik, J. W. Faust, H. F. Gray, and R. F. Greene, "Study of the Etch-Stop Mechanism in Silicon", J. Electrochem. Soc., 129, 2051 (1982).

7) E. D. Palik, H. F. Gray, and P. B. Klein, "A Raman Study of Etching Silicon in Aqueous KOH", J. Electrochem. Soc., 130, 956 (1983).

8) H. Seidel and L. Csepregi, "Three-Dimensional Structuring of Silicon for Sensor Applications", Sensors and Acuators, 4, 455 (1983).

9) L. Csepregi, K. Kühl, R. Nießl, and H. Seidel, "Technologie dünngeätzter Siliziumfolien im Hinblick auf monolithisch integrierbare Sensoren", BMFT-Forschungsbericht T 84-209, Fachinformationszentrum Karlsruhe, 1984.

10) N. F. Raley, Y. Sugiyama, and T. van Duzer, "(100) Silicon Etch-Rate Dependence on Boron Concentration in Ethylenediamine-Pyrocatechol-Water Solutions", J. Electrochem. Soc., 131, 161 (1984).

11) E. D. Palik, V. M. Bermudez, and O. J. Glembocki, "Ellipsometric Study of the Etch-Stop Mechanism in Heavily Doped Silicon", J. Electrochem. Soc., 132, 135 (1985).

12) O. J. Glembocki, R. E. Stahlbush, and M. Tomkiewicz, "Bias-Dependent Etching of Silicon in Aqueous KOH", J. Electrochem. Soc., 132, 145 (1985).

13) H. Seidel and L. Csepregi, "Etch-Stop Mechanism of Highly Boron-Doped Silicon Layers in Alkaline Solutions", Abstract 595, p. 839, The Electrochemical Society Extended Abstracts, Toronto, Canada, May 12 - 17, 1985.

14) X. Wu, Q. Wu, and W. H. Ko, "A Study on Deep Etching of Silicon Using EPW", p. 291, Proc. of the International Conference on Solid-State Sensors and Actuators - Transducers '85, Philadelphia, PA, June 11 - 14, 1985.

15) A. A. Volfson and V. K. Subashiev, "Fundamental absorption edge of silicon heavily doped with donor or acceptor impurities", Sov. Phys. Semicond., 1, 327 (1967).

Study of Electrochemical Etch-Stop for High-Precision Thickness Control of Silicon Membranes

BEN KLOECK, SCOTT D. COLLINS, NICO F. DE ROOIJ, AND ROSEMARY L. SMITH

Abstract—A method is described to control the thickness of single-crystal silicon membranes, fabricated by wet anisotropic etching. The technique of an electrochemical etch-stop on an epitaxial layer is used to yield better thickness control over the silicon membranes (± 0.2 μm s.d.) and hence improve the reproducibility of piezoresistive pressure sensors. Although the electrochemical etch-stop is not new, this paper reports the characterization and evaluation of the technique as a fabrication process. The output characteristics of piezoresistive pressure sensors fabricated using the electrochemical etch-stop technique are compared with previously fabricated pressure sensors not utilizing accurate control over membrane thickness. The benefits of the etch-stop approach become apparent when reductions in the pressure-sensitivity variations are considered. Without etch-stop, the sensitivity on one wafer varied by a factor of two from one sensor to the other. With etch-stop, the pressure sensitivity of devices fabricated on the same wafer can be controlled to within ± 4 percent s.d.

I. INTRODUCTION

A LARGE class of sensors and actuators use micro-machined silicon structures as sensing or active elements. The miniaturization of these devices to an ever smaller scale increases the requirements for dimensional control of micromachined components such as membranes, cantilever beams, suspended masses, etc. One important method that has been employed in the fabrication of microstructures is the electrochemical etch-stop at a reverse-biased p-n junction. This technique was first proposed by Waggener in 1970 [1] and has since been successfully applied to the fabrication of several different microsensor structures [2]–[5]. However, despite its increasing use, a comprehensive characterization of the etch-stop technique has never been presented.

The etch-stop technique combines the well known anodic passivation characteristics of silicon [6]–[9] with a reverse-bias p-n junction to provide a large etching selectivity of p-type silicon over n-type in anisotropic etches such as KOH and ethylenediamine/pyrocatechol (EDP). The dimensional definition of micromachined structures

Manuscript received November 23, 1987; revised November 17, 1988. This work was supported by the Hasler Foundation and by the Committee for the Promotion of Applied Scientific Research, Switzerland.

B. Kloeck, S. D. Collins, and N. F. deRooij are with the University of Neuchâtel, Institute of Microtechnology, Neuchâtel, Switzerland.

R. L. Smith is with the Department of Electrical Engineering and Computer Science, University of California, Davis CA 95616.

IEEE Log Number 8826318.

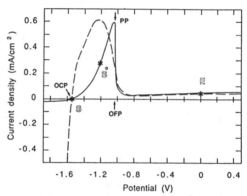

Fig. 1. The electrochemical I/V characteristics of p- (——) and n-type (– – – –) $\langle 100 \rangle$ oriented silicon, in 40-percent KOH aqueous solution at 60°C at a sweep rate of 1 mV/s. PP is the passivation potential of p-type silicon and OFP is the oxide formation potential of n-type silicon. $\mathbb{S}$ and $\mathbb{S}^*$ are working points for the substrate and $\mathbb{E}$ is the working point for the epitaxial layer during four-electrode etching.

can then be precisely controlled by taking advantage of current silicon diode fabrication technologies. The result is that etching can be stopped at a well defined p-n junction. To achieve this, a positive voltage is applied directly to the n-type silicon via an ohmic electrical contact while the electrical contact to the p-type silicon is accomplished via the etch solution with an appropriate counter electrode. Under sufficiently anodic biases (see I/V curves in Fig. 1) silicon passivates as a result of anodic oxide formation and etching stops. Since the majority of the potential drop is across the reverse-biased p-n junction, the p-type silicon remains essentially at open circuit potential (OCP) and etches. However, with the complete removal of the p-type silicon, the diode is destroyed and the n-type silicon becomes directly exposed to the etch solution. The positive potential applied to the n-type silicon then passivates it and etching terminates. Many silicon microstructures may be fabricated this way by selectively etching away p-type silicon and leaving the n-type silicon passivated. Definition of the microstructure morphology is precisely determined by the definition of the n-type silicon sections under anodic bias.

In this paper, a characterization of the electrochemical etch-stop in KOH is presented. Emphasis will be placed on the evaluation of several different electrochemical fabrication methodologies, as well as the pertinent process

Reprinted from *IEEE Trans. Electron Devices*, vol. 36, no. 4, pp. 663–669, April 1989.

parameters that control the physical nature of the etch-stop. The advantages of increased micro-structural control are demonstrated by applying the electrochemical etch-stop to the fabrication of a thin silicon membrane for use in a piezoresistive pressure sensor. The device characteristics and lot reproducibilies of these pressure sensors are compared with other devices fabricated without the aid of the electrochemical etch-stop.

II. EXPERIMENTAL

A. General

All experiments and pressure sensors used boron-doped (2×10^{15} at/cm^3), 3-in silicon wafers with an approximately 10-μm-thick n-type epitaxial layer deposited on the top surface. The epitaxial layer was phosphorous doped at 5 $\Omega \cdot$ cm. To etch the membranes, the wafers (both for the electrochemical studies and for the pressure transducers) were mounted in a plexiglass/stainless steel chuck using Perbunan 65 O-rings to seal the sample from the solution. The chuck was immersed in a 40 percent KOH solution at 60 $\pm$ 0.5°C (see Figs. 2 and 3). Electrochemical etching was accomplished using an IBM EC/225 Voltammetric Analyzer with a Saturated Calomel Electrode (SCE) and a platinum counter-electrode. Unless specified otherwise, all electrochemical potentials are referenced to the SCE at 60°C (0.253 V with respect to the NHE [10]). Membrane thicknesses were measured mechanically using a CaryCompar (Le Locle, Switzerland) that electronically reads the displacement of an inductive stylus. The surface roughness was recorded by means of a Talystep profilometer.

B. Fabrication of Test Devices

For the electrochemical studies, the n-type epitaxial layer surface was degenerately doped and a thick thermal oxide layer (1.5 μm) was grown to serve as a KOH etch mask for the back surface. After oxide patterning, the epitaxial layer was etched away except for a 5 by 5 array of 3 mm $\times$ 3 mm squares, thus forming 25 mesas of p-type silicon with the n-type epilayer on top of it. This was done to isolate the 25 diodes from each other and to have the possibility to make an electrical contact to the p-substrate. In the oxide on the back (p-type substrate), openings of 2 mm $\times$ 2 mm were aligned to the mesas to define the future geometry of the silicon membranes. An aluminum layer was deposited by e-beam evaporation and patterned to contact the epitaxial layer areas and the substrate. The aluminum was then alloyed at 450°C for 15 min. For some experiments, the wafer was diced into individual diodes.

C. Fabrication of Pressure Sensors

For the fabrication of the pressure sensors, the epitaxial wafers were thermally oxidized to 1.5 μm. The top oxide was removed and a new oxide was grown and patterned for boron implantation of the piezoresistors into the n-type epitaxial layer. Electrical connections were made using standard aluminum metallization. The design and

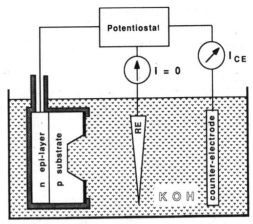

Fig. 2. Standard three-electrode system for etch-stop. RE is the SCE reference electrode.

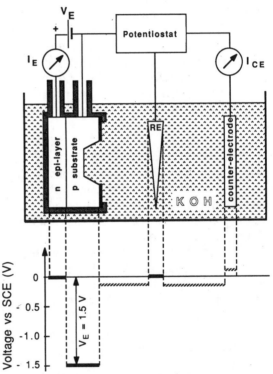

Fig. 3. Four-electrode electrochemical etch-stop configuration and voltage distribution with respect to the SCE reference electrode (RE). The standard three-electrode configuration of Fig. 2 is used with the addition of a potential V_E between the epitaxial layer and the substrate to externally maintain the substrate at etching potentials.

position of the resistors with respect to the silicon membrane were optimized with the modeling program SEN-SIM [11]. In the back-surface oxide, 1050 μm $\times$ 1050 μm square openings were patterned and etched. Alignment of the openings to the top resistors was accomplished by means of a double-sided wafer aligner. The size of the openings was calculated so that the final membrane geometry after etching was 520 μm $\times$ 520 μm.

Additional processing steps to provide direct electrical contact to the n-type epitaxial layer for the electrochemical etch-stop could be avoided by using the metallization of the sensor itself. The diode formed by the p$^+$ diffusion for the piezoresistors and the n-type substrate is in forward-bias mode when a positive voltage is applied. Thus, it forms no barrier for the etch-stop current. The voltage drop across this diode was merely added to the passivation potential that was applied. To provide electrical contacts to the p-substrate, two holes of 5 mm × 5 mm were etched through the epitaxial layer and aluminum was deposited and alloyed. Alternatively boron was diffused through the epitaxial layer and covered with aluminum.

III. RESULTS AND DISCUSSION

In the next sections, some practical process-relevant parameters of commonly used electrochemical etch-stop configurations are discussed in detail and a modified setup, the four-electrode configuration, is proposed for improved fabrication facility. The etch-stop technology will be evaluated by a detailed characterization of process parameters such as surface smoothness, membrane thickness reproducibility, and the influence of the applied passivation voltage. Finally, the suitability of electrochemical etch-stop for process automation will be demonstrated by its application to the fabrication of a specific device, a piezoresistive pressure sensor with a thin silicon membrane.

A. Etch-Stop Characterization

1) Etch-Stop Methodologies: In its simplest embodiment, the required passivating potential can be applied between the epitaxial silicon and a single inert metal electrode in the solution. Since any potential sufficiently anodic of the oxide formation potential (OFP = −1.080 V at 60°C [9]) will passivate the n-type silicon (see Fig. 1), any sufficiently large reverse bias will assure an etch-stop. However, the solution potential is ill defined and current dependent, which results in the lack of precise control over the fabrication parameters. Therefore, the use of a single metal electrode is not a viable process technique.

A three-electrode configuration, as shown in Fig. 2, overcomes the aforementioned limitations of a two-electrode configuration and is the preferred electrochemical arrangement. A potentiostat, in conjunction with a reference electrode, e.g., an SCE, and an inert counter-electrode, maintains a constant and reproducible solution potential with respect to the reference electrode. In order to establish the required voltage between the working electrode, i.e., the silicon wafer, and the SCE, the voltage between the working electrode and the SCE is measured and compared to the required voltage, and the current through the counter-electrode is adjusted so as to nullify the readings. In this way, the reference electrode remains currentless (high impedance input) and its interface potential with the solution is stable. In reality, the working electrode is grounded, but all electrochemical potentials are referenced to SCE.

By using three electrodes, the n-type epitaxial layer is kept at a well defined passivation potential, but an additional shortcoming of both the two- and the three-electrode configuration is that the substrate potential is not under direct electrical control. In the ideal case, it "floats" to OCP and is etched. In practice, however, the diode junction can be short-circuited, and since little current is required to passivate silicon [9], even a high-resistance short circuit will prevent the p-substrate from etching. The substrate potential, which in that case is defined by the total impedance of all electrical connections, will move to the right in Fig. 1, and if it becomes more anodic than the PP for p-type silicon (−1.04 V [9]), the substrate will be passivated. Common sources of p-n junction shorts are point defects in the junction and leakage at the border of the wafer, since the diode covers the entire surface of the wafer, about 45 cm². Thus, for practical application of etch-stop, a three-electrode configuration may not be sufficient.

Rather than to impose more stringent requirements on the epitaxial layer fabrication, it was preferred to modify the etch-stop arrangement so as to deal effectively with these problems. The new configuration uses four electrodes: an electrical connection to the substrate is added, as shown in Fig. 3. Whereas in the three-electrode method the potentiostat was connected to the n-type epitaxial layer, it is now connected to the p-substrate. A suitable voltage is applied directly to the substrate to assure that it maintains a controlled etching potential, close to the OCP of p-type silicon (−1.5 V to SCE [9]), independent of the magnitude or variations in the junction shorts. A second voltage supply V_E is connected between the epitaxial layer and the substrate to bias the epitaxial layer at passivating potentials, i.e., more anodic than −1.08 V with respect to SCE (the OFP for n-type silicon [9]), and thus more than +0.4 V with respect to the substrate. In practice V_E is set at about 1.5 V, which puts the epitaxial layer at 0 V with respect to SCE. In Fig. 3(b) a diagram of the potential distribution is sketched for the four-electrode arrangement. All potentials so applied are "hard" low-impedance connections, and the effects of unpredictable and fluctuating impedance pathways are eliminated. This method gave excellent results even with reverse-diode leakage currents as large as 10 mA per wafer from local p-n junction defects. It is noted here that, compared to the three-electrode method, the four-electrode configuration does not improve the surface quality of the etch-stopped membranes, but it allows etch-stop to be successful on every wafer, no matter how poor the junction quality is.

2) Etch-Stop Process Parameters: In this section the smoothness of etch-stopped surfaces and the thickness reproducibility of the membranes will be presented. These are major requirements for the mass production of high-quality micromechanical devices. Furthermore, the influence of the space-charge region (SCR) on etch-stop will be investigated in order to find out if the applied reverse bias can be used to vary the obtained membrane thickness according to the fabrication needs. Indeed, the width of the SCR is a function of the reverse-bias potential. Variations in the reverse-bias voltage could affect the thickness of the resulting membranes if the etch-stop is SCR dependent.

Smoothness of the membranes is important to increase the yield strength and the fatigue resistance. Fig. 4 shows a cross-sectional view of a cleaved silicon membrane fabricated by using etch-stop. The approximately 10-μm-thick membrane is seen at the bottom of the picture and two anisotropically etched ⟨111⟩ planes to the back and right. Despite an unpolished initial back-side wafer surface, the smoothness of the etch-stopped membrane surface approaches that of the extremely smooth ⟨111⟩ planes. Fig. 5 shows a profilometer scan of the etch-stopped surface and the initial surface from which the membrane was etched. The initial surface roughness of about 5 μm was smoothed to less than ±0.1 μm after etch-stop. This is a significant improvement over conventional timed etches, which at best demonstrate a steady-state limit roughness of 0.4 to 0.5 μm, even for initially polished wafers of less than 0.02-μm roughness [12].

For mass production, all devices should have identical mechanical properties. Hence, the reproducibility of the membrane thickness is another important fabrication feature. In order to study the thickness reproducibility across a wafer, 52 membranes were fabricated, equally spaced over a 3-in wafer. After etch-stop by means of the four-electrode method at a reverse diode bias of 1.5 V, the thickness of the membranes was measured. Fig. 6 shows a histogram of the measurements obtained for the membranes of one wafer. The membrane thickness was 10.7 ± 0.2 μm s.d. This distribution reflects the thickness variation of epitaxial silicon for the wafers used in these experiments, which is about 0.2 μm. Between different wafers, the thickness of the epitaxial layer can vary more than 0.5 μm, yielding a larger membrane thickness variation. Fig. 7 shows an SEM photograph off a cleaved membrane. The photo was taken while a reverse-bias potential was applied to the diode junction, so that electrons impinging on the p-n SCR were absorbed by electron-hole recombination. Hence, the SCR appears as a black line on the picture and is seen to coincide exactly with the etch-stopped membrane surface. This confirms that etching is stopped at the diode junction. Therefore, the etch-stop reproducibility is shown to be limited only by the reproducibility of the epitaxial layer growth process.

The SCR width can be an important process parameter for membrane thickness control if etch-stop is SCR dependent. To determine how the SCR affects etch-stop, membranes were fabricated by applying different epitaxial layer potentials. The resultant thickness was compared with the SCR width W, which is calculated by (1) for abrupt junctions [13]

$$W = \sqrt{\frac{2\epsilon_s}{q}\left(\frac{N_A + N_D}{N_A N_D}\right)(V_{bi} + V_E)} \quad (1)$$

where N_A is the substrate boron concentration ($N_A = 2 \times 10^{15}$ at/cm³), N_D the epitaxial layer phosphorus concentration ($N_D = 10^{15}$ at/cm³), ϵ_s the permittivity of silicon

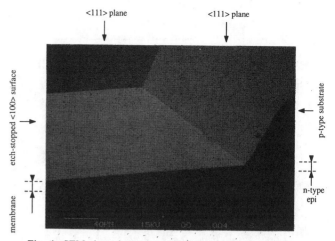

Fig. 4. SEM photograph of an etch-stopped membrane surface.

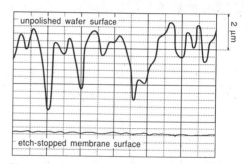

Fig. 5. Profilometer scan of an unpolished silicon wafer surface (upper) and of an etch-stopped membrane (lower).

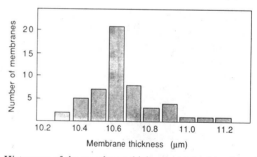

Fig. 6. Histogram of the membrane thickness obtained by four-electrode etch-stop.

(1.0536 10⁻¹² F/cm [13]), V_{bi} the built-in diode potential (0.7 V), and V_E the externally applied bias voltage. For the above dopant concentrations, the width of the p-type side of the SCR can be reduced to

$$L_p = 1/3 \cdot W = 0.47 \; 10^{-6} \cdot (0.7 + V_E)^{0.5}. \quad (2)$$

An array of 25 electrically isolated diodes, reverse biased at 8 different voltages, were etch-stopped on a single wafer. Voltages ranged from −0.80 to +4.00 V. The substrate was held at a potential of −1.5 V, so that bias voltages between 0.7 and 5.5 V were established across the p-n junction. According to (2), L_p ranged from 0.55 to 1.17 μm. The difference, 0.6 μm, should be reflected in the membrane thickness if any SCR influence is involved in etch-stop chemistry. In Fig. 8 the thickness of

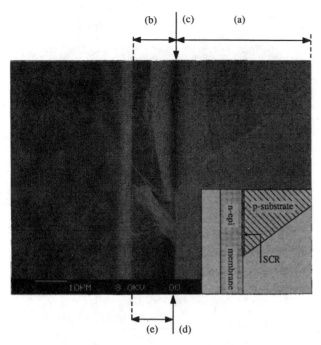

Fig. 7. Photograph of a cleaved membrane, showing that etching is stopped at the diode SCR. The inset indicates the relevant features of the picture; other lines are caused by imperfect cleavage of the sample. (a) p-type substrate. (b) n-type epitaxial layer. (c) Space-charge region. (d) Etch-stopped membrane surface. (e) Membrane.

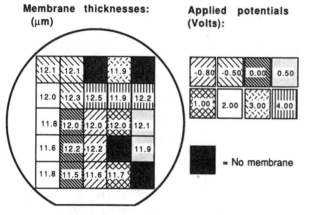

Fig. 8. Correlation investigation between membrane thickness and applied passivation potential: spatial distribution of the measurements.

each of the membranes is shown. The left-hand side of the figure schematically represents the wafer location of the membranes and their thicknesses; the shadings give the applied voltage on each membrane, as indicated on the right-hand side. The average thickness of 21 samples was 12.0 μm and the standard deviation was 0.25 μm. Fig. 9 plots the membrane thicknesses and the calculated L_p as a function of the potential. No obvious correlation between the membrane thickness and the SCR was observed. If, nevertheless, the SCR does have an effect, then it is negligible compared to other influences, such as epi-layer thickness. Thus, it may be concluded that the thickness of the membranes is not influenced by the magnitude of the applied reverse-bias voltage.

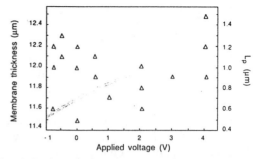

Fig. 9. Correlation investigation between membrane thickness and applied passivation potential: measured membrane thicknesses (Δ) and calculated L_p (p-type substrate side of the SCR) as a function of the applied voltage.

3) Etch-Stop Indicators: During a fabrication run it is useful to have an indicator that allows one to follow the process evolution and that automatically signals its completion. The current through the system is very well suited to perform this task. Fig. 10 shows current waveforms recorded as a function of time during four-electrode membrane etching. Curve *a* represents the ideal case where the reverse-biased diode current is negligible. The working point during substrate etching is indicated by *S* in Fig. 1. The counter-electrode current (I_{CE} in Fig. 3) is zero, since the substrate is kept at OCP. As soon as the etch front reaches the n-type epitaxial layer, the current is seen to increase and form a peak. As a fabrication tool this peak signals the onset of the etch stop. The existence and shape of the current peak follow naturally from the two-dimensional nucleation kinetics of oxide passivation [14] and from the wafer thickness taper and surface roughness. A complete explanation of this phenomenon is given elsewhere [15]. When the etch-stop process is finished across the entire wafer, a new steady-state current is reached at point *E* in Fig. 1, the working point for the passivated n-epitaxial layer, corresponding to a current through the counter-electrode of about 0.1 mA/cm^2. In practice, the final passivation current is higher, 0.2 mA/cm^2 in Fig. 10, curve *a* (j_{ox}), because of the interaction between the two voltages applied by the potentiostat and V_E. As the current reflects the magnitude of the etch rate of the passivation oxide in KOH, the steady-state etch rate of the passivated n-type epitaxial layer (E_p) can be calculated by Faradays law. In (3) it is assumed that all the current on a passivated membrane is used for the tetravalent oxidation of silicon (Si^0 to Si^{+4})

$$E_p = (j_{ox} \cdot A \cdot AMU)/(4 \cdot e \cdot \gamma_s) \qquad (3)$$

where j_{ox} is the measured current density, A is the atomic weight of silicon (28), AMU the atomic mass unit (1.6606 10^{-27} kg), e the elementary charge (1.60207 10^{-19} C), and γ_s the specific gravity of silicon (2.33). For a current density of 0.2 mA/cm^2, the etch rate is calculated to be 37 Å/min. Thus, leaving the wafer in the KOH will hardly etch the membranes anymore and there is no danger of overetching or piercing the membranes.

For wafers where the diode leakage current is not negligible, a typical simultaneous recording of counter-elec-

114

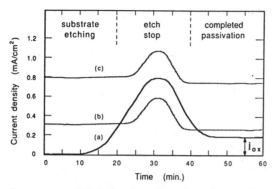

Fig. 10. Currents recorded during four-electrode etch-stop; Curve a—Counter-electrode current of a wafer with minimal diode leakage. j_{ox} is the current through the passivation oxide. Curve b—Counter-electrode current of a wafer with a large diode leakage current. Curve c—Epitaxial layer current recorded simultaneously with curve b.

TABLE I
PIEZORESISTIVE PRESSURE SENSOR SPECIFICATIONS

Chip dimensions:	1.1 mm x 3 mm x 0.4 mm
Membrane dimensions:	520 μm x 520 μm x 10 μm
Piezoresistors:	4 x 2 kΩ
Pressure sensitivity:	15 mV/V.bar
Output offset:	-5 to 5 mV/V
Non-linearity:	less than 0.1% to 400 mbar
Membrane rupture:	more than 20 bar
Temperature coefficients:	
of sensitivity:	-0.15 %/°C
of offset:	1.1 mbar/°C

trode current and epitaxial layer current (I_{CE} and I_E in Fig. 3) is shown in Fig. 10, curves b and c, respectively. The above discussion still holds, except that the counter-electrode current during substrate etching is no longer zero. Even though the externally applied voltage to the substrate is still -1.5 V, a steady-state situation is established across the network of interdependent interfacial and bulk impedances such that the actual potential at the exposed substrate surface is more anodic than OCP. The working point is now $\mathbb{S}^*$ in Fig. 1 and there is current flowing through the substrate–KOH interface. The substrate etch rate is slightly lower than in the ideal case, 14 μm/h at -1.2 V as opposed to 16 μm/h at OCP [9], but the etch process is not inhibited as long as the substrate surface potential stays sufficiently cathodic of the PP. If the diode leakage current is not too high, the voltage drop in the structure can be compensated for by adjusting the external substrate potential applied by the potentiostat to values that are more negative than -1.5 V, until the counter-electrode current is tuned to zero and the etch rate is again maximal. The shapes of curves b and c are essentially identical; the difference is the part of the diode leakage current that does not flow through the solution but goes through the substrate back to the negative lead of voltage supply V_E. Three-inch wafers with a total diode leakage current as high as 10 mA were etched successfully with the four-electrode etch-stop.

The aforementioned properties make the four-electrode etch-stop technique extremely suitable for integration in an automatic fabrication control system: any wafer can be processed, no matter how poor its diode quality is; the presence of the current peak allows automatic end point detection and the overetch protection makes it an uncritical, and thus easy to handle, but still very well controlled process step.

B. Application of Etch-Stop for Pressure Sensor Fabrication

The four-electrode etch-stop was used to control the thickness of membranes for piezoresistive pressure sensors that were developed to be mounted in the sidewall of

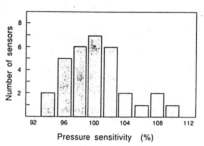

Fig. 11. Histogram of the pressure sensitivity of 32 pressure sensors from one wafer.

a heart catheter [16]. Table I summarizes some specifications of the sensor. It is obvious that the fabrication cost of individual sensors can be reduced if their output characteristics are reproducible. Therefore, the influence of the four-electrode etch-stop fabrication method on the reproducibility of the pressure sensitivity is examined in this section.

For the type of sensors described in the experimental section, the sensitivity (S) to a differential pressure (P_{diff}) is known to be a square function of the ratio membrane side (a) to thickness (h):

$$S = \frac{V_{out}}{V_{in} \cdot P_{diff}} \alpha \frac{a^2}{h^2}. \qquad (4)$$

Computer simulations with SENSIM showed that the thickness standard deviation of 0.2 μm (2 percent), as was measured previously, causes a variation in pressure sensitivity of 4 percent, confirming thus the square relationship of (4). The expected sensitivity variation is higher, however, because etch-stop implies also a variation of the membrane side length if the taper of the wafer is taken into account. In regions where the wafer is thicker, etching will continue longer until the epitaxial layer is reached; the pyramidal structure formed by the $\langle 111 \rangle$ oriented sidewalls of the holes will be deeper and thus the resulting membrane area smaller.

To determine the pressure sensitivity variation experimentally, pressure sensors were fabricated by means of four-electrode etch-stop, and their output characteristics were measured on the wafer, i.e., before the wafer was diced into individual sensors. This was done to avoid occasional influences of mechanical tensions introduced by encapsulation. The output voltage was recorded at room temperature for a pressure range from 0 to 300 mbar. Fig. 11 shows a histogram of the sensitivity of 32 sensors, reltive to the average value. The standard deviation of the measured pressure sensitivities was only 4 percent. This is an important improvement compared to previous generations of piezoresistive pressure sensors, where the membranes were fabricated without electrochemical etch-stop and the sensitivity on one wafer varied by a factor of two from one sensor to the other [17].

IV. CONCLUSION

The method of electrochemical etch-stop on an epitaxial layer was modified by controlling the potentials of both the epitaxial layer and the silicon substrate. All wafers that were etched with this four-electrode technique showed a successful etch-stop, including wafers with a large diode leakage current. In the latter case, the substrate was passivated when no substrate contact was provided (i.e., with two- or three-electrode methods), and no etching occurred.

The performance of four-electrode etch-stop was characterized in detail. It was shown that the surface of etch-stopped membranes is very smooth (± 0.1 μm pp) and that their thickness can be controlled to within ± 0.2 μm s.d. The influence of the diode SCR width on etch-stop was investigated, and it was concluded that it has no relevant influence on the membrane thickness.

It was demonstrated that four-electrode etch-stop is an easy-to-handle, yet powerful fabrication tool, since almost any wafer can be etched, the process can be automized if the current is used as an automatic end-point detector and there is no danger for overetching of the membranes.

As an example, a four-electrode etch-stop was used to fabricate membranes for piezoresistive pressure sensors. The pressure sensitivity of the devices could be controlled to ± 4 percent s.d., without any individual device adjustment.

ACKNOWLEDGMENT

The piezoresistive structures are fabricated at FAVAG Microelectronic S.A., Bevaix, Switzerland. We especially appreciated the stimulating discussions with A. Muller, A. Kayal, and S. Ansermet of this company.

REFERENCES

[1] H. A. Waggener, "Electrochemically controlled thinning of silicon," *Bell Syst. Tech. J.*, vol. 49, no. 3, p. 473, 1970.
[2] P. M. Sarro and A. W. van Herwaarden, "Silicon cantilever beams fabricated by electrochemically controlled etching for sensor applications," *J. Electrochem. Soc.*, vol. 133, no. 8, p. 1724, 1986.
[3] T. N. Jackson, M. A. Tischler, and K. D. Wise, "An electrochemical p-n junction etch-stop for the formation of silicon microstructures," *IEEE Electron Device Lett.*, vol. EDL-2, p. 44, 1981.
[4] M. Hirata, S. Suwazono, and H. Tanigawa, "Diaphragm thickness control in silicon pressure sensors using an anodic oxidation etch-stop," *J. Electrochem. Soc.*, vol. 134, no. 8A, p. 2037, 1987.
[5] M. Hirata, K. Suzuki and H. Tanigawa, "Silicon diaphragm pressure sensors fabricated by anodic oxidation etch-stop," *Sensors and Actuators*, vol. 13, no. 1, p. 63, 1988.
[6] E. D. Palik, J. W. Faust, H. F. Grey, and R. F. Green, "Study of the etch-stop mechanism in silicon," *J. Electrochem. Soc.*, vol. 129, no. 9, p. 2051, 1982.
[7] J. W. Faust and E. D. Palik, "Study of the orientation dependent etching and initial anodization of Si in aqueous KOH," *J. Electrochem. Soc.*, vol. 130, no. 6, p. 1413, 1983.
[8] O. J. Glembocki, R. E. Stahlbush, and M. Tomkiewicz, "Bias-dependent etching of silicon in aqueous KOH," *J. Electrochem. Soc.*, vol. 132, no. 1, p. 145, 1985.
[9] R. L. Smith, B. Kloeck, N. F. de Rooij, and S. D. Collins, "The potential dependence of silicon anisotropic etching in KOH at 60°C," *J. Electroanal. Chem.*, vol. 238, p. 103, 1987.
[10] A. J. Bard and L. R. Faulkner, *Electrochemical Methods.* New York: Wiley, 1980.
[11] K. W. Lee, "Modeling and simulation of solid state pressure sensors," Tech. Rep. 156, Dept. of Electrical Computer Eng., Univ. of Michigan, 1982.
[12] P. Meakin and J. M. Deutch, "The formation of surfaces by diffusion limited annihilation," *J. Chem. Phys.*, vol. 85, no. 4, p. 2320, 1986.
[13] S. M. Sze, *Physics of Semiconductor Devices*, 2nd ed. New York: Wiley, 1981, p. 74.
[14] J. A. Harrison and H. R. Thirsk, *Electroanalytical Chemistry*, vol. 5, A. J. Bard, Ed. New York: Dekker, 1971, p. 67.
[15] R. L. Smith, B. Kloeck, and S. D. Collins, "Mechanism of anodic passivation on ⟨111⟩ silicon in KOH," *J. Electrochem. Soc.*, to be published.
[16] B. Kloeck, B. Stauffer, D. C. Scott, and N. F. de Rooij, "Optimized design and fabrication methods for miniature piezoresistive pressure sensors," in *Capteurs 86 Conf. Proc.* (Paris), 1986, p. 74.
[17] B. Stauffer and B. Kloeck, Internal Rep. IMT 157 EC 04/85, Institute of Microtechnology, Univ. of Neuchâtel, 1985.

Surface micromachining for microsensors and microactuators

Roger T. Howe

Berkeley Sensor & Actuator Center, Department of Electrical Engineering and Computer Sciences and the Electronics Research Laboratory, University of California, Berkeley, California 94720

(Received 13 June 1988; accepted 24 August 1988)

Micromechanical structures can be made by selectively etching sacrificial layers from a multilayer sandwich of patterned thin films. This paper reviews this technology, termed surface micromachining, with an emphasis on polysilicon microstructures. Micromechanical characteristics of thin-film microstructures critically depend on the average residual stress in the film, as well as on the stress variation in the direction of deposition. The stress in low-pressure chemical vapor deposition polysilicon varies with deposition temperature, doping, and annealing cycles. Applications of surface micromachining to fabricate beams, plates, sealed cavities, and linear and rotary bearings are discussed.

I. INTRODUCTION

Orientation-dependent silicon etches, such as KOH and ethylene–diamine–pyrocatechol (EDP), have been successfully used for fabricating micromechanical structures from crystalline silicon or etch-resistant thin films.[1,2] Various etch-stop techniques have enabled the commercial production of silicon diaphragm pressure sensors and cantilever-beam accelerometers by backside etching of the silicon substrate (substrate micromachining).[3,4] An alternative micromachining process is to etch selectively a sacrificial layer underlying an etch-resistant thin film, a technology which has been termed "surface micromachining."[5] Originally demonstrated with metal films in the 1960s,[6] this process has been applied to fabricate static and dynamic micromechanical structures from low-pressure chemical vapor deposition (LPCVD) polysilicon,[7] LPCVD silicon nitride,[8,9] and polyimide[10,11] films.

This paper reviews the basic surface micromachining process, with emphasis on polysilicon microstructures. Extensions of the process to fabricate sealed cavities and micromechanical bearings are discussed. The mechanical characteristics of thin-film microstructures critically depend on the average residual stress in the film and the stress variation in the direction of film growth or deposition, as well as on the Young's modulus, Poisson's ratio, and other mechanical properties of the film.[12] After discussing the use of surface microstructures as sensitive probes of these thin-film properties, selected microsensor and microactuator applications of surface micromachining are reviewed.

II. SURFACE MICROMACHINING PROCESSES

The simple process of surface micromachining is illustrated in Fig. 1. A sacrificial layer is deposited on the silicon substrate, which may have been coated first with an isolation layer. Windows are opened in the sacrificial layer and the microstructural thin film is deposited and etched. Selective etching of the sacrificial layer leaves a free-standing micromechanical structure. The technique is applicable to combinations of thin films and lateral dimensions where the sacrificial layer can be etched without significant etching or attack of the microstructure, the isolation layer, or the substrate.

Polysilicon surface microstructures, such as beams and plates, are made by etching an underlying oxide film using various forms of HF as the micromachining etchant.[13–15] Hydrofluoric acid etches CVD phosphosilicate glass (PSG) much faster than thermal or undoped CVD oxides, making it attractive as the sacrificial layer.[15] A limitation on the process is the attack by buffered HF or HF vapors on $POCl_3$-doped polysilicon films, especially when they are deposited on oxides containing phosphorus.[16] LPCVD silicon nitride etches much more slowly in HF than oxide films, especially when deposited with a silicon-rich composition,[17] making it a desirable isolation film.[15] A further limitation on polysilicon surface structures is that large-area structures tend to deflect and attach to the substrate or isolation layer after the final rinsing step, a phenomenon that may be related to surface tension or residual contamination.[16,18]

Silicon nitride micromechanical beams can be fabricated by etching a polysilicon spacer layer in KOH, with the silicon substrated protected by an oxide or silicon nitride isolation layer.[8,9] If a window is opened in the isolation region underneath the silicon nitride film, the KOH etches a cavity in the substrate after removing the polysilicon sacrificial layer.[9] Polyimide surface microstructures can be made by selectively etching an aluminum sacrificial layer.[10,11] By using multiple coats of spun-on polyimide, 30-μm-thick suspend-

Sacrificial/Spacer Layer   Microstructure Layer   Freestanding Microstructure

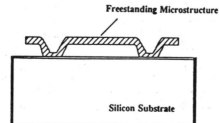

Isolation layer

Silicon Substrate   Silicon Substrate

FIG. 1. Surface microstructure fabrication process.

Reprinted with permission from *J. Vac. Sci. Technol. B*, vol. 6, no. 6, pp. 1809–1813, November/December 1988.

ed plates are possible.[11] Moreover, composite polyimide plates can be made depositing and patterning a metal film between coats.[19]

Sealed polysilicon diaphragms are made by defining thin regions of sacrificial oxide at the perimeter of the structure.[18,20] Thermal oxidation of the polysilicon and silicon substrate (reactive sealing) or CVD of oxide or nitride films seals the narrow openings after removal of the sacrificial oxide, as shown in Fig. 2. The cavity is sealed under vacuum in the case of reactive sealing, due to reaction of oxygen trapped inside the cavity, and also if the sealing films are deposited at low pressure. This process has also been used to seal LPCVD silicon nitride diaphragms using a plasma enhanced chemical vapor deposition (PECVD) silicon nitride sealing film.[9] Recently, deposition of an epitaxial silicon film has been shown to yield a hermetic seal of a surface cavity formed by selective etching of epitaxial and heavily doped silicon layers.[21] The residual hydrogen in the cavity after sealing is diffused through the epitaxial layer by high-temperature annealing in a nitrogen ambient.[21]

Multiple depositions of sacrificial and structural films greatly extend the range of structures which can be made.[22–25] Figure 3 shows the cross section of a polysilicon bearing prior to removal of the PSG sacrificial layers.[22,23] After releasing the structure in HF, the second polysilicon film is free to rotate or translate, depending on the layout, around the first polysilicon layer. The key step in this process is the definition of a constraining flange in the second polysilicon layer by first partially undercutting the first polysilicon layer and then depositing a second PSG spacer layer. Figure 4 is a scanning electron micrograph (SEM) of this flange structure, illustrating the imperfect undercoverage by the second PSG film. The second polysilicon film contains a void, due to the excessive depth of the undercut. In order to achieve better undercoverage, allowing a thinner sacrificial layer for a tighter bearing tolerance, LPCVD silicon nitride can be used as the second spacer layer at the expense of a longer micromachining etch.[26] Figure 5 is a SEM of the cross section of a linear flange bearing incorporating a 200-nm-thick Si$_3$N$_4$ film as the second spacer layer prior to micromachining in hydrofluoric acid.[26] The overhang of the 1.5-μm-thick first polysilicon layer, coated with the thin Si$_3$N$_4$ film, is faintly visible in Fig. 5. The capability of surface micromachining to fabricate microstructures with freedom to rotate or translate has stimulated research into the electromechanical drive of such structures.[27,28]

III. THIN-FILM MECHANICAL PROPERTIES

A central question for the design of surface-micromachined sensing or actuating structures is the mechanical

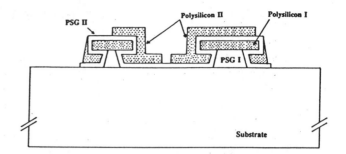

FIG. 3. Cross section of surface-micromachined flange bearing (Refs. 22 and 23).

properties of the structural thin films. In particular, as-deposited thin films are typically in a state of residual strain, due to either mismatch in the thermal expansion coefficient with the substrate or film nucleation and growth processes.[29] Constrained microstructures can buckle from residual compression or crack from residual tension. The mechanical response of structures is affected by the residual strain, even if the structure does not fail. Furthermore, if the residual strain varies in the direction of film growth, the resulting built-in bending moment will warp released structures, such as cantilever beams.

Because of their sensitivity to residual strain, surface microstructures are useful for making in situ measurements of residual strain. Average compressive strain can be measured by the relaxation of an undercut film edge[30,31] or by measuring the critical buckling length of clamped–clamped beams.[32] Tensile strain can be measured by a series of rings which are constrained to the substrate at two points on a diameter and spanned orthogonally by a clamped–clamped beam.[33] After removal of the sacrificial layer, tensile strain in the ring places the spanning beam in compression; the critical buckling length of the beam can be related to the average strain.

The average residual strain in LPCVD polysilicon films depends on deposition conditions, doping, and annealing cycles.[31,32,34,35] Residual strain in films deposited at 620–640 °C is found to be compressive, with a typical value of -3×10^{-3}. Annealing at temperatures over ~1000 °C is found to reduce the magnitude of the compressive strain in these large-grained films, as well as its variation through the thickness of the film.[32,36] By reducing the deposition temperature to the 570–610 °C range, fine-grained polysilicon films are obtained. Undoped films deposited at 580 °C on thermal oxide have an initial strain of $\sim-5\times10^{-3}$, similar to that of large-grained polysilicon. However, annealing at 600–1050 °C yields near-zero or tensile strain films.[37,38] As-

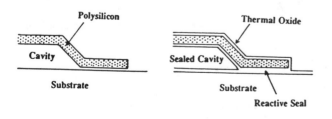

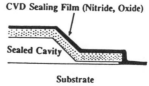

FIG. 2. Sealing processes (Refs. 18 and 20).

118

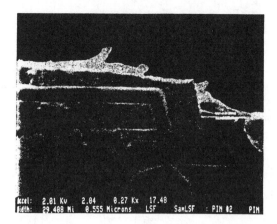

FIG. 4. SEM of polysilicon flange bearing using 500-nm-thick LPCVD PSG sacrificial layer (Ref. 22).

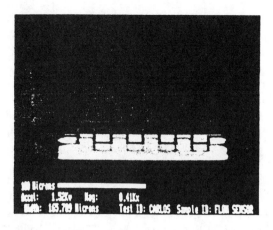

FIG. 6. SEM of 1.5-μm-thick polysilicon microbridge offset 3.5 μm from the substrate (Ref. 49).

deposited, undoped films deposited on PSG at 605 °C are found to have very low compressive, or tensile strain.[39]

LPCVD Si_3N_4 is highly tensile, which limits its microstructual applications. However, silicon-rich LPCVD Si_xN_y has a reduced tensile strain or even compressive strain, depending on the ratio x/y.[40] Spin-cast polyimide films are also in tension, with a strain of ~ 0.01.[41]

Thin-film surface microstructures cannot be designed without knowledge of the basic mechanical properties of the film. Again, micromechanical structures can be used as probes to measure the *in situ* mechanical properties. Young's modulus can be found from the resonant frequencies of beams[33,42] or from load-deflection experiments on membranes, if the Poisson's ratio of the film is known.[43-45] For LPCVD polysilicon and silicon nitride, mechanical properties will be likely to depend on deposition, doping, and annealing conditions. The characterization of mechanical properties is incomplete, although a database of careful measurements is beginning to accumulate.[12,33,34,43-45] For resonant microstructures, fatigue and internal friction are critical.[46] Finally, the static and dynamic friction and wear properties of thin films will be essential for the development of surface-micromachined bearings.[24,28]

IV. SELECTED APPLICATIONS

Polysilicon surface microstructures are thermally isolated from the substrate by the air gap shown in Fig. 1. As a result, resistively heated polysilicon microbridges are sensitive anemometers.[47-49] Figure 6 is a SEM of a 1.5-μm-thick *in situ* phosphorus-doped polysilicon microbridge which is suspended 3.5 μm above the substrate for thermal isolation.[49] The meander layout of the microbridge serves to relieve the residual compression in the film and avoid buckling. By measuring steady-state thermal losses in such structures, the thermal conductivity of heavily phosphorus-doped polysilicon was found to be 0.33 W cm^{-1} K^{-1}, a factor of 4 less than that of crystalline silicon.[50]

The SEM in Fig. 7 illustrates the step coverage by the second polysilicon film for the flange bearing shown in Fig. 5.[26] The extremely small gap (200 nm) between the two polysilicon layers is clear in this SEM. One application of this process is to define the mechanical boundary condition at one end of a microbridge fabricated from the second polysilicon film. The flange bearing permits only lateral displacement of the end of the polysilicon microbridge, enabling the buckled shape of the microbridge under a compressive load

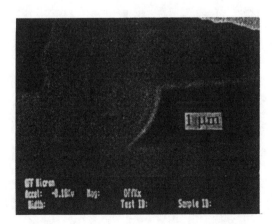

FIG. 5. SEM of polysilicon flange bearing using 200-nm-thick LPCVD Si_3N_4 sacrificial layer (Ref. 26).

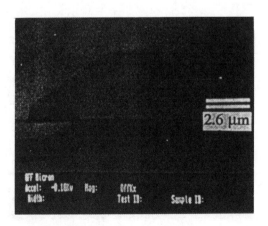

FIG. 7. SEM of polysilicon flange used for bridge-slider test structure (Ref. 26).

applied by a microprobe to be treated analytically. A SEM of this test structure is shown in Fig. 8, in which the end of the microbridge has become jammed in the linear flange bearing after being displaced by a microprobe.[26] Fracture of the microbridge eventually occurs at some lateral displacement, allowing the fracture strain of the polysilicon film to be calculated.[26] For as-deposited, *in situ* phosphorus-doped polysilicon films, the fracture strain is 0.017 ± 0.009. Annealing at 1000 °C for 1 h reduces the fracture strain to $(9 \pm 1) \times 10^{-3}$.

Figure 9 is a schematic illustration of a polyimide microfloating-element shear-stress sensor.[11] Fluid flow over the surface of the chip produces a shear stress due to the velocity gradient near the surface. The shear stress displaces the floating element, which is restrained by four tethers located perpendicular to the cross section of Fig. 9. The displacement is detected by means of the differential coupling of an ac signal to a pair of electrodes on the substrate by a metal film embedded in the floating element. Two matched metal–semiconductor field effect transistor (MOSFET's) are fabricated on the sensor chip to buffer the high-impedance electrodes.

This device is fabricated by selectively etching a 3-μm-thick aluminum sacrificial layer in a mixture of phosphoric–acetic–nitric acids in water from underneath the 32-μm-thick polyimide floating element, which contains a 30-nm-thick chromium conductor embedded 1 μm above the bottom surface of the element. A floating element planar to within < 1 μm is achieved by using tensile-strained evaporated chromium film, which matches the tensile strain in the polyimide film and minimizes the built-in bending moment in the plate.[19] Figure 10 is a SEM of a $500 \times 500 \mu$m floating-element shear sensor which is suspended by 1-mm-long, ~5-μm-wide, and 32-μm-thick polyimide tethers. The polyimide is patterned in an O_2 plasma in a parallel plate reactor, using an aluminum nonerodible etch mask. As shown in Fig. 10, polyimide is left in the field region in order to present a smoother surface to the flow. This device shows promise for the direct measurement of the mean and time-varying shear stress in turbulent boundary layers.[11]

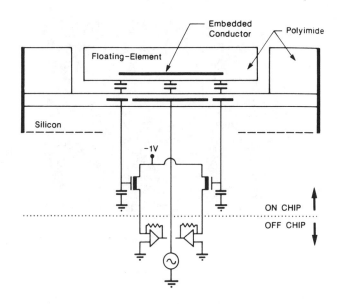

FIG. 9. Schematic cross section of polyimide microfloating-element shear sensor (Ref. 11).

V. CONCLUSIONS

Surface micromachining is a rapidly developing technology which extends the range of micromechanical materials and structures which can be fabricated using planar technology. If a suitable combination of structural, sacrificial, and isolation layers and a highly selective etchant are identified, then a simple fabrication sequence can be used to make micromechanical beams, plates, and other structures. If necessary, apertures can be opened in structural layers to accelerate the removal of sacrificial layers in order to minimize the etching or attack of structural or isolation films. The lateral dimensions of surface microstructures are defined by planar lithography and etching, and their thickness and offset from the substrate are defined by thin-film deposition or growth processes. Surface microstructures can be sealed through oxidation (polysilicon structures with no isolation layer) or by the CVD of films. The step-coverage characteristics of

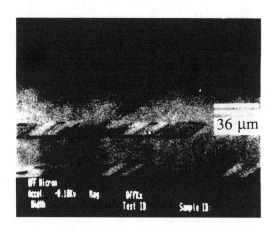

FIG. 8. SEM of microbridge with linear flange bearing at one end (Ref. 26).

FIG. 10. SEM of polyimide microfloating element with four tethers (dimensions in text) (Refs. 11 and 19).

CVD processes can be exploited to fabricate three-dimensional surface-micromachined bearings and other structures. Finally, surface micromachining is a single-sided process, in contrast to the double-sided processes necessary to fabricate silicon diaphragms using orientation-dependent silicon etching.

Since surface microstructures are fabricated from thin films, the understanding and control of their mechanical properties are crucial to successful sensor and actuator applications. Additional research is needed to establish the materials science of controlling residual stress and the Young's modulus and other mechanical properties of CVD films, such as polysilicon and silicon nitride. New thin-film materials may be needed to meet wear and abrasion requirements for microactuating structures.

On-chip electronics can be advantageous for the sensing or control of micromechanical structures. Integration of surface micromachining and electronic fabrication sequences typically requires significant process development and compromises between microstructural and electronic performance.[11,15,51] It has been demonstrated recently that polysilicon microbridges can be fabricated in a standard complementary metal–oxide semiconductor process, although an electrical isolation layer cannot be incorporated.[52]

Combining surface and substrate micromachining processes further extends the capabilities of microstructure fabrication. Recent examples of such combined processes include a thermally isolated polysilicon/silicon nitride microstructure[53] and a polysilicon one-way microvalve.[54]

ACKNOWLEDGMENTS

The author thanks Y.-C. Tai, L.-S. Fan, and Professor R. S. Muller of the University of California, Berkeley and Professor M. A. Schmidt of MIT for supplying SEM's. He acknowledges support by an NSF Presidential Young Investigator Award and by the Berkeley Sensor and Actuator Center, an NSF/Industry/University Cooperative Research Center.

[1]G. Kaminsky, J. Vac. Sci. Technol. B **3**, 1015 (1985).

[2]K. E. Petersen, Proc. IEEE **70**, 420 (1982).

[3]N. Raley, Y. Sugiyama, and T. Van Duzer, J. Electrochem. Soc. **131**, 161 (1984).

[4]B. Kloeck and N. F. de Rooij, in *4th International Conference on Solid-State Sensors and Actuators (Transducers' 87)*, Tokyo, Japan (IEE Japan, Tokyo, 1987), p. 116.

[5]Suggested by P. W. Barth, 1985.

[6]H. C. Nathanson, W. E. Newell, R. A. Wickstrom, and J. R. Davis, Jr., IEEE Trans. Electron Devices **14**, 117 (1967).

[7]R. T. Howe, in *Polysilicon Films and Interfaces*, edited by C. Y. Wong, C. V. Thompson, and K.-N. Ting (Materials Research Society, Pittsburgh, 1988), Vol. 106, p. 213.

[8]H. Guckel, D. K. Showers, D. W. Burns, C. R. Rutigliano, and C. G. Nesler, in *IEEE Solid-State Sensors Workshop* (IEEE, New York, 1986), p. 111.

[9]S. Sugiyama, T. Suzuki, K. Kawahata, K. Shimaoka, M. Takigawa *et al.*, in *IEEE International Electron Devices Meeting* (IEEE, New York, 1986), p. 184.

[10]G. Blackburn and J. Janata, in *Electrochemical Society Spring Meeting* (Electrochemical Society, Pennington, NJ, 1982), Vol. 82-1, p. 196.

[11]M. A. Schmidt, R. T. Howe, S. D. Senturia, and J. H. Haritonidis, IEEE Trans. Electron Devices **35**, 750 (1988).

[12]S. D. Senturia in Ref. 4, p. 11.

[13]R. T. Howe and R. S. Muller, in Ref. 10, Vol. 82-1, p. 184.

[14]R. T. Howe and R. S. Muller, J. Electrochem. Soc. **130**, 1420 (1983).

[15]R. T. Howe, in *Micromachining and Micropackaging of Transducers*, edited by C. D. Fung, P. W. Cheung, W. H. Ko, and D. G. Fleming (Elsevier, New York, 1985), p. 169.

[16]T. A. Lober and R. T. Howe, in *IEEE Solid-State Sensor and Actuator Workshop* (IEEE, New York, 1988), p. 59.

[17]Y.-C. Tai, UC Berkeley (personal communication).

[18]H. Guckel, D. W. Burns, C. R. Rutigliano, D. K. Showers, and J. Uglon, in Ref. 4, p. 277.

[19]M. A. Schmidt, R. T. Howe, S. D. Senturia, and J. H. Haritonidis, in *IEEE Micro Robots and Teleoperators Workshop* (IEEE, New York, 1987), p. 32.

[20]H. Guckel and D. W. Burns, in Ref. 9, p. 176.

[21]K. Ikeda, H. Kuwayama, T. Kobayashi, T. Watanabe, T. Nishikawa *et al.*, in *7th IEE Japan Sensor Symposium* (IEE Japan, Tokyo, 1988), p. 193.

[22]L.-S. Fan, Y.-C. Tai, and R. S. Muller, in Ref. 4, p. 849.

[23]L.-S. Fan, Y.-C. Tai, and R. S. Muller, IEEE Trans. Electron Devices **35**, 724 (1988).

[24]R. S. Muller, in Ref. 21, p. 7.

[25]M. Mehregany, K. J. Gabriel, and W. S. N. Trimmer, IEEE Trans. Electron Devices **35**, 719 (1988).

[26]Y.-C. Tai and R. S. Muller, in Ref. 16, p. 88.

[27]W. S. N. Trimmer and K. J. Gabriel, Sensors and Actuators **11**, 189 (1987).

[28]S. F. Bart, T. A. Lober, R. T. Howe, J. H. Lang, and M. F. Schlecht, Sensors and Actuators **14**, 269 (1988).

[29]R. W. Hoffman, in *Physics of Nonmetallic Thin Films*, edited by C. H. S. Dupay and A. Cachard (Plenum, New York, 1976), p. 273.

[30]P. G. Borden, J. Appl. Phys. **36**, 829 (1980).

[31]R. T. Howe and R. S. Muller, J. Appl. Phys. **54**, 4674 (1983).

[32]H. Guckel, T. Randazzo, and D. W. Burns, J. Appl. Phys. **57**, 1671 (1985).

[33]H. Guckel, D. W. Burns, H. A. C. Tilman, D. W. DeRoo, and C. R. Rutigliano, in Ref. 16, p. 96.

[34]S. P. Murarka and T. F. Retajczyk, J. Appl. Phys. **54**, 2069 (1983).

[35]M. S. Choi and E. W. Hearn, J. Electrochem. Soc. **131**, 2443 (1984).

[36]R. T. Howe and R. S. Muller, Sensors and Actuators **4**, 447 (1983).

[37]H. Guckel, D. W. Burns, C. C. G. Visser, H. A. C. Tilman, and D. DeRoo, IEEE Trans. Electron Devices **35**, 800 (1988).

[38]H. Guckel, D. W. Burns, H. A. C. Tilman, C. C. G. Visser, D. W. DeRoo *et al.*, in Ref. 16, p. 51.

[39]L.-S. Fan and R. S. Muller, in Ref. 16, p. 55.

[40]M. Sekimoto, H. Yoshirara, and T. Ohkubo, J. Vac. Sci. Technol. **21**, 1017 (1982).

[41]M. Mehrengany, R. T. Howe, and S. D. Senturia, J. Appl. Phys. **62**, 3579 (1987).

[42]K. E. Petersen and C. R. Guarnieri, J. Appl. Phys. **50**, 6761 (1979).

[43]E. I. Bromley, J. N. Randall, D. C. Flanders, and R. W. Mountain, J. Vac. Sci. Technol. B **1**, 1364 (1983).

[44]M. G. Allen, M. Mehregany, R. T. Howe, and S. D. Senturia, Appl. Phys. Lett. **51**, 241 (1987).

[45]O. Tabata, K. Kawahata, S. Sugiyama, H. Inagaki, and I. Igarashi, in Ref. 21, p. 173.

[46]R. T. Howe, in Ref. 4, p. 843.

[47]Y.-C. Tai, R. S. Muller, and R. T. Howe, in *3rd International Conference on Solid-State Sensors and Actuators (Transducers'85)* (IEEE, New York, 1985), p. 354.

[48]Y.-C. Tai, C. H. Mastrangelo, and R. S. Muller, in Ref. 4, p. 360.

[49]C. H. Mastrangelo and R. S. Muller, in Ref. 16, p. 43.

[50]Y.-C. Tai, C. H. Mastrangelo, and R. S. Muller, in *IEEE International Electron Devices Meeting* (IEEE, New York, 1987), p. 278.

[51]S. Sugiyama, K. Kawahata, M. Abe, H. Funabushi, and I. Igarashi, in Ref. 4, p. 444.

[52]M. Parameswaran, H. P. Baltes, and A. M. Robinson, in Ref. 16, p. 148.

[53]M. A. Huff and R. T. Howe, in Ref. 16, p. 47.

[54]S. Shoji and M. Esashi, in Ref. 21, p. 217.

FABRICATION TECHNIQUES FOR INTEGRATED
SENSOR MICROSTRUCTURES

H. Guckel, D. W. Burns

Wisconsin Center for Applied Microelectronics
Department of Electrical and Computer Engineering
University of Wisconsin, Madison, Wisconsin

INTRODUCTION

Research efforts on micromechanical sensors at the Wisconsin Center for Applied Microelectronics have addressed two fundamental issues: a significant miniaturation of mechanical devices and IC-compatible construction techniques. The justification for this research direction is found in a strong interest in sensor systems or multiple transducers and data extraction circuitry on a single chip.

A suitable test vehicle for the multitude of possible mechanical sensors is the pressure transducer. This device is mechanically a pillbox which is vacuum sealed if absolute pressures are to be measured. Applied pressures distort the pillbox. These geometric changes are sensed by strain sensitive resistors which decorate the transducer body or, in some cases, by capacitive techniques. Both sensing systems have been considered. The piezoresistive pressure read-out was chosen because it is easier to implement in the early development stages.

Commercial pressure transducers are normally fabricated by using a single crystal silicon diaphragm which is decorated with diffused resistors. The diaphragm size is typically somewhere near 0.250" x 0.250" x 0.001". This structure is fabricated by backetching the silicon wafer. Additional IC processing after diaphragm etching is difficult and thickness reduction of undoped silicon plates to, say, 0.0001" is not feasible. The etched silicon wafer is typically attached to a pyrex substrate to complete the pillbox structure. The mechanical device involves

therefore at least two materials. Differential thermal expansion is a problem and becomes very significant if one considers that the total expected resistance change is on the order of 1%. Improved processing sequences for microminiaturized pressure transducers must therefore involve the production of very thin diaphragms with excellent thickness control and should ideally produce an all-silicon pillbox.

A possible approach to an improved sensor technology involves the following:
(1) a silicon substrate which may already contain electronic processing details and
(2) a patterned, sacrificial layer which defines the interior of the pillbox and
(3) a patterned, deposited layer which together with the substrate forms the pillbox.

The removal of the sacrificial layer via lateral etching and a suitable batch sealing technique will yield pillboxes of suitable size and, in particular, of accurately controlled thickness since deposition thicknesses can be monitored closely. The sensing of pillbox geometry can be accomplished by decorating the sealed device with a patterned piezoresistive layer. This approach has been pursued for several material systems. The most promising results have been achieved by using a silicon substrate of arbitrary orientation, thermally oxidizing and patterning for interior cavity definition, covering with polysilicon and patterning, lateral etching, sealing and applying polysilicon strain sensitive resistors. Preliminary results have yielded functional pressure transducers of good quality. This input has now been used to investigate the possibility for a fully designed, easily manufactured component which will eventually be used with on-chip CMOS operational amplifiers.

Reprinted from *Rec. of the IEEE Int. Electron Devices Meeting*, 1986, pp. 176–179.

Design and Construction of Cavity

The initial pillbox structures, as stated earlier, involve a thermal post oxide for cavity interior definition. This technique produces a cavity which has a raised lip. Since typical device dimensions involve diaphragm sizes of 125 micron x 125 micron x 2 micron, strain variations due to the non-planar nature of the device are not desireable. This difficulty can be eliminated by employing isoplanar processing. Thus, wafer preparation starts with a thin, thermal strain relief oxide which is covered with 400Å of stoichiometric, LPCVD silicon nitride. The first mask is used to remove the nitride over the interior region of the pillbox. An oxidation to 7500Å, a strip and a re-oxidation produces a planar surface with 7500Å silicon dioxide regions. The second mask removes the nitride and strain oxide over the sealing region. In this region the polysilicon layer will contact the substrate. However, this region is also used to allow access for hydrofluoric acid for lateral etching of the internal oxide. The compromise, easy HF access and good polysilicon adhesion, is made by using many small contact areas which use roughly 50% of the surrounding oxide region.

The next processing step is that of polysilicon deposition. The deposition technique is always LPCVD from 100% silane. However, the deposition conditions have been changed signficantly from previous work (1). The argument for this involves the basic properties of polysilicon films and an improved understanding of plate deflection theory. Polysilicon films always exhibit compressive strain fields. This field in effect reduces the stiffness of a fully supported plate and therefore has a major effect on pillbox design. Calculations and experiments have been used to identify this property of deposited films as a first order input into plate deflection theory (2). This implies that variation in strain field over a wafer surface produces design conditions which cannot possibly lead to devices with identical pressure sensitivity. Consequently, the emphasis in polysilicon deposition techniques must be that of a film which has a nominal strain field value which does not change more than 1% over the wafer surface.

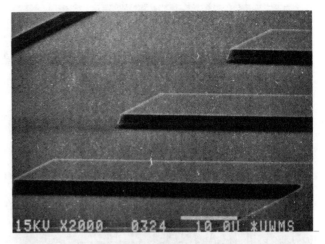

Figure 1. Free-standing cantilevers fabricated from fine-grain LPCVD polysilicon with low and high temperature annealing show very good uniformity. Although relatively high values of residual strain exist in the film (i.e. 0.1%), the strain variation in the direction of film thickness is minimal. A closer look at the top surface of the film (bottom) reveals a texture very similar to the single crystal substrate.

This can be obtained by using fine-grained polysilicon. Deposition conditions are adjusted so that the film deposits at the amorphous-polysilicon boundary near 580°C (3). The film is stabilized by low and high temperature annealing. Typically a strain level of 0.1% results with a crystal structure as indicated in fig. 1 (4). Mask #3 is used to define the polysilicon structure for reactive ion etching in an NF_3-O_2 plasma. Polysilicon retention is adjusted in such a way that all electronic wiring is on polysilicon rather than on the substrate. The argument for this is the steep polysilicon flank which makes the edge coverage difficult.

Etching of the polysilicon layer exposes the etch channel region. This procedure is followed by lateral etching. Typical etch times are 2 hours for a 125 x 125 $micron^2$ device. However, overetching is allowed because the etch is self-limiting.

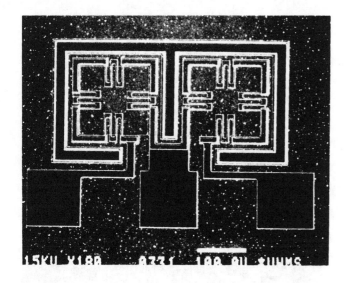

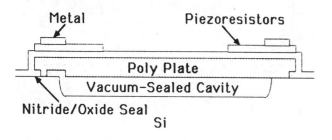

Fig. 2. An SEM of a completed device shows the active transducer on the left and the inactive device on the right. Three metal pads provide electrical connections to the strain-sensitive resistors. The resistors are placed in a highly symmetric configuration around the plate so that variation in pressure sensitivity with alignment error is minimized. The bottom illustration depicts the cross-section of the active sensor.

The basic concept in the sealing process, reactive sealing, involves on one hand the size of the etch channels relative to the cavity and on the other hand the supply of sealing material. The respective geometries are designed so that etch channel dimensions are always small relative to the cavity. Hence, in an oxidizing ambient, etch channel closure occurs and interrupts the supply of oxygen to the cavity interior. Any trapped oxygen continues to react and thereby produces a vacuum ambient.

The actual sealing process consists of a low temperature oxidation which is followed by an LPCVD silicon nitride deposition. The resistor layer, an LPCVD polysilicon film, is applied over the nitride. All three materials combine to form an excellent seal.

The construction of the sensing structure starts with a boron blanket implant. The implant is selected to yield a pre-determined temperature coefficient of resistance. This procedure differs significantly from diffused gages. Positive as well as negative TCR's can be realized and coefficients which are as small as 50 ppm/°C have been achieved. Resistor positioning on the diaphragm is again very different than for single crystal gages. It is possible to design fully active bridges. However, for the contemplated extraction circuitry a simple voltage divider with one active element is sufficient. It was stated earlier that the over-all diaphragm size is typically 125 micron x 125 micron. Resistor placement and geometry will therefore require insensitivity to typical misalignment errors as well as carefully controlled linewidth control. This has been achieved by constructing a voltage divider which involves first of all an active and an inactive pillbox. The inactive pillbox is simply an identical device which retains the internal silicon dioxide post. Resistor patterns on both pillboxes are identical. Four resistors which are connected in series are located on each box. The resistors meander and are turned around at the neutral stress points of the diaphragm. The layout is self-compensating for small alignment errors. A photoresist mask defines the heavy implanted regions for contacts and also reduces the resistance in non-strain sensitive areas such as outside the plate perimeter and at the neutral stress turn-around points. Polysilicon definition via the fifth mask and RIE etching is followed by a short 800°C oxidation which passivates the sensing structures. A contact opening mask is used to define oxide removal regions. Metal, 98% Al - 1% Si - 1% Cu is sputtered and defined via mask #7 to complete the device shown in figure 2.

Electrical tests for the device are still somewhat incomplete. However, resistance values of 20 kΩ with near zero TCR have been achieved. Contact resistance of 1.6Ω with 1.2Ω standard deviation has been measured and is encouraging. Typical gage factors are near 25 and vacuum transducers for one atmosphere service behave as expected. They show 1.2% full scale sensitivity, and are linear.

REFERENCES

(1)   H. Guckel and D. W. Burns, "Planar Processed Polysilicon Sealed Cavities for Pressure Transducer Arrays," Technical Digest, 1984 IEEE IEDM, pp. 223-225.

(2)   H. Guckel, D. W. Burns and C. R. Rutigliano, "Design and Construction Techniques for Planar Polysilicon Pressure Transducers with Piezoresistive Readout,"

IEEE Solid-State Sensors Workshop, Hilton Head Island, S. C., June 1986.

(3)   G. Harbeke, L. Krausbauer, E. F. Steigmeier, A. E. Widmer, H. F. Kuppert and G. Neugebauer, "Growth and Physical Properties of LPCVD Polycrystalline Silicon Films," J. Electrochem. Soc., March 1984, pp 675-682.

(4)   H. Guckel, T. Randazzo and D. W. Burns, "A Simple Technique for the Determination of Mechanical Strain in Thin Films with Applications to Polysilicon," J. Appl. Phys. 57(5), 1 Mar 1985, p. 1671-1675.

SURFACE MICROMACHINING OF POLYIMIDE/METAL COMPOSITES
FOR A SHEAR-STRESS SENSOR

Martin A. Schmidt, Roger T. Howe, Stephen D. Senturia
Microsystems Technology Laboratories
Massachusetts Institute of Technology
Cambridge, MA 02139

and

Joseph H. Haritonidis
Turbulence Research Laboratory
Massachusetts Institute of Technology
Cambridge, MA 02139

ABSTRACT

Surface micromachined polyimide/metal composites are studied for use as a shear-stress sensor. The role of residual stress in control of the shape of a released structure is reported. Several polyimide chemistries and two metals (aluminum and chrome) are used. The large compressive stress in evaporated aluminum makes it unattractive as a metal in this application. Evaporated chrome is tensile as deposited and matches the tensile stress in polyimide closely enough to make fabrication of floating-element sensors possible.

INTRODUCTION

Micromechanical components fabricated using surface and/or bulk micromachining are very sensitive to the mechanical properties of the constituent materials [1]. Residual stress can cause loss of adhesion and dimensional changes in patterned structures upon release. Also, non-uniform stress distributions through a film can cause substantial out-of-plane warpage. One example of where these problems have been encountered is in surface micromachining of polysilicon [2,3]. Residual compressive stresses have caused buckling of bridges and curvature of cantilevers, requiring the development of high temperature annealing procedures to reduce these stresses.

We have investigated polyimide and polyimide/metal composites as a surface micromachinable material for a floating-element shear-stress sensor [4]. This sensor is intended for use in turbulent boundary layer research where control of the shape of the microfabricated structure is critical. Excessive out-of-plane warpage due to residual stress non-uniformities will make this sensor unusable. We report here our findings on the fabrication of polyimide and polyimide/metal composite microstructures and the role of residual stresses.

FLOATING-ELEMENT SHEAR-STRESS SENSOR

Figure 1 illustrates a simplified process top-view and cross-section for a surface-micromachined floating-element shear-stress sensor. A square element and four tethers are formed out of a polyimide film deposited over a patterned aluminum spacer. The spacer is wet etched to release the square element from the substrate, leaving it anchored by the four tethers. The released element will displace laterally as shown due to the shear stress generated by a flowing fluid. Residual tensile stress in the polyimide keeps the element suspended above the substrate and determines the mechanical response to an applied load [5]. Figure 2 is a SEM photograph of a fabricated structure. The 8 µm thick polyimide element is 1 mm x 1 mm square with support tethers of 1 mm x 10 µm and is suspended above the substrate by 3 µm. The polyimide has been removed from the field region in this sample for clarity.

A differential capacitor readout scheme has been developed to transduce motion of the floating element. Figure 3 shows the readout scheme which requires the incorporation of a thin conducting shield plate in the polyimide element and three electrodes on the surface of the chip. Lateral motion of the element causes a change in the capacitive coupling between the central drive electrode and the two symmetrically placed sense electrodes. The sense electrodes are connected to the gates of a pair of matched depletion-mode MOSFETs to buffer the signal. The differential signal is then measured off-chip using a pair of transresistance amplifiers.

DEPOSIT AND PATTERN SPACER MATERIAL

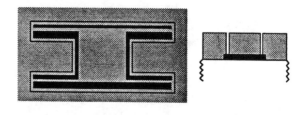

DEPOSIT AND PATTERN POLYIMIDE

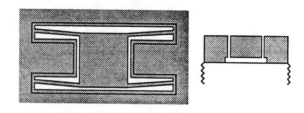

**WET ETCH SPACER MATERIAL
ELEMENT SHOWN DEFLECTED UNDER SHEAR-STRESS LOAD**

Fig. 1   Floating element process cross-section.

Reprinted from *Rec. of the IEEE Micro Robots and Teleoperators Workshop*, 4 pages, 1987.

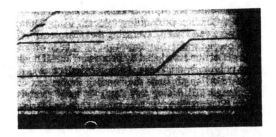

Fig. 2   SEM photogragh of a polyimide floating element.

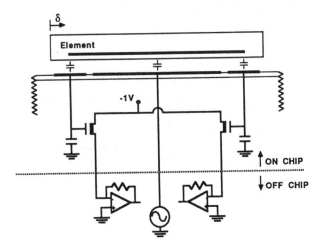

Fig. 3   Differential readout circuit.

We have investigated two polyimide chemistries, DuPont 2545 and 2555 and two thin metals, aluminum and chrome. The thin conductor metals are chosen for their excellent adhesion to polyimide. The polyimide thickness has ranged from 5 μm to 35 μm and the metal thickness from 300 Å to 1000 Å.

RESULTS AND DISCUSSION

The first experiments using this technology concentrated on fabricating polyimide structures without metal. Polyimide exists in a state of residual tensile stress as deposited [5]. Upon release, a polyimide structure will contract against its supports to release this stress. If there are any stress non-uniformities through the film, this will lead to bending of the structure out of the plane of the wafer. Close investigation of the structure in Figure 2 indicates a slight concave down curvature of the polyimide element with the center bowed up by 30 μm. We have found that both polyimide chemistries will exhibit warpage, but the direction of the curvature is dependent on the chemistry. The curvature is a strong function of the thickness of polyimide, such that increasing the polyimide thickness will decrease the curvature. It is possible to make the element shown in Figure 2 with less than 1 μm of total departure from flatness by increasing the thickness to 15-20 μm.

Fabrication of this sensor has required development of polyimide/metal microstructures. Particular attention has been paid to the sensitivity of this structure to residual stresses in the polyimide and metal. For turbulent boundary layer measurements, we would like the total out-of-plane warpage to be less than 1 μm. We will first discuss the general process sequence for fabricating these structures and then report on the particular results obtained.

FABRICATION

Figure 4 illustrates a cross-section through the process of fabricating a composite surface-micromachined cantilever. In making a sensor, we would begin with a substrate that contains the MOSFETs and elecrodes with a deposited oxide/polyimide passivation layer. A 3 μm thick aluminum spacer is evaporated and patterned. One coat of polyimide (1 μm thickness) is spun-cast on the wafer and partially cured at an intermediate temperature. The partial curing will improve the adhesion to the subsequent polyimide layers. A thin conductor is evaporated and patterned to form the shield plate. The remaining polyimide is spun-cast in multiple coats to the desired thickness. The polyimide is then cured at 400 °C for 45 minutes in nitrogen. A 3000 Å aluminum layer is evaporated and patterned as a non-erodible etch mask for the polyimide. The polyimide is etched using an O_2 plasma in a parallel plate reactor. The final step is release of the element in a aluminum wet etch consisting of phosphoric-acetic-nitric acid and water.

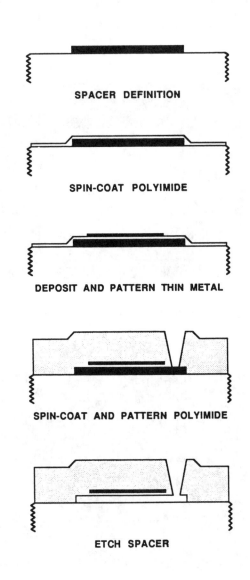

SPACER DEFINITION

SPIN-COAT POLYIMIDE

DEPOSIT AND PATTERN THIN METAL

SPIN-COAT AND PATTERN POLYIMIDE

ETCH SPACER

Fig. 4   Composite structure process cross-section.

Merging a thin conductor into the polyimide sustantially alters the shape of the structure due to the residual stress of the metal. Aluminum evaporated in both an electron beam and filament system has shown a very large compressive stress. Wafer curvature measurements on thicker films indicate stresses of 0.4-1.0 GPa. This high compressive stress manifests itself in two ways in the fabrication of composite elements. First, we have found that the stress can produce delamination of the aluminum from the underlying polyimide layer during processing. Figure 5 is a microphotograph of a 300 Å aluminum film embedded in polyimide. The high compressive stress has caused the aluminum to peel from the lower polyimide as evidenced by the "rippling" of the edge of the metal. Secondly, because of the high tensile stress of the polyimide and the asymmetric placement of the aluminum, the released elements will curl up substantially. Increasing the polyimide thickness will reduce the amount of curvature, but over the range of polyimide thicknesses investigated, it was not possible to make structures flat enough for use.

Chrome evaporated in an electron beam system is under tensile stress. The tensile stress is of sufficient magnitude to cause microcracking when deposited on polyimide. Figure 6 is a microphotograph showing the cracking of a 300 Å chrome layer on top of polyimide. The chrome remains a good conductor in spite of the microcracks. Released structures fabricated with this chrome also curve up, indicating that the tensile stress in the metal is less than the tensile stress in the polyimide, producing a net bending moment. However, the degree of curvature is substantially less than with aluminum, making the fabrication of a flat floating element possible.

Cantilevers have been fabricated out of the composite polyimide/chrome films and the tip deflection as a function of polyimide and chrome thicknesses studied. Table 1 presents a summary of cantilever tip deflection measurements for three samples as a function of cantilever length using DuPont 2545. These data were collected by averaging the measurements of several cantilevers on one wafer. Similar results have been obtained using 2555. The deflection is reduced by decreasing metal thickness and increasing polyimide thickness. Also, the deflection is dependent on the square of length since the bending moment is distributed along the length [6]. Using this data, we have established that functioning floating element shear sensors can be fabricated using 300 Å of chrome in 30 μm of polyimide and restricting the size of the element to 500 μm x 500 μm.

CONCLUSIONS

The development of a floating-element shear-stress sensor has required that a great deal of attention be paid to the mechanical properties of the materials used. In particular, shape control of microfabricated elements has necessitated an understanding of the role of residual stresses in a composite structure. We have investigated the effects of various polyimide/metal systems. A system that provides us with a functional floating-element sensor was identified.

ACKNOWLEDGEMENTS

Support for Martin Schmidt to work on this project was provided through a 3M Sensor Fellowship. Samples were fabricated in the Microelectronics Laboratory of the MIT Center for Materials Science and Engineering, which is supported in part by the National Science Foundation under Contract DMR-84-18718. The authors are indepted to Mehran Mehregany, Mark Allen, and Herb Neuhaus for valued discussions and assistance in sample fabrication.

Fig. 5   Thin aluminum plate in polyimide (1 mm x 1 mm).

SAMPLE		CANTILEVER DEFLECTION		
Polyimide Thickness	Chrome Thickness	Cantilever Length		
		1000 μm	500 μm	250 μm
18 μm	300 Å	99.5 μm	23.75 μm	6.5 μm
16 μm	600 Å	145 μm	42 μm	9 μm
32 μm	300 Å	22.25 μm	3.75 μm	0 μm

Table 1   Cantilever tip deflection.

Fig. 6   Microcracks in thin chrome layer.

REFERENCES

[1] S.D. Senturia, "Microfabricated Structures for the Measurement of Mechanical Properties and Adhesion of Thin Films", Proc. Transducers '87, Tokyo, 1987, pp. 11-16.

[2] R.T. Howe, "Polycrystalline Silicon Microstructures", in Micromachining and Micropackaging of Transducers, C.D. Fung, P.W. Cheung, W.H. Ko, and D.G. Fleming, eds., Amsterdam, Elsevier, 1985, pp. 169-187.

[3] H. Guckel, D.W. Burns, C.R. Rutigliano, D.K. Showers, and J. Uglow, "Fine Grained Polysilicon and Its Application to Planar Pressure Transducers", Proc. Transducers '87, Tokyo, 1987, pp. 277-282.

[4] M.A. Schmidt, R.T. Howe, S.D. Senturia, and J.H. Haritonidis, "A Micromachined Floating-Element Shear Sensor", Proc. Transducers '87, Tokyo, 1987, pp. 383-386.

[5] M.G. Allen, M. Mehregany, R.T. Howe, and S.D. Senturia, "Microfabricated Structures for the in-situ measurement of residual stress, Young's modulus, and ultimate strain of thin films", Appl. Phys. Lett., Vol. 51, July 1987, pp. 241-243.

[6] J.M. Gere and S.P. Timoshenko, Mechanics of Materials, PWS, Boston, 1984.

MICRO-DIAPHRAGM PRESSURE SENSOR

S. Sugiyama, T. Suzuki, K. Kawahata, K. Shimaoka

M. Takigawa and I. Igarashi

Toyota Central Research & Development Labs., Inc.

Nagakute, Aichi 480-11, Japan

ABSTRACT

A micro-diaphragm pressure sensor with silicon nitride diaphragm of 80 μm × 80 μm was fabricated by applying micromachining technique. The main feature is that it is a complete planar type pressure sensor formed by single-side processing solely on the top surface of (100) silicon wafer. The diaphragm and a reference pressure chamber are formed by undercut-etching of the interface between the diaphragm and the silicon substrate. The diaphragm thickness is 1.4 μm. Polysilicon piezoresistors are arranged to form a bridge on the micro-diaphragm. The output sensitivity is obtained more than 100 μV/V with respect to a pressure of 100 kPa. The temperature coefficient of the sensitivity is approximately −0.13 %/°C at the temperature range between −50°C and 150°C. The advantage is its suitability for development of integrated sensors, multiple sensors and sensor arrays.

INTRODUCTION

Piezoresistive silicon pressure sensors have been widely used for automotive, industrial and biomedical pressure sensing applications [1,2]. Usually, the silicon diaphragm dimensions are more than 1×1 mm^2 with a thickness of approximately 5-20 μm. These diaphragms are formed by using anisotropic silicon etching technique. Four piezoresistors are formed either by boron diffusion or ion implantation on top of the diaphragms. The pressure sensors have the advantages of low manufacturing cost due to application of the same batch processes as those for integrated circuit and excellent stability and mechanical properties of silicon [3]. However, the disadvantages are the difficulty in making micro-size diaphragm less than 100 μm, and yield losses in diaphragm etching because of utilization of both-side alignment and backside-etching technique.

Recently, micromachining which utilizes anisotropic etching of single-crystal silicon and forming of thin film is expected to contribute to the development of new functional devices for sensors and actuators [4-6]. This paper reports a micro-diaphragm pressure sensor with Si$_3$N$_4$ diaphragm of 80 μm × 80 μm, fabricated by applying the micromachining. This eliminates cumbersome both-side alignment steps which are necessary to align the resistor pattern with respect to the backside-etched diaphragm. The main feature is that it is a complete planar type pressure sensor with the diaphragm formed by single-side processing solely on the top surface of the silicon wafer. In this paper, the process to form the micro-diaphragm pressure sensor, and the experimental results will be described.

SENSOR PROCESSING

Figure 1 shows an idea of micro-diaphragm formation process. The substrate is (100) plane silicon. The diaphragm and a cavity as the reference pressure chamber are formed by undercut-etching of the interface between the diaphragm and the silicon substrate. The interlayer is formed as the etch-channel in the interface between the substrate and the diaphragm. Silicon nitride (Si$_3$N$_4$) layer is deposited on the interlayer as the diaphragm. The undercut is made by anisotropic silicon etching using alkali etchant through a hole opened in the Si$_3$N$_4$ layer to remove the interlayer and a part of the substrate. Figure 2 shows undercut-etching characteristics for various interlayers, thermal-SiO$_2$, phospho-silicate glass (PSG) and polysilicon. Each interlayer is formed to have a 2000 Å thick. When the polysilicon interlayer is utilized, the undercut-etch rate becomes about 100 times as fast as in the case of the thermal-SiO$_2$. After the entire polysilicon interlayer is removed, undercut-etching of the substrate practically stops. Therefore, the micro-diaphragm, with the shape corresponding to the shape of disappeared polysilicon interlayer, is formed, and a precisely-shaped cavity like a pyramid, with walls of four {111} crystalline planes, as a reference pressure chamber is hollowed out.

Figure 3 shows the cross section of the micro-diaphragm pressure sensor. The starting material is (100), n-type silicon wafer with a resistivity of 5 Ωcm. The first Si$_3$N$_4$ layer with a thickness of 500 Å is deposited by LPCVD. A square-shaped window of 80 μm × 80 μm is then photolithographically defined and opened in the Si$_3$N$_4$ layer. An etch channel is formed by means of 1500 Å thick polysilicon interlayer, deposited by LPCVD. The polysilicon channels are photolithographically defined and formed to cover the square-shaped

Reprinted from *Rec. of the IEEE Int. Electron Devices Meeting*, 1986, pp. 184–187.

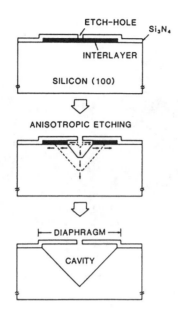

Fig. 1 An idea of micro-diaphragm formation
        process.

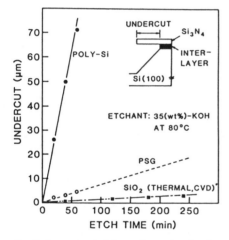

Fig. 2 Undercut-etching characteristics for
        various interlayers.

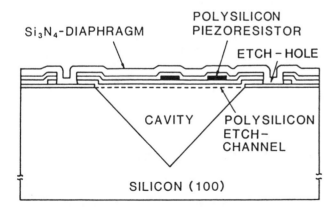

Fig. 3 Cross section of the micro-diaphragm
        pressure sensor.

window and to have a projecting area at each side
or corner of the square. This is followed by a
2000 Å thick second LPCVD-Si_3N_4 deposition. On the
second Si_3N_4 layer, a 2000 Å thick polysilicon is
deposited by LPCVD. Boron ion implantation is then
carried out with a dose of 1×10^{16} cm^{-2} at 30 keV
and activated at 900°C for 30 minutes. The 5 μm
wide polysilicon piezoresistors are then photolithographically defined and formed by reactive ion
etching. This is followed by a 2000 Å thick third
LPCVD-Si_3N_4 deposition to cover the polysilicon
piezoresistors from alkali etchant. Etch-holes are
formed such as to penetrate the second and the
third Si_3N_4 layers to reach the polysilicon channels. An anisotropic etchant of KOH is poured onto
the substrate through the etch-holes. The entirety
of the polysilicon channels are removed and then
the silicon substrate just under the diaphragm is
anisotropically etched through the etch-holes,
thereby forming the movable diaphragm and the
cavity as the reference pressure chamber.
Electrical contact holes are then defined and
opened to the piezoresistors. An aluminum layer
is deposited by vacuum evaporation, and photolithographically formed to make lead wires and electrodes. On top of these layers, 1 μm thick silicon
nitride is deposited to seal the etch-holes by
plasma CVD. Approximately 1.4 μm of total nitride
diaphragm is formed during these steps. The pressure inside the cavity is less than 0.3 Torr.
Hence, the pressure sensor can measure absolute
pressures.

Figure 4 shows an SEM photograph of a micro-
diaphragm cross section. The precisely shaped
cavity like a pyramid, with walls of {111} crystalline planes, is shown. Also, Fig. 5 shows SEM
photographs of close-up cross section of a micro-
diaphragm edge. The relationship between the diaphragm, a cavity and etch-holes is clearly shown
in (a). The narrow disappeared etch-channel of
1500 Å wide and the completely sealed etch-hole of
8 μm diameter are clearly shown in (b). Figure 6
shows a top view photomicrograph of a micro-
diaphragm. The 80 μm × 80 μm diaphragm, two polysilicon piezoresistors and four etch-holes with
8 μm diameter are clearly discernible.

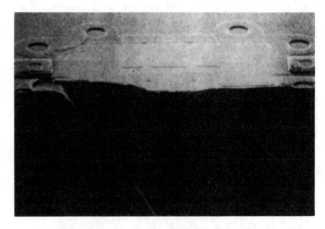

Fig. 4 Cross sectional SEM photograph of
        a micro-diaphragm pressure sensor.

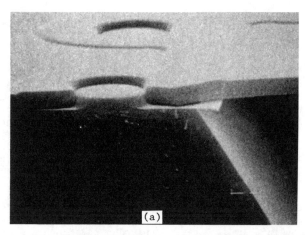

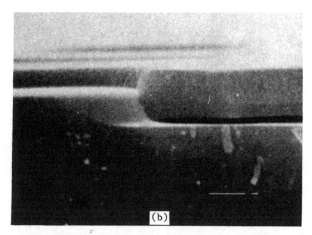

Fig. 5 Close-up cross sectional SEM photographs of a micro-diaphram edge. Relationship between
the diaphragm, a cavity and etch-holes is clearly shown (a). A completely sealed
etch-hole and a etch-channel (b).

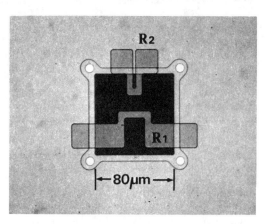

rig. 6 Top view photomicrograph of an 80 μm ×
80 μm micro-diaphragm pressure sensor.

Fig. 7 Resistance change of the two piezoresistors
on an 80 μm × 80 μm micro-diaphragm as
a function of pressure.

RESULTS AND DISCUSSION

Figure 7 shows the resistance change of the two individual piezoresistors of an 80 μm × 80 μm micro-diaphragm pressure sensor as a function of pressure. The central resistor R_1 is opposite in sign to the side resistor R_2. The bridge output voltage Vout as a function of pressure for the micro-diaphragm pressure sensor is shown in Fig. 8. A non-linearity of less than 0.5 % of the full scale is obtained. The pressure sensitivity is obtained approximately 130 μV/V with respect to a pressure of 100 kPa. The pressure sensitivity is rather low than that of a single-crystal silicon pressure sensor. The reduced pressure sensitivity of the micro-diaphragm pressure sensor is conjectured to be caused by the small crystalline grain size, the random orientation in crystalline direction and misarrangement of the polysilicon piezoresistors. This will be improved by recrystallization of the polysilicon piezoresistors using laser process and correct arrangment of the piezoresistors [7].

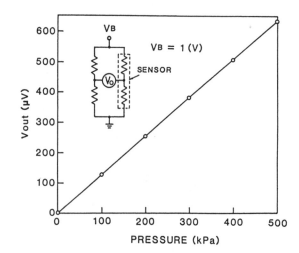

Fig. 8 Output voltage Vout as a function
of pressure for an 80 μm × 80 μm
micro-diaphragm pressure sensor.

The temperature dependence of the sensitivity is shown in Fig. 9. The temperature coefficient of the sensitivity is approximately -0.13 %/°C at the temperature range between -50°C and 150°C. This is similar to usual single-crystal silicon pressure sensors and an acceptable value for many applications [8].

The polysilicon piezoresistors are coated by the laminated silicon nitride film. Hence each piezoresistor is isolated from the other, thereby preventing leakage current due to a rise in temperature as in the case of P-N junction isolation. This enables the micro-diaphragm pressure sensor to be operated stably up to a high-temperature range. From the experiments carried out it has been confirmed that the pressure sensor is possible to operate at the temperature range of 300°C.

Figure 10 shows the stability for repetitive pressure applications of the micro-diaphragm pressure sensor. The repeatedly applied pressure is 700 kPa. The repeated period is 2.5 seconds. The sensitivity change is less than ±0.4 % at the repetitive pressure of 10^4 cycles. This assures that the etch-holes have been sealed airtight.

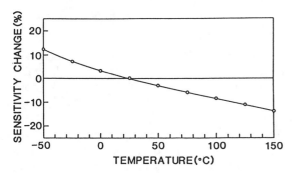

Fig. 9 Temperature dependence of pressure sensitivity for the micro-diaphragm pressure sensor.

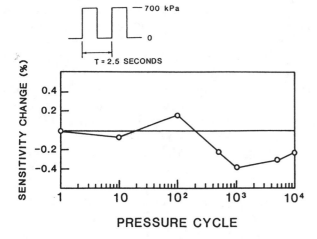

Fig. 10 Stability for repetitive pressure applications of a micro-diaphragm pressure sensor.

CONCLUSION

An 80 μm × 80 μm silicon nitride micro-diaphragm pressure sensor with polysilicon piezoresistors was fabricated by applying micromachining technique. The main feature is that it is a complete planar type pressure sensor. The diaphragm and a reference pressure chamber are formed by undercut-etching of the interface between the diaphragm and the silicon substrate. It is possible to conduct all the wafer processing steps solely on the top surface of the substrate, and to fabricate the pressure sensor by what is called single-side processing. This eliminates cumbersome both-side alignment steps which are necessary to align the resistor pattern with respect to the backside-etched diaphragm. The pressure sensitivity is obtained approximately 130 μV/V with respect to a pressure of 100 kPa. The temperature coefficient of the sensitivity is approximately -0.13 %/°C at the temperature range between -50°C and 150°C. The polysilicon piezoresistors are coated by laminated silicon nitride film. It has been confirmed that the micro-diaphragm pressure sensor is possible to operate at the temperature range of 300°C. The advantage is its suitability for development of integrated sensors, multiple sensors and sensor arrays.

REFERENCES

(1) W. G. Wolber and K. D. Wise, "Sensor Development in the Microcomputer Age", IEEE Trans. Electron Devices, ED-26 (1979), pp.1864-1874.
(2) W. H. Ko, J. Hynecek and S. F. Boettcher, "Development of a Miniature Pressure Transducer for Biomedical Applications", IEEE Trans. Electron Devices, ED-26 (1979), pp.1896-1905.
(3) S. Sugiyama, M. Takigawa and I. Igarashi, "Integrated Piezoresistive Pressure Sensor with Both Voltage and Frequency Output", Sensors and Actuators, 4 (1983), pp.113-120.
(4) K. E. Petersen, "Silicon as a Mechanical Material", Proceedings of the IEEE, 70 (1982), pp.420-457.
(5) E. Bassons, "Fabrication of Novel Three-Dimensional Microstructures by the Anisotropic Etching of (100) and (110) Silicon", IEEE Trans. Electron Devices, ED-25 (1978), pp.1178-1184.
(6) K. E. Petersen, "Silicon Sensor Technologies", Technical Digest, 1985, IEEE IEDM, pp.2-7.
(7) H. Guckel and D. W. Burns, "Laser-Recrystallized Piezoresistive Micro-Diaphragm Sensor", Digest of Technical Papers, 1985, 3rd Int. Conf. on Solid-State Sensors and Actuators, pp.182-185.
(8) S. Sugiyama, H. Funabashi, S. Yamashita, M. Takigawa and I. Igarashi, "Nonlinear Temperature Characteristics in Silicon Piezoresistive Pressure Sensors", Proc. the 5th Sensor Symposium, Tsukuba, Ibaragi, Japan (1985), pp.103-107.

Integrated Movable Micromechanical Structures for Sensors and Actuators

LONG-SHENG FAN, MEMBER, IEEE, YU-CHONG TAI, AND RICHARD S. MULLER, FELLOW, IEEE

Abstract—Movable pin joints, gears, springs, cranks, and slider structures with dimensions measured in micrometers have been fabricated using silicon microfabrication technology. These micromechanical structures, which have important transducer applications, are batch-fabricated in an IC-compatible process. The movable mechanical elements are built on layers that are later removed so that they are freed for translation and rotation. A new undercut-and-refill technique that makes use of the high surface mobility of silicon atoms undergoing chemical vapor deposition is used to refill undercut regions in order to form restraining flanges. Typical element sizes and masses are measured in millionths of a meter and billionths of a gram. The process provides the tiny structures in an assembled form, avoiding the nearly impossible challenge of handling such small elements individually.

I. INTRODUCTION

THE UNPRECEDENTED growth of integrated-circuit technology and computing techniques has made sophisticated data processing accurate, economical, and widely available. Today's electronic systems are capable of dealing with large numbers of physical input and output variables, but the transducers that provide interfaces between the electrical and physical world are in many cases outmoded and dependent on awkward hybrid-fabrication techniques. Many of the materials and processes used to produce integrated microcircuits, however, can be employed in new ways to produce microsensors and actuators. These structures complement the IC process and provide a means to produce new electronic systems.

Thus far, micromechanical transducer structures such as cantilevers, bridges, and diaphragms have been fabricated with IC-compatible processes for various useful applications. These structures, however, contain only bendable joints, a severe limitation on mechanical design capabilities for many applications. Microstructures with rotatable joints, sliding and translating members, and mechanical-energy storage elements would provide the basis for a more general micromechanical transducer-system design. Because such structures add important degrees of freedom to designers, we have investigated techniques to fabricate them using IC-based microfabrication processes [1]. Rotatable silicon elements, made using IC technol-

ogy, have also been reported by Gabriel *et al*. [2]. The new mechanical elements use polysilicon thin-film technology combined with techniques that we describe in this paper. An important advantage of the procedures described is that they provide mechanical structures containing more than one part in a preassembled form; this avoids individually handling the very tiny structures. The initial demonstration of the technique to make these structures employs polysilicon as the structural material and phosphosilicate glass for the sacrificial layer. Other materials may, however, be used in place of these, provided that they are compatible with the overall process.

II. STRUCTURES AND PROCESSES

A. Fixed-Axle Pin Joints

A pin joint is composed of an axle around which a member (rotor) is free to rotate. Movement along the axle by the rotor is constrained by flanges. Fig. 1 shows the cross section and top view of a pin joint fixed to the substrate that has been fabricated using polycrystalline silicon. The rotor, axle, and flange are all made of polysilicon that has been deposited by a low-pressure chemical-vapor-deposition (LPCVD) process on top of a silicon substrate. The pin joint is produced using a double-polysilicon process and a phosphosilicate glass (PSG) sacrificial layer in a three-mask process as indicated in Fig. 2. In this process, openings are first made by dry etching a composite layer of polysilicon on PSG deposited by sequential LPCVD processes. Another PSG layer is deposited over the entire structure including the edges of the circular openings. Photolithography steps are then used to expose bare silicon at specific locations so that a subsequent deposition of polysilicon will anchor to the silicon substrate at these desired places. After depositing and patterning the second polysilicon layer for axles and flanges, all previously deposited PSG layers are removed in buffered hydrofluoric acid (BHF). The remaining polysilicon layers form the pin-joint structure. The rotor is free to rotate when the PSG layer is removed by BHF. An SEM photo (Fig. 3) shows a completed pin joint of this type.

B. Self-Constraining Pin Joints

A rotating-joint structure that provides several new possibilities for mechanical design can be built using only a small variation on the process described above. To differentiate joints of this type from the fixed-axle pin joints

Manuscript received October 1, 1987; revised January 19, 1988. The Berkeley Sensor and Actuator Center is an NSF/Industry/University Cooperative Research Center. This work was partially supported by the U.S. Army Harry Diamond Laboratory.

The authors are with the Berkeley Sensor and Actuator Center, Department of Electrical Engineering and Computer Sciences, and the Electronics Research Laboratory, University of California, Berkeley, CA 94720.

IEEE Log Number 8820452.

Reprinted from *IEEE Trans. Electron Devices*, vol. 35, no. 6, pp. 724–730, June 1988.

134

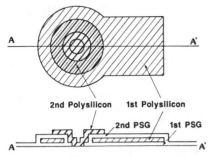

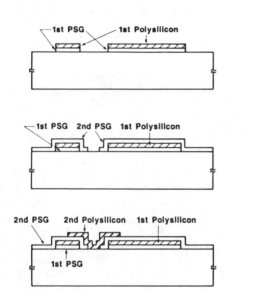

Fig. 1. Top view and cross section of a polysilicon micromechanical pin joint.

Fig. 2. Fabrication process for the anchored pin joint shown in Fig. 1.

Fig. 3. SEM photograph of anchored pin joint. The outer radius of the flange connected to the axle is 25 μm.

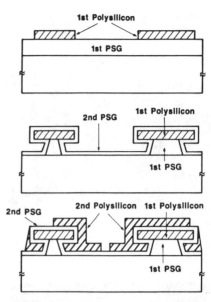

Fig. 4. Fabrication process for the self-constraining joint.

described above, we call these structures *self-constraining joints*. Self-constraining joints can, for example, allow for rotation while, at the same time, permitting translation across the silicon surface. These joints need to have a flange on the axle underneath the rotor to keep it in place. The axle can either be fixed to the substrate or else left free to translate across its surface.

Self-constraining joints are produced by a double polysilicon process with a PSG sacrificial layer. An undercut-and-refill technique is introduced to position the second-layer polysilicon both over and under the axle formed of first-layer polysilicon. Fig. 4 outlines the process for these joints, which are produced using two masks. In this process, after the PSG and polysilicon layers have been deposited by LPCVD, the polysilicon is patterned by dry etching. The next step is to use a timed etch of the first PSG layer to undercut the polysilicon. An optional mask may be used if only selected regions are to be undercut. Another PSG layer is then deposited. A second polysilicon layer that fills in the undercut regions is the patterned to produce axles and flanges. After this, all PSG layers are removed in a buffered hydrofluoric-acid solution. The remaining polysilicon layers form the self-constraining joint structure as shown in the SEM photograph of Fig. 5. The interleaved polysilicon layers, evident in Fig. 4, can be made because of the high surface mobility of silicon atoms during the LPCVD process. This permits the undercut regions to refill so that restraining flanges remain over and under the first-layer polysilicon.

For more complex structures, both pin joints and self-constrained joints can be made in the same process. Fig. 6 shows four-joint crank structures fabricated using both types of joints in a three-mask process (four, if the optional undercut mask is used). In Fig. 6, the joints at both ends of the central element are self-constraining but they are freed from the substrate. The other two joints are pinned to the substrate. Note that, except for the fixed joints, the entire structure has moved from its original position, indicated by a darkened pattern on the silicon substrate in Fig. 6. Using a surface profiler (Alpha-Step 200), we have found that in the darkened pattern there is a pit that is roughly 100 nm deep. This pit appears to be caused by enhanced etching of the silicon surface by the BHF under the polysilicon moving elements. The enhanced etching may, in turn, result from localized stress in this region. Further research to test this hypothesis is underway.

Fig. 5. SEM photograph of a self-constraining joint. The rotor is attached to a hub that turns in a collar projecting from a stationary polysilicon surface. There is a retention flange on the hub below the collar.

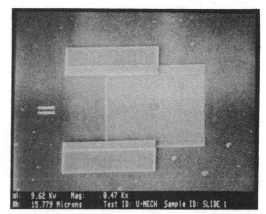

Fig. 7. Square slider with two edges restricted by flanges. One side of the slider is 100 μm and the central opening is a square of 10 μm.

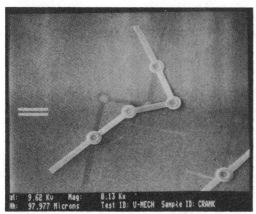

Fig. 6. SEM photograph of a four-joint crank having a central arm held by self-constraining joints that are free to translate. The original position of the crank is indicated by the darker pattern. All crank arms are 150 μm in length.

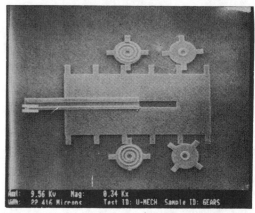

Fig. 8. Gear and slider combination. The slider is 210 by 100 μm. The toothed edges mesh with four gears, two of which have flat spiral springs attached.

C. Flanged Structures

The procedures carried out to make the two types of pin joints can be employed to produce other mechanical structures. For example, the three-mask pin-joint process can be used directly to fabricate the square slider shown in Fig. 7. The slider has a polysilicon moving element that is constrained by flanges along two of its edges so that only translational movement in one dimension is allowed. In the gear-slider combination of Fig. 8 (produced by the pin-joint process with four masks), the slider has a guide at its center to constrain movement to one dimension. Fig. 9 shows a crank-slot combination that requires five masks to produce. The slot element, formed by first-layer polysilicon, is pinned to the substrate by another element made of second-layer polysilicon so that both translational and rotational movements can take place. The joint for the crank-slot in Fig. 9 is self-constraining.

D. Design Variations

The versatility of the techniques described permits many potentially useful variations in design. As an example, Fig. 10 is a slider that can be thought of as a *dual* to the structure in Fig. 7. The flanges holding the parts together

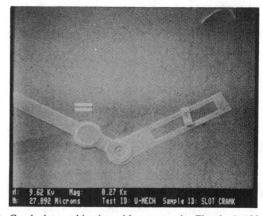

Fig. 9. Crank-slot combination with a center pin. The slot is 130 μm long and 20 μm wide. The diameters of the two joints are each 50 μm.

are on opposite members in each of these two versions of the mechanical slider. In the same sense, Fig. 11 shows a four-joint crank that is *dual* to the one in Fig. 6. All four joints in this crank are made using a self-constraining process; the two end joints are fixed and the two center joints can translate.

E. Micromechanical Bushings

The undercut-and-refill technique is useful for other applications also. When combined with pin joints, this tech-

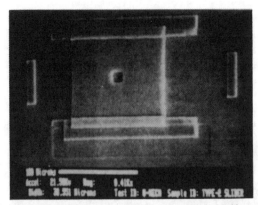

Fig. 10. Slider structure with outer edges guided by self-constraining joints. Stops limit the extent of lateral motion by the slider.

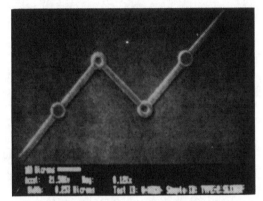

Fig. 11. Four-joint crank made with self-constraining joints.

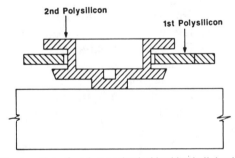

Fig. 12. Cross section of a micromechanical bushing built by the self-constraining-joint process.

nique permits the fabrication of bushings that can be used, for example, to elevate a rotor away from the silicon surface. This can greatly reduce frictional forces, especially if the bushing elements are coated with or made from another material such as silicon nitride that may provide better wear properties. Fig. 12 shows a cross section for a bushing produced using the self-constraining-joint process described above and one extra mask for anchoring to the substrate (three masks in all).

F. Polysilicon Springs

Mechanical energy storage is important in many systems, and therefore it is very desirable to be able to fabricate micromechanical springs. These elements can also be produced using the process described above. The flat spiral spring attached to a pin joint, shown in Fig. 13, is

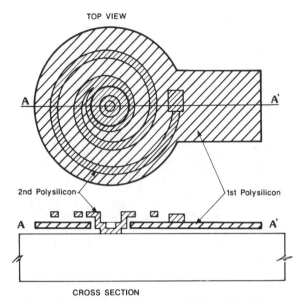

Fig. 13. Top view and cross section of flat spiral spring fixed on one end to an axle.

made of second-layer polysilicon and connected on one end to the axle of a pin joint. The other end is attached to a movable disc made of first-layer polysilicon. The spring, produced using four masks, returns the disc to its original position after it is displaced. Figs. 14 and 15 show SEM photographs of restraining springs connecting rotors to pin-joint axles. Both springs are made of 2-μm-wide second-layer polysilicon. Shown in Fig. 15 is a beam spring that has an appreciably larger spring constant than does the flat spiral spring of Fig. 14.

G. Processing Details

In the foregoing, we have described the essential techniques for the *in situ* fabrication of assembled micro structures. The achievable dimensions for the finished elements depend on the lithography and processing steps used, but they can be roughly estimated at ten or fewer micrometers. Any of the microstructures can be fabricated separately using fewer than four masks. Six masks are needed to build all of the structures in the same run. When all are produced at one time there is an unavoidable loss in element precision because of the extra processing steps.

To illustrate the processes more completely, we describe in fuller detail the steps used to fabricate all structures on the same chip. First, a 1.5-μm-thick phosphorus-doped (8 wt.%) LPCVD silicon-dioxide layer is deposited at 450°C on a (100) silicon substrate. Photolithography and the first mask are used to open selected areas on the substrate where the first-layer polysilicon is to be anchored. Undoped LPCVD polysilicon, 1.5 μm-thick, is then deposited at 630°C and patterned with a second mask in a CCl_4 plasma. The third mask is used to define the undercut regions for self-constraining and bushing structures. Buffered HF etching creates a 2-μm undercut. Next, a 0.5-μm-thick phosphorus-doped (8 wt.%) LPCVD sil-

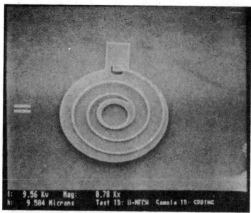

Fig. 14. SEM photograph of the spring-axle structure shown in Fig. 13. The 2.5-revolution spiral spring is made of 2-μm-wide second-layer polysilicon. Its inner end is fixed to an axle 10 μm in diameter, and its outer end is connected to a movable arm.

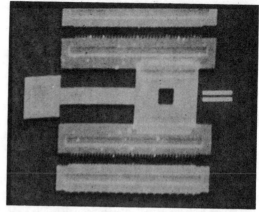

Fig. 16. Slider and bridge structure. One end of the bridge is anchored to the substrate and the other is attached to a slider. The central square opening is 20 μm on a side and the square slider is 80 μm on a side.

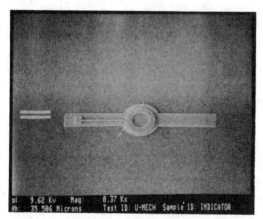

Fig. 15. SEM photograph of a beam spring attached to a central axle. The beam is 60 μm long and 2 μm wide.

icon-dioxide sacrificial layer is deposited at 450°C. The fourth and fifth masks are used to pattern the silicon-dioxide layer to anchor the spring element to the substrate and the rotor, respectively. Undoped LPCVD polysilicon, 1.0 μm thick, is then deposited at 630°C, defined and patterned using the sixth mask, and eteched in a CCl$_4$ plasma. Prolonged etching (and therefore thick resist films) are required to remove completely any residue of polysilicon from the regions near to topographic steps. For shorter etching times, a more isotropic plasma such as SF$_6$ might be used. A 1-h annealing step in nitrogen at 1000°C is used to reduce stress in the polysilicon. To release the structures from the oxide required 6 h of etching in a 5:1 buffered HF solution.

III. STUDY OF MECHANICAL PROPERTIES

An important use for these structures is to carry out research on the micromechanical properties of materials. This research is especially necessary since many of the materials have thus far been applied exclusively for electronics. One means for obtaining useful data is to carry out visual inspection of high-speed magnified video-tape images that show the response of dynamically actuated elements. Analysis of these data will permit studies of frictional behavior, damping, fatique limits, and of fundamental properties, such as Young's modulus, Poisson's ratio, and the orientational dependences of these parameters.

In general, a residual stress is found in LPCVD polysilicon after deposition. Previous papers [3], [4] have described useful ways to study uniform stress distributions. The stress distribution in the direction of polysilicon film growth is, however, very likely not to be uniform and therefore to induce a bending moment across the films. The bending moment is of special concern in cases where flatness is important.

Using slider structures (Figs. 7 and 10), we expect to be able to separate the effects of uniform and bending stresses in thin-film mechanical structures. Fig. 16 shows the top view of a slider and a bridging beam. The outer edge of the flanges are defined with teeth to act as measuring scales. One end of the beam is anchored on the silicon substrate, and the other end is connected to a self-constraining slider. Since the slider allows translational movement, the compressive-stress component in the polysilicon beam can be released after freeing the whole structure, while the bending moment will be left in the beam. Detailed analytical study of such structures to determine both the compressive stress and the bending moment in deposited polysilicon is underway.

IV. FRACTURE-STRENGTH STUDY

A flat spiral spring made of second-layer polysilicon is used to restrain a pin joint. Within its fracture limit, the spring can return the structure to its original position after it has been moved. Experiments have been done on these spring structures to estimate the lateral fracture stength of the polysilicon. A simplified mechanical analysis provides the basis for this estimate. For a 2 μm-wide 1-μm-thick spring extending 2.5 revolutions with inner radius $r_1 = 10$ μm connected to the central axle and outer radius $r_2 = 30$ μm connected to one arm, fractures occur at deflections of roughly 300°. For a spiral spring of width h and thickness t, assuming that adjacent turns do not come into contact, the energy stored is [5]

$$U = \int_0^l \frac{M^2 \, ds}{2 E_Y I} \qquad (1)$$

where M is the bending moment, E_Y is Young's modulus, I is the moment of inertia of the spring cross section, ds is the length of a small element of the spring, and l is the total length of the spiral. The spiral is generated using the equation $r = a\theta$ where θ is the polar-coordinate angle and a is a design constant chosen for a particular spring size. The bending moment can be shown [5] to be constant along the length of the spring. The angular deflection Φ is

$$\Phi = \frac{\partial U}{\partial M} = \int_0^l \frac{M}{E_Y I} \, ds = \frac{M}{E_Y I} \frac{r_2^2 - r_1^2}{2a}. \qquad (2)$$

The moment of inertia of the rectangular cross section is $I = th^3/12$ and the maximum stress in the spring σ_{max} is $Mh/2I$. Using these equivalents in (2), we can express the maximum stress in terms of the angular deflection Φ.

$$\sigma_{max} = \frac{E_Y h a}{r_2^2 - r_1^2} \Phi. \qquad (3)$$

The spiral springs were unwound using microprobes until they fractured. The bending moment in the spring loaded in this manner is constant along its length. Spring fractures occurred in all cases at deflection angles of $300 \pm 30°$. The fractures were observed in one or several locations and typically more then 20 μm away from the attachment points. The spiral springs (of the type shown in Fig. 13), have inner and outer radii of $r_1 = 10$ μm and $r_2 = 30$ μm, respectively. Other parameters are: thickness $t = 1$ μm, width $h = 2$ μm, and spiral constant $a = 1.27$ μm. Using the observed value of Φ_{fract} at fracture ($300°$ or 5.24 radians) in (3), we calculate a fracture strength σ_{fract} that is 1.7 percent times Young's modulus E_Y for thin-film polysilicon. At least two simplifying assumptions underlie the conclusions made above: 1) that the spring motion is entirely in the horizontal plane (neglecting possible vertical motions that would relax stress), and 2) that the spring has sufficient turns to be treated as ideal [6]. Other studies, still in progress, indicate a slightly lower fracture strength (in the order of 1.3–1.4 percent of E_Y). For a perspective on our results, we note that the highest reported fracture strength for single-crystal silicon is 2.6 percent times E_Y [7]. We expect that values for E_Y will depend on the deposition conditions for the film and on the direction of the stress relative to the growth direction.

To estimate E_Y for our polysilicon films, we make use of Johnson's analysis [8], X-ray diffraction studies showing the distribution of crystalline orientations in our films, and published orientation-dependent elastic constants for single-crystal silicon [9]. Ignoring grain-boundary effects, we estimate that E_Y for our films is 169 GPa and the fracture strength is in the 2 to 3 GPa range. For comparison, Guckel and co-workers have published a value

of E_Y for polycrystalline silicon of 22.2 Mpsi (153 GPa) [10].

Polysilicon Material Studies

An analysis to be published will detail the procedures sketched in the previous paragraph; we provide here only a few features of our studies of polycrystalline silicon to clarify the discussion. Using X-ray diffraction, we have found that the undoped LPCVD polysilicon films grown on PSG at 630°C have a preferred orientation that is generally in the (110) direction normal to the substrate. Annealing these films in nitrogen fosters grain growth but does not change their orientation. The polycrystalline-film orientation is actually described in terms of a distribution function derived from analysis of the X-ray diffraction data. This distribution function is used to calculate the effective film properties in terms of single-crystal parameters [8].

V. CONCLUSIONS

We have described a technique to build micro-scale movable mechanical pin joints, springs, gears, cranks, and sliders using a silicon microfabrication process. The ability of LPCVD polysilicon to fill undercut regions has been utilized in this research to build new structures including rotating and translating joints, bushings, and sliding elements.

The initial demonstration of the technology has employed polycrystalline silicon for the movable-joint members, but the process is not limited to using this material. The structural members might possibly be made from metals, alloys, and dielectric materials provided that these materials can be freed from their supporting substrate by selective etching of sacrificial (dissolvable) materials. Construction of these new elements gives rise to the need for research on mechanical parameters and properties for design. The process to produce these elements points up the need for further studies of sacrificial-layer etching, LPCVD growth, and remnant stresses in microfabricated systems. At the same time the realization of these structures brings new focus on the brightening prospects for producing microminiature prime movers [11].

The movable micromechanical structures can be batch-fabricated into multi-element preassembled mechanisms on a single substrate, or, if desired, they can be freed entirely from their host substrate to be assembled as separate elements. The potential uses for this new technology include the production of miniature ratchets, micro-positioning elements, mechanical logic, tuning elements, optical shutters, micro-valves, micro-pumps, and other mechanisms that have numberless applications in the macroscopic world. The method promises unheralded precision in the construction of miniature mechanical parts and systems with routine control at micrometer dimensions. Their manufacture in the world of micromechanics opens important avenues for further research and development.

ACKNOWLEDGMENT

We thank Prof. G. Johnson for valuable discussion on the characterization of polycrystalline films, and K. Voros, R. Hamilton, and the staff of the Berkeley Microfabrication Laboratory for help in experiments and fabrication.

REFERENCES

[1] L. S. Fan, Y. C. Tai, and R. S. Muller, "Pin-joints, springs, cranks, gears, and other novel micromechanical structures," in *Tech. Dig. 4th Int. Conf. Solid-State Sensors and Actuators* (Tokyo, June 1987), pp. 849–852 (U.S. patent pending).

[2] K. J. Gabriel, W. S. N. Trimmer, and M. Mehregany, "Micro gear and turbines etched from silicon," in *Tech. Dig. 4th Int. Conf. Solid-State Sensors and Actuators* (Tokyo, June 1987), pp. 853–856.

[3] R. T. Howe and R. S. Muller, "Stress in polycrystalline and amorphous silicon thin films," *J. Appl. Phys.*, vol. 54, pp. 4674–4675, Aug. 1983.

[4] H. Guckel, T. Randazzo, and D. W. Burns, "A simple technique for determination of mechanical strain in thin films with applications to polysilicon," *J. Appl. Phys.*, vol. 57, pp. 1671–1675, Mar. 1, 1985.

[5] S. Timoshenko, *Strength of Materials*, 3rd ed.   Princeton, NJ: Van Nostrand, 1955.

[6] R. P. Kroon and C. C. Davenport, "Spiral springs with small number of turns," *J. Franklin Inst.*, vol. 225, p. 171, 1938.

[7] G. L. Pearson, W. T. Read, and W. L. Feldmann, "Deformation and fracture of small silicon crystals," *Acta Metallurgica*, vol. 5, pp. 181–191, Apr. 1957.

[8] G. C. Johnson, "Acoustoelastic response of polycrystalline aggregates exhibiting transverse isotropy," *J. Nondestructive Evaluation*, vol. 3, pp. 1–8, 1982.

[9] H. J. McSkimin, W. L. Bond, E. Buehler, and G. K. Teal, "Measurement of the elastic constants of silicon single crystals and their thermal coefficients," *Phys. Rev.* vol. 83, p. 1080, 1951.

[10] H. Guckel, D. W. Burns, C. R. Rutigliano, D. K. Showers, and J. Uglow, "Fine grained polysilicon and its application to planar pressure transducers," in *Tech. Dig. 4th Int. Conf. Solid-State Sensors and Actuators* (Tokyo, June 1987), pp. 277–282.

[11] R. P. Feynman, "There's plenty of room at the bottom," in *Miniaturization*, H. D. Gilbert, Ed.   New York: Reinhold, 1961, pp. 282–296.

LIGA PROCESS: SENSOR CONSTRUCTION TECHNIQUES VIA X-RAY LITHOGRAPHY

W. Ehrfeld [1], F. Götz [2], D. Münchmeyer [1], W. Schelb [1], D. Schmidt [1]

[1]   Kernforschungszentrum Karlsruhe GmbH,
Institut für Kernverfahrenstechnik, P.O. Box 3640,
D-7500 Karlsruhe 1, Federal Republic of Germany

[2]   Steag AG, P.O. Box 10 37 62, D-4300 Essen 1,
Federal Republic of Germany

Abstract

A large variety of microsensors and components for microactuators can be fabricated by the so-called LIGA method which is based on deep-etch X-ray lithography, electroforming and molding processes (in German: Lithographie, Galvanoformung, Abformung). This microfabrication technique allows to generate devices with minimal lateral dimensions in the micrometer range and structural heights of several hundred micrometers from metallic and plastic materials. In contrast to orientation dependent etching of monocrystalline silicon there are no restrictions in the cross-sectional shape of the microstructures. Various concepts of sensors for measuring vibration, acceleration, position, spectral distribution, radiation, composition of mixtures etc are presented.

Introduction

Technological advances in lithography continue to drive the level of integration and the reduction in minimum feature size of semiconductor products. In the laboratory scale, optical lithography which is by far the most important technique in present production lines has reached the 0.5 μm range required for the fabrication of 16 Mbit memory chips. For mass production of 64 Mbit and even more advanced devices, X-ray lithography is expected to become the preferred technique [1]. It is capable to generate minimum dimensions in the 0.2 μm range.

Besides semiconductor devices, a large number of other microstructure products may be fabricated by means of X-ray lithography. The so-called LIGA process which is based on deep-etch X-ray lithography with high quantum energy synchrotron radiation, electroforming and molding processes (in German: Lithographie, Galvanoformung, Abformung) allows to produce microstructures with lateral dimensions in the micrometer range and structural heights of several hundred micrometers from a variety of materials [2, 3]. Its application potential covers microelectronics, sensors, microoptics, micromechanics and biotechnology [4].

This paper gives a short overview of the LIGA process and its application potential for fabricating microsensors.

Microfabrication Process

The principle of the LIGA method is evident from the process sequence shown in Figure 1. A polymeric material (resist) which changes its dissolution rate in a liquid solvent (developer) under high-energy irradiation, is exposed through an X-ray mask to highly intensive parallel X-rays. The radiation source is an electron synchrotron or an electron storage ring which, at present, is the only possibility to generate the highly collimated photon flux in the spectral range required for precise deep-etch X-ray lithography in thick resist layers. If one considers a thickness between 10 and 1000 μm for the resist layers to be structured, the optimum critical wavelength of the synchrotron radiation source ranges from some 0.1 to 1 nm for typical resist materials.

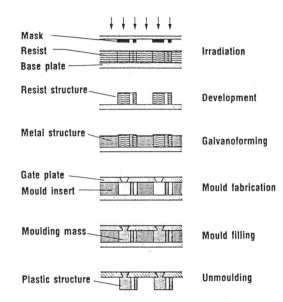

Fig. 1: Schematic representation of a process sequence of the LIGA method for mass fabrication of micro-devices.

In the next step, the resist structure is used as a template in an electroforming process where metal is deposited onto the electrically conductive substrate. In this way, a complementary metallic structure is obtained which can be either the final microstructure product or can be used as a microtool (mold insert) for multiple reproduction by means of a molding process.

The micromolding process has been optimized using methacrylate based casting resins with a special internal mold release agent. In the process sequence shown in Fig. 1 the mold material is introduced into the mold cavities through the holes of a gate plate. This plate which has a formlocking connection with the polymeric microstructures after hardening of the resin serves as an electrode in a second electroforming process for generating secondary metallic microstructures. It has been demonstrated by many experiments that, in spite of an aspect ratio of about 100 and minimum lateral dimensions of only some micrometers, a yield of approximately 100 % can be obtained in the micromolding process.

The secondary structures are perfect copies of the primary structures. Consequently, mass production of plastic and metallic microstructures should be feasible without a continuous utilization of a synchrotron radiation source which is only necessary for fabricating mold inserts.

Sensors

The development of modern sensors aims at miniaturization, more complex or array structures and integration with electronic signal conditioning. Furthermore, the expenditure for adjustment, trimming and replacement should be minimized and, accordingly, the manufacturing tolerances should be as small as possible. In this respect,

Reprinted from *Rec. of the IEEE Solid-State Sensor and Actuator Workshop*, 1988, pp. 1–4.

the LIGA method has a number of advantages which will be illustrated by a description of several sensor devices in the following.

Measurement of vibration and acceleration:

A LIGA configuration for measuring vibration or acceleration is shown schematically in Figure 2. By means of deep-etch X-ray lithography a template is generated which allows to manufacture a spring plate and a rigid stationary electrode directly upon a microelectronic circuit or, correspondingly, a multitude of such arrangements upon a completely processed silicon wafer. The change in capacity and the corresponding voltage change, respectively, may change directly the electric potential of the gate electrode of a MOSFET circuit

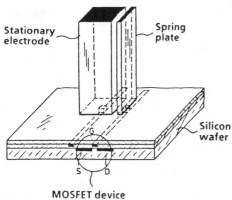

Fig. 2: Schematic representation of a LIGA sensor for measuring vibration or acceleration.

Compared to orientation dependent etching of monocrystalline silicon which is a well-known and proven technique for fabricating similar integrated sensors for vibration measurement, the LIGA technique should have several advantages. In particular, the fabrication of the electronic circuits can be fully completed prior to the fabrication of the sensor structure. Therefore, no mixing between micromechanical and microelectronic process steps occurs and a separate optimization is possible for the different fabrication procedures. Since the sensor element is arranged vertically upon the electronic circuit, the space requirement is minimized. This should be a further advantage compared to silicon micromechanics where the spring plates are usually fabricated in a plane parallel to the surface of the wafer. Nevertheless, the LIGA method allows also to produce cantilever structures which are arranged parallel to the wafer surface. Moreover, a two-axis sensor arrangement can be simply realized.

Measurement of position, displacement and magnetic fields:

For measuring position, displacement, small distances or changes of a magnetic field, sensor devices using inductive circuits are used in a large number of configurations. The LIGA method allows to generate small coils or complex coil arrays from materials with high electrical conductivity as well as microstructures with high aspect ratio from ferroelectric materials. Because of the large structural height attainable by the LIGA process, a low electrical resistance can be realized for small coils, which is an obvious advantage compared to corresponding configurations generated by thin film techniques whereas, compared to thick film techniques, much smaller dimensions and tolerances can be obtained.

Figure 3 shows a scanning electron micrograph of a helical structure generated by X-ray lithography. It is used for fabricating two intermeshing coils by electrodeposition which serve as the transmitting and receiving unit, respectively.

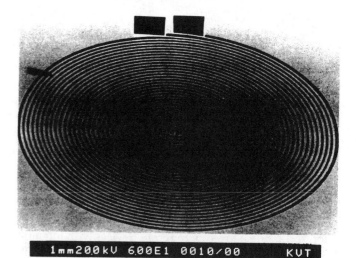

1mm 200kV 600E1 0010/00   KVT

Fig. 3: Scanning electron micrograph of a resist template for fabricating two intermeshing microcoils.

Microoptical devices and spectrometry:

In the field of microoptics, the LIGA process could be used for fabricating waveguides, gratings, small prisms and small cylindrical lenses, zone plates, polarizers and spectral filters for the infrared range, spatial filters, modulators and many other optical devices and structures. More or less arbitrarily, one may discriminate between two basic configurations. One is represented by a perforated, self-supporting membrane of plate structure which is put into the path of rays and where the optical effect is determined by the special pattern of perforation. The other configuration may be characterized by the fact that optical structures are generated on a stable substrate where the path of rays is parallel to the surface of the substrate and partially determined by a waveguide structure. In the following, two special examples of these basic configurations will be described.

Figure 4 illustrates the principle of a small simultaneous spectrometer. It has a Rowland configuration with a curved reflective grating and comprises a slab waveguide consisting of three layers of transparent X-ray resist with matched refractive indices. In spite of the relatively low resolution of this microspectrometer, it might be a cheap and reliable sensor for photometric analysis of mixtures, chromatic testing, optical film thickness measurement, etc. In the field of communication technology, this device can be applied as a low-cost wavelength division multiplexer/demultiplexer.

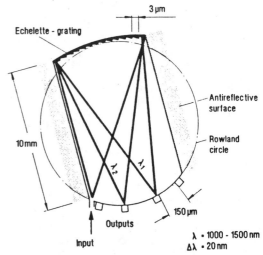

Fig. 4: Schematic representation of a simultaneous microspectrometer.

142

In the field of infrared spectroscopy, bandpass or cut-off filters can be realized as a so-called resonant mesh, i.e. by a metallic membrane with a multitude of cross-shaped, Y-shaped or circular openings whose dimensions and distances correspond to the wavelength to be transmitted [5, 6]. Such self-supporting membranes with the desired configuration of the openings can be fabricated by X-ray lithography and electroforming as demonstrated by Figure 5.

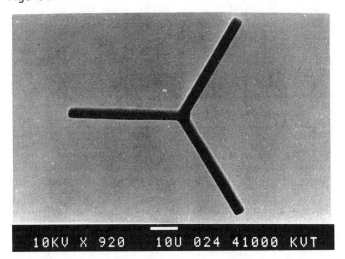

Fig. 5: Scanning electron micrograph of a Y-shaped opening with a slit width of 5 μm in a nickel membrane with a thickness of 300 μm.

Further development work deals with the fabrication of highly precise supporting and positioning elements for fiber optical sensors and with lithographic generation of waveguide elements from transparent X-ray resists.

Fluidic devices:

The first development work on the LIGA process aimed at the fabrication of micron-sized slit-shaped nozzles for uranium isotope separation as shown in Figure 6. Consequently, other flow devices and, in particular, fluidic sensors for measuring position, gas density, composition of mixtures etc. can be produced. Because of the small dimensions, a favorably low response time should be obtainable and such devices would be applicable even under conditions where microelectronic devices are difficult to operate, e. g. at a high radiation level. Further flow devices might be columns for chromatography, micro-cooling systems based on the Joule-Thompson effect, anemometric sensors, sieves for determining the size distribution of particles, micro-valves etc.

Measurement of radiation:

Concepts have been worked out for fabricating microchannel plates and arrays of secondary electron multipliers. In contrast to the standard processes for producing channel plates, the tolerances of the dimensions can be reduced and the positions can be exactly determined. As a result, single channels or groups of channels can be matched directly to other discrete microstructures at the input or output of a micro-channel plate. In the case of miniaturized electron multiplier arrays, the dynodes can be arranged in a ring-shaped or any other curved configuration required in an experiment.

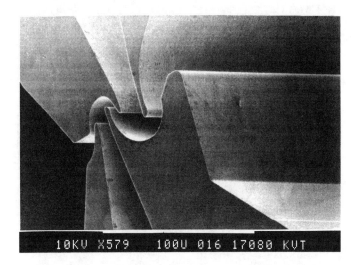

Fig. 6: Scanning electron micrograph of a double deflecting separation nozzle electroformed from nickel. The minimum slit width of the curved nozzle is 3 μm, the slit length is about 300 μm.

Electrical and Optical Microconnectors

The continuous increase of logic functions in microprocessor chips and the decrease in critical dimensions results in a corresponding demand for high density electrical connections with subminiaturized dimensions. A similar trend will probably be observable in the near future in optical communication technology, where the interconnection of optical waveguides, e. g. in the case of multiple monomode fiber connectors, requires micromechanical fabrication methods with submicron tolerances. Analogous requirements exist in modern sensor technology when sensor arrays have to be connected to VLSI chips.

Figure 7 shows an enlarged view of a part of a multi-pin plug with 100 poles per cm. The device is equipped with stable integrated guide pins which are fabricated in the same process sequence and ensure a simple insertion of the plugs.

Actuators

Since the LIGA method obviously allows to fabricate precise micromechanical components from various materials, its potential range of application also includes the fabrication of microacturators. Consequently, the current development work aims at the production of simple mechanical parts like gear wheels, bearings, joint couplings etc. and tests of various materials provided for fabrication of such devices. Moreover, feasibility studies are beginning which deal with the utilization of the LIGA process for fabricating micromotors and complete microactuator and positioning systems.

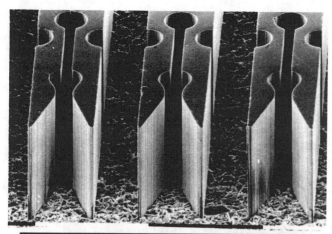

0.1mm20.0kV 4.24E2 0004/00   KVT

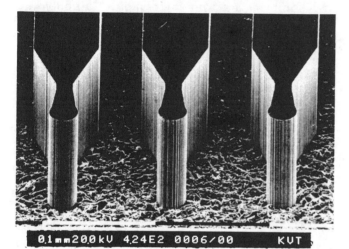

0.1mm20.0kV 4.24E2 0006/00   KVT

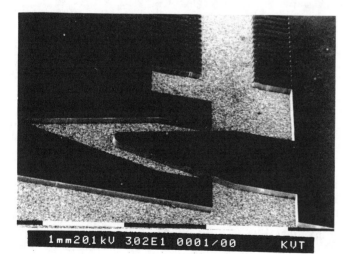

1mm20.1kV 3.02E1 0001/00   KVT

Fig. 7: Scanning electron micrographs of a multi-pin plug with 100 poles per cm.

Conclusions

The development work on the LIGA process is carried out by the Karlsruhe Nuclear Research Center in cooperation with the German industrial companies Steag AG, Essen, and Degussa AG, Frankfurt. The authors would like to emphasize that the feasibility of the process has been proven in the laboratory-scale only and large efforts are still necessary for commercialization. However, it is to be expected that this new microfabrication method will be superior to other processes if, in mass production of microstructures, specific requirements are imposed on the spatial resolution, the aspect ratio, the structural height, the parallelism of the structure walls and, in particular, if an unrestricted design of the cross-sectional shape and an optimum selection of the material required for the various microdevices has to be ensured.

These arguments should be important, above all, in the field of sensor technology where a precise mass fabrication of micromechanical, microoptical and microelectronic elements should result in favourable cost and where a trend exists in producing more and more smart and complex sensor elements. Moreover, the LIGA process should be also applicable in the field of microactuators. Thus, in connection with other microfabrication methods and semiconductor technology, the LIGA process might contribute to the future production of complete autonomous microsystems.

References

[1]   A.D. Wilson, X-Ray Lithography: Can it be Justified?, Solid State Technology, Vol. 29, pp. 249-255, May 1986.

[2]   E.W. Becker, W. Ehrfeld, P. Hagmann, A. Maner, D. Münchmeyer: Fabrication of microstructures with high aspect ratios and great structural heights by synchrotron radiation lithography, galvanoforming, and plastic moulding (LIGA process), Microelectronic Engineering, Vol. 4, pp. 35-56, 1986.

[3]   W. Ehrfeld and E.W. Becker, Das LIGA-Verfahren zur Herstellung von Mikrostrukturkörpern mit großem Aspektverhältnis und großer Strukturhöhe, KfK-Nachrichten, Vol. 19, No. 4, pp. 167-179, 1987.

[4]   W. Ehrfeld, P. Bley, F. Götz, P. Hagmann, A. Maner, J. Mohr, H.O. Moser, D. Münchmeyer, W. Schelb, D. Schmidt, E.W. Becker, Fabrication of Microstructures Using the LIGA Process, Micro Robots and Teleoperators Workshop, Hyannis, Massachusetts, Nov. 9-11, 1987, Proc. IEEE Catalog Number 87 TH 02404-8.

[5]   F. Keilmann, Infrared High-Pass Filter with High Contrast, International Journal of Infrared and Millimeter Waves, Vol. 2, No. 2, pp. 259-271, 1981.

[6]   H.-P. Gemünd, Filter für den submm-Bereich, Kleinheubacher Berichte, Vol. 29, pp. 501-505, 1986.

INTEGRATION OF MULTI-MICROELECTRODE AND INTERFACE CIRCUITS BY SILICON PLANAR AND THREE-DIMENSIONAL FABRICATION TECHNOLOGY

KOURO TAKAHASHI and TADAYUKI MATSUO

Department of Electronic Engineering, Tohoku University, Sendai (Japan)

(Received March 8, 1983; in revised form June 1, 1983; accepted July 29, 1983)

Abstract

A multi-microelectrode for simultaneous recording of single-unit action potentials is a useful device for studying the organization and function of neural systems.

We have fabricated a silicon probe which integrates a multi-microelectrode and interface circuits (preamplifier and analog switches) on a silicon chip by silicon planar and three-dimensional fabrication technology. This electrode has the following advantages: (1) it is easy to arrange the location of recording sites simply by changing photomasks; (2) crosstalk between the multi-electrodes can be reduced by the shielding effect of the silicon substrate; (3) integration of microelectrodes and preamplifiers on the same silicon chip eliminates the undesirable effects of stray lead capacitance; (4) analog switches serve as selectors or multiplexers for a parallel to series conversion of multichannel signals of neural activities.

In this paper, the design and fabrication processes of multi-microelectrode and interface circuits are described. In particular, the characteristics of polysilicon electrodes, design of a low-noise MOSFET for the preamplifier and silicon three-dimensional processes for the electrode probe are considered.

1. Introduction

The use of microelectrodes for recording action potentials generated in the central nervous system is an important technique for studying living systems at the cellular level. However, in the fabrication of conventional microelectrodes it is very difficult to get precision and reproducibility of the tip size, the probe shape, the thickness of insulator and the electrode impedance level. A new type of microelectrode has been developed using IC technology [1 - 3]. This type of microelectrode has a number of advantages: (1) many electrodes, with good uniformity of the probe shape and the impedance level, are made at once on a silicon substrate by batch fabrication; (2) the area, shape and layout of the recording sites can be determined by the photomask; (3) a multi-electrode structure can be fabricated using the same process as for a single electrode; (4) noise pickup due to stray capacitance and crosstalk between electrodes on a silicon substrate are eliminated because of the shielding effect of the substrate.

At the same time, the undesirable effects of stray capacitances in the connections between the electrode bonding pads and the preamplifiers are eliminated. One way of avoiding this effect is to locate buffer amplifiers close to the recording sites. Wise *et al.* reported a low capacitance multielectrode using the hybrid approach [4]. It is preferable to integrate both the electrode and the preamplifier on the same chip.

We have tried to achieve this by a completely monolithic approach. The integrated multi-microelectrode array reported here consists of gold/polysilicon multi-electrode, silicon-gate MOSFET preamplifiers and analog switches.

2. Design of integrated multi-electrodes

Figure 1 shows the top view and cross-section of the integrated multielectrode. The silicon substrate is shaped like a probe in order that its tip can be smoothly inserted into nerve tissue. The electrode site for recording action potentials is located on the tip of the probe. In this structure, the silicon substrate serves not only as a probe for the electrode but also as a substrate for the integrated circuits. For a multi-electrode chip, the inter-electrode coupling due to stray capacitances is greatly reduced, since the recording electrodes are formed on the grounded silicon substrate.

In order to fabricate this integrated multi-electrode, it is important to consider carefully the selection of the electrode materials and the process sequences.

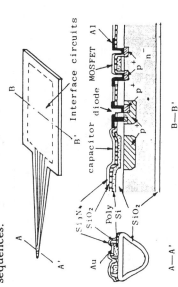

Fig. 1. Top view and cross-section (A–A' and B–B') of the fabricated integrated microelectrode.

2.1. Electrode probe shaping

The electrode carrier must be so sharp that it can be inserted into tissue with little damage. However, at the same time, it must have sufficient mechanical strength to endure the stress of insertion. In addition, the top side of the probe must be flat in order to define the recording sites and lead wires of the electrode by photolithography.

We use a chemical etching technique to shape the probe. Anisotropic and isotropic etching are available for the formation of an electrode substrate. For the top side etching of the silicon wafer, the anisotropic etchant APW (ethylenediamine, pyrocatechole and water) is used [5, 6]. Since this selectively etches (100)-oriented silicon except the area masked with the SiO_2 layer, side etching is negligible and the etched shape is defined exactly by the mask pattern. HF-HNO_3 isotropic etchant [7] (HF:HNO_3:Hac = 12: 1:2 by volume) is used for the back-side etching of the silicon wafer. This isotropic etching makes the electrode probe thin and gradually taper to its tip.

After shaping the probe, its surface is covered with an SiO_2 film except at the recording sites. Since the SiO_2 film isolates the probe from electrolyte, an artefact induced by photocarrier generation at the interface of the bare silicon probe and electrolyte can be avoided.

2.2. Recording electrode materials

The characteristics of recording electrode materials required for the integrated multi-microelectrode are: (1) ability to withstand high-temperature (over 900 °C) processes; (2) ability to define the pattern of electrodes by photolithography; (3) chemically and electrically stable in solution. Hence we have considered polysilicon and gold as suitable recording electrode materials.

Polysilicon is deposited in a layer 8000 Å thick on the electrode probe by CVD (chemical vapour deposition) of SiH_4 at 600 °C in N_2 carrier gas and is doped with boron in order to reduce its sheet resistance. After patterning of the recording electrodes by photolithography, the SiO_2 and Si_3N_4 films are coated by CVD to thicknesses of 2000 Å and 1000 Å respectively for electrical isolation. The Si_3N_4 film exhibits excellent electrical insulation and chemical stability in electrolyte.

However, polysilicon is not suitable as a recording site material because it is not stable in the electrolyte, which leads to an increase of electrode impedance due to the growth of natural silicon dioxide on the surface, and because of generation of a noise voltage by light. This light artefact arises because the potential barrier at the recording site–electrolyte interface decreases due to an increasing photo-induced carrier concentration in the polysilicon space-charge region. Therefore, it is necessary to coat the polysilicon site with some metal. Gold is suitable for this purpose, because of its chemical stability and its easy patterning. The gold coating procedure will be explained in Section 3.

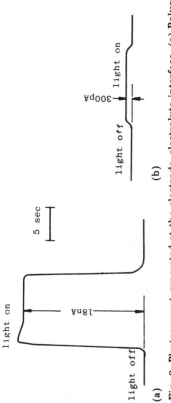

Fig. 2. Photocurrent generated at the electrode–electrolyte interface. (a) Polycrystalline silicon, (b) gold-deposited recording site.

Figure 2 shows the difference of carrier generation by light between the polysilicon and the gold-deposited recording site. The measurement is carried out under constant illumination conditions with a tungsten lamp. The gold-deposited site can reduce photo noise by about one sixtieth.

Figure 3 shows a photomicrograph of the polysilicon multi-microelectrode with gold deposited sites.

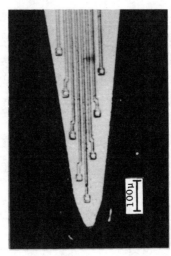

Fig. 3. Photomicrograph of the recording sites of the electrode probe.

2.3. Interface circuits

The amplitudes of action potentials measured by extracellular recording are from ten to several hundred microvolts and the impedance of the microelectrode is high. Therefore, the preamplifier recording these signals has to have high input impedance (more than 10MΩ) and a low noise characteristic. We use a MOSFET for this purpose, because it has very high input impedance and is suitable for integrated circuits. However, conventional MOSFETs exhibit relatively large $1/f$ noise in the low-frequency region compared with other active devices. It is therefore necessary to design and fabricate a low-noise MOSFET; we have carried out some preliminary experiments on MOSFET noise.

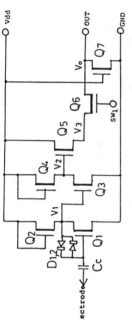

Fig. 6. Preamplifier circuit and analog switch of the integrated microelectrode.

size with $L = 10$ μm and $W = 1000$ μm was chosen. The gate bias voltage of Q_1 is set automatically at the operating point by the coupling capacitor and the diodes, and is independent of the polarization potential at the electrode site–electrolyte interface. As the voltage across the junction of the diode is nearly zero, the incremental resistance of the diode is so large (more than 100 MΩ) that the degradation of the input impedance is negligible.

The coupling capacitor C_c and the incremental resistance R_d of the diodes set the lower frequency limit f_l of the preamplifier, while the electrode impedance Z_e, the series capacitance of C_c and the input capacitance C_i of the preamplifier set the high frequency cut-off f_h. With $C_c = 30$ pF, $C_i = 15$ pF, $R_d = 100$ MΩ and $Z_e = 1$ MΩ at 1 kHz, we obtain $f_l = 53$ Hz and $f_h = 13$ kHz. This bandwidth agrees with that needed for extracellular recording (100 Hz to 10 kHz).

The analog MOSFET switch Q_6 connected to the output is for selecting or multiplexing signals recorded by the multi-electrode. The gate control signal for multiplexing is applied to its gate terminal by the other timing circuit. Q_6 is designed from the switching frequency, which is more than 160 kHz in order to multiplex eight channel signals of single unit-activities. The switching time of Q_6 is about 0.3 μs with a capacitance load of 100 pF.

3. Fabrication of an integrated multi-microelectrode

The three-dimensional process is necessary for the fabrication of an integrated multi-microelectrode in addition to silicon planar technology, and these two technologies have to be compatible with each other. The main fabrication steps for an integrated multi-microelectrode are: (1) interface circuits, (2) electrode probe, (3) packaging. The details of each of these procedures are described below.

3.1. Interface circuits

The interface circuits are constructed by self-aligning p-channel silicon gate MOSFETs. The fabrication steps before metallization are shown in Fig. 7:

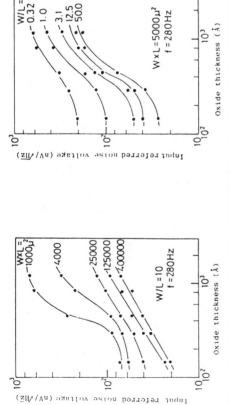

Fig. 4. Input referred noise voltage of MOSFET as a function of gate oxide thickness and gate area.

Fig. 5. Input referred noise voltage of MOSFET as a function of gate oxide thickness and W/L.

In general, the MOSFET noise voltage depends not only on surface states in the Si–SiO$_2$ interface but also on the gate geometry [8 - 10]. The surface states in the Si–SiO$_2$ interface depend on the gate insulator fabrication process (thermal oxidation, which is a well-established technique). Hence, we examined the relation between the noise voltage and the gate geometry of the MOSFET. We prepared many samples of polysilicon gate MOSFETs with different gate dimensions and measured their equivalent noise voltages referred to input as a function of gate insulator thickness and gate area. The gate insulator of all the samples tested is an SiO$_2$ film thermally oxidized at 1100 °C. Gate length (L) varied from 10 to 200 μm, gate width (W) from 500 to 2000 μm and the SiO$_2$ film thickness is in the range 140 to 1200 Å. Equivalent noise voltages are measured at 280 Hz under conditions of constant drain current density per gate width.

The experimental results are shown in Figs. 4 and 5. From these results, it is clear that a large gate area, a high ratio of W to L and a reduction of gate insulator thickness are recommended for low noise MOSFETs, provided there is a constant density of surface states. On the other hand, the required noise level of a preamplifier for extracellular recording has to be less than the thermal noise of a metal microelectrode, which is generally from 10 to 20 μV$_{pp}$ in the range from d.c. to 10 kHz bandwidth. Using these experimental results, we can design the gate geometry of the MOSFET as follows: insulator thickness < 300 Å, $W/L > 10$ and gate area > 4000 μm^2.

Figure 6 shows a preamplifier circuit with an analog switch for the integrated microelectrode, which is composed entirely of MOSFETs. Since noise in the preamplifier circuit is dominated by noise in the input device, the input-stage MOSFET Q_1 has to have low noise. From the above result, a gate

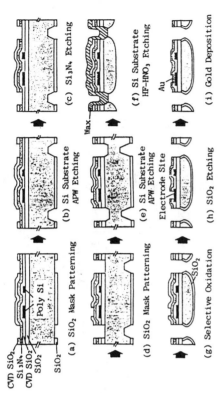

CVD SiO₂
Si₃N₄
CVD SiO₂
SiO₂
SiO₂

Poly Si

Electrode Site

Wax

Au

SiO₂

(a) SiO₂ Mask Patterning   (b) Si Substrate APW Etching   (c) Si₃N₄ Etching

(d) SiO₂ Mask Patterning   (e) Si Substrate APW Etching   (f) Si Substrate HF–HNO₃ Etching

(g) Selective Oxidation   (h) SiO₂ Etching   (i) Gold Deposition

Fig. 8. Processing steps for fabrication of an electrode probe and recording sites.

The fabrication steps of the electrode probe are shown in Fig. 8:

(a) opening into SiO_2 for silicon APW etching;

(b) silicon APW etching of the back side to a depth of 100 μm;

(c) Si_3N_4 etching;

(d) opening into SiO_2 for silicon etching of the top side and recording sites;

(e) silicon APW etching of both sides to a depth of 50 μm and 150 μm, respectively;

(f) silicon substrate $HF–HNO_3$ etching of the back side only (the top side is protected from etching by a wax mask);

(g) selective oxidation of the chemically etched surface of silicon;

(h) opening of recording sites (SiO_2 etching with Si_3N_4 mask after Si_3N_4 etching with SiO_2 mask);

(i) gold deposition on recording sites and aluminium deposition for the interconnections in the interface circuits.

In step (d), the SiO_2 patterning for opening the recording sites is prepared before silicon substrate etching. The recording sites can then be opened without photoresist mask in step (h). In step (f), the electrode probes are separated, but the top side of the wafer is protected from etching by a wax mask. If the silicon substrate etching is stopped when the silicon groove for separating the electrodes is etched away, the thickness of the electrode probe is equal to the depth of the silicon groove formed in step (e). However, since the SiO_2 mask on the back side is needle-shaped, the probes are shaped with a gradual reduction of volume toward the tip by side etching of the isotropic etchant. In selective oxidation (h), the bare surface of the etched silicon substrate is oxidized entirely, but the interface circuits and the electrode sites covered with Si_3N_4 film are not oxidized. In step (i), gold is deposited over the entire surface of the wafer. Then, by dipping the wafer in an ultrasonic cleaner, the gold deposited on the SiO_2 or Si_3N_4 films is removed, but that

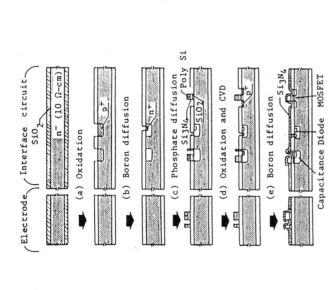

Electrode   Interface circuit

SiO₂

n⁻ (10 Ω·cm)

(a) Oxidation

p⁺

(b) Boron diffusion

n⁺

(c) Phosphate diffusion

Si₃N₄   Poly Si
SiO₂

(d) Oxidation and CVD

p⁺
Si₃N₄

(e) Boron diffusion

Capacitance Diode   MOSFET

(f) Si₃N₄ CVD

Fig. 7. Processing steps for fabrication of interface circuit and microelectrode before the definition of an electrode probe and recording sites (cross-sections correspond to A–A' and B–B' in Fig. 1).

(a) oxidation of n^- silicon wafer (10 $\Omega\cdot cm$, ⟨100⟩, 300 μm thick);

(b) boron diffusion for diode;

(c) phosphate diffusion for diode;

(d) oxidation for MOSFET gate (300 Å), Si_3N_4 CVD for gate passivation (300 Å) and polysilicon CVD for MOSFET gate electrode (8000 Å);

(e) boron diffusion for doping into source, drain and polysilicon;

(f) SiO_2, Si_3N_4 CVD for passivation and masking of selective oxidation.

As shown in step (d), MOSFET gate electrodes and recording electrode sites are defined at the same time by polysilicon CVD. In step (f), the integrated electrode is passivated entirely by the Si_3N_4 CVD film. Selective oxidation is explained in the next section.

3.2. Electrode probe and recording site definition

When the electrode probe is defined by chemical etching, the silicon wafer has three-dimensional structure. Once the silicon wafer is processed like this, it is difficult to coat uniformly with photoresist on the silicon probe surface and to define patterns by photolithography. Because of this restriction, the electrode probe has to be defined after the interface circuit processes.

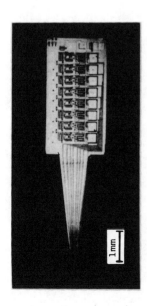

Fig. 9. Photograph of the integrated multi-microelectrode with eight microelectrodes and interface circuits.

on the polysilicon recording sites is not. This technique is based on the characteristics of gold; it adheres weakly to SiO_2 or Si_3N_4 but strongly to silicon, because it reacts with silicon and its interface changes to a eutectic as a result of annealing or high vacuum deposition.

Figure 9 shows an integrated multi-microelectrode fabricated with eight recording sites and eight channel interface circuits. The chip size is 2×8 mm² and the tip dimension of the probe is about 30 μm thick and 100 μm wide. The recording site area is 10×10 μm².

3.3. Packaging

The integrated microelectrode is mounted on a ceramic substrate and the connecting wires are attached on bonding pads by ultrasonic wire bonding. It is encapsulated in black wax, except at the tip of the probe, in order to preserve the interface circuits from light artefacts or mechanical and chemical damage.

4. Experimental results

The electrode impedance of the microelectrode is about 20 MΩ at 1 kHz, which compares well with the value for metal microelectrodes for extracellular recording of single-unit activities.

The preamplifier has about 26 dB voltage gain with a supply voltage of −9V and its bandwidth is from 100 Hz to 100 kHz. As previously mentioned, the high frequency cut-off is set by the electrode impedance and the series capacitance of C_c and C_i. However, the electrode impedance in the electrolyte is usually a function of frequency and is almost inversely proportional to it. Hence, the over-all frequency characteristics of the preamplifier are very little affected by the microelectrode impedance.

The input impedance of the preamplifier (53 MΩ) is calculated from the low-frequency cut-off (100 Hz) and the coupling capacitance C_c (30 pF). This value is reasonable for the microelectrode preamplifier.

On the other hand, the noise voltage referred to input is 40 to 50 μV_{pp}, which is larger than the designed level (less than 20 μV_{pp}). A possible cause is

an increase of surface states in the Si–SiO₂ interface of the MOSFET caused by the electrode probe fabrication process. Hydrogen annealing is effective in reducing surface states and improves the MOSFET noise performance. In this case, it would be also necessary to anneal the integrated electrodes in hydrogen gas in the final step of the electrode probe fabrication process.

The interelectrode crosstalk on the electrode chip itself is measured by dipping the recording sites into physiological saline and applying an a.c. signal through a reference electrode. From this measurement the crosstalk observed at the output of the preamplifier is less than −40 dB at 1 kHz between channels. This is negligible for identifying single-unit activity.

The insertion of the integrated electrode into the cortex of a cat is easy because the probe has a sharp point defined by the silicon chemical etching.

5. Conclusion

This paper has described the design, fabrication and characteristics of a new integrated multi-microelectrode array with MOSFET interface circuits. Multi-microelectrodes have been realized by extended polysilicon gates and their recording sites have been covered with a thin gold film. In comparison with a conventional metal microelectrode, this new microprobe structure possesses less parasitic capacitance from electrode to ground and less inter-electrode coupling. The frequency bandwidth and noise characteristics are comparable to those of a conventional microelectrode.

The present study has proved the feasibility of a new approach to extracellular microelectrodes.

Acknowledgements

This work was supported in part by the Japanese Ministry of Education, Science and Culture, under a Grant-in Aid for Scientific Research 57460127, 1982.

References

1 K. D. Wise, J. B. Angell and A. Starr, An integrated-circuit approach to extracellular microelectrodes, *IEEE Trans. Bio-Medical Engineering, BME-17* (3) (1970) 238 - 246.
2 P. Bergveld, Development, operation, and application of the ion-sensitive field-effect transistor as a tool for electrophysiology, *IEEE Trans. Bio-Medical Engineering, BME-19* (1972) 342 - 351.
3 A. Starr, K. D. Wise and J. Csongradi, An evaluation of photoengraved microelectrodes for extracellular single-unit recording, *IEEE Trans. Bio-Medical Engineering, BME-20* (1973) 291 - 293.
4 K. D. Wise and J. B. Angell, A low-capacitance multielectrode probe for use in extracellular neurophysiology, *IEEE Trans. Bio-Medical Engineering, BME-22* (3) (1975) 212 - 219.

5 E. Bassous and E. F. Baran, The fabrication of high precision nozzles by the anisotropic etching of (100) silicon, *J. Electrochem. Soc.*, *125* (8) (August 1978) 1321 - 1327.
6 A. Reisman, M. Berkenblit, S. A. Chan, F. B. Kaufman and D. C. Green, The controlled etching of Si in catalyzed ethylenediamine-pyrocatechol-water solutions, *J. Electrochem. Soc.*, *126* (8) (August 1978) 1406 - 1415.
7 Y. Ohta, M. Esashi and T. Matsuo, Multielectrode fabrication for simultaneous recording of nerve impulses using IC techniques, *J.J.M.E.*, *19* (2) (April 1981) 106 - 113.
8 S. Christesson, I. Lundström and C. Svensson, Low frequency noise in MOS transistors − 1, *Solid-State Electronics*, *11* (1968) 797 - 812.
9 R. S. Ronen, Low-frequency 1/f noise in MOSFETs, *RCA Rev.*, *34* (June 1973) 280 - 307.
10 H. Katto, Y. Kamigaki and Y. Itoh, MOSFETs with reduced low frequency 1/f noise, *Proc. 6th Conf. on Solid State Devices, Tokyo, 1974*, pp. 243 - 248.

A High-Yield IC-Compatible Multichannel Recording Array

KHALIL NAJAFI, KENSALL D. WISE, SENIOR MEMBER, AND TOHRU MOCHIZUKI

Abstract–This paper reports the development of a multielectrode recording array for use in studies of information processing in the central nervous system and in the closed-loop control of neural prostheses. The probe utilizes a silicon supporting carrier which is defined using a deep boron diffusion and an anisotropic etch stop. This substrate supports an array of polysilicon or tantalum thin-film conductors insulated above and below with silicon nitride and silicon dioxide. Typical probe dimensions include a length of 3 mm, shank width of 50 μm, and a thickness of 15 μm. These structures are capable of simultaneous high-amplitude multichannel recording of neural activity in the cortex. The probe fabrication process requires only four masks and is single-sided using wafers of normal thickness, resulting in yields which exceed 80 percent. The process is also compatible with the inclusion of on-chip MOS circuitry for signal amplification and multiplexing. A complete ten-channel signal processor which requires only three external probe leads is being developed.

I. INTRODUCTION

THE USE of microelectrodes [1]-[3] to record the activity of single neurons and groups of neurons has been the principal technique for studying the nervous system for many years. Using these electrodes, much has been learned about the operation of single neurons and about the functional organization of some neural structures–principally in the sensory and motor areas. Relatively little, however, has been learned about how neurons join together in circuits to process information. Part of the reason for this slow progress is certainly the tremendous complexity of many neural systems; however, an important factor has also been the rather severe limitations imposed by conventional microelectrodes, which exhibit great variability in their electrical impedance levels and physical shapes and permit recording from only a single point in tissue at a time. These limitations typically preclude any chance of testing hypotheses regarding neural interactions as circuits. Thus improved electrode structures are critically needed as a tool for studying information processing in the nervous system. Additionally, such electrodes are also needed for eventual application in a number of neural prostheses now under development [4]. These include auditory prostheses and a variety of devices which employ electrical stimulation for the activation of paralyzed limbs. While these devices are presently open loop, it is generally recognized that to achieve reasonable levels of performance they will have to condition their drive levels on physiological responses from the body. Hence, closed-loop control via the use of implantable recording electrode arrays will be necessary.

An electrode array capable of recording simultaneously from many points 20–100 μm apart in depth would provide an important advance in instrumentation for neurophysiology and for future neural prostheses, particularly if the array were chronically implantable. Such an array must meet a number of challenging requirements, however. It should possess closely controlled physical dimensions and electrical characteristics, and since the exposed recording sites are typically only a few micrometers on a side, probe geometries must be controlled to within a micrometer or better. The array should also be small enough to approach cells closely with a minimum of damage to the tissue. This requires an appropriate probe shape with overall dimensions measured in tens of micrometers or less. The probe materials must be biocompatible, physically strong, and compatible with a reproducible high-yield fabrication process. Finally, when chronically implanted, the array should free float in the cortex; thus the number of output leads must be minimized both to allow practical fabrication and to minimize their tethering effects.

Over the past fifteen years, several attempts to develop multielectrode probe structures based on solid-state process technology have been reported [5]-[8]. The resulting probes have been shown capable of recording the activity of single neurons and of neural populations, but have never been available in sufficient quantities to significantly influence neurophysiology. The limited availability of such structures has been due primarily to their difficult fabrication sequences, which have typically required double-sided processing of wafers less than 100 μm thick combined with precise chemical etching of selected areas. This paper reports a probe structure capable of meeting the requirements for chronic single-unit recording and which uses a simple four-mask single-sided process on wafers of normal thickness. The fabrication process is also compatible with the incorporation of on-chip circuitry to ease the problems of packaging and lead attachment in high-impedance multichannel recording arrays.

II. PROBE STRUCTURE AND FABRICATION

Fig. 1 shows the structure of the multichannel probe being developed. A silicon substrate supports an array of thin-film conductors which are insulated above and below by deposited

Manuscript received November 15, 1984; revised January 10, 1985. This work was supported by the National Institutes of Health under Contract NIH-NINCDS-N01-NS-1-2384.

K. Najafi and K. D. Wise are with the Solid-State Electronics Laboratory, Department of Electrical Engineering and Computer Science, University of Michigan, Ann Arbor, MI 48109.

T. Mochizuki is on leave with the Solid-State Electronics Laboratory, Department of Electrical Engineering and Computer Science, University of Michigan, Ann Arbor, MI 48109. He is with the Semiconductor Device Engineering Laboratory, Toshiba Corporation, Kawasaki, 210 Japan.

Reprinted from *IEEE Trans. Electron Devices*, vol. ED-32, no. 7, pp. 1206–1211, July 1985.

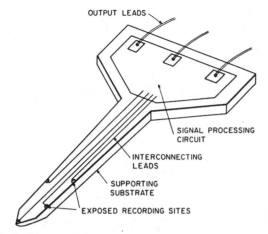

Fig. 1. Diagram of the multichannel recording array structure.

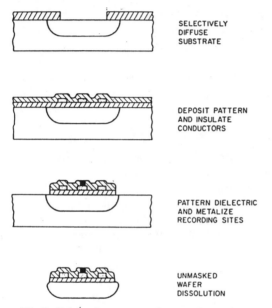

Fig. 2. Fabrication sequence for a passive probe.

dielectric layers. Openings in the upper dielectric are used to define recording sites at the ends of the conductors. On-chip circuitry is used to reduce the impedance levels on the electrode channels (which are capacitive and typically equivalent to about 10 MΩ at 1 kHz), amplify the neural signals (which are 0–250 μV in amplitude with a bandwidth of about 100 Hz–5 kHz), and multiplex the signals onto a single output lead. Operating from a single 5-V supply, the probe will thus require only three external leads.

Fig. 2 shows a summary of the process developed for a passive probe (without on-chip electronics). Fabrication begins with a p-type $\langle 100 \rangle$ silicon wafer of standard thickness and doping. The wafer is first oxidized and the oxide is patterned to define the intended probe areas. Next, these areas are subjected to a deep boron diffusion (15 h at 1175°C) to heavily dope the probe substrate. The masking oxide is then stripped and a combination of thermal oxide (300 nm), CVD silicon nitride (300 nm), and CVD silicon dioxide (800 nm) is deposited to form the lower dielectric. The relative proportions of nitride and oxide in this dielectric are important to achieve a composite insulator whose thermal expansion coefficient approxi-

mately matches that of the silicon, so that the resulting probes will not be warped.

Conductors of tantalum or polysilicon are next deposited and patterned, followed by the deposition of the upper oxide-nitride-oxide dielectrics. When tantalum is used, a thin (50 nm) nitride layer is deposited prior to the first oxide to prevent oxidation of the tantalum during the CVD process. The upper insulators are now patterned using a plasma process to open the recording sites and bonding areas. With the masking resist still in place, the exposed conductor areas are ion milled to remove any surface oxide, and gold is inlayed in the exposed areas. Lift-off is then used to remove the gold from everywhere except these regions. The gold recording sites are thus self-aligned. The field dielectrics outside the intended probe areas are now removed using a plasma etch. Finally, the wafer is thinned from the back in an isotropic etch (10-percent HF, 90-percent nitric acid) and is subjected to an unmasked etch in ethylene diamine-pyrocatechol (EDP) [9] to separate the individual probes. The EDP etch is known to stop when the boron concentration in the silicon exceeds a level of about 5×10^{19} cm^{-3}. This final etch thus dissolves the wafer and stops on the p$^+$ probe substrate. It does not attack any of the other materials used. The completed probe chips are removed from the etch, ready for lead attachment and mounting.

This process is capable of high yields (above 80 percent based on optical inspection), results in very small structures, and requires only single-sided processing on wafers of normal thickness. All etching steps are highly selective and self-stopping. Probe features can be controlled to within 1 μm or better. The finished silicon substrates can be as thick as 15–20 μm and of arbitrary two-dimensional shape. These substrates are strong yet flexible. For applications demanding greater stiffness, the probes can be withdrawn from the final etch before a complete etch stop is achieved, leaving a self-aligned support rib backing the shank. This rib forms naturally and results in no degradation of shank lateral dimensions since these are still controlled by the boron layer and the top-side portion of the final etch.

Fig. 3 shows some of the passive probes fabricated to date. Typical shank lengths have been 1.5–3 mm, with thicknesses of 8–15 μm. Shank widths are normally tapered from a width of 25 μm or less near the tip to 75–100 μm near the base. Fig. 4 shows the tip of a typical multielectrode probe, while Fig. 5 shows the cross section of a probe substrate. The lower probe edge is rounded as would be expected from an isotropic diffusion. The upper edge, however, is also rounded. This is thought to be due to the outdiffusion (segregation) of boron into the masking oxide as it diffuses laterally, suppressing the actual boron concentration near the surface. Fig. 6 shows substrate shape at the tip of a probe where tissue penetration is achieved. The rounded nature of the probe edges can be controlled to a considerable extent by the diffusion schedule employed and is thought to help in minimizing tissue damage.

Fig. 7 shows a probe mounted for acute (short-term) studies and a typical neuronal discharge recorded in gerbil cerebellar cortex. These probes penetrate the pia arachnoid layer covering the brain quite easily with minimal dimpling of the cortical surface. We have never broken a probe in such a procedure (in

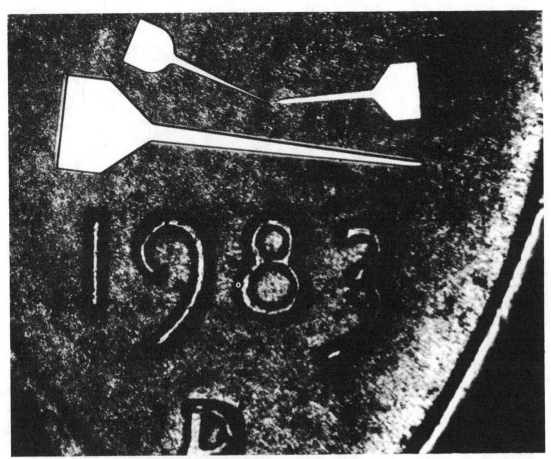

Fig. 3. Examples of probe substrates realized using the current process. The smaller probes are 1.8 mm long overall, with widths which taper to less than 20 μm near the tips. The substrates are compared with the date on a Lincoln cent.

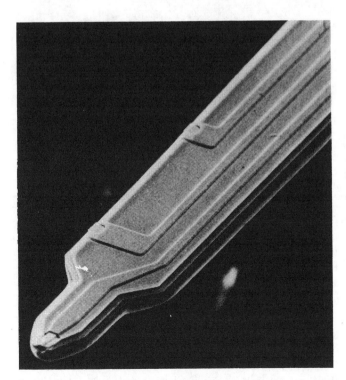

Fig. 4. SEM view of the tip of a multichannel recording array. The interconnect lines shown are 8 μm wide.

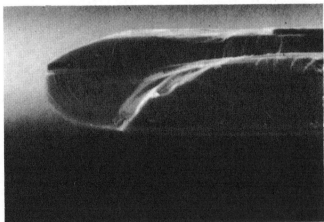

Fig. 5. Cross section of a probe substrate. The substrate was broken to allow examination of its profile. The probe thickness here is about 15 μm.

several hundred penetrations) so that the strength of these substrates is certainly adequate. In addition, not only do sites near the tip record neural signals readily, sites back along the shank also record high-amplitude neural activity. Thus these probes have shown that multichannel simultaneous recording in depth is possible and practical. Lead attachment and mounting pro-

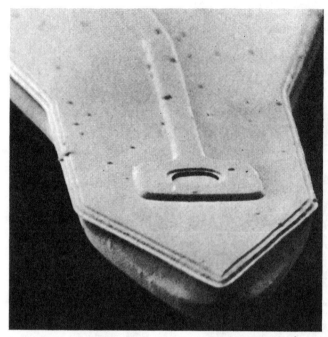

Fig. 6. SEM view of the tip of a typical probe. The rounding of the substrate edges, controlled by using a deep boron diffusion, is important in minimizing tissue damage by the array. The metalization width shown is 8 μm.

cedures have been developed for both acute and chronic (long-term) applications.

The probe materials are all thought to be biocompatible; however, for the chronic case, a number of packaging problems remain. Central among these is the encapsulation of the probes and their output leads. The polysilicon–oxide–nitride lead structure on the probe itself has been chosen so that the transition from conductor to insulator is intimate and avoids the adhesion/interfacial problems sometimes associated with dissimilar materials. Recording impedance levels on these probes have been stable for over 250 h in saline soak tests, with eventual leakage developing on the off-chip leads but not, to date, on the prove itself. Over the circuitry and the output leads, additional coatings will certainly be required. Work to develop such coatings is in progress, with most current efforts focusing on polyimide and on parylene. While much work remains, the packaging of sensor chips for chronic implantation is the subject of efforts in a number of laboratories, and the solutions to these problems promise a considerably broader impact on the overall sensor community.

III. INTERFACE CONSIDERATIONS

Careful attention to interface requirements is necessary to achieve the desired performance from the final probe structure. Fig. 8 shows an electrical equivalent circuit of the recording electrode–electrolyte interface and the amplifier input stage. The series double-layer of the metal–electrolyte surface [10] is represented by C_e, while R_e represents the real part of this surface impedance. The capacitance C_s represents shunt electrode coupling to the substrate and solution in parallel with the input capacitance of the recording amplifier. The resistance R_s represents any shunt leakage resistance from the recording electrode or the amplifier input stage.

(a)

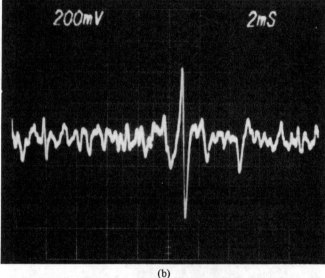

(b)

Fig. 7. (a) Photograph of a mounted multielectrode probe suitable for use in acute studies. (b) Neural activity recorded from a single cell in gerbil cerebellar cortex using a multichannel probe. The neural signals shown are about 200 μV before amplification.

Since the open-circuit dc potential between the solution and the metal electrode is somewhat unstable (variations of as much as 50 mV are observed for a gold surface in saline), it is important that the interface gain be near zero at dc and near unity in the passband. The permissible gain of eventual on-chip amplifiers will be limited by this baseline instability and by any input offset associated with the amplifiers themselves, since both of these quantities are large compared with the biological signal levels.

In order to achieve satisfactory ac recording performance C_e must be much greater than C_s. For a typical recording electrode, C_e = 10–20 pF, while C_s is about 0.8 pF. Thus in the passband, the ac gain is about 0.95, which is satisfactory for

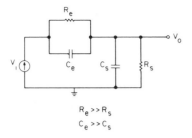

$$R_e \gg R_s$$
$$C_e \gg C_s$$

Fig. 8. Equivalent circuit of a typical microelectrode–electrolyte recording interface.

these applications, especially since it should be highly uniform from channel to channel. The silicon substrate acts as a ground plane under the electrodes and virtually eliminates crosstalk in the array [5]. The low-frequency bandwidth corner is set by R_s and C_e, and for $C_e = 10$ pF an $R_s > 160$ MΩ is required to set the corner frequency at <100 Hz.

The dc value of R_e is determined by the polarization behavior of the electrode recording surface. Noble metals polarize quite readily, and for gold in saline [7] the equivalent polarization resistance is about 5 MΩ·cm². For a 50-μm² recording surface this scales to a dc resistance of $R_e = 10^{13}$ Ω. Thus any R_s less then 10^{11} Ω should result in a dc gain near zero and stabilize the dc baseline level adequately.

While the probe process permits any metal which is not attacked in EDP to be deposited in the recording sites, gold is an appropriate choice because it is free from any surface oxide (ensuring a low ac impedance), polarizes easily, and is biocompatible. Therefore, by properly designing the interface electronics, an acceptable gain characteristic can be achieved and the electrode interface itself can provide the needed ac coupling. This avoids the problems involved in ac coupling the amplifier itself at these frequencies, signal levels, and required die areas. (A per-channel amplifier area of about 0.1 mm² or less is needed.) The major challenge in this interface thus becomes that of setting R_s in the appropriate range. Candidate approaches include using the resistance of a properly biased input protection diode or periodically resetting the input electrode line to ground via an input switch during a nonsampled time period.

IV. ON-CHIP CIRCUIT COMPATIBILITY

The probe structure and fabrication process have been designed to maintain compatibility with the use of on-chip circuitry for signal processing. The process puts no constraints on the starting substrate material aside from its ⟨100⟩ orientation, which is standard, and the substrate resistivity can be chosen to optimize device performance. The principal challenge to on-chip circuit realization lies in achieving compatibility with the very heavy boron diffusion used to control the shape of the probe substrate. The boron must be completely masked out of the circuit area without sacrificing shape control. This can be accomplished by boron doping only the perimeter of the rear portion of the silicon carrier, leaving the center (circuit) area (see Fig. 1) unaltered in its resistivity. Since the upper surface of the probe is protected by dielectrics, the probe shape is defined as before, with EDP etching from the back

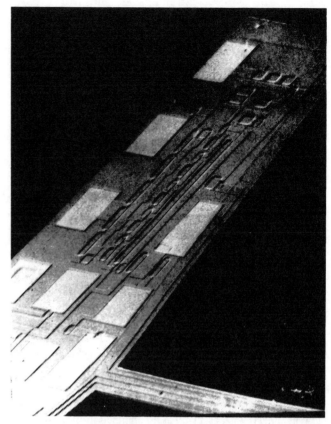

Fig. 9. SEM view of the rear portion of a probe containing a variety of NMOS test devices. This probe has established the compatibility of the probe process with the use of on-chip electronics, and full multichannel signal processor is being developed for use with chronic arrays.

and around the sides of the substrate. The boron-rich silicon perimeter prevents undercutting where it might otherwise occur. Since the active circuitry utilizes only the first few micrometers of material into the silicon, it is a relatively simple matter to withdraw the substrates from the EDP etch after a stop is achieved on the shanks and before excessive etching has occurred in the circuit areas. An etch stop there is really not needed.

A variety of masking dielectrics have been used to successfully mask the heavy boron diffusion, including 1 μm of silicon dioxide as well as various silicon dioxide/silicon nitride combinations. An NMOS test chip containing three recording electrodes, several different transistor geometries, process test structures, and a ring oscillator was fabricated in order to examine circuit compatibility. This active probe was defined in the usual manner using a deep boron diffusion. Care was taken in subsequent processing to minimize autodoping during the growth of the pad (sacrifice) and gate oxides. Control wafers were also processed in which the boron diffusion was not used. Fig. 9 shows one of these active probes. All of the devices on this test probe are fully functional, and the device thresholds on the wafers where the deep boron diffusion was used were not significantly different from those on the control wafers (i.e., within 0.2 V).

A block diagram of the circuitry being developed for use on these probes is shown in Fig. 10. Per-channel amplifiers are used to provide a gain of 100. The outputs of the ten channels

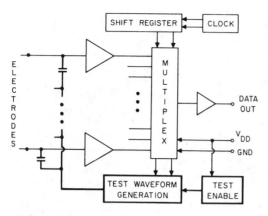

Fig. 10. Block diagram of the implantable on-chip circuitry being developed for multichannel multiplexed recording arrays. The circuitry used to provide self-test capability is shown in bold lines.

are multiplexed onto a common output data line using an analog multiplexer and broadband output buffer as shown. An on-chip clock is used to drive a two-phase dynamic shift register [11] which controls the multiplexer. The clock frequency is nominally 200 kHz, providing a per-channel bandwidth of about 6 kHz. A synchronization pulse is inserted as an additional channel to allow external regeneration of the on-chip clock and subsequent demultiplexing. The external circuitry has been developed and is capable of handling neural signals as low as 20 μV with as many as 40 channels multiplexed onto a single line. The on-chip circuitry has been designed to allow the electrode impedance levels to be tested *in vitro* or *in vivo* on demand [12]. The test mode is selected by temporarily pulsing the VDD line from 5 to 8 V. This pulse latches the test enable circuit into the test mode and gates a 1-kHz signal to the input lines via 60 fF coupling capacitors. The resulting input signal levels are amplified and multiplexed out in the usual way and provide an external indication of the electrode impedance levels, which are at present the most reliable known indicators of channel integrity.

The on-chip circuitry for a ten-channel probe has been designed and is being fabricated. While circuit details will be published separately, the active circuit area required is about 1.3 mm^2, with a power dissipation of 5 mW at 5 V. This chip design is based on a double-poly triply-implanted E/D NMOS process using 6 μm features. The inclusion of self-test capability requires an increase in die area of about 7 percent for the probe and an increase in power dissipation of about 12 percent. It does not affect the number of leads required by the probe, however, and provides a direct way of monitoring the integrity of the recording channels in both acute and chronic recording situations.

V. CONCLUSIONS

A multielectrode recording array has been developed that is suitable for use in studies of cellular physiology and for eventual application in the closed-loop control of neural prostheses. The probe permits simultaneous multichannel recording in depth in the cortex with minimal tissue disturbance. The passive probe process requires only four masks and permits substrates as small as 20 μm in width and 8–15 μm in thickness to

be realized reproducibly and with high yield. The process is compatible with the addition of on-chip circuitry for signal amplification and multiplexing, and such circuitry is being developed.

While substantial progress has been made in the development of the active probe structure desired, some important challenges remain. The implantable electronics must be integrated on the probe and shown to perform satisfactorily, and adequate encapsulation must be developed for these structures. This application is undoubtedly one of the most difficult ever faced in the area of IC packaging, and the results should be of interest in a wide range of other applications. Finally, there is a tremendous amount of physiology still to be learned which directly influences the usefulness of the arrays. With the successful development of these probe structures, however, the stage is at least set to address these questions in a meaningful way.

ACKNOWLEDGMENT

The authors would like to thank Dr. F. T. Hambrecht of the Neural Prosthesis Program, National Institutes of Health, for his interest and encouragement in this work. The many valuable ideas and assistance provided by K. L. Drake in probe fabrication, mounting, and testing are also gratefully acknowledged. Prof. D. J. Anderson and Prof. S. L. BeMent have played an important role in this program, assuming responsibility for electrode *in vitro* and *in vivo* testing. Their many contributions are much appreciated.

REFERENCES

[1] K. Frank and M. C. Becker, "Microelectrodes for recording and stimulation," in *Physical Techniques in Biological Research*, vol. 5, W. L. Nastuk, Ed. New York: Academic, 1964.
[2] R. C. Gesteland, B. Howland, J. Y. Lettvin, and W. H. Pitts, "Comments on microelectrodes," *Proc. IEEE*, vol. 47, pp. 1856–1862, Nov. 1959.
[3] D. A. Robinson, "The electrical properties of metal microelectrodes," *Proc. IEEE*, vol. 56, pp. 1065–1071, June 1968.
[4] F. T. Hambrecht and J. B. Reswick, Eds., *Functional Electrical Stimulation: Applications in Neural Prostheses*. New York: Marcel Dekker, 1977.
[5] K. D. Wise, J. B. Angell, and A. Starr, "An integrated circuit approach to extracellular microelectrodes," *IEEE Trans. Biomed. Engr.*, vol. 17, pp. 238–247, July 1970.
[6] A. Starr, K. D. Wise, and J. Csongradi, "An Evaluation of Photograved Microelectrodes for Extracellular Single-Unit Recording," *IEEE Trans. Biomed. Engr.*, vol. 20, pp. 291–293, July 1973.
[7] K. D. Wise and J. B. Angell, "A low-capacitance multielectrode probe for neurophysiology," *IEEE Trans. Biomed. Eng.*, vol. BME-22, pp. 212–219, May 1975.
[8] K. Takahashi and T. Matsuo, "Integration of multi-microelectrode and interface circuits by silicon planar and three-dimensional fabrication technology," *Sensors and Actuators*, vol. 5, pp. 89–99, Jan. 1984.
[9] R. M. Finne and D. L. Klein, "A water-amine-complexing agent system for etching silicon," *J. Electrochem. Soc.*, vol. 14, pp. 965–970, Sept. 1967.
[10] P. Delahay, *Double Layer and Electrode Kinetics*. New York: Wiley-Interscience, 1965.
[11] N. Koike, I. Takemoto, K. Satoh, S. Hanamura, S. Nagahara, and M. Kubo, "MOS area sensor: Design considerations and performance of an n-p-n structure 484 × 384-element color MOS imager," *IEEE Trans. Electron Devices*, vol. ED-27, pp. 1676–1682, 1980.
[12] K. D. Wise and K. Najafi, "A micromachined integrated sensor with on-chip self-test capability," in *Dig. IEEE Solid-State Sensor Conf.*, pp. 12–16, June 1984.

An Integrated Sensor for Electrochemical Measurements

R. L. SMITH, MEMBER, IEEE, AND D. C. SCOTT

Abstract—A method for the fabrication of a completely integrated solid-state electrochemical sensor which combines a minature liquid junction reference electrode with a CMOS ISFET is presented. The reference electrode is fabricated by preferentially etching silicon to form a porous silicon frit. The CMOS process provides electrical encapsulation of the ISFET. The performance of the reference electrode and CMOS ISFET as an integrated sensor is demonstrated.

INTRODUCTION

MUCH of the recent literature on new electrochemical sensors has centered on the application or development of chemically sensitive semiconductor devices with ISFET's (ion sensitive field effect transistors) enjoying a predominate role [1]–[5]. The solid-state construction of these devices makes them attractive for biochemical sensing for several reasons. 1) They are batch fabricated in silicon, using planar integrated circuit technology, which reduces individual cost and variations in device characteristics. 2) They offer the possibility of "smart" multisensors, allowing for on-site signal processing. 3) They are rugged, yet very small, which is an ideal combination for *in vivo* applications. Despite these advantageous characteristics, the development of commercially available sensors has been hampered by some very basic technological problems.

The deleterious effects of moisture on integrated circuits and electrical components is generally known, and to intentionally immerse an electronic package into an aqueous environment requires special attention to the encapsulation. In the past this has been accomplished by depositing Si_3N_4 or Al_2O_3 over the active FET regions and encapsulating the remainder of the silicon chip with organic encapsulants such as epoxies, polyimides, etc. [6]–[8]. Since all organic encapsulants hydrate to some extent, rather thick layers are required to obtain a satisfactory isolation from solution. Also, their use as encapsulants is usually very inconvenient, with the encapsulation often performed by hand. Recently, attempts to circumvent the encapsulation problem have been reported. Matsuo has demonstrated a unique solution to the problem of

Manuscript received April 30, 1985; revised August 25, 1985. This work was supported in part by the Swiss National Science Foundation and the Whitaker Foundation.

The authors are with the Centre Suisse d'Electronique et de Microtechnique, Neuchatel, Switzerland.

IEEE Log Number 8406340.

encapsulation by micromachining the silicon sensors into the shape of small needles and depositing 50–150 nm Si_3N_4 entirely around the tip of the needle [9], [10]. This allows most of the encapsulation to be disposed of at the wafer level with an electrical insulator known for its hydration resistance. However, in addition to the fabrication difficulties associated with this structure, the integrity of the Si_3N_4 film becomes suspect over the large areas required for complete coverage of the sensor.

Another serious problem that has restricted the use of ISFET's as a viable biomedical sensor has been the unavailability of a suitable solid-state reference electrode. A true solid-state reference electrode would comprise a material whose interfacial potential remained invariant under changing type and concentration of aqueous electrolyte. A few attempts to find such an electrochemical material have been presented with varying degrees of success [11], [12]. Since the interfacial potential is fundamentally limited by interfacial exchange currents, it appears unlikely that such a material will ever exist [13]. It is precisely this difficulty that has forced the majority of investigators to rely on the conventional types of reference electrodes and miniaturize them for compatibility with the ISFET. As expected, such miniaturizations are subject to tradeoffs. As an example, to miniaturize the conventional liquid junction reference electrode, smaller internal reference solution volumes are required. However, to prevent diffusional losses, which alter the reference potential, higher liquid junction impedances are necessitated. Also, fabrication of a conventional liquid junction type reference electrode is usually not process compatible with planar fabrication techniques and therefore not easily integrated with the ISFET [14]. Thin film metal depositions are process compatible and several investigations have involved the use of a thin film Ag/AgCl electrode which is either in direct contact with a test solution of constant Cl^- concentration [15], [16] or is covered with a polymer saturated with a solution of Cl^- [17]. However, both types are severely restricted by the variance of the reference electrode potential to changing concentration of its primary ion.

Presented in this paper is a planar process compatible liquid junction reference electrode, integrated on wafer with a CMOS ISFET. The result is a completely integrated electrochemical sensor with an electrical isolation from solution that significantly reduces the obligatory encapsulation area.

Reprinted from *IEEE Trans. Biomed. Eng.*, vol. BME-33, no. 2, pp. 83–90, February 1986.

157

EXPERIMENTAL

Shown in Fig. 1 is a cross-sectional diagram of the CMOS ISFET with integrated reference electrode design. In this application the ISFET is used as one half of the electrochemical sensor with a porous silicon membrane forming the liquid junction to electrically connect the internal reference electrode solution, KCl, to the external test solution and ISFET.

A. Fabrication

1) ISFET Fabrication: The ISFET's were fabricated according to a standard metal gate CMOS process. The substrate material was n-type, 4–6 $\Omega \cdot$ cm, $\langle 100 \rangle$ orient, 4 in diameter silicon wafers with 20 ± 0.5 mil thickness. Three individual chip designs were fabricated in each wafer: an n-channel ISFET/MOSFET pair, a p-channel ISFET/MOSFET pair, and a chip containing only the contact hole for the reference electrode. Only the reference electrode chip and the n-channel FET's are considered in this paper. The dimensions of each chip are approximately 1.6×1.2 mm. The FET's have channel lengths of 15 μm, W/L = 30, and a dual dielectric, composed of 500 ± 50 Å of silicon dioxide and 1100 ± 100 Å of LPCVD silicon nitride. No threshold setting implants were performed.

The CMOS process produced p-wells and drain/source diffusions with junction depths and sheet resistivity of 14 μm/1800 Ω per square and 3 μm/15 Ω per square, respectively.

2) Reference Electrode Fabrication: After the ISFET had been fabricated, the reference electrodes were added by anisotropically etching holes into the back side of the wafer using a Si_3N_4 mask and a 40–44 percent KOH etchant with well-known etching properties [18], [19]. For versatility, no etch stop mechanism was used to predetermine the depth of the KOH etch. Instead, the holes were etched at 120°C until the approximate depth was reached and then the etch rate reduced by lowering the temperature to between 70 and 90°C until the desired membrane thickness was reached, 10–70 μm. The ISFET circuitry was protected from the KOH etchant by means of a stainless steel holder designed to prevent the KOH solution from contacting the front ISFET circuitry during the etching process.

After the membranes were formed, aluminum was evaporated on the front side of the wafer to provide electrical connections to the silicon membranes and electrical contact made to the aluminum with silver epoxy. The silicon membranes were then anodized to porous silicon in 49 percent HF at current densities between 20 and 100 mA/cm^2 using either a Pt or Ir cathode and constant illumination from an IR filtered quartz iodide lamp. The intensity of the light source was not measured, but held constant for all the membranes fabricated. Electrical isolation of the aluminum was accomplished by either covering the aluminum with Apiezon wax or placing the wafer in a holder which exposed the back side of the wafer to the concentrated HF, but kept the front side of the wafer with aluminum on it immersed in a pH 7 buffer. The buffer

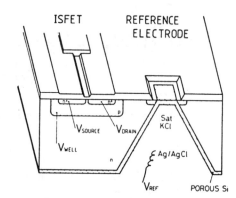

Fig. 1. A cross-sectional diagram of the CMOS ISFET and integrated reference electrode (IRE).

served as a etch stop to prevent further anodization of the silicon and protect the ISFET circuitry from HF once the silicon pores had completely penetrated the wafer. After formation of the porous silicon, the wax was removed and the bonding pads patterned from the aluminum.

The remainder of the IRE fabrication was performed on the individual chips after dicing, mounting, wire-bonding, and encapsulating the wire-bonds with epoxy (Shell Epon 825 and Jeffamine D-230). The substrates were designed to permit access to the back of the silicon chip after mounting. The reference electrode holes were filled with saturated KCl and sealed with a glass cover slide. The volume of the internal reference solution was approximately 1–2 μl. In all cases the electrical contact to the internal reference solution was by means of a Ag/AgCl redox couple. Two methods were used to prepare the redox couple: 1) silver wires (0.025 mm o.d.) were anodized electrically in 1 M HCl at current densities of 1.0 mA/cm^2, or 2) the cover slides were coated with thin films of Ti (20.0 nm) and Ag (1 μm) by thermal evaporation, and the Ag anodized chemically in a 1 : 1 : 10 $HCl : H_2O_2 : H_2O$ solution. For impedance measurements the reference electrodes were mounted on the tip of a glass capillary tube and back filled with at least 100 μl of saturated KCl.

B. Testing

All chemicals used were either reagent or electronic grade. Deionized water of 18 MΩ was used throughout testing and fabrication. Except where noted, all experiments were performed at a thermostated 25.0 ± 0.1°C. All pH measurements were made using an Orion 601 ion analyzer or Metrohm 654 pH meter, both with combination pH electrode. Impedance measurements were performed in a triple electrode configuration with a platinum or iridium counter electrode in conjuction with an IBM Voltametric Analyzer or Wavetech function generator and Kiethley 610 electrometer. ISFET characteristics were obtained with an HP 4145 Semiconductor Parameter Analyzer. Except where noted, all electrochemical potential measurements were made against the saturated calomel electrode (SCE). Electrochemical measurements on the ISFET were made in a feedback mode at a drain to source current of 100 μA using a Keithley 617 programmable

Fig. 2. Photomicrograph showing the CMOS ISFET and IRE sensor design. The active gate of the ISFET is the horizontal "u" at the lower right corner of the photo. An equivalent MOSFET is located immediately above the ISFET. To the left of the ISFET is the integrated reference electrode. The small square in the lower middle of the chip is the porous silicon frit. The anisotropically etched silicon hole is immediately beneath the frit and occupies a square whose boundaries are roughly coincident with the right, left, and bottom scribe lanes. The total sensor chip dimensions are 1.6 by 2.5 mm.

electrometer to measure the reference electrode (IRE or SCE) voltage.

RESULTS AND DISCUSSIONS

A photomicrograph of the completed sensor is shown in Fig. 2. The active gate region of the ISFET is found in the lower portion of the photograph and a metal gate FET used for comparisons in the upper portion. Both FET's are n-channel and located in a p-diffusion indicated by the surrounding rectangle. The porous silicon frit is seen as the square to the left of the ISFET gate. The internal reference solution reservoir is not visible from the front of the wafer, but is located immediately below the porous silicon frit, taking advantage of the normally unused silicon "real estate" on the back. Since the only part of the reference electrode that occupies silicon surface area is the porous silicon membrane, the size of the sensor could be considerably reduced by placing the reference electrode frit immediately adjacent to the ISFET gate but still outside the p-well diffusion. However, for this study, the distance between the ISFET and reference electrode was intentionally large to facilitate separations for testing each half of the electrochemical sensor individually, as well as together as a complete integrated sensor [20].

A. Integrated Reference Electrode (IRE)

Because the reference electrode liquid junction is of primary concern in determining the electrode stability and drift, the structure and electrochemical properties of the porous silicon also become important for use in the IRE. Ideally, the properties of the IRE liquid junction may be controlled by the fabrication parameters of the porous silicon, i.e., anodization current, illumination, HF concentration, silicon doping, etc. The literature contains varied reports as to the mechanism of formation and structure of the porous silicon. Reports of porous silicon pore diameters ranging anywhere between 1 nm and 1 μm, and a silicon content as low as 10 percent have been published [19], [23]–[26]. However, we found predictability and reproducibility of the porous silicon membranes to be poor. As a result, electrode properties that rely upon the chemical and structural properties of the porous silicon, i.e., drift, impedance, etc., were not easily controlled, and therefore, no attempt to provide a detailed correlation between the two is given here. Instead, general properties and trends that effect the reference electrode behavior will be given with selected examples to demonstrate a typical response.

The eventual destiny of porous silicon is SiO_2, silicon being thermodynamically unstable with respect to its oxide. Although the geometry and pore size may differ, this is not unlike the porous glass junctions used in some standard reference electrodes [27], [28], and it is not surprising that the two types of electrodes share similar properties. The pore diameter of the porous silicon used in the IRE was typically less than 50 nm. Shown in Fig. 3 is a scanning electromicrograph of a silicon chip that was cleaved through the IRE frit. The picture clearly shows the

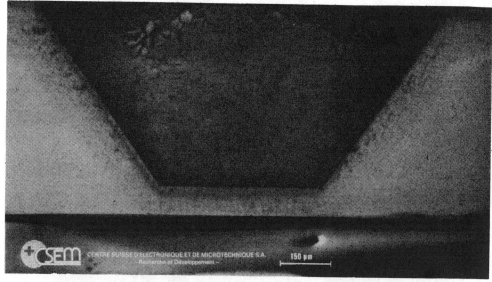

Fig. 3. Scanning electron micrograph showing a cross-sectional view of the integrated reference electrode cleaved through the porous silicon membrane. The side walls of the etched silicon holes are determined by the KOH anisotropic etching properties, and follow the 111 crystal planes (54° with respect to the surface). The porous silicon is clearly visible as a difference in texture of the silicon along the walls of the reference electrode hole. The membrane is located at the bottom of the photo.

TABLE I
THE IMPEDANCE CHARACTERISTICS OF THE INTEGRATED REFERENCE ELECTRODE OF FIG. 4. THE MEMBRANE WAS 50 μm THICK AND HAD AN AREA OF APPROXIMATELY 90 mil². THE POROUS SILICON WAS FORMED AT A CURRENT DENSITY OF 20 mA/cm² IN 49 PERCENT HF.

Frequency (Hz)	Impedance (kΩ)
0	211.0
50	179.8
100	169.8
500	151.0
1000	134.5
5000	91.0
10000	80.6

geometries formed by the anisotropic KOH etch, as well as the depth and texture of the porous silicon membrane located at the bottom of the photo.

As is the case for all liquid junction electrodes, the integrated reference electrode drifted considerably when first placed in solution. However, after 1–10 h, a stable potential of −45 to −56 mV versus the SCE was obtained. Thereafter, the drift remained linear to a first-order approximation at values between 10 μV and 2 mV per day depending upon the porosity and impedance of the porous silicon. After initial hydration, the potential drift was always positive and roughly inversely proportional to the junction impedance which ranged from about 800 to 10 kΩ. Table I shows the impedance of a typical IRE. The drift of this electrode was less than 20 μV per day with 1 μl of internal reference solution. It was assumed that the source of the long-term drift was diffusional Cl$^-$ losses from the internal reference solution. If the integrated reference electrode was stored in 3 M KCl, the potential remained within 2–3 mV of its orignal value for over two

weeks, regardless of its liquid junction impedance. Only one reference electrode remaining in solution (3 M KCl) for longer than two weeks was tested; however, it remained stable for over one month before testing was terminated. On top of the steady long-term drift, all electrodes tested showed a 4–8 h oscillation in potential of a few hundred microvolts, peak to peak, which appeared to be independent of the junction impedance. The cause of this oscillation was not discovered. The IRE potential was sufficiently immune to external noise to be tested without electrical shielding in most cases, even though relatively high liquid junction impedances were present. All reference electrodes tested showed a linear current voltage response between ± 0.700 V, indicating that a direct solution contact between internal reference solution and external test solution (pH 7 buffer, composed of 20 mM phosphate in saline) existed. Since silicon is directly exposed to solution, the maximum impedance the IRE could have is limited by the interfacial impedance of the silicon–solution interface which varies with the electroactive species present, and is generally not ohmic.

The IRE was insensitive to changes in KCl concentrations between 10^{-3} and 10^{-1} M in a background electrolyte of either 0.15 M NH$_4$Cl or 0.15 M KNO$_3$ buffered to pH 6 or 7 with phosphate. However, the IRE showed a small sensitivity to pH. The response generally increased with increasing impedance of the porous silicon, obtaining an almost Nerstian response when the limiting impedance of the silicon interface was reached, i.e., no silicon pores. The pH sensitivity of an electrode with impedance of 200 kΩ is shown in Fig. 4. This was approximately the range of impedance that showed the best all-round electrode properties for the IRE. The pH response of the IRE is not surprising when one considers that most oxides demon-

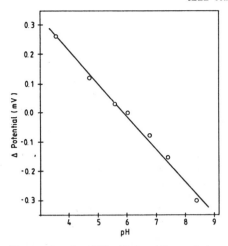

Fig. 4. The pH response of an IRE with best all-round electrode properties. The potential is given versus an arbitrary reference, and reflects only the relative changes in potential. The porous silicon was formed at an anodizing current of 20 mA/cm^2 and with a membrane area of 90 mil^2 and a membrane thickness of 50 μm. The impedance is given in Table I.

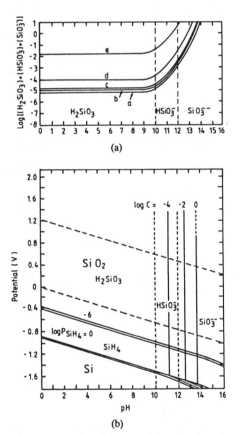

Fig. 5. (a) shows the solubility of silica at 25°C as a function of pH. a) Quartz, b) crystobalite, c) tridymite, d) vitereous silica, and e) amorphous silica. (b) shows the approximate potential–pH relationship for silicon/quartz. Diagrams show the areas of immunity, passivation, and corrosions for silicon. After Pourbaix et al. [31].

strate some degree of response to H$^+$ [29], [30]. In addition to the pH response shown, some irreversible changes of at most a few millivolts occurred when the electrodes were exposed to solutions with a pH 10 or greater. The potential usually showed a large and erratic shift, followed by a 1–3 h drift to approximately the original potential when the IRE was returned to a lower pH.

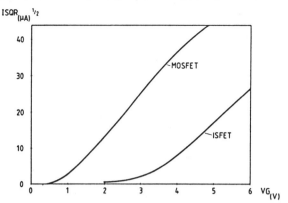

Fig. 6. Electrical characteristics of the CMOS ISFET and MOSFET. I_d versus gate voltage, V_g, of the MOSFET and ISFET with a drain to source voltage, V_{ds}, of 1.5 V. The difference in the gain of the two devices reflects the series resistance of the long drain and source diffusions of the ISFET. The ISFET characteristics were measured using an SCE.

It is assumed, considering the solubility of silicon oxide shown in Fig. 5, that the pores of the porous silicon fill with a gel of hydrated SiO$_2$, and that exposure to high pH values dissolves this gel. The long times required to regain its original potential reflect the time required to replenish the hydrated SiO$_2$ and reestablish the original diffusion potential.

B. CMOS ISFET with Electrical Encapsulation

The electrical characteristics of the metal gate FET and ISFET are compared in Fig. 6. The threshold shifts between the MOSFET and ISFET gates are typical of ISFET's with Si$_3$N$_4$ gate insulators [21], [22]. The lower gain of the ISFET is due to the long diffusion leads, which lower the effective drain to source potential seen at the channel by an amount equal to 2 × ($R_d I_{ds}$). The reverse bias current of the p-well to n-substrate is shown in Fig. 7. Measurements were performed in the dark. Because of the large surface area of the diode, the magnitude of this current increases with ambient illumination to approximately 200 nA.

The problem of encapsulation for the ISFET is reduced with the use of a CMOS design. This design takes advantage of a standard CMOS process to electrically isolate the active circuitry from the solution. Generally, the CMOS structure allows for the electrical isolation of integrated circuit components sharing the same silicon substrate by using the high impedance characteristic of a reversed biased diode between the substrate and the individual circuit. This electrical isolation technique can equally be applied to the encapsulation or electrical isolation problem of the ISFET. Use of the CMOS structure as an encapsulation aid was first proposed by Harrow in 1978 [32], and later again by Harame [15], [16]. Harame's proposed electrical test circuitry is shown in Fig. 8. In this arrangement the solution is in direct contact with the silicon substrate and essentially floats between ground and the solution potential. The precise potential is determined by the relative impedances of the solution/substrate and the reverse bias diode. The obvious advantage of this type of circuitry is that the obligatory area of encapsulation is

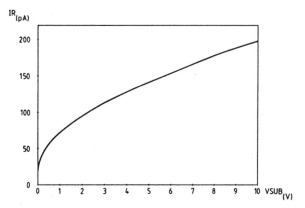

Fig. 7. Reverse bias characteristics of the p-well to substrate diode. Measurements were taken in the dark.

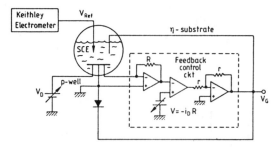

Fig. 8. CMOS ISFET and test circuitry proposed and·used by Harame *et al.* [15], [16]. In this arrangement the silicon substrate is exposed to the test solution and the electrical isolation of the ISFET is accomplished through the high impedance of the p-well/substrate reversed bias diode.

limited to the external electrical connections, i.e., wire-bonds, and leaves the remainder of the chip exposed to the solution.

However, the exposure of silicon to solution presents complications, particularly when the CMOS encapsulation scheme is used in conjunction with the integrated reference electrode. For example, the test circuitry presented by Harame is completely unsuited for incorporation with an integrated reference electrode. By electrically floating the silicon substrate, the reference electrode is forced to carry the current of the substrate/p-well reverse biased diode which can be as high as a few hundred nanoamperes under illumination. This current has two ramifications. 1) Since the sensor size is of primary importance, the current may alter or consume species necessary for the maintenance of the reference electrode potential of a miniature reference electrode. 2) Even when enough electroactive material is available to the reference electrode, the small size will present extremely high current densities. As an example, for a Ag/AgCl area the size of a normal bonding pad (4 × 4 mils) and a photo-induced reverse bias current of 100 nA, current densities as high as 1 mA/cm^2 will be forced through the reference electrode. This situation is not conducive to stable reference potentials, particularly when one considers that the ionic exchange current at Ag/AgCl is more than an order of magnitude lower than this.

A more desirable circuit arrangement would be to maintain the reference electrode and the silicon substrate at the same potential by means of a hard electrical connection to the substrate, thereby assuring that the reverse bias current is supplied by an external source and not forced through the reference·electrode. However, even this arrangement is insufficient when very small currents affect the reference electrode, as is the case for the integrated reference electrode presented here. Interfacial potentials of chemical or photogenerated origin between the reference electrode and the silicon substrate will constitute a photoelectrochemical cell capable of driving small currents through the reference electrode, even if the substrate is maintained at the same potential as that applied to the reference electrode. This is a particular problem when a reactive substance, such as silicon, is involved as one half of the electrochemical cell. The solution to this problem

is to use the measuring circuit shown in Fig. 9. Here, the silicon substrate is used to drive the solution potential to the desired value via a feedback circuit. The reference electrode is required to merely measure the solution potential and not to drive it. It may, then, be connected to a high impedance electrometer. This triple electrode technique is commonly used in electrochemistry to prevent large currents from passing through the reference electrode, and it works extremely well for the IRE/ISFET. As long as the silicon substrate does not degrade significantly from its use as the auxiliary electrode, the electrochemical behavior of the silicon/solution interface is of relatively little importance. After one month of testing in solution, there was neither noticeable degradation of the silicon surface nor deterioration in the electrical response of the IS-FET.

C. *Integrated Sensor*

The true test for the behavior of the CMOS encapsulated ISFET and IRE is their combined response as a complete electrochemical sensor. Since the Si_3N_4 deposited over the ISFET gate is known to have a pH response, we have measured the CMOS ISFET response to pH, although membranes selective to other ions could have been used. The electrical characteristics of the ISFET in different pH buffers are shown in Fig. 10. The electrical response of the complete sensor, i.e., ISFET and IRE combination, is essentially identical to response of the ISFET using a standard macro reference electrode. The only difference between the two sensors is the constant shift in threshold of about 50 mV, which reflects the difference in the standard potential of the SCE versus the saturated Ag/AgCl reference electrode. In Fig. 11 the pH response of the CMOS ISFET using the IRE is shown compared with the pH response of the same ISFET using a saturated calomel electrode as reference. As can be seen, the pH sensitivity of the ISFET with the SCE and IRE was 52.1 mV/pH and 49.6 mV/pH, respectively, with no visible deviations from linearity [16], [33] having correlation factors of 0.9998 and 0.9989. The slight difference in slope between the titration using the IRE and SCE could be attributed to the pH response to this particular IRE. However, since the pH response of the IRE is also linear with pH, the resulting contribution to the ISFET response remains linear. The pH titrations shown in the figure were per-

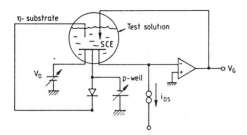

Fig. 9. Test circuitry used for the testing of the ISFET and IRE integrated sensor. The silicon is used to drive the solution potential to maintain the drain to source current at 100 μA. The solution potential is measured as V_{ref} using an SCE or the IRE. V_d was usually held constant at 1.5 V.

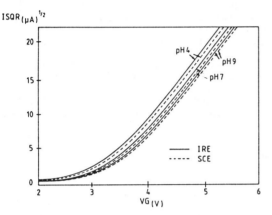

Fig. 10. Comparison of the CMOS ISFET electrical characteristics in various commercial pH buffers using the IRE (—) or SCE (- -). Drain to source voltage was 1.5 V.

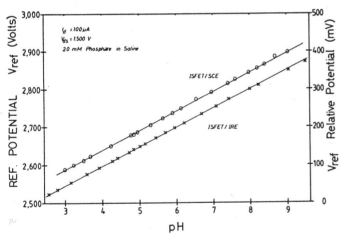

Fig. 11. The pH titration of the CMOS ISFET using either an SCE, (0), or IRE, (X), for reference. The titrations were performed in a constant 20 mM phosphate buffer in saline at 25°C. The pH sensitive Si_3N_4 response is the negative of the solution potential, V_{ref}.

formed by addition of concentrated HCl or NaOH to a constant concentration of 20 mM phosphate buffer in saline. If the pH titrations were done by changing the type of buffer, as is sometimes done when using standard commercially available buffers, the ISFET shows non-Nerstian behavior with large and varied drifts from buffer to buffer. This behavior is typical of so-called "blocked" interfaces showing a mixed potential [13], [34].

The CMOS ISFET's operated with SCE, as well as the integrated sensor, have long-term drift characteristics typical of standard nMOS ISFET's with silicon nitride sensitive layers, i.e., 1 mV/h at 25°C [10], [22]. Tests are now underway to compare these devices with integrated sensors employing aluminum oxide sensitive layers, which typically have drift values an order of magnitude lower [10].

CONCLUSION

The fabrication and testing of a minature liquid junction reference electrode, fabricated in a silicon wafer, and operated with a CMOS encapsulated ISFET, has been presented. The behavior of the integrated sensor was shown to be essentially identical to that of the ISFET operated with a standard macro reference electrode. The combination of CMOS encapsulation and the integrated reference electrode functioned extremely well. However, process control of the porous silicon was lacking. It is clear that better understanding of the formation mechanism and structure of porous silicon is needed before an acceptable degree of reproducibility is achieved.

ACKNOWLEDGMENT

The authors would like to thank Dr. A. Grisel of the CSEM for supporting this project, Dr. M. Koudelka (CSEM) for her assistance with Ag/AgCl thin film formation, Mostek Corporation for mask generation, and Solid State Scientific for the fabrication of the ISFET's used in this study.

REFERENCES

[1] O. F. Chan and M. H. White, "Characterization of surface and buried channel ion sensitive field effect transistors (ISFET's)," in *Proc. IEEE Int. Electron Devices Meet.*, Washington, DC, Dec. 5-7, 1983, p. 651.
[2] P. Bergveld and N. F. De Rooij, "The history of chemically sensitive semiconductor devices," *Sensors and Actuators*, vol. 1, p. 5, 1981.
[3] B. A. McKinley, J. Saffle, W. S. Jordan, J. Janata, S. D. Moss, and D. R. Westenskow, "*In vivo* continuous monitoring of K$^+$ in animals using ISFET probes," *Med. Instrum.*, vol. 14, p. 93, 1980.
[4] Y. Ohta, S. Shoji, M. Esashi, and T. Matsuo, "Prototype sodium and potassium sensitive micro ISFET's," *Sensors and Actuators*, vol. 2, p. 387, 1982.
[5] K. Shimada, M. Yano, K. Shibatani, Y. Komoto, M. Esashi, and T. Matsuo, "Application of catheter-tip ISFET for continuous *in vivo* measurements," *Med. Biol. Eng. Comput.*, vol. 18, p. 741, 1980.
[6] A. Sibbald, P. D. Whalley, and A. K. Covington, "A miniature flow-through cell with a four-function ChemFET integrated circuit for simultaneous measurements of potassium, hydrogen, calcium, and sodium ions," *Anal. Chim. Acta*, vol. 159, pp. 47–62, 1984.
[7] J. Harrow, "Medical applications of ion sensitive field effect transistors," *Trans. Amer. Soc. Artif. Intern. Organs*, vol. 27, p. 31, 1981.
[8] N. J. Ho, J. Kratochvil, G. F. Blackburn, and J. Janata, "Encapsulation of polymeric membrane-based ion-selective field effect transistors," *Sensors and Actuators*, vol. 4, p. 413, 1983.
[9] M. Esashi and T. Matsuo, "Integrated micro multi ion sensor using field effect of semiconductor," *IEEE Trans. Biomed. Eng.*, vol. BME-25, p. 184, 1978.
[10] T. Matsuo and M. Esashi, "Methods of ISFET fabrication," *Sensors and Actuators*, vol. 1, p. 77, 1981.
[11] T. Matsuo and M. Esashi, "Characteristics of parylene gate ISFET" (Extended Abstr.), *153rd Soc. Meet. ECS*, vol. 78-1, p. 83, May 1978.
[12] M. Yano, K. Shimoto, K. Shibalami, and T. Makimoto, U.S. Patent 4 269 682, May 26, 1981.
[13] S. Collins and J. Janata, "A critical evaluation of the mechanism of the potential response of antigenic polymer membranes to the corresponding antiserum," *Anal. Chim. Acta*, vol. 136, p. 93, 1982.

[14] P. A. Comte and J. Janata, "A field effect transistor as a solid state reference electrode," *Anal. Chim. Acta*, vol. 101, p. 247, 1978.

[15] D. Harame, J. Shott, J. Plummer, and J. Meindl, "An implantable ion sensor transducer," in *Proc. IEEE Int. Electron Devices Meet.*, Washington, DC, Dec. 1981, p. 467.

[16] D. L. Harame, J. D. Shott, L. Bousse, and J. D. Meindl, "Implantable ion-sensitive transistors," in *Proc. IEEE Biomed. Eng.*, Los Angeles, CA, Sept. 15–17, 1984.

[17] M. Koudelka, "Caracterisation d'une electrode Ag/AgCl miniaturisee," in *Proc. Journees d'Electrochimie*, Florence, Italy, May 28–31, 1985.

[18] J. W. Faust, Jr. and E. D. Palik, "Study of the orientation dependent etching and initial anodization of Si in aqueous KOH," *J. Electrochem. Soc.*, vol. 130, p. 1413, 1983.

[19] K. E. Petersen, "Silicon as a mechanical material," *Proc. IEEE*, vol. 70, p. 420, 1982.

[20] R. L. Smith and D. C. Scott, "A solid state miniature reference electrode," in *Proc. IEEE Symp. Biosensors*, vol. 61, Los Angeles, CA, Sept. 15–17, 1984.

[21] S. D. Moss, C. C. Johnson, and J. Janata, "Hydrogen, calcium, and potassium ion-sensitive FET transducers: A preliminary report," *IEEE Trans. Biomed. Eng.*, vol. BME-25, p. 49, 1978.

[22] J. Janata and R. J. Huber, "Ion sensitive field effect transistors," *Ion Select. Electrode Rev.*, vol. 1, p. 31, 1979.

[23] T. Unagami, "Formation mechanism of porous silicon layer by anodization in HF solution," *J. Electrochem. Soc.*, vol. 127, p. 476, 1980.

[24] G. Bomchil, R. Herino, K. Barla, and J. C. Pfister, "Pore size distribution in porous silicon studied by adsorption isotherms," *J. Electrochem. Soc.*, vol. 130, p. 1611, 1983.

[25] Y. Arita and Y. Sunohara, "Formation and properties of porous silicon film," *J. Electrochem. Soc.*, vol. 124, p. 285, 1977.

[26] T. Unagami and M. Seki, "Structure of porous silicon layer and heat-treatment effect," *J. Electrochem. Soc.*, vol. 125, p. 1339, 1978.

[27] A. K. Covington, "Reference electrodes," in *Ion Selective Electrodes*, D. A. Durst, Ed., NBS Special Publ. 314, 1969, p. 107.

[28] R. G. Bates, *Determination of pH. Theory and Practice.* New York: Wiley, 1973.

[29] K. Vijh, *Electrochemistry of Metals and Semiconductors.* New York: Dekker, 1979.

[30] I. Lauks, M. F. Yuen, and T. Dietz, "Electrically free-standing IrO_x thin film electrodes for high temperature, corrosive enviroment pH sensing," *Sensors and Actuators*, vol. 4, p. 375, 1983.

[31] J. Van Muylder, J. Besson, W. Kunz, and M. Pourbaix, in *Atlas d'Equilibres Electrochimiques*, M. Pourbaix, Ed. Paris, France: Gauthier–Villars, 1963.

[32] J. Harrow, personal communication, Dep. Bioeng., Univ. Utah, Salt Lake City, 1978.

[33] L. Bousse, N. De Rooij, and P. Bergveld, "Operation of chemically sensitive field-effect sensors as a function of the insulator–electrolyte interface," *IEEE Trans. Electron Devices*, vol. ED-30, p. 1263, 1983.

[34] R. P. Buck, "Kinetics and drift of gate voltages for electrolyte-bathed chemically sensitive semiconductor devices," *IEEE Trans. Electron Devices*, vol. ED-29, p. 108, 1982.

MOS Integrated Silicon Pressure Sensor

HIROSHI TANIGAWA, TSUTOMU ISHIHARA, MASAKI HIRATA, AND KENICHIRO SUZUKI

Abstract—An MOS integrated silicon-diaphragm pressure sensor has been developed. It contains two piezoresistors in a half-bridge circuit, and a new simple signal-conditioning circuit with a single NMOS operational amplifier. The negative temperature coefficient of the pressure sensitivity at the half-bridge is compensated for by a positive coefficient of the variable-gain amplifier with a temperature-sensitive integrated feedback resistor. The sensor was fabricated using the standard IC process, except for the thin diaphragm formation using the $N_2H_4 \cdot H_2O$ anisotropic etchant. The silicon wafer was electrostatically adhered to the glass plate to minimize induced stress. The −1750 ppm/°C temperature coefficient of sensitivity at the half-bridge was compensated for to less than +190 ppm/°C at the amplifier output in the 0–70°C range. A less than 20-mV thermal-output offset shift was also obtained after 26-dB amplification in the same temperature range.

I. Introduction

A SILICON-DIAPHRAGM pressure sensor [1] typically consists of a thin silicon diaphragm as an elastic material [2] and piezoresistive gauge resistors made by diffusing impurities into the diaphragm. Because a silicon single crystal has superior elastic characteristics, virtually no creep and no hysteresis occur, even if static pressure is applied. The pressure sensitivity or the gauge factor for silicon piezoresistors is many times larger than that of thin-metal-film resistors. This is due to a large piezoresistive-coefficient value [3]. The full-scale output voltage (a few hundred millivolts) obtained from bridge-connected piezoresistors, however, is not large enough to connect the output directly to A/D converters. Moreover, the sensitivity temperature coefficient for piezoresistive sensors is larger than that for conventional sensors with metal strain gauges. Thus a signal conditioner with a gain and a thermal-compensation circuit is necessary in pressure-sensing systems. Recently, piezoresistive pressure sensors integrated with an amplifier have been reported [4]–[9]. Signal-conditioning circuits were fabricated using bipolar monolithic technologies [4], [5], [7]. Ring-oscillator circuits whose frequencies were controlled by pressure-sensitive resistors were also constructed using bipolar I^2L technologies [8], [9].

A pressure sensor is typically combined with many functional peripheral circuits such as temperature-compensation circuits (span and offset), A/D converters, bus interfaces, and digitally controlled circuits. If they are fabricated on the same sensor die using silicon IC technologies, small and light-weight pressure sensors can be supplied with low prices compared to conventional metal-diaphragm sensors [10] having bonded metal gauges. They can respond to spreading demand for sensors in microcomputer-based systems. Pressure sensors will be widely used in such application fields as automobiles, medical instruments, and process equipment. To realize this highly integrated sensor, MOS technologies appear suitable due to their high packing density and low power consumption. No piezoresistive pressure sensor integrated with an MOS fabricated conditioner, however, has been reported.

This paper describes such an MOS integrated silicon-diaphragm pressure sensor [11] containing a new simple circuit with a single operational amplifier. First, the device operation and the circuit configuration will be presented. Next, the device-fabrication technologies peculiar to a pressure sensor and the experimentally obtained characteristics will be introduced.

II. Device Operation and Circuit Configuration

Due to the piezoresistive effect, the stress applied to a semiconductor resistor changes its resistance according to the following [12]:

$$\Delta R/R = \pi_l \sigma_l + \pi_t \sigma_t. \tag{1}$$

Here, ΔR is the resistance variation and R is the initial resistance with no applied stress. Stresses applied to the resistor in a longitudinal and a transverse direction are σ_l and σ_t, respectively, and π_l and π_t are piezoresistive coefficients in a longitudinal and a transverse direction, respectively. They depend on the resistor arrangement in crystallographic structures. For p-type diffused resistors, R_1 and R_2, arranged in ⟨110⟩ direction on an n-type silicon square diaphragm with (100) surface orientation, as shown in Fig. 1, the coefficients are approximately denoted [12] as

$$\pi_l = -\pi_t = \pi_{44}/2. \tag{2}$$

Here, π_{44} is the piezoresistive coefficient defined along silicon crystallographic axes. Then, resistance variations are simply expressed as

$$\Delta R_1/R_1 = -\Delta R_2/R_2 = \pi_{44}(\sigma_{1Y} - \sigma_{1X})/2 \tag{3}$$

where σ_{1X} and σ_{1Y} are X and Y directional stresses induced to resistor R_1. They are proportional to the fluid pressure P applied to the diaphragm. When connecting R_1 and R_2 in a half-bridge circuit, and exciting the bridge by V_{EXC}, output voltage V_o and its pressure sensitivity S are expressed as follows:

$$V_o = \pi_{44}(\sigma_{1Y} - \sigma_{1X}) V_{EXC}/4 \tag{4}$$

$$S = (1/V_{EXC}) (\partial V_o/\partial P) = (\pi_{44}/4) \partial(\sigma_{1Y} - \sigma_{1X})/\partial P. \tag{5}$$

Furthermore, the pressure sensitivity temperature coefficient

Manuscript received November 15, 1984; revised January 24, 1985.

The authors are with the NEC corporation, 4-1-1, Miyazaki, Miyamae-ku, Kawasaki, Kanagawa, Japan.

Reprinted from *IEEE Trans. Electron Devices*, vol. ED-32, no. 7, pp. 1191–1195, July 1985.

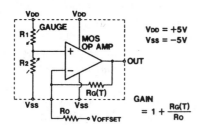

Fig. 1. Piezoresistor arrangement. Two p-type resistors R_1 and R_2 are diffused in the $\langle 110 \rangle$ direction on an n-type silicon square diaphragm with (100) surface orientation.

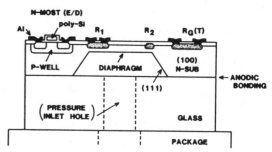

Fig. 2. Integrated silicon pressure-sensor circuit. The half-bridge output voltage is amplified by a noninverting MOS operational amplifier circuit. The feedback resistor $R_G(T)$ on the thick surrounding rim is fabricated by the same process as that for piezoresistors R_1 and R_2 at the diaphragm edges.

is

$$(1/S)(\partial S/\partial T) = (1/\pi_{44})(\partial \pi_{44}/\partial T). \qquad (6)$$

For the resistor arrangement as shown in Fig. 1, a negative temperature coefficeint is obtained because of the negative value of $\partial \pi_{44}/\partial T$ [12]. Thus sensitivity decreases with an operating temperature increase. When designing the integrated pressure sensor, the negative temperature coefficient must be electrically compensated for by the signal conditioner.

A signal conditioner has been widely used to amplify the small output voltage generated by a bridge circuit with thin-metal-film strain gauges. It consists of an instrumentation amplifier with three operational amplifiers to obtain a stable high gain and high common-mode rejection ratio. Although the silicon-diaphragm pressure sensor has a higher sensitivity than that of the metal-diaphragm sensor with bonded metal gauges, a signal conditioner is also necessary, because the pressure sensitivity and the offset voltage at zero supply pressure are sensitive to the operating temperature. In the integrated sensor, however, there is no extraneous common-mode noise into the transmission lines, because the sensor and the signal conditioner are positioned closely together on the same silicon die. Consequently, the instrumentation amplifier technology is not followed when designing the integrated signal conditioner. The circuit design is instead concentrated on temperature compensation to improve sensing characteristics and also on a simple configuration for easy production.

The integrated-sensor circuit is shown in Fig. 2. The p-type piezoresistors (R_1 and R_2) embedded at the diaphragm edges (shown in Fig. 1) are connected to the half-bridge circuit. Their resistances are changed by the applied pressure; one is increased and the other is decreased. The bridge output voltage is amplified by the MOS operational amplifier with a non-inverting feedback circuit. A pressure-independent resistor R_G, fabricated by the same process as that used for the piezoresistors, is used in the feedback circuit. A gain-setting re-

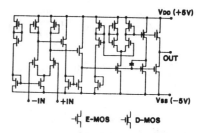

Fig. 3. NMOS operational amplifier circuit. It follows the conventional MOS amplifier with enhancement/depletion gain stages.

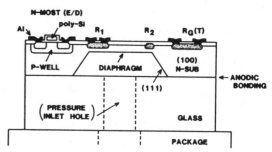

Fig. 4. Pressure-sensor cross-sectional diagram. Piezoresistors are embedded on the diaphragm. The conditioner circuit with a temperature-sensitive feedback resistor is made on the thick rim to be insensitive to the applied pressure.

sistor R_O is externally connected to the inverting input and biased at V_{OFFSET} for the offset compensation.

Specific characteristics of the proposed sensor are as follows:

1) The piezo-gauge-factor (π_{44}) temperature coefficient (negative) is equal to the R_G resistance temperature coefficient (positive) [12], when setting the impurity concentration for the p-type resistors (R_1, R_2, and R_G) to 3×10^{18} or 1×10^{20} cm^{-3}. Then the negative temperature coefficient for the pressure sensitivity, coming from π_{44}, can be compensated for by the positive temperature coefficient for the variable amplifier gain due to R_G.

2) Noninverting amplifier gain $1 + R_G/R_O$ is independent from the piezoresistor values.

3) Simpler amplifier-input-offset compensation and lower power consumption are accomplished than when using a bipolar instrumentation amplifier combined with an active four-resistor Wheatstone bridge (full bridge).

4) The design tolerance in determining all the resistors connected to the amplifier inputs is increased because there is no input-offset current at the MOS input stage.

The operational amplifier circuit is shown in Fig. 3. It consists of a differential-input stage, a differential-to-single-ended converter stage, a cascode stage, an output stage, and a phase-compensation circuit. The circuit follows a conventional MOS operational amplifier design [13] using enhancement/depletion gain stages and was designed to have minimum process-parameter dependence.

III. DEVICE FABRICATION

A cross-sectional view for the fabricated device is shown in Fig. 4. The starting material is an n-type silicon substrate with (100) surface orientation. Two piezoresistors with 3 ×

10^{18} cm^{-3} surface-impurity concentration were fabricated using boron ion implantation. One of them (R_1) is parallel to and the other (R_2) is perpendicular to the ⟨110⟩ diaphragm-edge orientation. Each piezoresistor consists of two parallel resistors (100 μm long and 20 μm wide) to prevent pressure-sensitivity decrease [1] due to a finite resistor length in the steep induced-stress region. Other peripheral components including R_G are fabricated during the same implantation process as that for the piezoresistors in the thick-rim area surrounding the diaphragm. Thus they are insensitive to pressure applied to the diaphragm. The NMOS operational amplifier in a p-well was fabricated using the conventional silicon IC process.

A thin diaphragm with 1-mm^2 area size was formed using the hydrazine-water ($N_2H_4 \cdot H_2O$) anisotropic etchant. SiO_2 deposited on the front surface and patterned SiO_2 on the back surface acted as etch-protect masks. The etching was carried out at 90°C in the reflux system with an etching rate of about 1.7 μm/min. The diaphragm thickness is controlled by the etching time to be about 30 μm.

The fabricated wafer was adhered on the borosilicate glass plate (Corning 7740) using anodic bonding [14]. Because the glass has a thermal-expansion coefficient nearly equal to that of silicon, it minimizes any unwanted stress due to thermal-coefficient mismatch between the silicon and the package header. The glass plate having 1-mm$^\phi$ pressure-inlet holes was prepared for differential- and gauge-pressure measurements. After the glass plate and the silicon wafer were assembled in the alignment jig, the glass-plate holes were aligned to the anisotropically etched cavity on the silicon back surface. They were electrostatically adhered by heating the assembly to 450°C, and applying 450 V across the silicon wafer and the glass plate for at least 15 min in a vacuum chamber to prevent being oxidized. Similarly, the glass plate with no holes was prepared for absolute-pressure-measuring sensors. In this case, the vacuum reference pressure was easily enclosed in each etched cavity when adhering the silicon with the glass plate in the vacuum chamber.

The glass-adhered silicon wafer was diced and individual dies were mounted on TO-5 headers with a pressure-inlet tube at the center. Finally, the cap with the pressure-inlet tube was hermetically sealed to the header.

Fig. 5 shows a die top-view photograph and cross-sectional view. A single die contains the earlier-described signal conditioner on a quarter-die area. It also contains the conventional instrumentation amplifier conditioner for the full-bridge circuit, as shown in Fig. 6. Therefore, the two integrated pressure-sensor types can be characterized using a single die. Fig. 7 shows the mounted die before the cap sealing.

IV. CHARACTERISTICS AND DISCUSSIONS

A 50-dB open-loop gain and 500-kHz unity-gain bandwidth were experimentally obtained for the operational amplifier only. A 5-mV mean-input-offset voltage with 3-mV standard deviation was also achieved. The amplifier operated at ±5-V supply voltage with 16-mW power consumption. The amplifier gain at room temperature was set to 26 dB by R_O for the pressure-sensor characterization.

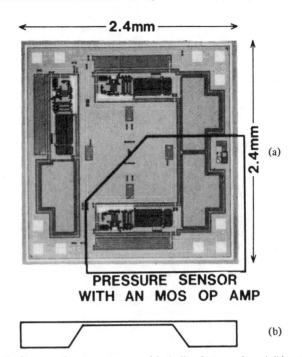

Fig. 5. Integrated-pressure sensor: (a) A die photograph and (b) a die cross-sectional view. The integrated sensor, as shown in Fig. 2, occupies a quarter die.

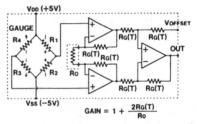

$$GAIN = 1 + \frac{2R_G(T)}{R_O}$$

Fig. 6. Integrated-pressure sensor with a conventional instrumentation amplifier conditioner for a full-bridge circuit. It is also integrated on a single die, as shown in Fig. 5, as well as the proposed sensor circuit with a single operational amplifier.

Fig. 7. Mounted die photograph. A silicon pellet bonded to the glass plate is mounted on a TO-5 header with a pressure-inlet tube.

(a)

(b)

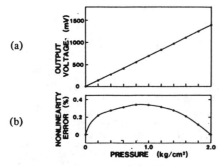

Fig. 8. Pressure-detection characteristics: (a) The pressure–voltage characteristics and (b) the terminal linearity error. They are measured at the amplifier output.

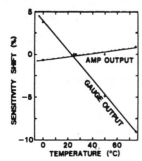

Fig. 9. Temperature characteristics of sensitivity. The negative temperature coefficient, as large as −1750 ppm/°C at the gauge-section (half-bridge circuit) output, is compensated for to less than +190 ppm/°C at the amplifier output.

Fig. 8 shows output voltage versus applied pressure curves. The pressure sensitivity, measured at the amplifier output terminal, was 700 mV/kg/cm². This value corresponds to the 35-mV/kg/cm² sensitivity in the piezoresistor bridge section for ±5-V excitation. The measured diaphragm thickness was about 35 μm for this device. This sensitivity agreed well with the simulation results using the Pressure Sensor/Transducer Simulator (PRESENTS) [15], based on the finite-element method, considering the sensitivity degradation caused by the finite-length resistor, the silicon elastic anisotropy, and the tapered-rim-supported condition (more weakly supported than the built-in-edge condition).

The terminal-linearity error at the amplifier output was less than 0.3 percent of full scale for the 2-kg/cm² range, as shown in Fig. 8. The experimentally obtained hysteresis was less than 0.1 percent, which was limited by the standard-pressure source used as a reference.

Typical temperature characteristics of sensitivity are shown in Fig. 9. The −1750-ppm/°C sensitivity temperature coefficient at the bridge was compensated for to less than +190 ppm/°C at the amplifier output in the 0–70°C range. A slight over-compensation occurred due to the positive R_G resistance temperature coefficient caused by a lower impurity concentration than the desired 3×10^{18} cm⁻³ concentration. It was shown that the temperature compensation of sensitivity was successfully achieved using the variable-gain amplifier.

An alternate compensation approach involves exciting the piezoresistor bridge circuit by a constant current source [16]. In this case, the π_{44} temperature coefficient can be compen-

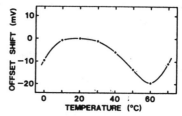

Fig. 10. Temperature characteristics of the offset at the amplifier output.

sated for by the temperature coefficient of the bridge constituting resistors themselves if the impurity concentration is set to an appropriate value such as 3×10^{18} cm⁻³. Because current-source integration is necessary in addition to amplifier integration, however, it is not as attractive for the monolithic integrated pressure sensor as the approach reported here. A single circuit amplifies and compensates the signal, thereby decreasing circuit complexity and power consumption using our approach. Therefore, it is preferable for integrated sensors.

Fig. 10 shows the temperature characteristics of the offset at the amplifier output. A less than 20-mV overall thermal-output-offset shift was obtained. The precise offset-generating mechanisms are not clear at present.

V. CONCLUSION

An MOS integrated pressure sensor has been developed. The temperature dependence of sensitivity in the piezoresistor bridge was well compensated for by using variable-gain amplifier with temperature-sensitive on-chip feedback resistor. The amplifier circuit is also applicable to other sensors which require thermal compensation and low power consumption.

ACKNOWLEDGMENT

The authors would like to express their gratitude to Dr. H. Shiraki for his encouragement. They would also like to acknowledge the contributions of S. Suwazono in the device fabrication.

REFERENCES

[1] S. K. Clark and K. D. Wise, "Pressure sensitivity in anisotropically etched thin-diaphragm pressure sensor," *IEEE Trans. Electron Devices*, vol. ED-26, pp. 1887–1896, Dec. 1979.
[2] K. E. Petersen, "Silicon as a mechanical material," *Proc. IEEE*, vol. 70, pp. 420–457, May 1982.
[3] O. N. Tufte, P. W. Chapman, and D. Long, "Silicon diffused-element piezoresistive diaphragms," *J. Appl. Phys.*, vol. 33, pp. 3322–3327, Nov. 1962.
[4] J. M. Borky and K. D. Wise, "Integrated signal conditioning for silicon pressure sensors," *IEEE Trans. Electron Devices*, vol. ED-26, pp. 1906–1910, Dec. 1979.
[5] R. E. Bicking, "A piezoresistive integrated pressure transducer," in *Proc. 3rd Int. Conf. Automat. Electron.*, pp. 21–26, 1981.
[6] K. Yamada, M. Nishihara, and R. Kanzawa, "A piezoresistive integrated pressure sensor," in *Proc. Solid-State Transducers '83*, pp. 113–114, May 1983.
[7] S. Sugiyama, M. Takigawa, and I. Igarashi, "Integrated piezoresistive pressure sensor with both voltage and frequency output," in *Proc. Solid-State Transducers '83*, pp. 115–116, May 1983.
[8] A. P. Dorey and P. J. French, "Frequency output piezoresistive pressure sensors," in *Proc. Solid-State Transducers '83*, p. 140, May 1983.

[9] H. Reichl and H. J. Hwang, "Frequency-analog sensors in I^2L-technique," in *Proc. Solid-State Transducers '83*, pp. 86–87, May 1983.

[10] J. McDermott, "Sensors and transducers," *Electron. Des. News*, pp. 122–142, Mar. 20, 1980.

[11] H. Tanigawa, T. Ishihara, M. Hirata, and K. Suzuki, "MOS integrated silicon pressure sensor," in *Proc. Custom IC Conf.*, pp. 91–95, May 1984.

[12] A. D. Kurtz and C. L. Gravel, "Semiconductor transducers using transverse and shear piezoresistance," in *Proc. 22nd ISA Conf.*, no. P4-1-PHYMMID-67, Sept. 1967.

[13] T. Ishihara, T. Enomoto, M. Yasumoto, and T. Aizawa, "High-speed NMOS operational amplifier fabricated using VLSI technology," *Electron. Lett.*, vol. 18, pp. 159–161, Feb. 1981.

[14] G. Wallis and D. I. Pomerantz, "Field assisted glass-metal sealing," *J. Appl. Phys.*, vol. 40, pp. 3946–3949, Sept. 1969.

[15] K. Suzuki, H. Tanigawa, M. Hirata, and T. Ishihara, "Analyses on silicon diaphragm pressure sensors," *IECE Japan*, Paper Tech. Group Electron Devices, ED84-56, Aug. 1984.

[16] J. Bryzek, "A new generation of high accuracy pressure transmitters employing a novel temperature compensation technique," *WESCON Conf. Rec.*, vol. 25, pp. 15.4.1–15.4.10., 1981.

HIGH-RESOLUTION SILICON PRESSURE IMAGER
WITH CMOS PROCESSING CIRCUITS

Susumu Sugiyama, Ken Kawahata, Masaaki Abe, Hirobumi Funabashi
and Isemi Igarashi

Toyota Central Research & Development Laboratories, Inc.
41-1, Yokomichi, Nagakute, Nagakute-cho, Aichi-gun
Aichi-ken, 480-11 Japan

ABSTRACT

A 32 × 32 (1k)-element silicon pressure imager with CMOS processing circuits using IC process combined with Si micromachining is described. The imager is organized of an X-Y matrix array of micro-diaphragm pressure cells with spacing of 250 μm. Polysilicon piezoresistors are arranged to form a full bridge on the each individual 100 μm × 100 μm silicon nitride diaphragm. CMOS electronics is formed around the array on the chip. Silicon chip size is 10 mm × 10 mm. Clock frequency is 4 MHz. The array is addressed as a memory, with an access time of less than 16 μsec for a cell and a readout time of less than 16 msec a frame. The output signals from 1024 cells are taken out as a serial analog waveform.

INTRODUCTION

Recently, silicon-based sensors are rapidly developed, utilizing silicon IC technology for automotive electronics, industrial automation, robotics and medical electronics [1]-[4].

In robotics particularly, tactile sensors having high resolution of two-dimension are needed to give the robot gripper an analog sense of touch for precise tasks of parts indentification, surface texture measurement, gauging and parts orientation. However, the requirments for tactile imager are not easily met with by conventional sensor fabrication techniques, and practical sophisticated pressure imager which rivals the performance of optical image sensor is not available yet. Micro-diaphragm pressure sensor fabricated by single-side processing solely on the top surface of the silicon wafer has been developed [5]. The sensor structure is a complete planar type, and is suitable for combining with MOS devices on the same chip. This paper reports a 32 × 32 (1k) element silicon pressure imager with CMOS processing circuits. The imager offers high resolution and high stability for two-dimensional pressure distribution. In this paper, the sensor structure, fabrication and readout system will be described.

STRUCTURE AND FABRICATION

Figure 1 shows a cross section of the proposed pressure imaging cell. A micro-diaphragm pressure sensor is formed by single-side processing in the each individual cell. The reference pressure chamber is formed by undercut-etching of the interlayer between the silicon nitride diaphragm and the silicon substrate. The center of the diaphragm is supported with a small spot on the silicon substrate. Polysilicon piezoresistors are arranged to form a full bridge on the diaphragm. The signal processing circuits are formed to the conventional double well CMOS struct.

The pressure imaging cells are fabricated in array form. Figure 2 shows the photomicrograph of a part of the array of the pressure imaging cells. Power source, ground, signal lines and Y-colums are run horizontally, and X-rows are run vertically on the chip. Four polysilicon piezoresistors are arranged in full bridge. CMOS analog switches, an NMOS power switch and CMOS logic circuits are included in the each individual cell. The pressure imager is formed 32 × 32 matrix organized array of pressure imaging cells with a spacing of 250 μm and CMOS processing circuits formed around the array.

Fabrication of the pressure imager was carried out using combination of 3 μm design rule CMOS process and silicon micromachining. The starting matrial is 10 Ω-cm n-type (100) silicon, 75 mm in diameter and 400 μm thickness. After oxidation of a 50 nm thick buffer oxide, n-well and p-well are formed by light ion implantation of phosphorus and boron. A 150 nm thick silicon nitride layer is deposited by LPCVD as a mask of LOCOS. Reactive ion etching (RIE) of the silicon nitride is used to defined the active device regions on the surface of the chip. After channel

Reprinted with permission from *Transducers '87, Rec. of the 4th Int. Conf. on Solid-State Sensors and Actuators*, 1987, pp. 444–447.
Copyright © 1987 by the Institute of Electrical Engineers of Japan.

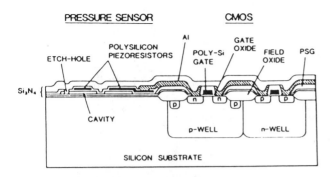

Fig. 1. Cross section of one cell of the silicon pressure imager.

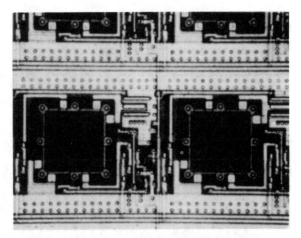

Fig. 2. Photomicrograph of a part of the array of pressure imaging cells.

stop boron ion implantation, a 1 μm thick LOCOS is carried out at 1100 °C in wet. After the nitride and the buffer oxide are removed, a 50 nm gate oxide is formed at 1000 °C in dry O_2. A threshold-adjusting is carried out using boron ion implantation with dose of a 7×10^{12} cm^{-2} at 30 keV. A gate polysilicon of 450 nm thick is deposited by LPCVD. After patterning the gate polysilicon, the PMOS source/drain regions are formed by boron ion implantation with dose of 3×10^{15} cm^{-2} at 30 keV, and NMOS source/drain regions are formed by asenic ion implantation with dose of 7×10^{15} cm^{-2} at 110 keV, and activated at 800 °C for 30 minutes in N_2. After 50 nm thick first silicon nitride deposition, a 200 nm thick polysilicon etch-channel is deposited by LPCVD. The polysilicon etch-channel is patterned as 100 μm × 100 μm square having projecting areas at each side and corner of the square. An 800 nm thick phosphosilicate-glass (PSG) is deposited by LPCVD to protect underlying devices from sodium contamination. After contact-holes are opend, PSG is reflowed to round the corner of contact-holes at 1000 °C. This is followed by a 100 nm thick second silicon nitride layer deposition. On the second nitride layer, a 200 nm thick polysilicon is deposited by LPCVD. Boron ion implantation is then carried out with a dose of 1×10^{16} cm^{-2} at 30 keV and activated at 900 °C for 30 minutes. Polysilicon piezoresistors are patterned by RIE. This is followed by LPCVD deposition of a 300 nm thickness third silicon nitride layer to cover the polysilicon piezoresistors from alkali etchant. Etch-holes are opened such as to penetrate the second and the third silicon nitride layers to reach the polysilicon etch-channel by RIE. The entirely of polysilicon etch-channel is removed through the etch-holes by solution of 50 wt% KOH, thereby forming the movable diaphragm and the cavity as the

reference pressure chamber. Electrical contact holes are then opened. An 800 nm thick aluminum layer is sputtered to form interconnection and bonding pads. After delineation, the aluminum layer is sintered at 400 °C for 30 minutes. A 1 μm thick final silicon nitride is deposited to seal the etch-holes and passivate the surface over the chip by plasma CVD. The pressure inside the cavity is less than 0.3 Torr. Hence, the pressure sensor can measure absolute pressure. Finally, windows of the bonding pads are opend by RIE, the wafer process of the pressure imager is completed. Figure 3 shows photograph of the completed 32 × 32 (1k) pressure imager. The overall chip size is 10 mm × 10 mm.

READOUT SYSTEM

The main features of this approach to pressure imaging are two-line readout system from full bridge of piezoresistors in the each individual cell, and array exciting system to low power consumption in large scale integration of piezoresistive pressure sensors.

Equivalent circuit of the array is shown in Fig. 4. Four polysilicon piezoresistors are arranged in full bridge, therefore two-line system is necessary for sensing signal readout. Two CMOS analog switches (SX_1, SX_2) to readout the sensing signal from the piezoresistive bridge, an NMOS power switch (Q_1) to excite the piezoresistive bridge and CMOS logic circuits (NAND and NOT) are included in the each individual cell. The array is addressed as a memory. Since only the NMOS switch (Q_1) selected from an X-row and a Y-colum is turned on in the array matrix, the power consumption required is equivalent to the value of the power consumption only for one cell in spite

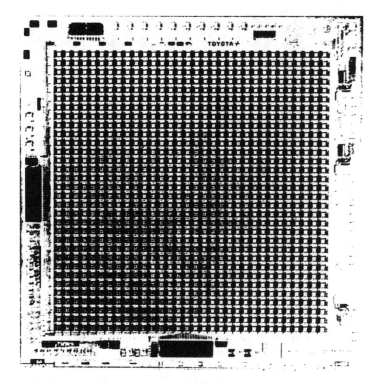

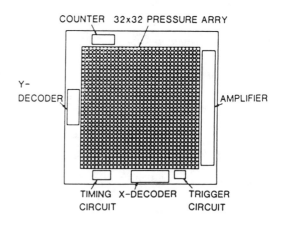

Fig. 3. Photograph of the silicon pressure imager with CMOS processing circuits.

of large scale integration. Signal readout from piezoresistive bridge is transmitted by the CMOS analog switches (SX_1, SX_2) which are turned on from selected X-row. Also, signal readout lines are selectivily conneced to analog output by CMOS analog switches (SY_1, SY_2) selected from a Y-colum to get higher rate of signal transmitting speed.

Figure 5 shows the overall organization of the signal processing electronics. A timing circuit, a 10-bit counter, an X-decoder, a Y-decoder, a serial/random selector, a trigger circuit, three operational amplifiers and other interface circuits are formed by CMOS circuits around the array on the chip.

Figure 6 shows the timing waveforms for outputs of the pressure imager. Expected cell outputs change just after rising of edges of the scan pulses. The sensing signals from 1024 cells are taken out as a serial waveform to the analog output terminal. The output pressure sensitivity of 100 $mV/g/mm^2$ is obtained. Typical clock frequency is 4 MHz. The scan of the array can be asynchronous pulse to the clock. A scan pulse of 60 kHz is typically provided, and an access time to a cell is 16 μsec, therefore the readout time is 16 msec (60 Hz) for a frame. Also, random access readout can be asynchronously made by external 10-bit parallel code. Table I shows the specification of the above proposed silicon pressure imager. One possible package assembly for the final pressure imager is shown in Fig. 7.

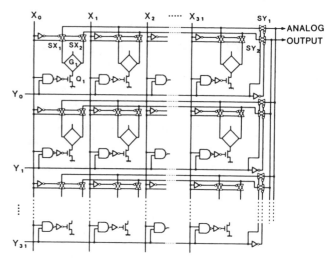

Fig. 4. Equivalent circuit of pressure imaging array.

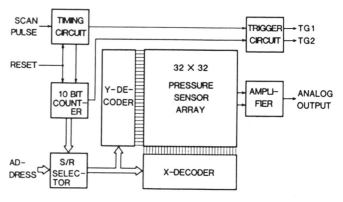

Fig. 5. Electrical block diagram of the pressure imager.

CONCLUSION

A 32 × 32 (1k)-element silicon pressure imager with CMOS processing circuits was fabricated for the detection of precise two-dimensional pressure distributions. The pressure imager consists of an X-Y matrix organized array of micro pressure sensing cells with a spacing of 250 μm and CMOS processing circuits. The sensing signals from 1024 cells are taken out as a serial analog waveform. Output sensitivity is designed to be 100 mV/g/mm^2. The readout time is 16 msec (60 Hz) for a frame.

The interpretation of data from the imager to permit image recognition of pressure/force is an area which will require considerable work. The realization of the pressure imager with high two-dimensional resolution will promote tactile sensor application for precision robotics.

ACKNOWLEDGMENT

The authors wish to thank Dr. M. Takigawa and Dr. T. Takeuchi for suggestion and support in integrated sensor design and fabrication.

REFERENCES

[1] K. E. Petersen, "Silicon Sensor Techologies", Technical Digest, IEEE IEDM, (Washington, Dc), Dec. 1985, pp. 2-7.

[2] H. Guckel and D. W. Burns, "Laser-Recrystallized Piezoresisteve Micro-Diaphragm Sensor", Digest of Technical Papers, 3rd Int. Conf. on Solid-State Sensors and Actuators, pp. 182-185, (Philadelphia), Jun, 1985.

[3] K. Chun and K. D. Wise, "A High-Performance Silicon Tactile Imager Based on a Capacitive Cell", IEEE Trans. Electron Devices, vol. ED-32, No 7, pp. 1196-1201, 1985.

[4] D. L. Polla, R.S. Muller and R. M. White, "Integraed Multisensor Chip", IEEE Electron Device Letters, vol. EDL-7, No. 4, pp. 254-256, 1986.

[5] S. Sugiyama, T. Suzuki, K. Kawahata, K. Shimaoka, M. Takigawa and I. Igarashi, "Micro-Diaphragm Pressure Sensor", Technical Digest, IEEE IEDM, (Los Angeles), Dec. 1986, pp. 184-187.

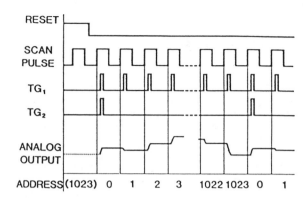

Fig. 6. Timing waveforms for outputs of the pressure imager.

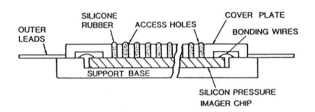

Fig. 7. Cross section of the proposed package for the pressure imager.

Table I
Typical specification of the silicon pressure imager.

NUMBER OF CELLS	1024 (32 x 32)
CELL SPACING	250 μm
DIAPHRAGM	100 μm x 100 μm
CHIP SIZE	10 mm x 10 mm
IC TECHNOLOGY	3 μm - CMOS
NUMBER OF FETS	ABOUT 16,000
SENSITIVITY	100 mV/g/mm^2 (MIN)
READOUT RATE	16 msec (TYP)
CLOCK FREQUENCY	4 MHz (TYP)
SUPPLY VOLTAGE	5 V D.C.
POWER CONSUMPTION	ABOUT 50 mW

Resonant-Microbridge Vapor Sensor

ROGER T. HOWE, MEMBER, IEEE, AND RICHARD S. MULLER, SENIOR MEMBER, IEEE

Abstract—A novel integrated vapor sensor is described that incorporates a polycrystalline silicon microbridge coated with a thin polymer film. The microbridge is resonated electrostatically and its vibration is detected capacitively using an integrated NMOS circuit. Vapor uptake by the polymer increases the mass-loading on the microbridge, thereby perturbing the first resonant frequency of the microbridge. In the prototype device, a 150-nm-thick layer of negative photoresist coats a 153-μm-long 1.35-μm-thick polycrystalline silicon microbridge. The phase between the excitation and output voltages at resonance is monitored as the sensor output signal. Exposure to saturated xylene vapor produces a phase shift of $-8°$ with a response time of less than 7 min.

I. INTRODUCTION

INTEGRATED SENSORS, in which the sensor is fabricated using silicon planar technology, offer advantages of miniaturization, integrated signal-processing electronics, multisensing capability, and the economies of batch fabrication [1], [2]. In particular, an integrated sensor for detecting organic vapors is attractive in comparison with existing analytical techniques, which require bulky and expensive instrumentation. Integrated vapor sensors that detect charge transfer from the bulk to a surface layer (gate-controlled diode and ion-sensitive FET) have been reported [3], [4], and a miniature gas chromatograph has been built on a silicon wafer [5].

An alternative means of detecting vapors is through their interaction with polymer films. The physical properties of a polymer film change upon exposure to an organic vapor. A chemically compatible organic vapor is both adsorbed onto the surface and absorbed into the bulk of the polymer, increasing the mass and thickness and altering the mechanical and electrical properties of the polymer [6]. This phenomenon, referred to simply as "sorption," is exploited in surface-acoustic-wave (SAW) vapor sensors, where the velocity of the elastic wave is perturbed by vapor uptake in a polymer film coating the substrate [7], [8].

In this paper, we describe the resonant-microbridge va-

Manuscript received May 15, 1985; revised November 12, 1985. This work was supported by the National Science Foundation under Grants ECS-78-21854 and ECS-81-20562, and by the University of California MICRO program. R. T. Howe was supported by a California Fellowship in Microelectronics and an IBM Fellowship.

R. T. Howe was with the Department of Electrical Engineering and Computer Sciences and the Electronics Research Laboratory, University of California, Berkeley, CA 94720. He is now with the Department of Electrical Engineering and Computer Science, Massachusetts Institute of Technology, Cambridge, MA 02139.

R. S. Muller is with the Department of Electrical Engineering and Computer Sciences and the Electronics Research Laboratory, University of California, Berkeley, CA 94720.

IEEE Log Number 8607552.

por sensor, a novel device that also utilizes vapor sorption by a polymer film. The mechanical element of the sensor is a polymer-coated polycrystalline silicon (poly-Si) microbridge. Underlying electrodes are used to electrostatically excite and capacitively detect the vibration of the microbridge. An integrated NMOS circuit is employed in converting the vibration into an electrical signal. Vapors are sorbed by the polymer film, increasing the mass-loading on the microbridge and thereby shifting its first resonant frequency [9]. It is not essential to use a microbridge as the resonant microstructure; the concept as first described utilized a polymer-coated cantilever beam [10].

We begin by describing the structure and fabrication of the prototype device, which is the first integrated-sensor application of poly-Si microstructure technology [11], [12]. The electromechanical frequency response of the driven microbridge must be modeled in order to design the structure and predict the sensitivity to changes in the polymer film mass. The analysis follows the treatment by Nathanson *et al.* of the resonant gate transistor, which was the initial demonstration of an integrated resonant microstructure [13]. Finally, experimental measurements of frequency response and sensitivity of the prototype are compared with theoretical predictions.

II. DEVICE STRUCTURE AND FABRICATION

Fig. 1 is a schematic cross section of the prototype resonant-microbridge vapor sensor. The microbridge is resonated by means of a dc bias voltage V_P and a sinusoidal drive voltage v_d applied to a diffused electrode beneath the center of the grounded microbridge. Vibration of the microbridge causes the air-gap capacitance C_s to be time-varying, generating a current $i_s = V_S \, dC_s/dt$. The derivative of the microbridge deflection is detected since C_s is proportional to the deflection; the desirability of this vibration-detection scheme will be clarified in the analysis of the electromechanical frequency response.

Measurement of the sense current i_s is indicated symbolically in Fig. 1. It is convenient to convert i_s into a voltage signal, amplify it, and buffer it using the on-chip NMOS circuit shown in Fig. 2. Neglecting parasitic and device capacitances, i_s produces a voltage $v_s \approx i_s/g_{m1}$ at the gate of transistor M_3. The circuit includes an external voltage source V_X that is necessary for biasing the amplifier if the enhancement and depletion threshold voltages deviate from their design values. Transistors M_4, M_5, and M_6 constitute a standard NMOS active load for transistor M_3 that minimizes gain degradation due to the body effect [14]. A source-follower stage buffers the output voltage.

Reprinted from *IEEE Trans. Electron Devices*, vol. ED-33, no. 4, pp. 499–506, April 1986.

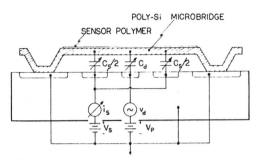

Fig. 1. Schematic cross section of resonant-microbridge vapor sensor.

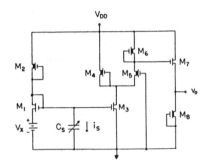

Fig. 2. NMOS vibration-detection circuit.

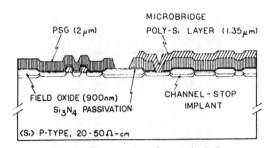

Fig. 3. Cross section after masks 1–6.

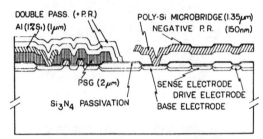

Fig. 4. Cross section of prototype resonant-microbridge vapor sensor.

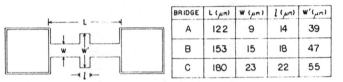

BRIDGE	L (μm)	W (μm)	l (μm)	W' (μm)
A	122	9	14	39
B	153	15	18	47
C	180	23	22	55

Fig. 5. Poly-Si microbridge dimensions.

The sensor fabrication process includes the essential steps for poly-Si microbridge and NMOS transistor processing, as well as additional steps needed to ensure compatibility [15]. In the following, we outline the fabrication sequence; a detailed description is given elsewhere [16]. The first masking steps are indistinguishable from the conventional poly-Si gate, depletion-load NMOS process using local-oxidation (LOCOS) isolation technology. After annealing the source–drain implant, which also dopes the active areas for the drive and sense capacitor electrodes, a 100-nm-thick layer of Si_3N_4 is deposited by low-pressure chemical vapor deposition (CVD). The purpose of this layer is to provide HF-resistant passivation for the drive and sense capacitor electrodes.

Phosphosilicate glass (7-percent phosphorus by weight, 2 μm thick) is next deposited on the wafer using atmospheric CVD. After densification at 1050°C, the surface is damaged by a low-energy argon implant. Windows are opened in the PSG at the bases of the microbridge. The PSG window edges are tapered at about 40° because the damaged layer etches more rapidly in buffered HF [17]. The Si_3N_4 passivation layer must be removed from the base windows by plasma etching. After etching the initial oxide from the base windows, a 1.35-μm-thick layer of poly-Si is deposited by low-pressure CVD. Internal stress in the poly-Si film is reduced by annealing at 1050°C in nitrogen for 20 min [18]. A trench, necessary for protecting the NMOS transistors from attack during the microbridge-undercutting etch, is then etched in the PSG film around the microbridge. The cross section of a portion of the device after this step is illustrated in Fig. 3.

NMOS circuit fabrication is completed with the etching of contact windows, surface-state anneal, metallization with sputtered Al/1-percent Si, and sintering. Since the transistors are covered with a film of Si_3N_4, the forming gas anneal is performed at 800°C before metallization [19]. It remains to protect the area outside the trench from buffered-HF attack. Nested layers of plasma-hardened AZ-1350J photoresist that are postbaked at 225°C are used in the prototype process. The plasma-hardening step consists of a brief exposure to CF_4/4-percent O_2 plasma to crosslink the surface of the resist and prevent flow during the high-temperature postbake [20]. This approach provides about 40 min of protection against buffered HF, after which the resist layers lose adhesion to the Si_3N_4 passivation layer at the perimeter of the trench.

Negative photoresist (Kodak 747 packaged as MN-50) is patterned on top of the microbridge in the final masking step as the sensor polymer for the prototype. We use a 150-nm-thick layer of this material for convenience in order to demonstrate the operation of the resonant-microbridge vapor sensor. Finally, the PSG island on which the microbridge poly-Si layer lies in Fig. 3 is etched in buffered HF. Since etching progress cannot be monitored visually, the etch time is estimated from previous experiments. Thirty minutes is sufficient to undercut 23-μm-wide poly-Si layers, for a 2-μm-thick PSG spacer layer. Fig. 4 illustrates the final cross section of the prototype.

Several microbridge designs are fabricated in order to investigate the electromechanical frequency response. Fig. 5 lists their lateral dimensions; the thickness is 1.35 μm and the substrate offset is 2 μm for the prototype designs. The center section of the microbridge is widened in order to increase the electrostatic driving force. To investigate the effect of apertures on the sharpness of the

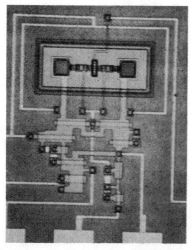

Fig. 6. Optical micrograph of sensor with type B apertured microbridge.

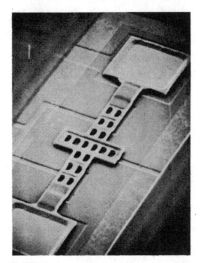

Fig. 7. SEM of type B microbridge with apertures.

mechanical resonance, each design is fabricated with and without apertures. An optical micrograph of the resonant-microbridge vapor sensor incorporating an apertured type B poly-Si microbridge is shown in Fig. 6; Fig. 7 is a scanning-electron micrograph of this microbridge.

III. ELECTROMECHANICAL FREQUENCY RESPONSE

In order to design this sensor and predict its performance, a first-order model is needed for the electromechanical frequency response of a driven microbridge with integrated NMOS vibration-detection circuit. In this section, we first consider the mechanical resonance and then analyze the electrostatic excitation and capacitive pickup of the vibration. Finally, we analyze the spurious output signal due to coupling between the drive and sense electrodes through the substrate resistance.

As a starting point, the deflection of a uniform bridge induced by a distributed sinusoidal driving force can be found by classical methods. The force is nonzero only on a segment of length l centered at the midpoint. Using d'Alembert's principle, the mechanical transfer function $M(j\omega) = \overline{Y}(j\omega)/F(j\omega)$ relating the average deflection of

the center section to the total applied force is [16], [21]

$$M(j\omega) = \frac{a^2 K^{-1}}{1 - (\omega/\omega_1)^2 + j(\omega/Q\omega_1)} \quad (1)$$

where ω_1 is the first resonant frequency, K is the effective spring constant, Q is the quality factor, and a is the mean of the first normal mode $X_1(x)$ over the center section. Although (1) is derived for a uniform bridge, the result is identical to that of a general mass-spring-dashpot resonator, suggesting that it can describe the resonance of the complex microbridge shown in Fig. 5 if appropriate values of ω_1, K, and Q are substituted.

The first resonant frequency of the microbridge is estimated using Rayleigh's method [22]. Neglecting the contribution of the polymer film to the elastic stored energy, we find that

$$\omega_1^2 = (Et^2/12L^4) \frac{\int_0^1 W(\theta) \, (d^2X_1/d\theta^2)^2 \, d\theta}{\int_0^1 \rho(\theta) \, W(\theta) \, X_1^2 \, d\theta} \quad (2)$$

where E is Young's modulus, t is the thickness of the poly-Si microbridge, and $\theta = x/L$. The density ρ varies along the microbridge because of apertures and the patterning of the polymer film. A value for the effective spring constant is found from the resonant frequency and the mass M of the microbridge, since by definition $K = \omega_1^2 M$. Finally, the quality factor at atmospheric pressure is limited by energy loss due to pumping air from beneath the microbridge, for the typical case where the substrate offset d satisfies the inequality $d < W/3$ [23]. By extending the first-order model for a micromechanical cantilever beam, an approximation can be found for the quality factor for an unapertured microbridge with widened center section [16]

$$Q^{-1} = \mu(E\rho)^{-1/2} \, t^{-2} d^{-3}[(W(L-l)/2)^2 + (W'l)^2/2] \quad (3)$$

where μ is the viscosity of air. Since the air under an apertured microbridge can escape more easily, we expect a sharper resonance and a higher quality factor than predicted by (3).

Having analyzed the mechanical resonance, we now turn to the electrostatic forcing of the resonance and the operation of the integrated vibration-detection circuit. The voltage sources v_d and V_P exert a force on the microbridge. The force is found by differentiating the electrostatic stored energy in the drive capacitor with respect to the bridge-substrate gap d and collecting terms at the frequency of the drive voltage. For small deflections, the result can be linearized, yielding in phasor notation

$$F(j\omega) = (\overline{C}_d d^{-1}) \, V_P V_d(j\omega) - (\overline{C}_d d^{-2}) \, V_P^2 \overline{Y}(j\omega) \quad (4)$$

where $\overline{C}_d$ is the mean air-gap drive capacitance (Fig. 1). By substituting for $F(j\omega)$ in terms of the mechanical transfer function $M(j\omega)$ and the average deflection

$\overline{Y}(j\omega)$, we can derive the electromechanical transfer function $N(j\omega) = \overline{Y}(j\omega)/V_d(j\omega)$

$$N(j\omega) = \frac{a^2 \overline{C}_d V_P d^{-1} K^{-1}}{[1 - (\omega/\omega_1)^2 + g] + j(\omega/Q\omega_1)}. \quad (5)$$

The parameter $g = a^2 \overline{C}_d V_P^2 d^{-2} K^{-1}$ in (5) reduces the peak in $N(j\omega)$ at the mechanical resonance.

The first step in analyzing the vibration-detection scheme is to evaluate the current i_s induced by motion of the microbridge. For small deflections, the sense capacitance C_s is given by

$$C_s = \overline{C}_s[1 - (\beta/d)\overline{y}] \quad (6)$$

where β is the ratio of the average deflection over the sense electrode to the average deflection over the drive electrode. Since we detect the derivative of the sense capacitance, as discussed above, the phasor sense current is

$$I_s(j\omega) = -j\omega V_S \overline{C}_s(\beta/d) \overline{Y}(j\omega). \quad (7)$$

The advantage of detecting dC_s/dt is the $-90°$ phase shift between i_s and $\overline{y}$. At mechanical resonance, (5) shows that $\overline{y}$ lags v_d by approximately $-90°$. Consequently, a feedback amplifier converting i_s into v_d need only invert the signal to satisfy the phase requirement for self-sustained oscillation.

We proceed to find the vibration-induced voltage v_s at the gate of transistor M_3 in the NMOS circuit shown in Fig. 2. The small-signal resistance in parallel with C_s is [24]

$$r_s = g_{m1}^{-1} \| g_{mb2}^{-1} \approx g_{m1}^{-1}. \quad (8)$$

The input capacitance C_i of the NMOS circuit is approximately the sum of C_{gs3}, the Miller capacitance C_M, and the depletion capacitance of the sense electrode C_s^d. Therefore, the vibration-induced voltage is

$$V_s(j\omega) = -r_s I_s(j\omega)/(1 + j(\omega/\omega_a)) \quad (9)$$

where $\omega_a = (r_s C_i)^{-1}$.

The drive voltage also causes a spurious feedthrough signal v_f at the gate of M_3 by means of capacitive coupling through the substrate. Fig. 8 shows the equivalent circuit for analyzing the feedthrough. The substrate resistance R_B and the depletion capacitance of the drive electrode diffusion C_d^d couple v_d to the sense electrode. Straightforward circuit analysis yields

$$V_f(j\omega) = \frac{-C_s^d C_i^{-1} \omega^2 (\omega_a \omega_b)^{-1} V_d(j\omega)}{(1 + j\omega/\omega_a)(1 + j\omega/\omega_b)} \quad (10)$$

where we have used the fact that $\omega_b = (R_B C_d^d)^{-1} \gg \omega_a$ [16].

We define the NMOS amplifier gain between the gate of M_3 and the output voltage $V_o(j\omega)$ indicated on Fig. 2 as $A_v(j\omega)$. Therefore, the overall transfer function from drive to output voltage can be found from (5), (7), (9), and (10)

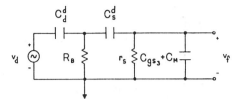

Fig. 8. Small-signal circuit for feedthrough calculation.

$$T(j\omega) = \frac{A_v(j\omega)}{(1 + j\omega/\omega_a)} \left\{ \frac{-j\omega\Gamma}{[1 - (\omega/\omega_1)^2 + g] + j\omega/Q\omega_1} \right.$$
$$\left. + \frac{-C_s^d C_i^{-1} \omega^2 (\omega_a \omega_b)^{-1}}{1 + j\omega/\omega_b} \right\}. \quad (11)$$

The parameter Γ in (11) is given by

$$\Gamma = a^2 \beta V_P V_S \overline{C}_d \overline{C}_s r_s d^{-2} K^{-1}. \quad (12)$$

It is clear from (11) that feedthrough contributes a term to $|T(j\omega)|$ that increases with frequency, obscuring the signal from the mechanical resonance of the microbridge. The above analysis assumes small dynamic deflections and that the voltage V_P does not cause significant static deflection of the microbridge [13], [16].

IV. THEORY OF SENSOR RESPONSE

The process of vapor penetration and dispersal in a polymer is referred to by the generalized term "sorption," since the vapor molecules are adsorbed onto the polymer surface as well as absorbed into the bulk of the polymer. Separation of adsorption and absorption is difficult as the polymer surface is ill-defined due to microporosity [6]. Sorption of organic vapors by polymers often involves large vapor uptake and consequent swelling of the polymer, in marked contrast to the familiar adsorption of gases onto crystalline surfaces.

A qualitative model based on the theory of polymer solutions [6] indicates that the solubility parameter difference between the polymer and vapor determines the amount of sorption. The solubility parameter δ is defined as $(\Delta \overline{E}_v/\overline{V})^{1/2}$, where $\Delta \overline{E}_v$ is the molar energy of vaporization and $\overline{V}$ is the molar volume [25]. Although not directly measurable for polymers, δ can be estimated from monomer properties. Solubility parameters range from 16 $(J \cdot m^{-3})^{1/2}$ for nonpolar polymers such as polyethylene to 31.5 $(J \cdot m^{-3})^{1/2}$ for highly polar polymers such as polyacrylonitrile [26]. Polymer sorption exhibits a limited selectivity among classes of organic vapors, since little sorption occurs when $|\delta_2 - \delta_1| > 6 (J \cdot m^{-3})^{1/2}$, where δ_1 and δ_2 are the solubility parameters of the polymer and vapor. However, significant uptake (over 50 percent of the sorption for the case of matched solubility parameters) occurs for $|\delta_2 - \delta_1| < 4 (J \cdot m^{-3})^{1/2}$ [16].

Sorption isotherms (equilibrium uptake as a function of vapor partial pressure) exhibit hysteresis if the ambient temperature T_a is less than the glass-transition temperature of the polymer T_g [6]. For such glassy polymers ($T_a < T_g$), irreversible chain-segment relaxation occurs during sorption, especially for large vapor uptakes. Rubbery polymers, for which $T_a > T_g$, exhibit little or no hysteresis, since the polymer chain segments have sufficient thermal energy to accommodate reversibly the sorbed vapor molecules.

The transient uptake of vapor is qualitatively different in rubbery and glassy polymers. Whereas penetrant dispersal in rubbery polymers follows Fick's law with large diffusivities that typically increase exponentially with sorbed vapor concentration, the process is much more complicated below the glass-transition temperature [27], [28]. A practical problem for sensor applications is that the rate of sorption is orders of magnitude slower in glassy polymers [16]. Clearly, a rubbery polymer is desirable for vapor sensing applications; unfortunately, the base polymer in negative photoresist is cyclized poly-isoprene, which is glassy at room temperature.

The resonant frequency shift resulting from vapor uptake by the polymer film can be calculated from (2). For the typical case where the polymer film is much thinner than the poly-Si microbridge, the normalized sensitivity of the first resonant frequency to a change $\Delta\sigma_f$ in the polymer film surface density is [16]

$$S_{f_1} = \frac{\Delta f_1/f_1}{\Delta\sigma_f/\sigma_f} \approx \left(-\frac{1}{2}\right)(\sigma_f/\sigma_p)\frac{\int_0^1 W_f(\theta)\,X_1^2(\theta)\,d\theta}{\int_0^1 W(\theta)\,X_1^2(\theta)\,d\theta}$$

$$(13)$$

in which $\theta = x/L$, σ_p is the poly-Si microbridge surface density, and W_f is the width of the sensor polymer film. Alignment tolerances generally require that $W_f(\theta) < W(\theta)$ and apertures may necessitate that $W_f(\theta) = 0$ for some values of θ. As a result, the ratio of integrals in (13) is always less than unity.

The resonant-microbridge vapor sensor can be operated in a closed-loop oscillator mode, which is desirable because the output signal is a shift in the frequency of oscillation. The associated feedback amplifier need only invert the output voltage of the detection circuit to satisfy the phase condition for oscillation, as discussed above. However, a large feedthrough signal and excessive $1/f$ noise in the NMOS transistors make this mode of operation impractical for the prototype [16].

Therefore, we investigate the sensitivity of the device by operating it in an open-loop mode, with the frequency of the drive voltage fixed at the nominal resonant frequency. As illustrated in Fig. 9, vapor uptake by the polymer film shifts the phase curve of the transfer function. The phase change observed at frequency f_1 is related to Δf_1 through the slope of the phase curve

$$\Delta\phi_1 = -\left.\frac{\partial\phi}{\partial f}\right|_{f_1}\Delta f_1. \qquad (14)$$

Therefore, the phase sensitivity $S_{\phi_1} = \Delta\phi_1/(\Delta\sigma_f/\sigma_f)$ is related to the frequency sensitivity by

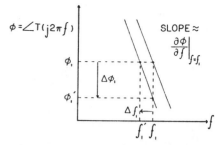

Fig. 9. Effect of vapor sorption on transfer function phase near resonance.

$$S_{\phi_1} = -\left.\frac{\partial\phi}{\partial f}\right|_{f_1}S_{f_1}f_1. \qquad (15)$$

Thus, a microbridge design must have both a large frequency sensitivity (13) and a steep phase curve at resonance for maximum phase sensitivity.

V. EXPERIMENTAL FREQUENCY RESPONSE

The transfer function is measured using an HP-3570A Network Analyzer and an HP-3330B Automatic Synthesizer. A test amplifier consisting of transistors M_3 through M_8 is fabricated to allow measurement of $A_v(j\omega)$. Small-signal parameters of the NMOS transistors are found from curve tracer measurements using the first-order MOS theory [16].

Several parameters are needed to apply the frequency response theory. The Young's modulus E of poly-Si is found by substituting the first normal mode for a uniform bridge into (2). Measured values of ω_1 for unapertured microbridges can be fit to within ± 10 percent with $E = 4 \times 10^{10}$ N $\cdot$ m^{-2}. Agreement with measurement is also good for the type A and type B apertured microbridges; however, the estimate is poor for the type C apertured microbridge. A possible explanation is that the first normal mode of the type C apertured microbridge, which has 44 widely distributed apertures [29], differs markedly from that of a uniform bridge. A second parameter in (11) is the quality factor of the resonance. For unapertured microbridges, Q can be found directly from (3). No attempt is made to predict the quality factors of the apertured microbridges.

The feedthrough expression in (10) requires a value for the substrate resistance, R_B. Since the resistivity of the test wafers is only specified as between 20 and 50 $\Omega \cdot$ cm, R_B is treated as an adjustable parameter. We determine R_B by equating the theoretical and measured $|T(j\omega)|$ at a frequency well beyond the mechanical resonance. This approach is verified by calculating the resistivity from R_B, using a simple trapezoidal model for the substrate resistance; in all cases, the inferred resistivities are within the specified range.

Fig. 10 compares the theoretical model of (11) with the experimental transfer function for the type B unapertured microbridge. The substrate resistance is the only empirical parameter for the theoretical curve. Although the general shapes of the amplitude and phase curves are correct, the agreement is not exact. Measurements on devices of

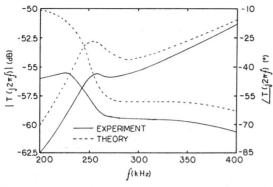

Fig. 10. Experimental and theoretical transfer functions for type B microbridge without apertures.

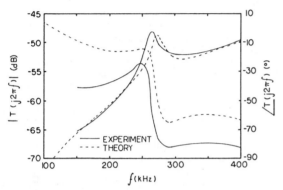

Fig. 11. Experimental and theoretical transfer functions for apertured type B microbridge.

this type indicate a variation of several decibels in the resonant peak; the experimental curve shown in Fig. 10 is typical. Clearly, the mechanical resonance is obscured by the large feedthrough signal. Numerical agreement between measured and calculated quality factors for unapertured microbridges is excellent [16].

In order to predict the frequency response of apertured microbridges, we substitute the measured quality factor into (11). Fig. 11 shows the comparison of (11) with the measured transfer function for the type B apertured microbridge. Agreement is good for the amplitude curve, although the first resonant frequency given by (2) is somewhat high. From these results and those for the types A and C microbridges, we conclude that the linear transfer function theory describes the essential features of the magnitude and phase responses [16], [29].

VI. EXPERIMENTAL SENSOR RESPONSE

The apertured type B microbridge is selected for investigation of the sensor response since (15) indicates that it has the largest phase sensitivity of the prototype designs. Substituting the dimensions and densities of the poly-Si and negative photoresist layers for this design into (13), we find that $S_{f_1} = -5.5 \times 10^{-3}$. From the measured phase slope and resonant frequency (Fig. 11), (15) predicts a phase sensitivity $S_{\phi_1} \approx -4°$. To demonstrate the sensor response, we select an appropriate organic vapor to give the maximum output signal. The organic solvent xylene is used, since its solubility parameter closely matches that

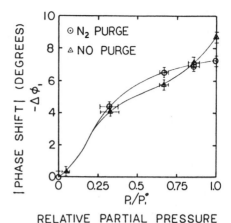

Fig. 12. Equilibrium phase shift as a function of relative xylene partial pressure, with and without purging in nitrogen after each measurement.

of the base polymer of negative photoresist, cyclized polyisoprene [25].

A range of partial pressures of xylene vapor is obtained by mixing N_2 and saturated xylene vapor (in N_2) in a gas proportioner. The phase of the transfer function at the nominal resonant frequency of the test device is monitored using an HP-3570 Network Analyzer. Equilibrium sensor response curves are measured both with and without flushing with N_2 after each measurement at a given xylene partial pressure. Fig. 12 is a plot of the sensor response $\Delta\phi_1$ as a function of relative partial pressure p_1/p_1^o, where p_1^o is the saturated vapor pressure at room temperature. Horizontal error bars reflect the measurement uncertainty in the gas proportioner whereas the vertical error bars represent the phase-measurement noise. As expected, the response exhibits hysteresis, since negative photoresist is glassy at room temperature. The measured phase shift at saturation and the slope of the phase curve in Fig. 11 imply that the resonant frequency shift is $\Delta f_1 \approx -2.7$ kHz. In addition, the calculated phase sensitivity implies that $\Delta\sigma_f/\sigma_f \approx 2$ for saturated xylene vapor. Such large vapor uptake and consequent swelling of the polymer film are not unusual for polymers and vapors with similar solubility parameters [25].

The transient response of $\Delta\phi_1$ for changes in xylene vapor pressure between zero and saturation is plotted in Fig. 13. By convention, the phase shift is normalized by the equilibrium value $\Delta\phi_1(\infty)$ and the abscissa is $t^{1/2}$. The overshoot observed in the absorption curve is not typical of the mass increase in polymer films during sorption. However, this phenomenon is observed in the elastic properties of polymers during sorption, suggesting that the polymer elasticity may also affect the sensor response [16]. Equilibrium is reached in less than 7 min for changes in vapor pressure between zero and saturation. In contrast, nearly 20 min are needed before the phase stabilizes when the xylene partial pressure is increased from zero to 32 percent of saturation [16]. This result is consistent with the general observation that penetrant mobility increases with increasing vapor uptake [27].

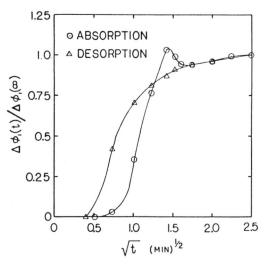

Fig. 13. Transient phase shift for changes in xylene partial pressure from zero to saturation.

VII. CONCLUSIONS

We have demonstrated a highly sensitive silicon integrated vapor sensor that incorporates a resonant poly-Si microbridge. In order to design the resonant microbridge and to predict the sensor response, a linear first-order theory for the electromechanical transfer function is developed. Measurements of the transfer function for three microbridge designs, with and without apertures, show that the theory explains the important features of the frequency response, including the feedthrough signal. This theory is used to predict the resonant frequency shift and the phase shift due to vapor sorption. The measured response of the prototype is consistent with equilibrium and transient sorption processes in glassy polymers. The prototype exhibits a responsivity of -0.3 Hz/ppm when exposed to saturated xylene vapor, comparing favorably with that estimated for the thin-membrane-SAW vapor sensor operating at 30 MHz [9], [16]. Frequency domain noise ultimately determines the minimum detectable vapor concentration. However, the excessive $1/f$ noise in the prototype NMOS amplifier precludes any measurement of the noise limits of this sensor.

Several aspects of the transfer function and sensor response theory have implications for improved sensor performance. High substrate resistance leads to excessive feedthrough in the prototype. An obvious solution is to use a high-resistivity epitaxial layer on a low-resistivity substrate [16]. The simplest way to reduce the $1/f$ noise in the MOS transistors is to remove the Si_3N_4 passivation layer from the transistors. Operation of the sensor in the closed-loop mode with a frequency-shift output signal should then be feasible. Finally, the theory of polymer sorption indicates that rubbery polymers are more attractive for vapor-sensing applications because of reduced hysteresis and faster transient response. The limited selectivity of the sorption process can be enhanced by use of pattern-recognition techniques to interpret the response of a multisensor chip incorporating several resonant-microbridge sensors with different polymer films [8], [16].

ACKNOWLEDGMENT

The authors would like to thank Prof. R. M. White, Prof. P. K. Ko, G. L. Halac, and J. I. Goicolea for helpful discussions and the staff of the Berkeley Microfabrication Laboratory for technical assistance.

REFERENCES

[1] W. G. Wolber and K. D. Wise, "Sensor development in the microcomputer age," *IEEE Trans. Electron Devices*, vol. ED-26, pp. 1864–1874, Dec. 1979.
[2] K. E. Petersen, "Silicon as a mechanical material," *Proc. IEEE*, vol. 70, pp. 420–457, May 1982.
[3] C.-C. Wen, R. C. Chen, and J. N. Zemel, "Gate-controlled diodes for ionic concentration measurement," *IEEE Trans. Electron Devices*, vol. ED-26, pp. 1945–1951, Dec. 1979.
[4] J. N. Zemel, "Ion-sensitive field affect transistors and related devices," *Anal. Chem.*, vol. 47, pp. 255A–265A, Feb. 1975.
[5] S. C. Terry, J. H. Jerman, and J. B. Angell, "A gas chromatograph air analyzer fabricated on a silicon wafer," *IEEE Trans. Electron Devices*, vol. ED-26, pp. 1880–1886, Dec. 1979.
[6] D. Machin and C. E. Rogers, "Sorption," in *Encyclopedia of Polymer Science and Technology*. New York: McGraw-Hill, vol. 12, 1970, pp. 679–700.
[7] H. Wohltjen and R. Dessy, "Surface acoustic wave probe for chemical analysis I–III," *Anal. Chem.*, vol. 51, pp. 1458–1475, Aug. 1979.
[8] C. T. Chuang, R. M. White, and J. J. Bernstein, "A thin-membrane-surface-acoustic-wave vapor-sensing device," *IEEE Electron Device Lett.*, vol. EDL-3, pp. 145–147, June 1982.
[9] R. T. Howe and R. S. Muller, "Integrated resonant-microbridge vapor sensor," in *IEDM Tech. Dig.* (San Francisco, CA), pp. 213–216, Dec. 1984.
[10] R. T. Howe, "An evaluation of surface-acoustic-wave and cantilever-beam oscillators for integrated organic-vapor sensors," M.S. thesis, Dept. of Electrical Engineering and Computer Sciences, Univ. of California, Berkeley, Oct. 1981.
[11] R. T. Howe and R. S. Muller, "Polcrystalline silicon micromechanical beams," in *Extended Abstracts, Electrochem. Soc. Meet.* (Montreal, Quebec, Canada), vol. 82-1, pp. 184–185, May 1982.
[12] R. T. Howe and R. S. Muller, "Polycrystalline silicon micromechanical beams," *J. Electrochem. Soc.*, vol. 130, pp. 1420–1423, June 1983.
[13] H. C. Nathanson, W. E. Newell, R. A. Wickstrom, and J. R. Davis, "The resonant-gate transistor," *IEEE Trans. Electron Devices*, vol. ED-14, pp. 117–133, Mar. 1967.
[14] B. J. Hosticka, R. W. Brodersen, and P. R. Gray, "MOS sampled data recursive filters using switched capacitor integrators," *IEEE J. Solid-State Circuits*, vol. SC-12, pp. 600–608, Dec. 1977.
[15] R. T. Howe and R. S. Muller, "Resonant polysilicon microbridge with integrated NMOS detection circuitry," in *Extended Abstracts, Electrochem. Soc. Meet.* (New Orleans, LA), vol. 84-2, pp. 892–893, Oct. 1984.
[16] R. T. Howe, "Integrated silicon electromechanical vapor sensor," Ph.D. dissertation, Dept. of Electrical Engineering and Computer Sciences, Univ. of California, Berkeley, Dec. 1984.
[17] J. C. North, T. E. McGahan, D. W. Rice, and A. C. Adams, "Tapered windows in phosphorus-doped SiO_2 by ion implantation," *IEEE Trans. Electron Devices*, vol. ED-25, pp. 809–812, July 1978.
[18] R. T. Howe and R. S. Muller, "Polycrystalline and amorphous silicon micromechanical beams: annealing and mechanical properties," *Sensors and Actuators*, vol. 4, pp. 447–454, 1983.
[19] B. Deal, "Drift (CMOS), Q_m, Q_f," in *Final Report, Wafer Reliability Assessment Workshop* (Lake Tahoe, CA), pp. 89–98, Oct. 1982.
[20] W. H.-L. Ma, "Plasma resist image stabilization technique (PRIST)," in *IEDM Tech. Dig.*, pp. 574–575, Dec. 1980.
[21] S. Timoshenko, *Vibration Problems in Engineering*. New York: Van Nostrand, 1955, pp. 245–251.
[22] C. M. Harris and C. E. Crede, *Shock and Vibration Handbook*, vol. 1. New York: McGraw-Hill, 1961, pp. 7–18.
[23] W. E. Newell, "Miniaturization of tuning forks," *Science*, vol. 161, pp. 1320–1326, Sept. 27, 1968.
[24] P. R. Gray and R. G. Meyer, *Analysis and Design of Analog Integrated Circuits*, 2nd ed. New York: Wiley, 1984, pp. 703–764.

[25] F. Rodriguez, *Principles of Polymer Systems*, 2nd ed.   New York: McGraw-Hill, 1982, pp. 26-29.

[26] J. Brandrup and E. H. Immergut, Eds., *Polymer Handbook*, 2nd ed. New York: Wiley-Interscience, 1975, pp. IV-347-IV-359.

[27] H. Fujita, "Diffusion," in *Encyclopedia of Polymer Science and Technology*, vol. 5.   New York: McGraw-Hill, 1966, pp. 65-82.

[28] C. E. Rogers, J. R. Semancik, and S. Kapur, "Transport process in polymers," in *Structure and Properties of Polymer Films*, R. W. Lenz and R. S. Stein, Eds.   New York: Plenum, 1973, pp. 297-319.

[29] R. T. Howe and R. S. Muller, "Frequency response of polycrystalline silicon microbridges," in *Tech. Dig. IEEE Int. Conf. on Solid-State Sensors and Actuators* (Philadelphia, PA), pp. 101-104, June 1985.

Integrated Multisensor Chip

D. L. Ľ. POLLA, MEMBER, IEEE, RICHARD S. MULLER, SENIOR MEMBER, IEEE, AND RICHARD M. WHITE, FELLOW, IEEE

Abstract—A multipurpose integrated-sensor chip has been fabricated for the simultaneous measurement of physical and chemical variables. The multipurpose chip which measures 8×9 mm^2 contains conventional MOS devices for signal conditioning, array accessing, and output buffering along with the following on-chip sensors: a gas-flow sensor, an infrared-sensing array, a chemical-reaction sensor, cantilever-beam accelerometers, surface-acoustic-wave (SAW) vapor sensors, a tactile sensor array, and an infrared charge-coupled device imager. The multisensing functions of this chip utilize both the pyroelectric and piezoelectric effects in ZnO thin films. Fabrication of the chip is carried out using a conventional 3-μm Si NMOS process combined with Si micromachining techniques. Compatible fabrication technology and sensor properties are described.

I. INTRODUCTION

A MULTIPURPOSE integrated-sensor chip that responds simultaneously to several different physical and chemical variables has been fabricated. The multipurpose chip contains conventional MOS devices for signal conditioning along with the following on-chip sensors: a gas-flow sensor, an infrared-sensing array, a chemical-reaction sensor, cantilever-beam accelerometers, surface-acoustic-wave (SAW) vapor sensors, a tactile-sensor array, and an infrared charge-coupled device imager. A photograph of this multipurpose sensor chip which measures 8×9 mm^2 is shown in Fig. 1. The intended function of each portion of the chip's area is indicated.

The multisensing functions of this chip utilize both the pyroelectric and piezoelectric effect in 1.0-μm-thick c-axis oriented ZnO thin films, fabricated in conjunction with conventional Si planar processing. In addition, micromechanical structures, such as cantilever accelerometer beams and supported thin membranes of low thermal mass, are also formed through Si micromachining.

II. FABRICATION

Fabrication of the integrated multisensor chip was carried out in an 11-mask process in the University of California, Berkeley, EECS Microelectronics Laboratory using 3-μm design rules. The major steps in the fabrication sequence are summarized in Fig. 2. The starting material is 35 $\Omega \cdot$cm p-type (100) silicon, 5 cm in diameter and 325 μm thick. After an initial oxidation to form a 65-nm gate oxide, a 100-nm-thick layer of silicon nitride is deposited by LPCVD techniques.

Anisotropic dry etching of the silicon nitride is used to

Manuscript received November 18, 1985; revised January 28, 1986. This work was supported in part by the National Science Foundation under Grant ECS 81-20562 and in part by the State of California MICRO program.

The authors are with the Department of Electrical Engineering and Computer Sciences and the Electronics Research Laboratory, University of California, Berkeley, CA 94720.

IEEE Log Number 8607947.

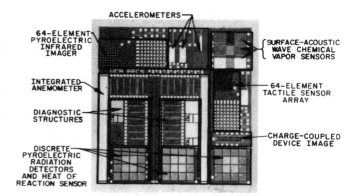

Fig. 1. Die photograph of multisensor chip measuring 8×9 mm^2 with intended function identified for each region of the chip.

define the active device regions on the front side of the chip in the first photolithographic step. Typical NMOS technology, using a boron channel stop and local oxidation, is employed to fabricate the sense amplifier. A backside heater is formed by the boron ion implantation.

In order to avoid loading the pyroelectric signal with bias circuitry, input MOSFET's are depletion type, obtained through a threshold-adjusting implant of arsenic (2×10^{12} cm^{-2} dose), at an energy of 125 keV. Enhancement-mode MOSFET's are formed using a threshold-adjusting implant of boron (5×10^{11} cm^{-2} dose).

Polycrystalline silicon is deposited next using LPCVD at 635°C to form the NMOS gates, backside-capacitor electrode, and certain amplifier resistive loads. After patterning, the source/drain region and underlying on-chip sensor heater are formed by phosphorus ion implantation (3×10^{15} cm^{-2} dose), at 150 keV. A 0.4-μm CVD oxide, doped with phosphorus, is then deposited to encapsulate the MOS transistors. The phosphorus-doped glass, in addition to protecting underlying devices from sodium contamination, protects the NMOS transistors during subsequent ZnO sputtering and is critical for controlling NMOS threshold voltages.

Thin-film (1.0 μm) capacitors of ZnO are next formed by RF planar-magnetron sputtering at 200 W for 30 min at 10 mtorr using a 50-percent Ar 50-percent V_2 mixture. The substrate-target distance is 4 cm and the substrate temperature is maintained at 230°C. Patterning of the ZnO films is carried out using an etch consisting of acetic acid:phosphoric acid:water (1:1:30). A second layer of 0.4-μm-thick CVD oxide is then deposited over the entire wafer to insulate the ZnO film. Etching of contact holes is then carried out, followed by aluminum-silicide sputtering to form interconnection and bonding pads. After delineation of the aluminum-silicide layer, specific sensor metallizations (Pt, SnO, or Cr) are

Reprinted from *IEEE Electron Device Lett.*, vol. EDL-7, no. 4, pp. 254–256, April 1986.

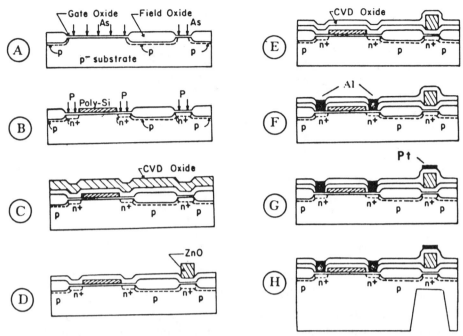

Fig. 2. Major processing step in the fabrication of the multisensor chip: (*A*) initial oxidation, channel-stop implantation, LOCOS, and threshold implantations; (*B*) poly-Si deposition and delineation, self-aligned source and drain ion implantations; (*C*) CVD SiO₂ dielectric encapsulation; (*D*) ZnO deposition and definition; (*E*) CVD SiO₂ dielectric encapsulation; (*F*) Al-Si sputtering; (*G*) Sensor metallizations (Pt, SnO or Cr); and (*H*) EDP anisotropic backside etching.

deposited over the entire wafer with subsequent lift-off delineations.

The backside Si is anisotropically etched as the last step of the multisensor fabrication process to minimize the underlying thermal mass of pyroelectric-based sensor structures. Backside Si anisotropic etching is then carried out for 220 min in a solution of ethylenediamine:pyrocatecol:pyrazine:water (250 ml:80 g:1.5 g:80 ml) heated to 100°C. This etchant removes approximately 300 μm of silicon, leaving a membrane 25 μm thick. Since the etch rate is about 1.4 μm per minute, timing of the etch step provides adequate control of the membrane thickness. The front side of the wafer is protected during EDP etching through O-ring mounting the Si wafer on a teflon holder.

A final backside masking step is carried out to define the regions over which an EDP etch-through of the wafer (necessary to fabricate integrated accelerometers) is to take place. The top-side surface containing the finished circuitry and sensor structures is coated with black wax and placed in contact with a glass cover slide. The wafer is then sealed around its edges with Dow Corning 732 RTV, the etch-through is completed, and the RTV and black wax are removed. At this point, we applied polymeric coatings to some wafers to provide sensitivity to selected vapors. Although not demonstrated in our fabrication run, a low-temperature protective CVD oxide could also be applied to encapsulate devices for use in adverse environments (before laying down the polymer films).

A cross section of the pyroelectric chemical-reaction sensor is shown in Fig. 3. This structure measuring 450 × 450 μm² consists of seven layers supported on a 25-μm-thick Si

substrate. Each of the nine die obtained per wafer was mounted in a 64-pin dual in-line ceramic package.

III. SENSOR FUNCTIONS

The multisensing chip utilizes the pyroelectric effect in the following sensing devices: gas-flow sensor, infrared-detector array, chemical-reaction sensor, and charge-coupled device imager.

In the gas-flow sensor [1], a thin-film on-chip polycrystalline-silicon resistor or a phosphorus ion-implanted resistor serves as a heater, and two symmetrically placed ZnO strips are cooled differentially, depending on the velocity of the stream of gas flowing over the surface of the sensor. The pyroelectrically determined temperature difference is a measure of the flow velocity. For nitrogen flowing parallel to the chip surface and perpendicular to the ZnO thin-film strips, the output voltage is observed to vary approximately linearly with flow less than 2 m/s.

The pyroelectric infrared-detector array [2] consists of 1-μm-thick ZnO capacitors in an 8 × 8 pattern with each sensing element measuring 70 × 70 μm² and spaced on 140-μm centers (dimensions commonly used in photovoltaic detectors). The array is supported by a 25-μm-thick Si membrane. The backside electrode is formed of polycrystalline-silicon while the top-side electrode is formed of either SnO or Cr. Infrared radiation absorbed by SnO-coated or Cr-coated ZnO regions produces heating, and hence a pyroelectric voltage. The measured black-body responsivity at $T = 300$ K and $f = 24$ Hz is 4.3×10^4 V·W⁻¹ with a corresponding measured detectivity $D = 3.1 \times 10^7$ cm·Hz·W⁻¹.

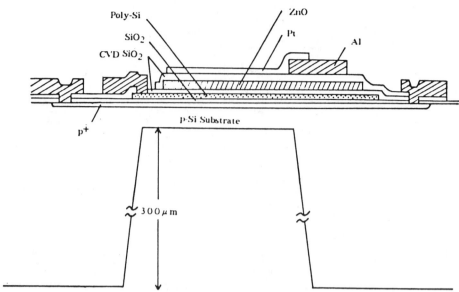

Fig. 3.   Cross section of the chemical-reaction sensor.

The chemical-reaction sensor [3] detects the heat exchanged in the reaction of gas molecules on the overlying surface. The sensor in this multichip configuration measures 450×450 μm^2, and is composed of a 0.2-μm-thick film of Pt supported on an etched Si membrane coated with the pyroelectric ZnO thin film. Heat liberated in the chemisorption of CO gas on the Pt thin-film surface induces a surface charge through the pyroelectric effect in the ZnO capacitor. The charge is transduced directly to an on-chip MOSFET amplifier. Preliminary experiments show a peak output voltage of 22 mV when 0.1 mtorr of carbon-monoxide gas is introduced over a clean Pt surface. Temperature variations as low as 18 μK have been detected as a result of this chemical reaction.

A 12-element, three-phase, infrared charge-coupled device imager is based on the pyroelectric detection of chopped incident radiation, scanned parallel to the CCD channel. A dc voltage applied across the ZnO capacitor establishes a filled electron well immediately below the detector structure. The modulated pyroelectric charge produced on the surface of the ZnO capacitors is added to the thermally generated charge in the potential wells of adjacent MOS capacitor structures. The modulated charge is transferred to a nearby CCD channel. A complete characterization of this imager is presently underway.

Sensors which utilize the piezoelectric effect in ZnO include: a tactile-sensing array, an SAW chemical-vapor sensor, and an integrated microbeam accelerometer.

The integrated ZnO tactile sensing array [4] is designed for high-resolution robotics applications. An 8×8 array of ZnO capacitors measuring 70×70 μm^2 is used to detect force applied normal to the chip surface. The piezoelectric-induced charge responsive to an applied pressure is coupled to individual MOSFET amplifiers. The measured response is 5.4 mV/g, linear with loading up to 120 g. The measured piezoelectric coefficient d_{33} of an individual tactile cell is 14.4 nC/N.

The SAW chemical vapor sensor contains an absorptive polymer (Shipley AZ 1450J photoresist) supported on a thinned-membrane region located in the propagation path of the SAW. Absorption of vapors from the surrounding atmo-

sphere results in changes in the density, thickness, and stiffness of the polymer film. As a result, the phase velocity of the SAW changes and the characteristic oscillation frequency shifts. Response at the 70 parts-per-billion level has been observed [5].

Integrated microbeam accelerometers [6] are based on the detection of strains in miniature cantilever beams. The beams are composite structures consisting of Si, SiO_2, ZnO, SiO_2, and Al, anisotropically etched on three sides. Time-varying strains due to a changes in velocity induce a piezoelectric charge that is detected by on-chip n-channel MOSFET amplifiers. Typical measured output voltages are 5 mV for 100-g acceleration [6].

Although the simultaneous measurement of the seven physical and chemical variables has no immediate application for a single sensing system, the multichip sensor demonstrates: 1) the possibility of carrying out multifunction sensing on one integrated-circuit chip; 2) the versatility of ZnO thin-film technology for IC sensing applications; and 3) the possibility that low processing cost per function can be attained if a sensor chip is produced using IC-production techniques.

REFERENCES

[1] D. L. Polla, R. S. Muller, and R. M. White, "Monolithic integrated zinc-oxide on silicon pyroelectric anemometer," in *Proc. 1983 IEEE Int. Electron Devices Meeting* (Washington, DC), Dec. 1983, Abstr. 28.4, pp. 639–642.
[2] D. L. Polla, R. S. Muller, and R. M. White, "Fully integrated ZnO on silicon pyroelectric infrared detector array," in *Proc. IEEE Int. Electron Devices Meeting* (San Francisco, CA), Dec. 1984, Abstr. 14.4.
[3] D. L. Polla, R. M. White, and R. S. Muller, "Integrated chemical-reaction sensor," presented at the 3rd Int. Conf. Solid-State Sensors and Actuators, Philadelphia, PA, June 1985.
[4] D. L. Polla, W. T. Chang, R. S. Muller, and R. M. White, "Integrated zinc oxide-on-silicon tactile sensor array," in *Proc. IEEE Int. Electron Devices Meeting* (Washington, DC), Dec. 1985, Abstr. 5.7.
[5] C. T. Chuang, R. M. White, and J. J. Bernstein, "A thin-membrane surface-acoustic-wave vapor-sensing device," *IEEE Electron Device Lett.*, vol. EDL-3, pp. 145–148, June 1982.
[6] P. Chen, R. S. Muller, R. D. Jolly, G. L. Halac, R. M. White, A. P. Andrews, T. C. Lim, and M. E. Motamedi, "Integrated silicon microbeam PI-FET accelerometer," *IEEE Trans. Electron Devices*, vol. ED-29, pp. 27–33, Jan. 1982.

MICROSENSOR PACKAGING AND SYSTEM PARTITIONING*

STEPHEN D. SENTURIA and ROSEMARY L. SMITH

Microsystems Technology Laboratories, Department of Electrical Engineering, Massachusetts Institute of Technology, Cambridge MA 02139 (U.S.A.)

(Received April 9, 1987; accepted November 26, 1987)

Abstract

The possibility of using microfabrication technology for creating integrated sensors and actuators also brings with it new design problems. This paper addresses three of these design problems: system partitioning, package design and process optimization in the presence of the technological constraints that arise when one wishes to build both electronic components and sensor structures as part of the same device.

1. Introduction

The purpose of this paper is to address some general issues involving the use of microfabrication technologies to create new types of measurement devices. The tone is informal and descriptive, under the assumption that regular technical papers provide good examples of how various investigators choose to confront the issues raised here.

We begin with some definitions. *Transducers*, as classically defined, are devices that convert one form of energy to another. In the present context, it is more useful to consider such devices as elements that convert a physical (or chemical) variable into an electrical quantity, regardless of whether the energy is obtained from the physical system or from energy sources associated with the transducer. Transducers can be used for measurement, for actuation and for display. This paper emphasizes issues associated with measurement, although there are obvious parallels in the other areas.

A *microsensor* is a measurement transducer made with techniques of *microfabrication*. Some of these techniques are well established in the integrated circuit industry; others are specific to microsensors (microfabrication methods are discussed further in Section 3). There is nothing particularly new about microsensors. Examples are well known (see Fig. 1). The photodiode converts incident optical energy into electrical energy, and is an example of a direct energy-conversion transducer. The photoconductor

works differently: incident light energy changes the relative populations of electrons in various quantum states, thus changing the conductivity of the element. The energy in the electric circuit, however, is supplied from an external source. The phototransistor shares features of both other devices. The base current in the transistor is supplied by electron–hole generation from incident light, but most of the energy in the collector circuit is supplied by an external power source. These latter two transducers are examples of *parametric* microsensors, in which the physical variable modifies a parameter of the sensor element, which is then measured or detected through the element's behavior in an electrical circuit.

(a)   (b)   (c)

Fig. 1. Examples of microsensors: (a) photodiode; (b) photoconductor; (c) phototransistor.

2. System issues

The title of this paper suggests that 'Microsensor Packaging and System Partitioning' is a discipline that has well-understood principles, and that examples will readily illustrate how these principles impact on any given engineering design problem. Unfortunately, such a suggestion would be imprecise, at best. While there have been some excellent publications on specific techniques used for microsensor packaging [1 - 3], most attempts to elucidate principles have produced more controversy than agreement. The nub of the controversy is the so-called 'smart sensor', which merges sophisticated electronic data processing with the microsensor. The reader is hereby alerted that while some of the assertions presented here are based on technical judgments that are quantitatively defendable, some are also, to a certain extent, based on strongly felt opinions of the authors, and may not be universally accepted. *Caveat emptor.*

Measurement systems have a great deal of *modularity.* This is illustrated schematically in Fig. 2. Three modules are identified. The transducer's function has already been discussed. The schematic 'packaging' boundary suggests that transducer packaging presents special problems: some of the transducer requires environmental access while the rest may require protection from the interface (the packaging issue is discussed further in Section 4). The *interface circuit* supplies excitation to the transducer (if needed), accepts the response and performs additional functions such as amplification, linearization, or data conversion. The *data system* provides overall control, and accepts the data for subsequent use. The various components communicate with one another over highly-standardized interconnections,

---

*Paper presented at the 14th Automotive Materials Conference, Ann Arbor, MI, U.S.A., November 19, 1986.

resides *everywhere* in the system; it is a system issue. Conceptually, modularity aids in calibration because the functional performance of each module can be independently discovered, optimized and compensated. On the other hand, there are examples where cost advantages are achieved by allowing one portion of the system (such as the data system) to implement a compensating correction for the calibration of individual transducers. Thus, in considering a microsensor design, there is a system-level decision to be made: whether to trim individual devices to standard calibrations (the modular approach), or to use the interface circuit or data system to compensate for device-to-device variations (the system approach). Both approaches are used.

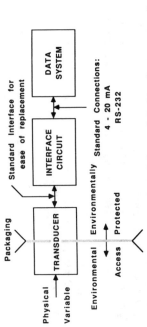

Fig. 2. Schematic illustration of the modularity of measurement systems.

such as the 4-20 mA or RS-232C interfaces between data system and interface circuit. Furthermore, the design of tranducers is often made to provide for easy replacement; hence, the interface circuits are designed to connect to standardized types of transducers. Among the benefits of this modularity is the fact that transducers, interfaces and data systems can all be designed and optimized separately. Microsensors that merge the transducer with other parts of the measurement system do not have such modularity. The implications of this are discussed in Section 3.

Measurement systems must be *calibrated*. Figure 3 illustrates a highly simplified piezoresistive bridge that could be used to measure pressure. Two of the resistors are presumed to be pressure dependent, the other two are not. (This can be achieved by fabricating two of the resistors in a thinned diaphragm portion of the microsensor, which is allowed to deform under pressure, straining the resistors and changing their values.) The excitation voltage $x(t)$ is supplied by the interface circuit, and the output voltage $y(t)$ is returned to the interface circuit. It is seen that the relation between the inferred pressure $p(t)$ and the output $y(t)$ is non-linear. The calibration expresses the accuracy with which the inferred pressure reflects the actual pressure applied to the diaphragm.

Where does the calibration reside? It resides in the precision and repeatability with which the resistors and the deformable diaphragm are manufactured, in the stability of these components, in the extent to which the response $y(t)$ can be made independent of all other physical effects, notably temperature and package-induced stresses, in the accuracy with which the excitation waveform $x(t)$ is produced, and in the accuracy of the amplification and data-conversion portions of the system. In summary, the calibration

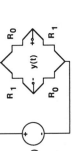

Fig. 3. A piezoresistive bridge. $R_1 = R_0(1 + \alpha P(T))$. Parametric input-output relation: $P(t) = 2y(t)/(\alpha x(t) - \alpha y(t))$.

3. Microfabrication

Microfabrication refers to the collection of techniques used by the electronics industry for the manufacture of integrated circuits. The success of microfabrication is immense, combining the economies of batch fabrication with the dimensional precision of photolithography. A variety of materials are compatible with batch-fabrication techniques, and these materials can be deposited and patterned in may ways. In addition, and of great significance for the microsensors field, the technologies of microfabrication are needed by, and are therefore supported by, the electronics industry. Thus, it is not necessary for the sensor industry to provide the capital development costs for most of the process technologies.

Standard integrated circuit processing techniques have been used to make a variety of devices that function as sensors. Table 1 lists types of microsensor devices that are already well established. In many cases, these are simply microfabricated versions of existing 'macro' transducers. In other cases, notably the charge-coupled optical imaging devices, there is no corresponding macro-device. Careful examination of the Table, however, shows that the well-established devices are those for which the packaging problem is most readily solved. Heat and magnetic fields readily penetrate standard encapsulation materials, and hermetic window technologies for optical devices have a long history. Encapsulation technologies for pressure microsensors are in commercial use, but in many cases, packaging artifacts limit either accuracy or drift specifications. Chemical microsensors are not yet able to take full advantage of the microfabrication technologies, in part, because of packaging limitations.

In addition to new fabrication capabilities for existing types of sensors, microfabrication offers the promise of new types of devices, based on unique properties of microelectronic devices (such as the charge-coupled device or carrier-domain magnetometer devices), on new fabrication capabilities such as micromachining [4, 5], or on the promise of being able to merge signal conditioning and signal processing with the primary sensing device (the 'smart sensor'). *Micromachining* refers to a set of special deposition and/or

TABLE 1
Electronic components and their uses as sensors

Element		Sensor use
Resistors	Metal	Thermometers
	Polysilicon	Strain gages
	Semiconductors	Magnetic sensors
		Photoconductors
Diodes	Schottky	Light and radiation
	p-n junctions	Temperature
Capacitors	Between conductors	Position sensors
		Dielectric properties
	MOS capacitors	Charge-coupled devices
		Work-function changes
Transistors	MOS	Electrochemical potential
		Charge sensors
	Bipolar	Temperature
		Light
		Magnetic fields
Electronic function modules for signal processing	Differential amplifiers	
	Digital logic	

economic justifications for building smart sensors, but in the many public discussions of this topic, only the attractiveness has been emphasized; the hazards have often been ignored.

The capabilities of microfabrication, with their potential for new and improved microsensors, create a set of *partitioning* decisions for the designer. The *system* must be partitioned between the batch-fabricated microsensor and the rest of the system. The *process technology* must be partitioned between standard process steps that are readily available in the integrated circuit industry, and non-standard process steps required for the specific microsensor design. Both packaging and the need to optimize process technology impose *constraints* on these partitioning and design decisions. The rest of the paper explores these issues, starting with packaging.

4. Packaging

As used here, *packaging* refers to first-level packaging, *i.e.*, enough encapsulation to permit device handling, performance evaluation and actual use in at least some applications. The packaging problem, as stated earlier, is that the details of the package affect every level of microsensor design, including how the measurement system is set up, how it partitioned and how the microsensor part of the system is designed. Therefore, *it is necessary to design the microsensor and the package AT THE SAME TIME*. This surprisingly simple suggestion often meets with great opposition. One problem is that 'packaging people' are usually not the same as 'sensor people', and getting them to work together can be difficult. Further, because package design can be expensive, there is a reluctance to commit effort without some evidence that the microsensor will work. Nevertheless, it is the authors' opinion that without a package design, even a temporary, simple package design, effort spent on microsensor development can be a mistake. The cost of fabrication of a microsensor does not depend heavily on details of the layout. Therefore, one might as well use a layout that can be successfully tested in a package. By putting some effort into the packaging problem early in the design, unrealistic designs that cannot be packaged are avoided.

The approach involving simultaneous package and sensor design is described with reference to Fig. 4. The design sequence is conceived at four levels: partitioning, specification of interfaces, design specifications and detailed design. Iteration at all levels is assumed. Each of the levels is now discussed:

(a) Partitioning

For a microsensor measurement system, the *partitioning* decision addresses how much of the system is to be merged into the batch-fabricated microsensor part (the 'chip'), and how much is to be 'off-chip'. The position taken here is very simple:

MINIMIZE THE ON-CHIP PART OF THE SYSTEM

etch processes with which mechanically complex structures can be fabricated either in or on planar substrates. Diaphragms, cantilevers, moveable capacitor plates and through-substrate holes are among the types of structures that are readily created. The *smart sensor* is an attractive idea, at least at first. One imagines a batch-fabricated device that performs an entire measurement and presents an output signal in a form readily accepted by a microprocessor. This is the promise; but there are problems.

The principal difficulty with the smart-sensor concept is the loss of modularity. Because the design of microfabricated parts must be done in a monolithic fashion, every detail of the device must be designed at once. It is no longer possible to have the interface expert work independently of the sensor expert. Furthermore, because the production specification for a smart-sensor consists of a mask set and a process description, any change in *either* the sensor or the interface design requires a completely new mask set. Fabrication errors in *either* the sensor *or* the interface can ruin both when they are fabricated together. Of perhaps greater significance is the fact that the optimization of process sequences for electronic components may not be compatible with the optimum process with which to fabricate the transducer. By trying to create a merged design, the quality of both types of components may be compromised. Finally, the sensor package must be designed along with the microsensor. Thus, the design overhead in a non-modular merged 'smart' sensor can be very large. There may still be good

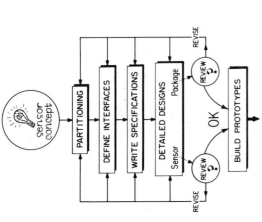

Fig. 4. Schematic illustration of the sensor/package design process.

This approach is useful for several reasons. It forces the designer to decide which of the many *possible* functional parts (such as extra resistors, transistors for amplification, switches for multiplexing) are actually *essential* for successful performance of the microsensor's task. It also forces the designer to address questions of operating environment, control of parasitic responses and overall system architecture early in the design. Adding functionality may increase the process complexity (hence the cost), may reduce the yield (hence increase the cost) and increase the size or packaging complexity (and hence the cost) of the device. The only justification for such increased costs is a documentable performance benefit, either at the sensor level or at the overall system level. There are correctly partitioned examples where the microsensor consists of a simple set of electrodes with no added functionality, and other correctly partitioned examples where the microsensor contains a sensing element plus transistor amplification and trimmable resistors for adjustment of temperature compensation and calibration. The idea is to avoid seduction by complexity just because highly-merged 'smart' sensors are *possible*; they should also be *required* by the application at hand before the 'smart' sensor-designer builds them.

The partitioning of the microdielectrometry system [6, 7] provides a useful example. The microdielectrometer is intended for measurements of dielectric constant and conductance in very insulating materials (see a system schematic in Fig. 5). It is based on a pair of interdigitated electrodes, one of which is driven, and the other of which collects charge through the medium under test. Because of the intended high-impedance application, beyond the capabilities of a simple passive electrode pair, the sense electrode must be physically close to the first stage of amplification; otherwise the

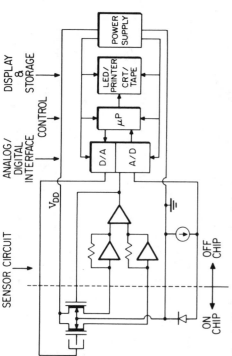

Fig. 5. The system partitioning of the microdielectrometer.

leakage currents in wiring insulation would limit the measurement. Hence, the device as conceived requires one transistor. Given that one is required, a second is also required to allow an accurate differential measurement to be made that provides temperature and pressure compensation and cancels out process-induced variations. There are many examples where this minimal device set, a property-dependent element and a matched pair of transistors (see Fig. 6), is appropriate. In the case of the microdielectrometer, an additional diode is added to the chip to provide a temperature measurement capability. This adds no process complexity, no additional area and greatly enhances the performance of the sensor. Hence it is justified under the minimalist approach. However, the next stage of analog amplification in Fig. 5, which *could* have been added with relatively modest cost, is *not* included on chip because *it is not necessary*. A benefit of this decision was the discovery (after the fact) that the basic sensor of Fig. 6 would operate

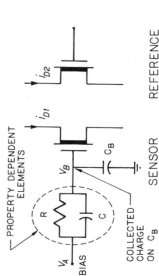

Fig. 6. The canonical minimum 'smart' sensor: a property-dependent element and a matched pair of transistors.

188

successfully at temperatures much higher than would have been possible had the analog electronics been added to the chip.

(b) Define system interfaces

A microsensor has a variety of interfaces: an electronic interface to the measurement system; a mechanical interface with the environment (with attendant chemical, thermal and pressure/stress characteristics); a cabling or interconnect requirement with the measurement system; materials requirements for chemical stability, thermal stability, mechanical properties and, in some cases, biocompatibility. During the process of defining these various interfaces, it is useful to list possible electrical, chemical, mechanical and thermal parasitic effects that can arise either in the microsensor itself or in its package. It is often possible to make relatively modest design changes to eliminate serious parasitic effects. The best time to find them is before the first prototype is built.

(c) Design specifications

As the various system requirements and interfaces become clear, possible combinations of microsensor and package can be evaluated for their success in dealing with each of the requirements and possible parasitic effects. Out of this process, it becomes possible to be specific about how the microsensor should be built and how it should be packaged. Ideas can be made explicit on issues such as on-chip device count and type, circuit requirements and expected nominal system performance, layout-related issues such as chip size, special structures (such as diaphragms) and location of pin-outs, ideas on passivation, a package concept with suggested materials, overall package dimensions and assembly concepts, package fabrication and assembly concepts, overall package dimensions and provision for connection of the microsensor to the rest of the system.

During this phase of the design, it is also important to deal with test, acceptance and calibration issues. Since it may be impossible to test microsensors for measurement performance until they are packaged, some attention must be given to how to accept sensors prior to packaging. Equally important is a package acceptance criterion, since in many cases, package costs exceed sensor costs. Decisions on how the measurement is to be calibrated may result in additional elements, such as trimmable resistors. Such elements should be added before the final mask design is complete.

(d) Detailed design

The final design phase creates the detailed design. For the microsensor, it consists of a process flow and a set of masks (process issues are discussed in Section 5). For the package, it consists of a fabrication process and drawings for the various parts and for the tooling needed to make the parts. The assembly must also be specified in terms of process, intermediate acceptance tests and procedures for calibration, trim and final test.

This may all sound straightforward but in fact, it rarely is. The authors are not aware of any commercial part that went smoothly through the four design stages outlined above, with a final microsensor and package design emerging together. The microdielectrometer is a good example. The first device, while admittedly a research vehicle, had a ghastly bonding-pad placement that prevented effective packaging. Had the packaging been considered earlier, time and money would have been saved. Indeed, the authors' experience suggests that the weak link in the design process is that too little attention is paid to the package until the microsensor design is so far along that changes become difficult. With suitable attention to packaging issues early enough in the design, a more balanced optimization of both parts of the microsensor/package combination becomes possible.

5. Technology constraints

(a) Material and process selection

The design process described in Section 4 assumed implicitly that the transition from desired specifications to a successful process could be readily accomplished. This Section examines some of the constraints and difficulties that are encountered in selecting the various processes and materials that ultimately become the microsensor fabrication sequence [8].

The microsensor geometry and structure determine many of the material and/or process requirements. For example, if insulator thicknesses greater than about 1 μm are required, sputtering must be used. Conformal step coverage requires chemical vapor deposition. Small lateral feature sizes (<1 μm) may require advanced patterning techniques, such as X-ray or electron-beam lithography. If continuous films less than about 100 Å thick are required, they must be created thermally because deposited films of this thickness are not usually integral.

Beyond the issue of selecting the individual process technologies to create the desired structure, there are several critical levels of *interaction among process steps*. As process steps are assembled into a process sequence, the effects of subsequent processes on the results of the earlier steps must be carefully considered. At high temperatures ($\sim$900 °C), dopant redistribution can occur. Chemical compositional and morphological changes can occur at lower temperatures ($\sim$600 °C). Interfacial properties can be modified at temperatures as low as 300 °C, and can also be affected by exposure to ultraviolet light and soft X-rays, which can happen during plasma etching or metallization. Finally, exposure to wet chemicals during cleaning and etching can affect various materials.

The semiconductor process-modeling program SUPREM is very useful for examining what happens in the semiconductor part of a microsensor. However, the effects of process steps on layers deposited on the semiconductor, particularly on their mechanical properties, cannot presently be predicted in general. Furthermore, there is no automatic way to find an optimum selection of process steps to build any particular structure. An example is useful [8].

189

Figure 7(a) shows a structure in which gold is deposited on silicon dioxide over a step etched in silicon nitride. The process sequence involves, first, growing the oxide on a silicon wafer, then depositing and patterning the silicon nitride. The first critical decision concerns the patterning of the silicon nitride. A plasma etch should be used, because the other available etch method (hot phosphoric acid) requires a deposited oxide mask, and during mask removal, there would be an unacceptable undercut of the oxide, which could not be successfully covered by the deposited gold (see Fig. 7(b)). Having decided on nitride patterning, the next step is gold deposition. In this case, because gold does not adhere well to oxide, an adhesion layer of chrome is evaporated beneath the gold. The adhesion layer and gold must be sequentially deposited under vacuum. Otherwise, the chrome may oxidize and no longer adhere to gold. The presence of the adhesion layer dictates that if the gold is to be patterned, the patterning should be done with lift-off rather than wet etching, because the wet etchant for the gold would attack and underetch the chrome adhesion layer.

Process designers are accustomed to dealing with the type of process-step interaction described above. However, because microsensors may involve structural elements such as diaphragms or beams in addition to electronic components, it may turn out that the process for making the electronics cannot be optimized without sacrificing the quality of the structural element, and that the process for the structural element cannot be optimized without sacrificing the performance of the electronic components. The process and performance compromises inherent in merging electronics with microsensor structures must be carefully considered. Just as was true for the packaging decision, the mutual constraints on process optimization argue for keeping the microsensor process as simple as possible. Avoid added complexity.

(b) Technology partitioning

There is partitioning decision in specifying the process sequence that is just as critical as the system-partitioning decision discussed earlier. Many of the process steps are readily available as part of standard integrated-circuit processes. As a result, the relatively low cost, ready availability and established reliability of these process steps makes them particularly attractive for use in microsensor design. In many microsensors, however, some non-standard process steps are needed, either because of mechanical structural elements, novel materials or unusual packaging requirements. Therefore, the

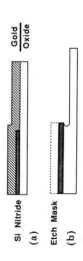

Si Nitride / Gold / Oxide / Etch Mask

(a)

(b)

Fig. 7. Interaction among process steps: (a) desired structure; (b) oxide undercut that would result from removal of wet-etch mask.

overall process sequence must be partitioned into steps done prior to, during, or after a set of standard processes. In doing this partitioning, it must be recognized that simultaneous optimization of processes for electronic and other elements may not be possible. In addition, special handling requirements for mechanical structures such as beams and diaphragms must be considered. These structural elements may be fragile, may be sufficiently non-planar to affect processing and may be sensitive to residual stresses that can be modified by other process steps.

An example of technology partitioning is shown in Fig. 8 [9]. The structure in Fig. 8(a) consists of a chemically-sensitive field-effect transistor (CHEMFET) on a chip that also contains a porous silicon diaphragm, which is created by hydroflouric acid anodization of a silicon diaphragm. The technology partitioning is illustrated by the fact that the silicon diaphragm is first created using anisotropic etch techniques, then the underside of the diaphragm is doped. These steps are done before the standard processing. During the standard processing, the CHEMFETs are built in a diffused dopant well, and during the step in which this well is created, dopant is also diffused into the top side of the diaphragm connecting up with the previous bottom diffusion. The anodization of this fully-doped diaphragm is then done after the standard processes, exposing only the back side of the wafer to HF, resulting in the final device structure.

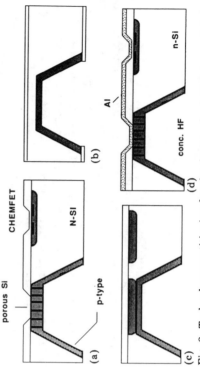

Fig. 8. Technology partitioning for the device of ref. 9: (a) complete structure; (b) pre-circuit processing steps (anisotropic etching and a deep diffusion); (c) during-circuit processing (form transistor well and diffuse through remaining part of diaphragm); (d) post-circuit processing (anodize to form porous silicon).

6. Conclusions

Microfabrication creates many new opportunities for microsensors. However, a loss of modularity results from the merging of the basic microsensor with electronics, and this poses new problems of system and tech-

190

nology partitioning. Seduction by the promise of total system integration (the 'smart sensor') should be avoided without careful analysis. Functionality should be added to a basic sensor structure only on the basis of demonstrated and defendable improvements in cost or performance. Design the first-level package and the microsensor together, and compatibly. If both electronic and structural components must be merged in a design, be prepared for some difficult compromises when attempting to optimize process sequences. Finally, and more optimistically, real advances in microsensor technology should be forthcoming in the next few years. The technological capabilities are just beginning to be tapped. There is unlimited opportunity for innovation.

References

1 C. D. Fung, P. W. Cheung, W. H. Ko and D. G. Fleming (eds.), *Micromachining and Micropackaging of Transducers*, Elsevier, Amsterdam, 1985.

2 W. H. Ko and T. Spear, Packaging of implantable electronics: past, present and future developments, in W. H. Ko, J. Mugica and A. Ripart (eds.), *Implantable Sensors for Closed-Loop Prosthetic Systems*, Futura Publishing Company, Mount Kisco, New York, 1985, pp. 259 - 304.

3 L. Bowman and J. D. Meindl, The packaging of implantable integrated sensors, *IEEE Trans. Biomed. Eng., BME-33*, (1986) 248 - 255.

4 K. E. Petersen, Silicon as a mechanical material, *Proc. IEEE, 70* (1982) 420 - 457.

5 R. T. Howe, Polycrystalline silicon microstructures, in C. D. Fung, P. W. Cheung, W. H. Ko and D. G. Fleming (eds.), *Micromachining and Micropackaging of Transducers*, Elsevier, Amsterdam, 1985, pp. 169 - 187.

6 N. F. Sheppard, Jr., D. R. Day, H. L. Lee and S. D. Senturia, Microdielectrometry, *Sensors and Actuators, 2* (1982) 263 - 274.

7 S. D. Senturia and D. R. Day, Packaging considerations for the microdielectrometer and related chemical sensors, in C. D. Fung, P. W. Cheung, W. H. Ko and D. G. Fleming (eds.), *Micromachining and Micropackaging of Transducers*, Elsevier, Amsterdam, 1985, pp. 29 - 39.

8 R. L. Smith, Technology constraints on microsensor design, in S. D. Senturia, R. T. Howe and R. L. Smith, *Microsensors*, Massachusetts Institute of Technology, Class Notes for Special Course 6.77S, 1986 (unpublished).

9 R. L. Smith and D. C. Scott, An integrated sensor for electrochemical measurements, *IEE Trans. Biomed. Eng. BME-33* (1986) 83 - 90.

Micromachined Packaging for Chemical Microsensors

R. L. SMITH AND S. D. COLLINS

Abstract—A review of the critical issues involved in chemical microsensor packaging and encapsulation is made, and a hybrid solution is presented. In the design approach, the microsensor is divided into two principal physical parts, an electrode and electronics-containing substrate, and a micromachined membrane package. The fabrication and, hence, the resultant performance of each part is independently optimizeable. The final microsensor is constructed by, first, binding the two parts together at the wafer level, followed by die separation, and then lead attachment. The micromachined membrane holders are then filled with liquid membranes to yield functioning sensors. A calcium ion sensor fabricated by this method is demonstrated.

I. INTRODUCTION

SPECIFIC, and often independently addressed, areas of chemical microsensor packaging and encapsulation that require the attention of the designer are: 1) electronic isolation of active devices from solution, 2) lead attachment and encapsulation, and 3) membrane attachment and isolation. In addition to these, there remains the reference electrode. Several solutions have been found for each individual problem area, but these often result in added complexity to the FET fabrication process, making the process noncompatible with standard IC processing techniques. This means that the addition of circuitry to achieve the ever sought after multisensor chip becomes exorbitant in development and manufacturing cost. This then removes the low-cost disposable feature that would make this device desirable.

A brief review of the techniques and materials that have been employed in CHEMFET and other chemical microsensor packaging and encapsulation is given below. A combination of some of these techniques with silicon micromachining is then described, which presents a hybrid solution to microchemical sensor packaging and encapsulation, and which requires minimal deviation from standard processing techniques for the FET's or any associated circuitry.

II. REVIEW

A. Electronic Isolation

The electrodes, FET's, and any other electronic devices that are on the same chemical sensor substrate must be electrically isolated from the surrounding conductive solution in order to operate properly. The CHEMFET exemplifies the requirement for electronic isolation. The basis of operation of any FET is that a field is induced in the channel region that controls the conductivity between the source and drain. Since the gate of a CHEMFET includes the surrounding solution, isolation from the solution is as important to proper functioning as is the isolation of MOS components in circuits, from one another and from the environment. In addition, the surrounding solution is of changing chemistry, which is sensed by a change in gate potential. All interfaces with the solution will also have a characteristic potential that can alter interfacial processes, i.e., exchange currents, that occur there. Since FET fabrication is most readily accomplished using planar technologies, CHEMFET fabrication processes have traditionally employed solid-state coatings with low water and ion permeability, such as silicon nitride and aluminum oxide, on the uppermost surface. Chemical vapor deposition (CVD) and patterning techniques for high-quality materials have been developed and perfected by the IC industry. However, after separation of the individual die from the silicon wafers, the substrate becomes exposed at the sides of the die. This poses an encapsulation problem for the sensor package design.

Isolation of the exposed substrate can be accomplished by encapsulation with an insulating organic, such as epoxy resin [1]–[4]. One approach, developed by Matsuo [5], [6], was to micromachine the sensor wafer into needles, attached together at one end like the teeth of a comb. These structures were then coated on all sides by CVD silicon nitride. Processing of these devices is severely complicated by the nonplanar and very fragile structure. More recently, the application of diode isolation and SOI techniques to the problem of substrate isolation have been employed [7]–[9]. These techniques have been developed for circuits and therefore exist as standard processes. The combination of one of these isolation technologies and a top surface barrier coating appears to be the most successful and the most promising method of electronic isolation and enables a part of the packaging to be accomplished on wafer, by solid-state materials and methods.

B. Input/Output Lead Wires

Chemical microsensors, especially active devices such as the CHEMFET, require electrical signal and power supplying leads, which communicate between the device in solution and the outer "dry" data acquisition environment. Therefore, a means by which leads can be attached to the chip and their electrical isolation from one another are necessary. In addition, encapsulation of the leads from water and ions in the solution is required to prevent deleterious corrosion. Several approaches to this problem

Manuscript received October 13, 1987; revised January 21, 1988.

R. L. Smith was with the Massachusetts Institute of Technology, Cambridge, MA 02139.

IEEE Log Number 8820664.

Reprinted from *IEEE Trans. Electron Devices*, vol. 35, no. 6, pp. 787–792, June 1988.

have been explored. Most often, commercially available bonding techniques have been employed such as wire bonding [1]–[7] and tape automated bonding (TAB) [10], [11]. The physical bonding of the leads to the chip has not been a significant problem. Rather, the difficulty lies in the geometrical puzzle of how to encapsulate the bonds and lead wires without covering the sensitive gate region. The bonding pads for CHEMFET's are most conveniently placed along one edge of the chip, as far as possible from the active gate area. With this configuration, the bonds can be coated with an epoxy or other viscous liquid coating, which is subsequently cured, without coating the gates.

There are no suitable commercially available chip carriers or cables for these devices. Printed circuit cards and dual lumen catheters are among the hand-fashioned chip carriers and cabling that have been employed. TAB bonding can reduce this problem; however, failure of the adhesion layer with long-term exposure to ionic aqueous solutions makes commercially available Kapton® tapes inadequate for encapsulation [10].

Special methods of lead attachment and encapsulation have been proposed and tested. The gate regions can be protected with photolithographically patterned materials, such as Riston® [10], prior to wire bonding and encapsulation, and later removed. This technique requires some special processing techniques, such as mechanical grinding, but they can be performed at the wafer level. A totally different approach is to create "back-side contacts," which involves the etching of wafer via holes [12], diffusing dopants through the entire substrate by thermal gradient [13], [14], or employing SOI fabrication techniques [9]. These methods place the lead attachment and encapsulation problem on the back of the chip, which need not be exposed to solution if a flow cell is employed [9], [15]—a very attractive alternative. However, this method has significant limitations: 1) combination of back-side contacts *and* on-chip electronics (substrate) isolation involves very complex processing; 2) the number and placement of i/o leads is limited by the large space required per contact and/or poor spacial resolution capabilities; 3) the difficulties encountered in making IC fabrication compatible with back-side contact formation; and 4) it does not solve the problem for *in vivo* sensors where chip and leads are immersed in electrolyte.

It is the authors' opinion that the method of choice for lead attachment for a multichemical sensor chip will be one of the more standard methods, either wire or TAB bonding, with custom-made chip carriers and custom "tape" materials and monolithic cable fabrication.

C. Membranes

Generally speaking, electro-chemical microsensors either employ selectively sensitive materials as transducers of chemical energy to a sensed potential, or they measure the rate of a reaction, represented by the current. CHEMFET's are made chemically sensitive to a specific ion or other chemical species by attaching a sensing membrane material in series with the gate insulator. The chemically established membrane potential is effectively in series with any applied gate bias and is therefore sensed in the same manner as a change in the gate voltage of the FET. Several solid-state membrane materials exist, such as LaF_3, $AgCl$, and Si_3N_4, which establish potentials selectively to fluoride, chloride, and hydrogen ions, respectively. Many of the solid-state materials that are used as hydrogen-ion-sensitive membranes are also insulators and excellent diffusion barriers to water and ions. They are often incorporated as the uppermost layer of the FET gate insulator and as an encapsulant. They can be integrated into the FET fabrication at the wafer level. However, for sensing most other chemical species, organic membranes are employed.

There exists a host of different organic membrane systems, selective to a wide variety of ions [1], [4], [15], [16]. These materials are attached to the FET gate insulator by physio-chemical adhesion. Any electrical shunt path, either vertical or horizontal, through or around the membrane–solution potential generating interface, will diminish the potential sensed by the FET. Horizontal shunts between membrane-covered FET's will create mutually dependent sensors and diminish their sensitivity. Therefore, the membrane integrity (no holes), adhesion, and isolation from other sensing gates are crucial to proper operation. Many solutions to this problem have been presented. Hand-painted epoxy wells and silanization of surfaces was the original approach [4]. It soon became evident that this was not a commercially viable technique. The Riston® masking technique described earlier for lead encapsulation has also been applied to membrane-well fabrication [10].

Microfabricated meshes [16] have been made in spun-cast polyimide films, which improved the adhesion of subsequently solvent cast polymeric membrane materials. Alternatively, membranes may be formed by spin casting [17] a plasticized polymeric matrix onto a wafer containing FET's, or thin-film electrodes, and locally doping the organic film with an ion-sensitive material, e.g., ionophores.

All of the techniques mentioned so far preclude the use of liquid-membrane materials and the possible fabrication of the classical ISE in miniature, i.e., membrane/filling solution/redox couple/metal/amplifier. This structure is desirable because each interface in this electrochemical system is thermodynamically well defined and therefore can be fabricated (presumably) in such a way that it is "well behaved" with respect to drift and reproducibility. This is not the case for the CHEMFET, which has a semiconductor–insulator–ionic solution or semiconductor–insulator–membrane–ionic solution gate structure. The coupling between ionic and electronic conduction in these systems is unknown. The insulator–membrane interface is blocked [18], i.e., impermeable to charge transfer, and therefore is not thermodynamically well defined. It is possible that instabilities in CHEMFET behavior, i.e., drift, emanate from this interface. Modification of the CHEMFET structure to include an inner reference solution and redox couple between the membrane and gate insulator would result in a highly improved chemical microsensor.

This approach has recently been pursued by several investigators. For example, a device very similar in concept was fabricated nearly a decade ago by Come and Janata, who laboriously pasted individual capillary tips over the gate regions of CHEMFET's, epoxied them in place, and then filled them with a pH buffer solution, entirely by hand, to create a reference FET [3]. Prohaska [19], [20] has surface micromachined silicon nitride microchambers to form thin-film nearly planar electrochemical cells.

Micromachining of chambers in Pyrex glass plates, by etching and laser drilling [21]–[23] and then bonding these structures to the sensor-containing substrate is a more recent approach. Pyrex has been chosen as the membrane-holding material because it can be hermetically sealed to silicon substrates by field-assisted bonding. The process of field-assisted bonding employs high fields and temperature in order that a sufficiently large anodizing current may flow across the silicon substrate–glass interface to chemically bond the glass to the substrate [24], [25]. The intention is to bond these structures over thin-film electrodes, e.g, silver/silver chloride, which will provide a stable redox couple to a filling solution containing chloride. The advantages of micromachined membrane holders are:

1) They can be fabricated independently from the electrodes and any electronics, and the two joined together at the wafer level.

2) These and other structures can be micromachined into three-dimensional forms of great dimensional and functional variety, e.g., to include flow and fill channels.

3) The stacking of the micromachined substrate over the sensor substrate adds significant vertical dimension to the sensing region such that wire bond encapsulation is no longer a difficulty.

There are, however, pitfalls in the to-date proposed methods of miromachined membrane-holder fabrication and attachment. The use of field-assisted bonding to attach these structures to the sensor substrate containing thin-film electrodes and/or FET's poses the following technical difficulties:

1) The high fields and temperatures required for bonding are very detrimental to thin-film electrode materials such as silver and to MOSFET devices without gate protection.

2) The bonding technique requires a conductive plane on the substrate surface that comes in direct contact with the glass. This conductive layer (doped polysilicon [21]–[23]) will short together adjacent sensors, unless those regions that are left exposed are somehow isolated.

3) The bonding technique requires an extremely planar surface, < 1000-Å steps, which is more planar than a silicon wafer surface after the fabrication of integrated circuits by standard processing techniques (which include planarization steps). This means that the electrodes and any other devices or circuits on the substrates need be fabricated by other than traditional means. For this reason and because of the high field requirements, these structures have not yet successfully employed FET's at the sensing site.

Anodic bonding is not the only means of attachment and hermetic sealing of the sensing surface is not necessary if that surface has been coated with a moisture barrier, solid-state film such as silicon nitride. Also, hermetic sealing of the surface does not provide encapsulation of bonded leads nor of the substrate and its electronics. It does have the advantage of providing excellently adherent parts. Micromachined substrates can be bonded to silicon substrates by other means, including the attachment by organic adhesives, thermoplastics, polyimide film, glass frits, and reflow oxides. These methods do not require high temperatures or high fields, and do not require highly planarized surfaces. Therefore, all the advantages of the micromachined cavities can be utilized, and in addition, 1) FET's can be placed at the sensing site for at site impedance transformation and resultant improved signal-to-noise ratio and 2) the use of materials other than Pyrex glass as the micromachined substrate can be employed.

III. The Micromachined Package

The most attractive micromachinable material for microsensor packaging is silicon [26]. Fine geometrical control in all three dimensions is possible with the use of anisotropic etchants. Etch-stopping techniques exist for anisotropic and isotropic etchants. The structural possibilities tease the imagination. For example, one can combine porous silicon membranes with anisotropically etched cavities to produce a micro reference electrode [7]. Flow channels and valves can also be incorporated into the machined substrates to aid in filling and for flow analysis. Therefore, silicon was the material of choice for the micromachined package described here.

The design presented here was meant to incorporate what the authors believe have been the most successful techniques applied to the previously described problem areas of chemical microsensor packaging and encapsulation, with a multisensor chip application in mind. One important aspect of this design is that the fabrication of the sensor and electronics are sufficiently standard that foundry services can be utilized, at least up to the final contact hole via formation. Electronic isolation is achieved by diode isolation. Either the sensing FET's and all other on-chip circuitry are of a single type, i.e., nMOS or pMOS, and placed in the opposite-type well, or a twin-tub CMOS technology can be employed. The FET's can have polysilicon gates, as long as the gates are or can be electrically floated [27], [28]. The top surface of the sensor chip is completely coated with silicon nitride, or other encapsulating, solid-state layer, except for the bonding pads. This layer and final metallization represent a deviation from most VLSI processes and require some process development to ensure desired electronic operating characteristics. The addition of thin-film silver electrodes over the gate regions of the FET's would be a last additional step in order to make the microfabricated ISE-plus-amplifier described earlier.

A. Fabrication

The packaging of the sensor began with a 4-in wafer containing approximately 2500, 1.46×1.87 mm die. Every third die contains a single n-MOS CHEMFET in a p-well and an aluminum-gate MOSFET of identical structure. The wafers were processed commercially (Solid State Scientific) following a metal-gate CMOS process, with dual dielectric over the gates, comprised of 500 ± 50 Å of silicon dioxide and 1100 ± 100 Å of LPCVD silicon nitride. The detailed design, fabrication, and testing of the CHEMFET's used here has been previously reported [7]. The silicon nitride proved to be an excellent passivation layer, with leakage current less than 20 pA over an exposed area greater than 1 mm^2. The mask layout for the micromachined package was designed, and dimensions were assigned in accordance with the sensor wafer layout. The cavity and bonding area patterns were photolithographically transferred onto both sides of an oxidized (100)-orientation 2-in-diameter double-side polished silicon wafer with the aid of an infrared aligner. The oxide was removed from the patterned areas in hydrofluoric acid. The wafer was then placed in KOH at 60°C where exposed silicon was anisotropically removed from both sides of the wafer. Etching was terminated when the pyramidal pits forming on either side met one another approximately midway through the wafer. A sketch of the resultant structure, taken in cross section through the CHEMFET region, is shown in Fig. 1.

The micromachined cavities were positioned over the CHEMFET gate and the large bonding openings were positioned over the bonding pad area. The latter were positioned such that the borders of the opening were just at the edge of the scribe lanes. With this configuration, after the attachment of the micromachined substrate, individual sensor die-plus-membrane-holders could be separated with a diamond diesaw.

The under-side of the micromachined wafer, which would be attached to the sensor wafer, was then coated with epoxy. This was accomplished by applying a thin film of epoxy (Shell Epon 825 and Jeffamine D-230) onto a glass slide, placing the under-side of the machined wafer onto the epoxy-coated slide and then gently pulling the wafer and slide apart. Although this method worked well, more controllable techniques such as screen printing, spray coating or photopatterning, and other materials could be employed. The machined 2-in wafer and 1/4 of the 4-in sensor wafer were then aligned with respect to one another under a microscope with an x-y-z positioning stage and a vacuum pickup arm. A manual contact-type wafer aligner can readily be used for this procedure, with the machined wafer replacing what is normally the mask.

When aligned, the two wafers were brought into contact and left at room temperature to partially cure for 12

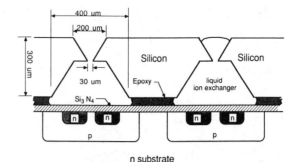

Fig. 1. A cross-sectional sketch of the micromachined package: the positioning of the membrane chambers with respect to the underlying CHEMFET gates, and approximate dimensions.

h, and then completely cured at 80°C for another 8 h. A photomicrograph of the adhered machined wafer and the sensor wafer is shown in Fig. 2, which also clearly shows the bonding area and membrane cavities. The sandwiched structure is then diced, individual die are glued to a printed circuit card, and aluminum wires were wedge bonded to the bonding pads on the chip and to the copper leads of the PC card. The wires were then coated in epoxy, which was dispensed from a needle. The machined substrate is approximately 300 μm thick and provides an excellent barrier to the flow of epoxy into the gate region. This second application of epoxy was then fully cured.

B. Electrochemical Testing

The microsensors were loaded with a liquid ion exchanger for serum calcium ion (Orion membrane number 9825). The membrane choice was dictated by its availability and clinical interest. The chamber was filled with the liquid membrane by positioning the liquid over the chamber opening while applying vacuum to evacuate the air inside the chamber. Upon release of the vacuum, the liquid fills the chamber. An Orion barrel electrode was charged with the same membrane material and tested along with the microsensor for comparison. Both sensors were titrated with CaCl$_2$ in a constant background electrolyte of 0.2 M KCl. All potentials are referenced to a Saturated Calomel Electrode (SCE). The CHEMFET was operated in a feedback mode [1], [7] and all measurements were made at room temperature, 24°C.

C. Results and Discussion

The responses of several microsensors and the Orion macro electrode to calcium ion concentration is shown in Fig. 3. The Orion electrode gives a linear response to Ca^{++} in the range from 0.01 to 0.1 M, with a slope equal to the theoretical value of 30 mV/pH. Although the microsensors show a slightly lower sensitivity (27 mV/pH) than the Orion macro electrode, their response is very reproducible. The microsensors gave an identical response to repeated titration after 24-h immersion in a solution containing 0.1-M CaCl$_2$ and 0.2-M KCl.

The microsensor drift reached a steady-state value of less than 0.1 mV/h after the first hour of exposure to

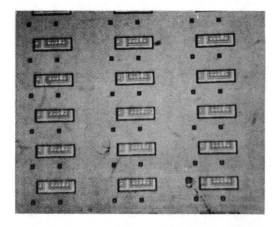

Fig. 2. A photomicrograph of the aligned and epoxied together, micro-machined package and CHEMFET wafer, with exposed bonding pad regions. Bonding pads 1 and 5 are the source and drain of the CHEMFET and pad 2 is the p-well contact. Bonding pads 3, 4, and 5 are the source, gate, and drain of the MOSFET, respectively.

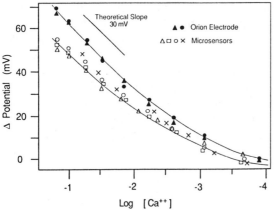

Fig. 3. The potential response of the Orion macro electrode (filled symbols) and the microsensor (open symbols) to calcium ion titration. Each symbol represents a single titration.

solution. This level of drift was maintained for the next 5 days, during which time it continued to respond to additions of $CaCl_2$. The device sensitivity was not checked by titration after 30 h of operation. After 5 days, the device no longer responded to $CaCl_2$ and a sudden shift in gate potential of approximately 10 mV was noted. It is assumed that the membrane material had either dissolved into solution, or had otherwise developed an electrolytic shunt.

It was noted that these microsensors were sensitive to rigorous movement, such as shaking, but their gate potentials always returned to their original value (within 1 mV), *even after repeated removal from solution and biasing*. This is highly unusual for unshielded FET sensors and may mean that the silicon package provides some electrostatic shielding. The conductivity of the package may also explain the slightly lower sensitivity of the microsensor relative to the macroelectrode, i.e., an electrical shunt path across the membrane through the conductive silicon package may exist. This is possible since the package demonstrated here has no insulating layer, other than native oxide, on the inner walls of the membrane chamber. This can be remedied by oxidation and/or the application of LPCVD silicon nitride after micromachining the package.

A micro ion sensor with micromachined membrane containing cavities attached to and positioned over a CHEMFET has been demonstrated. Several aspects of this design, combined together, differentiate it from previously described microsensor structures: 1) The membrane isolating and positioning cavities are micromachined in silicon and as such can be made of smaller dimensions and with greater three dimensional flexibility than can be presently achieved in glass or in thin films. 2) Attachment of the micromachined cavity containing wafer to the FET containing wafer is achieved without anodic bonding and hence without the need to planarize the substrate nor electrostatically protect the FET's. 3) The FET is placed directly below the membrane and thereby provides at site impedance transformation and consequently improved signal-to-noise ratio.

Further development in this direction is the construction of the micro-miniature ISE configuration described earlier, over a FET. The assembly of this package is the same, with the following differences: a thin film of Ag is deposited and patterned over the CHEMFET gate, which are consequently chloridized, and the cavities are filled with a chloride-ion-saturated hydrogel prior to the addition of a polymeric ion selective membrane. These structures are currently being fabricated. They are expected to perform even better with respect to stability and sensitivity than the membrane-coated CHEMFET presented here.

ACKNOWLEDGMENT

The authors would like to thank M. Schmidt of the Department of Electrical Engineering and Computer Science, MIT, for so generously volunteering to assist us in mask generation and fabrication of the micromachined package.

REFERENCES

[1] J. Janata and R. J. Huber, "Chemically sensitive field effect transistors," in *Ion Selective Electrodes in Analytical Chemistry*, H. Freiser, Ed. New York: Plenum, 1980, pp. 107–174.
[2] M. Esashi and T. Matsuo, "Integrated micro multi ion sensor using

field effect of semiconductor,'' *IEEE Trans. Biomed. Eng.*, vol. BME-25, p. 184, 1978.

[3] P. A. Comte and J. Janata, ''A field effect transistor as a solid state reference electrode,'' *Anal. Chim. Acta.*, vol. 101, p. 247, 1978.

[4] P. T. McBride, J. Janata, P. A. Comte, S. D. Moss, and C. C. Johnson, ''Ion-selective field effect transistors with polymeric membranes,'' *Anal. Chim. Acta.*, vol. 101, pp. 239–245, 1978.

[5] A. Shimada, M. Yano, K. Shibatani, Y. Komoto, M. Esashi, and T. Matsuo, ''Application of cather-tip ISFET for continuous in vivo measurements,'' *Med. Biol. Eng. Comput.*, vol. 18, p. 741, 1980.

[6] T. Matsuo and M. Esashi, ''Methods of ISFET fabrication,'' *Sensors and Actuators*, vol. 1, p. 77, 1981.

[7] R. L. Smith and D. C. Scott, ''An integrated sensor for electrochemical measurements,'' *IEEE Trans. Biomed. Eng.*, vol. BME-33, p. 83, 1986.

[8] D. Harame, J. D. Shott, J. Plummer, and J. D. Meindl, ''An implantable ion sensing transducer,'' in *IEDM Tech. Dig.*, p. 46, 1981.

[9] T. Sakai, H. Hiraki, and S. Uno, ''Ion sensitive FET with a silicon-insulator-silicon structure,'' in *Tech. Dig. 4th Int. Conf. Solid-State Sensors and Actuators* (Tokyo, June 6–12, 1987), pp. 711–714.

[10] N. J. Ho, J. Kratochvil, G. F. Blackburn, and J. Janata, ''Encapsulation of polymeric membrane-based ion-selective field effect transistors,'' *Sensors and Actuators*, vol. 4, p. 413, 1983.

[11] S. D. Senturia and D. R. Day, ''An approach to chemical microsensor packaging,'' in *Tech. Dig. Int. Conf. Solid-State Sensors and Actuators* (Philadelphia, PA, June 1985), pp. 198–201.

[12] H.-H. van den Vlekkert, B. Kloeck, D. Prongue, J. Berthoud, B. Hu, N. F. de Rooij, E. Gilli, and P. de Crousaz, ''A pH-ISFET and an integrated pH-pressure sensor with back-side contacts,'' in *Tech. Dig. 4th Int. Conf. Solid-State Sensors and Actuators* (Tokyo, June 6–12, 1987), pp. 726–729.

[13] H. E. Cline and T. R. Antony, ''On the thermomigration of liquid wires,'' *J. Appl. Phys.*, vol. 49, pp. 2777–2786, 1978.

[14] C. C. Wen, T. C. Chen, and J. N. Zemel, ''Gate-controlled diodes for ionic concentration measurement,'' *IEEE Trans. Electron Devices*, vol. ED-26, p. 1945, 1979.

[15] A. Sibbald, P. D. Whalley, and A. K. Covington, ''A miniature flow-through cell with a four-function ChemFET integrated circuit for simultaneous measurements of potassium, hydrogen, calcium and sodium ions,'' *Anal. Chim. Acta*, vol. 159, pp. 47–62, 1984.

[16] G. Blackburn and J. Janata, ''The suspended mesh ion selective field effect transistor,'' *J. Electrochem. Soc.*, vol. 129, p. 2580, 1982.

[17] U. Oesch and W. Simon, ''Opportunities of planar ISE membrane technology,'' in *Tech. Dig. 4th Int. Conf. Solid-State Sensors and Actuators* (Tokyo, June 6–12, 1987), pp. 755–759.

[18] R. P. Buck, ''Kinetics and drift of gate voltage for electrolyte-bathed chemically sensitive semiconductor devices,'' *IEEE Trans. Electron Devices*, vol. ED-29, p. 108, 1982.

[19] O. J. Prohaska, ''New developments in miniaturized electrochemical sensors,'' in *Tech. Dig. Int. Conf. Solid-State Sensors and Actuators* (Philadelphia, PA, June 1985), pp. 402–401.

[20] O. J. Prohaska et al., ''Multiple chamber-type probe for biomedical application,'' in *Tech. Dig. Transducers '87* (Tokyo, June 1987), pp. 812–815.

[21] L. Bousse, F. Schwager, L. Bowman, and J. D. Meindl, ''A new encapsulation technique for microelectrodes and ISFETs,'' in *Proc. 2nd Int. Meeting Chemical Sensors* (Bordeaux, France, July 1986), pp. 499–501.

[22] H. Blennemann, L. Bousse, L. Bowman, and J. D. Meindl, ''Silicon chemical sensors with microencapsulation of ion-selective membranes,'' in *Tech. Dig. 4th Int. Conf. Solid-State Sensors and Actuators*, (Tokyo, June 6–12, 1987), pp. 723–725.

[23] M. Decroux, H.-H. van den Vlekkert, and N. F. de Rooij, ''Glass encapsulation of Chemfets: A simultaneous solution for Chemfet packaging and ion-selective membrane fixation,'' in *Proc. 2nd Int. Meeting on Chemical Sensors* (Bordeaux, France, July 1986), pp. 403–406; also in *Tech. Dig. 4th Int. Conf. Solid-State Sensors and Actuators* (Tokyo, June 6–12, 1987) pp. 730–733.

[24] M. P. Borom, ''Electron-microprobe study of field assisted bonding of glass to metals,'' *J. Amer. Ceramics Soc.*, vol. 56, pp. 254–257, 1973.

[25] A. D. Kurtz, J. R. Mallon, and H. Bernstein, ''A solid state bonding and packaging technique for integrated sensor transducers,'' ISA ASI 73246, pp. 229–238, 1973.

[26] K. E. Petersen, ''Silicon as a mechanical material,'' *Proc. IEEE*, vol. 70, p. 420, 1982.

[27] R. L. Smith, J. Janata, and R. J. Huber, ''Electrostatically protected ion sensitive field effect transistors,'' *Sensors and Actuators*, vol. 5, p. 126, 1984.

[28] L. Bousse, J. Shott, and J. D. Meindl, ''A process for the combined fabrication of ion sensors and CMOS circuits,'' *IEEE Electron Device Lett.*, vol. EDL-9, no. 1, pp. 44–46, 1988.

BONDING TECHNIQUES FOR MICROSENSORS

W. H. Ko, J. T. Suminto and G. J. Yeh

Introduction

Bonding of one substrate to another substrate has become one of the important steps in the fabrication of microsensors. In integrated circuit (IC) packaging technology, the technique is used to bond the IC die to the leadframe and for the final seal on hermetic packages. The attachment of the die to a package substrate serves the purpose of providing a mechanical support, a thermal path, and sometimes an electrical contact. Hermetic packages are used to protect or isolate the IC chip from a hostile environment. The materials used for die bonding and package sealing include: metal, eutectic solders [1,2,3], epoxies [2,4,5], polyimides [5], ceramic and low temperature glasses [4,5]. As an illustration, when fabricating an ultrasonic delay line, the most crucial part of the process is the bonding of the piezoelectric transducer to the delay medium. In order to minimize the mismatch of sonic load impedance it is necessary to bond the transducer onto a delay medium with an extremely thin and uniform adhesive layer to optimize the performance of the device. Epoxy has been used for this bonding purpose, but the bond should be thinner than 0.1 μm. This method requires elaborate procedures in a very clean environment [6]. When using soft-metal bonds such as the well known indium bond, a thicker bond is permissible because of the closer match in their sonic impedance. Thermocompression bonding techniques [7] may then be used to bond the indium-coated surfaces together, but the elevated temperature tends to produce stress due to differential thermal expansion when the bond is returned to room temperature. This shortcoming can be overcome by room temperature compression metallic bonding [8,9]. This bonding is performed by applying high pressure on the 'fresh' indium-coated or Al/Au-coated transducer and substrate, i.e. performed in the vacuum evaporator immediately after coating. In the fabrication of silicon sensors, e.g. pressure sensors, ISFET's, solar cells etc., since the silicon chip will be used in exposed, hostile, and potentially abrasive environments, it will often be necessary to use mounting techniques substantially different from the usual IC packaging methods. Electrostatic bonding of glass or other ceramic materials to metal, silicon or gallium arsenide [14-35] can fulfill many of the requirements for bonding and mounting micromechanical structures.

The techniques which have been used for bonding a transducer to a substrate or for the final seal in hermetic packages are listed as follows:

1. Eutectic Bonding [1,2,3]
2. Epoxy Bonding [4,5]
3. Polyimide Bonding [5]
4. Nonuniform Press Bonding [6]
5. Thermocompression Metallic Bonding [7]
6. Room Temperature Compression Metallic Bonding [8,9]
7. Ultrasonic Welding [10,11]
8. Seam Welding [12]
9. Laser Welding [13]
10. Electrostatic Bonding [14-35]
11. Low Temperature Glass Bonding [36,37]

In our laboratory, eutectic bonding, electrostatic bonding and low temperature glass bonding techniques have been developed in the fabrication of piezoresistive [3] or capacitive [22] pressure sensors, and have given very promising results. These techniques are discussed below.

Eutectic Bonding

In a 2-component phase diagram where there is either partial or no solid solubility between the components, there is a eutectic point, corresponding to the composition of the lowest melting temperature. In the Sn/Pb system, for example, the temperature of the eutectic point is 183°C and the eutectic composition is 61.9% Sn and 38.1% Pb by weight. An examination of the phase diagram for an Si/Au system reveals the eutectic temperature is 363°C and the eutectic composition is 97.1% Au and 2.85% Si by weight; no solid solubility exists between the components.

In the die attachment process, heat is transmitted to the package from the heater block. The die can be heated by resistance heating, hot gas, hot collet, or infrared. The gold from the package and the silicon from the chip are in intimate contact. As the temperature increases, the gold atoms start to diffuse rapidly into the silicon. The gold atoms diffuse interstitially, but most of them terminate in a silicon site where a vacancy existed. When enough gold has diffused to equal the eutectic composition, a very thin liquid layer forms at the interface. As the temperature continues to rise above the eutectic temperature, a larger volume of eutectic alloy is formed. The eutectic alloy formation will continue until one of the two reacting materials is used up. The Si-Au eutectic contains 97% Au; therefore, the limiting factor here is the gold available from the substrate.

The die bond can be made using a Au-backed silicon die, a bare silicon die, or either type of die with a preform. The advantage of using a Au-backed die is that the gold backing, if deposited soon enough after wafer fabrication, will prevent the silicon from oxidizing. If the silicon has had a chance to oxidize, it is difficult to form the Au-Si eutectic bond. Some mechanical motion is required during the chip bonding process in order to break through any silicon dioxide that may have formed on the back of the chip. Once the oxide is broken down, the eutectic formation proceeds.

It is common practice to use a small preform about 0.0015 in. thick, with a composition very close to the eutectic, to start the reaction during die attachment. This piece is placed under the die before applying pressure. The preform melts at the die attachment temperature and provides a wet surface to facilitate the bonding process. The preforms usually consist of pure gold (Au) or gold-silicon (Au/Si) on gold backed dies. Lead-indium-silver (Pb/In/Ag) and lead-tin (Pb/Sn) can be used on

Reprinted with permission from *Micromachining and Micropackaging of Transducers*, 1985. Copyright © 1985 by Elsevier Science Publishers.

nickel backed dies. Special alloys also used include gold-tin (Au/Sn) and gold-germanium (Au/Ge), where their properties are useful. These alloys are selected because of their good thermal and electrical conductivity and their resistance to corrosion.

Eutectic bonding using preform has been used in our laboratory for the fabrication of a miniature pressure transducer [3]. The completed transducer structure is illustrated in Fig. 1. A silicon diaphragm, with a diffused piezoresistive bridge, is bonded to a silicon substrate which is already etched into a big crater to form a reference chamber. The bonding is performed by sandwiching a 80%Au-20%Sn preform about 1 mil thick between the two substrates. A holder, as shown in Fig. 2, keeps the assembly in position and a pressure of about 150 psi is applied. This assembly is then put into a vacuum chamber and gradually heated to 280°C and then cooled to room temperature. The whole heating cycle takes 4 hours, one hour to raise the temperature and three hours to decrease it. This sealing had been proven hermetic. Table-1 shows a typical short term performance of these piezoresistive pressure transducers. Experiments were carried out to determine the source of long-term baseline drift [3]. The result is summarized in Fig. 3. The diaphragm without sealing material (80% Au / 20% Sn) showed very good stability, with or without experiencing the heating cycle of the sealing process. Diaphragms with a metal preform sealed on, but without the second diaphragm bonded on the back, showed a very large negative drift, while sealed transducers showed a large positive drift. The rate of drift decreases exponentially with time. Several sets of samples were run with similar results. From this it was suggested that the long-term baseline drift is caused by stresses in the alloy material which relax with time due to creep in the alloy. The cause of the initial stress is the unequal thermal expansion of the silicon and the Au-Sn alloy when cooled from sealing temperature (280°C) to room temperature.

199

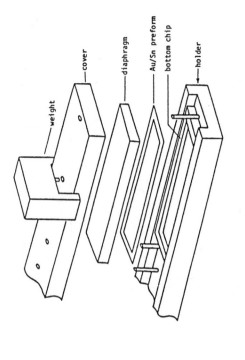

Fig. 2 Holder for the eutectic bonding of PS-6 piezoresistive pressure sensor.

Table 1 Thermal Expansion Characteristics of Silicon and #7740 Pyrex Glass

$X = \Delta L/L \ (25°C)$ 　 $\alpha = \Delta L(T)/L(T)/\Delta T$

TEMP (°C)	SILICON		7740 PYREX (#1)		7740 PYREX (#2)	
	X_{si} (ppm)	α_{si} (ppm/°C)	X_{glass} (ppm)	α_{glass} (ppm/°C)	X_{glass} (ppm)	α_{glass} (ppm/°C)
-50	-165	---	-236	---	-210	---
-25	-117	1.92	-160	3.04	-145	2.6
0	-62	2.2	-80	3.2	-70	3.0
25	0	2.48	0	3.2	0	2.8
50	68	2.72	80	3.2	75	3.0
100	217	2.98	243	3.26	225	3.0
150	381	3.28	406	3.26	380	3.1
200	556	3.5	571	3.3	535	3.1
250	739	3.66	735	3.28	690	3.1
300	928	3.78	900	3.3	845	3.1
350	1121	3.86	1066	3.32	1010	3.3
400	1319	3.96	1232	3.32	1180	3.4
450	1520	4.02	1398	3.32	1360	3.6
500	1725	4.1	1620	4.44	1570	4.2
550	1935	4.2	1970	7.0	1940	7.4

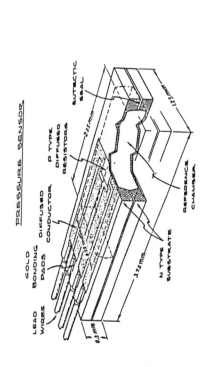

Fig. 1 The structure of PS-6 piezoresistive pressure sensor.

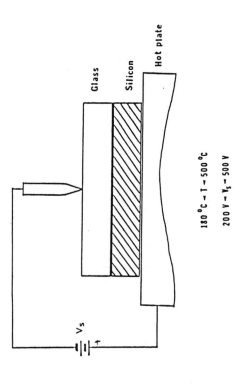

Glass

Silicon

Hot plate

V_s

$180\,^\circ C < T < 500\,^\circ C$

$200\,V < V_s < 500\,V$

Fig. 4 The schematical set-up for electrostatic bonding of silicon to glass.

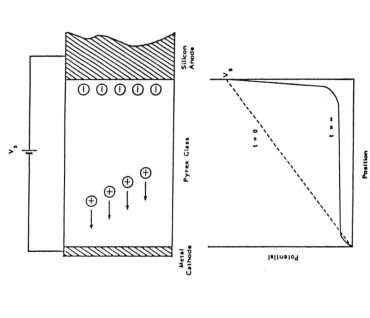

V_s

Metal Cathode

Pyrex Glass

Silicon Anode

$t = 0$

$t = \infty$

V_s

Potential

Position

Fig. 5 Initial and final equilibrium potential distributions across the glass during electrostatic bonding.

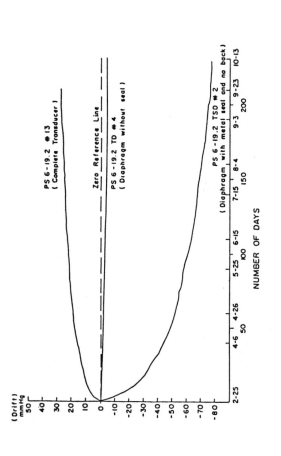

(Drift) mmHg

PS 6-19.2 #13 (Complete Transducer)

Zero Reference Line

PS 6-19.2 TD #4 (Diaphragm without seal)

PS 6-19.2 TSD #2 (Diaphragm with metal seal and no back)

NUMBER OF DAYS

Fig. 3 Long-term drift of PS-6 sensor diaphragm and assembled transducer with Au-Sn alloy seal.

Electrostatic Bonding

Electrostatic bonding is an important means of encapsulating sensors at the chip or wafer level. It allows the silicon chip to be hermetically bonded to a glass support chip, effectively capping one side of the silicon and shielding it from the surrounding environment.

The electrostatic bonding process can be accomplished on a hot plate in atmosphere or vacuum at temperatures between 180° and $500^\circ C$ (well below the softening point of the Pyrex glass). Electrostatic attraction between the glass and silicon pieces serves to pull the two into intimate contact, thus eliminating the need for applying mechanical pressure to the wafers.

The bonding set-up is shown schematically in Fig. 4. The polished Pyrex glass is placed against the polished surface of the silicon. A cathode electrode is held against the outer surface of the glass wafer and the whole assembly is heated on a hot plate, which also serves as an anode, to approximately $450^\circ C$. A 200-1000 volt potential is then applied between the electrodes. At the elevated temperature, the electric potential between the two wafers causes them to be pulled into close contact and they bond almost instantly.

The electrostatic attraction between the glass and silicon wafers is developed as follows. At elevated temperatures (yet below the softening point of Pyrex), the positive sodium ions in the glass become quite mobile and they are attracted to the negative electrode on the glass surface where they are neutralized. The more permanently bound negative ions in the glass are left, forming a space charge layer in the glass adjacent to

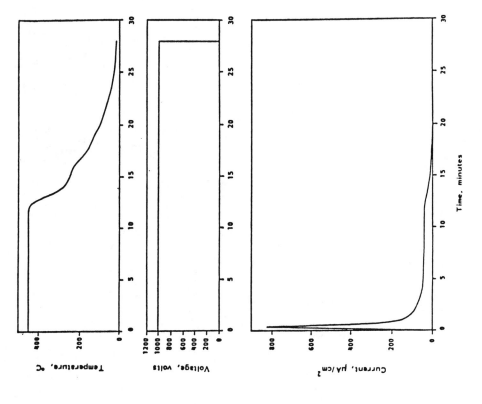

the silicon surface. The time-varying potential distribution is shown in Fig. 5 as a function of position in the glass plate. After the Na$^+$ ions have drifted toward the cathode, most of the potential drop in the glass occurs at the surface next to the silicon. The two wafers then act as a parallel plate capacitor with most of the potential being dropped across the several micron wide air gap between them. The resulting E-field between the surfaces serves to pull them into contact with a force of approximately 350 psi for E= 3 x 10^{-6} V/cm.

Once the wafers are in contact, almost all of the applied potential is dropped across the space charge layer in the glass. The extremely high fields which develop in that region transport oxygen out of the glass to bond with the silicon surface. The seal appears to be chemical in nature, possibly a very thin layer of grown SiO_2.

The temperature, voltage, and current density during the bonding process is recorded as shown in Fig. 6. During the bonding, the temperature and applied voltage is kept constant. A pulse of current will occur when the voltage is switched on, indicating a drift of sodium ions. Very soon a space charge region is built up, and the bonding occurs. Looking through the glass, the bonded region will become a dark grey color; when this region expands throughout the whole wafer, the bonding complete. Though the bonding is irreversible, in common practice the voltage is kept on while the bonded sample is cooling down. After it reaches near room temperature, the applied voltage is turned off.

The bonding process can be used to seal various insulating glasses to matching metals, alloys or semiconductor materials. Of particular interest are a variety of borosilicate glasses, e.g. Corning #7740 and #7070. Seals were also made to soda lime #0080, potash soda lead #0120. aluminosilicate #1720, fused silica and fiber optics [14]. Strong seals have also been made to ceramic Cer-VIT (Owens Illinois) [14] and beta-Alumina [34]. Among metals and alloys whose thermal properties match those of commercial glasses, seals have successfully been made to [14,17] tantalum, titanium, Kovar, Niromet 44, Al, Fe-Ni-Co alloy , and to the semiconductors silicon [21-31], e.g. in silicon pressure transducers and solar cells, and gallium arsenide [34].

The main requirements for the materials to be bonded are:

(1) The glass must be slightly conductive when heated to temperatures well below its softening point.

(2) The metal used must not inject mobile ions into the glass. As we mentioned previously, the space charge region is responsible for the bonding. However, the metallic anode could provide the glass with positive ions which would be transported across the Na-depleted region by the high electric field and compensate the charge of the immobile negative ions. Those ions whose mobilities are orders of magnitude lower than that of Na$^+$ will not affect the build up of the space charge region. But, those ions whose mobilities are comparable to or higher than that of sodium will replace the sodium ions so that the conductivity of the boundary layer will not be reduced. Under this condition, therefore, a space charge region will not be generated and no bond will be formed. This situation apparently exists when trying to bond silver metal to glass.

Fig. 6 The temperature, voltage, and current density profiles during the electrostatic bonding process.

(3) The surface roughness of both the glass and metal should be less than 1 micron rms, since the deformation of either the glass or the metal is very limited. Also, the surface should be free from dust or other contamination. A silicon surface with thermally grown oxide less than 4600 Å thick can also be bonded to Pyrex #7740. Fig. 7 shows the current flow during the electrostatic bonding of silicon with different oxide thicknesses to Pyrex #7740 at 500°C at an applied voltage of 100U V. The oxide was thermally grown at 1000°C. This shows that if the thickness of the oxide exceeds 4600 Å the current flow becomes negligibly small and the bonding will fail. Only those with an oxide thickness less than 2000 Å will form a good bond.

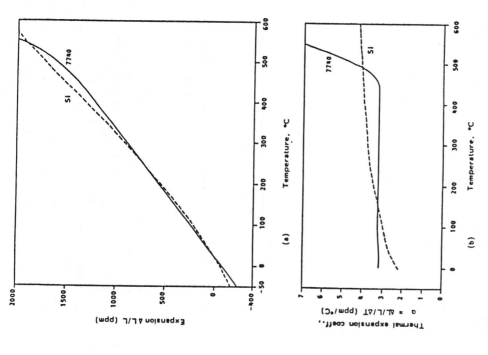

Fig. 8 Thermal expansion properties of #7740 Pyrex glass and single crystal silicon.

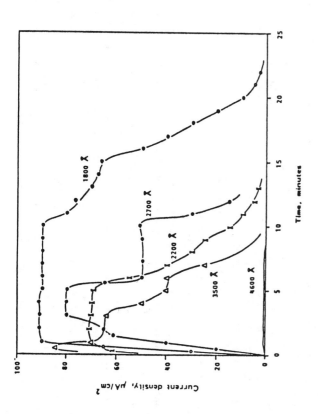

Fig. 7 Current density profiles during the bonding of glass to silicon with various thickness of glass on it.

(4) The thermal expansion coefficients of the two materials should be closely matched. This requirement results from the fact that the bond actually occurs at an elevated temperature, and a major mismatch of expansion coefficients will cause one of the materials to crack upon cooling. For this reason, Corning #7740 Pyrex glass is used for bonding to silicon. Fig. 8(a) shows the thermal expansion properties of Pyrex #7740 compared to silicon. It can be seen they behave similarly at lower temperatures. However, the thermal expansion of Pyrex #7740 deviates from silicon at temperatures greater than 300°C. The plot of thermal expansion coefficients, Fig. 8(b), shows a drastic increase of 7740 at temperatures greater than 450°C. Fig. 9 represents the results of a study to determine the residual stress in bonding silicon to one batch of #7740 Pyrex glass. The residue stress between the two materials vanishes at a bonding temperature approximately equal to 300°C. Therefore, in order to achieve a no-stress bond, a 300°C temperature is preferred. The thermal expansion of Pyrex #7740 may vary from batch to batch. Table-1 compares the thermal characteristic of Pyrex #7740 from two different batches. The derived bonding temperature may vary depending upon the #7740 glass used.

Bonding process parameters vary widely with application and materials. Temperature, voltage, current density, time and atmosphere are important. Bonding temperatures as low as 200°C have been used successfully. Values to 500-600°C are used for irregularly surfaced devices and rough glass. For silicon devices with aluminum metalization, temperatures cannot exceed 450°C. Higher temperatures can be used, but process advantages are reduced and operational problems such as handling of soft glass are introduced.

The voltage requirement depends upon temperature and glass type and thickness. Typically 1000 volts is used for 15-60 mil thick glasses at moderate temperature bonding. Fig. 10 from one of our studies shows that at least 450 volts is required in order to achieve a successful bonding of 60 mil thick pyrex to silicon at a temperature near 200°C. Current density varies in the same manner as voltage, typically being 1 mA/cm² but varies

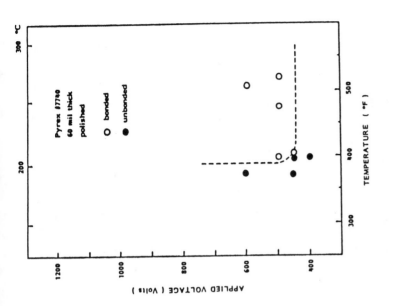

Fig. 10 Voltage and temperature requirements for a successful bonding of 60 mil thick Pyrex to silicon.

Laser heating techniques have also been used in electrostatic bonding [31,32]. This laser assisted electrostatic bonding process combines the application of an electric field with heating of pyrex glasses via a CW CO_2 laser exposure to produce bonds similar to those created by the standard electrostatic bonding process. Standard electrostatic bonding requires the application of high pressures and heating of both the silicon (or metal) and glass by contact with a hot cathode. The laser assisted electrostatic bonding process differs in that thermal energy is rapidly generated only in the glass by the absorption of CO_2 laser energy. Also, the addition of high mechanical pressure is not needed in the laser assisted process. The benefit of this approach is to maximize the glass softening while minimizing the time-temperature stress to the silicon (or metal).

In the bonding of GaAs-glass a special treatment should be performed prior to the bonding process [34]. Even in a reducing atmosphere of H_2 and N_2, the surface of GaAs will form a non-adherent oxide layer which prohibits the bonding. One probable mechanism is surface reaction between gallium and oxygen-containing complexes in the glass. An empirical solution to the problem was found by prebaking the glass in a reducing

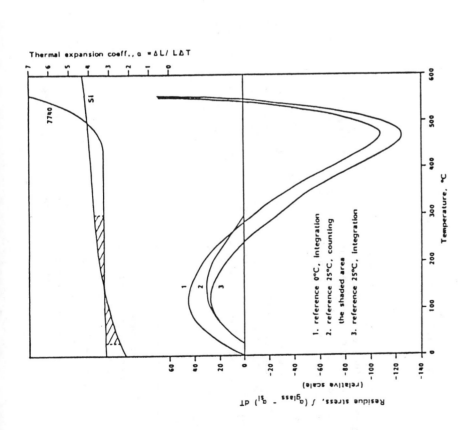

Fig. 9 The residual stress in silicon after it is bonded to #7740 Pyrex glass at different temperatures.

greatly from run to run. Process time is dependent on temperature, voltage, and the bonded area. Typically 5-10 minutes is enough. A longer time is required when using lower temperatures where ion mobility is lower.

An ambient atmosphere is adequate for bonding and may be preferable due to the presence of free oxygen. However, a special application may require a controlled atmosphere. Examples include bonding a semiconductor device with corrosion sensitive metallization and bonding where vacuum sealing is desired. Atmospheres that have been used successfully in silicon-glass bonding include air, nitrogen, forming gas, argon, and helium [28].

203

Table 2   Chemical Constituents of Bulk and Sputter-coated #7740 Pyrex Glass

Corning technical data:

	SiO$_2$	Al$_2$O$_3$	Na$_2$O	K$_2$O	B$_2$O$_3$
	81%	2%	4%	0.5%	13% wt.

Auger Analysis:

	Si	B	O
Sputtering target	22%	14%	64%
Sputtered Pyrex Film	37%	8%	55%
Quartz	35%	-	65%

Remarks:

1. Si LVV peak of pyrex 7740 sputtering target occurs at 74 eV which is closed to Si-O bond.

2. Si LVV peak of sputtered pyrex film occurs at 89 eV which is near to elemental Si

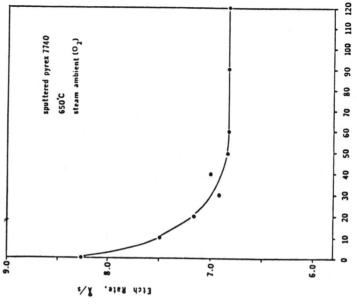

Fig. 12   Etch-rate of sputter coated Pyrex film in buffer HF after being annealed at 650°C in a steam ambient.

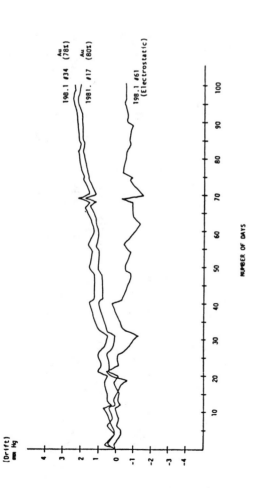

Fig. 11   Long-term stability of Au-Sn alloy sealed and electro-statically sealed PS-6 transducers.

atmosphere at 400°C for 15 hours. This procedure is believed to remove oxygen residues contained in the glass which are amenable to chemical reaction. The bonding temperature is 360°C with an applied voltage of 800 V. The charge build up takes 5-10 minutes and the entire operation is completed within 30 minutes. The glass used is Corning #0211.

Electrostatic bonding of a silicon piezoresistive pressure transducer (which is shown in the upper half of the structure in Fig. 1) to a Pyrex #7740 backpiece was studied. The results of this study are shown in Fig. 11. For the purpose of comparison, this figure also shows the pressure sensors sealed with metal alloy (80% Au / 20% Sn and 78% Au / 22% Sn) eutectic bonding. These devices were run for 100 days and long-term drift for this entire test was quite small. For devices sealed with the metal alloy, the maximum drift rate was 6 mmHg/month, while the electrostatically sealed devices had drift rates of less than 1 mmHg/month.

Electrostatic Bonding of Si to Si

Adopting the idea of silicon-glass bonding, silicon and silicon can also be electrostatic-bonded together via a thin layer of pyrex glass [33].

The surfaces of the silicon to be bonded should be polished. One of the silicon surfaces was coated with a thin Pyrex film. This film was sputter deposited on the silicon using an RF sputtering system. The target was a 5 in. dia. circular shaped Corning 7740 Pyrex glass plate bonded on

204

In order to have a successful bonding, the two members to be bonded should be in intimate contact over the whole bonding area. Since the applied voltage is low (50V), the electrostatic force which will pull the two members together is also small. Any small tilt of the bonding surface or small contamination will cause a bonding failure.

The silicon piezoresistive pressure transducer, which is shown in the upper half of the structure in Fig. 1, was electrostatically bonded on the silicon substrate via a 4 μm thick sputter-coated Pyrex film. The baseline drift of this device was studied and compared to the device bonded on Pyrex #7740 substrate. The result is shown in Fig. 14. Undoubtedly, the performance of the device bonded on a silicon substrate was better than that bonded on a Pyrex substrate. Because the silicon sensor and the silicon substrate were of the same material and had exactly the same thermal expansion, the thermal stress came only from the bonding medium, that is the sputter-coated Pyrex film. Since the film was very thin compared to the thickness of either the sensor or the substrate, the thermal stress exerted on the sensor's diaphragm was small.

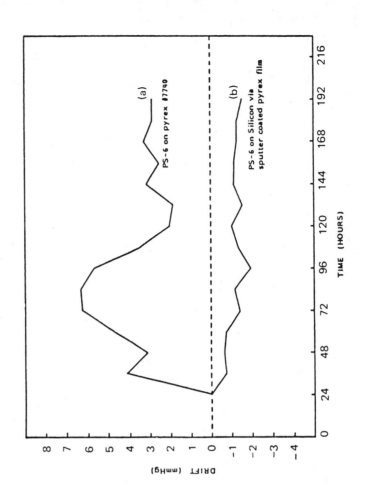

Fig. 14 Baseline drift of PS-6 piezoresistive pressure sensors (a) bonded on Pyrex substrate and (b) bonded on silicon substrate via a thin Pyrex film.

an Al target holder via a thin layer of DuPont 6838 silver epoxy. The assembly was then dried and cured in a 150°C oven. Before sputtering the vacuum chamber was pumped down to 1 X 10⁻⁵ torr. The sputtering was carried out in an 8 mtorr 10% oxygen in argon ambient and the power was set at 300W. Both the deposition rate and substrate holder were cooled by water flow. The deposition rate was about 40 Å/min. A 4 μm thickness of pyrex was deposited on the silicon. We found that a minimum glass thickness of approximately 3 μm was desirable in order to achieve a satisfactory bonding.

The as sputter-coated Pyrex film was analyzed using Auger Electro-spectroscopy (AES). A comparison of the compositions of the sputter-coated Pyrex and bulk Pyrex analyzed in AES is shown in Table-2. It shows that the sputter-coated film contained 37% Si instead of 22% Si as in the target material. The Si LVV peak of this sputtered pyrex was at 89eV, near elemental Si, instead of 74eV in the Si-0 bond. Therefore, the as sputter-coated pyrex was Si rich glass. It was also found that this film brokedown at approximately 20V which caused the failure of the electrostatic bonding. In order to restore the Si-0 bond, this sputter-coated Pyrex film had to be annealed at a high temperature. Fig. 12 shows the etch rate of this sputter-coated pyrex film in buffer HF after being annealed at 650°C in a steam ambient for various lengths of time. An annealing of about 1 hour proved to be sufficient.

In performing the electrostatic bonding, the Si wafer with coated Pyrex was placed on another bare surface Si. The two members were aligned in the desired orientation and held in position by the top electrode, as shown in Fig. 13. The negative electrode of the dc power supply was applied to the coated Si member. After the sandwich was temperature stabilized, a gradually increasing dc voltage was applied to the members. A maximum voltage of 50V is adequate for the bonding. An abrupt increase of the voltage would cause an arc at the edge of the bonding members and no bonding would occur. When the voltage reached 50V, a high current pulse would occur and then rapidly decrease to a constant low current of about 100μA/cm². The assembly was held in this condition for 5 min., after which the substrate heater was shut off so that the temperature would decrease to near room temperature. The voltage was then removed and the bonding operation was complete.

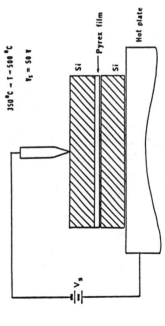

Fig. 13 The schematical set-up for electrostatic bonding of silicon to silicon via a sputter coated Pyrex film.

205

Low Temperature Glass Bonding

The advantage of electrostatic bonding is uniformity, but the electric field applied is detrimental to some field effect devices. Low melting point glass needs no electric field, but the thickness and uniformity is hard to control. In order to have the advantages of these two methods, sputtered low melting temperature glass was explored. Since the glass can be applied uniformly on the silicon surface, high voltage is not necessary.

materials which melt and flow at sealing temperature each time they are thermally processed. Devitrifying glasses are thermosetting materials which crystallize by surface nucleation in a time-temperature relationship to produce their specified properties. Once the sealing glass crystallizes, its thermal stability is improved because its softening point is not that of the original glass, but is that of the crystalline material which exhibits increased chemical durability compared to the original glass. These glasses used to be applied in four different ways: spraying, screen printing, extrusion and sedimentation. After the glass is applied, it has to be preglazed to remove the organic residues produced by vehicle and binder decomposition. The sandwich of substrate-glass-substrate is then heated to the sealing temperature and a weight of at least 1 psi is applied to the sandwich. Sealing cycles depend on the geometry of the seal and the composition of the solder glass used.

The thickness of the glass applied by spraying, screen printing, extrusion or sedimentation is very difficult to control, and it is almost impossible to deposit a thickness less than 2 μm because the particle size of the glass is about 0.2 μm. By coating the glass on a substrate using the sputtering system, the glass film can be uniformly controlled to below 1 μm, and sealing is still possible.

Corning 7593 devitrifying glass frit has a temperature coefficient close to that of silicon. It was chosen for the sealing of Si to Si in our experiment. The sputtering target was made as follows: A slurry (which is a mixture of Corning 7593 glass frit and the carrier of amyl acetate with 12% nitro-cellulose by weight) about 60 mil thick was painted on the Al target holder. This assembly was dried and cured at about 150°C. To prevent cracking of the glass pie, the assembly was cooled slowly to room temperature.

The sputtering was performed in an RF sputtering system. Both the target and the substrate holder were water cooled. Pure oxygen was used as the sputtering gas. The vacuum chamber was pumped down to about 1×10^{-5} torr before introducing the oxygen. The pressure during the sputtering was kept at 8 microns, and a power of 300 W was applied. The deposition rate was about 50 Å/min. The sputtered glass does not need to be glazed. Two pieces of silicon coated with 8000 Å thickness of glass were put face to face and a pressure about 1 psi was applied to keep the assembly in position and help the glass flow at the sealing temperature. The assembly was put into a 650°C furnace with oxygen ambient, as shown in Fig. 11, and cured for 30 minutes. The resulting bond strength was excellent, and it has been proved hermetic under He leak detection.

Conclusion

For miniature pressure transducers, the major problems are packaging and long-term stability. Previous studies [3] show that the major cause of long-term drift in the device is not related to the sensor design or processing, but rather to the assembly and packaging of the device. In particular, thermal stresses developed by thermal expansion during the sealing process appear to be the major cause of baseline drift. Therefore, choosing the right technique and the right material to seal the device is very important.

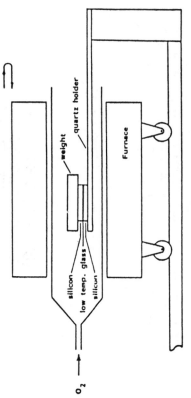

Fig. 15   Furnace used for bonding silicon to silicon via a sputter coated low melting temperature glass.

Phosphosilicate glass (PSG) films which have a relatively low melting point have been used for silicon to silicon wafer bonding [36]. A SiO2 film 1-2 μm thick was thermally grown on the wafers. A PSG layer containing approximately 10% P_2O_5 was formed on the SiO2 surfaces by exposing the wafers to a mixture of phosphorous oxychloride ($POCl_3$) vapor and O2 gas at 900°C. The wafers were then aligned and clamped together inside a quartz vacuum-chuck assembly which forced the wafer surfaces into intimate contact by connecting them to a vacuum pump throughout the bonding process. The bonding assembly was introduced into a furnace at 1100°C and heated for 30 minutes, then slowly cooled to room temperature. Fusion of the PSG layers took place readily and the resulting bond strength was excellent provided the wafer surfaces were clean and reasonably flat. However, the sealing temperature (1100°C) is still considered to be high. Corning Glass Work has introduced a series of glass frit (#75xx) suitable for hermetic low temperature sealing in a variety of applications [37]. The sealing temperature of these sealing glass frits ranges from 415°C to 650°C, and their thermal expansion coefficient ranges from 2 to 5 times that of the thermal expansion coefficient of the silicon. Technology Glass Corporation has also introduced a suspension of high purity, low sodium, ultrafine glass powder in isopropyl alcohol [38]. It is designed for silicon to silicon wafer bonding. The thermal expansion coefficient is about 3 times of that of silicon, and the sealing temperature is 510-540°C.

The glass for hermetic low temperature sealing can be divided into two types: vitreous and devitrifying. Vitreous glasses are thermoplastic

Eutectic bonding and electrostatic bonding have been developed to seal piezoresistive pressure transducers. The transducer utilizing the electrostatic bonding technique showed a more rigid seal with better long-term stability.

Electrostatic bonding of glass to silicon with oxide can also be achieved if the oxide thickness does not exceed 2000 Å. Since thermal stress developed by thermal expansion during the sealing process appears to be the major cause of baseline drift, the thermal expansion of Pyrex #7740 and silicon were studied and compared. The study showed that there is a bonding temperature at which the residual stress is minimal. Therefore, electrostatic bonding at the minimum stress temperature of 300°C is preferred.

Electrostatic bonding of silicon to silicon via a thin sputter-coated Pyrex #7740 film can also be achieved. The as sputter-coated pyrex film has to be annealed in order to restore the Si-O bond and to achieve a successful seal. The piezoresistive pressure transducer bonded on silicon via a sputter-coated Pyrex film shows smaller baseline drift than that of the device bonded on Pyrex #7740 substrate, because the silicon transducer and the silicon substrate have exactly the same thermal expansion coefficient, and the only thermal stress comes from the very thin Pyrex film which only exerts a small stress on the sensor's diaphragm.

Low temperature glass bonding does not need a high electric field which is detrimental to some field effect devices. Utilizing the RF sputtering technique, a thin and uniform low temperature glass film can be coated on a silicon surface. Two glass-coated silicon wafers can be sealed together by simply heating the assembly to the melting point of the glass which is quite low. A thin glass film of about 1 micrometer is enough to achieve sealing. This seal has been proved hermetic.

Acknowledgement

This work was partly supported by NIH grant no. RR02024 the Semiconductor Chemical Transducer Resource and no. RR00857 the Biomedical Electronics Resource. The assistance of members of the Electronics Design Center staff is also gratefully acknowledged.

References

[1] Valero, L., "The Fundamental of Eutectic Die Attach", Semiconductor International, May 1984, p.236-241.

[2] Harper, C. A., ed., Handbook of Thick Film Hybrid Microelectronics, p.8-32 - 8-39.

[3] Ko, W. H., Hynecek, J., Boettcher, S. F., "Development of a Miniature Pressure Transducer for Biomedical Applications", IEEE Trans. on Electron Dev., ED-26, 12, p.1896-1905 (1979).

[4] Singer, P. H., "Die Bonding and Package Sealing Materials", Semiconductor International, Dec. 1983, p.62-65.

[5] Bolger, J. C., and Mooney, C. T., "Die Attach in Hi-Rel P-Dips: Polyimides or Low Chloride Epoxies ?", IEEE 1984 Electronics Components Conf., p.63-67.

[6] Papadakis, E. P., "Nonuniform Pressure Device for Bonding Thin Slabs to Substrates", J. Adhesion, 3, p.181-194 (1971).

[7] Butts, A., and van Duzee, G. R., "Low Temperature Bonding of Silver", Silver in Industry, Addicks, L., ed., New York: Reinhold, 1940.

[8] Sittic, E. K., and Cook, H. D., "A Method for Preparing and Bonding Ultrasonic Transducer Used in High Frequency Digital Delay Lines", Proc. IEEE (Letters), p.1375-1376 (Aug. 1968).

[9] Knox, J. D., "A Room Temperature Non-Indium Metallic Bond Tested by Welding Acoustic Shear-Wave Transducer to Paratellurite", RCA Rev., 34, p.369-372 (1973).

[10] Larson III, J. D., and Winslow, D. K., "Ultrasonically Welded Piezoelectric Transducers", IEEE Trans. on Sonics and Ultrasonics, SU-18, 18, p.112-152 (1971).

[11] Noguchi, T., Fukukita, H., and Fukomoto, A., "Simple Ultrasonic Welding Method for Bonding Thickness Mode Transducers", IEEE Trans. on Sonics and Ultrasonics, SU-21, 1, p.55-56 (1974).

[12] Integrated Circuit Engineering Corp., Integrated Circuit Fabrication, 2nd ed., p.5-21 (1979).

[13] Krishnaswamy, H. N., and Boccelli, V. E., "Micro-circuit Flatpack Sealing by Laser Welding", SAMPE Quarterly, p.360-368 (July 1977).

[14] Wallis, G., and Pomerantz, D. I., "Field Assisted Glass-Metal Sealing", J. of Appl. Phys., 40, p.3946-3949 (1969).

[15] De Nee, P. B., "Low Energy Metal-Glass Bonding", J. Appl. Phys. 40, p.5396-5397 (1969).

[16] Wallis, G., "Direct Current Polarization During Field-Assisted Glass-Metal Sealing", J. Am. Cer. Soc., 53, p.563-567 (1970).

[17] Wallis, G., Dorsey, J., and Beckett, J., "Field Assisted of Glass to Fe-Ni-Co Alloy", Cer. Bull., 50 (12), p.958-961 (1971).

[18] Borom, M. P., "Electron-Microprobe Study of Field-Assisted Bonding of Glasses to Metals", J. Am. Cer. Soc., 56, p.254-257 (1973).

[19] Kim. C., and Tomozawa, M., "Glass-Metal Reaction in AC Electric Field", J. Am. Cer. Soc., 59, p.321-324 (1976).

[20] Brownlow, J. M., "Glass Related Effects in Field-Assisted Glass-Metal Bonding", IBM Rep., RC 7101 (May 1978).

[21] Pomerantz, D. I., "Anodic Bonding", U. S. Pat. 3,397,278, Aug. 13, 1968.

[22] Ko, W. H., Bao, M. H., and Hong, Y. D., "A High-Sensitivity Integrated-Circuit Capacitive Pressure Transducer", IEEE Trans. on Electron Dev., ED-29, p.48-56 (1982).

[23] Sander, C., Knutti, J., and Meindl, J., "A Monolithic Capacitive Pressure Sensor with Pulse-Period Output", IEEE Trans. on Electron Dev., ED-27, p.927-930 (1980).

[24] Lee, L. S., and Wise, K. D., "A Batch-Fabricated Silicon Capacitive Pressure Transducer with Low Temperature Sensitivity", IEEE Trans. on Electron Dev., ED-29, p.42-48 (1982).

[25] Petersen, K. E., "Fabrication of an Integrated Planar Silicon Ink Jet Structure", IEEE Trans. on Electron Dev., ED-26, p.1918-1920 (1979).

[26] Roylance, L. M., and Angell, J. B., "A Batch-Fabricated Silicon Accelerometer", IEEE Trans. on Electron Dev., ED-26, p.1911-1917

[27] Younger, P. R., "Hermetic Glass Sealing by Electrostatic Bonding",
J. Non-Crystalline Solids, 38 & 39, p.909-914 (1980).

[28] Minucci, J. A., Kirkpatrick, A. R., and Kreisman, W. S., "Integral
Glass Sheet Encapsulation for Terrestial Panel Applications", 12th
IEEE Photovoltaic Specialist Conf., p.309 (1976).

[29] Younger, P. R., Kreisman, W. S., and Landis, G. A., "Terrestial
Solar Arrays with Integral Glass Construction", 13th IEEE
Photovoltaic Spec. Conf., p.729-732 (1978).

[30] Kirkpatrick, A. R., Kreisman, W. S., and Minnuci, J. A., "Status of
Electrostatically Bonded Integral Covers for Silicon Solar Cells",
IEEE Photovoltaic Spec. Conf., p.573-576.

[31] Engelkrout, D. W., Day, A. C., and Horne, W. E., "Current Research
in Adhesiveless Bonding of Cover Glasses to Solar Cells", 16th IEEE
Photovoltaic Spec. Conf., p.108-114 (1982).

[32] Horne, W. E., "Electro-Optically Assited Bonding", U. S. Pat.
4,294,602, Oct. 13, 1981.

[33] Brooks, A. D., and Donovan, R. P., "Low Temperature Electrostatic
Si-to-Si Seals Using Sputtered Borosilicate Glass", J. Electrochem.
Soc., 119, p.545-546 (1972).

[34] Hok, B., Dubon, C., and Ovren, C., "Anodic Bonding of GaAs to
Glass", to be published.

[35] Dunn, B., "Field Assisted Bonding of Beta-Alumina to Metals", J.
Am. Cer. Soc., 62, p.545-547 (1979).

[36] Bassous, E., "Fabrication of Novel Three Dimensional Microstructure
by Anisotropic Etching of (100) and (110) Silicon", IEEE Trans. on
Electron Dev., ED-25, p.1178-1185 (1978).

[37] Sealing Glass, Corning Tech. Publ. (1981).

[38] TG-120 Si-to-Si Sealing Glass, Technol. Glass Corp., Tech. Publ..

Authors:

W. H. Ko, J. T. Suminto, and G. J. Yeh

Electrical Engineering and Applied Physics Department
and Electronics Design Center
Case Western Reserve University
Cleveland, Ohio 44106

208

SILICON FUSION BONDING FOR PRESSURE SENSORS

Kurt Petersen, Phillip Barth, John Poydock,
Joe Brown, Joseph Mallon Jr., Janusz Bryzek

NovaSensor
1055 Mission Court
Fremont, CA 94539
(415) 490-9100

ABSTRACT

Two novel processes for fabricating silicon piezo-resistive pressure sensors are presented in this paper. The chips described here are used to demonstrate an important new silicon/silicon bonding technique, Silicon Fusion Bonding (SFB). Using this technique, single crystal silicon wafers can be reliably bonded with near-perfect interfaces without the use of intermediate layers. Pressure trans-ducers fabricated with SFB exhibit greatly improved perfor-mance over devices made with conventional processes. SFB is also applicable to many other micromechanical structures.

INTRODUCTION

Micromachining technology is often severely con-strained 1) because of the structural limitations imposed by the traditional backside etching process and 2) because few practical methods have yet been demonstrated for true silicon/silicon direct bonding. Devices fabricated using techniques such as thermomigration of aluminum, eutectic bonding, anodic bonding to thin films of pyrex, intermed-iate glass frits and "glues" suffer from thermal expansion mismatches, fatigue and creep of the bonding layer, com-plex and difficult assembly methods, unreliable bonds and/or expensive processes. In addition, none of these processes provide the performance and versatility required for advanced micromechanical applications. For example, the (backside) cavity sidewalls slope outward from the diaphragm in anisotropically etched pressure sensor chips, forcing the overall chip area to be substantially larger than the active sensing diaphragm area. Another problem area is that narrow, hermetic gaps between two single crystal wafers are difficult to fabricate with precision because of the intermediate layers required. Finally, the most exciting class of new micromechanical structures, which have recently been developed, are actually thin film movable structures that are incompatible with current silicon/silicon bonding techniques.

Workers have addressed these problem by fabricating beams [1], diaphragms [2], and (recently) more complex structures such as mechanical springs and levers [3] from polycrystalline silicon using sacrificial-spacer etching. The silicon/silicon bonding technique described here has been previously discussed in several papers [4, 5, 6, 7]. SFB (Silicon Fusion Bonding) is a very powerful process for creating a wide range of **all single crystal** mechanical structures, that can replace polysilicon in many of the currently used sacrificial-etching methods. In this paper, several pressure sensor designs will be presented that illustrate the practical implications of this technology. Another paper at this meeting describes an acceleration sensor realized with the same processing techniques [8].

FABRICATION OF LOW PRESSURE SENSORS

One important application of silicon/silicon bonded sensors is ultra-miniature catheter-tip transducer chips for in vivo pressure measurements. A typical chip currently in use for this product is shown as Design 1 in Figure 1. This device has dimensions of 2400 μm by 1200 μm by 175 μm thick. It has a 650 μm square active diaphragm which is 8.0 μm thick. Diaphragm thickness is controlled to within 0.5 μm by an etch-stop process. The silicon wafer is anodically bonded to a glass constraint wafer which is only 125 μm thick. Much of the processing of this product requires lithography, etching, bonding, and general handling of these extremely thin wafers. A piezoresistive half-bridge is located on two edges of the diaphragm for a pressure sensitivity of 18 μV/V/mmHg. Linearity of these chips is better than about 1.2 % of the full scale output. These chips have been in production at NovaSensor for about a year and a half.

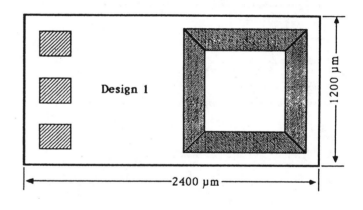

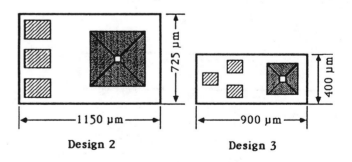

Figure 1   The three designs shown schematically here represent three generations of cathetar-tip sensors fabricated at NovaSensor. Design 1 uses a "conventional" silicon pressure sensor process sequence. Designs 2 and 3 use the SFB process.

Reprinted from *Rec. of the IEEE Solid-State Sensor and Actuator Workshop*, 1988, pp. 144–147.

A new generation of silicon fusion bonded (SFB) chips makes it possible to fabricate much smaller chips with equivalent or better overall performance. Schematic drawings of these chips are shown in Figure 1 and the fabrication procedure is outlined in Figure 2. The bottom, constraint substrate is first anisotropically etched with a square hole which has the desired dimensions of the diaphragm. In the most recent generation of chips, the bottom wafer has a (standard) thickness of 525 μm and the diaphragm is 250 μm square, so the anisotropic etch forms a pyramidal hole with a depth of about 175 μm. At the same time the pattern for these holes is defined, an alignment pattern is produced on the backside of the wafer in a double-sided aligner. Next, the etched constraint wafer is SFB bonded to a top wafer consisting of a p-type substrate with an n-type epi layer. The thickness of the epi layer corresponds to the required thickness of the final diaphragm for the sensor.

The bulk of the second wafer is removed by a controlled-etch process, leaving a bonded-on single crystal layer of silicon which forms the sensor diaphragm. Next, resistors are ion-implanted, contact vias are etched, and metal is deposited and etched. These patterns are aligned

to the buried cavity using a double-sided mask aligner, referenced to the marks previously patterned on the backside of the wafer at the same time as the cavity. All these operations have a high yield because they are performed on wafers which are standard thickness. In the final step, the constraint wafer is ground and polished back to the desired thickness of the device; about 140 μm. The bottom "tip" of the anisotropically etched cavity is truncated during this polishing operation, thereby exposing the backside of the diaphragm for a gauge pressure measurement.

In an optional configuration, the initial cavity need not be etched to completion. Figures 3 and 4 show SEM's of a device with a sealed reference cavity only 30 μm deep. This design can be directly compared to the polysilicon sacrificial-layer pressure sensor techniques currently under investigation [2].

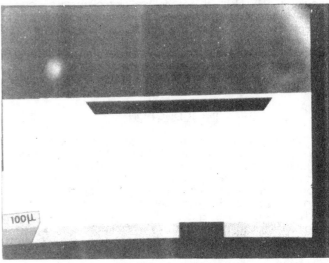

Figure 3 SEM cross-section of a test structure consisting of a 30 μm deep cavity anisotropically etched in the constraint wafer, and a 6 μm thick, 400 μm wide diaphragm suspended over the reference cavity. The fabrication procedure follows the general sequence outlined in Figure 1.

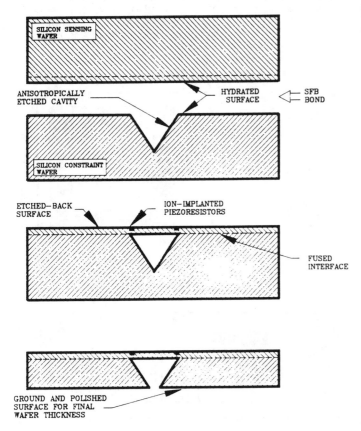

Figure 2 Fabrication process of SFB bonded low pressure sensor suitable for ultra-miniature cathetar-tip applications.

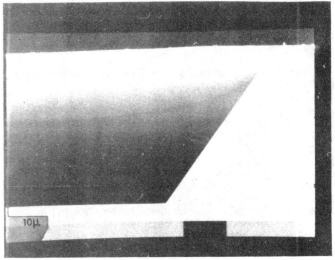

Figure 4 Magnified SEM cross-section of the bond region, reference cavity, and 6μm diaphragm in Figure 2.

Despite the fact that all dimensions of this chip are about half those of the conventional chip described above, the pressure sensitivity of the SFB chip is identical to the larger device and its linearity is actually improved to 0.5% of full scale. The implementation of SFB technology increases wafer yield because the wafers are not as thin (fragile) during most wafer processing steps. In addition, of course, the small size of these chips is advantageous since each wafer contains almost 16,000 chips.

A comparison of conventional and SFB technology (as applied to ultra-miniature catheter-tip transducers) is shown in Figure 5. For the same diaphragm dimensions and the same overall thickness of the chip, the SFB device is almost 50% smaller. In this special application, chip size is critical. SFB fabrication techniques make it possible to realize extremely small chip dimensions, which may permit this catheter-tip pressure sensor to be used with less risk to the patient in many situations.

FABRICATION OF HIGH PRESSURE SENSORS

High pressure absolute sensors have been constructed in a modified SFB process to create much thicker diaphragms and to provide a larger mass of silicon in the constraint wafer, thereby optimizing materials compatibility and minimizing thermal mismatch problems.

First, the oxidized constraint wafer is aligned and exposed on both sides. The bottom surface has a "marker" pattern which will be used as an alignment reference later in the process in a manner similar to the low pressure sensor. The top surface has a round cavity pattern which will correspond to the diameter of the sensing diaphragm. After the oxide is wet etched, the silicon is plasma etched in CF_4 or anisotropically etched in KOH or EDP on both sides of the wafer simultaneously to a depth of about 10 μm.

Next, all photoresist and oxide layers are stripped from the constraint wafer and the top surface of this wafer (the surface with the round depression) is bonded to another n-type wafer at 1100°C. This operation is illustrated in the top part of Figure 6. After bonding, the top wafer is mechanically ground and polished back to a thickness corresponding to the desired pressure range. For a 900 um diaphragm with a thickness of 200 μm, this chip will result in a pressure sensor with a full scale output of 130 mV at 4000 psi. Linearity is exceptional at about 0.2% of full scale.

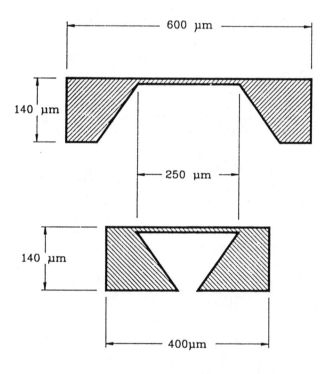

Figure 5   Comparison of miniature, low pressure silicon sensors fabricated using a "conventional" process, and the SFB process described here. For the same diaphragm dimensions and design groundrules, the SFB process results in a chip which is 50% smaller than the conventional process.

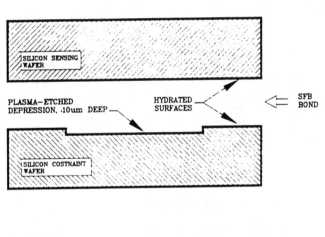

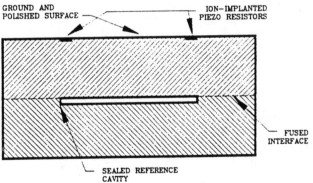

Figure 6   Fabrication process for SFB bonded high pressure sensors suitable for pressure ranges from 1,000 psi to 10,000 psi.

211

Figure 7 SEM cross-section of a silicon/silicon bonded reference cavity of the type used in the high pressure sensor. The narrow cavity is only 2 µm deep in this figure. After plasma-etching the shallow cavity, the top wafer is SFB bonded as outlined in Figure 6.

Finally, an insulator is deposited, piezoresistors are implanted and annealed, contact holes are opened, and metal is deposited and etched. This completes the process for the pressure sensor. No back-side etching and no anodic bonding are required. In addition, 100% of the bulk of the chip is single crystal silicon. A monolithic single crystal silicon sensor chip provides important performance advantages in temperature coefficient of offset. The SFB chip, a total silicon thickness of over 600 µm, exhibits a very low temperature coefficient of bridge offset of about 0.3%/100°C of full scale output. A similar chip made with conventional backside etching and anodic glass/silicon bonding for the constraint, exhibits a typical TC of zero higher by a factor of 3.

CONCLUSION

The successful development of these two products shows that SFB bonding can be applied on a commercial basis. Not only are the processes employed here simpler than those currently used to build conventional pressure transducers for the same applications, but the yields are higher, the costs are lower, and the chip performance is improved compared to conventional technologies.

Beyond the two devices shown here, the potential of SFB bonding in other micromechanical structures is enormous. This process eliminates most of the disadvantages of previous silicon/silicon bonding methods. It is hermetic; it does not require any intermediate bonding layers; it accurately preserves any pattern previously etched in either or both bonded wafers; it has a high yield strength, as much as double that of anodic bonds; it can be used to create vacuum reference cavities; it can be used in place of sacrificial layer technology in many applications; it eliminates thermal expansion and Young's modulus mismatches in bonded wafer structures. The interface itself is a true single-crystal/single-crystal boundary. We have demonstrated, for example, that pn junctions with low leakage and sharp break-down voltages can be formed by bonding a wafer with a p-type diffusion to an n-type wafer.

We have only begun to exploit this extremely powerful new technology. During the next few years, silicon fusion bonding will revolutionize the field of silicon microstructures and will have a vast impact on high performance silicon microsensors.

ACKNOWLEDGEMENTS

The authors would like to acknowledge the important contributions of Ted Vermeulen, Rose Scimeca, Van Nguyen, and Terry Cookson in the development of SFB for silicon sensor applications.

REFERENCES

1] R.T. Howe and R.S. Muller, **Integrated Resonant-Microbridge Vapor Sensor**, International Electron Devices Meeting, December 1984, pg. 381.

2] H. Guckel, D.W. Burns, C.R. Rutigliano, D.K. Showers, and, J. Uglow, **Fine Grained Polysilicon and Its Application to Planar Pressure Transducers**, International Conference on Solid-State Sensors and Actuators, June 1987, pg. 277.

3] L.S. Fan, Y.C. Tai, R.S. Muller, **Pin Joints, Gears, Springs, Cranks, and Other Novel Micromechanical Structures**, International Conference on Solid-State Sensors and Actuators, June 1987, pg. 843.

4] J. Lasky, S. Stiffler, F. White, and J. Abernathey, **Silicon-on-Insulator (SOI) by Bonding and Etch-back**, International Electron Devices Meeting, December 1985, pg. 684.

5] H. Ohashi, K. Furukawa, M. Atsuta, A. Nakagawa, and K. Imamura, **Study of Si-Wafer Directly Bonded Interface Effect on Power Device Characteristics**, International Electron Devices Meeting, December 1987, pg. 678.

6] L. Tenerz, and B. Hok, **Silicon Microcavities Fabricated with a New Technique**, Electronics Letters, 22, pg. 615, (1986).

7] J. Ohura, T. Tsukakoshi, K. Fukuda, M. Shimbo, and H. Ohashi, **A Dielectrically Isolated Photodiode Array by Silicon-Wafer Direct Bonding**, IEEE Electron Device Letters, EDL-8, pg. 454, (1987).

8] P.W. Barth, F. Pourahmadi, R. Mayer, J. Poydock, K. Petersen, **A Monolithic Silicon Accelerometer with Integral Air Damping and Over Range Protection**, Solid-State Sensors Workshop, Hilton Head, S.C., 1988.

Part 3
Microsensor Transduction Principles

SOME of the basic transduction principles that have been applied to microsensors are described in this section. Many further examples of how these principles are applied can be found throughout the volume.

Many microsensors are fabricated using techniques that are the same as or only slightly modified from those used to make microelectronic devices and integrated circuits. Therefore, it is natural to begin a review of transduction principles with an overview of how microelectronic devices themselves might function as transducers. This subject is important for two reasons: (1) it may be possible to use an existing device structure as a microsensor; and (2) the sensitivity of electronic components to their environment affects the extent to which electronic components can be added to primary transducers in order to create what have been termed *smart sensors*. The list of basic microelectronic devices is brief: resistors (whether made from metals, polysilicon, or through diffusions in silicon), capacitors (either having a parallel-plate or interdigitated geometry), diodes, bipolar transistors, field-effect transistors (FET's), and circuits created from suitable combinations of these components.

Resistors are used for the sensing of temperature, chemical species (adsorbed or absorbed), magnetic fields, and, when combined with deformable structures, strain—hence any physical effect that can create a strain. Examples in this section of resistors used as thermal sensors are found in the papers by Hocker, Johnson, Higashi, and Bohrer; by Petersen, Brown, and Renken; by Stemme; and by Tai and Muller. Also, pairs of metal lines can serve as thermocouples, and these too, have application for thermal sensing, as illustrated in the paper by Choi and Wise. The critical issue for thermal sensors is the creation of suitable thermal-isolation structures, and all of the papers of Section 3.2 demonstrate useful approaches to this problem. Chemically sensitive resistors are discussed in papers by Chang and Hicks; and by Heiland and Kohl. In these cases, the critical issues are the selection of the sensing material, its stability, selectivity, and processing requirements. In addition, contacts to the chemically sensitive resistor material, and the possibly deleterious effects of moisture are important considerations. Finally, because there are so many examples elsewhere in this reprint collection of microsensors for strain based on piezoresistance, this particular topic is not covered in this section.

Capacitors are used both for position sensing (plate spacing), and for a variety of applications in which charge-imaging between the plates can be exploited, including capacitors combined with piezoelectrics, with dielectric materials subjected to a varying potential, or with electrolytes. In the special case of the metal-oxide-semiconductor (MOS) capacitor, one of the plates (the gate) is accessible to a variety of environments while the other plate is actually in the semiconductor. The charge on the semiconductor plate can act as the channel of a MOSFET, providing a remarkably effective transduction of induced charge. Indeed, in the case of the chemically sensitive MOSFET, the electrochemical potential difference between the conductors affects the charge on the plates and hence the conductance of the FET channel. The role of the MOS structure in microsensors is explored in the paper by Senturia.

Diodes are used for sensing of light and of temperature. The fundamental temperature dependence of p-n junctions also appears in bipolar-transistor characteristics, and has been exploited in a particularly effective way in the integrated-circuit temperature sensor described in the paper by Timko.

Magnetic fields cause electric currents to deviate in direction through the Lorentz force. This gives rise to the Hall effect, to magnetoresistance, and is further exploited in a wide variety of special transistor structures in which magnetic fields steer currents within the device. An example of this latter type is found in the paper by Goicolea and Muller. Other examples are reviewed in the paper by Baltes and Popović in Section 5.3.

One of the ways of enhancing the sensing capabilities of microelectronic devices is to add sensitive films to the structure. Examples of chemically sensitive resistors have already been discussed. Piezoelectric and/or pyroelectric films can create charge when subjected to variations in strain or temperature. When combined with a suitable capacitive readout circuit, such as that at the gate of a MOSFET, direct transduction of strain or temperature change is accomplished. Examples of these principles are illustrated in the paper by Polla, Muller, and White. Another example is the adsorption or absorption of vapors by a polymer film, as discussed in the paper by Grate *et al*. In this case, the effect of the absorption is to modify the mass of the film, which can be sensed with acoustic-wave devices.

Mass loading of structural parts affects their mechanical stiffness, and, hence, their wave-propagation characteristics. Three types of acoustic-wave devices are described here: the surface acoustic-wave device (in a paper by Wohltjen), the Lamb-wave device (in the paper by Wenzel and White), and the bulk shear-wave device (in the paper by Martin, Ricco, and Hughes). These devices can be used either as passive sensors, or as the frequency-determining components in a resonant circuit. In this latter case, shifts in resonant frequency become indicators of the variable being sensed. Other effects, such as liquid viscosity, can affect the wave-propagation properties of these devices. Because of the complexity of these interactions, ultrasonic sensing is characterized by careful mechanical modeling.

THE ROLE OF THE MOS STRUCTURE IN INTEGRATED SENSORS*

STEPHEN D. SENTURIA

Department of Electrical Engineering and Computer Science, and Center for Materials Science and Engineering, Massachusetts Institute of Technology, Cambridge, MA 02139 (U.S.A.)

Abstract

The basic properties of the Metal–Oxide–Semiconductor (MOS) structure are reviewed with attention to the separation of electrochemical effects from charge-imaging effects. Two basic classes of sensors are identified: chemical sensors, which respond to changes in the chemical potential of a charged species within the gate structure, and charge sensors, which combine the charge-imaging property of the MOS structure with a property-dependent coupling element. The chemical sensors discussed include ion-sensitive field-effect devices, ion-controlled diodes and transition-metal-gate gas sensors. Charge sensors include charge-coupled imaging devices, charge-flow transistors, piezoelectric-gate strain sensors and membrane and/or beam-based capacitors for pressure and acceleration measurement.

1. Introduction

The term 'integrated sensor' suggests the joining, within an integrated circuit, of the primary measurement function of a sensor with signal-conditioning, or, to use Middelhoek's term [1], signal-modification functions. Where the integrated circuit technology involves the MOS structure, the joining together of measurement and signal modification provides a rich field for sensor applications. This paper reviews the ways in which the intrinsic physics of the MOS structure can be combined with the amplification and charge-transfer capabilities of MOS-based electronic devices to build truly integrated sensors.

The list of sensor devices that incorporate MOS structures is very large. The ion-sensitive field-effect transistor (ISFET) [2], the closely related ion-controlled diode (ICD) [3] and the palladium-gate hydrogen sensor (PdFET) [4] exploit the electrochemical equilibrium that exists between the

semiconductor and the gate material. Charge-flow transistors (CFT) [5, 6], piezoelectric MOS strain transducers [7, 8] and MOS accelerometers [9, 10] exploit the steady-state charge-imaging properties of the MOS structure. The transient charge-imaging properties of the MOS structure lead to a variety of charge-transfer devices, such as charge-coupled devices (CCD) used either for signal processing [11, 12] or optical imaging [13, 14]. Finally, any sensor that can be fabricated with a process compatible with basic MOS technology can incorporate MOS integrated circuits as part of the sensor to perform such tasks as differential amplification, multiplexing and feedback control of sensor bias.

Section 2 of this paper contains a review of fundamental MOS device physics with particular emphasis on those aspects that can affect sensor design. Section 3 then reviews current sensor technology, highlighting the relation between the specific sensor and the underlying physical principles of operation.

2. MOS physics: the electrochemical circuit

This section reviews the salient features of MOS device physics that affect sensor design. Additional details can be found in standard texts [15, 16] and in the comprehensive review of chemically sensitive field-effect devices by Janata and Huber [17].

Figure 1 shows a schematic diagram of all of the components of the MOS structure, organized to emphasize the fact that this structure forms a kind of electrochemical 'circuit'. The basic building blocks are a gate material (the 'M' of 'MOS' because of its historical identification as a metal), an insulator layer (the 'O', representing silicon dioxide) and a semiconductor

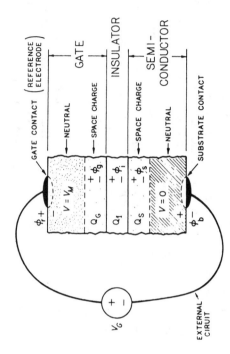

Fig. 1. The electrochemical circuit of an MOS device.

*Based on an Invited Paper presented at Solid-State Transducers 83, Delft, The Netherlands, May 31 - June 3, 1983.

substrate (the 'S'). In the case of electrolyte-based sensors, the 'gate' is a solution, and may also include a chemically sensitive membrane between the solution and insulator.

The electrically neutral bulk region of the semiconductor is taken as the reference point for the internal electrostatic potentials everywhere in the circuit ($v = 0$). Near the semiconductor–insulator interface, however, there can be a space-charge layer, having charge per unit area Q_S, and across which a potential drop ϕ_s appears. Thus, the semiconductor–insulator interface has an electrostatic potential of ϕ_s relative to the semiconductor neutral bulk. The insulator itself can have charges within it, denoted by Q_I (per unit area), and a potential drop across it denoted by ϕ_i. The electrostatic potential within the neutral region of the gate is denoted by v_M. Between the neutral gate region and the gate–insulator interface, there can be a space-charge layer, having charge per unit area Q_G, and across which a potential drop ϕ_g appears. To complete the circuit, Fig. 1 shows a contact to the semiconductor substrate, a voltage source which applies an external voltage of v_G to the gate relative to the substrate, and a contact to the gate material (equivalent to the reference electrode in an electrolyte system). For reasons explained below, the contact to the semiconductor and the contact to the gate will each have a potential drop, denoted by ϕ_b and ϕ_r, respectively.

A. The work function, or chemical potential

The key point in understanding the electrochemical circuit as presented in Fig. 1 is the path for electron exchange provided by the substrate-contact-to-gate-contact part of the circuit. It is generally true that materials in contact with each other so that both particles and energy can be exchanged will, at thermal equilibrium, have the same temperature T and the same electrochemical potential $\bar{\mu}$. The electrochemical potential consists of two components, the chemical potential (or, for the case of electrons, the work function), μ, and the electrostatic potential, denoted by v in Fig. 1, but often referred to by other symbols. For electrons, it is convenient to express μ as equal to $-q\phi$, where ϕ denotes the work function potential in volts (e.g., 2.1 V for aluminium relative to vacuum). With this notation, we can write

$$\bar{\mu} = -q\phi - qv \qquad (1)$$

If two materials, denoted 1 and 2, are in thermal equilibrium, then equating their electrochemical potentials requires that an electrostatic potential difference appear between the two materials which is the negative of their chemical-potential difference:

$$v_2 - v_1 = -(\phi_2 - \phi_1) \qquad (2)$$

For example, the potential differences ϕ_b and ϕ_r in Fig. 1 represent the contact potentials that result from the difference in semiconductor-to-contact-metal work functions, and contact-metal-to-gate work functions, respectively.

Equation (2) also applies to ionic species that can exchange between two materials. Thus, in the case of a solution in contact with a chemically sensitive membrane, at equilibrium, there is a potential difference between the solution and interior of the membrane that depends on the chemical potential difference of the species between solution and membrane [18]. This leads to the well-known Nernstian dependence of potential difference on the log of ion concentration.

Returning to Fig. 1, in the absence of any applied voltage ($v_G = 0$), and because of the path for electron equilibration provided by the external circuit, we can apply eqn. (2) to the gate as material 2 and the semiconductor as material 1 (in which $v = 0$), and obtain

$$v_M = -\phi_{ms} \qquad (3)$$

where ϕ_{ms} is the gate-to-semiconductor work function difference, expressed in volts. This equilibrium potential difference is intrinsic to the MOS structure. An alternative description of the same effect, more appropriate to the case of an electrolyte gate, is that at equilibrium,

$$v_M = \phi_{b,EQ} + \phi_{r,EQ} \qquad (4)$$

where the 'EQ' denotes equilibrium conditions. That is, for an electrolyte case without a membrane, v_M corresponds to the inner potential of the solution. (The case with a membrane is discussed in Section 3.B.) These observations lead to the first sensor-related results:

(1) Any chemical change that affects the electron work function of the gate material modifies v_M, the bulk gate potential relative to the bulk semiconductor potential.

(2) The specific location of this change in potential is at the gate contact (ϕ_r). All other potential drops in the external circuit remain unchanged.

When a non-zero external voltage v_G is applied, there can be transient currents as the charges Q_G and Q_S (and, perhaps, Q_I) adjust toward their steady-state values. Under d.c. conditions, however, assuming that the insulator is thick enough to block conduction, there is no current in the circuit. Therefore, the various contact potentials, ϕ_b and ϕ_r, have the same values they had in thermal equilibrium. As a result, the only effect of the applied bias is simply to increase v_M:

$$v_M = -\phi_{ms} + v_G \qquad (5)$$

In Section 3, we will examine how these results appear in the operation of ISFET, ICD and PdFET devices.

B. Steady-state charge imaging

Turning again to Fig. 1, Gauss's Law requires that

$$Q_G + Q_I + Q_S = 0 \qquad (6)$$

This fundamental constraint is referred to in this paper as the *charge-imaging* property of the MOS structure. Furthermore, Kirchhoff's Voltage Law requires that

$$v_M = \phi_g + \phi_i + \phi_s \tag{7}$$

These two constraints provide the basis for complete analysis of the behaviour of the MOS device. The problem reduces to understanding the relation between the structure of each component of the space-charge layer and the corresponding potential drop across it for each of the types of materials present in the structure: metal, insulator, semiconductor or electrolyte. We consider first quasistatic steady-state conditions, in which applied voltage variations are sufficiently slow to permit Q_G and Q_S to follow with their d.c. steady-state values.

The metal gate is the simplest. Because of the high density of electron states at the Fermi level in a metal, gate charge layers of very high charge per unit area can be induced with negligibly small surface potential drops ϕ_g. Furthermore, the induced charge at a metal surface has a very small screening length (Debye length), typically fractions of nm. Therefore, the metal-gate-to-insulator interface is modelled as having a sheet of charge at the metal interface with $\phi_g = 0$. The charge on the metal gate, Q_G, takes on whatever value is required to satisfy eqns. (6) and (7); *i.e.*, the metal 'images' the rest of the charges in the structure.

The insulator may have non-zero charge distributed in it due to imperfections at either interface, to entrapped ions or to trapped electrons or holes. Generally, these charges change only slowly with time, if at all. For this discussion, the charge distribution within the insulator will be taken as fixed in time. The issue of changes in insulator charge is addressed briefly in Section 3.A.

Analysis of the effect of insulator charge is most easily treated by considering a special case called 'flat-band' conditions, in which the external voltage v_G is adjusted to whatever value is required to reduce the space-charge in the semiconductor Q_S and the corresponding semiconductor surface potential ϕ_s to zero. We denote this value of v_G by the special notation v_{FB}. It is evident that to reach flat-band conditions, the voltage v_G must, first, cancel the intrinsic potential difference established by the gate-to-semiconductor work function difference (eqn. (3)), and, secondly, be sufficient to set up a gate charge Q_G that exactly images all of the charge in the insulator, Q_I. The amount of excess voltage required to set up this image charge depends on the details of the charge distribution within the insulator (*i.e.*, how far from the gate it is located), on the distribution of the image charge within the gate and, for an electrolyte gate, on possible ion-exchange and other chemical reactions between an electrolyte and the insulator surface [18].

We have already remarked that for a metal gate, the charge is modelled as a sheet charge at the insulator interface. For an electrolyte gate, however, this charge is distributed between surface sites and a complex double layer.

One approach to analysing such a structure is to solve the Poisson–Boltzmann equation in each region of the double layer using an assumed model for the surface-charge chemistry [19 - 21]. This approach is of significant value in understanding specific experimental results, but has the disadvantage of obscuring an important physical insight within the depths of a complex calculation. In this paper, an alternative description is used which lumps the detailed electrochemistry into two parameters describing the charge distribution. The first parameter is the total charge in each layer (*i.e.*, Q_G, Q_I, etc). The second parameter is the *charge centroid*, which is the charge-weighted average distance of the charge from the interface [22]. The advantage of this approach is that the effects of changes in the shape of a charge distribution, for example, due to a change in Debye length, can be readily visualized.

Referring to Fig. 2, the total insulator charge Q_I is related to its charge density $\rho_I(x)$ by

$$Q_I = \int_0^{t_i} \rho_I(x)\, dx \tag{8}$$

where t_i is the insulator thickness. The charge centroid of the insulator charge distribution, measured from the gate interface, and denoted by x_i, is defined as

$$x_i = \frac{1}{Q_I} \int_0^{t_i} x\rho_I(x)\, dx \tag{9}$$

The corresponding integrals over the charge distribution within the gate space-charge layer that define Q_G and the gate charge centroid, x_g, have as their limits the gate–oxide interface and the edge of the neutral bulk of the gate (where $\rho_G = 0$). Finally, if ϵ_g and ϵ_i are the dielectric permittivities of the gate material and insulator, respectively, then the potential difference ϕ_0 that must exist between the bulk gate and the semiconductor interface so that all of the insulator charge is imaged by gate charge is given by

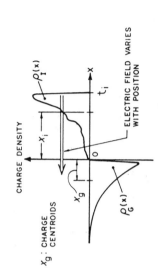

Fig. 2. Charge imaging between insulator and gate charge at flat band.

217

$$\phi_0 = -\left[\frac{x_g}{\epsilon_g} + \frac{x_i}{\epsilon_i}\right] Q_1 \tag{10}$$

It will be convenient to identify the effective capacitance associated with flat band by C_1, defined by:

$$\frac{1}{C_1} = \left[\frac{x_g}{\epsilon_g} + \frac{x_i}{\epsilon_i}\right] \tag{11}$$

With this definition, we can write the flat-band voltage as

$$v_{FB} = \phi_{ms} - \frac{Q_1}{C_1} \tag{12}$$

Note that for the commonly cited case of a metal gate ($x_g = 0$) and all of the insulator charge located at the semiconductor interface ($x_i = t_i$), then we obtain the familiar result

$$v_{FB} = \phi_{ms} - \frac{Q_1}{C_0} \tag{13}$$

where $C_0 = \epsilon_i/t_i$. In general, however, C_1 is not the same as C_0. Note further, that for the case of an electrolyte gate, ϕ_0 may depend on an ionic species concentration through the potential required to control the insulator surface charge set up by ion-exchange interactions with binding sites at the insulator surface.

Having now determined v_{FB}, we note that any applied voltage v_G that differs from v_{FB} must go into producing non-zero image charge in the semiconductor. Since v_{FB} includes the effect of imaging the insulator charge in the gate, and since we can model the insulator as a linear dielectric medium, the effect of $v_G \neq v_{FB}$ is to add a uniform electric field to that in the insulator at flat band. This is accomplished by changing Q_G and Q_S by equal but opposite amounts. The normal use of MOS structures in sensing devices involves producing that sign of Q_S that corresponds to, first, depletion of the semiconductor surface of mobile carriers, and, secondly, the creation of an inversion layer of mobile carriers opposite in sign to the original substrate doping. For the purposes of the present discussion, it is satisfactory to represent the semiconductor space charge as consisting of two terms: (1) fixed depletion-layer bulk charge Q_B (of the order of 10^{-8} coulomb/cm^2, depending on the square root of the substrate doping), which assumes its d.c. steady-state value when the surface potential reaches $2\phi_F$ (of the order of 0.6 V, and logarithmically dependent on substrate doping): (2) mobile inversion charge, Q_N, which is located within a thin sheet near the semiconductor surface, and which in d.c. steady-state conditions appears in significant amounts only after Q_S exceeds the steady-state value of Q_B. Furthermore, it is acceptable for present purposes to approximate ϕ_s as having the value $2\phi_F$ when Q_N is non-zero, and to assume that the

interface state density is small. More accurate models can be formulated [15, 16].

With this model of the semiconductor charge, but keeping track of the finite charge centroid within the gate, we can analyse the conditions that occur for $v_G \neq v_{FB}$. The charge imaging condition requires that

$$Q_G - Q_{G,FB} = -Q_S \tag{14}$$

where 'FB' denotes flat-band conditions (see Fig. 3). Taking account of the electric field in the oxide and in the gate space-charge layer set up by Q_S yields

$$(\phi_g + \phi_i) - \phi_0 = -\frac{1}{C_2} Q_S \tag{15}$$

where we define a second effective capacitance, C_2, as

$$\frac{1}{C_2} = \frac{x_g}{\epsilon_g} + \frac{t_i}{\epsilon_i} \tag{16}$$

If we assume that $\phi_s = 2\phi_F$, and use the result that

$$Q_S = Q_B + Q_N \tag{17}$$

we find the mobile inversion charge to be

$$Q_N = -C_2(v_G - \phi_{ms} - 2\phi_F - \phi_0) - Q_B \tag{18}$$

This can be expressed in the familiar form

$$Q_N = -C_2(v_G - V_T) \tag{19}$$

where V_T is the threshold voltage, defined as

$$V_T = \phi_{ms} + 2\phi_F - \frac{Q_B}{C_2} + \phi_0 \tag{20}$$

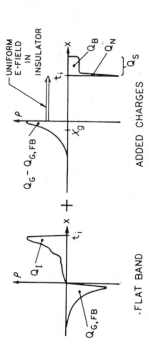

Fig. 3. Schematic illustration of additional charge imaging away from flat-band conditions.

218

where $\phi_0 = -Q_1/C_1$. Note that these expressions differ from the usually quoted forms in which both C_1 and C_2 are approximated simply by the oxide capacitance C_0. This leads us to additional results:

(3) The gate charge centroid affects both V_T (eqn. (20)) and the mobile charge density (eqn. (19)) through the C_2 term.

(4) An oxide charge centroid that differs from t_i results in different scale factors for Q_B and Q_1 (through ϕ_0) in V_T.

The charge centroid is small for metal gates, but may be important for electrolyte gates, particularly near the point of zero surface charge where the entire gate charge density must be supplied by the diffuse Gouy–Chapman layer. Charge-centroid effects can thus produce both an ionic-strength dependence and an over-all bias dependence in both the threshold voltage and mobile charge density.

C. Transient charge imaging

Equation (19) defines the amount of inversion charge that can be imaged under d.c. steady-state conditions beneath a gate to which an external voltage of v_G has been applied. However, unless the region beneath the gate is directly connected to a source of inversion charge, such as the source diffusion of a MOSFET, it takes time to develop d.c. steady-state conditions. When a voltage in excess of V_T is first applied to an isolated MOS structure (one not connected to a source of inversion charge), a condition called deep depletion results, in which the depletion layer penetrates farther into the semiconductor than under steady-state conditions, and the surface potential ϕ_s exceeds $2\phi_F$. Because of the increased depletion depth and absence of inversion charge, the effective capacitance of the structure is smaller than C_2. Hence, there is less total charge on the MOS structure than when steady-state conditions are reached. Thermal generation of hole-electron pairs both in the depletion layer and at the semiconductor interface provides a source of minority carriers, which collect under the gate, gradually collapsing the depletion layer toward its steady-state depth and reducing ϕ_s toward $2\phi_F$, and, at the same time, increasing the incremental capacitance of the structure toward C_2. The characteristic time to achieve d.c. steady-state conditions is called the capacitance recovery lifetime, τ, and can vary from fractions of a second to many seconds. On a time scale that is short compared to τ, the MOS device still obeys the imaging constraints (eqns. (6) and (7)), but the mobile inversion charge Q_N is smaller than that in eqn. (19). We refer to operation in this domain as exploiting the transient charge-imaging characteristics of the MOS structure.

Equation (19) defines what is often called the 'capacity' of a charge-storage 'well' beneath an MOS gate. A partially filled well, then, is an MOS structure which has been biased into deep depletion, and which has collected an amount of mobile charge less than its capacity. This mobile charge can come from thermal generation (background effects), from photogeneration of carriers (leading to imaging devices) or via charge transfer from an adjacent potential well. Figure 4 illustrates how the potentials on adjacent

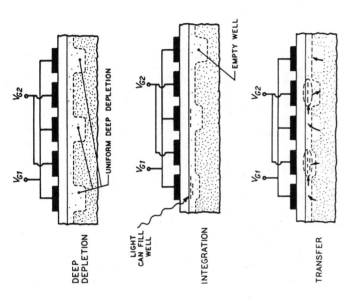

Fig. 4. Schematic illustration of CCD operation. During deep depletion and integration periods, $v_{G_1} > v_{G_2} > v_{G_2}$. During transfer, gate potentials are cycled so that $v_{G_2} > v_{G_1}$.

gates in a charge-coupled device can isolate a well from a diffused source or drain, and how cycling the gate potentials from one well to the next. While the concept of well capacity is useful, accurate models of charge transfer devices must include effects not considered here, specifically, lateral gradients in the surface potential which accelerate the mobile charge and the dynamics of trapped charges in the wells [23].

D. MOS electronics

Except for the case of charge-coupled devices and related charge-transfer devices, which intentionally isolate MOS structures from sources of inversion charge in order to operate in the transient charge-imaging mode, the semiconductor surface can be considered in contact with a source of inversion charge, and, hence, can be modelled in the quasistatic steady state. Figure 5 illustrates the cross-section of a standard MOSFET, with the source diffusion providing contact to the channel of inversion charge beneath the gate, and shows the corresponding current–voltage characteristics. The detailed expressions for the MOSFET operation are not essential to this paper. For illustration, however, it is convenient to consider an approximate

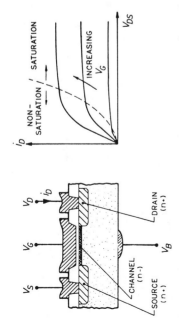

Fig. 5. Schematic MOSFET cross section and current–voltage characteristic.

expression for operation in the non-saturation region, $|v_{DS}| < |v_{GS} - V_T|$ and assuming no substrate bias, so that v_{GS} equals the applied gate-to-substrate voltage v_G (15):

$$i_D = \frac{\mu_0 C_2 W}{L}\left[v_{DS}(v_G - V_T) - \frac{v_{DS}^2}{2}\right] \qquad (21)$$

where μ_0 is the inversion charge mobility, W is the channel width and L is the channel length. This expression presumes that the mobility is not field dependent and assumes that the effective capacitances C_1 (within V_T) and C_2 are selected to have values appropriate to the actual charge distribution in the gate.

An equivalent way to think of eqn. (21) is in terms of charge imaging. The equation can be rewritten in the form

$$i_D = \left[\frac{\mu_0 W v_{DS}}{L}\right] Q_C \qquad (22)$$

where Q_C is the average channel charge-per-unit-area:

$$Q_C = C_2(v_G - V_T - v_{DS}/2) \qquad (23)$$

At fixed v_{DS}, therefore, the drain current varies with channel charge. Any mechanism that changes Q_C can be detected. These include changes in v_G, (the usual amplification mode), changes in V_T (the ISFET or the PdFET) or changes in Q_C itself, as in the CFT and in the MOS mechanical transducers. The MOSFET structure also provides the basis of operation of the ICD, as discussed in Section 3.B. below.

An important issue concerns matching of transistors for differential measurements. MOS sensors often exploit two nominally matched transistors, one as primary sensor and the other as a reference device (see Section 3.C for examples). For the present discussion, it must be noted that unless the devices have identical threshold voltages, insulator thicknesses, areas,

channel mobilities and gate charge centroids, their i–v characteristics will not be identical. It is possible to achieve excellent control of W and L with photolithography. However, if the two devices receive different gate materials, then it is possible, first, that gate insulator thicknesses may differ slightly because of processing steps, secondly, that insulator charge distribution and/or channel mobilities may differ because of process induced charges and interface states and, thirdly, that the centroid of the gate charge may be different in the two devices, hence producing different values of V_T and C_2. Therefore, the common practice of using feedback bias circuits to bring the two transistors to the same drain current may nevertheless introduce errors in inferred values of V_T, in the case of chemical sensors, or of Q_C, in the case of charge sensors.

3. MOS sensors: an overview

A. The MOS hydrogen sensor

MOS devices that use palladium as their gate metal exhibit a sensitivity to hydrogen. First reported [4] and recently reviewed [24, 25] by Lundström et al., the device structure has attracted attention from many groups. There appear to be two distinct mechanisms responsible for the operation of the device; first, a work-function change when hydrogen adsorbs on the palladium surface and also when hydrogen dissolves in the metal and, secondly, a charge-imaging effect attributed to a hydrogen-dependent dipole layer at the metal–insulator interface. The existence of an interface effect has been shown by Lundström and DiStefano [26], but Armgarth and Nylander showed that the interface effect could account only for a long-term instability in the hydrogen response, but not for the entire hydrogen response [27]. In the context of the discussion of Section 2 above, it should be recognized that gases adsorbed on or dissolved in metals can affect the work function for electrons, either by creating a dipole layer at the metal–ambient interface or by affecting the intrinsic chemical potential of the alloy formed when the gas dissolves. These changes affect the external electrochemical circuit (the electron exchange path), and therefore appear in ϕ_{ms} in eqn. (12). Dipole layers at the metal–insulator interface involve ion exchange between the metal and the insulator, and properly should be attributed to the charge-imaging term (ϕ_0 of eqn. (10), or Q_1/C_1 of eqn. (12)). The bulk potential of electrons within the metal, v_M, is not affected by insulator dipole layers; however, the dipole layer will affect how much charge appears in the semiconductor for a given v_M.

Extensions to other transition-metal gates and other gases have been reported [24, 25, 28, 29]. Poteat and Lalevic [28] show that a silicon nitride cap over the silicon dioxide gate insulator does not affect hydrogen sensitivity, and observe a definite but slow response to CO. Krey et al., however, assert that CO sensitivity depends on holes in the gate metal allowing the CO to reach the metal–insulator interface [29]. However, the fact that their work was done in 55% relative humidity raises the possibility that the mechanism for their CO effect is metal-catalysed production of hydrogen from chemisorbed moisture groups on the oxide surface. Further work is required to clarify this issue.

For thick gate insulators (50 - 100 nm), there is no significant change in semiconductor interface state density produced by hydrogen [28]. However, Keramati and Zemel report that when the insulator is thinned to the 2 - 3 nm range, a definite interface state effect is observed [30]. Such thin-dielectric devices actually resemble Schottky diodes more than the MOS structures analysed here, as discussed by Ruths et al. [31].

B. The ISFET and the ICD

Both the ion-sensitive FET (ISFET) and ion-controlled diode (ICD) sense changes in mobile semiconductor charge Q_N produced by the combined effects of electrochemical equilibrium in the external circuit and charge imaging across the gate insulator. In the case of the ISFET, Q_N is monitored via the FET drain current (eqn. (17)), whereas in the ICD, the capacitance of a source-to-substrate diode is monitored. As Q_N changes from zero (depletion) to a substantial non-zero value (strong inversion), the effective capacitance between the source and substrate increases because of the added area of the channel region, which for large Q_N becomes part of the source–substrate junction. More detailed models, in which the distributed R–C nature of the channel region in weak inversion is considered, have been published [3, 32]. In either device, the essential issues are the electrochemical and charge-imaging properties of the electrolyte–insulator–semiconductor system. The discussion below applies equally to ISFET and ICD devices.

Figure 6 illustrates the two types of structures used in ion-sensitive devices. When no membrane is used (Fig. 6(a)), the surface of the gate insulator functions as the chemically sensitive 'membrane'. SiO_2 was used initially [2], but a variety of stability problems involving possible hydration, microcracking and the introduction of semiconductor interface states [17, 33] have led to investigations of the use of other insulators deposited over thermal SiO_2, such as silicon nitride [34 - 37], aluminium oxide [20, 36] and tantalum oxide [38], all of which yield nearly ideal Nernstian pH responses. With all these materials, ion exchange reactions between the electrolyte and the insulator surface modify the charge imaging in the semiconductor (i.e., Q_N) for a given bulk electrolyte potential v_M. It should be emphasized that electrolyte–insulator interactions do not modify v_M, which is fixed by the external electrochemical circuit via the reference electrode (Fig. 1).

When an ion-exchange membrane is used (Fig. 6(b)), such as valinomycin in PVC to achieve potassium sensitivity [39], the description of physical mechanism changes slightly. Because particle exchange occurs across the solution–membrane interface, there will be a membrane–electrolyte potential ϕ_m which depends on the electrolyte concentration of the exchanging species. Furthermore, provided that the membrane is impregnated with electrolyte and is at least several Debye lengths thick, the bulk or inner potential of the membrane relative to the semiconductor bulk becomes identified with the v_M of Fig. 1, except that v_M is now equal to $\phi_b + \phi_r + \phi_m$. However, the charge–voltage characteristics of the *membrane–insulator* interface must still be considered when evaluating charge-imaging effects in field-effect devices (i.e., C_2 in eqns. (19) and (20)). This particular issue has not yet received as careful attention in the literature as have other aspects of ISFET and ICD physics.

The diverse applications of membrane-containing ISFET devices have been reviewed in ref. 17, including potassium, calcium, chloride, cyanide and iodide ions. The use of pH-sensitive devices in conjunction with membranes that modify the pH within the membrane to achieve response to penicillin [40] or semipermeable membranes to achieve response to CO_2 [41] have also been reported, as has the proposal of antigen–antibody interactions as a basis for immunoassay devices [17]. With the increased breadth of applications, there is a need for continued study of the membrane–insulator interface, particularly if the membranes are sufficiently thin that the membrane–solution and membrane–insulator space-charge layers touch, in which case the definition of v_M becomes complicated. The electrochemical aspects of the thin membrane have been discussed by Zemel et al. [42].

One interesting approach to decoupling these effects, using, in this case, a screen-printed and fired thick film of pH glass as the membrane, was to interpose a metal gate layer between the insulator and the membrane, assuring sheet-charge characteristics at that interface [43]. In this approach, the FET is functioning strictly as a potentiometric sensor of changes in v_M, i.e., as a charge-sensing device (see 3.C); all of the electrochemistry is confined to the membrane–electrolyte system. This approach also provides flexibility in device layout, in that the membrane need not be directly over the channel, a feature which may improve prospects for robust packaging of the devices.

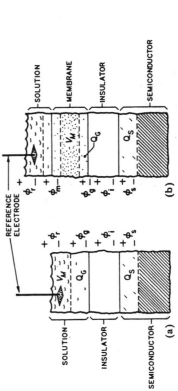

Fig. 6. Electrolyte-based structures (a) without membrane, (b) with ion-exchange membrane.

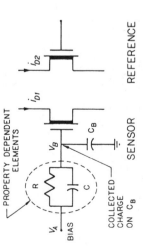

Fig. 8. Illustrations of (a) piezoelectric and (b) displacement-between-electrode capacitors that can be used in the charge sensor of Fig. 7 for strain, pressure and acceleration measurement.

An example of the ways in which MOS electronics can be combined with property-sensitive coupling elements to achieve the benefits of integration is illustrated by the floating-gate charge-flow transistor (Fig. 9). Here the property-dependent element is configured out of a pair of interdigitated electrodes over field oxide, connected to each other either by a thin film (Fig. 10) of sensor material, or, in the case of its use for chemical reaction monitoring, by the bulk sample of reaction medium [45]. The thin-film case will be described briefly here.

An equivalent circuit for the coupling element is shown in Fig. 11. The capacitance C_X represents stray capacitance through the ambient; the distributed R-C transmission line, with sheet resistance R_S, sheet capacitance C_S and sheet-to-substrate capacitance C_T, is the property-dependent portion of the coupling element. For example, if the thin film is hydrated aluminium oxide, the sheet resistance becomes an exponential function of dew point in the ambient [46]. Calculation of the sinusoidal-steady-state transfer function is straightforward, and for a given device and electrode geometry, yields calibration curves of the type illustrated in Fig. 12 for two different values of C_S. The magnitude of the transfer function is plotted against the phase, with the product of sheet resistance and frequency being the parameter. For fixed frequency of operation, changes in sheet resistance are detected by changes in both amplitude and phase of the transfer function. The feedback circuitry with which such detection is accomplished is illustrated in Fig. 13, showing the on-chip reference FET as well as the CFT. The feedback circuit forces the two drain currents to be equal. Because the two FETs are identical, the voltage applied to the reference gate by the feedback circuit must equal the voltage that appears on the CFT floating gate by charge-flow from the driven gate through the thin film. The fact that no steady-state current

C. Charge-sensing devices

Figure 7 illustrates schematically how MOSFETs are combined with property-dependent elements to build sensors that are based on steady-state charge imaging within the MOS structure. A bias v_A (which may be zero for a piezoelectric sensor) is applied to one side of a property-dependent element, represented as a lumped parallel-RC circuit for simplicity. The other side is connected to the FET gate, and, hence, is loaded by the gate-to-substrate capacitance, C_B. In effect, a voltage divider is set up between the RC element and C_B. The incremental charge (hence, the incremental voltage) appearing on C_B is imaged in Q_N, and thus can be monitored by measuring the effect on drain current under suitable drain bias. Usually, a second identical on-chip FET is provided as a reference for use in feedback circuits. An example is discussed later in this section.

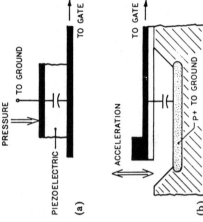

Fig. 7. A generic charge sensor, with a property-dependent R-C element coupling a bias source to the gate-to-substrate capacitance C_B. Also shown is a reference FET, nominally identical to the sensor FET, for use as a reference element in feedback circuits.

Figure 8 illustrates the variety of structures that can be used as property-dependent elements in charge-sensing devices. Figure 8(a) illustrates a piezoelectric strain or pressure sensor [8, 44] which can be connected to the FET gate. Here, the charge that appears on C_B is obtained from the piezoelectric effect within the coupling element. Figure 8(b) illustrates one of many possible structures, including membranes, cantilever beams and multiply-supported beams, that can be used to build capacitors that are either position-, strain- or pressure-sensitive because of mechanical changes in electrode spacing. An MOS-based example of such a device is the cantilever-beam accelerometer of Petersen et al. [9]. Membrane-based MOS pressure sensors should also be feasible. A variant that combines both piezoelectric strain sensing with a cantilever-beam assembly has been reported by Chen et al. [10]. In this structure, the beam provides an acceleration-dependent strain in the piezoelectric element that is deposited on the beam, and the piezoelectric charge is sensed. (Charge-coupled devices and related MOS optical imaging structures also fall in the general category of MOS charge sensors, but will not be discussed further here. Commercial products and numerous published reviews are available [23].)

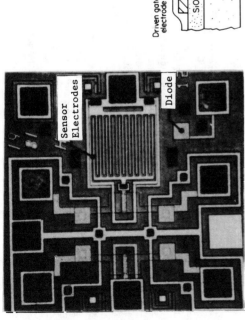

Fig. 9. Top view of a floating-gate charge-flow transistor (CFT), including an on-chip diode temperature sensor.

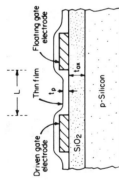

Fig. 10. Cross section of the electrode region of the CFT when used in conjunction with a thin-film sensor material.

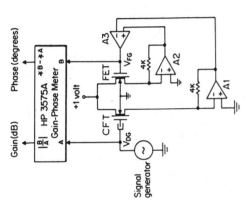

Fig. 11. Equivalent circuit for the thin-film use of the CFT.

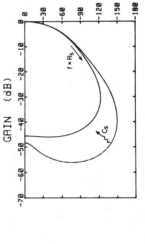

Fig. 12. Magnitude–phase plot of the transfer function for the thin-film CFT.

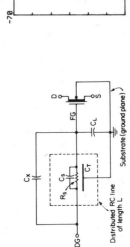

Fig. 13. Feedback circuit and measurement system for CFT sensors.

4. Summary and conclusion

The physical mechanisms underlying the operation of MOS sensing devices have been reviewed, with specific attention to clarification of the role of electrochemical equilibrium in the external circuit and charge imaging across the MOS gate insulator. The diversity of applications of the MOS structure for sensing has been explored, and specific detailed issues involving charge-imaging effects and the gate-insulator interface have been raised. It is clear that the future of MOS sensing devices will be rich in both breadth of use and the performance advantages that come with the integrability of feedback and signal-conditioning elements with the basic sensor. However, problems with choice of sensor materials, with stability and with packaging, as with all sensors, and specific problems with understanding the interface chemistry involved in the chemical sensors continue as the dominant areas for future work.

Acknowledgements

The preparation of this paper required review of a vast literature. Two undergraduate assistants, Valerie LeMay and Linda Mougel, did much of the ground work in locating, classifying and summarizing this literature, and their contribution is gratefully acknowledged. The author also wishes to thank Professor Charles Sodini for his critical reading of the manuscript. This work was supported in part by the National Science Foundation under grant ECS-8114781.

is required in sensing the sheet resistance makes the device capable of measuring very high resistances, up to 10^{16} Ω/□ at 1 Hz. The use of a pair of FETs in this fashion compensates for temperature variations, pressure variation and process variations. Finally, an example of the flexibility of the MOS environment for combining sensors is illustrated in Fig. 9, in which an on-chip diode serves as a local temperature sensor [47]. This additional measurement device can be added at minimal incremental cost, and is particularly useful for chemical reaction monitoring applications where temperatures as high as 300 °C are encountered.

References

1 S. Middelhoek and D. J. W. Noorlag, Three-dimensional representation of input and output transducers, *Sensors and Actuators*, 2 (1981/82) 29 - 41.

2 P. Bergveld, Development of an ion-sensitive solid-state device for neurophysiological measurements, *IEEE Trans. Biomedical Eng., BME-17* (1970) 70 - 71.

3 C.-C. Wen, T. C. Chen and J. M. Zemel, Gate-controlled diodes for ionic concentration measurement, *IEEE Trans. Electron Devices, ED-26* (1979) 1945 - 1951.

4 I. Lundström, M. S. Shivaraman, C. S. Svensson and L. Lundkvist, Hydrogen-sensitive MOS field-effect transistor, *Applied Phys. Letters, 26* (1975) 55 - 57.

5 S. D. Senturia, C. M. Sechen and J. A. Wishneusky, The charge-flow transistor : a new MOS device, *Applied Phys. Letters, 30* (1977) 106 - 108.

6 S. L. Garverick and S. D. Senturia, An MOS device for a.c. measurement of surface impedance with application to moisture monitoring, *IEEE Trans. Electron Devices, ED-29* (1982) 90 - 94.

7 R. S. Muller and J. Conragan, Transducer action in a metal–insulator–piezoelectric–semiconductor triode, *Applied Phys. Letters, 6* (1965) 83 - 85.

8 K. W. Yeh and R. S. Muller, Piezoelectric DMOS strain transducers, *Applied Phys. Letters, 29* (1976) 521 - 522.

9 K. E. Petersen, A. Shartel and N. F. Raley, Micromechanical accelerometer integrated with MOS detection circuitry, *IEEE Trans. Electron Devices, ED-29* (1982) 23 - 27.

10 P.-L. Chen, R. S. Muller, R. D. Jolly, G. L. Halac, R. M. White, A. P. Andrews, T. C. Lim and M. E. Motamedi, Integrated silicon microbeam PI-FET accelerometer, *IEEE Trans. Electron Devices, ED-29* (1982) 27 - 33.

11 W. S. Boyle and G. E. Smith, Charge-coupled semiconductor devices, *Bell System Tech. J., 49* (1970) 587 - 593.

12 G. F. Amelio, M. F. Tompsett and G. E. Smith, Experimental verification of the charge coupled device concept, *Bll System Tech. J., 49* (1970) 593 - 600.

13 G. F. Amelio, W. J. Bertram, Jr., and M. F. Tompsett, Charge-coupled imaging devices: design considerations, *IEEE Trans. Electron Devices, ED-18* (1971) 986 - 992.

14 W. F. Tompsett, G. F. Amelio, W. J. Bertram, Jr., R. R. Buckley, W. J. McNamara, J. C. Miukkelsen and D. A. Scaler, Charge-coupled imaging devices: experimental results, *IEEE Trans. Electron Devices, ED-18* (1971) 992 - 996.

15 S. Sze, *Physics of Semiconductor Devices*, 2nd Edition, Wiley, New York, 1981, Chapters 7 and 8.

16 E. H. Nicollian and J. R. Brews, *MOS Physics and Technology*, Wiley, New York, 1982.

17 J. Janata and R. J. Huber, Chemically sensitive field-effect transistors, in H. Freiser (ed.), *Ion-Selective Electrodes in Analytical Chemistry*, Plenum Press, New York, 1980, volume 2, pp. 107 - 175.

18 R. P. Buck, Potential-generating processes at interfaces; from electrolyte/metal and electrolyte/membrane to electrolyte/semiconductor, in P. W. Cheung *et al.* (eds.), *Theory, Design, and Biomedical Applications of Solid State Chemical Sensors*, CRC Press, West Palm Beach, Florida, 1978, pp. 3 - 36.

19 W. M. Siu and R. S. Cobbold, Basic properties of the electrolyte–SiO₂–Si system: physical and theoretical aspects, *IEEE Trans. Electron Devices, ED-26* (1979) 1805 - 1815.

20 C. D. Fung, P. W. Cheung and W. H. Ko, Electrolyte–insulator–semiconductor field-effect-transistor, *IEEE Tech. Digest of IEDM*, (1980) 689 - 692.

21 I. Lauks, Polarizable electrodes, *Sensors and Actuators, 1* (1981) 261.

22 S. D. Senturia, J. Rubinstein, S. J. Azoury and D. Adler, *J. Applied Phys., 52* (1981) 3663 - 3666.

23 R. Melen and D. Buss (eds.), *Charge-Coupled Devices: Technology and Applications*, IEEE Press, New York, 1977.

24 I. Lundström, Hydrogen sensitive MOS-structures, part 1: principles and applications, *Sensors and Actuators, 1* (1981) 403 - 426.

25 I. Lundström and D. Söderberg, Hydrogen sensitive MOS-structures, part 2: characterization, *Sensors and Actuators, 2* (1981/82) 105 - 138.

26 I. Lundström and T. DiStefano, Hydrogen induced interfacial polarization at Pd–SiO₂ interfaces, *Surface Science, 59* (1976) 23 - 32.

27 M. Armgarth and C. Nylander, A stable hydrogen-sensitive Pd gate metal-oxide–semiconductor capacitor, *Applied Phys. Letters, 39* (1981) 91 - 92.

28 T. L. Poteat and B. Lalevic, Transition metal-gate MOS gaseous detectors, *IEEE Trans. Electron Devices, ED-29* (1982) 123 - 129.

29 D. Krey, K. Dobos and G. Zimmer, An integrated CO-sensitive MOS transistor, *Sensors and Actuators, 3* (1982/83) 169 - 177.

30 B. Keramati and J. N. Zemel, Pd–thin-SiO₂–Si diode. I. Isothermal variation of H₂ induced interfacial trapping states, *J. Applied Phys., 53* (1982) 1091 - 1099; and II. Theoretical modelling and the H₂ response, *ibid.*, pp. 1100 - 1109.

31 P. F. Ruths, S. Ashok, S. J. Fonash and J. M. Ruths, A study of Pd/Si MIS Schottky barrier diode hydrogen detector, *IEEE Trans. Electron Devices, ED-28* (1981) 1003 - 1009.

32 G.-C. Chern and J. N. Zemel, Temperature dependence of the gate-controlled portion of ion-controlled diodes, *IEEE Trans. Electron Devices, ED-29* (1982) 115 - 123.

33 P. R. Barabash and R. S. C. Cobbold, Dependence of interface states properties of electrolyte–SiO₂–Si structures on pH, *IEEE Trans. Electron Devices, ED-29* (1982) 102 - 108.

34 T. Matsuo and K. Wise, An integrated field-effect electrode for biophysical recording, *IEEE Trans. Biomedical Elec., BME-21* (1974) 485 - 487.

35 R. M. Cohen, R. J. Huber, J. Janata, R. W. Ure, Jr. and S. D. Moss, A study of insulator materials used in ISFET gates, *Thin Solid Films, 53* (1978) 169 - 173.

36 H. Abe, M. Esashi and T. Matsuo, ISFETs using inorganic gate thin films, *IEEE Trans. Electron Devices, ED-26* (1979) 1939 - 1944.

37 I. R. Lauks and J. N. Zemel, The Si₃N₄/Si ion-sensitive semiconductor electrode, *IEEE Trans. Electron Devices, ED-26* (1979) 1959 - 1964.

38 T. Matsuo and M. Esashi, Methods of ISFET fabrication, *Sensors and Actuators, 1* (1981) 77 - 96.

39 S. D. Moss, J. Janata and C. C. Johnson, Potassium ion-sensitive field effect transistor, *Anal. Chem., 47* (1975) 2238 - 2243.

40 S. Caras and J. Janata, Field effect transistor sensitive to penicillin, *Anal. Chem., 52* (1980) 1935 - 1937.

41 K. Shimada, M. Yano, K. Shibatani, Y. Komoto, M. Esashi and T. Matsuo, Application of catheter-tip ISFET for continuous *in vivo* measurement, *Med. Bio. Eng. Comput., 18* (1980) 741 - 745.

42 J. N. Zemel, B. Keramati, C. W. Spivak and A. D'amico, Non-FET chemical sensors, *Sensors and Actuators, 1* (1981) 427 - 473.

43 M. A. Afromowitz and S. S. Yee, Fabrication of pH sensitive implantable electrode by thin-film hybrid technology, *J. Bioeng., 1* (1977) 55 - 60.

44 R. G. Swartz and J. D. Plummer, Integrated silicon–PVF₂ acoustic transducer arrays, *IEEE Trans. Electron Devices, ED-26* (1979) 1921 - 1931.

45 N. F. Sheppard, D. R. Day, H. L. Lee and S. D. Senturia, Microdielectrometry, *Sensors and Actuators, 2* (1982) 263 - 274.

46 T. M. Davidson and S. D. Senturia, The moisture dependence of the electrical sheet resistance of aluminum oxide thin film with application to integrated moisture sensors, *Proc. IEEE International Reliability Physics Symposium, San Diego, CA, April, 1982*. pp. 249 - 252.

47 S. D. Senturia, N. F. Sheppard, Jr., H. L. Lee and S. B. Marshall, Cure monitoring and control with combined dielectric/temperature probes, *Proc. 28th National SAMPE Symposium, Anaheim, CA, April, 1983*, pp. 851 - 861.

224

A Two-Terminal IC Temperature Transducer

MICHAEL P. TIMKO, MEMBER, IEEE

Abstract—A monolithic IC temperature transducer with an operating temperature range of −125°C to +200°C has been designed, fabricated, and tested. The two-terminal device, which is fabricated using laser trimmed thin-film-on-silicon technology, is a calibrated temperature-dependent current source with an average output impedence of 10 MΩ over the 3.5-V to 30-V range of input voltage. Overall absolute accuracies of ±0.5°C from −75°C to +150°C have been achieved on a scale of 1 μA/K under optimum operating conditions.

I. Introduction

SILICON bipolar transistors have a property which can be exploited to produce a voltage that is directly proportional to absolute temperature (PTAT). If the transistors are designed so as to minimize other temperature effects, this voltage will be a highly accurate representation of the temperature over a wide range. A temperature sensor, however, must have several other important properties to be truly flexible and widely useful. A list of specifications for all temperature transducers includes the following:

1) accuracy versus temperature (the predictability of the output parameter);

2) linearity (the constancy of the change in output parameter per degree over the temperature range);

3) interchangeability (absolute calibration);

4) signal level;

5) remote sensing capability (the maximum allowable distance between the sensor and supporting electronics and the number and quality of connections necessary);

6) temperature range;

7) cost [including support electronics to allow for deficiencies in 2)–5)].

While an IC temperature transducer probably cannot be as accurate as, for instance, a platinum resistance thermometer and the temperature range is limited to about −125°C to +200°C, the flexibility of monolithic integration allows the designer to include many features that no other medium can achieve.

Previous IC designs using the PTAT voltage specifically to sense temperature concentrated on amplifying and buffering the voltage over a limited temperature range with only rough calibration [1]. In contrast, the emphasis in this new design is the best achievable accuracy over the widest possible temperature range. This goal has reached a point never before obtained in an integrated circuit by combining a unique circuit configuration with the latest laser-trimmed thin-film resistor technology to produce a calibrated PTAT *current*. The fact that the output parameter is current, and that the sensor

is relatively insensitive to the voltage across it, makes the device easy to use, even over long wires.

II. Temperature Dependence

The temperature sensitive parameter of interest is the difference between the V_{be}'s of two transistors operating at a constant ratio of collector current densities. This can be derived from the equation for emitter current density [2]:

$$J_e = \frac{1}{\alpha} J_s (e^{q V_{be}/kT} - 1) \tag{1}$$

where

J_e emitter current density;
J_s emitter saturation current density;
α common base current gain;
V_{be} base emitter voltage;
q charge on an electron;
k Boltzmann's constant:

and

T absolute temperature.

If we assume $J_e \gg J_s$, then for two transistors at current densities J_{e1} and J_{e2}:

$$V_T = V_{be1} - V_{be2} = \frac{kT}{q} \ln \frac{\alpha_1 J_{e1} J_{s2}}{\alpha_2 J_{e2} J_{s1}}. \tag{2}$$

For V_T to be proportional to absolute temperature, the logarithmic term must be constant. For well-matched transistors $\alpha_1 = \alpha_2$ and $J_{s1} = J_{s2}$. (In the circuit, as long as these parameters remain in constant ratio over temperature, the resulting error will be removed by the calibration trim.)

If J_{e1}/J_{e2} is a constant r, then

$$V_T = T \frac{k}{q} \ln r. \tag{3}$$

There are several ways to achieve the desired ratio of current densities. Two identical transistors operated at two different collector currents, held at a constant ratio, could be used. However, a more practical method is to operate two transistors of unequal emitter areas at equal collector currents. In order to have control over this ratio, the larger transistor emitter can be made up of a number of parallel emitters similar to that of the smaller transistor. The area ratio is then accurately equal to the ratio of the number of emitters. (In the schematic diagrams in this paper, a number immediately adjacent to an emitter refers to the number of similar emitters in that transistor relative to other transistors of the same type, n-p-n or p-n-p.)

Fig. 1 is an elementary application of the above principle. Neglecting for the moment the variation of I_c with V_{ce} and the

Manuscript received May 29, 1976; revised August 4, 1976.

The author is with the Semiconductor Division, Analog Devices, Wilmington, MA 01887.

Reprinted from *IEEE J. Solid-State Circuits*, vol. SC-11, no. 6, pp. 784–788, December 1976.

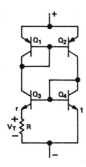

Fig. 1. Elementary example of a PTAT generating circuit.

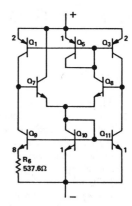

Fig. 2. Simplified schematic of the new circuit showing the major components contributing to the dc operation.

effects of base current, this circuit will conduct a current proportional to absolute temperature for any voltage greater than two V_{be} across the device. Transistors Q_1 and Q_2 constrain the collector currents of Q_3 and Q_4 to be equal. Since the emitter area ratio of Q_3 and Q_4 is r, the difference in their V_{be}'s is V_T in (3). For $r = 8$,

$$V_T = \frac{kT}{q} \ln 8 = T \times 0.1792 \text{ mV/K} \tag{4}$$

or 53.44 mV at 25°C.

This voltage, when amplified and buffered, would make a useful output parameter. But, since V_T is impressed across resistor r, the current drawn by each side of the circuit is

$$I = \frac{V_T}{R} = \frac{kT}{q} \frac{\ln r}{R}. \tag{5}$$

If R has a temperature coefficient of zero, then the total current drawn by the circuit is also proportional to absolute temperature. Under the original assumptions for this elementary circuit, the supply current would be independent of input voltage (above two V_{be}'s) and the output impedence would be infinite.

Naturally, the circuit of Fig. 1 is insufficient for good results when made with real components. The Early effect and finite β would reduce the output impedance drastically, the circuit would not start up dependably, and once started, it would probably oscillate under some otherwise useful conditions.

III. THE NEW CIRCUIT: DC ACCURACY

Fig. 2 is a simplified schematic of the new circuit in which the idealized principle of Fig. 1 can be approached with real components. Transistors Q_1, Q_3, Q_9, and Q_{11} correspond to the four transistors of Fig. 1. These transistors make up what will be referred to as the PTAT core since they define the temperature dependence of the circuit.

The collector currents of Q_1 and Q_3 are equal assuming only equal emitter areas and equal voltages at the bases of Q_7 and Q_8. The collector currents of Q_9 and Q_{11} are forced to be equal to each other and to those of Q_1 and Q_3 by the combined action of Q_7 and Q_8.

Bias current for Q_7 and Q_8 is provided by diode Q_{10} such that no nonPTAT component is added to the total supply current. The terminal current of this configuration is then three times the current in each branch or

$$I = 3 \frac{kT}{q} \frac{\ln 8}{R_6}. \tag{6}$$

Note that the value for R_6 on the schematic results in a scale factor of 1 μA/K.

Transistor Q_5, which has half the emitter area of Q_1 or Q_3, operates at about half the current available from Q_{10}. As a result, the collector currents of Q_7 and Q_8 are nearly equal. The Early effect still causes errors between the two diodes and their associated transistors, creating an offset between Q_7 and Q_8. Since these errors are reduced from those of Fig. 1 by n-p-n β (typically 400), they are insignificant except at temperature extremes.

Diode Q_{10} is the key to the success of this circuit configuration. Its importance lies in the fact that it forces all currents to $V-$ that are not in the Q_1, Q_3, Q_9, Q_{11} loop to sum to a PTAT current. Later, it is shown that even the high-temperature leakage currents can be measured in this way to extend the upper limit of operation.

In order to have good accuracy over temperature, R_6 must have a low temperature coefficient (TC). To put it another way, any TC in R_6 will show up as a TC in the absolute error of the sensor. This fact precludes the use of diffused resistors which have TC's of $\simeq 2000$ ppm. SiCr thin-film resistors, however, can be made with consistent TC's of -30 to -50 ppm/°C. The effects of the aluminum interconnect and emitter resistance TC's are of the same order and were experimentally adjusted for optimum results. The inclusion of thin film also allows these devices to be laser trimmed at the wafer level to obtain excellent absolute accuracy as well as PTAT performance. Exact calibration allows for direct interchangeability of the finished devices.

IV. THE COMPLETE CIRCUIT

Fig. 3 is a complete schematic of the new circuit showing the additional components necessary to make the chip more closely approximate the ideal.

The absolute calibration of this device is accomplished by laser trimming R_5 and R_6 at the wafer level. Resistor R_5 was added to allow the initial current of the device to be either increased or decreased as needed so that the nominal design value could be the desired target value. The average amount

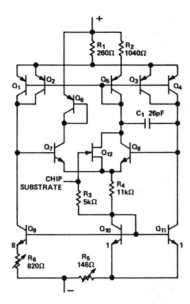

Fig. 3. Complete schematic of the two-terminal temperature transducer.

of trimming needed is thereby minimized and the yield is improved. The effective resistance for determining output current from V_T is now $R_6 - 2R_5$ since the emitter currents of Q_{10} and Q_{11} through R_5 cancel part of the voltage drop across R_6.

Q_{12} provides a small current to start the device when power is applied. Note that this current finds its way to V^- through Q_{10} without contributing to output error.

As shown in the schematic, the gate of Q_{12} (the chip substrate) is not connected to V^- as is normally the case in monolithic IC's. This unusual configuration is important in helping to extend the temperature range of the device well beyond the point at which leakage currents become significant. The single largest source of junction leakage current is the pocket containing C_1 and transistors Q_1 through Q_5. This current does not upset the balance of the PTAT core but, if simply brought out to V^-, would add an error to the result. By connecting this and all other substrate currents to the positive side of Q_{10}, these currents are measured by the circuit. The leakage currents eventually accounted for the high temperature limit of the circuit by upsetting the balance of Q_7 and Q_8. At some point (about 210°C) the collector current of Q_8 is reduced to zero and the device no longer functions. Instead of a direct connection between the substrate and Q_{10}, R_3 has been inserted to isolate certain substrate capacitances which otherwise would have a detrimental effect on the frequency stability of the circuit.

Since the substrate of this chip is not electrically connected to either output lead, special packaging considerations are required. The prototype devices were assembled in TO-52 metal transistor packages with an aluminum oxide substrate between the chip and the header. A special temperature sensor package would be necessary to realize the full capabilities of this device.

Resistor R_1 and R_2 were added to provide emitter degeneration and therefore reduce the base width modulation effects between Q_5 and the quad of lateral p-n-p's.

The combination of R_4 and C_1 determines the frequency compensation for the main feedback loop. The values shown, which were those used in the prototypes, somewhat overcompensate the circuit.

Transistor Q_6 balances the collector voltages of Q_7 and Q_8. More importantly, it provides protection for the device if the applied voltage is inadvertently reversed.

The simplicity of this design is a major factor in extending its useful temperature range. Additional components could be added to refine the accuracy at the cost of more parasitic leakage and a reduced upper temperature limit. The overall accuracy of the transducer compares very favorably with conventional sensors which are more difficult and less convenient to use.

V. LAYOUT

In this device, the layout design is as important as the circuit design. Fig. 4 is a chip photograph of a prototype device. While on-chip dissipation is not a major problem, the critical temperature sensing transistors Q_9, Q_{10}, and Q_{11} are located at the opposite end of the chip from the main power consuming components Q_1, Q_2, Q_3, Q_4, Q_7, and Q_8. Note that Q_9 has eight large rectangular emitters with interdigitated base stripes. Q_{11} is made in the same way with one identical emitter. (The two unconnected emitters were included for experimentation with other emitter ratios and current densities.) Note that the base box and isolation pocket of Q_{11} are identical to those of Q_9. This provides first-order cancellation of leakage currents through these junctions at high temperatures.

Q_{10} (which also has an extra unconnected emitter) has an additional interesting design feature. In an ordinary diode-connected transistor, the combination of falling V_{be} and rising collector resistance produces saturation problems at high temperatures. Q_{10} is designed to avoid this as shown in Fig. 5. The majority of the collector resistance of a monolithic n-p-n transistor is in the epi between the buried layer and the

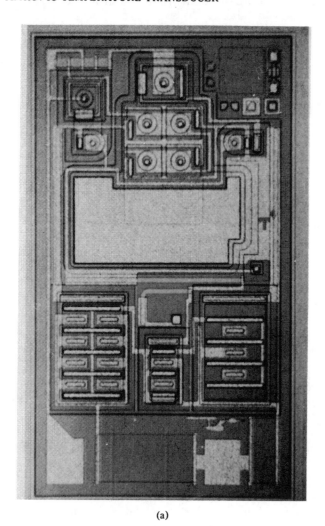

(a)

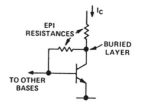

Fig. 5. Detail of diode Q_{10}.

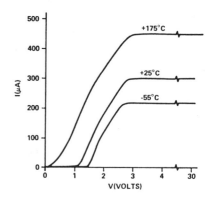

Fig. 6. *V–I* plot of a typical device at three temperatures.

(b)

Fig. 4 (a) Photograph of the 37 × 62-mil chip. (b) Key to Fig. 4(a) showing the location of all components.

surface collector contact. By making two surface contacts on opposite ends of the transistor and using them as a Kelvin

contact, the voltage of buried layer is maintained near the base voltage over the whole temperature range. The voltage at the terminal into which the collector current flows changes with epi resistance, but this voltage is not critical to the accuracy of the circuit. While not as critical, Q_5 is also Kelvin-connected for the same reason.

Transistors Q_1 through Q_4 are wired as a cross-connected quad [3] and carefully matched in geometry and parasitics to insure faithful tracking over temperature.

VI. PERFORMANCE

Fig. 6 shows the *V–I* characteristic for voltage from zero through the operating region. While this particular device enters its linear region at about 2.7 V, the minimum guaranteeable operating voltage, taking into account process variables, is 3.5 V. The output impedance varies over the active region from 5 MΩ at 5 V to over 20 MΩ at 15 V and above. Actually, the output current changes no more than about 2–3 μA over the whole useful voltage range. Maintaining the voltage across the device within a volt of any operating point keeps this error below 0.2°C.

A supply voltage of 5 V was chosen as a convenient level at which to trim the devices. All measurements for absolute accuracy were also made with 5 V across the device.

As mentioned earlier, these devices are adjusted to a scale factor of 1 μA/K by the laser wafer trim process discussed elsewhere [4]. The temperature of the wafer is carefully measured during the trim procedure and each device is then trimmed to indicate that temperature. This procedure presently yields a spread of results after assembly and burn-in such that about half the devices are accurate to within ±1°C at 25°C.

Fig. 7 shows the absolute error of three different devices between −125°C and +200°C. These devices were selected for good calibration at 25°C in order to emphasize the relative

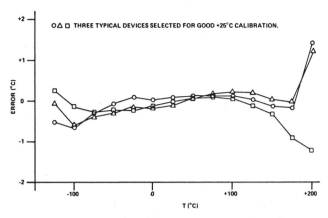

Fig. 7. Absolute error versus temperature.

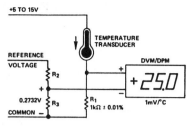

Fig. 8. Typical application as a centigrade thermometer.

linearity. The measurements were done by mounting the devices under test in an aluminum block suspended in a temperature chamber into which a calibrated thermocouple was also inserted. A direct comparison was made between the thermocouple and the devices when the block had settled near the desired temperature. The figure shows accuracy within ±0.5°C from -75°C to +150°C on all three units. This nonlinearity over this range compares favorably with the only other IC temperature transducer available which specifies a nonlinearity of 1.8°C over the -55°C to +125°C range. Thermocouples, the most widely used sensors, have a standard limit of error of ±0.8°C over a similar temperature range.

At the extremes of temperature, the inherent linearity is degraded but some devices remain useful for another 50°C on each end. At high temperatures, junction leakages, saturation resistances, and high-level injection effects eventually come into play. Most devices are useful at +175°C and a selected few will operate at +200°C. At the low end, β degradation and increasing current in Q_{12} cause the errors. We found it very difficult to evaluate the devices below -125°C because the test fixture would not settle long enough to do the measurements. The only other point we could get was to immerse the device and the thermocouple into liquid nitrogen. Most of the devices still functioned but the output impedance was seriously degraded and the accuracy at 5 V was no better than ±5°C. Early devices which were placed on 200°C powered burn-in have shown only a few tenths of a degree C change in over 2000 h after a 100-h initial burn-in period.

VII. APPLICATION

The current mode operation of this device lends itself to numerous applications. Fig. 8 is probably the simplest and most straightforward way to use the two-terminal temperature transducer as a centigrade thermometer. Resistor R_1 converts the current to an accurate voltage. Voltage divider R_2 and R_3 provides the Kelvin to centigrade conversion using whatever low TC reference is available. (In most cases, the reference and power supply are available from the meter.) For a Fahrenheit thermometer, the resistors are simply changed for a different scale factor and offset. For a Kelvin thermometer (or for centigrade with digital conversion from Kelvin), R_2, R_3, and the reference are not needed. The meter would then be connected directly across R_1. (The TC's of R_1 and the meter are, of course, critical in this circuit.) The current mode output also makes this device uniquely useful for creating a variety of linearly or functionally temperature dependent signals to use to compensate for known temperature coefficients.

VIII. CONCLUSION

The two-terminal IC temperature transducer described in this paper develops a total supply current that is directly proportional to absolute temperature by relying on a fundamental property of bipolar transistors. The inherent linearity, insensitivity to operating voltage, and laser trim calibration make the device uniquely useful and applicable among the other sensors presently available.

REFERENCES

[1] R. C. Dobkin, "Monolithic temperature transducer," in *ISSCC Digest Tech. Papers*, pp. 126–127, Feb. 1974.
[2] P. E. Gray *et al.*, *Physical Electronics and Circuit Models of Transistors*. New York: Wiley, 1964.
[3] M. A. Maidique, "A high-precision monolithic super-beta operational amplifier," *IEEE J. Solid-State Circuits*, vol. SC-7, pp. 480–487, Dec. 1972.
[4] P. R. Holloway, "A high-yield, second generation 10-bit monolithic DAC," to be published in the *IEEE J. Solid-State Circuits*, Apr. 1977.

A SILICON MAGNETO-COUPLER USING

A CARRIER-DOMAIN MAGNETOMETER

J.I. GOICOLEA and R. S. MULLER

*Department of Electrical Engineering and Computer Sciences
and the Electronics Research Laboratory
University of California, Berkeley 94720*

ABSTRACT

A new *IC* isolation coupler based upon a sensitive magnetometer is described. The circuit is functionally similar to an opto-coupler in that a signal can be transferred to a dc-isolated circuit. In contrast to the opto-coupler, however, both transmitter and receiver for the magneto-coupler are built using a linear bipolar *IC* process. A fully integrated sense amplifier has been fabricated with two carrier-domain magnetometer sensors to produce a signal coupler having 250 kHz bandwidth.

INTRODUCTION

Electrical isolation is desirable in many electronic applications to protect circuits from large voltage spikes (such as are encountered in industrial controls), to avoid ground loops that significantly increase system noise, and to link signals between circuits having large offsets in dc-voltage levels. Opto-couplers, in which a light-emitting diode and an optical sensor are employed, are frequently used as isolators, but opto-couplers require hybrid fabrication techniques. The magneto-coupler provides an isolator in a circuit that can be fabricated using conventional *IC* processing. The magneto-coupler is made possible because of the wide dynamic range and linear response of a new type of carrier-domain magnetometer [1].

DESCRIPTION

The magneto-coupler consists of a magnetic-field source and a detector, separated by a dielectric medium. The field is generated by current in a flat coil, patterned in the *IC* metalization layer and dielectrically isolated from the rest of the circuit by a layer of silicon dioxide (the field oxide, 1μm in thickness) as shown in Figure 1 [1]. The coil has a pitch of 20μm and produces a magnetic field of 0.3 G/mA at the magnetic sensor. The field produced for a given current drive varies inversely with the metal pitch, and could, therefore, be increased by about an order of magnitude without difficulty.

Underneath the coil are two carrier-domain magnetometers (CDM), measuring $100 \times 800 \ \mu$m^2 [1,2].

The CDM is an *npnp* device that uses two minority-carrier domains (beams) that are deflected by a magnetic field. The deflection mechanism is coupled to a regenerative structure that enhances the primary effect, producing a very high sensitivity (we have built devices to detect fields as low as 30 mG, much smaller than the earth's magnetic field of 700 mG).

The *npnp* structure (figs 2,3) can be viewed as merged *npn* and *pnp* transistors which share a common base-collector junction. During operation this junction is always reverse-biased. The *npn* transistor is biased by a current source at its emitter, and both transistors are in the forward-active region of operation. Electrons are injected into the base of the *npn* transistor (layer 1) and collected in the *pnp* base (layer 3) which is, at the same time, the *npn* collector. Similarly, holes are injected from the *pnp* emitter (layer 4) and collected in the *npn* base (layer 2). Due to the resistive voltage drops in the bases (as a result of x- directed currents), carrier injection is localized over a very small portion of the base-emitter junction, concentrating the current into two narrow beams (domains). The first domain consisting of electrons, and the second composed of holes. Detailed analysis[1] of the currents and voltages within layers (2) and (3) indicates that the injection point of each domain along the x−direction tends to coincide with the point at which the other domain is collected.

When a y−directed magnetic field is present, the Lorentz force $q\vec{v} \times \vec{B}$ deflects both domains in the same direction (along x) due to the opposite polarities and velocities of the carriers. The tendency of each domain to start at the point where the other one is collected creates a regenerative process that shifts the injection point along the x−direction, greatly multiplying (from 10 to 100 times) the original deflection. This shift is translated into a differential current ($I_{C1}−I_{C2}$ in Figure 3) that constitutes the output signal. The hole current in the CDMs is limited to 3-4 mA and is supplied through front contacts to avoid voltage drops in the substrate.

Two CDM sensors are implemented below the coil to cancel offset and to compensate for the presence of unintended magnetic fields. The signal is cou-

Reprinted from *Rec. of the IEEE Int. Electron Devices Meeting*, 1985, pp. 276-279.

pled in opposite phase to the two sensors; external fields, in contrast, are detected in-phase in both CDM sensors. Figure 4 shows that this subtractive signal detection very effectively reduces sensitivity to external fields.

The on-chip amplifier, shown schematically in Figure 5, consists essentially of a current-to-voltage converter. A common-base stage is loaded with a Wilson current mirror that converts the differential signal to single-ended format. The output is buffered through a common-emitter stage. The biasing circuit includes a current source and a negative voltage supply for the emitter and base regions of the CDMs. The magneto-coupler was fabricated on a commercial production line, using a 9 μm epitaxial linear bipolar process.

RESULTS

As shown in Figure 6, the measured bandwidth of the magnetic coupler is 250 kHz. The upper frequency is consistent with the predicted transit time through the 8 μm base width of the vertical *pnp* transistor making up the CDM. Regeneration in the device reduces the bandwidth approximately to the frequency corresponding to the transit time divided by the regenerative gain which is about 30 in the structures tested. Bipolar processes for faster devices would yield significantly greater bandwidth. The coupler output depends linearly on the current in the flat coil and responds to dc signals, making it suitable for analog applications. Measurements shown in Figures 7 and 8 demonstrate the linearity. The measured minimum detectable input current in the coupler is approximately 250 μA.

Sensitivity to external fields is reduced by a factor of 10 when using the differential configuration explained above. Higher rejection ratios can be obtained by separately trimming the bias currents (and therefore adjusting the sensitivity) of each CDM.

CONCLUSION

The magneto-coupler isolator has demonstrated:

(1) the use of a magnetic field produced by an on-chip coil to couple signals to sensitive magnetic detectors,

(2) compatible fabrication (with no extra steps) of the magnetic sensors, transmitting coil and analog bipolar circuitry.

Since the magnetic coupler is a monolithic *IC*, it has inherent advantages over alternative isolators such as opto-couplers which demand the separate use of a light-emitting element and a detector (forcing a hybrid realization).

Acknowledgements: The cooperation and courtesy of Dr. G. B. Hocker and Honeywell Inc. is gratefully acknowledged. The experimental circuits were designed and laid out at UC, Berkeley and fabricated at Honeywell, Colorado Springs, Colorado. This research was supported in part by the National Science Foundation (Grant NSF-ECS-81-20562, Dr. F. Betz, monitor) and in part by the State of California MICRO program.

REFERENCES

[1] J.I. Goicolea, R. S. Muller and J.E. Smith, A Highly Sensitive Silicon Carrier-Domain Magnetometer, *Sensors and Actuators*, 5(1984) 147-167

[2] J. I. Goicolea, R. S. Muller and J. E. Smith, "An Integrable Silicon Carrier-Domain Magnetometer with Temperature Compensation," *Tech. Digest: 1985 IEEE Int. Conference on Solid-State Sensors and Actuators*, 300-303, Philadelphia, PA (June 11-14, 1985).

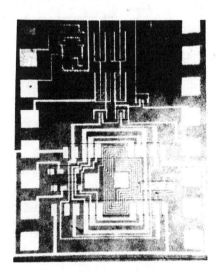

Figure 1. Photograph of a CDM magneto-coupler. The cross-connected CDMs are above and below the flat coil in the lower half of the chip.

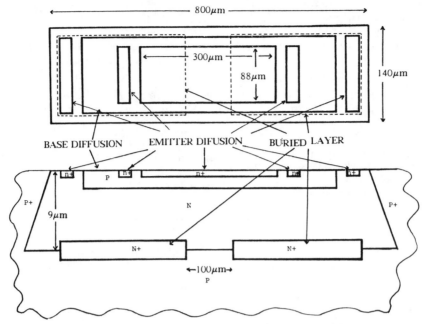

Figure 2. Lay out and cross section of the carrier-domain magnetometer.

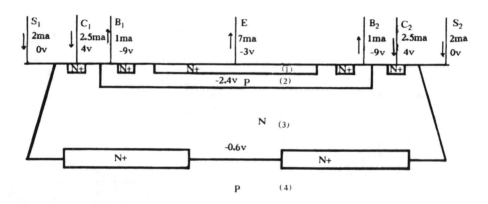

Figure 3. Biasing and operation of the CDM. The output signal is the difference $I_{C1} - I_{C2}$. Typical bias values are: $I_E = 7\text{mA}$, $V_{B1} = V_{B2} = -9\text{V}$, $V_{C1} = V_{C2} = 4\text{V}$, $V_{S1} = V_{S2} = 0\text{V}$.

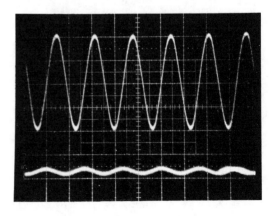

Figure 4. Sensitivity reduction of coupler to extraneous magnetic fields; the top trace is measured directly on a single sensor, the bottom trace (about 10% as large) is an output with differential detection.

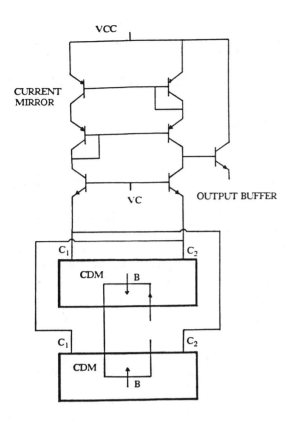

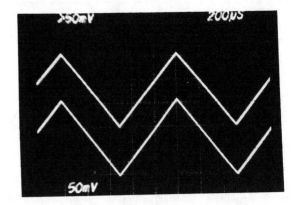

Figure 7. Response of the magneto-coupler to a 1 kHz triangular current waveform applied to the flat coil. The lower trace shows the current through the coil (20 mA per division). The upper trace is the coupler output processed by an on-chip current-to-voltage converter.

Figure 5. Circuit diagram for the magneto-coupler. The bias circuitry for the CDMs has been omitted for clarity.

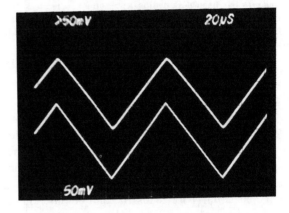

Figure 8. Magneto-coupler response as in Fig. 7 with a 10 kHz triangular current-source input.

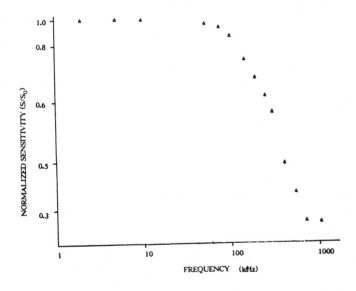

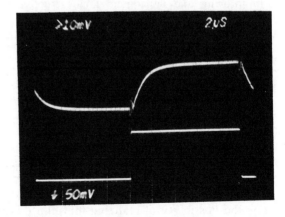

Figure 6. Relative sensitivity as a function of frequency of the signal applied to the magneto-coupler.

Figure 9. Magneto-coupler response as in Fig. 7 with a 50 kHz square-wave, current-source input.

A Silicon-Thermopile-Based Infrared Sensing Array for Use in Automated Manufacturing

IL HYUN CHOI AND KENSALL D. WISE, SENIOR MEMBER, IEEE

Abstract—This paper describes a new low-cost infrared detector array that has been realized using standard silicon MOS process technology and micromachining. This array uses thermopiles as infrared detecting elements and multiple layers of silicon oxide and silicon nitride for diaphragm windows measuring 0.4 mm × 0.7 mm × 1.3 μm. Each thermopile consists of 40 polysilicon–gold thermocouples. A high fill factor for this array structure has been achieved by using the boron etch-stop technique to provide 20-μm thick silicon support rims. The array shows a response time of less than 10 ms, a responsivity of 12 V/ W, and a broad-band input spectral sensitivity. The process is compatible with silicon MOS devices, and a 16 × 2 staggered array with on-chip multiplexers has been designed for applications in process monitoring. The array theoretically achieves an NETD of 0.9°C and an MRTD of 1.4°C at a spatial frequency of 0.2 Hz/mrad in a typical imaging system.

I. INTRODUCTION

THERMAL DETECTORS have been widely used for infrared detection [1]–[4]. They operate at room temperature, respond to a wide range of infrared radiation, and are inexpensive. These unique properties make them suitable for various tasks that cannot be fulfilled by photon-type detectors and are motivating their continued development, particularly for industrial and commercial applications where inexpensive but reliable detectors are required. Moreover, detector arrays can expand the usefulness of these detectors in numerous areas, including process monitoring and noncontact nondestructive testing, by providing area thermal information with a higher sensitivity and data-acquisition speed than single detectors.

The availability of thermal-detector arrays has previously been limited to pyroelectric detectors [5]–[8]. Attempts to realize thermopile detector arrays using conventional shadow-masking techniques result in arrays that are too large in chip size, slow in speed, and low in array packing density to be practical. In addition, if the number of array elements is large, signal readout can be a serious problem due to the low signal levels and large number of leads involved.

Recently, the possibility of realizing thin-film thermopile-based detectors and electronic circuits on the same

Manuscript received April 19, 1985; revised July 25, 1985. This work was supported in part by the Air Force of Scientific Research under Contract F49620-82-C0089 and by the Semiconductor Research Corporation under Contract 84-01-045.

The authors are with the Solid-State Electronics Laboratory, University of Michigan, Ann Arbor, MI 48109.

IEEE Log Number 8405799.

silicon substrate has been suggested [9], [10]. However, such structures have not been previously demonstrated nor have the associated performance limitations been fully explored. This paper describes a high-performance thermopile-based detector that is compatible with a conventional silicon IC process and examines the performance limitations of silicon thermopile arrays containing on-chip readout circuitry.

II. SINGLE THERMOPILE DETECTORS

A. Basic Theory

The basic structure of a silicon thermopile detector is shown in Fig. 1. The thermopile consists of a series of thermocouples, with the hot junctions supported on a thin dielectric window in the wafer and the cold junctions supported over the silicon rim. Incident radiation heats the hot junctions, which are thermally isolated from the rim, and the thermocouples convert the temperature rise into an electrical output voltage. The thermopile output is then a sum of voltage generated from each thermocouple since the thermocouples are serially interconnected.

The temperature difference between the hot junctions and cold junctions can be conveniently thought of as generated by two components of radiation: radiation incident on the hot-junction region that covers the center of the window (coated with IR-absorbing blacks) and radiation incident between the blacked area and the rim. Then

$$\Delta T = \Delta T_1 + \Delta T_2$$

$$= \frac{\eta_1 Q_{in} A_1}{G_t(r_1)} + \eta_2 Q_{in} \int_{r_1}^{r_2} \frac{2\pi r}{G_t(r)} \, dr \quad (1)$$

where $G_t(r)$ is the total thermal conductance of the structure, η_1, η_2 are absorptivity, Q_{in} is the power density of incident radiation, A_1 is the hot junction area, and r_1, r_2 are the inner and outer radii of hot and cold junctions, respectively. For simplicity, it is assumed that the radiative and thermal conduction losses through air around the detector are negligible and thus that all heat flow is lateral through the thermocouples and diaphragm to the rim. Then the total thermal conductance of the structure $G_t(r)$ becomes a sum of thermal conductance of the thermocouple leads $G_{lead}(r)$ and the thermal conductance of the diaphragm $G_{diap}(r)$ as follows:

$$G_{lead}(r) = N \sum_i \frac{\sigma_i a_i}{r_2 - r} \quad (2)$$

Reprinted from *IEEE Trans. Electron Devices*, vol. ED-33, no. 1, pp. 72–79, January 1986.

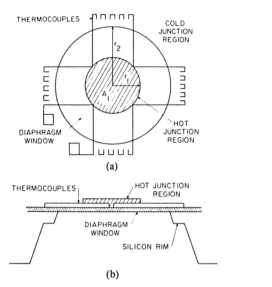

Fig. 1. Basic structure of a silicon thermopile. (a) Top view, (b) cross section.

$$G_{\text{diap}}(r) = \frac{2\pi t \sigma_{\text{diap}}}{\ln (r_2/r)} \qquad (3)$$

where a_i is the cross-sectional area of each thermocouple lead, N is the total number of thermocouples, σ_i, σ_{diap} are thermal conductivity of each thin-film material, and t is the thickness of the diaphragm. Since the induced voltage across the junctions is $\Delta V = N(\alpha_1 - \alpha_2) \Delta T$, where α_1 and α_2 are the thermoelectric powers of the thermocouple materials, respectively, the responsivity (defined by $R_v = \Delta V/(Q_{in}A_t)$) becomes

$$R_v = N(\alpha_1 - \alpha_2) \left(\frac{\eta_1}{G_t(r_1)} \frac{A_1}{A_t} + \eta_2 \frac{1}{A_t} \int_{r_1}^{r_2} \frac{2\pi r}{G_t(r)} \, dr \right) \qquad (4)$$

where A_t is the total active area. If the second term of (5) is negligible compared with the first and the active area A_t is taken as the hot-junction area A_1, as is usually the case, then

$$R_v = \eta_1 N(\alpha_1 - \alpha_2)/G_t(r_1). \qquad (5)$$

As the above equations show, we should choose thermocouple materials having a large difference in their thermoelectric power, diaphragm materials with a low thermal conductivity, and a radiation absorber with an absorptivity of near unity to achieve high responsivity. Typically, semiconductors show a high Seebeck effect, and as the impurity concentration decreases, their thermoelectric power increases [11]. However, the resistance of the thermopile limits the minimum detectable signal level through its thermal noise. This noise can be reduced by increasing the impurity concentration in the semiconductor and since this reduces the responsivity, it is necessary to compromise for optimal performance. A material such as highly doped polysilicon [12] is near optimum for the thermocouples in this sense.

Detectivity takes into account the responsivity and detector noise simultaneously. For the present thermopile where thermal noise is dominant, the dc detectivity normalized by detector size and noise bandwidth can be expressed as

$$D^* = \eta_1 \frac{(\alpha_1 - \alpha_2)}{G_t(r_1)} \left(\frac{NA_1}{4kTR_e} \right)^{1/2} \qquad (6)$$

where k is Boltzmann's constant, T is the ambient temperature, and R_e is the electrical resistance of each thermocouple in the thermopile.

Response time is also an important performance measure for a detector. For this type of detector, response time is limited by the thermal time constant as

$$\tau = C_t/G_t \qquad (7)$$

where

$$C_t = \sum_i v_i \rho_i H_i \qquad (8)$$

and v_i is the volume of ith-layer material, ρ_i is the density of ith-layer material, and H_i is the volume-specific heat of ith-layer material. A higher thermal conductance assures a faster detector response; however, it will reduce the induced temperature difference across the junctions and degrade responsivity. Another approach to improve speed is to reduce the thermal capacitance of the structure by choosing window materials with a low density and specific heat or by utilizing thinner layers.

B. Design Considerations

A major concern in designing a thermopile detector is what materials are to be used for the thermocouples. Doped polysilicon and gold have been chosen here. Polysilicon shows a high Seebeck effect [12], [13], and although the resistance of such devices is typically rather high, it can be kept in a reasonable range if the doping concentration is high enough and the appropriate polysilicon line dimensions are used. Various materials having a high thermoelectric power with the opposite sign can be used for the second thermocouple material; however, the material must either be protected from the silicon anisotropic etching solution (which is used to form the supporting diaphragm) or not attacked by it. Gold has been chosen as the second material because it is not attacked by ethylenediamine-pyrocatechol-water (EDP), a silicon anisotropic etchant. Since this material has a negligible Seebeck coefficient, the overall thermoelectric power of the detector is determined by the polysilicon lines.

Another important aspect of the detector is the diaphragm. Thin-film dielectric materials such as silicon dioxide and silicon nitride exhibit a low thermal conductivity, which is desirable for diaphragm materials. For our prototype structure, the contribution of these dielectric layers to the total thermal resistance is much less important than that of the thermopile leads. However, any significant mismatch in the thermal expansion coefficients between the silicon substrate and the thin-film window materials will result in a warped diaphragm after its formation. In order that thermal matching may be achieved, stacked layers of silicon dioxide and silicon nitride are used. Compared with the silicon substrate, silicon dioxide (thermal or pyrolytic CVD) is in compressive stress while high-temperature silicon nitride is in tensile stress. Thus the overall tension of the dielectric diaphragm can be controlled by appropriately choosing the material thickness and deposition conditions.

To form the thin diaphragm using silicon anisotropic etching, it is necessary to pattern both sides of the wafer. Diaphragm window size variations inevitably arise due to wafer thickness variations and will cause considerable variations in the detector sensitivity. To avoid such problems, the boron etch stop [13], [14] has been used. A diffused layer on the front of the wafer is used to define the window edge and, as a result, window size is not susceptible to small variations in wafer thickness. In addition, this method provides significant tolerance to front–back mask misalignments.

Responsivity and detectivity are strongly dependent on the geometry of diaphragm windows and thermocouples. Fig. 2(a) and (b) shows the calculated theoretical voltage responsivity and specific detectivity with respect to diaphragm radius for three different linewidths and numbers of thermocouples. Responsivity increases almost linearly as the diaphragm radius increases when the hot junction area is fixed. Meanwhile detectivity also increases, but at a much reduced rate (approximately in a square-root fashion). This is because detector resistance rises in proportion to the diaphragm radius, and thus detector thermal noise increases according to the square-root law. However, the maximum detector size will be limited by the detector response time. As Fig. 2(c) shows, the thermal time constant rises rapidly with respect to the diaphragm radius and also depends on the thermocouple linewidth and number of couples. In addition, a large semiconductor thermopile may result in an excessively high device resistance and make low-noise preamplifier design difficult or impractical. Thus, thermopile geometry should be carefully chosen, considering response time and available preamplifiers as well as sensitivity.

C. Detector Fabrication and Characterization

A fabricated silicon thermopile detector is shown in Fig. 3. The chip measures 3.6 mm $\times$ 3.6 mm. The diameter of the diaphragm window is 1.6 mm, and there are 32 polysilicon–gold thermocouples per window. Each thermocouple line is 16 μm wide. The diaphragm consists of SiO_2 and Si_3N_4 layers with a total thickness of 1.3 μm. The silicon rim realized by the boron etch stop is approximately 20 μm thick.

The starting point for detector fabrication is an n-type 1-$\Omega \cdot$ cm $\langle 100 \rangle$-oriented silicon wafer, which is oxidized to obtain a 1-μm-thick silicon dioxide layer for use as a diffusion mask. After photolithography on both sides of the wafer (diffusion patterns on the front side and alignment patterns on the back side), boron is diffused into the Si substrate from a solid source for 20 h at 1150°C to define the thin rim area. All of this oxide is then removed,

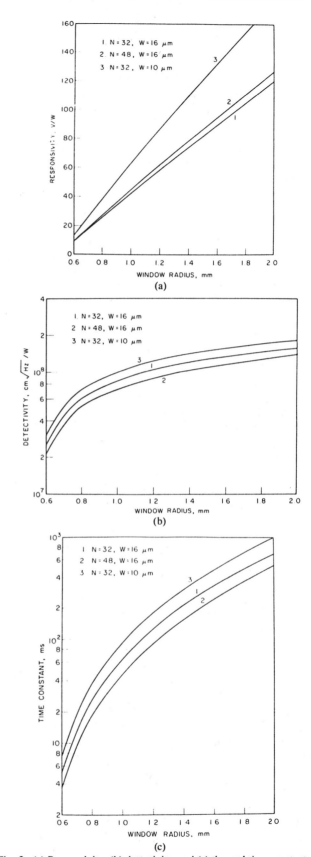

Fig. 2. (a) Responsivity, (b) detectivity, and (c) thermal time constant versus diaphragm radius for various line widths (W, in micrometers) and thermocouple numbers (N). For these calculations, the blacked area of the window was held constant at 0.79 mm², η_1 was assumed to be 0.98, the net thermoelectric power of the polysilicon–gold couples was chosen to be 184 μV/°C, and the thickness of the dielectric diaphragm and the gold and polysilicon leads was 1.3, 0.3, and 0.8 μm, respectively.

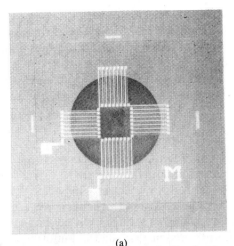

(a)

(b)

Fig. 3. A fabricated thin-film thermopile detector. The chip size is 3.6 mm × 3.6 mm and the diaphragm diameter is 1.6 mm. The line width measures 16 μm. (a) top view, (b) bottom view.

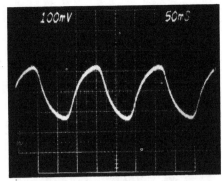

Fig. 4. A typical output waveform in response to chopped radiation from a 500 K blackbody radiation source. $Q_{in} = 0.29$ mW/cm² (after 1000 × amplication).

TABLE I
MEASURED PERFORMANCE OF SINGLE THERMOPILES

Parameter	Conditions	Measured	Values	Units
Type of Samples		n-type (Phos. doped)	p-type (B doped)	
Resistance	25°C	73 - 82	220 - 280	kΩ
Responsivity	(500°K,DC)	20 - 25	52 - 56	V/W
Detectivity	(500°K,DC,1 Hz)	5.0 - 6.0 x 10⁷	6.8 -7.3 x 10⁷	cm √Hz/W
Time Constant	500°K B.B. source	25	25	msec

Fig. 4 shows an output signal waveform in response to chopped radiation from a 500 K blackbody radiation source. The performance characteristics of the device are listed in Table I. The measured sheet resistance for p-type polysilicon leads was distributed between 135 and 170 Ω/□, while that for n-type samples was in the range from 46 to 52 Ω/□. The p-type thermopiles exhibit about twice the responsivity of the n-type thermopiles, while their detectivities are similar. This is due to the fact that the p-type leads generate more thermal noise even though they show a higher Seebeck effect as a result of the lower doping levels.

The effect of ambient temperature on responsivity and device resistance is shown in Fig. 5. Responsivity is almost flat over the temperature range from −5 to 80°C. Device resistance decreases linearly as temperature rises, with a temperature coefficient of about −200 ppm/°C for p-type thermopiles and −70 ppm/°C for n-type thermopiles. This device can inherently handle large incident radiation levels and has a wide dynamic input power range. The minimum detectable input power, which is limited by the detector noise, is about 1.2×10^{-9} W/$\sqrt{Hz}$ for the p-type samples and about 1.6×10^{-9} W/$\sqrt{Hz}$ for the n-type samples. These levels correspond to input power densities of 1.5×10^{-5} W/cm² and 2.0×10^{-5} W/cm², respectively. The maximum input power, which is limited by the heat-handling capacity of the thermopile structure, is rather high since the silicon rims dissipate heat efficiently and the dielectric diaphragm is not susceptible to thermal shock. These have handled input densities of 5 W/cm² without any damage or degradation. The detector responsivity is linear with respect to measured incident power density over the range from 10^{-4} to 10^{-2} W/cm² (as limited by the source used), and we expect that the total linear range is substantially wider.

300 nm of new thermal SiO₂ is grown at 1000°C, and 200-nm-thick Si₃N₄ and 800-nm-thick SiO₂ are deposited at 770°C by an atmospheric CVD (APCVD) process to form the window dielectric. Approximately 800 nm of polysilicon is pyrolytically deposited at 670°C, patterned, and doped with a suitable impurity (boron or phosphorus). After an additional CVD oxide (250 nm) is deposited and contact windows are opened, gold-on-chrome (300 nm in total) is evaporated and defined as interconnects. Finally, diaphragm windows are patterned on the back side and the wafer is immersed in EDP for 4 to 5 h. Chip separation is simultaneously performed during this EDP etching step.

These delicate-looking devices have shown a suprisingly high yield. Over 90 percent of the thermopile detector chips batch-fabricated in a typical 2-in wafer were defect free. They are also rugged enough to survive ultrasonic vibration and 1-m drop tests. The separated chips were mounted in TO-5 packages and bismuth blacks were applied over the window areas. The detectors were then bonded and hermetically sealed in an argon atmosphere. KBr windows were used to transmit the infrared radiation onto the detectors.

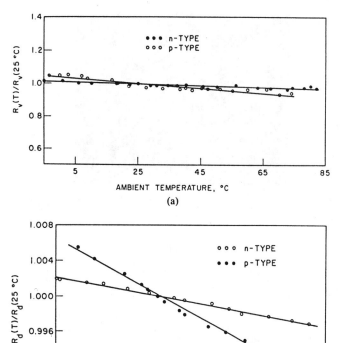

Fig. 5. (a) Detector responsivity and (b) detector resistance versus ambient temperature.

located only about 1 mm from the hot junction area; heat loss through the surrounding gas (argon) to the mount is thought to have been significant. If this loss is eliminated, the responsivity would be improved by about 13 percent according to our calculation, bringing about experimental results more closely in line with the theoretical results shown in Fig. 2.

III. LINEAR DETECTOR ARRAYS

For many applications, a detector array is required since it can provide area thermal imaging with faster speed and better focal-plane sensitivity than a single detector and can simplify mechanical scanning complexity. Various array and scanning schemes [16] have been developed to effectively acquire two-dimensional thermal information. However, an ideal array should provide uniform sensitivity, high spatial resolution, and on-chip readout capability. In consideration of these requirements, the feasibility of realizing an array of thermopiles containing on-chip readout circuits has been investigated.

A. Diaphragm Window Arrays

Several arrays of diaphragm windows are shown in Fig. 6. Each array consists of two linear window subarrays, and each subarray has 30 window elements. The two linear subarrays are staggered (vertically offset) by 250 μm with respect to each other. This staggered structure can be used to improve the vertical resolution of the array by delaying the readout time of the offset subarray when the detectors are scanned horizontally.

The diaphragm window element again consists of SiO_2 and Si_3N_4 in a sandwich structure with dimensions of 0.4 mm $\times$ 0.8 mm $\times$ 1.3 μm. The center-to-center spacing between the windows is 500 μm, and the rim formed by the boron etch-stop is about 20 μm thick. The main functions of this Si rim are to heat-sink the cold junctions and to provide mechanical support for the array structure. It is noteworthy that single-crystal silicon has a typical thermal conductivity of 0.84 W/cm $\cdot$ °C.

The ruggedness and yield of these thin diaphragms are rather high. Approximately 98 percent of the diaphragm windows in a typical 2-in wafer were defect free. This result assures the practical feasibility of this approach. As far as window packing density is concerned, it can be increased as much as required by reducing the dimensions of the windows and rims; however, the sensitivity of the thermopile elements sets the lower limit on the window size and thus an upper limit on packing density for a given application.

Blacks also play an important role in determining detector performance; the detector output almost doubles when blacks are added, as compared to a non-black device. To examine the amount of reflected radiation on the detector surface, blacks were coated on the back side of some samples and the front side was irradiated. The detector output was then compared with that for the samples with blacks on the front (irradiated) side. The magnitude of the output signal for the back-coated samples is about a factor of 2 smaller than for the front-coated samples, indicating that approximately 50 percent of the incident radiation is reflected at the non-blacked surface.

The measured response time ($\tau_{63\%}$) was 25 ms for these blacked detectors, independent of the resistivity type. This indicates that the impurity type and doping level of the polysilicon lines are not important in determining the response time, unlike the detector responsivity. However, the response time was roughly doubled when the blacks were added.

Detector performance could be further improved by optimizing the structure and the thermocouple materials. For example, use of tantalum instead of gold could help to enhance the device responsivity since tantalum has a significantly lower thermal conductivity than gold and is not attacked by EDP. Better mounting to reduce heat conduction through the surrounding gas could also improve the responsivity. The mounting surface for these detectors was

B. On-Chip Circuit Compatibility

Polysilicon-gate PMOS was chosen to test the compatibility of an MOS process with the thermopile process described above. The use of PMOS minimizes the number of process steps and the number of masks needed. Using PMOS, the field dielectric, polysilicon, and metal layers are shared between the circuitry and the detectors.

The fabrication steps are essentially the same as for a

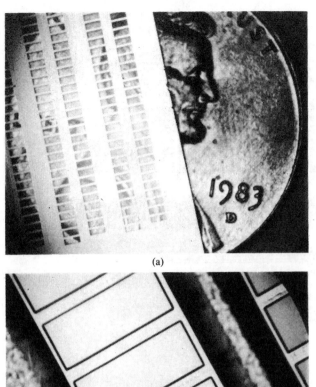

(a)

(b)

Fig. 6. Thin diaphragm array. Each window consists of multiple layers of SiO$_2$ and Si$_3$N$_4$, measuring 0.4 mm × 0.7 mm × 1.3 μm. (a) Top view, (b) Bottom view.

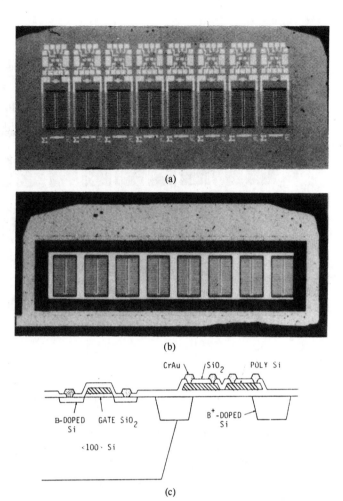

(a)

(b)

(c)

Fig. 7. A prototype thermopile (8 elements) linear array with on-chip PMOS test devices. The chip size is 5.5 mm × 2.5 mm. (a) Top view, (b) bottom view, (c) cross section.

single thermopile described above except for an additional step to define source–gate–drain regions for enhancement PMOS devices. The 850-nm-thick gate oxide was thermally grown at 1000°C in an O$_2$ atmosphere. The total number of masks used in the combined process is seven, with three masks shared between the thermopile and MOS devices. This is only one mask more than required for the single thermopiles.

A fabricated prototype array with on-chip PMOS test devices is shown in Fig. 7. The array consists of 8 thermopiles and some simple devices that are used to test the process compatibility with conventional silicon-gate PMOS. The dimension of each window is 0.4 × 0.7 mm^2 and the center-to-center spacing between elements is 500 μm. The number of thermocouples per element is 40.

The completed detector chip was blackened on the back side and mounted in a package. The sample was tested using a 500-K blackbody radiation source. In testing, the back side of the chip was irradiated to reduce reflection at the surface and to eliminate unwanted irradiation to the PMOS devices.

TABLE II
MEASURED PERFORMANCE OF ARRAY ELEMENTS

Parameter	Conditions	Measured Values
Material		B-doped Poly Si-Au
Detector Area		0.4mm x 0.7 mm
Diaphragm Thickness		1.3 μm
Resistance	25°C	42 - 47 kΩ
Noise Voltage	10 kHz B.W.	3 - 4 μV (rms)
Responsivity	(500°K,DC)	8.4 -12.6 V/W
Detectivity	(500°K,DC,1 Hz)	1.6 - 2.5 x 10^7 cm$\sqrt{\text{Hz}}$/W
Time Constant	63 % of V_{p-p}	5 - 10 msec

Typical characteristics of these detector elements are listed in Table II. The response time of the elements is about three times faster than for the larger single detectors, while the responsivity and resistance are lower. The improved time constant and device resistance are mainly due to the small rectangular window, which provides a low thermal capacitance (0.8 × 10^{-6} J/°C) and allows an appropriate geometric dimension (250 μm × 10 μm) for the polysilicon leads. This rectangular window geometry improves the vertical fill factor by 60 percent compared to a circular window geometry having the same area, although it degrades responsivity due to a reduced thermal resistance (1.18 × 10^4 °C/W). The detectivity nonuniformity of the present array has been as high as 20 percent. This

239

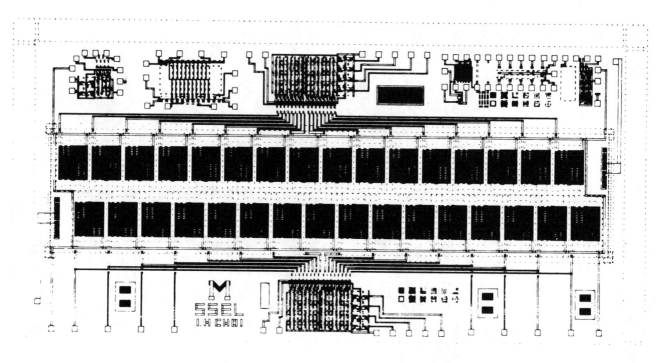

Fig. 8. Layout of a 16 × 2 thermopile linear array with on-chip multiplexing. The chip size is 11 mm × 5.5 mm, with 10-μm feature size.

nonuniformity is caused mainly by nonuniformity in the thickness of polysilicon leads deposited by the APCVD process. Use of an improved CVD process such as LPCVD could significantly improve the uniformity.

The enchancement PMOS transistors ($Z/L = 50/10$ in micrometers) included on the test chip showed typical I_D–V_{DS} characteristics, with a $V_T = -1.3$ V when the substrate was grounded. Transconductance was $g_m = 0.24$ m℧ at the bias point of $V_{DS} = -10$ V and $V_{GS} = -4.6$ V. These parameter values are in good agreement with basic MOS theory, assuming a $Q_{ss} = 1.3 \times 10^{-8}$ C/cm^2 and $\mu_p = 150$ cm^2/V · s. The breakdown voltage from the drain with the source and substrate grounded was about -44 V. These results indicate that PMOS devices with reasonable performance can be readily obtained by using the fabrication process for the silicon thermopile.

C. Staggered Linear Arrays with On-Chip Multiplexers

As the next step in this work, a thermopile array with on-chip multiplexers has been designed and is now being fabricated using a silicon-gate E-D PMOS process. A layout of the array is shown in Fig. 8. It consists of two staggered linear subarrays of thermopiles and two 16:1 analog multiplexers, one for each subarray. Each subarray contains 16 elements, each measuring 100 μm × 800 μm. The center-to-center spacing between the elements is 600 μm. MOSFET's with a large aspect ratio ($Z/L = 20$) are used for the analog switches to reduce their associated noise as well as channel resistance. Switching noise will be removed by external signal conditioning. A separate thermopile, polysilicon resistors, and diodes are also included on the chip to measure the cold-junction temperature and

TABLE III
SYSTEM PARAMETERS

Lens Clear Aperture D_0	50 mm
Lens Focal Length f_0	50 mm
Optical Transmission	0.8
Scanning Efficiency	0.8
Operating Temperature	300° K
Spectral Band	8–14 μm
Scanning Interval	3 sec/picture
Detectivity D*	3.5 x 10^7 cm$\sqrt{\text{Hz}}$/W
Detector Angular Subtense	16 mrad x 8 mrad
Number of Detectors	32
Total FOV	1024 mrad x 768 mrad
Reference B.W. of Electronics	5 KHz

the thermoelectric power of the polysilicon lines. The total chip size is 11 mm × 5.5 mm using 10-μm design rules.

In order to estimate the theoretical performance of an infrared imaging system using this array, noise equivalent temperature difference (NETD) and minimum resolvable temperature difference (MRTD) given by Lloyd [17] have been calculated. The system parameters used to calculate NETD and MRTD are listed in Table III. A typical Gaussian modulation transfer function (MTF) is assumed for the overall system. According to this calculation, the NETD is 0.9°C and the MRTD is about 1.4°C at a spatial frequency of 0.2 Hz/mrad, which implies that a standard target with a spatial period of 0.5 cm and a target-to-background temperature difference of 1.4°C is detectable at a distance of 1 m. This is adequate for many industrial applications. Meanwhile, where substantial temperature differences are being scanned, window size can be further decreased and a larger number of elements can be included in the array to improve spatial resolution. Although we have assumed a scanning image system, simple nonscanned mounting is also adequate for many other applications, for example in process monitoring.

IV. Conclusion

A broadband thermopile-based infrared detector array has been successfully realized using a conventional silicon IC process and micromachining technology. This array shows a fast response time with modest responsivity, operates over a wide ambient temperature range, and responds to a broad spectral signal range. In addition, since this array involves a diaphragm window structure with high packing density and good thermal insulation, this structure should be useful for other detector types such as bolometers, pyroelectric detectors, or gas-flow sensors. Although this thermopile array has exhibited process compatibility with silicon PMOS devices, CMOS or NMOS technology could also be used with a modest increase in detector process complexity. A 16 × 2 staggered linear array containing on-chip analog multiplexers has been designed and is being fabricated. It is expected that this array will be useful for various industrial and commercial applications, including process monitoring and noncontact testing where moderate resolution is acceptable. Since the array can operate over a wide ambient temperature range, applications in *in situ* process monitoring are of particular interest.

Acknowledgment

The authors would like to express their appreciation to Dr. R. Toth of the Dexter Research Center, Dexter, MI, and A. J. Meijer of Sensors, Inc., Ann Arbor, MI, for their contributions to detector packaging and testing, and F. J. Schauerte of the General Motors Research Laboratories, Warren, MI for his contributions to layout conversion.

References

[1] R. J. Keyes, "Recent Advances in optical and infrared detector technology," in *The Optical and Infrared Detectors, Topics in Applied Detectors*, R. J. Keyes, Ed. New York: Springer-Verlag, 1980, pp. 301–315.

[2] S. C. Chase, Jr., J. L. Engel, H. W. Eyerly, H. H. Kieffer, F. D. Palluconi, and D. Schofield, "Viking infrared thermal mapper," *Appl. Opt.*, vol. 17, no. 8, pp. 1243–1251, 1978.

[3] R. Tice and J. Euskirchen, "Live infrared fire imaging surveillance," *Opt. Spectra*, pp. 32–35, Sept. 1978.

[4] F. C. Gillett, E. L. Dereniak, and R. R. Joyce, "Detectors for infrared astronomy," *Opt. Eng.*, vol. 16, no. 6, pp. 544–550, 1977.

[5] E. H. Putley, R. Watton, and J. H. Ludlow, "Pyroelectric thermal imaging devices," *IEEE Trans. Sonics Ultrason.*, vol. SU-19, pp. 263–268, 1972.

[6] S. Iwasa, J. Gelpey, J. Marciniec, D. Marshall, W. White, D. Lamb, S. Liu, and D. Paffel, "Pyro/CCD direct signal injection theory and experiment," in *IEDM Tech. Dig.* (Washington, D.C.), pp. 522–525, Dec. 4–6, 1978.

[7] C. B. Roundy, "Pyroelectric self-scanning infrared detector arrays," *Appl. Opt.*, vol. 18, no. 7, pp. 943–945, 1979.

[8] D. L. Polla, R. S. Muller, and R. M. White, "Fully-integrated ZnO on silicon pyroelectric infrared detector array," in *IEDM Tech. Dig.*, (San Francisco), pp. 382–384, Dec. 9–12, 1984.

[9] C. Shibata, C. Kimura, and K. Mikami, "Far IR sensor with thermopile structure," in *Proc. 1st Sensor Symp.*, *IEE of Japan, Tech. Comm. on Electron Devices*, pp. 221–225, 1981.

[10] G. R. Lahiji and K. D. Wise, "A batch fabricated Si thermopile infrared detector," *IEEE Trans. Electron Devices*, vol. ED-29, pp. 14–22, 1982.

[11] G. Busch and H. Schade, "Thermoelectric effects," in *Lectures on Solid State Physics*. New York: Pergamon, 1976, pp. 371–379.

[12] G. R. Lahiji, "A monolithic thermopile detector fabricated using integrated-circuit technology," Solid-State Electronics Lab, Univ. of Michigan, *Tech. Rep.* no. 150, pp. 64–78, 1981.

[13] D. I. Jones, P. J. Le Comber, and W. E. Spear, "Thermoelectric power in phosphorus-doped amorphous silicon," *Phil. Mag.*, vol. 36, no. 3, pp. 541–551, 1977.

[14] A. Bohg, "Ethylene diamine–pyrocatechol–water mixture shows etching anomaly in boron-doped silicon," *J. Electrochem. Soc.*, vol. 118, pp. 401–402, 1971.

[15] N. F. Raley, Y. Sugiyama, and T. Van Duzer, "(100) Si etch-rate dependence on boron concentration in ethylenediamine pyrocatechol-water solution," *J. Electrochem. Soc.*, vol. 131, no. 1, pp. 161–171, 1980.

[16] W. L. Wolfe and G. J. Zissis, *The Infrared Handbook*. Ann Arbor, MI: The Infrared Information and Analysis (IRIA) Center, Environmental Research Institute of Michigan, 1978, ch. 10.

[17] J. M. Lloyd, *Thermal Imaging Systems*. New York: Plenum, 1975, pp. 166–211.

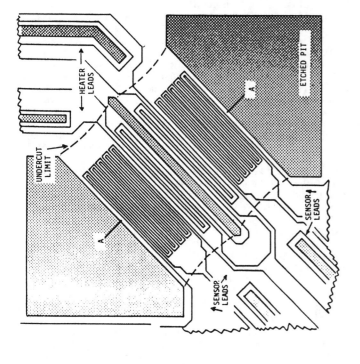

FIGURE 1: PLAN VIEW OF AIR FLOW SENSOR MICROSTRUCTURE.

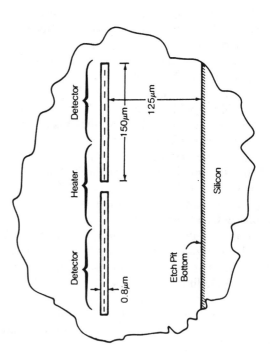

FIGURE 2: CROSS SECTION, A-A, OF AIR FLOW SENSOR MICROSTRUCTURE.

A MICROTRANSDUCER FOR AIR FLOW AND DIFFERENTIAL PRESSURE SENSING APPLICATIONS

G. B. Hocker, R. G. Johnson, R. E. Higashi, P. J. Bohrer

Introduction

The micromachining of small structures in silicon became an attractive approach to the development of new types of microtransducers with the appearance of papers by Bassous (1), by Bean (2), and by Peterson (3) in 1978. These papers outlined the principles and methods for precision anisotropic etching of single crystal silicon, and presented many examples of structures fabricated by these techniques. By means of innovations on this early work, we have developed techniques for the fabrication of precisely dimensioned dielectric film structures containing temperature sensitive resistors that can be thermally isolated from the chip. This development has enabled the attainment of locally high temperatures with very low input power and little heating of the chip, and has led to novel concepts of integrated thermal sensors and transducers that can be built with the low cost, thin film-batch processing characteristic of integrated circuit fabrication. There is, in addition, the option of incorporating active silicon circuit elements on the chip with the microtransducer, thus achieving a larger scale device integration with lower cost, lower device size, and easier applicability. These expectations are now well on the way to being realized in practical devices at Honeywell with the development of an airflow and differential pressure sensor using these techniques.

Microtransducer Design for Air Flow Sensing

The choice of materials for the transducer was strongly influenced by the objective of making the finished device easy to manufacture. For example, silicon nitride was chosen as the dielectric because it can be deposited to tightly controlled specifications as well as because it is an excellent insulator and passivating film. A nickel iron alloy was chosen for the resistor metallization because the deposition technology was well established. In addition, the alloy's higher resistivity than pure metals allows the delineation of resistors in the 500 to 1000 ohm range on a 150x350 micron area using 5 micron line widths while maintaining a substantial temperature coefficient of resistance of about .003 per degree centigrade (4). Standard (100) silicon was chosen for the substrate because of its desirable anisotropic etching properties and its suitability for later circuit integration with the transducer.

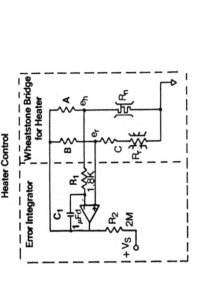
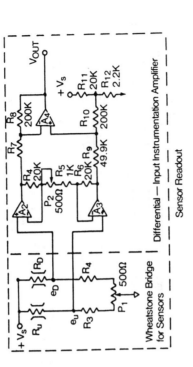

FIGURE 3: HEATER AND SENSOR CIRCUITS FOR THE AIR FLOW SENSOR FOR MASS FLOW.

The transducer consists of a pair of silicon nitride bridge microstructures, as shown in Figures 1 and 2, containing a central heater resistor divided equally between the two bridges, and two identical detector resistances placed adjacent to and symmetrically with respect to the heater resistance. In operation, air flows across the chip perpendicular to the long axes of the bridge structures, cooling the upstream detector and heating the downstream detector. The resulting temperature and corresponding resistance differential yields a circuit output measurement of the flow rate. The use of identical detector resistors eliminates zero point offsets due to interfering factors such as variations in ambient pressure and temperature. The bridge structures are less than one micron thick, and their small heat capacity provides a short thermal time constant, typically .005 second or less, and a high sensitivity to rapid flow changes.

The air space beneath the bridges thermally decouples the heater and detectors from the silicon to a large extent. Consequently, large heater temperature differences in the 100-200°C range relative to the silicon can be sustained by small power inputs. The thermal efficiency is typically 15°C per milliwatt of input power under no-flow conditions. The thin nitride film permits the detector resistors to be placed closely adjacent to the heater such that they can operate at about 60 per cent of the heater temperature elevation and can develop large temperature differentials under air flow conditions.

The etched cavity below the bridge pair is precisely limited at the sides by the etch-resistant (111) planes of the silicon, and at the bottom of the cavity and the ends of the bridges by accurate timing of the etch duration. The symmetry and effectiveness of the etched undercut of the bridges is maximized by orienting the axes of the bridges at 45 degrees to the <110> directions in the silicon.

Drive Circuit Operation

Figure 3 shows heater and sensor circuits used to operate the flow sensor. The sensor circuit is a conventional wheatstone bridge circuit. However, the heater circuit, as shown in Figure 3, is uniquely adapted to the flow sensor to provide an output proportional to mass flow, and to minimize errors due to ambient temperature changes. The heater circuit is designed to keep the heater temperature at a constant differential above the ambient air temperature under conditions of ambient temperature variation and air flow variation. The ambient temperature is sensed by a similar heat sunk resistor on the chip, and the chip temperature, which remains within about one degree of the ambient air, is a satisfactory approximation to the ambient.

This mode of heater operation also reduces, but does not eliminate, the effects of changes in air or gas composition which could alter the thermal conductance and thus otherwise change the operating temperature of the heater and detector resistances. However, extreme changes of thermal conductance, such as that encountered between air and helium, cause large differences in power needed to hold the heater at its constant temperature differential.

The error integrator is the active component in the heater circuit. It integrates the voltage differences seen on the wheatstone bridge, and changes the voltage to the heater to maintain the bridge circuit balance. However, because the temperature characteristic of the nickel iron alloy is not linear, ambient temperature changes do not produce proportional changes in the heater, R_h, and reference

243

resistor, Rr, when the heater is in operation at an elevated temperature. Consequently resistors A, B, and C must be chosen to compensate for this effect to keep the wheatstone bridge ratios equal at a given operating temperature differential. With proper choice of resistor values, the error in heater temperature can be kept less than 1.0 per cent of its differential above ambient over an ambient temperature range of -40C to +80C.

Chip Packaging and Application Considerations

The chip is fabricated with passivation techniques to permit its exposure to air flow environments over a long operating life. To prevent corrosion, gold pads and gold wire bonds are used, and all other surfaces are passivated with silicon nitride. The chip housing protects the microstructure from mechanical damage, and can provide the smooth flow channel that is desired for precise flow measurements. Mass flow measurements require calibrations that depend on channel size, of course, and a factor limiting the accuracy of mass flow measurement is the onset of turbulence, or velocity-dependent flow profiles across the channel. Therefore it is important to keep Reynolds numbers below the turbulent flow values, and it is desirable to have fully developed laminar flow at the chip for all channel sizes. The small channel sizes, which can be comparable to or less than chip dimensions, and the resulting high flow impedance facilitate the sensing of dynamic differential pressures, for example, between two rooms, across a duct flow orifice, or any application in which a small flow through the sensor is permissible. The detector microtransducer provides an output response proportional to the mass flow rate. This response can be calibrated to yield a differential pressure measurement in which the range and shape of the response curve can be determined by the dimensions that determine the flow impedance of the housing. In addition, the small channel dimensions that can be obtained provide a high impedance relative to the impedance of common pneumatic connecting lines. Consequently, it is possible to remotely locate the differential pressure sensor by using convenient connecting lines, thus reducing installation costs and, in some applications, reducing environmental stresses on the sensor.

The effects of dust coming through the flow line are minimized by the chip surface being parallel to the flow direction. In contemplated differential pressure sensing applications, a long life in typical industrial atmospheric environments can be assured by removing most of the dust by the use of a suitable filter. This approach is practical because of the low throughput rate, typically 20 to 50 cm3

per minute, made possible by the small channel dimensions and the high sensitivity of the detector. Accelerated life tests with heavy dust concentrations have demonstrated an equivalent life in normal industrial air in excess of 20 years with no dust accumulation on the chip and no deterioration of response.

Performance Characteristics

An outstanding property of the sensor is the large temperature differential that can be developed between the two detector resistances when using small air flows, thus making it practical to use a simple metal film in which the temperature coefficient is much smaller than the temperature coefficient of resistance of more complex diodes and thermistors. Figure 4 shows the temperature changes of the detector resistors in typical differential pressure measurement, and the resulting temperature differential between the upstream and downstream detector resistances. Figure 5 shows pressure differential characteristics at four ambient temperatures, and illustrates the temperature compensation that can be achieved over a broad operating temperature range. The sensor measures mass flow quite accurately under density changes caused by ambient temperature variations or ambient pressure changes. Under the more diverse conditions of gas species variation, with different molecular masses, thermal

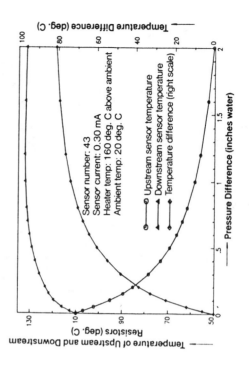

Sensor number: 43
Sensor current: 0.30 mA
Heater temp: 160 deg. C above ambient
Ambient temp: 20 deg. C

Upstream sensor temperature
Downstream sensor temperature
Temperature difference (right scale)

Temperature of Upstream and Downstream Resistors (deg. C)
Temperature Difference (deg. C)
Pressure Difference (inches water)

FIGURE 4: TEMPERATURE DIFFERENTIALS UNDER FLOW CONDITIONS IN A TYPICAL SMALL CHANNEL AS A FUNCTION OF PRESSURE DIFFERENCE BETWEEN CHANNEL PORTS.

conductivities, and mean free paths, the sensor likewise performs well, although with somewhat greater span changes for some gas types. Figure 6 shows mass flow response characteristics for air and CO_2 which, for identical responses, yield mass flow rate differences of 12 per cent. For minor air composition differences the errors are quite small.

Conclusions

We have developed a novel highly sensitive air or gas flow sensor which performs well as a mass flow sensor and differential pressure sensor. It is especially suited to applications in the low differential pressure range from 0 to 1.0" of water column. Without change of calibration, it provides a more accurate measure of mass flow rates for different gaseous species than other commercially available mass flow sensors. Because it can be fabricated by conventional thin film deposition and silicon processing techniques, it offers the possibility of lower cost and broader applications than present commercially available gas flow sensors.

References

1. Bassous, E., (1978). Fabrication of novel three dimensional microstructures by the anisotropic etching of (100) and (110) silicon. IEEE Transactions on Electron Devices, ED-25. No. 10. pp 1178-1185.

2. Bean, K.E., (1978). Anisotropic etching of silicon. IEEE Transactions on Electron Devices, ED-25. No. 10. pp 1185-1193.

3. Peterson, K. E., (1978). Dynamic micromechanics on silicon: Techniques and devices. IEEE Transactions on Electron Devices, ED-25. No. 10. pp 1241-1250.

4. Iron-nickel and related alloys of the Invar and Elinvar types, (1949). International Nickel Company. Inc., New York, NY.

All authors affiliated with
Honeywell Physical Sciences Center
10701 Lyndale Avenue South
Bloomington, Minnesota 55420

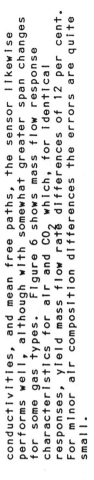
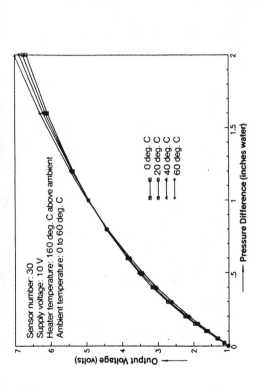

FIGURE 5: TEMPERATURE COMPENSATION ACHIEVED IN DIFFERENTIAL PRESSURE MEASUREMENT OVER 0 TO 60°C RANGE.

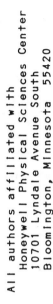
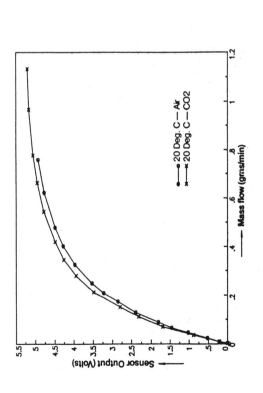

FIGURE 6: RESPONSE TO MASS FLOW IN A TYPICAL SMALL CHANNEL FOR AIR AND CO_2.

HIGH-PRECISION, HIGH-PERFORMANCE MASS-FLOW SENSOR WITH INTEGRATED LAMINAR FLOW MICRO-CHANNELS

Kurt Petersen and Joseph Brown
Transensory Devices, Inc., Fremont, CA

and

Wayne Renken
Innovus Inc., Milpitas, CA

ABSTRACT

In this presentation, a micromechanical sensor chip is described which consists of micromechanical bridges suspended over channels etched in silicon. The bridges are composed of two insulating layers between which a meandering thin film metal resistor is sandwiched. After the bridge structures and electrical contact pads are formed on the first wafer, a second wafer is prepared which contains etched channels identical to the first, but without the bridges. When the two wafers are aligned and bonded together, a very reproducible, well-defined "tube" is formed with high-precision bridge structures suspended across the center. Typical dimensions of the tubular micro-channel are 140 microns high by 540 microns wide. The bridges are 540 microns long and about 150 microns wide.

INTRODUCTION

Two aspects which are crucial for high-accuracy gas flow measurements include not only a well-characterized and reproducible transduction mechanism, but also a well-defined and repeatable gas flow geometry. The transduction mechanism itself is important, of course, because it plays a major role in determining resolution, linearity, accuracy, speed, and other performance characteristics. The geometry is equally important, however, since the flow rate is proportional to the cube of the tube radius through which the gas is flowing - small variations in tube diameter may result in large gas flow measurement errors. In addition, the size and shape of the flow channels also influence characteristics such as laminar versus turbulent flow, which in turn affect heat transfer parameters.

The techniques of silicon micromechanics can be exploited to provide both these functions.

The unique shape of the bridges, shown in Figure 4, has been designed to optimize strength, minimize thermal losses to the supporting substrate, maintain structural stability under high gaseous velocities, provide very high corrosion resistance, and simplify fabrication procedures. The thermal resistance from the bridge to the substrate has been

calculated to be greater than 10,000 C/W. This property makes it possible to operate the thin film resistive element at greatly reduced temperatures (25 C above ambient), while still decreasing the response time to the millisecond range.

Besides exhibiting response times 2 to 3 orders of magnitude faster than conventional mass flow sensors, the lower operating temperature permits the sensor to be used more reliably with reaction-prone gases. Furthermore, the materials of construction are exceptionally inert to those gases commonly used, for example, in the semiconductor processing industry.

SENSOR FABRICATION

Processing for the mass-flow sensor chip involves two wafers whose initial preparation is identical. First, p-type, [100] oriented, 100 mm wafers are heavily doped with boron to serve as an etch-stop layer. A 70 micron thick epitaxial layer is grown over the p+ boron layer, which will define the height of the flow channel. After an insulating layer is deposited on both wafers, the "cap wafer" is etched to form the channel regions, then coated with a silicon nitride corrosion-resistant film. The "bridge wafer," however, is now coated with a thin film metal layer with a high temperature coefficient of resistance to a thickness of about 0.2 microns. The metal film is etched into meandering resistor patterns with linewidths of about 7.0 microns and resistance values of about 1000 Ohms per bridge element. Another insulating thin film is deposited over the etched metal pattern to fully encapsulate the thin film resistors for maximum environmental protection. Next, via holes are etched through the second insulating layer, stopping at the resistor thin film, and a gold bonding film is deposited and patterned in the via holes. The bridges themselves are formed after the bridge pattern is etched completely through the two insulating layers and the exposed silicon in anisotropically etched down to the p+ etch-stop layer. To coat and protect the bare silicon itself, a final insulating layer is deposited over the wafer, then etched away from the bonding pad areas.

Reprinted from *Rec. of the 3rd Int. Conf. on Solid-State Sensors and Actuators*, 1985, pp. 361-363.

246

The two wafers are aligned, clamped, and bonded with an intermediate adhesive layer. Since the bridge elements are protected by the bonded "cap" wafer, ordinary sawing procedures can be employed to dice the wafer assembly and to expose the electrical contact pads.

SENSOR PERFORMANCE

For fifteen years the industry standard mass flow controller for gases in semiconductor fabrication applications has been based upon the tubular type of sensor shown in Figure 1. The silicon microsensor described here introduces a new horizon in performance, particularly when applied in a control system designed to take advantage of its novel attributes. In most key performance areas the differences are great:

* Measurement Precision - Process results which are related to flow control percision have long been observed to be at odds with published flow controller precision specifications. This is especially true of run-to-run reproducibility, which implies poor flow control calibration stability. Short-term inconsistencies can be caused by external temperature gradients or transients resulting from other parts of the system turning on and off, and flow controller specifications typically ignore these effects. The microsensor is far less sensitive to external influences because the sensing element is buried in the flow stream.

* Sensor Response - The sensor response to a transient in mass flow can be characterized approximately as a first-order exponential lag with a time constant which depends upon the sensor geometry and the heat transfer rate. Since the heat transfer rate varies with both flow rate and type of gas, it has a range of values for a given sensor, being about 0.5 to 5 milliseconds for the microsensor, and about 1 to 10 seconds for through-the-wall types. An attempt has been made in most commercial flow controllers to cancel a portion of the sensor lag with a phase lead filter in the signal conditioner, but the extreme variability of the lag precludes effective compensation. When excessive compensation is added, it leads to bizarre signal overshoots during fast transients.

* Flow Control Dynamics - Gas mass flow controllers using the microsensors are combined with fast electromagnetic modulating valves to extend the fast response characteristics from measurement to control. Closed-loop control to within 2% of flow following a set-point change is about 25 milliseconds for the microsensor compared with several seconds for traditional thermal flow controllers. The speed-up is important to applications such as sandwiched dielectrics, graded composition films, super-lattice structures, single-wafer processing, and many other processes that require precise dynamic control of the process atmosphere.

* Turn-on Surges - Traditional thermal mass flow controllers combine slow sensors with fast valves, an unfortunate combination that results in large flow overshoots during turn-on. These surges are attenuated by special "soft-start" circuits that ramp up the flow slowly over a period of seconds while the sensor stabilizes. The microsensor, being faster than the valve, is in full control of a quick turn-on transient and provides regulation with no overshoot.

* Sensor Temperature Rise - Thermal flow sensors must operate at some rise above stream temperature to function. Because the inside-out construction of the microsensor isolates it from external thermal noise, it can operate at a cooler temperature with the same signal/noise ratio. The microsensor is regulated to operate at a constant 25 C above ambient, while traditional sensors expose the stream to surface temperatures as high as 100 C above ambient. The lower temperatures of the microsensor avoid contaminant production in reactive gases.

* Sensor Reliability - Contamination has historically been the primary cause of flow controller failure in the semiconductor industry. The low temperature and inert surface materials used in the microsensor greatly reduce contaminant formation, thereby increasing the inherent reliability of each flow channel. In addition, the use of redundant sensors and a smart controller for dynamic calibration monitoring provides an improved recovery strategy in the event of problems.

* Packaging - The small size of the microsensor results in a number of system packaging advantages: major reductions in gas system size, closer placement to the process chamber, minimized dead volumes, faster atmosphere change dynamics, and greater freedom of design in equipment architecture.

The above features define the flow controller needed to support the next generation of semiconductor processing equipment. Silicon micromechanics provides the techniques, as well as the technical perspective to conceive and to realize such advanced, high performance structures.

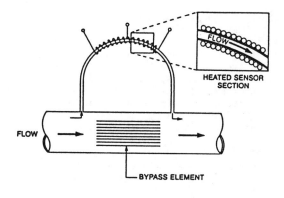

Figure 1   Conventional through-the-wall type thermal mass-flow sensor.

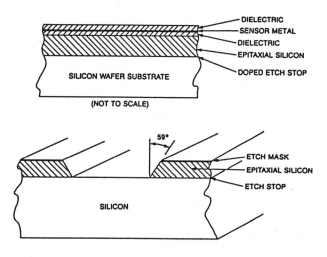

Figure 2   Cross-section of silicon wafer layers.

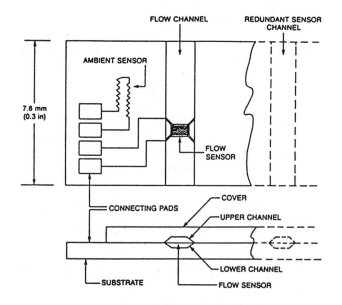

Figure 3   Configuration of the sensor and flow channel elements on the completed sensor chip.

Figure 4   Scanning electron micrograph of a sensor bridge element suspended over an etched channel. The black bar is 100 microns long.

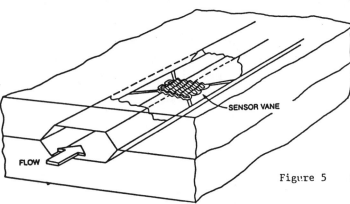

Figure 5   Cutaway drawing of a sensor bridge suspended in the middle of a flow channel, after wafer lamination.

248

A Monolithic Gas Flow Sensor with Polyimide as Thermal Insulator

GÖRAN N. STEMME

Abstract—A novel small monolithic gas flow sensor has been designed and fabricated by use of micromachining of silicon. Its operation is based on the cooling of an electrically heated mass by the gas flow, and detection of the mass's temperature by a diode. The small size, 0.4 mm by 0.3 mm by 30 μm, of the hot part of the sensor gives a fast thermal response (time constant 50 ms). By using polyimide as a thermal insulator a high gas flow sensitivity is achieved. The shape of the sensor will present very little obstruction to the gas flow and also makes it easy to mount.

I. INTRODUCTION

A VERY COMMON type of gas flow sensing with no moving parts is based on heat transfer by convective cooling between the sensor and the gas flow. A classical example of such a sensor is the hot-wire anemometer. However, the hotwire anemometer has several drawbacks: It is difficult to calibrate and requires rather complicated electronics. Furthermore, it is fragile and difficult to mount in the gas stream. To solve these problems, a number of gas flow sensors in silicon have been developed [1]–[5]. The silicon gas flow sensor to be presented here has several advantages over other gas flow sensors. It is extremely small. A special geometry and the small size of the sensor results in minimal disturbance of the gas flow and good measurement accuracy. The small mass provides a fast thermal response. Another advantage of this sensor is good thermal isolation between the hot part of the sensor and the supporting structure, thus increasing the sensitivity and further improving the thermal response time. The sensor has a mechanical structure that makes it easy to mount. By using standard silicon process technology, silicon micromachining, and batch-fabrication, it is possible to produce low-cost identical small gas flow sensors. It is also possible to integrate electronics with the sensor, in order to control the hot part temperature and to form the output signal.

II. SENSOR DESIGN

The gas flow sensor consists of three main parts shown schematically in Fig. 1 and with photographs in Fig. 2. It has a base plate with five electrical bonding pads that connect the sensor to external circuits. A silicon beam, extending from the base plate, protrudes through a hole in

Manuscript received December 2, 1985; revised March 21, 1986.
The author is with the Department of Solid State Electronics, Chalmers University of Technology, S-41296 Gothenburg, Sweden.
IEEE Log Number 8609979.

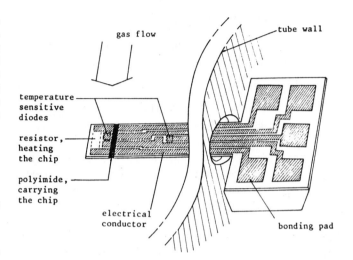

Fig. 1. The gas flow sensor with its three parts; the base plate (1.5 mm by 1.0 mm by 0.3 mm), the silicon beam (1.6 mm by 0.4 mm by 30 μm), and the chip (0.4 mm by 0.3 mm by 30 μm).

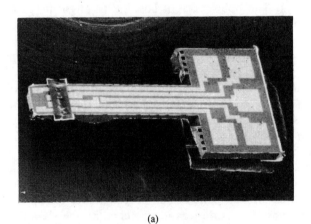

(a)

(b)

Fig. 2. SEM photographs of the (a) front side and (b) back side of the sensor, without and with the protective silicon frame.

Reprinted from *IEEE Trans. Electron Devices*, vol. ED-33, no. 10, pp. 1470–1474, October 1986.

the tube wall into the gas stream in which the flow is to be measured. In order to measure the beam temperature, a temperature-sensing diode is integrated on the silicon beam. At the far end of the silicon beam, four electrical conductors connect the active part of the sensor, a 0.3 mm by 0.4 mm by 30 μm chip, which is attached to the beam by a bridge of thermally insulating polyimide. This chip is heated electrically by dissipation in an integrated 100-Ω ion-implanted resistor. An ion-implanted diode similar to the one on the silicon beam is also integrated on the chip. The diode and the resistor on the chip, together with an external temperature regulator, control the chip temperature.

III. FABRICATION

The sensor was fabricated on a 2–6-$\Omega \cdot$ cm (100)-oriented n-type silicon wafer 225 μm thick and unpolished on the backside. The micromachining was achieved by using EDP (ethylenediamine, pyrocatechol, and water), an anisotropic silicon etch [6], [7]. The etch was held at 100°C. All masking of the silicon etch was done by silicon dioxide, which is very resistant to the EDP etch. After a 0.5-μm thermal oxide was grown, the first front side photolithographic process was performed, carefully aligned to the wafer flat, defining indicator holes that were etched through the wafer. The back side oxide was preserved to protect the back side of the wafer during silicon etching. The holes made it possible to align the pattern on the back of the wafer with the pattern on the front side. The back side pattern defined the area where the silicon was to be thinned down with the front side oxide preserved.

The first front side etch also resulted in an array of holes with different known depths shown in Fig. 3(a). These holes were used to determine the thickness of the membrane, shown in Fig. 3(b), which was etched next from the back side. Since the back side was unpolished, the oxide back side pattern had uneven edges. Due to the anisotropic etch characteristic of EDP and because the pattern was aligned parallel and perpendicular to the (110) directions, the oxide was underetched by the size of unevenness of the oxide edges, producing even edges of silicon. The thickness of the silicon membrance was controlled by inspection under a microscope, and by counting the number of front side depth references etched through. The etching of the membrane was not critical and resulted in a membrane thickness of 20–50 μm.

Next followed a series of standard process steps of photolithography and ion implantation forming the resistors, diodes, and the ohmic substrate contacts. The layout of the sensor is shown in Fig. 4. Chromium (40 nm) and gold (200 nm) were chosen as interconnection metals because of their resistance to the silicon etch used. As an intermediate step silicon dioxide was deposited to a thickness of approximately 1 μm on the back of the wafer. A short silicon etch (approx 1 h) on the front side finally defined the silicon beam and the active chip. The silicon etch used is orientation dependent, which means that it

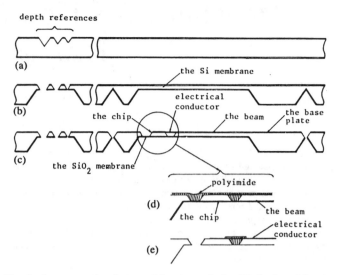

Fig. 3. A cross-sectional view of the sensor during fabrication. After the (a) first, (b) second, and (c) third silicon etch steps. (d) Shows the sensor after polyimide application and (e) after polyimide lithography and SiO_2 membrane etch.

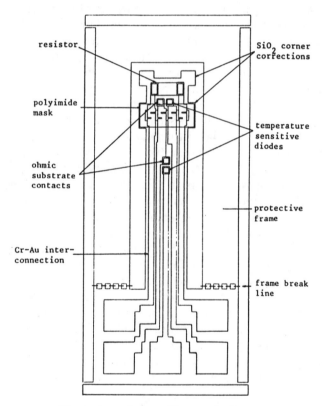

Fig. 4. Sensor layout with protective frame.

etches different crystallographic directions at different speeds. EDP etching of a (100) silicon wafer with a square hole in the silicon dioxide aligned with the (110) direction will produce a hole in the silicon with virtually no undercutting of the silicon dioxide due to the slow etch speed of the (111) directions that will constitute the side walls of the hole. A right-angle convex corner of silicon dioxide, however, similarily aligned, will when exposed to the etch present silicon directions that have a much faster etch speed. This results in an undercut of the silicon dioxide convex corners by the order of the hole depth. In order

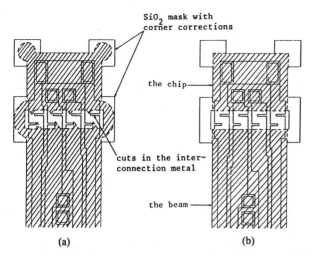

Fig. 5. Silicon etching using corner corrections in the silicon dioxide mask and a special metal pattern to form the chip and the beam (shaded areas represent silicon): (a) shows how the corner corrections and the metal bridges are under etched during the silicon etching and (b) is the final shape of the chip and the beam.

to make right-angle convex corners of silicon, the sensor's silicon dioxide etch mask had special corner corrections [6] everywhere a convex silicon corner was wanted on the silicon beam or on the chip as illustrated in Fig. 5. A special pattern in the metal layer was used together with the convex corner etch characteristic of EDP to create the undercut channel under the four chrome-gold conductors connecting the chip to the beam. The etch uncovered the oxide deposited on the back and resulted in the chip hanging on the thin oxide membrane. This oxide membrane had many cracks but still had the strength to carry the small chip as shown in Fig. 3(c). By using a microscope with a large working distance, it was easy to see when the chip and the beam had the desired form, without taking the wafers out of the etch.

Polyimide was spun on the front of the wafer at very low speed, filling the grooves and the space under the metal bridges, as illustrated in Fig. 3(d). Since the polyimide used is also a negative-working photoresist, a photolithographic step removed all polyimide except the one under (and over) the metal bridges, achieving the purpose of the polyimide, which was to join the beam to the chip and act as a thermal insulator. This is shown in Fig. 3(e).

Finally, the wafers were broken into individual components along the grooves that were etched through the wafers. The components had a frame of silicon to protect the active chip during handling. By breaking this frame, it was removed as a last step before the mounting of the sensor. See Fig. 2.

IV. THEORY

The heat transfer of a plate in a parallel flow [8] is generally represented by

$$Nu \equiv \frac{hL}{k_f} = C \cdot Re^n \qquad (1)$$

where Nu is the Nusselt number, L is the length of the

plate in the direction of the gas stream, h is the convective heat transfer coefficient of the fluid, k_f is the thermal conductivity of the fluid, Re is the Reynold's number, and C and n are constants depending on the physical properties of the fluid and the type of flow (laminar or turbulent). Reynold's number is proportional to the free stream velocity v by the relation

$$Re = \frac{\rho v L}{\mu} \qquad (2)$$

where ρ is the fluid mass density and μ is the viscosity of the fluid. Under steady-state conditions there is a balance between the input electrical power to the chip P_e and the sum of the losses due to convective cooling by the gas flow P_v and conduction to the supporting beam P_c.

$$P_e = P_v + P_c \qquad (3)$$

where

$$P_v = hA(T_{\text{chip}} - T_{\text{gas}}) \qquad (4)$$

$$P_c \sim k_s(T_{\text{chip}} - T_{\text{beam}}). \qquad (5)$$

P_v in (4) represents the loss due to convective cooling where A is the total area (both front and back sides of the chip) and $T_{\text{chip}} - T_{\text{gas}}$ is the temperature difference between the chip and the fluid. The conduction heat transfer P_c in (5), is proportional to the thermal conductivity of the supporting material k_s, and to the temperature difference between the chip and the beam. This results in an expression showing the relation between the electrical power dissipation and the flow velocity.

$$P_e = \frac{k_f CA(T_{\text{chip}} - T_{\text{gas}})}{L} \left(\frac{\rho v L}{\mu}\right)^n + P_c. \qquad (6)$$

A flat plate with a leading edge has, for all reasonable gas flow velocities and lengths L (i.e., $Re < 200\,000$), a laminar boundary layer, regardless of the nature of the gas flow [8]. This gives a value of $n = 0.5$. But since the sensor does not have a perfect leading edge and since the etched side of the sensor area may not be perfectly flat, a different value of n is possible.

If the temperature difference, $T_{\text{chip}} - T_{\text{gas}}$, between the chip and the flow is kept constant by using an external temperature regulator, and if it is assumed that the gas parameters (thermal conductivity, density, and viscosity) are constant (true for a limited gas temperature range), then it is possible to simplify (6) to

$$P_e = C_1 v^{0.5} + P_c \qquad (7)$$

where C_1 and P_c are constants. This equation can be used to obtain an expression for the gas flow sensitivity of the sensor

$$\frac{\frac{\partial P_e}{\partial v}}{P_e} = \frac{0.5 C_1 v^{-0.5}}{C_1 v^{0.5} + P_c}. \qquad (8)$$

If polyimide, instead of silicon, is used as supporting material between the chip and the beam, then the thermal conductivity loss P_c is reduced due to the lower thermal conductivity of polyimide (5). This will increase the gas flow sensitivity (8).

V. Measurements

The measurement were carried out with the sensors mounted in a tube of 5-mm inside diameter connected to the outlet, i.e., at atmospheric pressure, of a calibration flow meter. The sensor was mounted to provide a parallel flow over the front surface of the chip and with the tip of the sensor 1.6 mm from the tube wall, as illustrated in Fig. 1. The gas flow, air, was regulated at the inlet of the flow meter and the flow conditions were such that the chip was in the turbulent core of the flow where the velocity profile is relatively flat. This, together with the flow meter and knowing the tube cross area, made it possible to calculate the gas flow velocity at the sensor chip. Using a temperature feedback regulator control, the sensor chip was maintained at a constant elevated temperature while the power dissipated in the heater resistor was measured, i.e., the current and the voltage, for different gas flow velocities. The temperature characteristics of the sensor's temperature sensing diodes had been determined by use of an environmental chamber. The gas flow temperature was measured with a thermometer and with a diode while the sensor was unheated. Two types of sensors were used in order to investigate the improved performance of a sensor with a polyimide bridge: one type where the silicon between the beam and the chip was completely removed and replaced by polyimide, acting as supporting and insulating material, and a second type where approximately 50 percent of the silicon, between the chip and the beam, was left as supporting material.

In Fig. 6 the power dissipation versus flow velocity has been plotted for the two types of sensors, where the chips were regulated to 100°C above gas temperature (the gas temperature was approximately 23°C). Using (7), a least square fit, regarding P_c and C_1, to measured data was made. There is a good agreement in Fig. 6 to the square-law dependence assumed in (7). These calculations show also that the silicon bridge type of sensor has considerably, approximately 3.7 times, larger thermal conductive losses than the sensor with polyimide as bridging material, as shown in Fig. 6. An analysis shows that, since the polyimide thermal conductivity is approximately 2000 times smaller than for gold, some 90 percent of the thermal conductive losses, in the polyimide bridge sensor case, is lost through the metal part of the bridge. If then the temperature difference over the bridges and the bridge thermal resistances is considered, a thermal conductivity loss factor of 3.7 between the silicon bridge sensor and the polyimide bridge sensor seems reasonable. Due to the poor insulation between the chip and the beam of the silicon bridge sensor, the heating of the chip will also result in heating of the silicon beam to a temperature of some 92°C (measured by the diode integrated on the beam). This is to be compared to a beam temperature rise to 36°C

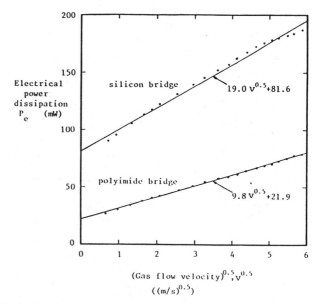

Fig. 6. Electrical power dissipation versus the square root of the gas flow velocity for the polyimide bridge sensor and the silicon bridge sensor. The least square fits to the data points are also plotted. The sensors were regulated to a temperature of 100°C above gas temperature.

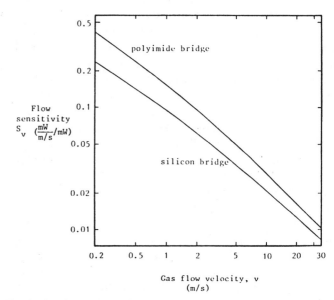

Fig. 7. Gas flow sensitivity versus gas flow velocity for the polyimide bridge sensor and the silicon bridge sensor, based on the least square fits of Fig. 6.

for the polyimide bridge sensor. The cooling of the heated beam in addition to the cooling of the heated sensor chip may be included in (6) as an increase in the cooling area, thus giving the silicon bridge sensor a larger equivalent value of C_1 than for the polyimide bridge sensor.

The sensitivity S_v for the two types of sensors, defined as: $(\partial P_e / \partial v)/P_e$, is plotted in Fig. 7 using the least square fit values of P_c and C_1 from Fig. 6. These plots show that the polyimide bridge sensor has a higher sensitivity than the silicon bridge sensor (more than twice the sensitivity for low gas flow velocities), due to its smaller thermal conduction losses.

Measurements of speed were also made by letting the resistor power dissipation have the form of a pulse and

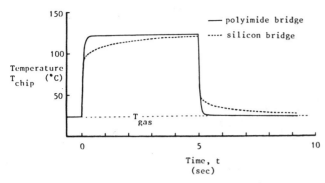

Fig. 8. Temperature responses from pulses of the electrical power dissipation for the polyimide bridge sensor and the silicon bridge sensor. The power applied in order to obtain the same temperature of 100°C above gas temperature was 130 and 45 mW for the silicon bridge sensor and the polyimide bridge sensor, respectively. The gas flow velocity was 1 m/s.

studying the temperature responses of the sensors. The pulse power to each sensor was adjusted to give the same temperature of approximately 100°C above gas temperature. Typical responses can be found in Fig. 8. The polyimide bridge type of sensor reached a steady-state value much faster than the silicon bridge sensor. When the pulse was applied, it took the polyimide bridge sensor 0.45 seconds to reach one degree centigrade from the steady-state value, while it took the silicon bridge sensor more than 15 seconds to reach the same temperature. Similarily, the cooling time, i.e. the time to reach one degree centigrade from gas temperature, was 0.45 seconds for the polyimide bridge sensor and 13 seconds for the silicon bridge sensor. The silicon bridge sensor has both a fast and a slow time constant, which gives the sensor a two-stage response, as shown in Fig. 8. The heating of the chip is represented by the fast time constant, while the slow time constant is due to the heating of the much larger beam mass. In the polyimide bridge sensor case, this beam heating is negligible, so the thermal response is determined by the fast time constant. A detailed analysis of the rising waveform of Fig. 8 yields a 50-ms time constant when fit with a simple exponential; however, final settling to within 1°C takes 0.45 s, which is longer than would be required for a purely exponential waveform. The fast switching time of the polyimide bridge sensor makes it suitable for use in a pulse-mode operation where the gas temperature is measured with the chip diode during the unheated period. Note that it is necessary to know the gas flow temperature in order to make compensation for fluctuations in gas temperature.

VI. Conclusions

A gas flow sensor fabricated by using micromachining techniques and silicon fabrication technology has been described. It presents very little obstruction to the gas flow, has a fast thermal response, a high sensitivity, is easily mountable, and can be mass produced using batch processing.

Acknowledgment

The author wishes to acknowledge Prof. J. T. Wallmark for many helpful discussions and for useful criticism of the manuscript.

References

[1] L. A. Rehn *et al.*, "Dual-element solid-state fluid flow sensor," *SAE Trans.*, vol. 89, pp. 705–710, 1980.
[2] J. H. Huysing, "Monolithic flow sensors, A survey," in *Proc. Vaste Stof Sensoren Conf.* (University of Delft, The Netherlands), pp. 39–48, Dec. 1980.
[3] A. F. P. van Putten and S. Middelhoek, "Integrated silicon anemometer," *Electron. Lett.*, vol. 10, pp. 425–426, Oct. 1974.
[4] R. W. M. van Riet and J. H. Huysing, "Integrated direction-sensitive flow-meter," *Electron. Lett.*, vol. 12, pp. 647–648, Nov. 1976.
[5] A. F. P. van Putten, "An integrated silicon double bridge anemometer," *Sensors and Actuators*, vol. 4, pp. 387–396, Nov. 1983.
[6] K. E. Bean, "Anisotropic etching of silicon," *IEEE Trans. Electron Devices*, vol. ED-25, pp. 1185–1193, Oct. 1978.
[7] K. E. Petersen, "Silicon as a mechanical material," *Proc. IEEE*, vol. 70, pp. 420–457, May 1982.
[8] J. R. Welty, *Engineering Heat Transfer*. New York: Wiley, 1978.

LIGHTLY DOPED POLYSILICON BRIDGE AS AN ANEMOMETER

Yu-Chong Tai and *Richard S. Muller*

Berkeley Integrated Sensor Center
An NSF/Industry/University Cooperative Research Center
Department of Electrical Engineering and Computer Science
and the Electronics Research Laboratory
University of California, Berkeley CA 94720

Abstract

A polysilicon bridge 200 μm long, 5 μm wide, and 1.5 μm thick having a lightly doped region 1.9 μm long at its center is used as a low-power anemometer. Experiments on the current-voltage characteristics and the flow-meter behavior are interpreted in terms of mechanical and electrical models. An analytical model of the electrical-conduction mechanism in polysilicon resistors at both low and high bias levels is derived to model performance. The heating power needed for the lightly doped polysilicon-bridge anemometer is only 2 mW for acceptable sensitivity at moderate flow rates.

I. Introduction

Conventional hot-wire anemometers made of metal filaments are usually bulky and power consuming. This research describes a new hot-wire polysilicon-bridge anemometer made by using a sacrificial-layer micromachining technique together with conventional silicon technology. The micromachining technique helps to reduce the size of the sensor; silicon technology provides precision and potential for economical production.

The feasibility of using polysilicon bridges as anemometers was demonstrated in our previous work[1]; however, the initial designs were constructed of heavily doped polysilicon and not optimal. We found that the uniformly doped polysilicon bridge has a strong thermal coupling to the silicon substrate. This coupling lowers the heating efficiency that at least 5 mW is needed for a signal in the order of a few milivolts.

In this paper we present an improved design, the new polysilicon-bridge structure shown in Fig. 1. This bridge is lightly doped at the center, but the rest parts remain heavily doped. The heavily doped sections provide thermal isolation but a low resistance path for electrical conduction. The polysilicon-bridge anemometer is 200 μm long, 5 μm wide, and 1.5 μm thick. The central lightly doped region is 1.9 μm long [Appendix A]. The offset of the polysilicon bridge from the substrate is 3 μm.

II. Fabrication

Figure 2 shows an SEM photograph of a fabricated polysilicon bridge. The bridge fabrication process starts with 4-inch silicon wafers coated with 3 μm-thick LPCVD phosphosilicate glass (PSG). This PSG is used both as the sacrificial layer and as the phosphorus-diffusion source. After the PSG deposition, a 40 nm thick LPCVD nitride layer is deposited and patterned. The nitride acts as a diffusion barrier for phosphorus and thus defines the lightly doped region at the center of the polysilicon bridge. This technique allows us to calibrate lateral diffusion in polycrystalline silicon [Appendix A], a process that has not previously been reported upon. A 1.5 μm thick, lightly doped polysilicon layer is then deposited and patterned. This is followed by a high-temperature phosphorus-diffusion step, which also anneals out intrinsic stress in the deposited polysilicon. Conventional aluminum metalization follows and a time-controlled PSG etch in 5:1 buffered hydrofluoric acid frees the bridge. The PSG etch is stopped when the polysilicon bridge is freed completely, while the two supports for the polysilicon bridge are only slightly undercut because of their large sizes. Oxygen plasma etch then cleans away the photoresist and completes the process.

III. Electrical Conduction in Polysilicon Resistors

Since ohmic power is used to heat the hot-wire polysilicon-bridge anemometer, an adequate current-voltage model is needed for theoretical study. Unfortunately, existing theories for the current-voltage behavior derived from carrier-trapping models [2-7] apply only to low-bias cases, but a polysilicon-bridge anemometer is always used in high-bias regions. Therefore, we derive a new model for application in the high bias regime in the following.

In order to consider the high-bias case, a close look at the biased band diagram around a grain boundary is necessary (Fig. 3). According to thermionic emission theory [8], The net current across the grain boundary is

$$I = I_0 \, exp(-\frac{E_{a0}}{kT}) \, [exp(\frac{qV_L}{kT}) - exp(-\frac{qV_R}{kT})], \qquad (1)$$

here I_0 is a proportional constant, and E_{a0} is the activation energy at zero bias. Equation (1) is applicable to any bias, but V_L and V_R must be expressed in term of the applied voltage first.

To express V_L and V_R, the carrier-trapping model is adopted, and uniform grain size and grain-boundary width, W_{gb}, are used. We first write that the sum of the voltage components, $V_L + V_{gb} + V_R$, equals the voltage drop across a single crystallite V_g,

$$V_g = \frac{V}{N_g} = V_L + V_{gb} + V_R, \qquad (2)$$

where V_L and V_R are the voltage drops across the left and the right depletion regions respectively, and V_{gb} is the voltage across the grain boundary. A second equation is that the charge trapped in the grain boundary equals the charge in the depletion regions.

$$N_t W_{gb} = N_B (\lambda_L + \lambda_R), \qquad (3)$$

where N_t is the grain-boundary trapped charge density, N_B is the doping concentration in the crystallites, and λ_L and λ_R are depletion widths at the left and right sides respectively. The trapped charges in the grain boundary are assumed to be uniformly distributed. The rest equations then come from the Poisson equation at a grain boundary with the assumption of abrupt junction. By integrating the Poisson equation twice, we derive the voltages defined in Fig. 3 in terms of λ_L and λ_R. Using Eqs. (2) and (3), we further eliminate λ_L and λ_R and write V_L, V_R, and V_{gb}

Reprinted with permission from *Transducers '87, Rec. of the 4th Int. Conf. on Solid-State Sensors and Actuators*, 1987, pp. 360–363.

$$V_L = \frac{1}{2}\xi V_g - \frac{1}{16 V_{B0}}\xi^2 V_g^2 \qquad (4.1)$$

$$V_R = \frac{1}{2}\xi V_g + \frac{1}{16 V_{B0}}\xi^2 V_g^2 \qquad (4.2)$$

$$V_{gb} = (1-\xi)V_g \ . \qquad (4.3)$$

V_{B0} and ξ are defined by

$$V_{B0} = \frac{qN_t^2 W_{gb}^2}{8\,\varepsilon_{Si}N_B}, \qquad \xi = \frac{\varepsilon_0 N_t}{(\varepsilon_{Si}N_B + \varepsilon_0 N_t)},$$

where ε_{si} is the permittivity of silicon, and ε_0 is the permittivity in vacuum. Substituting Eq. (4) into Eq. (1), we derive Eq. (5) which is applicable to any bias,

$$I = I_0 \exp(-\frac{E_{a0}}{kT})\exp(-\frac{q\xi^2 V_g^2}{16 V_{B0}kT})\times 2\,\sinh(\frac{q\xi V_g}{2kT}). \quad (5)$$

Two features of Eq. (5) are important. One resides in the existence of coefficient ξ which arises from considering voltage drops across the grain-boundary and the depletion regions. The voltage drop across the two depletion regions is this coefficient times the total grain-voltage drop -- ξV_g, which therefore appears in the argument for the hyperbolic sine in Eq. (5). Another feature is the extra exponential term involving $(\xi V_g)^2$. The voltage drop in the depletion regions is not equally shared between V_L and V_R; V_L is always less than V_R as shown in Eq. (4). The extra exponential term comes from the different term between V_L and V_R in Eqs. (4.1) and (4.2).

IV. Heat-Flow Model

For a hot-wire anemometer, the steady-state relationship between power generation and heat dissipation can be written [9]

$$Power = IV = K_{eff}(v)(T - T_a), \qquad (6)$$

where T is the thermistor temperature, T_a is the ambient temperature, v represents the flow velocity, and $K_{eff}(v)$ is an effective heat-dissipation constant. Since a boundary-layer flow will be established for a polysilicon bridge on a 3 by 3 mm^2 die in a laminar-flow wind tunnel, the heat-dissipation constant can be written

$$K_{eff} = K_0 + K_1\sqrt{v}, \qquad (7)$$

where K_0 consists of the thermal conduction of the bridge, the thermal conduction of the air, the thermal radiation, and the natural convection. These terms are assumed not to be a function of flow velocity. $K_1\sqrt{v}$ represents the boundary-layer forced convection. The $\sqrt{v}$ dependence is the characteristic of a laminar-flow forced convection. This dependence is the same as that used in [1,10].

V. Experimental Results and Theoretical Interpretation

In this section, we interpret the experimental results using Eqs. (5), (6), and (7). There are six parameters -- I_0, E_{a0}, ξ, V_{B0}, N_g, and K_0 -- to be determined (K_1 is to be determined afterwards). To do this, experiments in a closed no-flow hot-chuck system and in the open air are performed. The hot-chuck experiment is done by measuring the currents at a constant bias voltage with variable hot-chuck temperatures; The open-air experiment is done at room temperature by measuring the voltages with variable currents. Iterations then determine these unknowns so that they simultaneously fit the measured data. The procedure is as follows. We first choose a value for K_0. The temperature of the central region in the bridge is then calculated from Eq. (6)

$$T = \frac{IV}{K_0} + T_a, \qquad (8)$$

where T_a is the measured hot-chuck temperature, and zero flow velocity is used. With the calculated data, activation energies

with different applied voltage are extracted and compared to the theoretical values. The theoretical activation energy is derived from Eq. (5)

$$E_{a,eff} = -\frac{\partial lnI}{\partial(1/kT)}$$

$$\approx E_{a0} - \frac{q\xi V_g}{2} + \frac{q\xi^2 V_g^2}{16 V_{B0}} \quad if \quad \frac{\xi V_g}{2} \gg kT/q. \quad (9)$$

By fitting Eq. (9) to the extracted activation energies, we can obtain E_{a0}, ξ, V_{B0}, and N_g; however, ξ is calculated from Eq. (4) assuming that W_{gb} is 0.7 nm [7] and the crystallites are totally depleted.

Knowing E_{a0}, ξ, V_{B0}, and N_g, we use Eq. (5), (6), and (7) to further fit the V-I data measured in the open air. I_0 and K_0 are new fitting parameters at this stage. This new K_0 then is substituted back for the next iteration until both parameters converge. Table 1 lists all the final iterative values.

Table 1

Major Parameters		
Notation	Value	Remarks
E_{a0}	0.305 eV	Activation energy at zero bias.
ξ	0.954	Defined in Eq. (4).
V_{B0}	0.2 eV	Zero-biased grain-boundary potential barrier.
N_g	11.3	Effective number of grains.
I_0	3.15×10^{-3} A	Used in Eq. (5).
K_0	1.12×10^{-5} W/ °C	Used in Eq. (7).
K_1	6×10^{-8} W/ °C	Used in Eq. (7).
W_{gb}	0.7 nm	Grain-boundary width [7].
T_a	21 °C	Ambient temperature.

The final results are plotted in Figs. 4, 5, and 6. The solid curves in these figures are theoretical calculations from Eqs. (5), (6), and (7). In Fig. 6, the I-V characteristics in the open air show a negative incremental resistance region when the driving current is larger than 220 μA. In this region the increase in bridge temperature, due to the increase of driving current, reduces the voltage required for a given current because of the exponential terms involving temperature in Eq. (5).

The calibration of the polysilicon-bridge anemometer is performed in a wind tunnel calibrated by a commercial hot-wire anemometer (TSI 1650) [11]. A constant-current configuration is used to evaluate the anemometer theory. The voltage changes are induced by variations in the air flow velocity. The electrical responses due to step-velocity variations have exponential-like characteristics. Figure 7 shows the measured and theoretical differential signals due to velocity variations. Solid curves represent the theoretical calculations using K_1 as the fitting parameter. Note that the output characteristics fall quite close to each other even with different driving currents. In other words, the differential output voltage is insensitive to the driving currents.

The speed of response of the anemometer is also examined. Since the step response nearly follows an exponential characteristic, a time constant, τ, is therefore defined as the time needed for a 63.2% output change. Figure 8 shows the measured time constants at different step velocities. The specific driving current of this measurement is 350 μA. In Figure 8, τ^{-1} is plotted against the square root of the step velocities. Experimentally, we find that τ is inversely proportional to $\sqrt{v}$.

VI. Conclusions

A 200 µm long, 5 µm wide, and 1.5 µm thick polysilicon-bridge with an estimated 1.9 µm, lightly doped, central region has been fabricated and its application to an anemometer has been studied. Good thermal isolation to the substrate of the lightly doped region reduces the ohmic power needed for anemometer operation. As an example, less than 2 mW is needed for a differential signal readout of 30 mV at a flow velocity of 1.5 m/s. The inverse of the time constant is found to be a linear function of the square root of velocity. A new electrical conduction model of the polysilicon resistors is also presented. The electrical conduction model is applicable to any bias.

Appendix A: Lateral Diffusion of Phosphorus in Polysilicon

The measurement of lateral diffusion in polysilicon is described here. In general, the resistance of the polysilicon resistor R is

$$R = R_\square \frac{(L_{mask} - 2L_{LD})}{W} , \qquad (A.1)$$

where $R_\square$ is the sheet resistance, L_{mask} is the mask length, L_{LD} is the lateral diffusion length, and W is the width of the lightly doped region. From Eq. (A.1), if we plot R against L_{mask}/W, we obtain a straight line having an intersection with the axis where L_{mask} equals $2L_{LD}$. Experimentally, polysilicon resistors having a range of W and L_{mask} values are fabricated. These polysilicon resistors are 1.5 µm thick and the phosphorus-diffusion step is performed at 1000 °C for one hour. The resistances at low bias are measured and the results are shown in Fig. 9. The solid lines in Fig. 9 are least-mean-square fits to the data and the lateral diffusion lengths for different widths are extracted from their slopes. Table A1 lists the extracted values of L_{LD} for different widths.

Table A1

W (µm)	3	5	10	15	20
L_{LD} (µm)	4.226	4.310	4.116	4.045	3.552

From the table, we calculate an average lateral diffusion length of 4.05 µm. If we assume the doping profile has a complementary error function distribution, the lateral diffusion constant of phosphorus in LPCVD polysilicon is 2.56×10^{-12} cm^{2}s^{-1} at 1000 °C. Since the mask length is 10 µm for the polysilicon-bridge anemometer, the remaining length of the lightly doped region is 1.9 µm.

References

[1] Y.C. Tai, R.S. Muller, and R.T. Howe,"Polysilicon bridges for anemometer application", *Digest of Technical Papers, International Conference on Solid-State Sensors and Actuators*, Philadelphia PA, June 4-7, 1985, pp.354-357.

[2] G.J. Korsh and R.S. Muller,"Conduction properties of lightly doped polycrystalline silicon," *Solid-State Electron., 21*, 1978, pp. 1045-1051.

[3] J.Y.W. Seto,"The electrical properties of polycrystalline silicon," *J. Appl. Phys., 42*, 1971, pp. 5247-5254.

[4] T.I. Kamins,"Hall mobility in chemically deposited polycrystalline silicon," *J. Appl. Phys., 46*, 1971, pp. 4357-4365.

[5] N.C. Lu, L. Gerzberg, C. Lu, and J.D. Meindl,"Modeling and optimization of monolithic polycrystalline silicon resistors," *IEEE Trans. Electron Devices, ED-28*, July 1981, pp. 818-830.

[6] M.M. Mandurah, K.C. Saraswat, and T.I. Kamins,"A model for conduction in polycrystalline silicon -- part I: theory," *IEEE Trans. Electron Devices, ED-28*, Oct. 1981, pp.1163-1171.

[7] M.M. Mandurah, K.C. Saraswat, and T.I. Kamins,"A model for conduction in polycrystalline silicon -- part II: comparison of theory and experiment," *IEEE Trans. Electron Devices, ED-28*, Oct. 1981, pp.1171-1176.

[8] S.M. Sze, *Physics of Semiconductor Devices*, Wiley, New York, 2nd edition, 1981, p.256.

[9] E.D. Macklen, *Thermistors*, Electrical Publications Limited, 1979, p. 39.

[10] G.N. Stemme,"A Monolithic Gas Flow Sensor with Polyimide as Thermal Insulator," *IEEE Trans. Electron Devices, ED-33*, Oct. 1986, pp.1470-1474.

[11] D.L. Polla, R.S. Muller, and R.M. White,"Monolithic integrated zinc oxide on silicon pyroelectric anemometer,", *Technical Digest*, International Electron Devices Meeting(IEDM), Dec. 1983, pp. 639-641.

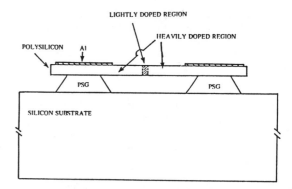

Figure 1   Cross section of a lightly doped polysilicon bridge.

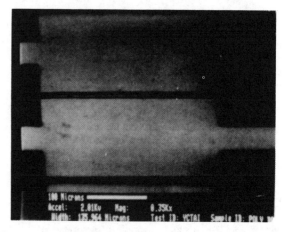

Figure 2   SEM photograph of a lightly doped polysilicon bridge. The bridge is 200 micrometer long, 5 micrometer wide, and 1.5 micrometer thick.

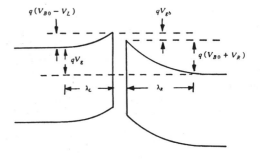

Figure 3   Band diagram at a grain boundary under bias.

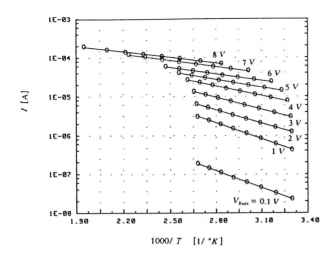

Figure 4   Experimental results of the current-temperature measurement in a no-flow hot chuck. The solid lines are L.M.S. fits.

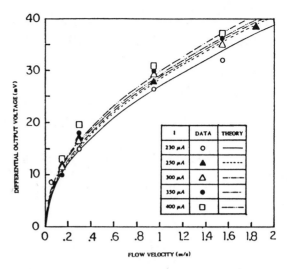

Figure 7   The differential-voltage flowmeter characteristic of the polysilicon-bridge anemometer.

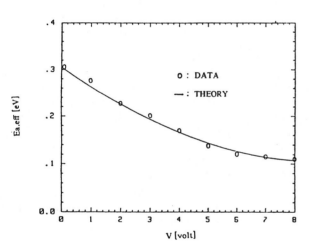

Figure 5   Experimental results of the activation energy versus bias voltage.

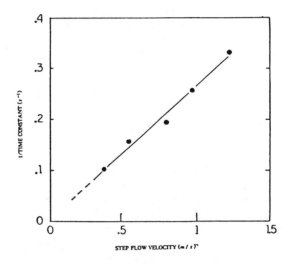

Figure 8   The time-constant measurement. The inverse of the measured time constant is plotted against the square root of the step flow velocity.

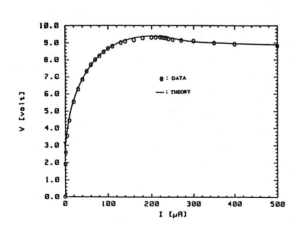

Figure 6   The current-voltage characteristic of the polysilicon-bridge anemometer in still air.

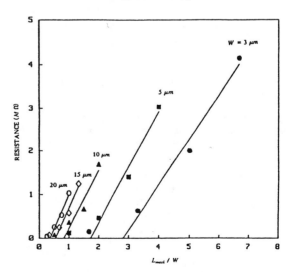

Figure 9   Experimental results of the low bias resistance versus L_{mask}/W.

Section 3.3: Sensitive Films

PYROELECTRIC PROPERTIES AND APPLICATIONS OF SPUTTERED ZINC-OXIDE THIN FILMS

D.L. Polla, R.S. Muller, and R.M. White

Department of Electrical Engineering and Computer Sciences
and the Electronics Research Laboratory
University of California, Berkeley, California 94720

ABSTRACT

Thin films of zinc oxide that are active piezoelectrics also exhibit a pyroelectricity comparable to that of bulk zinc oxide. The pyroelectric effect can be used in a number of sensors in which a temperature variation is related to the quantity sensed. A multi-sensor chip made to explore a number of these applications is described. Experimental performance data is given for an infrared detector array, mass fluid-flow sensor, and a chemical-reaction sensor.

INTRODUCTION

The use of sputtered zinc oxide (ZnO) films in SAW television filters is well known, and such films are beginning to find applications in integrated SAW sensors. In this paper we discuss sensor applications of the *pyroelectric* effect of thin-film zinc oxide, made by the same technique employed to produce *piezoelectric* zinc-oxide films. Monolithic pyroelectric sensors are now possible, and one could even consider making multifunction integrated sensors that employ both the pyroelectric and piezoelectric properties of zinc oxide.

PYROELECTRICITY IN ZINC OXIDE

The pyroelectric coefficient p^σ is defined as the differential change in polarization with temperature under constant stress (tensor σ) and low electric field

$$p^\sigma = \left. \left(\frac{\partial P}{\partial T} \right) \right|_{\sigma, E = 0} \qquad (1)$$

The measured pyroelectric coefficient as a function of temperature is shown in Figure 1. For a 1.0 μm-thick ZnO film with a resistivity of $2.9 \times 10^7 \Omega$-cm, the room-temperature value of p^σ is 1.4×10^{-9} C cm^{-2}K^{-1} [1] which can be compared to a value of 8×10^{-10} C cm^{-2}K^{-1} reported for bulk ZnO [2]. Since typical ZnO film thicknesses are approximately a factor of 100 smaller than the Debye screening length for films of this resistivity, extremely long charge retention times have been observed. By encapsulating the ZnO film in thin layers of *CVD* silicon dioxide, a characteristic e^{-1} charge decay time in excess of 32 days has been measured (Figure 2). For typical ZnO sensors measuring 70μm $\times$ 70μm $\times$ 1.0μm that were used for pyroelectric sensing in this work, a temperature change ΔT of 1 degree induces a pyroelectric voltage of 0.153 V for a 298 K ambient temperature.

Three monolithic pyroelectric sensors will now be described: infrared detector array, anemometer, and chemical-reaction sensor.

PYROELECTRIC INFRARED-DETECTOR ARRAY

The pyroelectric effect in thin films of ZnO is utilized in a 64-element infrared detector array [3], made in a compatible process with NMOS integrated circuits. Radiation absorbed by a thin-silicon membrane detector alters the internal polarization of the oriented pyroelectric film, and induces a surface charge that is detected by an on-chip MOSFET amplifier.

The fully-integrated structure consists of 1 μm-thick sputtered ZnO pyroelectric capacitors in an 8×8 array, with each sensing element measuring 70×70 μm^2 and encapsulated on both sides with

Reprinted from *Rec. of the IEEE Ultrasonics Symp.*, 1985, pp. 495–498.

CVD silicon dioxide. The back-side electrode is formed of phosphorus-doped CVD polycrystalline silicon. The top-side electrode (which is exposed to infrared radiation) is formed either by thin-film sputtering of tin-oxide or vacuum evaporation of chromium. The entire structure is supported on a 25μm-thick membrane of silicon (Figure 3).

In each sensor element, a one degree temperature change induces a measured pyroelectric charge of 74 pC, equivalent to 153 mV across the capacitor.

Each array element is connected directly to the gate of an n-channel, depletion-mode MOSFET amplifier with a 3 μm channel length and $W/L = 23$. The MOSFET drain is connected to an ion-implanted resistive load of 1 kΩ. This MOSFET stage has a gain of -10 dB and serves as a buffer to convert pyroelectric charge into an output voltage.

Figure 4 shows the response of an array element with a chromium electrode versus chopping frequency f using a calibrated radiant flux source (1.89×10^{-5}W cm^{-2}) at T=298 K. At f= 24 Hz, the voltage responsivity (output voltage/incident radiant power) is 4.3×10^4VW^{-1}. The measured blackbody detectivity D^* under these measurement conditions is $= 3.1 \times 10^7$cm$\sqrt{\text{Hz}}$ W^{-1}. This value is approximately a factor of 2 to 5 below that obtain by commercial *hybrid, room temperature* infrared detectors. The high-frequency slope associated with the data in Figure 4 indicates a composite detector time constant of 0.74 s.

The detector time constant was also measured by a focused helium-neon laser $\lambda = 0.6328\mu$m. The time constant associated with the measured rise- and fall-time is approximately 0.78 s, in excellent agreement with the high-frequency voltage responsivity data. Other characterization data are summarized in Table I.

PYROELECTRIC ANEMOMETER

Thin films of pyroelectric ZnO have been combined with NMOS planar technology to produce a fully integrated anemometer [1]. The anemometer is based on the differential cooling rate of two temperature sensors placed up- and down- stream from a heat source. The sensors consist of two μm-thick ZnO pyroelectric capacitors, encapsulated in silicon dioxide. Sense electrodes on the capacitors are connected directly to differentially coupled depletion-mode MOSFET pairs.

A resistive heater is located in the center of the chip between heat-sensitive zinc oxide capacitors at opposite chip edges. Fluid flows parallel to the chip surface, establishing a thermal boundary layer with increasing thickness in the direction of flow (Figure 5).

Flow measurements were carried out in a small (4 cm diameter) wind tunnel utilizing a N$_2$ gas source instrumented with a commercial hot- wire anemometer. Figure 6 shows the measured differential output voltage versus flow velocity with an overall circuit gain of 10.7. The output voltage was 184 mV (S/N ratio = 200) at a velocity of 2 ms^{-1}. A dc offset voltage of 15 mV, measured with no flow, has been subtracted out of the data in Figure 5. Further characterization results are summarized in Table II.

INTEGRATED CHEMICAL-REACTION SENSOR

Using silicon micromachining and pyroelectric thin-films, an integrated sensor capable of measuring heats-of-reaction has been built and tested. The sensor detects the heat exchanged in the chemical reaction of gas molecules on a solid surface. The sensor is composed of a thin film of metal supported on an etched silicon membrane coated with a one-μm -thick pyroelectric zinc-oxide thin film. Heat liberated in a chemical reaction at the surface metal film induces surface charge that is detected by an on-chip MOSFET amplifier.

The sensor has measured the heat liberated in the chemisorption of carbon-monoxide gas at the surface of platinum. For this application, the measured signal-to-noise ratio indicates sensitivity to temperature variations as small as 180μK.

Figure 7 shows the effect of introducing CO into a vacuum initially pumped to 2×10^{-8}Torr. For a constant CO partial pressure of 10^{-4} Torr, the maximum output voltage of 22 mV is reached after 15 s. The voltage decays subsequently to zero as CO molecules are chemisorbed on all available Pt bonding sites. From Figure 7, we can calculate the temperature change occurring in the ZnO thin film. Considering the maximum output signal of 22mV together with the ZnO capacitance and the MOSFET amplifier gain, we calculate the induced polarization charge as 5×10^{-14} C. Considering the pyroelectric coefficient measured for ZnO thin films [1], and the sensor area, the pyroelectric induced charge can be related to the change in temperature

$$\Delta T = \frac{Q_P}{p^{\sigma}A_d} = 18\text{mK} \qquad (2)$$

From the noise observed in this experiment, we calculate the lower limit for temperature sensitivity to be

$$\Delta T = \frac{C_d V_{s(min)}}{p^{\sigma}A_d} = 180\,\mu\text{K} \qquad (3)$$

where $V_{s(min)}$ is the output signal at which the signal-to-noise ratio is unity. Table III summarizes the properties of integrated chemical reaction sensor.

MULTI-SENSORS

An integrated multi-sensor chip that responds to several different physical and chemical variables has been fabricated [5]. The chip contains conventional MOS devices for signal conditioning along with the following on-chip ZnO pyroelectric sensors: a gas-flow sensor, an infrared-sensing array, infrared charge-coupled device imager, and a chemical-reaction sensor. Integrated sensors making use of the piezoelectric properties of the ZnO films have also been incorporated on this same chip. These include cantilever-beam accelerometers, surface-acoustic-wave (SAW) vapor sensors, and a tactile-sensor array. Although this chip which contains seven different sensors has not been designed for a specific multi-sensing application, the technology utilized in its fabrication is sufficiently flexible for use in a multi-sensing environment.

CONCLUSIONS

RF-planar-magnetron sputtered zinc oxide thin films, fabricated using the same parameters as used in films for ultrasonic devices exhibit substantial pyroelectric activity. Among the thermal sensors that have been investigated are a monolithic infrared-imaging array, a mass-flow gas sensor, and a chemical heat-of-reaction sensor. A multi-sensor chip that includes both piezoelectric and pyroelectric devices has been demonstrated.

Acknowledgements. Research supported in part by the National Science Foundation (grant ECS 81-20562, Dr. F. Betz, technical monitor), and in part by the State of California MICRO program. Helpful laboratory support by O. Lau is gratefully acknowledged.

References

[1] D.L. Polla, R.S. Muller, and R.M. White, "Monolithic Integrated Zinc-Oxide on Silicon Pyroelectric Anemometer," *IEEE Int. Electron Devices Meeting*, 639-642, Washington D.C. (December 1983).

[2] G. Heiland and H. Ibach, "Pyroelectricity of Zinc Oxide," Solid State Comm. 4, 353-356 (1966).

[3] D.L. Polla, R.S. Muller, and R.M. White, "Fully Integrated ZnO on Silicon Pyroelectric Infrared Detector Array," *IEEE Int. Electron Devices Meeting*, 382-384, Washington D.C. (December 1984).

[4] D.L. Polla, R.M. White, and R.S. Muller, "Integrated Chemical-Reaction Sensor," *IEEE Int. Conference on Solid-State Sensors and Actuators*, 33-35, Philadelphia, PA (June, 1985).

[5] D.L. Polla, R.S. Muller, and R.M. White, "Integrated Multi-Sensor Chip," to be published.

TABLE I: IR DETECTOR ARRAY PARAMETERS

Pyroelectric Coefficient	$1.4 \times 10^{-9} \mathrm{Ccm^{-2}K^{-1}}$
Dielectric constant	10.3
Detector Element Area	$4.9 \times 10^{-5} \mathrm{cm^2}$
ZnO Film Thickness	1.0 μm
Resistance	$2.9 \times 10^{12} \Omega$
Capacitance	447 fF
Voltage Responsivity	$4.3 \times 10^4 \mathrm{V\ W^{-1}}$
Noise-Equivalent Power (f=24 Hz)	$2.3 \times 10^{-10} \mathrm{WHz^{-1/2}}$
Time constant	0.76 s

TABLE II: PYROELECTRIC ANEMOMETER PARAMETERS

Pyroelectric Sensor Length	3990 μm
Pyroelectric Sensor Width	190 μm
ZnO Film Thickness	1.9 μm
Chip Temperature	311 K
Typical Responsivity (1m/s)	93 mV
Response time	1.4 s

TABLE III: CHEMICAL REACTION-SENSOR

Pyroelectric Coefficient	$1.4 \times 10^{-9} \mathrm{Ccm^{-2}K^{-1}}$
ZnO Film Thickness	1.0 μm
Sensor-Element Area	$4.9 \times 10^{-5} \mathrm{cm^2}$
Sensor-Element Resistance	$2.9 \times 10^{11} \Omega$
Sensor-Element Capacitance	450 fF
Typical Output Voltage	22 mV
Minimum Detectable Temperature	180 μK

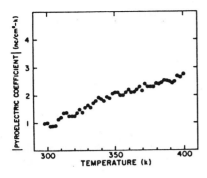

Fig. 1. Measured pyroelectric coefficient as a function of temperature for a 1.0μm-thick ZnO film.

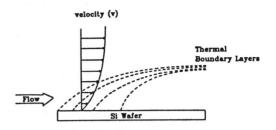

Fig. 5. Thermal boundary layer established over the surface of the silicon chip due to incident fluid flow of differing temperature than that of the chip.

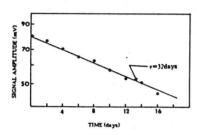

Fig. 2. Static charge decay time associated with a dielectrically encapsulated ZnO thin film.

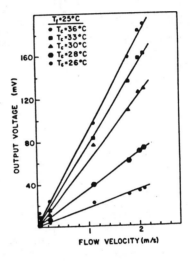

Fig. 6. Differential output voltage of the pyroelectric anemometer versus flow velocity measured at different chip temperatures.

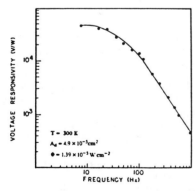

Fig. 3. Cross-section of a discrete infrared detector. The total membrane thickness is approximately 25μm.

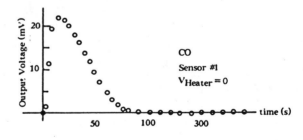

Fig. 7. Chemical reaction sensor voltage response to carbon monoxide.

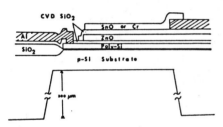

Fig. 4. Voltage responsivity versus frequency for a pyroelectric detector with a chromium sensor electrode.

TIN OXIDE MICROSENSORS ON THIN SILICON MEMBRANES

Shih-Chia Chang
David B. Hicks
Electronics Department
General Motors Research Laboratories
Warren, Michigan 48090-9056

Abstract

Tin oxide based microsensors with integrated polysilicon heaters were fabricated on thin silicon membranes (~2 μm thick) generated by anisotropic wet chemical etching using ethylene diamine-pyrocatechol (EDP) as the etchant. Good thermal isolation and lower power consumption were achieved. The sensing elements, tin oxide thin films, were prepared either by sputter-deposition or by metallo-organic deposition (MOD). The completed microsensors showed good response to alcohol vapor and good stability of the integrated polysilicon heaters.

Introduction

Semiconductor gas sensors generally have to be operated at elevated temperatures (150°C-700°C) to attain the speed and magnitude of response required for practical applications [1-3]. Consequently, for the development of portable gas sensors or integrated sensing systems with on-chip control/logic circuits, effective thermal isolation for the active sensing area becomes imperative in order to keep the power consumption low and to protect adjacent circuitry from extreme temperature generated by the sensor heater. One way of attaining low power consumption and good thermal isolation is by fabricating a sensing element with an integrated heater in thin film form on a suspended thin membrane. By doing this, both the heating mass and heat transport path are greatly reduced. Silicon membranes can be readily generated by an anisotropic wet chemical etching technique using either potassium hydroxide (KOH) or ethylene diamine-pyrocatechol (EDP) as the etchant [4,5]. Such a technique has been used to fabricate piezoresistive or piezocapacitive pressure transducers [6,7]. When using EDP, precise control of the membrane thickness can be achieved by using a heavily doped boron layer as the etch-stop.

The main purpose of this work is twofold: a) to fabricate tin oxide-based microsensors on micromachined thin membranes to attain good thermal isolation, and b) to evaluate sputter-deposited and metallo-organic deposited tin oxide thin films.

This paper describes the detailed processing procedures used to fabricate these microsensors and also presents the various sensor responses to reducing gaseous species (propylene, vapors of alcohol, ethyl ether, methoxyflurane and halothane).

Experimental

1. Device Fabrication

A five-mask process was used to fabricate tin oxide thin film microsensors on silicon membranes.

Most of the processing steps are similar to those used in our previous work [8]. The major additions and modifications adopted in the current fabrication process are described below:

a) The formation of a thin silicon membrane

A silicon membrane with a thickness of ~2 μm was formed by anisotropic chemical etching using EDP. EDP has a fast etch rate (~80 μm/h) in the Si <100> direction, a relatively slower etch rate (<10 μm/h) in the Si <111> direction, a very slow etch rate (<20 nm/h) for SiO_2, and a close-to-zero etch rate for heavily boron doped silicon. Thus, a heavily boron doped layer (doping concentration $\geq 5 \times 10^{19}$ cm^{-3}) generated by ion implantation (dosage $\sim 5 \times 10^{16}$ cm^{-2}, implant voltage ~200 keV) was used as the etch-stop.

b) The formation of the sensing element

Tin oxide thin films were prepared either by sputter-deposition [3] or by metallo-organic deposition (MOD) [9]. For the sputter-deposited samples, the sensing element processing steps are the same as those used in our previous work. For the MOD technique, an ink was prepared by dissolving tin (II) 2-ethylhexanoate in xylene. The ink was spun onto a silicon wafer, fired to form a 100 nm to 200 nm tin oxide film, and subsequently patterned either by reactive ion etching or by wet chemical etching.

c) The formation of metal interconnects

An aluminum (Al)/chromium (Cr) double layer with a thickness of 1 μm/50 nm was used to form metal interconnects for the tin oxide as well as the polysilicon heater. Chromium not only provides good ohmic contact for tin oxide, it also prevents electromigration of the interconnect metals at the heater contacts. Severe electromigration was observed if only Al was used as the contact metal, and would result in poor contact or premature contact failure. The processes used to form the MOD tin oxide metal interconnects had to be modified due to the high reactivity of the MOD films. The most commonly used process for metal interconnect fabrication consists of the following steps: a) Deposition of metal(s) on top of the material to be electroded, b) photolithographically defining the interconnect features into a photoresist layer, and c) transferring the defined feature into the underlying metal layer by wet chemical etching. However, when the above procedure is used, two problems are encountered:

(1) During wet chemical etching of Al, active hydrogen (H*) is produced at the metal/MOD tin

Reprinted from *Rec. of the IEEE Solid-State Sensors Workshop*, 1986.

oxide interface according to the following general equation:

$$Al + H(X) \longrightarrow Al(X) + H* \qquad (1)$$

where H(X) is an etchant composed of H_3PO_4, HNO_3, and CH_3COOH. The MOD tin oxide is readily reduced to tin by H*,

$$SnO_2 + 4 H* \longrightarrow Sn + 2H_2O \qquad (2)$$

and is removed by the etchant, resulting in a bad (or open) tin oxide contact.

(2) In photolithographic processes for the patterning of the metal interconnect feature, the weak chemical reaction between the hydroxide-based positive resist developer and Al produces active hydrogen

$$Al + 3 (OH) \longrightarrow Al(OH)_3 \longrightarrow AlO_3^- + H_3O^+ \qquad (3)$$

The reduction of SnO_x by the active hydrogen as expressed by Eq. 2 at the metal/SnO_x interface degrades the metal contact of MOD tin oxide film.

Two approaches were used to overcome the MOD tin oxide/metal contact problems: a) by using a lift-off technique [10]; b) by using an inverse metallization technique. In the lift-off technique, the MOD tin oxide thin film sensing element was prepared as described in the previous section, an opposite contrast metal interconnect feature was defined in a positive photoresist layer and Al/Cr was then deposited. The metal on top of the photoresist layer was subsequently lifted off by acetone. In this process, the co-presence of positive photoresist developer, MOD tin oxide and metal was avoided, and wet chemical etching of the metal interconnect was eliminated. In the inverse metallization scheme, the Al/Cr metal interconnect was fabricated prior to the deposition of the MOD film. The Al layer at the metal/SnO_x area was etched off, exposing the underlying Cr layer. The MOD tin oxide sensing element was then fabricated as described before. In both cases, the chemical reaction expressed in Eq. 2 was avoided and a reliable metal/MOD tin oxide contact was obtained. The schematic diagrams of the cross-sectional views of the inverse metallization structure and conventional metallization structure are shown in Fig. 1. A complete microsensor structure is shown schematically in Fig. 2.

Figures 3a and 3b are the SEM pictures of a finished tin oxide microsensor on a Si membrane, showing the top and cross-sectional views, respectively. The dimensions of the sensing element (the sensing element was a sputter-deposited tin oxide, in this particular case) are 25 μm x 25 μm.

2. **Results and Discussion**

The electrical resistance of tin oxide film strongly depends on its stoichiometric composition and can be effected by fabrication processes. For instance, in the reactive ion etching (RIE) process, the sample was immersed in an intense plasma containing reactive reducing gaseous species such as chlorine radicals. A decrease in sample resistance was observed after the RIE process. (The sample resistance was restored after a proper heat treatment in air.) Oxygen plasma exposure, on the other hand, tends to increase the tin oxide resistance due to its

strong oxidation strength. Hence, in fabricating tin oxide microsensors as well as other transition metal oxide-based sensors, the effects of microfabrication processes on the properties of the sensing materials (intrinsic sample conductance as well as gas sensitivity) should be carefully investigated. From our work on tin oxide microsensors, we found that although the sensor resistance can be affected by the RIE and oxygen plasma treatments, no discernible, detrimental effect on the gas sensitivity by such treatments was observed. In this work, the exposure of tin oxide films to an oxygen plasma was kept to a minimum or totally avoided.

The responses of the MOD tin oxide thin film microsensors to propylene and to vapors of ethyl alcohol, ethyl ether, methoxyflurane and halothane in air are shown in Fig. 4. The relative sensitivities of the microsensor to the organic vapors tested are, in decreasing order, ethyl alcohol > ethyl ether > methoxyflurane > halothane. This is congruous with the relative chemical activeness of the respective molecular species. The alcohol vapor sensitivity of the current microsensor is about one third of that given by a similar sensor that has not been subjected to micromachining processes. This indicates that the protective black wax coating on top of the sensing device during the EDP cavity etching process has a detrimental effect on the alcohol sensitivity of the tin oxide sensor.

The integrated sensor heater was a boron doped polysilicon layer with a thickness of ~5000 A. The sheet resistance of the heater versus temperature is shown in Fig. 5. In this experiment, a voltage of ~3.6 V and a current of ~40 mA were applied to the heater, which is equivalent to a power consumption of ~0.15 watt. From the R(T) curve of Fig. 5, the sensor temperature was estimated to be ~250°C.

Summary

Microsensors with tin oxide sensing elements and integrated polysilicon heaters were fabricated on thin silicon membranes generated by anisotropic wet chemical etching using ethylene diamine-pyrocatechol (EDP) as the etchant. The fabricated microsensors had good thermal isolation and low power consumption, and showed good response to organic vapors such as ethyl alcohol and ethyl ether.

Acknowledgements

We would like to thank M. Putty, J. Biafora and C. Puzio for their assistance in microfabrication, R. Laugal and A. Micheli for providing metallo-organic inks and W. Lange for taking the SEM pictures.

References

[1] S. R. Morrison, Sensors and Actuators, 2, 329 (1982).

[2] G. R. Heiland, Sensors and Actuators, 2, 343 (1982).

[3] S. C. Chang, IEEE Trans. Electron Devices, ED-26, 1875 (1979).

[4] K. E. Petersen, Proceedings of IEEE, 70, 420 (1982).

[5] N. F. Raley, Y. Sugiyama, and T. V. Duzer, J.
      Electrochem. Soc. 131, 161 (1984).

[6] S. K. Clark and K. D. Wise, IEEE Trans. Electron
      Devices, ED-26 1887, (1979).

[7] C. S. Sander, J. W. Knutti, J. D. Meindl, IEEE
      Trans. Electron Devices
      ED-17, 927, (1980).

[8] S. C. Chang and D. B. Hicks, GM Research
      Publication, GMR-4954, (1985).

[9] S. C. Chang and A. L. Micheli, GM Research
      Publication GMR-5212, (1985).

[10] M. Hatzakis, J. Electrochem. 116, 1033 (1969).

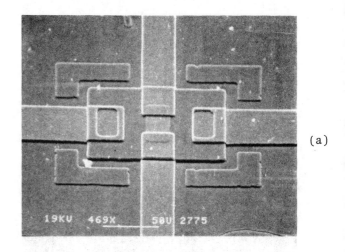

(a)

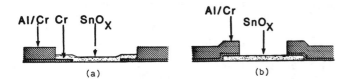

Fig. 1   Schematic diagrams of the cross-sectional
          view of the inverse metallization structure
          (a), and the conventional metallization
          structure (b).

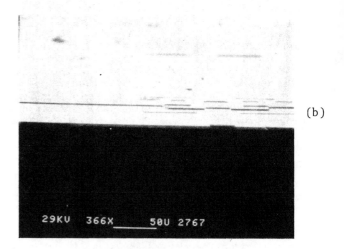

(b)

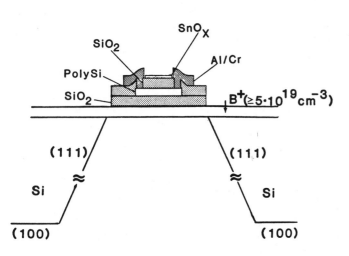

Fig. 2   Schematic diagram of a completed microsensor
          on a thin Si membrane. The Al/Cr metal
          contact on tin oxide is not shown.

Fig. 3   SEM pictures of a completed microsensor on a
          thin Si membrane (a) top view, (b) cross-
          sectional view.

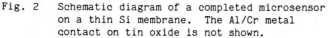

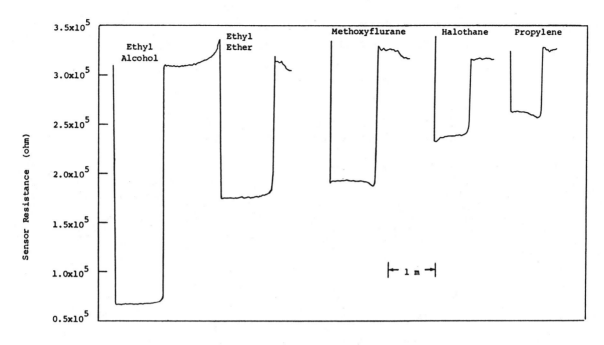

Fig. 4   Sensor responses to 200 ppm of ethyl alcohol,
ethyl ether, methoxyflurane, halothane and
propylene in air. Sensor material: MOD tin
oxide, sensor temperature = ~250°C.

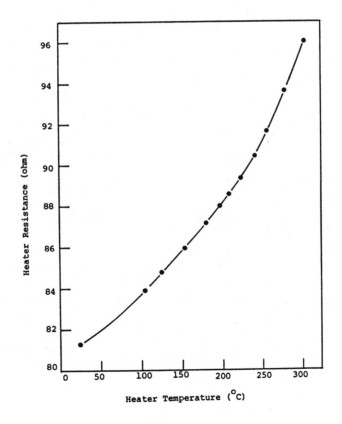

Fig. 5   Polysilicon heater resistance versus
temperature.

265

PROBLEMS AND POSSIBILITIES OF
OXIDIC AND ORGANIC SEMICONDUCTOR GAS SENSORS

G. Heiland and D. Kohl

2. Physikalisches Institut, Rheinisch-Westfälische
Technische Hochschule Aachen, D-5100 Aachen, F.R.G.

Aspects of gas sensing in air like selectivity, stability, contacts, and integration are discussed. Some applications are shown. Reducing gases are detected preferentially by means of n-type metal oxides, whereas p-type phthalocyanines (Pcs) can be used to sense oxidizing gases. A spatially separated catalyst is proposed for the Pcs allowing the observation of larger halogen containing molecules.

1. INTRODUCTION

Semiconducting gas sensors are mainly used for the detection of noxious or dangerous gases in the air and also for control of technical processes. The field has grown very fast during the last years. An excellent presentation of the state was obtained at the Meeting of Fukuoka (1). The intention of the present paper is to give not a review but only some comments. This also means an apology for not mentioning important contributions. Only homogeneous sensors are treated, where the conductance of a thin film, a sintered layer or a crystal is changed by the gas and measured with two electrodes (2) (3) (4) (20). An exception is made for the integrated gas sensors, using oxides.

2. METAL OXIDES

2.1 Principles and Examples

Preferentially SnO_2 and ZnO are used. Their n-type conductivity is due to an excess of metal (oxygen vacancies) or to the addition of foreign metal atoms acting as donors. In air these oxides are covered by chemisorbed oxygen binding electrons and decreasing the conductivity near the surface (5). Fig. 1 shows the scheme of a sensor and a typical time dependence of conductance : Reducing gases react with the adsorbed oxygen freeing electrons and thereby increasing the conductance. Also the lattice oxygen at the surface can be included in the reduction process, whereby donors (oxygen vacancies) are produced (6). In the presence of a small concentration of a reducing gas in air a steady state is reached characterized by a reduction process concurrent with a reoxidation by the air and by a continuous consumption of the gas by oxidation. After the end of the exposure the surface recovers by desorption of reaction products and reoxidation : The conductance decays again.

Fig. 2. shows, that acetic acid is detected by ZnO layers. The sensitivity reaches a maximum at

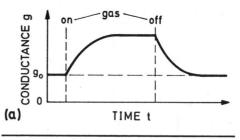

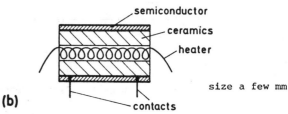

Fig. 1. Schematic drawings for a homogeneous semiconducting gas sensor : (a) Transient exposure to a gas increasing the conductance. (b) Example of a device. A ceramic tube serves as insulating support.

500°C. The decomposition of acetic acid has been studied under well defined conditions on SnO_2 single crystals by reactive scattering using a mass spectrometer (7). A similar temperature dependence has been found for other gases on SnO_2, whereby temperature and height of the maxima vary with the type of the gas (Fig. 3). The low temperature flank can be ascribed to the onset of a catalytic reaction, as proofed experimentally (8) (9), whereas the high temperature decay is due to the increasing speed of the surface reoxidation by the air. The reactions can be influenced by the addition of catalysts, e.g. noble metals. An increased temperature is required not only to get the reaction started, but also to obtain a sufficient speed of response including a fast and complete recovery.

Calibration curves of two commercial sensors are reproduced in Fig. 4. In moderately sintered samples the resistance can reside at barriers between grains. This model is not applicable for thin films and single crystals. At high temperatures also a variation of the bulk conductivity can take place by diffusion of donors (3). In many cases a clear discrimination between pure adsorption and subsurface phenomena appears difficult. An example is provided by the semiconducting oxides studied at high temperatures for the monitoring of combustion processes (11) (12).

Reprinted from *Rec. of the 3rd Int. Conf. on Solid-State Sensors and Actuators*, 1985, pp. 260–263.

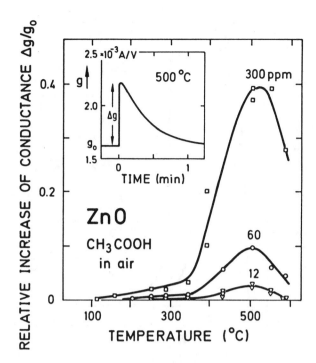

Fig. 2. Sintered layer without additions. Relative increase of conductance $\Delta g/g_0$ as a function of temperature. The insert shows the time dependence of the signal after a short injection of acetic acid vapor (7).

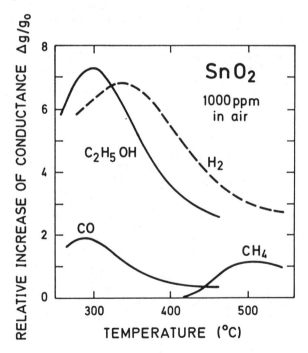

Fig. 3. Sintered layer, thickness 0.05 mm, 0.05 wt % Sb. Relative increase of conductance by reducing gases as a function of temperature. After N. Komori et al. (8)

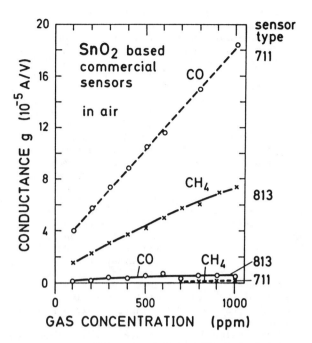

Fig. 4. Sintered layers with stabilizing admixtures like SiO_2. Conductance as a function of CO and CH_4 concentration. Type 711 probably uses a lower temperature and other catalytically active additions than type 813. After J. Watson (10).

2.2 Problems and Possibilities

The <u>sensitivity</u> is sufficient in many cases and shall not be discussed.

The <u>selectivity</u> depends on several parameters. The choice of temperature allows some discrimination (Fig. 3). Catalytically active metals (addition $\approx 1\%$) change position and height of the maxima and therewith the ratio of sensitivities (9). Also structure and thickness of the oxide layer exert an influence. E.g. a thick layer (0.5 mm) of SnO_2 at high temperature (520°C) is proposed for a methane sensor (8). The sensitivity $\Delta g/g_0$ for CH_4 exceeds that for H_2, CO and C_2H_5OH by a factor of ten and more, in contrast to the results with a thin layer (Fig. 3). An example for the difference of CO and CH_4 signals obtained with two commercial sensors is given in Fig. 4. Furthermore a modulation of temperature with periods of about one minute can be used for the sensing of CO in the presence of other reducing gases (4) (13).

Also filters are proposed. Carbon reduces the effect of NO_x (13), which, as a strongly oxidizing gas, counteracts the signal of reducing gases. For a multisensor system a charcoal filter is used (14). Good results are reported for the separation of H_2 and H_2S with zeolites (15). Besides the action as molecular sieves also a catalytic effect has to be considered. A high selectivity in favour of H_2 was observed on a sintered SnO_2 layer coated with an SiO_2 film (16). If several active gases are present, there signals may not sum up, but, an interaction might take place at the surface.

<u>Stability</u> also represents an important property of the sensors. From Fig. 2 and 3 it can be seen that the temperature has to be fairly constant to

avoid fluctuations of sensitivity. Also the humidity of the air has a disturbing influence. OH groups stick to the oxide surface and cause variations of the sensitivity either by a shift of surface conductivity or by catalytic activity, as considered for CO detection (3) (17). The oxide surface is open to the air. Therefore the calibration can be changed by chemical reaction and by adsorption. For sensors operating at a lower temperature the long-time stability is improved by transient heating in regular intervals. Also a periodic recalibration is used, e.g. after one week for the alcohol sensors of road side breath analysers. For commercial sensors a good stability during several years is reported.

The electrical contacts for measurements of the conductance require attention, too. In some experiments the change from Au to Pt contacts causes a remarkable variation of conductance and gas sensitivity (17). The contact resistance and its variation by gas reactions might dominate the observed conductance. More detailed studies to separate space charge effects at the contacts from the catalytic influence of the contact metals are desirable.

The integration of gas sensitive oxide films on Si-chips encounters a difficulty : Only lower temperatures are allowed. For an operation at higher temperatures a thermal insulating is needed (18). Another type of integrated device has been proposed: A thin film of SnO_2 forms the gate of a field effect transistor. The adsorbed oxygen produces a space charge layer penetrating the film and inducing a charge in the Si surface. Thus any variation in oxygen coverage by reaction with gases causes a signal (19).

3. ORGANIC SEMICONDUCTORS

As an example the phthalocyanines (Pcs) will be mentioned here. They are especially apt for gas sensing because of their thermal and chemical stability (20). Vapor deposition of thin films in vacuum is possible without decomposition of the Pc molecules. The temperature of permanent operation in air is limited to about 160°C by evaporation losses. The usually p-type conducting Pcs can be prepared n-type by purification in vacuum, but the oxygen of the air turns them p-type again and increases the conductance to an equlibrium value (21). Preferentially those gases are detected, which have a large electron affinity compared to oxygen. They increase the conductance of thin films beyond the oxygen induced level, e.g. halogens and NO_2.

About the selectivity not much is known so far. The model of key and lock seems promising in the case of large organic molecules and biosensors, but, it does not hold for the Pcs. Some halogen containing molecules produce no signal in the accessible temperature range up to 160°C, also with a catalyst at this temperature. However, a spatial separation of semiconductor and catalyst allows sufficiently high temperatures of the metal catalyst (21). The Pc film remains at lower temperature, but reacts with the fragments of the molecule. New results for the detection of CCl_4 are given in Fig. 5. As a new parameter the temperature of the catalyst can be varied in a wide range to obtain some discrimination of different gases.

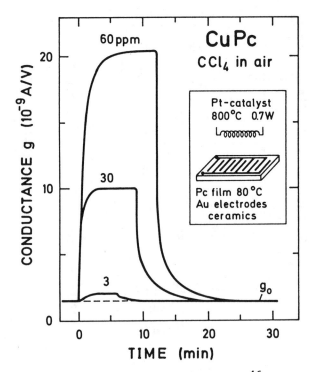

Fig. 5. Cu-Phthalocyanine film, about 10^{16} molecules/cm². Conductance during transient exposure to various concentrations of CCl_4 in air. No signal is observed if the catalyst is not heated. The insert shows the arrangement in a schematic way. Size a few mm. Both temperatures are stabilized. (21)

The stability requires an equilibrium with the oxygen of the air. After a change of temperature, especially after cleaning by a short heat treatment, this equilibrium is reached only slowly at a low temperature.

4. CONCLUSION

Semiconductor gas sensors offer a wide field for research. Compared to the perfect Si, the oxides and Pcs exhibit properties, which are not so well defined, e.g. stoichiometry (of oxides), purity and structure. For gas sensing not only an appropriate surface reaction is required, but also a "recoil" on the carrier density. Inspite of these difficulties there are already successful applications. For the solution of remaining problems and the realization of promising possibilities the two complimentary ways will contribute : The old approach by trial and error and the study of reactions by means of modern surface physics under well defined conditions (6) (7).

268

REFERENCES

(1) Proceedings of the International Meeting on Chemical Sensors, Fukuoka, Japan, T. Seiyama et al. (edts.), Kondansha Ltd, Tokyo, and Elsevier, Amsterdam, 1983.

(2) S.R. Morrison, Sensors and Actuators 2 (1982) 329.

(3) G. Heiland, Sensors and Actuators 2 (1982) 343.

(4) W. Hagen, R.E. Lambrich and J. Lagois, Festkörperprobleme, Vieweg, 23 (1983) 259.

(5) G. Heiland and H. Lüth, "Adsorption on Oxides", in The Chemical Physics of Solid Surfaces and Heterogeneous Catalysis, Vol. 3 B, D.A. King and D.P. Woodruff (edts.), Elsevier, Amsterdam (1983).

(6) G. Heiland and D. Kohl (1) p. 125.

(7) W. Thoren, D. Kohl and G. Heiland, ECOSS 7 (1985).

(8) N. Komori, S. Sakai and K. Komatsu (1) p. 57.

(9) N. Yamazoe, Y. Kurokawa and T. Seiyama, Sensors and Actuators 4 (1983) 283.

(10) J. Watson, Sensors and Actuators 5 (1984) 29.

(11) E.M. Logothetis and W.J. Kaiser, Sensors and Actuators 4 (1983) 333.

(12) K. Ihokura, K. Tanaka and N. Murakami, Sensors and Actuators 4 (1983) 607.

(13) Figaro Engin. Inc., TGS 203 Techn. Report.

(14) B. Bott and T. Jones, this Conference.

(15) G.N. Advani and A.G. Jordan, J. Electron. Mat. 9 (1980) 29.

(16) K. Fukui and K. Komatsu (1) p. 52.

(17) R. Lalauze, N. Bui and C. Pijolat, Sensors and Actuators 6 (1984) 119.

(18) Shih - Chia Chang, General Motors Research Publication GMR-4954 (1985).

(19) K. Dobos, D. Krey and G. Zimmer (1) p. 464.

(20) B. Bott and T.A. Jones, Sensors and Actuators 5 (1984) 43.

(21) H. Laurs and G. Heiland, publication in preparation.

THE USE OF PARTITION COEFFICIENTS AND SOLUBILITY PROPERTIES TO UNDERSTAND AND PREDICT SAW VAPOR SENSOR BEHAVIOR

J.W. Grate and A.W. Snow
U.S. Naval Research Laboratory
Washington, DC 20375-5000, USA

D.S. Ballantine, Jr.
Geo-Centers, Inc.
4710 Auth Place
Suitland, MD 20746, USA

H. Wohltjen
Microsensor Systems, Inc.
P.O. Box 90
Fairfax, VA 22030, USA

M.H. Abraham, R. A. McGill, P. Sasson
Department of Chemistry
University of Surrey
Guildford Surrey GU2 5XH, United Kingdom

ABSTRACT

Surface acoustic wave devices coated with a thin film of a stationary phase sense chemical vapors in the gas phase by detecting the mass of the vapor which distributes into the stationary phase. This distribution can be described by a partition coefficient. An equation is presented which allows partition coefficients to be calculated from SAW vapor sensor frequency shifts, and results are presented for nine vapors into SAW coating "fluoropolyol". Partition coefficients have also been determined independently by GLC and the results are in good agreement. The relationship between SAW frequency shifts and partition coefficients allows SAW sensor responses to be predicted if the partition coefficient has been measured by GLC or if it can be estimated by various correlation methods being developed.

INTRODUCTION

Surface acoustic wave (SAW) devices have been investigated by several groups for sensing chemical vapors in the gas phase[1-10]. The frequency of a SAW device in an oscillator circuit is measurably altered by small changes in mass or elastic modulus at the surface. Vapor sensitivity is typically achieved by coating the device surface with a thin film of a stationary phase which will selectively absorb and concentrate the target vapor. Vapor sorption increases the mass of the surface film and a shift in the oscillator frequency is observed. SAW devices have the potential to be adapted to a variety of gas phase analytical problems by strategic design or selection of coating material. Full realization of this potential will require methods to quantify, understand, and finally to predict the vapor/coating interactions responsible for vapor sorption.

Sorption of ambient vapor into the SAW device coating until equilibrium is reached represents a partitioning of the solute vapor between the gas phase and the stationary phase. This process is illustrated in Figure 1. The distribution can be quantified by a partition coefficient, K, which gives the ratio of the concentration of the vapor in the stationary phase, C_s, to the concentration of the vapor in the vapor phase, C_v (equation 1).

$$K = \frac{C_s}{C_v} \qquad (1)$$

An equation is derived herein which allows K to be calculated directly from observed SAW vapor sensor frequency shifts. This conversion provides a standardized method of normalizing empirical SAW data, and does so in a way that provides information about the vapor/coating equilibrium.

In this investigation, we examine the stationary phase referred to as "fluoropolyol", whose structure is shown in Figure 2. SAW vapor sensors coated with this soft, polymeric material have high sensitivity to certain toxic vapors[2]. Partition coefficients have been determined for nine vapors using SAW sensor frequency shifts. In addition, partition coefficients for the same vapors into fluoropolyol were determined independently by gas-liquid chromatographic (GLC) measurements. The values resulting from these two different techniques are in good agreement.

These results demonstrate that the mechanism of action of a coated SAW device is the same as that of GLC, i.e., reversible absorption of the vapor in the gas phase into the stationary phase. This confirms the solubility model for the interaction of vapors with the SAW coating; i.e., the solute vapor dissolves and distributes into the solvent stationary phase. Finally, the correlation between SAW vapor sensor responses and GLC partition coefficients creates a means for predicting SAW sensor behavior. Thus, if GLC partition coefficients are available from experimental measurement, or can be reliably predicted, then SAW vapor sensor responses can also be predicted.

Reprinted with permission from *Transducers '87, Rec. of the 4th Int. Conf. on Solid-State Sensors and Actuators*, 1987, pp. 579–582.

$$K = \frac{C_S}{C_V}$$

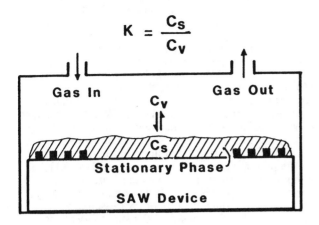

Figure 1. Distribution of vapor between gas phase and stationary phase.

Figure 2. Fluoropolyol structure.

EQUATION RELATING SAW FREQUENCY SHIFT TO THE PARTITION COEFFICIENT

The change in oscillator frequency observed when the mass on the surface of the SAW device increases can be described by equation (2).[6]

$$\Delta f = \frac{(k_1 + k_2) f_0^2 m}{A} \qquad (2)$$

Δf is the frequency change; m is the mass increase; k_1 and k_2 are material constants for the piezoelectric substrate; f_0 is the unperturbed resonant frequency of the device; and A is the active sensing area. It can be derived that the frequency shift (Δf_v) caused by the mass of vapor absorbed into the stationary phase coated on a SAW device is related to the partition coefficient (K) by equation (3).

$$\Delta f_v = \frac{\Delta f_S C_V K}{\rho} \qquad (3)$$

Δf_v is obtained in Hz when Δf_S (the frequency shift caused by the mass of the stationary phase applied to the device) is expressed in KHz, C_V (vapor concentration in the gas phase) is in g liter^{-1}, and ρ (stationary phase density) is in g mL^{-1}. The principal assumption inherent in equation (3) is that the SAW device functions as a mass sensor only, mechanical effects being negligible.[6] In addition, it is assumed that mass loading of the coating by vapor is low, since high mass loading would cause the coating density to change.

PARTITION COEFFICIENTS DETERMINED FROM SAW VAPOR SENSOR FREQUENCY SHIFTS

SAW vapor sensors were prepared by spray coating one delay line of a dual delay line SAW device with a dilute solution of fluoropolyol, as has been described previously[2]. The fluoropolyol coating causes the frequency of the device to change by an amount Δf_S, which provides a measure of the amount of coating material applied. These sensors were tested against vapors by alternately exposing them to clean air or a calibrated vapor stream using an automated vapor-generation instrument described in reference [11]. The change in the frequency observed when the gas over the sensor was changed from clean air to vapor gives the frequency shift caused by the vapor. For reliable measurements of K, this shift must be determined only after the sensor has reached a stable, equilibrium response level.

Partition coefficients were calculated from the observed frequency shifts (Δf_v) using equation (3). These values will be referred to as SAW partition coefficients and denoted K_{SAW}. Each sensor was exposed to each vapor at a minimum of two different concentrations (usually four, sometimes seven different concentrations) and at least four exposures at each concentration. A SAW partition coefficient was calculated for each vapor exposure, and these values were converted to logarithms. Then, all the log K_{SAW} values for each vapor on a particular sensor were averaged. Finally, the results from all sensors were averaged and are reported in Table 1. These results represents nearly 900 measurements of SAW frequency shifts.

Table 1. Log K_{SAW} and Log K_{GLC} Values

Vapors	Log K_{SAW}[a]	Log K_{GLC}[b]
DMMP	5.77	7.53[c]
Dimethylacetamide	5.65	7.29[c]
1-Butanol	3.24	3.66
2-Butanone	3.15	3.48[c]
Water	2.88	2.89
Diethyl Sulfide	2.78	2.54[d]
Toluene	2.69	2.64[c]
1,2-Dichloroethane	2.39	1.94[c]
Isooctane	1.97	1.22

[a] These values are averages from sensors described in the text.
[b] At 298°K.
[c] Determined from values measured at 333°K and corrected to 298°K with equation (4).
[d] Estimated value using a correlation equation.

Three sensors were examined in this study. One 158 MHz SAW device was coated with 207 KHz of fluoropolyol and tested against vapors one week later. Two 112 MHz devices were coated with 106 and 104 KHz, respectively, and were tested both one day and two months after coating. Reproducibility was good and aging appeared to have little effect.

Data from device to device showed some systematic variation, which could be most easily explained by errors in the measurement of the coating material applied. However, the actual source of the variation is not definitely known.

PARTITION COEFFICIENTS MEASURED BY GAS-LIQUID CHROMATOGRAPHY

Partition coefficients for a wide variety of solute vapors were determined by gas-liquid chromatography with fluoropolyol as the stationary phase, using methods described in reference 12. Partition coefficients as defined in equation (1) are actually identical to the Ostwald solubility coefficients usually denoted as L. We will use the symbol K, and refer to GLC partition coefficients as K_{GLC}.

GLC peaks on the fluoropolyol were generally broad, especially at 298°K. Therefore, 26 K_{GLC} values were determined at both 298°K and 333°K and the following correlation was found to hold:

$$\log K_{GLC} (298K) = -0.728 + 1.470 \log K_{GLC} (333K)$$

$$n = 26, \ sd = 0.156, \ r = 0.986 \qquad (4)$$

When retention times were too long to measure at 298°K, equation (4) and the measured value of K_{GLC} at 333°K were used to estimate the value of K_{GLC} at 298°K. Equation (4) shows that thermostatting a SAW sensor to fractions of a degree is not critical. However, changes of 5 to 10 degrees (e.g. due to varying ambient conditions) will become significant, especially for solutes with large K_{GLC}.

Log K_{GLC} and log K_{SAW} values for nine solute vapors are compared in Table 1, with the vapors in order of decreasing log K_{SAW}. All of the GLC values refer to 298°K, either by direct measurement, or via equation (4) as described above. In one additional case, diethyl sulfide, the log K_{GLC} value was estimated from various correlations we have constructed using solvatochromic parameters. With the exception of this estimated value, the order of decreasing log K values is identical for the SAW and GLC measurements. Indeed, there is good agreement between all but the highest Log K values, such that the SAW sensor frequency shifts could be estimated using K_{GLC} values and equation (3).

DISCUSSION

The experimental conditions for measuring partition coefficients with a SAW device are somewhat different than those for GLC measurements. SAW measurements, for instance, are carried out at finite vapor concentrations while the GLC measurement usually refers to infinite dilution. In addition, the SAW measurements reported here were conducted at ambient temperatures, while the GLC measurements were rigorously thermostated. Finally, the calculation of K_{SAW} assumes that the vapor causes the sensor to respond based on mass effects alone; if mechanical effects become significant for a particular vapor/coating interaction, then the calculated K_{SAW} will be altered proportionately.

One or more of the above factors may be responsible for differences in the precise values of log K_{SAW} and log K_{GLC} shown in Table 1.

The overall correlation observed clearly shows that partition coefficients are a useful concept for thinking about SAW sensor behavior (see also references 3 and 7). Indeed, the calculation of K_{SAW} values by equation (3) provides a standardized method of normalizing empirical SAW data which also provides information about the magnitude of the vapor/coating interaction. We have previously normalized our data by dividing the sensor response by the ppm of vapor in the gas phase and the KHz of coating. Normalization to yield a partition coefficient, K_{SAW} is very similar, and requires only that the vapor concentration be expressed in g liter^{-1}, and that the density of the stationary phase coating be factored out, according to equation (3).

The experimental correlation between K_{GLC} and K_{SAW} demonstrates that relative retention times for various vapors on a GLC column with a given stationary phase should be a reliable indicator of the relative sensitivity of a similarly coated SAW sensor to these vapors. This requires, of course, that the GLC measurement and SAW device operation be at the same temperature. On a more quantitative level, if absolute K_{GLC} values are determined, then estimates for actual SAW sensor frequency shifts can be made using equation (3). Finally, a clear relationship between SAW sensor responses and K_{GLC} values means that methods developed to predict K_{GLC} values will also be useful in predicting SAW sensor responses.

A simple examination of the order of the partition coefficients determined in this study illustrates the importance of solubility properties (Table 1). The lowest K values are those of isooctane, a solute which is not dipolar or polarizable, and which cannot accept or donate hydrogen bonds. Solutes which are more polarizable, such as dichloroethane, toluene, and diethyl sulfide have greater K values than isooctane. However, these solutes are still incapable of hydrogen bonding. The top of the list contains exclusively those solutes which can accept and/or donate hydrogen bonds. Vapor sorption is also influenced by the saturation vapor pressure, P^o_v, of the solute vapor, with lower P^o_v giving larger partition coefficients (for example, DMMP).

Solubility interactions can be placed on a more quantitative scale by the use of solvatochromic parameters (13). Such parameters are available for a wide range of vapors which may act as solutes in a vapor/coating interaction. Unfortunately, similar parameters are not yet available for a wide range of coating materials. The challenges, therefore, are to develop methods to characterize the solubility properties of stationary phases, and ultimately to be able to predict partition coefficients for any vapor with any characterized phase. This work is in progress, and equations of the general form shown in (5) are being used to predict log K_{GLC} values for fluoropolyol, and for various other stationary phases which have been useful as SAW device coatings[2].

$$\log K_{GLC} = \text{constant} + s\pi^* + a\alpha$$
$$+ b\beta + 1 \log L^{16} \qquad (5)$$

The variables π^*, α, β, and $\log L^{16}$ describe the solubility properties of the vapor. π^* is the dipolarity/polarizability; α is the hydrogen bond donor acidity; β is the hydrogen bond acceptor basicity; $\log L^{16}$ measures the tendency to partition into hexadecane. Coefficients s, a, b, and l describe the stationary phase. For example, b, as the coefficient for vapor β, measures the hydrogen bond donor acidity of the stationary phase. Once the coefficients have been determined for a particular stationary phase, then it will be possible to predict K values for all vapors for which π^*, α, β, and $\log L^{16}$ are known. Then, via equation (3), it will be possible to predict the responses of a SAW vapor sensor to these same vapors.

Finally, specific vapor/coating interactions may sometimes be encountered where mass effects alone do not adequately account for SAW sensor response. Mechanical effects will be implicated in such cases, but have been difficult to estimate thus far. Using equation (3) and a measured (GLC) or predicted (equation (5) K value, it is now possible to calculate the mass only SAW frequency shifts, and to estimate the mechanical frequency shift from the difference between the observed frequency shift and the mass-only frequency shift.

REFERENCES

(1)   H. Wohltjen and R.E. Dessy, Analytical Chemistry, 51(9), 1979, pp. 1458-1475.

(2)   D.S. Ballantine, Jr., S.L. Rose, J.W. Grate, and H. Wohltjen, Analytical Chemistry, 58, 1986, pp. 3058-3066.

(3)   H. Wohltjen, A. Snow, and D. Ballantine, Proc. of the Int. Conf. on Solid State Sensors and Actuators - Transducers '85, Philadelphia, PA, June 11-14, 1985, IEEE Cat. No. CH2127-9/85/0000-0066, pp. 66-70.

(4)   A.W. Barendsz, J.c. Vis, M.S. Nieuwenhuizen, E. Nieuwkoop, M.J. Vellekoop, W.J. Ghijsen, and A. Venema, Proc. of the 1985 IEEE Ultrasonics Symp.

(5)   S.J. Martin, K.S. Schweizer, S.S. Schwartz, and R.L. Dunshor, Proc. of the 1984 IEEE Ultrasonics Symp.

(6)   H. Wohltjen, Sensors and Actuators, 1984, pp. 307-325.

(7)   A. Snow, and H. Wohltjen, Analytical Chemistry, 56(8), 1984, pp 1411-1416.

(8)   A. Bryant, D.L. Lee, and J.F. Vetelino, Proc. IEEE 1981 Ultrasonics Symposium, pp 171-174.

(9)   A. D'Amico, A. Palma and E. Verona, Sensors and Actuators, 1982/1983, pp 31-39.

(10)   C.T. Chuang, R.M. White, J.J. Bernstein, IEEE Elect. Dev. Lett., EDL-3(6), 1982, pp 145-148.

(11)   J.W. Grate, D.S. Ballantine, Jr., and H. Wohltjen, Sensors and Actuators, 11, 1987, pp 173-188.

(12)   M.H. Abraham, P.L. Grellier, and R.A. McGill, J. Chem. Soc. Perkin Trans. I, in press.

(13)   M.J. Kamlet and R.W. Taft, Acta. Scand., Ser. B, B39, 1985, pp 611-628, and references therein.

Editors' Note: For readers seeking further information on this topic, the following reference has been provided by the authors for this Reprint Book.

J. W. Grate, A. Snow, D. S. Ballantine, Jr., H. Wohltjen, M. H. Abraham, R. A. McGill, and P. Sasson, *Anal. Chem.*, vol. 60, p. 869, 1988.

MECHANISM OF OPERATION AND DESIGN CONSIDERATIONS FOR SURFACE ACOUSTIC WAVE DEVICE VAPOUR SENSORS

HANK WOHLTJEN

Naval Research Laboratory, Chemistry Division, Code 6170, Washington, DC 20375 (U.S.A.)

(Received March 7, 1984; in revised form May 15, 1984; accepted May 21, 1984)

Abstract

Surface acoustic wave (SAW) devices offer many attractive features for application as vapour phase chemical microsensors. This paper describes the characteristics of SAW devices and techniques by which they can be employed as vapour sensors. The perturbation of SAW velocity by polymeric coating films is investigated both theoretically and experimentally. Highest sensitivity can be achieved when the device is used as the resonating element in a delay line oscillator circuit. A simple equation has been developed from theoretical considerations which offers reasonably accurate quantitative predictions of SAW device frequency shifts when subjected to a given mass loading. In this mode the SAW device behaves very like conventional bulk-wave quartz crystal microbalances except that the sensitivity can be several orders of magnitude higher and the device size can be several orders of magnitude smaller. Detection of mass changes of a few femtograms by a SAW device having a surface area of 10^{-4} cm^2 is theoretically possible.

Introduction

The phenomenon of waves which occur on the surface of solids was first described by Lord Rayleigh in 1885 [1]. Subsequent studies of surface waves were conducted primarily by seismologists who were interested in the propagation of mechanical waves at the earth's surface (*i.e.*, earthquakes). In 1965 White and Voltmer [2] developed the interdigital transducer which allows the convenient generation of surface waves in piezoelectric solids. This breakthrough precipitated a considerable amount of work in the application of surface acoustic wave (SAW) methods to radio-frequency signal processing. Here the small size, high performance and relative simplicity of SAW filters, delay lines and convolvers found many applications, particularly in electronic countermeasure systems for aerospace use [3, 4].

The first report of a sensor for chemical vapours based on SAW device technology appeared in 1979 [5, 6]. Both quartz and LiNbO$_3$ SAW devices

were evaluated for their performance as detectors in a gas chromatograph system when coated with sensitizing organic films. Investigations of SAW vapour sensing using polyvinyl chloride films have been reported by Fertsch, White and Muller [7, 8]. A comparison of bulk wave and surface wave device sensors for SO$_2$ was made by Bryant, Lee and Vetelino [9]. The surface wave device was found to detect less than 100 ppb of SO$_2$, which was at least an order of magnitude more sensitive than the bulk wave sensor. Recently a clever scheme for SAW vapour sensing using a thin membrane device has been reported by Chuang, White and Bernstein [10] and a hydrogen sensor using a palladium coated SAW device has been described by D'Amico, Palma and Verona [12].

SAW devices as chemical sensors are attractive because of their small size (*e.g.*, <0.1 cm^3), ruggedness, low cost, electronic output, sensitivity and adaptability to a wide variety of vapour phase analytical problems.

Rayleigh surface wave characteristics

The displacement of particles near the surface of a solid which is propagating a Rayleigh surface wave is shown in Fig. 1. The particle displacements have two components; a longitudinal component (*i.e.*, back and forth, parallel to the surface) and a shear vertical component (*i.e.*, up and down). The superposition of these two components results in surface particle trajectories which follow a retrograde elliptical path around their quiescent positions. As suggested by their name, these surface waves have most of their energy localized within one or two acoustic wavelengths of the surface. This permits easy and strong interaction with the medium adjacent to the surface. (Surface acoustic waves can also have an associated electric field if they are generated on a piezoelectric substrate.) Other particle displacements are possible (*e.g.*, shear horizontal rather than shear vertical), but such waves are not Rayleigh waves. Consideration of other wave types, *e.g.*, Love waves and Stonely waves, which are found in layered structures, is beyond the scope of this discussion.

The primary reason for focusing on Rayleigh waves is that they are generated quite readily in a variety of piezoelectric substrates using an interdigital transducer electrode (IDT). The interdigital electrode is fabricated

SURFACE PARTICLE DISPLACEMENT

UNSTRESSED

RAYLEIGH WAVE

Fig. 1. Deformation of the SAW substrate resulting from the propagation of a Rayleigh surface wave. The vertical displacement amplitude is typically about 10 Å.

Reprinted with permission from *Sensors and Actuators*, vol. 5, no. 4, pp. 307–325, 1984. Copyright © 1984 by Elsevier Sequoia.

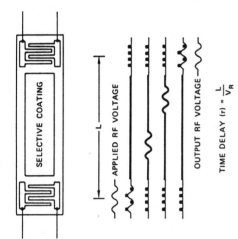

Fig. 2. Top view and side view of a SAW delay line propagating a Rayleigh wave. The vertical displacement of the wave is greatly exaggerated for clarity.

microlithographically from a thin film of metal (typically 1000 - 2000 Å thick) which is vacuum evaporated on to the polished piezoelectric substrate. These electrodes will generate a Rayleigh surface wave in the piezoelectric substrate if a radio frequency (RF) voltage is applied. The time-varying voltage will result in a synchronously varying deformation of the piezoelectric substrate and the subsequent generation of a propagating Rayleigh surface wave. The wavelength (and hence the operating frequency) of the Rayleigh wave is determined by the spacing between the electrodes. The electrical impedance of the IDT on a given substrate is determined by the number of electrodes and the electrode overlap length. This overlap length also determines the width of the acoustic beam that is produced. For a typical piezoelectric substrate such as quartz, which has a Rayleigh wave velocity (V_R) of about 3100 m/s, an operating frequency (f) of 31 MHz will result when the wavelength (λ) of the acoustic wave is 1×10^{-4} metres (100 micrometres) since $f = V_R/\lambda$. Thus the interdigital pattern required to operate at this frequency would typically consist of finger electrodes 25 micrometres wide separated by gaps of 25 micrometres. Practical microfabrication lithography limits the finger width in state-of-the-art devices to about 0.25 micrometres (*i.e.*, 3 GHz operating frequency). A lower frequency limit for practical SAW devices is about 10 MHz. Below this frequency the devices become too large (*e.g.*, greater than several cm² in area). The efficiency with which the RF voltage applied to the IDT couples to the mechanical deformation depends heavily on the piezoelectric substrate material used. Quartz is not as good in this respect as other materials (such as LiNbO₃) and as a result, requires more fingers on the IDT to achieve efficient power transfer. The operating bandwidth of an IDT is inversely related to the number of finger pairs (N) in the electrode. A typical IDT having 50 finger pairs and operating at 30 MHz (f_0) would have a bandwidth of about 0.6 MHz (*i.e.*, f_0/N), which means that the RF frequency would have to be 30.0 MHz ± 0.3 MHz for efficient power transfer to occur. Electrically, the IDTs appear as a capacitive load to the RF voltage source and a series matching inductance is often employed to improve the power transfer efficiency. Further information on the design and electrical characteristics of SAW IDTs can be found elsewhere [3, 11].

As previously mentioned, the SAW device substrate must be piezoelectric to permit the easy generation of a Rayleigh wave with an IDT. Ordinarily, polished single crystal slabs of quartz or lithium niobate are used because of their desirable SAW characteristics (low cost, good piezoelectric coupling, low temperature coefficient of delay, low acoustic loss, etc.). Although other materials are available, certain crystalline orientations are frequently chosen to optimize either the piezoelectric coupling coefficient or the temperature behaviour of the substrate in SAW applications. For example, in quartz the Rayleigh wave velocity of the *ST* cut exhibits a zero temperature coefficient of delay around room temperature, whereas the *YX* cut has a more efficient piezoelectric coupling coefficient but a non-zero temperature coefficient. Thus, the choice of substrate material depends somewhat on the intended application. The substrate surface which supports

the Rayleigh wave must be very smooth to prevent energy losses from wave scattering at imperfections. Optically polished surfaces are usually adequate. Surface acoustic waves offer many attractive possibilities in the design of miniature delay lines and resonators because the wavelength of sound in the piezoelectric substrate is about 10⁵ times smaller than that of an electromagnetic wave of the same frequency. Perhaps the simplest SAW device is a delay line (Fig. 2), which consists of an IDT at each end of an appropriate piezoelectric substrate. One IDT acts as a transmitter and the other as a receiver of acoustic energy which travels along the substrate surface. The propagating Rayleigh wave is free to interact with matter in contact with the delay line surface. This wave/matter interaction results in alteration of the wave's characteristics (amplitude, phase, velocity, harmonic content, etc.) and is the basis of SAW device chemical sensors.

Considering the SAW delay line in greater detail, one realizes that the delay time (τ) is directly related to the propagation path length (*i.e.*, the distance between IDT electrodes). A problem which deserves attention is encountered in the generation and detection of surface waves on delay lines. Namely, the simple IDTs described here are bidirectional and will permit a significant amount of energy to be reflected off the edge of the piezoelectric substrate immediately behind the electrode. This gives rise to an effect known as triple transit echo, which can result in anomalous signals appearing at the receiving IDT. Triple transit echo is ordinarily eliminated by the application of acoustic energy absorbers (*e.g.*, silicone adhesive) at the ends of the delay line. Alternatively the ends of the substrate can be cut at an angle to direct the reflection slightly off axis, thereby destroying its phase coherence.

Another problem can arise when a slight amount of surface wave energy is scattered within the substrate and converted into bulk acoustic waves, which then reflect off the bottom surface of the substrate and interfere with the surface wave on the top [13]. This effect is easily reduced by adding grooves or absorbing material to the bottom of the substrate to destroy the phase coherence of the bulk acoustic wave energy.

SAW delay line sensors are most easily configured to monitor either changes in the SAW amplitude or SAW velocity. Amplitude measurements are made by using an apparatus in which the Rayleigh wave is excited using an RF power source and the power at the receiving end of the SAW delay line is measured with an RF power meter. More elaborate balanced systems involving a reference device do, of course, make greater measurement precision possible. Measurements of SAW velocity changes are performed indirectly with extremely good precision by using the SAW delay line as a resonating element. The apparatus required to do this is shown in Fig. 3. An RF power amplifier is used to feed back energy from the output IDT to the input IDT, causing the system to oscillate at the IDT resonant frequency. Oscillations occur only when the amplifier gain is greater than the losses of the delay line. The resonant frequency of the device is altered by changes in the velocity of the Rayleigh wave and can be measured very accurately using a digital frequency counter. Experience has shown that the measurement of SAW velocity perturbations using the delay line oscillator offers vastly

superior precision compared to measurement of SAW amplitude perturbations. Thus, subsequent discussion will emphasize SAW delay line oscillator capabilities.

Theory of SAW interactions

(a) SAW amplitude perturbations

Once a Rayleigh surface wave has been generated by application of an RF voltage at the transmitting transducer, it can undergo several types of interactions which can significantly alter the energy content of the wave. In general, these lossy interactions will involve either direct absorption of the wave energy or redirection of the energy away from the receiving transducer. In either case the energy of the wave (and hence its amplitude) will be reduced at the receiving transducer and a corresponding reduction in the radio-frequency voltage will be observed. Losses in the SAW device substrate arising from scattering at imperfections, viscosity of the crystal and phonon–electron interactions are usually quite small (typically less than 1 db/cm at 100 MHz) and are quite constant with time and moderate temperature fluctuations (e.g., ±20 °C). Attenuation of the surface wave can also result if a conducting or semiconducting film is placed in contact with the vibrating piezoelectric crystal surface. Ordinarily a metallic film will cause an attenuation of a few decibels per centimetre at 100 MHz. Interactions of the electric field of the surface wave with charge carriers in thin semiconducting films can also result in attenuation of the wave amplitude. Surface wave energy is also lost to the ambient gas medium adjacent to the SAW device through coupling of the longitudinal component of the acoustic wave on the surface to a corresponding compressional acoustic wave in the gas [15]. Losses of less than 1 db/cm at 100 MHz are common.

The presence of a thin organic film on the SAW device surface can potentially cause attenuation of the wave through interaction with both the longitudinal and the vertical shear components of the Rayleigh wave [16]. Attenuation increases with the length of the SAW device and also with the inverse of the wavelength (i.e., operating frequency) in relation with oscillator circuit performance. If the adjacent medium is thick enough (i.e., its thickness is greater than λ), then compressional waves will be generated. This condition is easily met by ambient gases but for films which are very thin (less than $\lambda/100$) compressional waves cannot exist and can therefore be neglected. Thus, the attenuation produced by a thin organic coating film will be caused primarily through interactions with the shear component of the Rayleigh wave. In this situation the viscosity, elastic modulus and thickness of the organic coating will significantly affect the attenuation of the shear component. For organic films whose thickness is less than one per cent of the acoustic wavelength (e.g., 1 micrometre film on a 31 MHz SAW device) the observed attenuation is usually less than 3 db/cm [14]. This level of

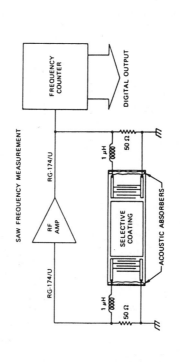

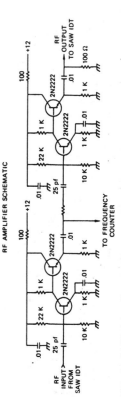

Fig. 3. SAW delay line oscillator system (top) and detailed RF amplifier schematic (bottom).

276

attenuation can be quite significant for SAW delay line oscillators since the loss introduced by the film makes it more difficult to sustain oscillation in the device.

(b) SAW velocity (frequency) perturbations

It was mentioned previously that the most precise measurements of changes in SAW velocity could be made by using the SAW delay line as the resonating element of an oscillator circuit. Determination of the resonant frequency with a precision better than 1 part in 10^7 is readily accomplished using a simple digital frequency counter. Stable oscillation in a SAW delay line is obtained when the oscillation frequency (ω) satisfies the following equation [17]:

$$\omega = (2n\pi - \phi_e)V_R/L \qquad (1)$$

where n = an integer, ϕ_e = electrical phase shift from IDTs and amplifier, V_R = SAW velocity and L = delay line path length between IDT centres.

From this relationship it is apparent that many resonant frequencies are possible depending on the value of n. In practice n is restricted to only a few values in a simple delay line because of bandwidth constraints imposed by the IDTs. It is possible to design SAW delay line oscillators which have only one resonant mode [18], but this is not essential for useful SAW sensors.

Ordinarily ϕ_e and L are constant (although a slight temperature dependence is sometimes observed). This means that changes in the resonant frequency of a SAW oscillator will occur if the SAW velocity is perturbed through an interaction with a film adjacent to the SAW device surface.

Auld [19] has derived an expression from a perturbation analysis of a SAW device coated with a non-conducting, isotropic overlay film which accurately describes the fractional change in wave velocity when the film is very thin (e.g., film thickness less than 1% of the wavelength). Auld's resulting equation is:

$$\frac{\Delta V_R}{V_R} = \frac{-V_R h}{4P_R}\left[\rho'|v_{Ry}|^2 + \left(\rho' - \left(\frac{4\mu'}{V_R^2}\right)\left(\frac{\lambda'+\mu'}{\lambda+2\mu}\right)\right)|v_{Rz}|^2\right]_{y=0} \qquad (2)$$

where h = film thickness (metres), V_R = Rayleigh wave velocity (m/s), ρ' = film mass density (kg/m^3), λ' and μ' = film Lamé constants (N/m^2), (λ' = Lamé const., μ' = shear modulus) and $(v_{Ry})_{y=0} = \surd/P_R$ and $(v_{Rz})_{y=0} = \surd/P_R$ are the normalized particle velocity components at the surface. This equation can be simplified by setting

$$C_1 = \frac{-V_R |v_{Ry}|^2_{y=0}}{4P_R} \qquad (3)$$

$$C_2 = \frac{-V_R |v_{Rz}|^2_{y=0}}{4P_R} \qquad (4)$$

Thus

$$\frac{\Delta V_R}{V_R} = (C_1 + C_2)h\rho' - C_2 h\left(\left(\frac{4\mu'}{V_R^2}\right)\left(\frac{\lambda'+\mu'}{\lambda'+2\mu'}\right)\right) \qquad (5)$$

However, the normalized particle velocity surface components $|v_{Ry}|^2/P_R$ and $|v_{Rz}|^2/P_R$ are frequency dependent and are tabulated by Auld for many SAW substrates in the form of a constant which must be multiplied by the wave frequency (f) in Hertz.

Clearly then $C_1 = k_1 f$ and $C_2 = k_2 f$ and

$$\frac{\Delta V_R}{V_R} = (k_1 + k_2)fh\rho' - k_2 fh\left(\frac{4\mu'}{V_R^2}\left(\frac{\lambda'+\mu'}{\lambda'+2\mu'}\right)\right) \qquad (6)$$

Values of k_1 and k_2 for several popular SAW substrate materials are provided in Table 1. It can also be shown that in the case of a SAW oscillator of quiescent frequency f:

$$\Delta V_R/V_R = \Delta f/f \qquad (7)$$

Here Δf is the change in resonant frequency due to a perturbation in the wave velocity by a thin overlay film. If the perturbation is small, then $f - \Delta f \cong f$ and

$$\Delta f = (k_1 + k_2)f^2 h\rho' - k_2 f^2 h\left(\frac{4\mu'}{V_R^2}\left(\frac{\lambda'+\mu'}{\lambda'+2\mu'}\right)\right) \qquad (8)$$

It should be noted that the first half of this equation is independent of the shear modulus (μ) and that the quantity $(\lambda'+\mu')/(\lambda'+2\mu')$ is in the range 0.5 to 1.0, with a value of 0.85 being typical for many polymeric materials. The shear modulus-dependent portion of eqn. (8) becomes negligible when $4\mu'/V_R^2$ is less than 100 (i.e., for 'soft' polymers). Thus it is possible to write

TABLE 1
Material constants for selected SAW substrates (derived from Auld [19])

Substrate	Rayleigh wave velocity V_R (m/s)	k_1 (m^2 s/kg)	k_2 (m^2 s/kg)
Y cut X propagating quartz	3159.3	-9.33×10^{-8}	-4.16×10^{-8}
Y cut Z propagating LiNbO$_3$	3487.7	-3.775×10^{-8}	-1.73×10^{-8}
Z cut X or Y propagating CdS	1702.2	-8.33×10^{-8}	-2.67×10^{-8}
Z cut X or Y ZnO	2639.4	-5.47×10^{-8}	-2.06×10^{-8}
Z cut X propagating Si	4921.2	-9.53×10^{-8}	-6.33×10^{-8}

a simplified relationship between the resonant frequency of a SAW oscillator and the coating film properties:

$$\Delta f = (k_1 + k_2)f^2 h\rho'$$ (9)

The quantity $h\rho'$ is simply the mass per unit area of the coating film. Small variations in either h or ρ will cause a corresponding variation in the oscillator frequency. Also noteworthy is the fact that the magnitude of the frequency change (Δf) produced by changes in mass per unit area (i.e., sensitivity) is highly frequency dependent. In the case of a 1 micrometre ($h = 10^{-6}$ m) thick polymer film ($\rho = 1000$ kg/m^3) coating, a YX quartz SAW oscillator ($k_1 = -9.33 \times 10^{-8}$, $k_2 = -4.16 \times 10^{-8}$ m^2 s/kg) resonating at 31 MHz, the simplified theory predicts a frequency shift (Δf) of -129.6 kHz.

Experimental

All of the observations reported here were made with a SAW device designed for operation at 31 MHz. The device was fabricated on an ST quartz substrate (1/2 inch wide and 2 inches long), one side of which was polished (Valpey-Fisher, Inc., Hopkinton, Mass.). The interdigital electrodes consisted of 50 finger pairs of 2000 Å gold on 200 Å chromium. Each finger was 25 micrometres wide and spaced 25 micrometres from the adjacent finger. The finger overlap length was 7250 micrometres. The IDTs were fabricated using standard microfabrication procedures employing an optical lithographic mask exposure of Shipley AZ-1350J photoresist followed by a chlorobenzene soak, development, metal evaporation and lift-off of the stencil and unused metal. The centre-to-centre spacing between the IDTs was 2.0 cm. A narrow bead of silicone adhesive was applied to each end of the device to eliminate triple transit echoes. Electrical connections to the IDTs were made using metal pressure clips. Poly(methyl methacrylate) (PMMA), cis poly(isoprene) (PIP), and poly(ethylene maleate) (PEM) were applied to the device by spin coating from solution at 1900 rpm for 1 minute, followed by a 5 minute bake at 110 °C to drive off residual solvent. Film thickness depended upon polymer solution concentration, solvent and spin speed. The film thickness was determined gravimetrically with microscope slides substituted for the SAW device. Spin coating of the microscope slides was performed under conditions identical to those for the SAW device. Measurement of the film mass together with knowledge of the polymer density and microscope slide area allowed a very simple but reliable determination of its thickness.

SAW frequency measurements were made using a simple two-transistor class A RF amplifier and an identical buffer amplifier to drive the FLUKE frequency counter. A circuit diagram for this apparatus is shown in Fig. 3. The resonant frequency of a clean SAW delay line oscillator was recorded with the resonant frequencies for the same device coated with progressively thicker films of PEM, PMMA and PIP. All measurements were obtained at room temperature and ambient atmospheric pressure.

Results and discussion

(a) SAW frequency measurements

The frequency shifts obtained from a 31 MHz ST-quartz SAW oscillator with various coating thicknesses of poly(methyl methacrylate), poly(isoprene) and poly(ethylene maleate) are shown in Figs. 4 to 6, respectively. These curves and the response predicted by eqn. (9) (for YX quartz) are dis-

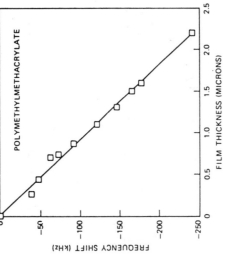

Fig. 4. SAW oscillator frequency shift produced by various thicknesses of poly(methyl methacrylate).

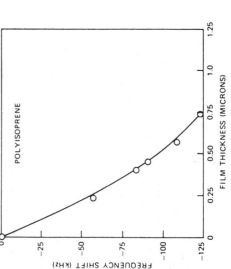

Fig. 5. SAW oscillator frequency shift produced by various thicknesses of cis-poly-(isoprene).

(i.e., smaller elastic modulus), although the exact cause is not known. Considering the number of assumptions and simplifications required to arrive at eqn. (9), the theoretical results are in very good agreement with experiment. The theoretical responses were obtained using material constants $(k_1 + k_2)$ for YX quartz, since constants for ST quartz were not available. It is believed that the constants for YX and ST quartz should be quite similar. The scatter observed in the frequency shift vs. film thickness data is primarily a result of errors in the gravimetric determination of film thickness. Frequency shifts produced by the film could be measured with a precision of 1 Hertz. Under typical conditions (ambient temperature and pressure) the baseline noise observed from the 31 MHz SAW oscillator was less than ±1 Hertz r.m.s. The sensitivity of the SAW oscillator can be appreciated by considering that a 1 micron thick film of PMMA produced a frequency shift greater than 100 kHz. This means that film thickness variations of about 0.1 Å are measurable.

The signal generated by a sensor for a given stimulus (i.e., the sensitivity) is not its only important characteristic. From the point of view of detectability, the signal-to-noise ratio is most important. In a SAW delay line oscillator signal, noise appears as random or drifting frequency fluctuations. Drift results from slow thermal changes in the SAW device substrate and electronic amplifier. The resonant frequency (ω) of a SAW delay line oscillator is given by eqn. (1) where the ratio (V_R/L) is equal to the reciprocal of the delay time ($1/\tau$). One can readily see that if the temperature of the substrate changes, then the delay time will change due to the linear thermal expansion coefficient of the substrate. This results in a temperature-dependent frequency shift. ST quartz is a material with a temperature coefficient of delay,

$$\frac{1}{\tau}\frac{\delta\tau}{\delta T},$$

which is nearly zero at room temperature, and is thus very desirable for sensor applications. Use of other substrates with non-zero temperature coefficients of delay (e.g., LiNbO$_3$) will result in greater temperature drift, which may require compensation with a reference device or oven. Also apparent is the fact that small changes in the phase shift of the electronic amplifier (ϕ_e) due to thermal variations in the components will show up as frequency drift. The relative change in frequency for a change in amplifier phase shift was shown by Ash to be [18]

$$\delta f/f = -\delta\phi_e/2\pi n \tag{10}$$

Here the factor $2\pi n$ is analogous to the Q of the oscillator and as the resonator Q increases (by increasing the delay line length and hence n), the effect of amplifier phase shift variations can be greatly reduced.

The random frequency noise encountered in SAW delay line oscillators results from Johnson-type thermal noise present at the input of the RF amplifier. Lewis has considered the problem of the short-term stability of a

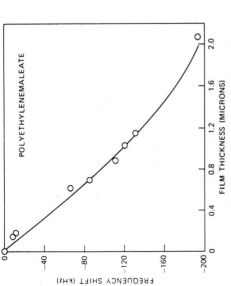

Fig. 6. SAW oscillator frequency shift produced by various thicknesses of poly(ethylene maleate).

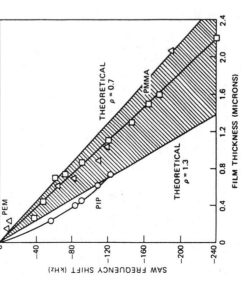

Fig. 7. SAW oscillator frequency shifts produced by films of poly(ethylene maleate) (△), cis-poly(isoprene) (○), and poly(methyl methacrylate) (□) compared with response predicted by eqn. (11). Shaded region shows the theoretical frequency shift assuming a film density (ρ) in the range of 0.7 to 1.2 g/cm³.

played on the same axis in Fig. 7. The responses obtained for PEM and PMMA are quite similar and lie within the envelope of theoretically predicted response for films whose densities are in the range 0.7 to 1.3 gm/cm³. The behaviour of the PIP is slightly different from that of the other two films studied and this could be related to its 'softer' viscoelastic properties

SAW delay line oscillator system [17] and has developed an expression describing the output power spectrum of such an oscillator:

$$S_{RF}(\omega) \simeq \frac{(2/\pi)kT(NF)G^2}{\left[\dfrac{2kT\omega_0}{P_0 Q}\right]^2 + \left[Q\dfrac{(\omega - \omega_0)}{\omega_0}\right]^2} \qquad (11)$$

where $S_{RF}(\omega)$ = power spectrum magnitude, k = Boltzmann's constant, T = absolute temperature, NF = amplifier noise figure, G = amplifier gain, ω_0 = SAW resonant frequency, Q = system quality factor and P_0 = output power.

For the maximum stability (i.e., minimum frequency noise) of a SAW oscillator it is desirable that the output power spectrum be small at all frequencies (ω) different from the resonant frequency (ω_0). Thus by inspection of eqn. (11), one readily sees that the most stable system will result when the amplifier noise figure (NF) is low, the amplifier gain (which is required to compensate for the delay line losses) is low and the output power (P_0) and delay line quality Q are high. Under these conditions the sharpest resonance will be obtained. The amplifier output power and noise figure can be considered as factors independent of the SAW delay line. The amplifier gain required to produce oscillation and the system Q are heavily dependent upon the losses experienced by the SAW delay line. A thick coating film on a SAW oscillator vapour sensor could have the result of absorbing a larger amount of energy from the device, which necessitates a correspondingly larger amplifier gain. Furthermore, the Q (which is classically defined as the ratio of energy stored to the energy dissipated per oscillation cycle) will also be greatly reduced. While a thicker coating film could have the ability to interact with more vapour and generate a greater signal, it could also have the effect of increasing the baseline noise. Thus it appears that for a given sensor, an optimal coating thickness will exist which provides a maximum signal-to-noise ratio.

From the preceding considerations it should be clear that careful attention must be given to the design of the SAW delay line and its RF amplifier for optimal performance to be achieved. The delay line should have low insertion loss. This can be accomplished by having a large number (e.g., 50) of pairs of interdigital electrodes and series inductors accurately tuned with them to achieve maximum power transfer with the RF amplifier. The RF amplifier itself should be operated with moderate gain (e.g., 20 db) and should have a low noise figure (e.g., 3 db at 100 MHz). Proper heat sinking of the RF amplifier will help to minimize thermally-induced electronic phase shift variations.

(b) SAW vapour sensor considerations

It has been shown that the SAW delay line oscillator is sensitive to both the mass per unit area of material present on the surface wave device and to the mechanical properties of the material (eqn. (8)). No direct sensitivity to a vapour is indicated. For this reason a coating film is required to convert

this mass-sensitive device into a vapour concentration-sensitive device. The coating, through either physisorption or chemisorption, must interact selectively with the vapour of interest to cause a change in mass or mechanical property of the coating. While the SAW oscillator is easily capable of detecting monolayer adsorption on to its surface, greater sensitivity can be obtained by using a thicker coating film through which the vapour can diffuse and interact with a greater number of sorption sites than on a smooth surface. Vapour diffusion rate and device response time are closely related. Clearly a film that is very dense will not permit easy diffusion and the device will respond very slowly.

For polymeric coatings rapid diffusion is most easily obtained if the polymer is highly amorphous and well above its glass transition temperature. For such a film the shear modulus is probably small enough to make its mechanical contribution to the SAW oscillator frequency a negligible quantity. Thus, for vapour-permeable films eqn. (9) will be the most accurate predictor of SAW frequency shift. Incremental changes in film density or thickness (i.e., mass per unit area) will result in a corresponding resonant frequency change. Coatings such as thin liquid films that can interact with a vapour will also work on the SAW device. However, the volatility of most liquids is high enough to cause a steady baseline drift due to evaporation. Coatings developed for bulk wave piezoelectric crystal vapour sensors will also work with the SAW device [21, 22].

The reversibility of the coating will depend entirely on the nature of the vapour/coating interaction. Interactions involving chemisorption will usually be quite selective but irreversible under normal conditions and will be most useful in dosimetry applications where an integral signal related to the concentration and time of exposure to a particular vapour is desired. The large dynamic range capability of the SAW oscillator and the low cost of the device permit irreversible interactions to be considered as a practical system. For reversible responses, interactions involving physisorption will be required. Unfortunately the relatively low energies of adsorption associated with physisorption processes could permit several different vapours with similar properties to interact with the coating film. Thus it is possible that the coating specificity and reversibility will be conflicting requirements. One possible solution to the specificity problem could involve the use of multiple sensors with different coatings to allow compensation for known interferences. Once again the attractive characteristics of the SAW device (i.e., very small size and low cost) make this approach practical.

(c) Ultimate performance limits

The sensitivity and size of SAW delay oscillator sensors is directly related to their resonant frequency. Equation (9) predicts that the frequency shift obtained for a given mass loading will increase with the square of the operating frequency. Thus, a device operating at 3 GHz will exhibit frequency shifts 10 000 times greater than a 30 MHz device for the same mass loading. However, it is expected that the baseline noise for the 3 GHz device

The fabrication of SAW devices to operate at frequencies up to 3 GHz poses no significant technical challenge. Optical lithographic techniques will easily produce devices operating around 300 MHz. For higher frequencies, electron beam or X-ray lithography systems would be required, but the process would still be very simple since only a single level of pattern masking is needed. Above 300 MHz the main challenge will be to develop suitable RF amplifiers and electrical connecting leads to permit stable device operation.

There are numerous piezoelectric materials that can be used as SAW device substrates. From an inspection of Table 1 it appears that quartz is a good choice, since the constants k_1 and k_2 are relatively large, resulting in greater sensitivity (eqn. (11)). It is also practical to consider thin piezoelectric films deposited on a silicon wafer [10]. Such an approach would allow many devices to be fabricated in one step, resulting in lower cost. Other SAW devices besides the delay line oscillator could also be useful as sensors. In particular, the SAW resonator device could offer some advantages over the delay line, particularly at higher frequencies [18].

The SAW vapour sensor is similar to the bulk wave piezoelectric quartz crystal vapour detector originally developed by King [20] and extensively investigated by Guilbault et al. [21, 22]. Both are mass sensitive and require a selective coating; both use a shift in resonant frequency as the indicating signal. However, there are several significant differences which deserve mention. First, the SAW device is capable of operating at frequencies at least two orders of magnitude higher than the bulk wave device and is therefore capable of much greater sensitivity. Secondly, the planar geometry of the SAW device allows one side to be rigidly mounted, making it even more rugged than the bulk wave devices. Thirdly, the device is intrinsically a micro sensor with its ability to occupy a volume as small as a few nanolitres. Finally, the SAW device allows multiple sensors to be fabricated close together on the same substrate. Thus, temperature drift compensation using a reference device can be achieved with great precision.

Equation (9) permits a direct comparison to be made between the sensitivities of bulk wave piezoelectric quartz crystal oscillator vapour sensors and SAW vapour sensors. Sauerbrey [23] was the first to develop an analytical expression that describes the frequency shift obtained from a bulk wave crystal oscillator under a given mass loading. King [20] reported that for quartz, the Sauerbrey equation is

$$\Delta f = 2.3 \times 10^6 f_0^2 (\Delta W/A) \qquad (12)$$

where f_0 is the quiescent resonant frequency of the oscillator (in MHz), ΔW is the change in mass of the crystal (in grams) and A is the area of the crystal (in cm²). Equation (9) can be reconstructed to fit the same form as the Sauerbrey equation if one realizes that the product of film thickness (h) and density (ρ) is simply the mass per unit area. For a YX quartz SAW delay line oscillator, the frequency shift resulting from a given mass loading can be described by

will also be somewhat higher, resulting in a signal-to-noise ratio which is less than 10 000 times greater. It is interesting to note that the frequency shift obtained for a given mass loading is independent of the length of the SAW delay line. The baseline noise, however, depends on the effective delay line oscillator Q, which is related to the length of the device. Thus, longer delay lines will produce higher signal-to-noise ratios not by increasing the signal but rather by decreasing the noise. The SAW delay line oscillators used in this study operating at 31 MHz had a delay line length which supported about 300 wavelengths of the Rayleigh wave between the IDT electrodes. The Q of these devices was high enough to provide frequency noise less than 1 Hz (i.e., stability of better than 1 part in 10^7). If one can maintain the same numbers of cycles (wavenumber) between the IDTs at other frequencies, then comparable stabilities should result. Thus the baseline noise is expected to increase linearly with frequency (i.e., stay at 1 part in 10^7), whereas the signal increases with the square of the frequency. The overall result is a signal-to-noise ratio which increases linearly with frequency.

The area occupied by the device is also frequency dependent. On ST quartz the IDT aperture width (the electrode overlap length) should be about 28 wavelengths [11] to provide a 50 ohm impedance frequently required by the RF amplifier. If a constant number of wavelengths is maintained, then the device length will also depend on the wavelength. Thus, the area of a SAW delay line with constant IDT impedance from device to device and constant number of wavelengths will vary inversely with the square of the operating frequency. The reduction of device area also has implications with regard to the minimum mass change that can be detected by a SAW delay line oscillator. Since the device is sensitive to mass per unit area, a reduction in device area results in a corresponding reduction in the minimum detectable mass change. A summary of estimated SAW device sensor performance at various operating frequencies is presented in Table 2. Detection of a few femtograms of material with a 3 GHz device having 10^{-4} cm² of active area seems theoretically possible.

TABLE 2
Estimated SAW device sensor performance

Frequency	Device area [1] (cm²)	Baseline noise [2] (Hz)	Minimum detectable mass change [3] (g)
30 MHz	1	3	3×10^{-9}
300 MHz	10^{-2}	30	3×10^{-12}
3 GHz	10^{-4}	300	3×10^{-15}

[1] Approximate area required to produce a delay line on ST Quartz with 50 ohm IDT impedance and a cavity number of wavelengths of 300.
[2] Assuming a SAW oscillator stability of 10^{-7}.
[3] Minimum mass change required to produce a signal plus noise-to-noise ratio of 2 ($S + N/N = 2$). Frequency shift based on eqn. (9).

$$\Delta f = 1.3 \times 10^6 f_0^2 (\Delta W/A) \qquad (13)$$

At first glance the bulk wave oscillator appears to be the most sensitive since it will produce the greater signal (Δf) for a given frequency and mass loading. However, bulk wave oscillators with fundamental mode resonant frequencies greater than 15 MHz are quite difficult to fabricate and very fragile in operation due to the extremely small crystal thickness required. SAW devices and their associated electronics are very easy to fabricate and operate at 300 MHz and, with carefully designed electronics, can go to 3 GHz. Thus, the SAW device offers the highest possible sensitivity due to its higher operating frequency.

A direct comparison between a 15 MHz bulk wave oscillator and a 300 MHz SAW oscillator shows that the SAW device should produce a frequency shift more than 200 times greater than that of the bulk wave device under identical mass loading (i.e., mass per unit area).

Conclusion

The effects of various coating film thicknesses on the velocity of propagating Rayleigh surface waves have been studied experimentally and compared with theory. For a given film loading, the measurement of SAW velocity with a delay line oscillator system affords the greatest sensitivity to film variations. If a vapour of interest perturbs the mass of the selective coating film by diffusing into the film and either chemisorbing or physisorbing, then the mass loading on the SAW device will increase, resulting in a wave velocity reduction and a corresponding downward shift in the delay line resonant frequency. A highly simplified analytical expression has been derived from theoretical considerations which predicted SAW delay line oscillator frequency shifts for a given mass loading. In spite of its many simplifying assumptions, reasonably good agreement with experimental data was obtained. Thus, the mechanism of operation of the SAW vapour sensor has been shown to be quite similar to that of the bulk wave piezoelectric quartz crystal vapour sensor. The significantly higher operating frequencies obtainable with the SAW device result in a sensitivity which is theoretically orders of magnitude greater than that of the bulk wave device. The high sensitivity of the SAW device, together with its low cost, small size, ruggedness and electronic output make it very attractive for application in a broad spectrum of vapour phase analytical problems. The ultimate success of these devices appears to be limited only by the performance of the selective coating films. In practical systems it may be necessary to compensate for interference vapours by using an array of devices with different coatings. The planar geometry and small size of the SAW vapour sensor make it an ideal vehicle in such a system.

Acknowledgement

The author wishes to thank Dr. N. L. Jarvis and Dr. A. Snow of the Naval Research Laboratory for many helpful discussions during the course of this work.

References

1 Lord Rayleigh, On waves propagated along the plane surface of an elastic solid, *Proc. London Math. Soc.*, *17* (1885) 4 - 11.
2 R. M. White and F. W. Voltmer, Direct piezoelectric coupling to surface elastic waves, *Appl. Phys. Lett.*, *7* (1965) 314 - 316.
3 R. M. White, Surface elastic waves, *Proc. IEEE*, *58* (1970) 1238 - 1276.
4 Special issue on surface acoustic wave devices and applications, *Proc. IEEE*, *64* (1976) 579 - 788.
5 H. Wohltjen, Methods of detection with surface acoustic wave and apparati therefore, *U.S. Patent 4 312 228* (1982).
6 H. Wohltjen and R. E. Dessy, Surface acoustic wave probe for chemical analysis I. Introduction and instrument description, *Anal. Chem.*, *51* (1979) 1458 - 1475.
7 M. T. Fertsch, A surface acoustic wave vapor sensing device, *Masters Thesis*, U.C. Berkeley (1980).
8 M. T. Fertsch, R. M. White and R. S. Muller, *Surface acoustic wave vapor sensing device, Device Research Conf., Ithaca, NY, June, 1980*.
9 A. Bryant, D. L. Lee and J. F. Vetelino, A surface acoustic wave gas detector, *Proc. IEEE Ultrasonics Symp., IEEE Cat., 81CH1689-9* (1981) 1735 - 1738.
10 C. T. Chuang, R. M. White and J. J. Bernstein, A thin membrane surface acoustic wave vapor sensing device, *IEEE Elec. Dev. Lett., EDL-3* (1982) 145 - 147.
11 D. P. Morgan, Surface acoustic wave devices and applications 1. Introductory review, *Ultrasonics*, *11* (1973) 121 - 131.
12 A. D'Amico, A. Palma and E. Verona, Palladium-surface acoustic wave interaction for hydrogen detection, *Appl. Phys. Lett.*, *41* (1982) 300 - 301.
13 R. S. Wagers, Spurious acoustic responses in SAW devices, *Proc. IEEE*, *64* (1976) 699 - 702.
14 H. Wohltjen, *U.S. Naval Research Laboratory Memorandum Report #5314* (1984).
15 A. J. Slobodnik, Attenuation of microwave acoustic surface waves due to gas loading, *J. Appl. Phys.*, *43* (1972) 2565 - 2568.
16 K. Dransfeld and E. Salzmann, Excitation, detection, and attenuation of high frequency elastic surface waves, in W. P. Mason and R. N. Thurston (eds.), *Physical Acoustics*, Vol. 7, ch. 4, Academic Press, New York, 1970, pp. 219 - 273.
17 M. F. Lewis, Surface acoustic wave devices and applications 6. Oscillators — the next successful surface acoustic wave device? *Ultrasonics*, *12* (1974) 115 - 123.
18 E. A. Ash, Fundamentals of signal processing devices, in A. A. Oliner (ed.), *Acoustic Surface Waves*, ch. 4, Springer-Verlag, New York, 1978, pp. 117 - 124.
19 B. A. Auld, *Acoustic Fields and Waves in Solids*, Vol. 2, Wiley-Interscience, New York, 1973, ch. 12.
20 W. H. King, Piezoelectric sorption detector, *Anal. Chem.*, *36* (1964) 1735 - 1739.
21 Y. Tomita, M. Ho and G. G. Guilbault, Detection of explosives with a coated piezoelectric quartz crystal, *Anal. Chem.*, *51* (1979) 1475 - 1478.
22 J. Hlavay and G. G. Guilbault, Applications of the piezoelectric crystal detector in analytical chemistry, *Anal. Chem.*, *49* (1977) 1890 - 1898.
23 G. Sauerbrey, Verwendung von Schwingquarzen zur Wägung dünner Schichten und zur Mikrowägung, *Z. Phys.*, *155* (1959) 206 - 222.

A Multisensor Employing an Ultrasonic Lamb-Wave Oscillator

STUART W. WENZEL AND RICHARD M. WHITE, FELLOW, IEEE

Abstract—We present initial experimental, analytical, and numerical evaluations of a new microsensor that employs ultrasonic Lamb waves propagating in a thin plate supported by a silicon die. Changes of oscillator frequency indicate magnitudes of the variables sensed. Because it is sensitive to many measurands, the device could operate as a microphone, biosensor, chemical vapor or gas detector, scale, pressure sensor, densitometer, radiometer, or thermometer. Lamb waves offer unique means for obtaining selective response and permit sensitive operation in the low-megahertz frequency range in vacuum, in a gas, or while immersed in a liquid.

I. INTRODUCTION

THE LAMB-WAVE sensor concept presented here is similar to ultrasonic oscillator sensors employing surface acoustic waves (SAW's) to measure pressure, acceleration, and chemical vapors and gases [1], [2]. Major benefits over the SAW accrue from the use of Lamb waves, propagating in a plate whose thickness is small compared with the ultrasonic wavelength. Notable are the convenience of operating in the low-megahertz frequency range, and the possibility of operation while immersed in a liquid. Both are results of employing the low-velocity, flexural zeroth-order Lamb wave. Another benefit is the high sensitivity of the Lamb-wave device to changed mass per unit area of surface, a feature that can be employed in biological and chemical sensors. The thin plate also has a low heat capacity and hence a rapid thermal response.

Most of the discussion that follows concerns a sensor whose key element is a planar sheet whose thickness is much smaller than the wavelength of the ultrasonic waves propagating in it. We might call this an "acoustically thin plate" or a "membrane." Strictly speaking, it is neither a "plate" nor a "membrane" as usually defined in the mechanical engineering literature. Unlike such plates, ours is so thin that its elastic response to deformation can be influenced by in-plane tension that develops during the initial fabrication process. And, since flexural Lamb waves can propagate on these structures, they are unlike membranes, which are formally regarded as being infinitely flexible. In spite of these formal differences, in what follows we will use for convenience of expression the terms plate and membrane interchangeably when referring to our structures.

Manuscript received October 15, 1987; revised January 4, 1988.

The authors are with the Berkeley Sensor and Actuator Center, Department of Electrical Engineering and Computer Sciences, and the Electronics Research Laboratory, University of California, Berkeley, CA 94720.

IEEE Log Number 8820449.

We first summarize the characteristics of Lamb-wave propagation in plates, and include the effect of in-plane tension on the wave velocity of the lowest order antisymmetric (A_0) mode. Next we consider how the velocity can be affected by external stimuli, together with ways the velocity can be measured in a sensor. Then, after describing the design and fabrication of our devices, we present initial sensor test results.

II. LAMB-WAVE CHARACTERISTICS

Lamb waves are elastic waves that propagate in plates of finite thickness. These waves are divided into symmetric and antisymmetric modes to indicate the symmetry, about the median plane of the plate, of the particle displacements associated with the waves. In a plate whose thickness is much larger than the wavelength, the lowest order symmetric and antisymmetric Lamb waves become very much like a SAW propagating on a semi-infinite medium. In very thin plates, the lowest order symmetric Lamb wave (usually labeled S_0) becomes dispersionless, and its velocity becomes considerably larger than that of a SAW propagating on an identical material. In thin plates, the lowest antisymmetric mode A_0 involves flexure, and its wave velocity decreases monotonically to zero as the plate is made vanishingly thin.

For Lamb-wave propagation in homogeneous, isotropic plates, the phase and group velocities of the lowest order symmetric mode (v_{ps} and v_{gs}) and the lowest order antisymmetric mode (v_{pa} and v_{ga}) are given by [3]

$$v_{ps} = \sqrt{\frac{E'}{\rho_p}}, \qquad v_{gs} = v_{ps} \tag{1}$$

and

$$v_{pa} = \sqrt{\omega d}\left(\frac{E'}{12\rho_p}\right)^{1/4}, \qquad v_{ga} = 2v_{pa}. \tag{2}$$

Here, $E' = E/(1 - v^2)$, E is Young's modulus, v is Poisson's ratio, ρ_p is the plate density, and d is the plate thickness.

We consider propagation of the A_0 mode and seek an expression for the phase velocity of this wave that includes the effects of in-plane tension, which may develop during membrane fabrication and affect the device operation. Knowledge of tension effects may add insight to the types of measurands that may be sensed.

Consider a time-harmonic wave, traveling in the x-di-

Reprinted from *IEEE Trans. Electron Devices*, vol. 35, no. 6, pp. 735–743, June 1988.

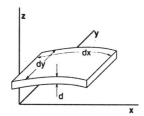

Fig. 1. Elastically isotropic, homogeneous plate assumed in the analysis of Lamb wave propagation.

rection in an isotropic plate, and let the vertical z-directed particle displacement be w. Assuming no variation with y, the z-component of force on an elemental piece of the membrane of thickness d, length dx, and width dy (Fig. 1) is composed of a stiffness term dF_{zS} and a tension term dF_{zT} [4], [5]

$$dF_{zS} = -\frac{E'd^3}{12}\frac{\partial^4 w}{\partial x^4}\, dx\, dy$$

$$dF_{zT} = T_x \frac{\partial^2 w}{\partial x^2}\, dx\, dy$$

where $T_x\, dy$ is the component of tension in the direction of propagation generated by forces along the boundaries of the membrane that are parallel to the y-direction. In this derivation we ignore the effect of this tension upon the Young's modulus of the membrane material.

The total z-directed force dF_z is

$$dF_z = dF_{zS} + dF_{zT} = -\frac{E'd^3}{12}\frac{\partial^4 w}{\partial x^4}\, dx\, dy + T_x \frac{\partial^2 w}{\partial x^2}\, dx\, dy.$$

Relating the force to the acceleration of the elemental membrane volume, we have

$$-\frac{E'd^3}{12}\frac{\partial^4 w}{\partial x^4}\, dx\, dy + T_x \frac{\partial^2 w}{\partial x^2}\, dx\, dy = M\, dx\, dy\, \frac{\partial^2 w}{\partial t^2}$$

or

$$\frac{\partial^2 w}{\partial t^2} = \frac{1}{M}\left(T_x \frac{\partial^2 w}{\partial x^2} - \frac{E'd^3}{12}\frac{\partial^4 w}{\partial x^4}\right) \quad (3)$$

where M is the mass per unit area and $M\, dx\, dy$ is the mass of the elemental volume. Solving (3) by separation of variables, we write

$$w(x, t) = \Psi(x)\, \Phi(t). \quad (4)$$

Substituting into (3) and rearranging gives

$$\frac{1}{\Phi}\frac{\partial^2 \Phi}{\partial t^2} = \frac{1}{M}\left(T_x \frac{1}{\Psi}\frac{\partial^2 \Psi}{\partial x^2} - \frac{E'd^3}{12}\frac{1}{\Psi}\frac{\partial^4 \Psi}{\partial x^4}\right). \quad (5)$$

For (5) to be valid for all x and t, the quantities on each side of the equation must be constants, and (5) becomes

$$\frac{1}{\Phi}\frac{\partial^2 \Phi}{\partial t^2} = -\omega^2 \quad (6a)$$

$$\frac{1}{M}\left(T_x \frac{1}{\Psi}\frac{\partial^2 \Psi}{\partial x^2} - \frac{E'd^3}{12}\frac{1}{\Psi}\frac{\partial^4 \Psi}{\partial x^4}\right) = -\omega^2. \quad (6b)$$

The solution to (6a) has the form $\Phi(t) = Ae^{\pm j\omega t}$. From (6b), Ψ has the form $\Psi = Be^{\gamma x}$. We will launch waves with interdigital transducers, so that $w(x, t)$ and Ψ are periodic in x with period $\lambda_n = P/n$ ($n = 1, 2, 3, \cdots$) where P is the IDT period. Note that n may be restricted to certain allowed values, obtained from spatial Fourier analysis of the transducer. For example, if one uses transducers having equal finger widths and gaps between fingers, and further drives the transducers differentially with respect to a ground plane on the membrane, the voltages and displacements of adjacent fingers will be 180° out of phase. Consequently, n will be restricted to odd integers ($n = 1, 3, 5, \cdots$). Thus, $\gamma = \pm jk_n$, $k_n = 2\pi n/P$, and (6b) can be solved for ω in terms of P and n

$$\omega_n = \frac{2\pi n}{P}\sqrt{\frac{1}{M}\left(T_x + \left[\frac{2\pi n}{P}\right]^2 \frac{E'}{12}d^3\right)}. \quad (7)$$

Equation (3) becomes, for a wave traveling in the $+x$-direction

$$w(x, t) = \Psi(x)\, \Phi(t) = C_n e^{j(\omega_n t - k_n x)}. \quad (8)$$

The phase velocity v_{pa_n} is

$$v_{pa_n} = \frac{\omega_n}{k_n} = \frac{\omega_n}{2\pi n/P} = \sqrt{\frac{1}{M}\left(T_x + \left[\frac{2\pi n}{P}\right]^2 \frac{E'}{12}d^3\right)} \quad (9)$$

which is valid for $d \ll \lambda_n = P/n$.

In the absence of tension ($T_x = 0$), (9) is equivalent to (2) as can be seen by substituting $\omega_n = 2\pi n v_{pa_n}/P$ for ω and $\rho_p = M/d$ in (2). For a very flexible material ($E' = 0$), (9) reduces to the phase velocity for flexural waves on an infinitely thin membrane [5]

$$v_{pa_n} = \sqrt{\frac{T_x}{M}}. \quad (10)$$

We can rewrite (9) in a more general form in order to apply it to composite membranes

$$v_{pa_n} = \frac{2\pi n}{P}\sqrt{\frac{B_n}{M}} \quad (11)$$

where

$$B_n = \left(\frac{P}{2\pi n}\right)^2 T_x + B_0 \quad (12)$$

is the effective bending stiffness of the plate including tension, and B_0 is the material stiffness contribution to this term ($B_0 = E'd^3/12$ for a homogeneous, isotropic plate).

For future reference, we include an expression for the group velocity v_{ga_n} of the A_0 mode

$$v_{ga_n} = \frac{d\omega_n}{dk_n} = v_{pa_n}\left(1 + \frac{v_{pa_n}^2(T_x = 0)}{v_{pa_n}^2}\right) \quad (13)$$

where $v_{pa_n}(T_x = 0) = 2\pi n/P\sqrt{B_0/M}$ is the phase velocity with no tension in the membrane. For zero tension, $v_{ga_n} = 2v_{pa_n}$, in agreement with the group velocity expression in (2). It can be seen from (9) and (13) that the effect of tension is to *increase* the phase velocity while *decreasing* the group velocity of the A_0 mode.

The phase velocity of the A_0 mode also depends upon the loading effect of any fluid that might contact the membrane. This loading is reactive (no attenuation due to radiated energy loss into the fluid) when $v_{pa_n} < v_f$ (v_f is the sound velocity in the liquid), and dissipative (radiated energy and wave attenuation) for $v_{pa_n} > v_f$ [6]. Using the radiation load method of Kurtze and Bolt [7], we write equations governing the phase velocity for these two cases:

$$\frac{\alpha}{\beta}\left(\frac{\alpha^2}{\beta^4} - 1\right)\mu\sqrt{1 - \beta^2} - 1 = 0, \qquad \text{for } v_{pa_n} < v_f$$
(14)

$$\alpha^6 - \beta^4\alpha^4 + \beta^6/4\mu^2(1 - 1/\beta^2)^2 = 0,$$
$$\text{for } v_{pa_n} > v_f.$$
(15)

In these equations, $\beta = v_{pa_n}/v_f$, $\alpha = \omega_n/\omega_0$, $\omega_0 = v_f^2\sqrt{M/B_n}$, and $\mu = \omega_0 M/\rho_f v_f$, where ρ_f is the fluid density. Substituting $\omega_n = 2\pi n v_{pa_n}/P$ and rearranging, (14) becomes

$$\frac{2\pi n}{\rho_f P}\left[\left(\frac{2\pi n}{P}\right)^2\frac{B_n}{v_{pa_n}^2} - M\right]\sqrt{1 - \left(\frac{v_{pa_n}}{v_f}\right)^2} - 1 = 0,$$
$$\text{for } v_{pa_n} < v_f$$
(16)

and (15) becomes

$$\left(\frac{2\pi n}{P}\right)^6 B_n^2 - \left(\frac{2\pi n}{P}\right)^4 MB_n v_{pa_n}^2 + \frac{\rho_f^2}{4[v_f^{-2} - v_{pa_n}^{-2}]^2} = 0,$$
$$\text{for } v_{pa_n} > v_f.$$
(17)

Therefore, the phase velocity of the A_0 mode for a plate having mass per unit area M and effective bending stiffness B_n is affected by both the density and the compressional wave velocity of the surrounding fluid.

We may obtain additional velocity information for these multilayer structures by using the program of Nassar and Adler [8], which does not presently include tension effects.

III. ULTRASONIC OSCILLATOR SENSORS

The fundamental phenomenon responsible for the operation of ultrasonic oscillator sensors is *elastic wave propagation along a path whose characteristics can be altered by a measurand*. Table I lists the factors that influence elastic wave velocity in a piezoelectric medium, together with illustrative ways of causing each sort of velocity change.

As with bulk and surface elastic waves, one may use Lamb waves for sensing by measuring the characteristics of passive delay lines, delay-line feedback oscillators, or individual transducers. In comparison with bulk waves

TABLE I
FACTORS THAT COULD INFLUENCE WAVE VELOCITY AND ILLUSTRATIVE
SOURCES OF EACH INFLUENCE

INFLUENCES	ILLUSTRATIONS
Change elastic stiffness	Temperature; bombardment; curing
Change density	Temperature; bombardment; curing
Change tension	E or H field; force; pressure; acceleration
Change thickness	Etching; deposition; adsorption
Change length	Temperature
Change piezoelectric stiffening	Illumination; dielectric loading
Load mode -- reactive	Liquid or gas loading; sorption
Load mode -- dissipative	Liquid or gas loading

and SAW's, Lamb waves provide a number of advantages for use in sensors:

1) As we have shown elsewhere [9], the calculated sensitivity of the phase velocity of Lamb waves to added mass or thickness (as in a chemical vapor sensor) should be at least an order of magnitude greater than that of a SAW operating with the same wavelength (assumed to be 141 μm in our examples).

2) The coupling of a piezoelectric film of a given thickness (zinc oxide in our device) is substantially higher for a Lamb wave on an acoustically thin membrane than for a SAW on a thick substrate.

3) Using the low-velocity A_0 Lamb mode permits a sensor of given wavelength to operate at a frequency much lower than a corresponding SAW sensor, simplifying and reducing the cost of the associated amplifier.

4) The velocity of the A_0 mode can, in practical devices, be made lower than the velocity of compressional waves in common liquids, permitting low-loss operation of an immersed sensor.

5) The heat capacity of a thin membrane can be quite low, permitting one to realize sensing elements that respond quickly to thermal energy. This is useful in radiometry, and could enable thermal-desorption spectroscopy to be employed in a vapor or gas sensor.

A Lamb-wave sensor shares with other acoustic sensors the problem of providing *selective* response. Conventional means for achieving selectivity with SAW sensors (such as use of a reference device and solubility-parameter matching in vapor sensors) could also be used with the Lamb-wave sensor. Moreover, some additional methods are available with these sensors. The velocities and particle motions of the symmetric and antisymmetric Lamb waves depend somewhat differently on the characteristics of the propagation medium. The A_0-mode velocity depends strongly on membrane thickness, but the lowest symmetric mode (S_0) velocity is independent of the thickness for thin membranes. Tension should affect the A_0 mode much more strongly than the S_0 mode. The S_0 mode in a very thin plate is nondispersive, but the A_0 mode is dispersive, permitting operation at several different frequencies (corresponding to integrally related wavelengths) with a single transducer structure. Finally, because of its small heat capacity, the thin membrane has a rapid thermal response, as shown by its response to chopped infrared radiation discussed below. This also means that thermal-desorption spectroscopy should be usable in chemical vapor membrane sensors for identifying sorbed species, since a deposited conducting film could heat the membrane quickly.

IV. Device Fabrication and Operation

A typical oscillator design is shown in Fig. 2. It consists of a Lamb-wave delay line (plate and piezoelectric transducing film) and a feedback amplifier (A). Planning ultimately to integrate the delay line with the amplifier (and possibly more auxiliary circuitry), we formed the thin plate in a silicon wafer by depositing LPCVD silicon nitride uniformly over a wafer and then etching away the silicon from beneath the nitride. Following formation of the thin plate, we sputtered zinc oxide onto an evaporated aluminum ground plane, and then formed interdigitated transducing electrodes on top of the zinc oxide.

The delay-line device on which most measurements were made consists of a 2-μm-thick silicon nitride plate, 0.3-μm-thick evaporated aluminum ground plane, 0.7-μm-thick ZnO, and 0.35-μm-thick aluminum interdigitated finger transducer electrodes. The transducers have 100-μm-period, 2.5-mm apertures, 50 fingers on each of the two transducers (split finger design to reduce reflections [10]), and contact pads off the membrane. The distance between the nearest edges of the transducers is 2.5 mm. The plate itself measures 3 mm $\times$ 9 mm, and the ends are angled to reduce coherent end reflections. Details of each fabrication step follow.

Silicon Nitride: LPCVD silicon nitride was deposited at 835°C in a 5-to-1 gas ratio of dichlorosilane and ammonia, to obtain a low-stress film suitable for fabricating membranes [11]. The deposition time was 5 h to obtain a 2-μm-thick film.

Membrane Release: Using a two-sided alignment technique [12], we patterned windows in the backside nitride by plasma etching. We then etched away the silicon (using the nitride as an etch mask) with ethylenediamine, pyrocatechol, and water (EDP) to leave a 2-μm-thick nitride membrane on the front side.

Zinc Oxide Deposition and Transducer Fabrication: We deposited 0.7 μm of ZnO by RF planar magnetron sputtering. Onto this we evaporated 0.35 μm of aluminum and used the marks generated during the two-sided alignment step to pattern interdigital transducers centered on the membranes. For etching the aluminum, we used a solution of KOH, $K_3Fe(CN)_6$, and water (1 g : 10 g : 100 ml), which does not etch ZnO [13].

Auxiliary Apparatus: The feedback amplifier used in the oscillator configuration of Fig. 2(a) consists of two cascaded LM733 differential video amplifiers. The device was driven untuned, and an amplifier gain of approximately 40 dB was necessary to sustain oscillation. We measured frequency directly from this device with a frequency counter. No reference oscillator was used for this initial testing. The short-term instability of the oscillator in still air was ± 1 Hz.

For measuring the frequency of minimum insertion loss, we applied RF pulses of 0.5–2-μs duration to one trans-

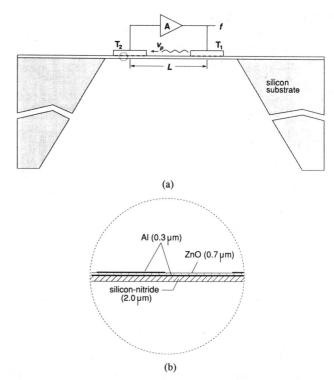

(a)

(b)

Fig. 2. Experimental Lamb-wave oscillator. (a) Schematic cross-sectional view of membrane device (drawn to exaggerated vertical scale) showing transducers on membrane and feedback amplifier. (b) Expanded view showing nitride membrane, aluminum ground plane, ZnO layer, and a transducer electrode. Note that widths of transducer electrodes are much larger than their separations from the ground plane, causing the electric fields beneath electrodes to lie mostly in the vertical direction.

ducer and tuned the RF frequency to maximize the received signal amplitude from the second transducer. The velocity of maximum transducer coupling is found by multiplying this frequency by the acoustic wavelength. Using the same apparatus we could measure group delay times and hence group velocity.

We measured the A_0 velocity of the maximum transducer coupling on the test device for $n = 1$, for comparison with computer results for zero tension. The measured value of v_{pa_1} is 474 m/s, compared to 422 m/s predicted with the Nassar–Adler program. From (11) and (12), using $v_{pa_1}(T_x = 0) = 422$ m/s, and $v_{pa_1} = 474$ m/s, the tension T_x is 528 N/m. To obtain this value we used $M = 11.34$ kg/m^3, found by summing the mass per unit area of the individual layers using density values from the literature [14], [15]. If we assume that this tension is generated by the 2-μm nitride layer *alone*, then the tensile stress in the nitride would be 2.6×10^8 N/m^2, which is reasonable for this nitride deposition process.

The measured group velocity v_{ga_1} is 833 m/s. Using (13), the theoretical value of group velocity should be 850 m/s, or 2 percent higher than the measured value; this result is well within the experimental measurement error. (Note that (2), which does not take tension into account, predicts that $v_{ga_1} = 2v_{pa_1} = 948$ m/s, 14 percent above the measured value.)

Fig. 3 shows the Lamb-wave oscillator sensor mounted in a test fixture that permits making coaxial connections

Fig. 3. View of device from above, showing 3 mm × 9 mm membrane at center with the silicon-nitride layer facing upward. Four coaxial feedthroughs connect, via gold bonding wires, the input and output interdigitated electrodes to external amplifiers.

Fig. 4. Side view of membrane device in its experimental mounting fixture, capped with a teflon block for test as a vapor and gas sensor. Inlet and outlet pipes are visible beneath the fixture.

to each set of transducing electrodes. Inlet and outlet pipes in the mounting fixture (Fig. 4) and a tight-fitting teflon cap also permit test of the device as a chemical vapor or gas sensor. Incidentally, as a demonstration of device ruggedness and the feasibility of spin-casting thin polymer films onto the membrane for chemical sensing, we spun the device in its holder (Fig. 4) on a photoresist spinner at 3700 rpm without damage.

V. EXPERIMENTAL, ANALYTICAL, AND NUMERICAL RESULTS

From the fundamental oscillation-condition equation [9]

$$f = \frac{\nu_p}{2\pi L} (2\pi N - \phi_E) \qquad (18)$$

(in which L is the acoustic path length, N is an integer, ϕ_E is the electronic phase shift of the amplifier, and ν_p is the phase velocity of the mode being observed), one can determine the change of oscillator frequency that will be caused by a change of phase velocity or separation of transducers. In order to determine the frequency shift caused by a specific measurand, we must find the change of velocity or separation caused by that measurand. For example, the Lamb-wave analog of the SAW chemical sensor would involve a change of mass loading as molecules of a vapor or gas were absorbed in a thin film that coated the plate. Increasing the thin plate's temperature would reduce elastic stiffness somewhat, and would also cause a change in tension because of thermal expansion. Heating both the plate and its surrounding silicon "frame" would affect both the velocity and the transducer separation.

Initial results for the device operating in the A_0 mode with $n = 1$ will now be presented. The measurands considered are arranged in the order described by White [16].

A. Acoustical

An air-borne sound wave impinging on the thin plate will modulate the net membrane tension, causing the phase velocity to become time varying and producing a modulation of the oscillator frequency. The sensor output will be a frequency-modulated voltage with information about the impinging sound wave carried in the sidebands [9], [17].

B. Biological and Chemical

Lamb-wave phase velocity can be altered by physical (mass and/or stiffness) changes that occur in a sensitive coating film on the membrane upon interaction with chemical or biological species [18], [19]. Our published computer simulations [9] showed that, at a particular wavelength, the A_0-mode device could be an order of magnitude more sensitive than the corresponding SAW device while operating at about one-tenth its frequency. The liquid loading test results reported below verify the ability of A_0-mode devices to operate while immersed; experimental study of A_0-mode biological and chemical sensors is in progress.

C. Mechanical

Because the Lamb-wave device responds to changes of membrane tension, loading of its surfaces, and changes of the dimensions of the transducers, a number of applications to mechanical measurements appear possible.

1) Force: A force could be applied to the membrane directly, increasing tension and causing an increase in oscillator frequency. It would be important to arrange the application of the force to avoid interfering with wavefront planarity, for example, by causing ultrasonic wave diffraction about a stylus contacting the thin nitride plate. In order to determine the response of the sensor of Fig. 3 to a force, we applied a known force on the silicon die, rather than the membrane, outside the area occupied by the membrane. Thus the die served as a sort of cantilever beam loaded by the force at a point beyond the thin plate. A pantograph was used to exert the force in a consistertly downward direction.

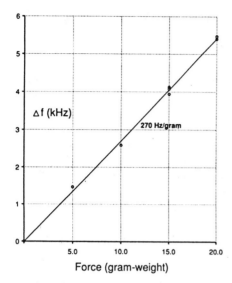

Fig. 5. Increase of membrane oscillator frequency as force is applied to the silicon surrounding the membrane at a point located approximately $\frac{3}{8}$-in to the right of the center of the membrane (as shown in Fig. 3).

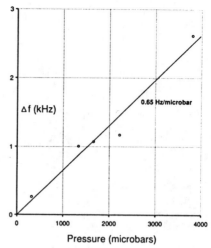

Fig. 6. Increase of membrane oscillator frequency upon application of differential pressure produced by a variable-height column of water above the membrane.

Fig. 5 shows the upward frequency shift as a function of force. Note the relatively linear dependence and the measured sensitivity, 270 Hz/gram weight. One expects that greater sensitivity could be obtained with proper design of the chip and the mounting structure. For measuring small forces, applying the force directly to the thin plate should be advantageous.

2) Pressure: Response to gas pressure can be realized in two ways. If the membrane is subjected to unequal pressures on its two sides, changes of membrane tension will cause an oscillator frequency change. If both sides of the membrane are subject to the same gas pressure, membrane tension will be constant, but loading of the propagating Lamb wave will depend upon the pressure.

To simulate application of a differential pressure to the membrane, we loaded the membrane with a column of deionized water of variable height. Results of this test are plotted in Fig. 6. Only the change of frequency with change of column height is significant here, because it was difficult to determine accurately the absolute height of the column owing to the depression of the membrane surface below that of the silicon die. As Fig. 6 shows, the variation of frequency with column height, and hence differential pressure, is linear, with a slope of 0.65 Hz/μbar (6.5 Hz/Pa). As the short-term instability of the water-loaded oscillator was about ± 1 Hz, the minimum detectable pressure change would have been approximately 4.6 μbar, if we adopt the convention of taking the minimum as the level for which the signal-to-noise ratio equals three. It is interesting to compare the Lamb-wave device sensitivity with that of a temperature-compensated SAW device [20] whose reported sensitivity was only 0.4 Hz/Pa.

The loading of a propagating A_0 mode by a gas can cause both resistive (dissipative) and reactive (velocity) effects. If the phase velocity of the unloaded membrane is smaller than the sound-wave velocity in the gas, only a reactive shift to a lower velocity will be caused by the gas loading. We will discuss this in the next section.

3) Mass, Density: Loading one or both sides of the membrane with a fluid can cause quite large velocity changes and oscillator frequency shifts. The analysis summarized below shows that the magnitude of the change depends primarily upon the density of the liquid; the sound velocity in the liquid has a somewhat smaller effect.

To determine the suitability of the membrane sensor for applications requiring immersion, we loaded an operating sensor with several different liquids. When we filled the 6-μl etch pit (above the membrane in Fig. 3) with de-ionized water, we observed three effects:

1) the frequency for optimal coupling (minimum insertion loss) of the interdigitated transducers decreased 36 percent;
2) the insertion loss of the membrane delay line increased 6 dB;
3) the delay-line oscillator continued to operate at a lower frequency when loaded on one or both sides.

We interpret observations 1 and 3 as resulting from a decrease in the A_0 mode velocity owing to loading by the liquid (whose velocity of sound, 1480 m/s, is much higher than the Lamb-wave velocity on the unloaded membrane). The velocity changes reported below are *much* larger than the transducer bandwidth, which in our device was only 2.6 percent. This is possible because the transducers, being located on the membrane itself, are loaded by the liquid and experience large shifts in their operating frequencies. This behavior permits these devices to have an unusually large dynamic range. Finally, we believe that observation 2 represents a 3-dB decrease in the efficiency of piezoelectric coupling at each of the two transducers owing to the increased mass loading of the membrane in the vicinity of the transducers.

The results of loading experiments with de-ionized water and methanol are listed and compared with expected values in Table II.

TABLE II
A_0-MODE PHASE-VELOCITY FOR FLUID LOADING ON ONE SIDE OF THE
COMPOSITE MEMBRANE OF FIG. 2

Fluid	v_{pa_1} (m/s)	
	Experiment	Theory
Air	473.7±0.4	--
De-ionized water	303.5±0.8	303.6
Methanol	322	322.3

(Relevant values are: Membrane: $M = 1.134 \times 10^{-2}$ kg/m²; $B_1 = 6.50 \times 10^{-7}$ Nm; $P = 100 \times 10^{-6}$ m. Air: $\rho_f = 1.21$ kg/m³; $v_f = 343$ m/s. De-ionized water: $\rho_f = 1 \times 10^3$ kg/m³; $v_f = 1480$ m/s. Methanol: $\rho_f = 791$ kg/m³; $v_f = 1103$ m/s. For air, the density was doubled to represent two-sided loading.)

The theoretical values are obtained from (16) and (17). In order to solve these equations, we needed the value of B_1 for our multilayer structure. For an isotropic homogeneous plate with no stress, $B_n = B_0 = E'd^3/12$. Expressions for the bending stiffness of multilayer structures under tension are much more complicated. Rather than using such expressions, we substituted the measured air-loaded phase velocity into (17) (using this equation because $v_{pa_1} > v_{f_{air}}$) and solved numerically for B_1. The values used were $v_{pa_1} = 473.7$ m/s, $v_f = 343$ m/s (air), $\rho_f = 2.42$ kg/m³ (two times the actual air density is used to represent two-sided loading), $M = 11.34 \times 10^{-3}$ kg/m², and $P = 100$ μm, which yielded $B_1 = 6.50 \times 10^{-7}$ Nm. We then used this value of B_1 in (16) and solved for the liquid-loaded phase velocities v_{pa_1}. As Table II shows, the agreement with theory is excellent. Numerically, we found that for the cases considered here, the velocity was about 20 times more sensitive to changes of density than to changes of sound velocity in the liquid.

Fig. 7 shows the frequency shifts, calculated from (16), expected when a membrane oscillator with 100-μm-period transducers is loaded on one side with a fluid whose density and velocity of sound vary incrementally from those of water. In these calculations, viscosity has not been taken into account and the temperature is assumed to be constant.

In a similar fashion, from (16) we calculated the oscillator frequency shifts caused by density changes in a gas that loads the membrane on both sides. At STP, this gas has the density (1.21 kg/m³) and sound velocity (343 m/s) of air. For this analysis we assumed ideal gas behavior, constant temperature, sound velocity independent of pressure, and zero viscosity. The transducer period was assumed to be 236.1 μm, while the other device dimensions remained unchanged. This yielded a calculated velocity in vacuum of 300 m/s, so that again $v_{pa_1} < v_f$. The results (Fig. 8) show linear behavior over a large range of gas densities. At the higher densities the response is quite nonlinear, but the density is calculable analytically using (16). Note that the effect upon wave velocity of compression of the membrane (change of thickness d) at the higher pressures was not taken into account.

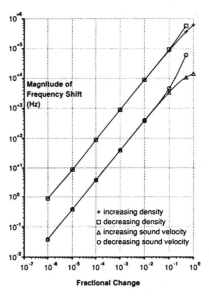

Fig. 7. Calculated membrane oscillator frequency shifts as functions of fractional change of density (with constant fluid sound speed, $v_f = 1480$ m/s) and sound velocity (with constant density, $\rho_f = 1 \times 10^3$ kg/m³) in a fluid. Frequency decreases (negative shift) as ρ_f increases (+) or as c_0 decreases (O).

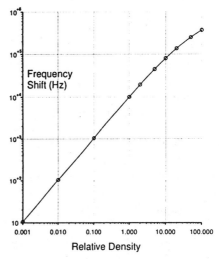

Fig. 8. Calculated frequency shift (from vacuum value) of membrane oscillator immersed in air at various densities. Density is given normalized to air density at STP (1.21 kg/m³). (See text for assumptions made.)

D. Thermal

We have observed sensor response to radiation from a warm body. Fig. 9 shows the experimental frequency shift produced by allowing radiation from a nearly black body (hot plate coated with Nextel Velvet coating #101-C10 black) to be incident upon the silicon nitride membrane. The observed shift in the quiescent oscillator frequency (4.67 MHz, at ambient temperature, 24.5°C) is a linear function of the temperature rise of the black body above ambient over about a 30-degree range; the sensitivity is about −17 Hz/°C. As the black-body temperature increases further, the oscillator frequency decreases superlinearly, as Fig. 9 shows. This behavior can be understood qualitatively as resulting from a transition from a

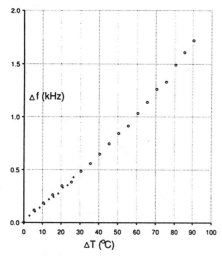

Fig. 9. Response of membrane oscillator to radiation from a nearby heated blackened plate. Ordinate: difference between plate temperature and ambient temperature (24.5°C). Abscissa: decrease of oscillator frequency from its initial value (4.67 MHz). Measurements were made on two different runs, the first over a 30–50°C range (+) and the second over a 30–120°C range (○).

linear relationship—valid for small temperature differences—to a higher power dependence valid for radiative heat transfer between surfaces at widely separated temperatures.

In an attempt to explain the transient heating test results, we calculated the steady-state temperature of a homogeneous membrane exchanging heat by radiation and convection. (Heat conduction in the film appears to have a minor effect.) A fundamental uncertainty is the emissivity, or absorptivity, of our nitride membrane: the membrane, backed by a reflective Al coating, is only 2 μm thick, while the peak wavelength for radiation from a black body at 400 K is near 7 μm. We obtain, for various assumed emissivities, a dependence of membrane temperature upon the hotplate-to-ambient temperature difference that is in excellent qualitative agreement with the frequency shift data of Fig. 9. The calculated temperatures may be too high, however, since the resulting relative frequency dependence, 10^{-5}/°C, is much smaller than one might expect to result from the temperature dependence of elastic stiffness, for instance. Initial measurements with steady-state heating show a more reasonable variation of about 1.8×10^{-4}/°C. The latter value suggests the need for use of a reference device in conjunction with an active device.

Because the membrane is quite thin, it equilibrates quickly. Experiments performed with a chopper between the heater and sensor showed that oscillator output sidebands resulting from heater-induced frequency shifts were evident even when the radiation was chopped at 100 Hz.

VI. CONCLUSIONS

We have described initial experimental, analytical, and numerical results on an ultrasonic oscillator sensor that employs Lamb waves propagating in a membrane whose thickness is a small fraction of a wavelength. We have shown that, for A_0-mode operation, this device responds to forces applied to the sensor die, differential pressure applied to the membrane, loading by a fluid on one or both its sides, and absorption of radiant energy. Analytical predictions and experimental results for liquid loading agree extremely well. Experimentally, we found that the device continues to oscillate while totally immersed; this ability, together with its calculated high sensitivity to the addition of small amounts of mass to the membrane, suggests its use for chemical and biological sensing. We also found agreement between experimental values of phase and group velocities and an analysis of the effect of intrinsic stress in the membrane, coupled with predictions obtained from a numerical analysis program.

The device is suited to use a tool or testbed for studying certain chemical processes, such as etching of films deposited on the membrane, or exothermic or endothermic reactions in liquids that contact the sensor. Because the elastic interactions between a liquid and a low-velocity A_0 mode can be purely reactive, only an evanescent disturbance is excited in the liquid. Accordingly, it appears that only very small liquid volumes need be used.

We believe that many inexpensive quasi-digital sensors can be based on this simple structure. With it one could realize an accelerometer, barometer, thermometer, vapor or gas sensor, and so on. With the addition of suitable electrodes or a ferromagnetic film, it could sense electric and magnetic fields as well. Because of its sensitivity to many different measurands, for selective response it will be necessary to design the device properly for a given application, or to use one or more active and reference sensors together. It appears that information obtained from the different responses of the several propagating modes and operating wavelengths of the device can be used to obtain precise information about individual measurands.

ACKNOWLEDGMENT

We gratefully acknowledge L. Field's adaptation of the Sekimoto LPCVD silicon nitride process for the Berkeley Microfabrication Facility, the demonstration by E. S. Kim and R. S. Muller that zinc oxide could be deposited succesfully on such nitride films, and discussions with R. T. Howe about resonant microsensors. We also appreciate collaborative work on Lamb waves with K. Uozumi, of Aoyama Gakuin University, Tokyo, and help with the acoustic wave analysis program provided by E. Adler, of McGill University.

REFERENCES

[1] *IEEE Trans. Ultrasonics, Ferroelectrics, Frequency Control, Special Issue on Acoustic Sensors*, vol. UFFC-34, no. 2, Mar. 1987.
[2] H. Wohltjen and R. Dessy, "Surface acoustic wave probe for chemical analysis. I. Introduction and instrument description; II. Gas chromatography detector; III. Thermomechanical polymer analyzer," *Anal. Chem.*, vol. 51, pp. 1458–1475, Aug. 1979.
[3] I. A. Viktorov, *Rayleigh and Lamb Waves.* New York: Plenum, 1967.
[4] S. Timoshenko and S. Woinowsky-Krieger, *Theory of Plates and Shells.* New York: McGraw-Hill, 1959.
[5] L. E. Kinsler, A. R. Frey, A. B. Coopens, and J. V. Sanders, *Fundamentals of Acoustics.* New York: Wiley, 1982.

[6] R. D. Watkins et al., "The attenuation of Lamb waves in the presence of a fluid," Ultrasonics, vol. 20, pp. 257–264, Nov. 1982.

[7] G. Kurtze and R. H. Bolt, "On the interaction between plate bending waves and their radiation load," Acustica, vol. 9, pp. 238–242, 1959.

[8] A. A. Nassar and E. L. Adler, "Propagation and electromechanical coupling to plate modes in piezoelectric composite membranes," in Proc. IEEE Ultrasonics Symp. (Atlanta, GA, 1983), pp. 369–372.

[9] R. M. White, P. J. Wicher, S. W. Wenzel, and E. T. Zellers, "Plate-mode ultrasonic oscillator sensors," IEEE Trans. Ultrasonics, Ferroelectrics, Frequency Control, vol. UFFC-34, no. 2, pp. 162–171, Mar. 1987.

[10] T. W. Bristol, "Analysis and design of surface acoustic wave transducers," in Proc. IEE Specialist Seminar Component Performance and Systems Applications of SAW Devices (Aviemore, Scotland, 1973), vol. 109, pp. 115–129.

[11] M. Sekimoto, H. Yoshihara, and T. Ohkubo, "Silicon nitride single-layer x-ray mask," J. Vac. Sci. Technol., vol. 21, pp. 1017–1022, Nov./Dec. 1982.

[12] R. M. White and S. W. Wenzel, "Inexpensive and accurate two-sided semiconductor wafer alignment," Sensors and Actuators, vol. 34, no. 4, pp. 391–395, 1988.

[13] G. L. Halac, "Fabrication of a piezoelectric accelerometer with integrated circuit technology," M.S. Research Rep., Univ. of California, Berkeley, June 1982.

[14] A. J. Slobodnik, Jr., E. D. Conway, and R. T. Delmonico, Microwave Acoustics Handbook, vol. 1A, Air Force Cambridge Research Labs., Bedford, MA, Oct. 1973.

[15] W. A. Fate, "High-temperature elastic moduli of polycrystalline silicon nitride," J. Appl. Phys., vol. 46, no. 6, pp. 2375–2377, June 1975.

[16] R. M. White, "A sensor classification scheme," IEEE Trans. Ultrasonics, Ferroelectrics, Frequency Control, vol. UFFC-34, no. 2, pp. 125–126, Mar. 1987.

[17] R. T. Howe, "Resonant microstructures," in Transducers '87 Dig., June 1987.

[18] H. Wohltjen, "Mechanism of operation and design considerations for surface acoustic wave device vapor sensors," Sensors and Actuators, vol. 5, pp. 307–325, June 1984.

[19] J. E. Roederer and G. J. Bastiaans, "Microgravimetric immunoassay with piezoelectric crystals," Anal. Chem., vol. 55, pp. 2333–2336, Dec. 1983.

[20] D. Hauden, "Miniaturized bulk and surface acoustic wave quartz oscillators used as sensors," IEEE Trans. Ultrasonics, Ferroelectronics, Frequency Control, vol. UFFC-34, no. 2, pp. 253–258, Mar. 1987.

ACOUSTIC WAVE DEVICES FOR SENSING IN LIQUIDS

S. J. Martin, A. J. Ricco, and R. C. Hughes
Sandia National Laboratories, Albuquerque, NM, USA

Abstract

Horizontally polarized shear waves (HPSWs) have been used for acoustic sensing of liquid viscosity and for measuring sub-monolayer mass changes accompanying electrode reactions. HPSWs are excited on a piezoelectric substrate using interdigital transducers in much the same fashion as the Rayleigh waves used in surface acoustic wave (SAW) sensors. HPSW sensors are shown to have several advantages compared to SAW devices and bulk crystal oscillators for liquid sensing.

Introduction

Surface acoustic wave (SAW) devices have received considerable attention for their versatility and extreme sensitivity as gas phase sensors [1-3]. These devices utilize a pair of interdigital transducers to excite and detect a Rayleigh wave which propagates across the intervening crystal surface. Typically, the surface is modified with a thin film capable of sorbing a particular species from the gaseous environment. Surface accumulation of the species results in perturbation of the wave velocity and/or attenuation of the Rayleigh wave. For reasons to be discussed shortly, high frequency (>100 MHz) acoustic wave devices have not been used for liquid sensing. Bastiaans [4] has demonstrated the use of low frequency (10 MHz) SAW devices for aqueous sensing. Bruckenstein [5] and Kanazawa [6] have used the bulk crystal oscillator, which also operates at relatively low frequencies, for sensing in liquids as well.

When used in an oscillator circuit, SAW devices and bulk crystal oscillators are extremely sensitive to changes in surface mass. Sorbed species move synchronously with surface particle displacements, causing a perturbation in frequency, Δf, related to the sorbed mass by the Sauerbrey equation [7]:

$$\Delta f/f_o = -c_m f_o \delta_m, \qquad (1)$$

in which c_m is the mass sensitivity factor of the device, f_o is the unperturbed operating frequency, and δ_m is the accumulated mass density (mass/area) of species on the surface. Factors other than mass-loading may contribute to frequency shifts, but in many cases these effects are insignificant. Wohltjen has pointed out that the sensitivity factor c_m in Eq. 1 is quite similar for SAW devices and bulk crystal oscillators [1]; the greater overall sensitivity of Rayleigh wave devices is largely a result of their 1 to 2 orders of magnitude higher operating frequency.

The polarization (direction of particle displacement) of an acoustic wave is unimportant in gas phase sensing. In liquid sensing, however, wave polarization direction is a crucial factor in determining the extent of wave damping by the medium. Displacements normal to the surface generate compressional waves in the liquid which dissipate wave energy, causing attenuation proportional to frequency [8]:

$$\alpha = c_\alpha f \qquad (2)$$

Displacements lying in the plane of the surface couple viscously to the adjacent liquid, causing much less attenuation.

The Rayleigh wave has two components of particle displacement, one parallel to the surface along the direction of wave propagation and the other normal to the surface. Generation of compressional waves by the surface-normal component of a Rayleigh wave on ST-quartz and in contact with water results in a calculated attenuation of 4.2 dB/cm-MHz. At low frequencies and over short path lengths, this attenuation is manageable: Bastiaans has demonstrated microgravimetric immunoassay using 10 MHz SAW devices, i.e., the weighing of antigens sorbed from aqueous solution onto an antibody-modified quartz surface [4]. Unfortunately, operation at higher frequencies where sensitivity is greater (Eq. 1) results in attenuation so severe that near-total extinction of the Rayleigh wave occurs.

Horizontally Polarized Shear Waves

Although liquid phase sensing using the Rayleigh wave is difficult, the interdigital transducer is capable of exciting other waves, one of which is better suited for such applications. Lau et al. demonstrated [9] that bulk waves can be generated by an interdigital transducer and propagated parallel to the surface when excited at the frequency f = v_0/d, where d is the transducer periodicity and v_0 is the velocity of the bulk wave propagating at zero angle to the surface. These waves, which propagate at 5060 m/s in ST-quartz (1.6 times the Rayleigh velocity), are called variously shallow bulk acoustic waves (SBAWs), surface-skimming bulk waves (SSBWs), and horizontally polarized shear waves (HPSWs). For liquid sensing applications, the essential feature of the HPSW is that it has only a single polarization direction, lying in the plane of the surface and normal to the propagation direction. The absence of a surface-normal component permits this wave to propagate efficiently at a solid/liquid interface, while the in-plane component allows wave-perturbing coupling with sorbed species.

In this paper, we describe some of the uses of acoustic wave devices utilizing the HPSW for sensing in liquid environments. Mass sensitivity equivalent to a fraction of a monolayer is demonstrated by experiments utilizing the HPSW device as a real-time electrode monitor: mass changes are correlated with cell current during electrodeposition reactions and during cyclic voltammetry. We show in addition that the HPSW responds to changes in liquid viscosity, providing the basis for a microviscometer.

Experimental Methods and Apparatus

HPSW devices for liquid sensing were constructed on ST-quartz. HPSWs are excited and detected by a pair of interdigital transducers separated by 6.5 mm and positioned on either side of a bottomless teflon cell containing ~50 μl of the liquid to be examined, as shown in Fig. 1. The polarization direction of the HPSW is indicated.

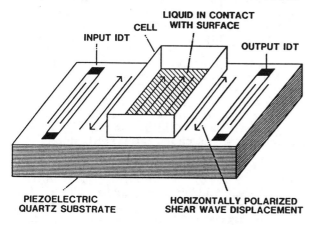

Fig. 1. Acoustic wave sensor for operation in liquids. Interdigital transducers on ST-quartz excite and detect the horizontally polarized shear wave which probes the solid/liquid interface.

The transducers, comprised of 50 finger-pairs having 32 μm periodicity, are patterned from 200 nm thick Cr/Au. The 6 mm high cell defines a wave-liquid interaction length 3.0 mm in the wave propagation direction and 4.1 mm transverse to it. A seal is obtained by holding the cell against the surface with slight pressure, resulting in 1-3 dB of propagation loss between transducers. Details of the RF measurements are given elsewhere. [3,10]

Measurements in Liquids

Two mechanisms for the perturbation of the HPSW propagation characteristics at the solid/liquid interface will be explored: (1) viscous coupling of the wave to the liquid; (2) mass loading of the wave by species sorbed from the solution. The experimental frequency response of the HPSW sensor of Fig. 1 is shown in Fig. 2. Notable is the signal level obtained with a bare surface (solid line) compared to that obtained with a cell containing water (dashed line).

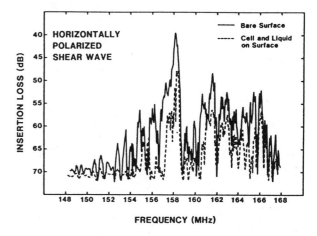

Fig. 2. Frequency response of HPSW device with and without liquid in wave path.

The liquid accounts for about 3.5 dB of loss, while the cell contributes the remainder. The device can be operated effectively in an oscillator loop at 158 MHz, at which frequency the mass sensitivity (*vide infra*) is comparable to that obtained with Rayleigh wave devices in gas phase sensing [3].

Measurement of Viscosity [10]

The shear polarization of the HPSW couples viscously to liquid on the device surface, causing a thin liquid layer adjacent to the surface to oscillate. The amplitude of liquid movement decays exponentially with distance from the surface with a decay length $\delta = \sqrt{2\eta/\rho\omega}$, in which η is liquid viscosity, ρ is liquid density, and ω is the angular frequency of HPSW excitation. The oscillating surface entrains a liquid thickness $\delta \approx 50$ nm for water at 158 MHz. Viscous coupling of the wave to the liquid results in a perturbation in the HPSW velocity and also introduces attenuation. The propagation loss incurred by the wave in propagating through the cell containing a liquid is given by:

$$L = A \sqrt{\omega \rho \eta} \quad \text{(dB)} \quad (3)$$

in which the prefactor A is approximated by $1.2\omega l/\rho_S v_o^2$; ρ_S is the substrate density and l is the liquid-solid interaction length defined by the cell (Fig. 1). To use the device as a microviscometer, the change in signal level at the output transducer is measured as liquid is placed in the cell; viscosity is calculated from Eq. 3.

Viscous loss was measured for liquids having viscosities between 0.3 and 1500 cP. Methanol, acetone, castor oil, ethylene glycol, and a number of water/glycerol mixtures provided a number of data points spanning the aforementioned range. Fig. 3 shows a plot of normalized power loss, $L/\sqrt{\rho}$, vs. $\sqrt{\eta}$ for the various liquids tested; η values are from the literature. Below ~80 cP, $L/\sqrt{\rho}$ varies linearly with $\sqrt{\eta}$ as predicted by Eq. 3. The upper limit of measurable viscosity is reached when the liquid relaxation time, which is proportional to viscosity, becomes as long as the period of wave excitation.[10] In this high viscosity regime, the liquid displays viscoelastic behavior, with loss attaining a constant value.

A unique aspect of the HPSW microviscometer is that it samples only a 50 nm thick liquid layer adjacent to a surface. In some systems,

surface viscosity may differ appreciably from bulk shear viscosity due to liquid-surface interactions.

Measurement of Mass Changes

In addition to measuring viscosity, the HPSW can be used to measure mass deposited from (or removed into) a liquid. In general, the perturbation in wave velocity due to surface mass accumulation depends both on changes in surface mass and stiffness properties. In many cases the mass contribution dominates, with the fractional velocity perturbation given by:

$$\Delta v/v_o = - c_m f_o \delta_m, \quad (4)$$

with variables defined as in the functionally equivalent Sauerbrey equation, Eq. 1. The mass sensitivity factor c_m is approximated by $2.5/\rho_S v_o = 2 \times 10^{-7}$ m^2-s/kg and is, in fact, within 20% of the Sauerbrey result. The signal represented by Eq. 4 adds to the (constant) shift arising from the entrainment of liquid through viscous coupling. Thus, the HPSW allows sensor operation at high frequencies (>100 MHz) where sensitivity is greatest, without the severe attenuation experienced by Rayleigh wave devices. Although they suffer relatively minor attenuation in liquids compared to SAW devices, bulk crystal oscillators cannot be constructed to operate at the high frequencies where sensitivity is greatest due to physical limitations on how thin the quartz substrate can be made.

To demonstrate the measurement of mass accumulated from solution, the electrodeposition of silver onto a palladium electrode (vacuum evaporated onto the quartz substrate) was examined. Since deposition rate is controlled by electrode current (galvanostatic control), direct calibration of the HPSW mass sensitivity is obtained by comparing the frequency shift with the charge passed for a plating process of known current efficiency. For the silver plating experiment, which has current efficiency near unity [11], aqueous electrolyte containing 0.3 M $[Ag(CN)_2]^-$ and 0.3 M free CN^- in a basic solution was added to the electrochemical cell; a Ag wire served as counter electrode. Current density was ~0.3 mA/cm^2. Fig. 4 shows the fractional frequency change vs. the charge passed. The HPSW monitor reveals an induction period, during which no significant change in surface mass occurs. This may be due to reductive dissolution of surface impurities and/or Pd hydride formation. Following the induction period, the frequency changes at a rate of 1.0 ppm/μC; assuming a deposition efficiency of 100%, this implies a mass

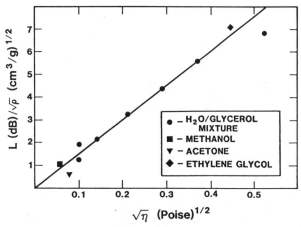

Fig. 3. Relative power loss between transducers due to viscous damping of the horizontally polarized shear wave as a function of liquid viscosity.

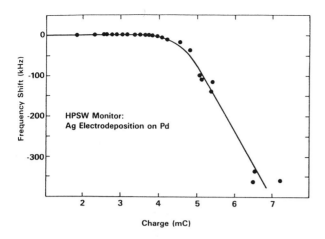

Fig. 4. Frequency shift of HPSW oscillator during electrodeposition of Ag onto a surface Pd electrode.

sensitivity of 1.6×10^{-7} m^2-s/kg, reasonably close to the calculated value.

Another illustration of electrode monitoring is provided by cyclic voltammetry. The results of this experiment, which has previously been performed using bulk crystal oscillators [5,6], are shown in Fig. 5. The working electrode is a vacuum evaporated gold film. Both current-potential and frequency shift-potential curves are shown for two potential cycles, taken in 0.5 M H$_2$SO$_4$. At approximately +0.66 V applied potential, a sudden increase in anodic current results from the formation of a single monolayer of oxide on the Au surface [5]. Concomitant with this oxide formation, a sudden decrease in frequency corresponding to the mass of this oxide monolayer is registered by the HPSW device. The HPSW monitor shows that the oxide gradually dissolves at less positive applied potentials. The 18 ppm frequency excursion corresponds to a change in surface coverage of 5×10^{15} oxygen atoms/cm^2. The noise level is on the order of 0.2 ppm, so resolution is a small fraction of a monolayer. Because this experiment used a two electrode cell, the features of the voltammogram differ markedly from those of the more usual three electrode voltammogram.

Conclusions

Acoustic wave devices using the horizontally polarized shear wave are effective sensors of fluid viscosity and allow highly accurate measurement of mass changes on a surface immersed in a liquid. The HPSW allows operation at high frequencies where sensitivity is greatest, without the losses associated with operation of Rayleigh wave devices in solution.

This work was performed at Sandia

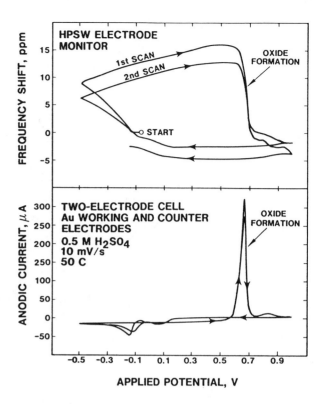

Fig. 5. Simultaneous measurement of surface mass change (top) and current flow while cycling potential applied to a surface Au electrode.

National Laboratories and supported by the U. S. Department of Energy under contract no. DE-ACO4-76DP00789.

References
1. H. Wohltjen and R. Dessy, Anal. Chem., 51 (1979) 1458-1475.
2. C. T. Chuang and R. M. White, Proc. IEEE Ultrasonics Symp. (1981) 159-162.
3. A. J. Ricco, S. J. Martin, and T. E. Zipperian, Sensors and Actuators, 8 (1985) 319-333.
4. J. E. Roederer and G. J. Bastiaans, Anal. Chem., 55 (1983), 2333-2336.
5. S. Bruckenstein and M. Shay, J. Electroanal. Chem., 188 (1985) 131-136.
6. R. Schumacher, G. Borges, and K. K. Kanazawa, Surface Science, 163 (1985) L621-6.
7. G. Sauerbrey, Z. Physik, 155 (1959) 206.
8. B. A. Auld, "Acoustic Fields and Waves in Solids", Vol. 2, John Wiley, New York (1973), pp. 283-286.
9. K. F. Lau, K. H. Yen, J. Z. Wilcox, and R. S. Kagiwada, Proc. Symp. on Frequency Control (1979), 388-395.
10. A. J. Ricco and S. J. Martin, Appl. Phys. Lett., submitted.
11. F. A. Lowenheim, "Electroplating", McGraw-Hill, New York (1978), pp. 257-263.

Part 4
Microsensor Circuit Interfaces

SENSORS are usually linked to signal processors through customized interface circuits whose design is determined, on the input side, by the transduction characteristics of the sensor and, on the output side, by the signal processor. A function that is often carried out by interface circuits is to correct for nonideal behavior of the sensing element; for example, for its nonlinear response, baseline drift, or zero-signal offset. Other functions include compensation for temperature, pressure, humidity, and other environmentally uncontrolled parameters.

Two papers are included in this section that deal with the behavior and performance of piezoresistive pressure sensors. Bryzek discusses the effects of temperature and signal levels on the performance of a resistive-bridge interface circuit. Suzuki, Ishihara, Hirata, and Tanigawa discuss nonlinear characteristics and compensation by judicious placement of the piezoresistive elements. One method of temperature compensation, for example, is to make a simultaneous measurement of the variable of interest and of temperature at the sensing site, thus building a primitive form of a multisensor. Multisensors, in turn, generally demand more extensive signal processing and conditioning than simple sensors. Multiplexing of signals to reduce lead count, and hence the overall size, may be required, and normalizing signal types and levels prior to data acquisition and processing is usually necessary. A paper by N. Najafi, Clayton, Baer, K. Najafi, and Wise presents a circuit architecture for interfacing with microsensor arrays.

In corrosive sensing environments, output signals may require amplification and buffering or modulation encoding to make them less sensitive to possible changes in the input–output lead impedance. In addition, circuits for implanted sensors, which need to be battery powered, must have low power consumption and long-term stability. Two papers, by Spencer, Fleischer, Barth, and Angell; and by Smith, Bowman, and Meindl, deal with these issues as they apply to resistive- and capacitive-type sensors, respectively.

Special consideration is required to sense small variations in a sensing capacitor. In many cases parasitic capacitance is comparable to or even greater than the sensed value. At-site measurement or the use of parasitic-insensitive methods is needed. The application of switched-capacitor techniques to this measurement problem is presented in the paper by Park and Wise. Another approach to the precise measurement of capacitance using a charge redistribution scheme while providing a very high sensitivity is described by Kung, Lee, and Howe.

AN ARCHITECTURE AND INTERFACE
FOR VLSI SENSORS

N. Najafi, K. W. Clayton, W. Baer, K. Najafi, and K. D. Wise

Center for Integrated Sensors and Circuits
University of Michigan
Ann Arbor, Michigan 48109-2122

ABSTRACT

A VLSI sensor interface has been designed for use in bus-organized sensor-driven process control systems where high accuracy and high reliability are important. The interface permits 12-bit digital sensor data to be communicated to the host processor over a bidirectional parallel data bus which includes parity checking. The sensor can be self-testing and employs digital compensation of cross-parameter sensitivities. The interface has been implemented using discrete commercial components and has been designed in monolithic form in 3μm single-metal double-poly CMOS technology. The chip has a die area of 10.8mm x 8.5mm before compaction and has a simulated power dissipation of 75mW. The on-chip microprocessor operates at 4MHz and the 12-bit ADC has a conversion time of 14μSec.

INTRODUCTION

As integrated transducers are combined with increasing amounts of on-chip or in-module circuitry, where to partition the electronic system and how much electronics to include with the "sensor" become major issues. Integrated sensors, and particularly those associated with automated manufacturing and process equipment control, are likely to evolve into smart peripherals, and the definition of appropriate sensor interface standards is currently the subject of at least three national committees. For process control applications, high accuracy and high reliability are important goals, suggesting the use of substantial amounts of on-chip circuitry. More and more sensors are also being implemented as multi-transducer arrays, requiring input as well as output data for their operation. For example, thermally-based flow devices require heater control as well as temperature readout, and more advanced devices require that the heater be externally adjustable. This implies a bidirectional sensor interface capable of enhanced control; however, there are no agreed-on standards, feature lists, or architectures at the present time for the high-end integrated sensors which will be needed for the 1990s and beyond.

This paper reports initial results from a research program which has a number of long-term goals, including 1) the development of improved architectures for integrated sensing systems, 2) understanding the implications of using increased amounts of on-chip electronics on sensors, 3) the development of standards for the sensor-system interface, 4) the development of high-performance, modular, generic interface circuits for use on sensors, and 5) the study of the barriers associated with sensor performance at the 12-bit level. The work is intended to address the problems associated with the development of a generic process controller for the 1990s and involves a number of fundamental issues in system partitioning, controller architecture, sensor function, and sensor testing/compensation. It is hoped that the work will serve to initiate discussions which will lead to the development of standards for sensors and their associated control systems.

This paper describes a possible organization for high-end sensors and associated interface protocols. The interface described is addressable, programmable, self-testing, compatible with a bidirectional digital sensor bus, and offers 12-bit accuracy using internally-stored compensation coefficients. The design is sufficiently flexible to allow upward-compatible sensor designs to be inserted in existing equipment without reprogramming the host system and will accommodate differing sensor features. Thus, it should be possible within this organization to automatically upgrade machine performance in the field while preserving the ability of the sensor manufacturer to provide innovative and unique product features.

SYSTEM ORGANIZATION

Figure 1 shows a block diagram of the overall sensor-driven control system. A host computer, acting as system controller, interfaces with up to 256 nodes over a standardized bidirectional digital bus. Each node can contain a number of sensors and/or actuators. It is expected that each actuator will effectively be in a system feedback loop with one or more sensors. The sensors represent the critical element in the control loop since overall system operation depends on their accuracy. If they are not accurate and reliable, then no matter how sophisticated the controller is, the machine will not perform adequately. Thus, it is expected that the use of sensors which cannot be tested from the host controller will become increasingly unacceptable as automated control systems evolve into the 1990s.

Figure 2 shows the bus developed for communication between the nodes and the host. This Michigan Parallel Standard (MPS) bus contains 16 lines, including eight bidirectional data lines (D0-D7), a parity line, four control lines (HHS, NBR, NHS, and NAL) for synchronizing message transfers, and three power lines. For the discrete implementation of this interface, ±12V supplies have been used, while the monolithic design uses ±5V supplies. Message transfer is initiated by the host when it addresses a specific node to request or transfer data. The first (address) byte is decoded by all nodes on application of HHS. The addressed node acknowledges by pulsing NHS high and sets NBR high, locking out other nodes for subsequent bus data. Subsequent message bytes are entered and acknowledged using HHS and NHS, respectively, as

Reprinted from *Rec. of the IEEE Solid-State Sensor and Actuator Workshop*, 1988, pp. 76–79.

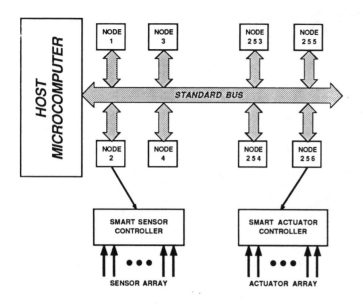

Fig. 1: Block diagram of a sensor-driven control system

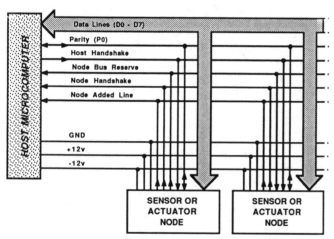

PARALLEL MICHIGAN PROVISIONAL STANDARAD (MPS) BUS

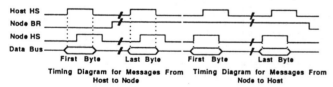

MESSAGE TRANSFER DIAGRAMS BETWEEN HOST AND NODES

Fig. 2: The Michigan Parallel Bus (MPS) and its message transfer diagrams.

indicated in Fig. 2. Messages are of variable length, with the bus released as NBR drops low. The NAL line can be used as a system interrupt in the event of situations requiring immediate action by the host or can be used to signal the insertion of a new sensor in the event of a replacement or addition to the system.

Figure 3 shows a block diagram of a sensing node for this "fifth-generation" sensor design. An array of transducer outputs are amplified, multiplexed, and converted to digital data under the control of a microprocessor-based microcontroller, which interfaces to the external sensor bus through a communication interface. The PROM contains the node address as well as information as to the sensor

compensation techniques and interface protocols. Thus, upon insertion in the host system, the node can deliver to the host information as to its features, the compensation techniques employed, and the compensation coefficients. This PROM is encoded at the time of sensor test and allows digital "trims" to be substituted for the traditional analog laser trims long associated with cross-parameter compensation and offset/slope control. Amplifier offsets and gains are set via this PROM as well and are internally compensated so that the analog signals remain in-range for the analog-to-digital converter (ADC). Actual precision compensation (e.g., for temperature) is performed by the host system, where much greater processing power is assumed. This organization attempts to implement as many system features associated with high accuracy and reliability as possible while maintaining the sensor electronics at a moderate level of complexity. It is possible that sixth-generation sensors will eventually do all compensation on-chip, thus simplifying the host system; however, if suitable standards can be developed for both the hardware and the software interface to the sensors, the use of data compensation in the host via standard modules should not present a serious

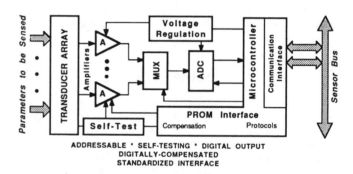

ADDRESSABLE • SELF-TESTING • DIGITAL OUTPUT
DIGITALLY-COMPENSATED
STANDARDIZED INTERFACE

Fig. 3: Block diagram of a VLSI fifth-generation integrated sensor

problem. In the event of a parity error, the host requests retransmission, and in the event of repeated errors, the node can be tested by the host and a particular sensor, or an entire node, can be removed from service. The goal is to be able to detect sensor drift prior to catestrophic failure, however, so that timely replacements can be made and equipment downtime can be avoided.

The sensors are sampled sequentially by the microcontroller so that their most recent values are always present in RAM; hence, the external host system is not required to wait for data conversion by the node, and sensor reads appear much like memory accesses.

DISCRETE SYSTEM IMPLEMENTATION

The above system organization has been implemented in discrete form using an IBM PC/XT as host. The sensor node was implemented on an Augat URG1 wirewrap panel and requires 22 commercial components. The microcontroller was realized using a National NSC-800 microprocessor. This present system is configured to measure reference pressure, chamber/unknown pressure, and temperature, and can be used as an automated sensor characterization system. The system noise level currently corresponds to about 11 bits, with temperature automatically controlled from -70°C to +205°C and pressure controlled from vacuum to 2000 Torr. Figure 4 shows representative signals on the system bus during a message transfer from a node to the host. System speed here is set by the host.

Figure 5 shows an example of digital compensation of the sensor data using PROM-based compensation coefficients. The commercial piezoresistive pressure sensor used for this example was measured from 50 Torr to 1000 Torr at over 2000 points. The uncompensated pressure nonlinearity over this range had a maximum of about 0.6 percent. Based on this data, compensation coefficients were automatically generated using a statistical analysis program for a fifth-order polynomial to convert the measured pressure response into true pressure. The compensated nonlinearity is reduced to less than 0.1 percent. Similar results appear possible for the compensation of temperature effects. Thus, it appears feasible to considerably enhance the performance of sensors using digital compensation (and its ability to handle nonlinear effects). Sensor stability then becomes a major concern and focus in the realization of high-accuracy devices. It is not yet known whether polynomial compensation can be used to reduce the

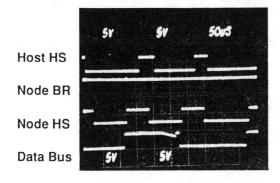

Fig. 4: Representative signals on the MPS system bus during a message transfer from a node to the host.

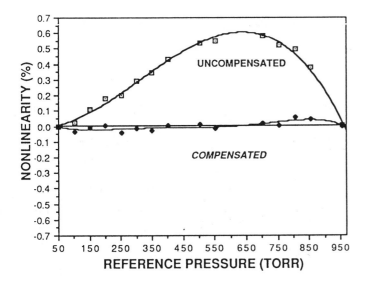

Fig. 5: Digital compensation of pressure nonlinearity for a commercial piezoresistive pressure sensor at 21.5°C.

required accuracy in the data converter so that the entire signal path can be handled using PROM-based compensation. If it can be, then substantial savings in testing and die area may be possible.

MONOLITHIC INTERFACE IMPLEMENTATION

Figure 6 shows a representation of the interface in more highly integrated form. The transducer array is implemented on one chip, which includes front-end circuitry and initial signal amplification. The sensor interface is realized as a single monolithic chip and interfaces with a commercial PROM such as a 2716. While in a practical (ca. 1994) application, the system might appear as a single monolithic chip (after all of the associated circuitry is optimized), at the present time a hybrid implementation appears more realistic using an off-chip PROM. Splitting the analog and digital portions of the interface chip into a hybrid might also be done for the near term.

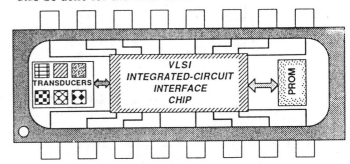

Fig. 6: A high-end hybrid (modular) integrated sensor for semiconductor process control.

Figure 7 shows a block diagram of the sensor interface chip as designed in monolithic form. The chip is realized in 3μm single-metal double-poly CMOS technology. Up to seven voltage-level sensor outputs are amplified and digitized under the control of a custom microprocessor. The chip also accepts eight digital (event) inputs as well as one pulse-rate (FM) sensor output. A 256-bit RAM stores the digitized sensor data, while a ROM stores the program code for the processor. The communication interface can be configured for either the MPS bus or for a serial bus; all other circuitry is independent of the bus configuration used. The processor employs a 12-bit internal bus, consistent with handling sensor data at the 12-bit level.

The microprocessor executes up to sixteen different commands from the host and implements 22 different instructions internally. All transducers can be read on command or only a single transducer can be interrogated. Compensation data can also be readout or self-test can be initiated on command. In addition, an 8-bit register can be loaded from the host for the control of internal flags or on-chip actuators.

Figure 8 summarizes the layout area of the chip in 3μm CMOS. The microprocessor and communication interface are custom implementations, while the amplifier and ADC were realized with the aid of a silicon compiler. Substantial compaction of the layout is thought to be possible as circuits are further optimized, and this optimization will be performed prior to initial integration of the interface. The present overall chip size is 10.3mm x 8.5mm. The use of a 2μm double-metal double-poly process would also be expected to decrease this die area by about 50%. As

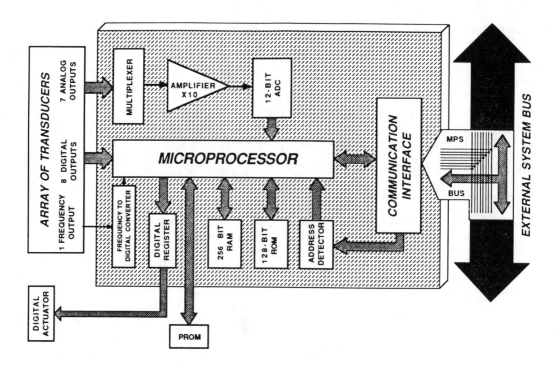

Fig. 7: Block diagram of the integrated sensor/actuator
interface chip.

noted in Fig. 9, the simulated power dissipation for the entire chip is 75mW. The amplifier has an open loop gain of 87dB, an output swing from -4.8V to +4.3V, a bandwidth of 2.3MHz, and a closed-loop gain of 20dB. The common-mode and power-supply rejection ratios are -71dB and -80dB, respectively. The analog-to-digital converter has a conversion time of 13.5μSec at 12 bits.

While the die area associated with the full monolithic implementation of the sensor interface is relatively large by present standards, by the early 1990s it will not appear so formidable as further feature size reductions, technology improvements, and circuit refinements reduce the die size and the associated chip cost. It appears likely that such interfaces will be viewed as increasingly attractive and practical for process control systems over the next few years as efforts continue to achieve higher accuracy and improved system reliability. While substantial challenges lie ahead in generic controller design, sensor testing, circuit development, and fundamental transducer design, it is hoped that this work will be useful as a starting point in the development of high-end sensors for process equipment control and perhaps for other applications as well.

ACKNOWLEDGMENTS

The authors would like to acknowledge the support of the Semiconductor Research Corporation for this work under Contract 87-MP-085. Mr. T. Huang contributed substantially to the layout of the monolithic sensor interface while S. Cho and J. Cowles assisted in the development of the process technology and in the design of the analog portions of the interface circuitry.

Microprocessor	5.6x6.1mm^2
ALU	1.2x2.6mm^2
Data Path	3.7x1.8mm^2
Control Unit	2.3x2.3mm^2
Analog-to-Digital Converter	4.2x4.7mm^2
Communication Interface	1.8x2.1mm^2
Frequency-to-Digital Converter	1.2x1.1mm^2
RAM	1.3x1.1mm^2
ROM	0.8x0.3mm^2
Operational Amplifier	0.8x0.3mm^2
Address Detector	0.6x0.4mm^2
Multiplexer	0.4x0.4mm^2

Figure 8: Layout areas for individual circuit blocks of
the interface chip

Active Area*	10.3x8.5mm^2
Power Dissipation	~75mW
Supply Voltage	±5V
Technology	3μm Single-Metal, Double-Poly CMOS
Amplifier Gain	20dB
ADC Conversion Time	13.5μsec, 12-Bits
Microprocessor Clock	4MHz (Simulated @ 2,4,8 MHz)
RAM Size	256 Bits
FIFO Size	144 Bits (18x8)

*Compaction and optimization of circuit blocks will reduce the
area in future designs considerably.

Fig. 9: Characteristics of the integrated interface chip.

MODELING PERFORMANCE OF PIEZORESISTIVE PRESSURE SENSORS

Dr. Janusz Bryzek

IC Sensors, Inc., 430 Persian Drive, Sunnyvale, CA 94089

ABSTRACT

In this paper the basic piezoresistive pressure sensor characteristics in a bridge configuration will be discussed. For the first time several unique modeling approaches developed by the author, both analog and digital, will be introduced. Test system configuration leading to the derivation of model components is outlined. An example of the span temperature error calculations with a single resistor calibration is shown.

1. INTRODUCTION

Piezoresistive IC pressure sensors offer a very simple electronic structure when compared to microprocessors or monolithic memories. Mechanical configuration is also very straightforward, as shown in Fig. 1. Due to the fact however that they perform mechanical to electrical conversion in the presence of the unavoidable high temperature sensitivity of silicon, electrical modeling of their performance with a high degree of accuracy is not a very simple task.

There are two major areas of applications for precision sensor modeling. The first group of applications includes temperature compensation wherein the values of the temperature compensating components are calculated as a result of the sensor test results. The second group of applications involves a direct pressure calculation or display based on a stored numerical sensor model and measured sensor outputs.

2. BASIC SENSOR CHARACTERISTICS

There are three primary variables, not effected by the sensor excitation source or the load resistances, which enable the modeling of sensor's performance in essentially every possible electrical configuration:
- Bridge output voltage representing primary pressure output; usually output voltage is divided between a zero pressure output voltage, independent of pressure, and a pressure dependent component.
- Normalized pressure sensitivity representing output voltage change with full scale pressure change, normalized to unity bridge voltage and

unity pressure, expressed often in [mV/V/psi].
- Bridge resistance, consisting of four arm resistances two of each increasing with pressure and the other two decreasing with pressure.

Each of these variables is influenced by several effects, out of which the most important are the following:
- differential pressure across the diaphragm
- operating temperature
- excitation voltage or current level

In order to build an appropriate model a number of measurements has to be performed. The minimum number of measurements is set by the sophistication level of the model. More accurate models require more measurements to be made.

If certain functions can be defined as a result of known correlations with some other variables

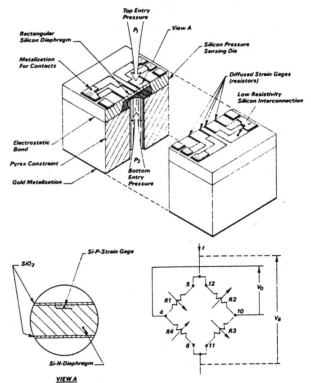

Fig. 1 Sensor structure and configuration.

Reprinted from *Rec. of the 3rd Int. Conf. on Solid-State Sensors and Actuators*, 1985, pp. 168–173.

(eg. temperature coefficient of pressure sensitivity and bridge resistance as a function of doping concentration) then the required number of measurements may be smaller.

The limit of possible modeling accuracy is set of course by the stability and thermal hysteresis of the sensor.

3. BASIC ANALOG MODEL

One of the simpler analog models of pressure sensors is shown in Fig. 2. This model consists of four elements for each temperature and each current level:
- resistance R representing bridge resistance of a perfectly balanced bridge,
- resistance r representing bridge unbalance (offset) at zero pressure,
- resistance q representing pressure sensitivity of the strain gages subjected to compression,
- resistance p representing pressure sensitivity of the strain gages subjected to tension.

Differentiation between p and q sensitivities is justified by the occurence of different stress levels in the areas of tension and compression on the diaphragm. For example, in the rectangular diaphragm the maximum stress at the edge has about two times higher amplitude (and different polarity) than the stress in the center.

Basic model discussed here is based on the following assumptions:
- bridge resistance is symmetrically distributed between left and right half-bridges,
- pressure sensitivity of strain gages increasing with pressure is the same in both half-bridges and the same is true for pressure sensitivity of

strain gages decreasing with pressure.

The accuracy of these assumptions is a function of sensor configuration and symmetry. For one of the designs, developed by IC Sensors and illustrated in Fig. 1, bridge configuration consists of two inverted half-bridges and the assumptions are sufficient for the basic temperature compensation applications. It should be noted however that the accuracy of zero pressure output (offset) modeling will be a function of bridge symmetry.

All four elements of the model can be derived from the results of simple measurements performed with a constant current source I for the sensor's excitation. Denoting V0 and V1 the output voltage and E0 and E1 the bridge voltage, both respectively at zero input pressure and full scale input pressure, the values of model's components can be found as follows:

$$R = (E0 - V0) / I$$
$$r = 2 V0 / I$$
$$q = (V1 - V0 + E1 - E0) / I$$
$$p = (V1 - V0 - E1 + E0) / I$$

This procedure should be repeated for each test temperature and for each test current as well as for each additional pressure needed to model pressure nonlinearity. Each of the four model's elements can be then defined as a respective function of these three variables. From this point on, any curve fitting technique can be used to make an interpolation for the other points.

For basic temperature compensation applications it is usually sufficient to know values of elements R, r, p, q only for two temperatures in order to calculate the required compensating components.

The selection of a closed bridge test configuration as opposed to a direct resistance measurements offers several major advantages:
1. The model will represent the distribution of voltage potentials and leakage currents across the diaphragm very close to a final operating distribution; since all strain gages are isolated from the silicon substrate through a p-n junction, this model configuration includes all parasitic effects that would not be included by a direct resistance measurement.
2. Excitation current can be turned-on for a long enough time before the measurements, to allow a proper warm-up of the sensor, without effecting the testing through-out.
3. Smaller number of data acquisition channels is required (two instead of four for each sensor) resulting in major savings on cabling and input modules for production systems supporting simultaneously several hundred sensors.
4. Better accuracy of measurements. Typical resistance change with full scale pressure change is in the range of 2% of bridge's arm resistance. Requirement for a pressure sensitivity modeling accuracy of only .1% creates a demand for a bridge's arm resistance measurement accuracy of 10 ppm in order to deliver combined 20 ppm accuracy from four measurements.

Voltage measurement mode in a closed bridge configuration offers a distinctive advantage over a resistance measurement mode: ability to measure directly a differential output voltage with a simultaneous rejection of a common mode signal. For the same modeling accuracy as above only .05% voltage measurement accuracy is required as long

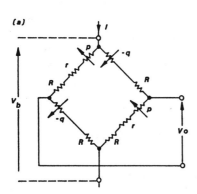

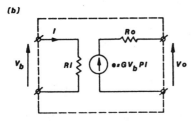

Fig.2 Basic model of pressure sensor
(a)bridge configuration,
(b)equivalent electrical circuit.

as a voltmeter has high enough common mode rejection.

5. Closed bridge configuration rejects a majority of common mode errors as first order temperature sensitivity and RFI interference simplifying requirements for stabilization of temperature and shielding.

Using the calculated model's components we can easily create the equivalent electrical circuit useful in modeling applications as shown in Fig. 2b. The circuit is analogous to an instrumentation amplifier model and it consists of three elements: input resistance Ri, output resistance Ro, and pressure controlled voltage source e. Their values can be easily calculated:

$$e = G\ V_b\ P_i \qquad \text{for voltage supply}$$
$$\text{or}\quad e = G\ R_i\ I\ P_i \qquad \text{for current supply}$$
$$R_i = R + (r + p - q)\ /\ 2$$
$$R_0 = R + (r + p - q)\ /\ 2$$

where G is normalized pressure sensitivity and Pi is input pressure.

This equivalent circuit does not include common mode components, as the effect of static pressure for example.

4. APPLICATION EXAMPLE

To demonstrate the effectiveness of the intoduced model we will evaluate the effect of single resistor calibration on temperature coefficient of full scale pressure span with no output load condition. There are two possible supply configurations for the sensor. With a constant voltage source (Fig. 3a), span calibrating resistor R should be connected in a series with the sensing bridge and when constant current source (Fig. 3b) is used the resistor R should be connected in parallel to the bridge. From electrical point of view both circuits are equivalent and the same value of resistor R is required to provide temperature compensation of span.

Insertion of the calubration resistor R introduces a loss of output span for given supply voltage or current. There are two different expres-

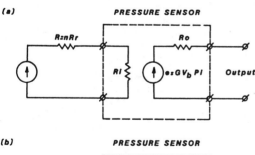

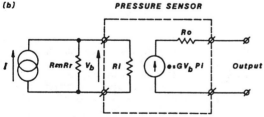

Fig.3 Span calibration resistor for constant voltage(a) and constant current (b) excitation.

sions for output span as a function of configuration. For a voltage mode the bridge voltage will

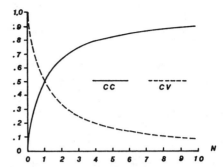

Fig.4 Loss of span as a function of calibrating resistor R expressed in multiples of bridge resistance Rr.

be defined as follows:
$$V_b = E\ R_i\ /\ (R + R_i)$$
and then span:
$$S = E\ R_i\ G\ P_i\ /\ (R + R_i)$$

For current mode the equations are modified as follows:
$$V_b = I\ R_i\ R\ /\ (R_i + R)$$
$$S = I\ R_i\ R\ G\ P_i\ /\ (R_i + R)$$

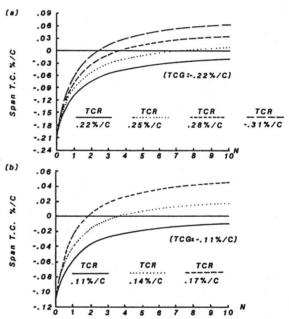

Fig.5 Effect of single resistor R calibration on span temperature coefficient. Resistor value expressed in multiples N of input resistance at reference temperature.

Denoting by Rr input resistance of the bridge at reference temperature and expressing calibrating resistor R as a multiple N of reference bridge resistance Rr to make a relationship more general in nature:

$$R = N\ R_r$$

we can now plot output span normalized to span without normalizing resistor R at the same excitation level, as shown in Fig. 4. For small values N a constant voltage (CV) mode is providing smaller loss of a signal. For high values N a constant current (CC) mode provides lower loss of a span.

305

Let's compare now temperature errors for both cases. To perform temperature errors evaluation we have to assume temperature coefficients of bridge resistance and pressure sensitivity. To compare errors for two diffusion processes we will use average temperature coefficients TC for high impedance and low impedance processes used at IC Sensors:

	High Impedance	Low Impedance
TC Ri:	A = .25% / C	.14% / C
TC G :	B = - .22% / C	- .11% / C

Bridge resistance and pressure sensitivity can be now related to temperature deviation dT from reference condition:

$$Ri = Rr (1 + dT A)$$
$$G = Go (1 + dT B)$$

Span temperature error can be then plotted as shown in Fig. 5. To show the effect of possible process variations a span error for several other possible temperature coefficients of bridge resistance is shown. Comparing these plots against eg. AAMI requirements of .1%/C maximum span error for blood pressure sensors it can be seen that the low impedance process meets the requirements for almost any calibration requirement (represented by a constant N) while the high impedance process requires at least 60% loss of output signal in the voltage mode to include process variations.

Another interesting aspect of these results, for higher values of N, is a small sensitivity of a span temperature error to process variations yielding a different temperature coefficient of bridge resistance. Also another trend is clear: lower value of N is required to achieve zero temperature coefficient of output span for a lower impedance process.

5. HIGH PERFORMANCE ANALOG MODEL

The model introduced formerly was assuming symmetry within a bridge. This allows minimization of the number of measurements to be made to create the model. The major drawback of that model was a relatively low accuracy of balancing the bridge with no pressure applied. The other drawback was the same expression for input and output resistance, decreasing modeling accuracy in such applications as calculation of shunt calibration for medical blood pressure sensors.

Major improvements of modeling accuracy can be achieved when the model consisting of four different resistors R1, R2, R3, R4, changing independently with pressure, temperature and excitation power level is used (Fig. 6). The equivalent electrical circuit will be the same as before, however the expressions for differential input and output resistance will be different:

$$Ri = (R1+R4) (R2 + R3) / (R1+R2+R3+R4)$$
$$RO - (R1+R2) (R3 + R4) / (R1+R2+R3+R4)$$

In order to calculate all four bridge resistors we need two times more measurements as before: at least four measurements for each pressure, temperature and supply voltage or current value. The author has developed a unique method allowing an efficient testing of sensors in closed bridge configuration yielding all model components with full accuracy of the system used. This test system uses two current sources I1 and I2 (Fig. 6) for sensor excitation. Two measurement cycles are employed.

During the first measurement cycle only the main current source I1 is used (I2 = 0). Five measurements are representing offset voltage V0 and four bridge arm voltages V1, V2, V3 and V4. During the second measurement cycle the additional cross current I2 is being injected across the bridge output and two additional voltage measurements across resistors R1 and R2, respectively V1c and V2c, are performed.

Calculation of bridge arm resistors is also split into two phases. During the first phase the set of ratiometric simultaneous equations is being solved, yielding values of resistors accurate within ratiometric accuracy of the voltmeter operating in several volt range. Ratiometric arrangement of equations improves the accuracy over a direct voltage approach. The set of simultaneous equations is very straightforward and easily converging:

$$R1 / R4 = V1 / V4$$
$$F2 / R3 = V2 / V4$$
$$R1 / R2 = Abs(V1-V1c) / Abs(V2-V2c)$$
$$(V1+V4) (R1+R2+R3+R4) = I1 (R1+R4) (R2+R3)$$

As it can be seen the value of crosscurrent I2 is not used directly in the equations, so requirement for the accuracy of this current is drastically reduced. It should stable between from the time of V1c measurement to the time of V2c measurement; typically this is equivalent to stability requirement over 20 ms time frame.

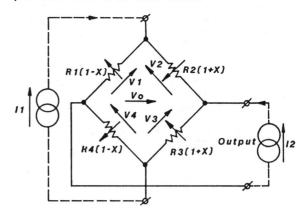

Fig. 6 High performance model of pressure sensor. Dash line indicates interconnection of current sources during testing.

When one would try to compare the derived value of bridge offset voltage based on calculated resistor values with the directly measured value V0, there would not be a perfect match. This is the effect of a lower absolute voltmeter accuracy at eg. 10 V range, where arm voltages are being measured, then at eg. 100 mV range, where the offset is measured.

Thus the second phase of calculations is now implemented. There was a single measurement left unused up to now: offset V0, creating a single degree of freedom. There are however four resistors that require correction. The author has decided to redistribute symmetrically values of all resistors (Fig. 6) in order to match calculated and measured offset values:

$$R1 = R1 (1 - x)$$
$$R2 = R2 (1 + x)$$

R3 = R3 (1 - x)
R4 = R4 (1 + x)

Using simple algebraic manipulation the correction value of x may be derived as:

$$x = (-T + \sqrt{Y}) / (2 Z)$$

where:

T = 2 I1 (R2R4 + R1R3) + V0 (R1-R2+R3-R4)
Z = (R2 R4 - R1 R3) I1
Y = T 2 - 4 Z (Z - (R1 + R2 + R3 + R4) V0)

Implementing the modeling approach outlined above with HP3456 voltmeter demonstrated consistently a 10 ppm differential measurement accuracy of resistors between a direct four wire technique and a modeling technique introduced here. In typical applications with 5V across the pressure sensor this is equivalent to 50 microvolts of offset error during the first modeling phase of calculation. The second phase of calculations corrects this error.

Using creative data acquisition techniques a system testing simultaneously 200 sensors was built based on a total of only 406 analog input channels with the effective modeling speed of 5 sensors per second per each test point.

6. SINGLE SEGMENT NUMERICAL MODEL

For some applications, already employing microprocessors with EEPROM, it may become more cost effective to perform temperature compensation in a digital form rather than in an old fashioned analog way using resistors as compensating components. For this type of application a simple single segment approximation of sensor characteristics will be used.

The output voltage V0 of a pressure sensor can be expressed as a combination of a zero pressure output e independent of pressure and a pressure related component generated as an effect of pressure sensitivity G. For numerical modeling we will use an independent linear approximation for temperature dependance of zero pressure output and for pressure sensitivity:

e = e0 (1 + dT A)
G = G0 (1 + dT B)

where e0, G0 are respectively normalized zero pressure output and pressure sensitivity at reference temperature, A, B are their respective average (over temperature range of interest) temperature coefficients and dT is a deviation of a temperature from a reference condition.

Output voltage V0 and bridge resistance R can now be expressed as follows:

V0 = (e + G Pi) Vb
R = R0 (1 + dT c)

and we can define now the value of temperature deviation dT as:

dT = (R - R0) / R0 C

Input pressure value now can be found after some straightforward algebraic operations as:

Pi = R0 C V0/Vb - e0 (R0 C + A (R - R0))
Pi = Pi / G0 [R0 C + B (R - R0)]

Expression for Pi allows for a direct pressure calculation based on two measurements: bridge output voltage V0 and bridge voltage Vb in a constant current excitation mode after performing a one time modeling phase during which six constants: A, B, C, e0, G0, R0 are being defined and stored.

The accuracy of the algorithm is primarily a function of temperature nonlinearities of a sensor itself and thus a selection of a temperature span over which compensation is performed is important. For IC Sensor devices a total error of 0.5% over 0C to 50C may be expected.

To maintain that error band over a military temperature range usually a three segment approximation would be required, with a total of 18 constants stored.

Another factor limiting pressure measurement accuracy is pressure sensitivity of bridge resistance, introducing bridge voltage change not only with temperature but also with pressure. This change results as the effect of not quite symmetrical strain gages sensitivity to pressure. For example a stress in the center of a rectangular diaphragm is about two times smaller than at the edge. If a sensor is designed to maximize the output with strain gages located in maximum compression and tension locations, then bridge resistance will be changing with applied pressure as a result of different pressure sensitivity of strain gages in tension and compression.

For a number of sensor designs this change is on the order of magnitude equivalent to 2 degree C tempertaure error. This translates, for IC Sensor devices, to about .07% error band.

7. POLYNOMIAL BASED NUMERICAL MODEL

Polynomial approximation of sensor characteristics allows the improvement of modeling accuracy within given temperature range over the single segment approach.

The second order approximation improves the accuracy about four times, especially for the lower temperature range. Modified expressions for the offset e, pressure sensitivity G and bridge resistance R will be then as follows:

e = e0 [1 + X1 dT + X2 (dT)²]
G = G0 [1 + Y1 dT + Y2 (dT)²]
R = R0 [1 + Z1 dT + Z2 (dT)²]

where as before E0, G0, R0 are respectively the reference values of offset, sensitivity and bridge resistance at reference temperature and X1 through Z2 are respective coefficients of a parabolic curve fitting into the test data.

It should be pointed here that a least mean square fit through at least four points pratically eliminates the possibility of a perfect fit through test points with a large deviation for other temperatures.

The temperature deviation dT can be found from the last expression as:

$$dT = [\sqrt{Z1\,2 - 4\,Z2\,(1 - R/R0)} - Z1] / 2\,Z2$$

Including that sensor output voltage V0 is a function of a zero pressure output e, pressure sensitivity G and bridge voltage Vb as it was above, the input pressure Pi can now be calculated as follows:

Pi = (V0/Vb - e) /G

As before the input pressure is calculated based on the output voltage and bridge voltage measurements with the assumption that nine constants are being defined during the temperature compensation cycle.

From the efficiency point of view the second order fit through the test data is essentially equivalent to the approach with two single segments. The number of stored coefficients with two segments would be 12 instead of 9 for a second order fit. If the reference point for both segments is made the same then required number of coefficients decreases to 9 and both approaches are comparable from efficiency point of view.

8. SUMMARY

Two analog and two digital modeling approaches of piezoresistive pressure sensor characteristics were presented in this paper. Presented approaches are not limited to pressure sensors and also can be used, with small modifications, to model other sensors as well, eg. mass air flow devices, accelaration sensors and others employing resistive bridge configuration.

The progress in technology will improve sensors performance in new generations of these devices. Due to a widespread low cost intelligence (micro-computers) it will change also a definition of "good" sensors: as long as characteristics are stable and correctable then a lower priced device will be a better one for a given application. It is then expected that proper modeling techniques will be playing an increasingly important factor in manufacturing cost effectiveness.

NONLINEAR ANALYSES ON CMOS INTEGRATED SILICON PRESSURE SENSOR

Kenichiro SUZUKI, Tsutomu ISHIHARA, Masaki HIRATA and Hiroshi TANIGAWA

NEC Corporation
1-1, Miyazaki, 4-chome, Miyamae-ku, Kawasaki, Japan

ABSTRACT

This paper reports theoretical and experimental results of analysis on nonlinear characteristics, for the newly designed CMOS integrated pressure sensor with square silicon diaphragm. It is shown that nonlinearity is caused by both the large diaphragm deflection effect and nonlinear piezoresistance effect of resistors. The optimum layout for piezoresistors to minimize their nonlinearity is also shown. The measured nonlinearity for the fabricated device quantitatively agrees well with the numerically analyzed nonlinearity.

I. INTRODUCTION

Silicon smart sensors, integrated with signal conditioners, are attractive for their high performance and for their interface use feasibility. Sensitivity and offset variations, as well as temperature dependence, can be corrected by on-chip peripheral circuits [1]. However, some characteristics, such as nonlinearity, are difficult to be compensated for by analog peripheral circuits. Therefore, a design approach to decrease the nonlinearity is important, as well as further integration of peripheral circuits.

Recently, it was reported that nonlinearity arose from both large deflection in the diaphragm [2] and higher order piezoresistive coefficients for the sensing element [3-6] in a silicon pressure sensor. K. Yamada et al. [6] described the application of higher order piezoresistive coefficients to a silicon circular diaphragm pressure sensor. The nonlinearity, however, has never been analyzed on a square silicon diaphragm pressure sensor, which can be more precisely and easily formed by using anisotropic etching techniques than a circular diaphragm pressure sensor.

In this paper, the notation for the nonlinear piezoresistance effect is first discussed in terms of the stress patterns on a square diaphragm pressure sensor and the higher order piezoresistive coefficients, which were newly measured. The nonlinearity for the newly designed CMOS integrated pressure sensor are then calculated, using the previously obtained notation, and results are compared with the measured nonlinearity. Last, design considerations to minimize nonlinearity for the pressure sensor were investigated.

II. A Notation for Nonlinear Piezoresistance Effect

A. Stress Analysis

Large diaphragm deflection induces nonlinear stress on the diaphragm, in response to the applied pressure. Nonlinearity is caused by the finite elongation in the central interior diaphragm plane. Considering this large deflection effect, approximate formulas for use in regard to a square diaphragm have been derived by using the strain-energy method [7]. These approximate formulas can be numerically solved by the successive approximate method. The results for a maximum deflection and for a maximum stress on the diaphragm are shown in Figs. 1 and 2, to compare with the values obtained from the small deflection theory (linear analysis). The former maximum deflection vertically occurs at the center of a square diaphragm, due to the applied pressure. In Fig. 1 the maximum deflection values are shown as a function of the dimensionless pressure, which is normalized by using half the diaphragm edge length "a", diaphragm thickness "h" and applied pressure "P". In the figure, the maximum deflection values, obtained from the small deflection theory, are also shown using a broken line. The latter maximum stress on the diaphragm is induced at the mid point on a square edge and perpendicularly to the ege, due to the applied pressure. In Fig. 2, two maximum stresses, labeled total stress in the figure, are shown under tension and compression conditions. Tension and compression are caused by applying pressure to the top and then to the bottom of the rigidly supported diaphragm membrane, respectively. The individual total stress consists of two components; bending stress and membrane stretching stress. In both tensive and compressive cases, the same membrane stress is induced, because the central interior plane in the diaphragm is stretched for either pressure. On the other hand, the bending stress magnitudes are the same, but the signs are opposite. Therefore, the tensive total stress becomes larger in magnitude than the compressive total stress. Figures 1 and 2 show that the deflection and the stress values, determined from the small deflection theory, are in proportion to the applied pressure, while their values, obtained from the large deflection theory, are not the same.

B. Higher Order Piezoresistive Coefficients

The nonlinear fractional resistance change in a piezoresistor embedded in the diaphragm has the general notation [4,6]

$$\frac{\triangle R}{R_o} = \sum_{i=1}^{n} \left(C_{l\,i}\ \sigma_l^{\,i} + C_{t\,i}\ \sigma_t^{\,i} \right) , \tag{1}$$

where σ_l and σ_t are longitudinal and transverse stresses, respectively, induced onto piezoresistors. $C_{l\,i}$ and $C_{t\,i}$ are longitudinal and transverse ith-order

Reprinted from *Rec. of the IEEE Int. Electron Devices Meeting*, 1985, pp. 137–140.

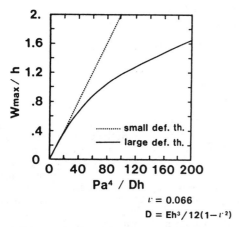

$\iota = 0.066$

$D = Eh^3 / 12(1 - \iota^2)$

Fig. 1 Dimensionless maximum deflection.

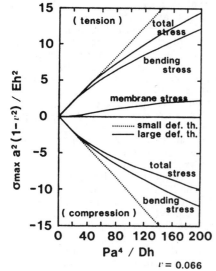

$\iota = 0.066$

Fig. 2 Dimensionless maximum stresses.

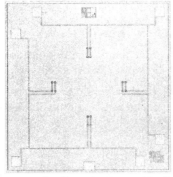

Fig. 3 Photograph of specimen used in Section II B.

Table 1 Higher order piezoresistive coefficients

C_{l1} ((cm²/dyn)⁻¹)	C_{l2} ((cm²/dyn)⁻²)	C_{l3} ((cm²/dyn)⁻³)
5.8×10^{-11}	1.4×10^{-21}	-1.5×10^{-31}
C_{t1} ((cm²/dyn)⁻¹)	C_{t2} ((cm²/dyn)⁻²)	C_{t3} ((cm²/dyn)⁻³)
-5.6×10^{-11}	3.2×10^{-21}	$0.0 (\ll 10^{-31})$

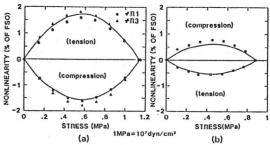

$1 \text{MPa} = 10^7 \text{dyn/cm}^2$

Fig. 4 Comparison between measured and calculated nonlinearities for two kinds of piezoresistors. (a) $R_\parallel$. (b) $R_\perp$.

(100) crystal orientation. Two kinds of piezoresistors are laid out on the diaphragm. One, denoted $R_\parallel$, is parallel to the <110> diaphragm edge. The other, denoted $R_\perp$, is perpendicular to the diaphragm edge. Each of them consists of two same size resistors embedded in parallel to decrease the finite resistor length effect [1]. The surface boron concentration for resistors is 3×10^{18} cm⁻³. By applying pressure in the 4 kg/cm² range to the top and then the bottom of the diaphragm in the specimen, fractional resistance changes for $R_\parallel$ and $R_\perp$, respectively, have been measured. It should be noted that each resistor is stretched when applying pressure to the top of the diaphragm, while it is compressed when applying pressure to the bottom.

From both measured data regarding fractional resistance change, $\Delta R/R_0$, and stresses, σ_l and σ_t, calculated under the large diaphragm deflection condition, the values for piezoresistive coefficients in Eq. (1), as shown in Table 1, have been obtained by the method of least squares. C_{l1} and C_{t1} in Table 1 are constants, as usually called piezoresistive coefficients. In addition, higher order piezoresistive coefficients are given in Table 1. Figs. 4(a) and (b) show nonlinearity values for $R_\parallel$ and $R_\perp$, as a function of transverse stress and as a function of longitudinal stress, respectively. In the figure, the dots and the solid lines are measured values and calculated values. From this data, a set of piezoresistive coefficients in Table 1 were found to closely reproduce nonlinearity for piezoresistors.

III. Nonlinearity for CMOS Integrated Pressure Sensor

As an integrated sensor consists of both sensing elements and peripheral circuits, nonlinear response from peripheral circuits, in addition to response from the sensing element itself, indicate the total device nonlinear characteristics. Nonlinearity for the

piezoresistive coefficients. In practice, the third-order polynomial approximation in Eq. (1) is sufficiently useful.

In order to obtain the values for C_{li} and C_{ti} ($i = 1, 2, 3$), specimens have been fabricated as shown in the photograph in Fig. 3. This pressure sensor has a 1mm×1mm square diaphragm, about 27μm thick with

fabricated CMOS integrated pressure sensor has been investigated, including measuring nonlinearity due to amplifier distortion.

Figure 5 is a photograph of the fabricated CMOS integrated pressure sensor with a square diaphragm. Piezoresistors are embedded near one edge of the diaphragm to minimize the variation in their impurity concentration. They are arranged in a Wheatstone bridge configuration with four active arms. Peripheral circuits are also laid out, more than 100 μm from the diaphragm edge, to prevent circuit characteristics from being degraded by induced stress. Peripheral circuits achieve signal amplification and temperature dependence compensation. If constant voltage is applied, the total nonlinearity for the bridge output can be calculated from the third-order polynomial approximation, as previously mentioned, and the following Eq. (2),

$$\frac{\triangle V}{V_{cc}} = \frac{\left\{1+\left(\frac{\triangle R}{R}\right)_{\perp}\right\}^2 - \left\{1+\left(\frac{\triangle R}{R}\right)_{\parallel}\right\}^2}{\left\{2+\left(\frac{\triangle R}{R}\right)_{\parallel}+\left(\frac{\triangle R}{R}\right)_{\perp}\right\}^2} \quad , \qquad (2)$$

where $\triangle V$ and V_{cc} are differential output voltage for the bridge and an applied supply voltage, respectively. The nonlinearities for the two kinds of resistors, $R_{\parallel}$ and $R_{\perp}$, and for the output from the device are shown in Fig. 6, where measured and calculated values are indicated by dots and lines, respectively.

From measurements on the on-chip peripheral circuits, nonlinearity due to amplifier distortion has been concluded to be less than 0.05 %. In Fig. 6, the calculated values closely agree with the measured values.

IV. Design Considerations

If the stress induced by the applied pressure is represented by the sum of a linear component σ_L and a nonlinear component σ_{NL}, Eq. (1) (n=3) can be approximately transformed to

$$\frac{\triangle R}{R_o} = (C_{l1}\,\sigma_{Ll}+C_{t1}\,\sigma_{Lt})+ (C_{l1}\,\sigma_{NLl}+C_{t1}\,\sigma_{NLt})$$
$$+ \sum_{i=2}^{3}\left(C_{li}\,\sigma_{Ll}^i + C_{ti}\,\sigma_{Lt}^i\right) \quad , \qquad (3)$$

because $\sigma_L \gg \sigma_{NL}$, $C_{l1} \gg C_{l2}$, C_{l3} and $C_{t1} \gg C_{t2}$, C_{t3}. In Eq. (3), the second term and the third term are related to the values due to the large diaphragm deflection effect and the resistors nonlinear piezoresistance effect, respectively. Figure 7 shows nonlinearity generated from the second term in Eq. (3) as functions of structure ratio a/h and applied pressure range values. The nonlinearity, induced at the mid point on the diaphragm, near the location where the resistors are laid out, is shown on the ordinate. In the figure, the broken lines also show the region where more than silicon rupture stress (2×10^9 dyn/cm²) occurs. From Fig. 7, it is shown that, the second term in Eq. (3) becomes serious for higher a/h values. Figure 8 shows two nonlinear effects calculated by using the second and the third term in Eq. (3), respectively, for the piezoresistors in the fabricated integrated pressure sensor (a/h≈18). The figure indicates that nonlinearity

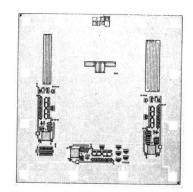

Fig. 5 CMOS integrated pressure sensor.

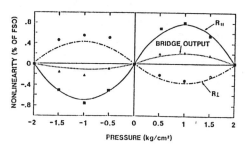

Fig. 6 Nonlinearity for integrated pressure sensor.

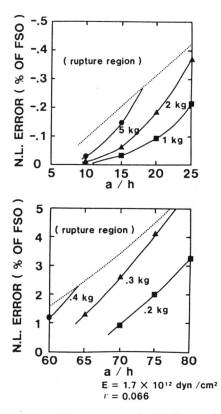

Fig. 7 Nonlinearity generated from the second term in Eq. (3), as a functions of structure ratio a/h and applied pressure range values.

for piezoresistor $R_\perp$ is caused by two different effects in the same order. For $R_{||}$, nonlinearity is mainly caused by the nonlinear piezoresistance effect.

As the two kinds of piezoresistors, $R_{||}$ and $R_\perp$, have opposing nonlinearity signs in any applied pressure, their nonlinearities counterbalance each other at the bridge output (See Fig. 6). For the fabricated integrated pressure sensor, however, this counterbalance has not been sufficient, because $R_{||}$ is much less linear than $R_\perp$. If the resistors layout is varied, the bridge output can be shown to become more linear. Figure 9 shows the maximum nonlinearity calculated for $R_\perp$ and $R_{||}$, laid out along the center line of a diaphragm, as a function of x/a; the ratio of the distance from the center of the diaphragm to half diaphragm edge length. In the figure, the a/h value and the resistor length are assumed to be 25 and zero, respectively. This figure shows that the nonlinearity for $R_{||}$, positioned near $x/a \approx 0.78$, is closest to zero, while the nonlinearity for $R_\perp$ near $x/a \approx 0.85$ is closest to zero. These are positions where the second and the third term in Eq. (3) most closely counterbalance each other. Figure 9 also shows the calculated maximum nonlinearity in bridge output as a function of the $R_{||}$ location, when $R_\perp$ is located at $x/a = 1.0$. From this figure, optimum layout location for $R_{||}$ is near $x/a = 0.82$.

V. CONCLUSION

Nonlinearity characteristics, for a CMOS integrated pressure sensor with a square diaphragm, were analyzed using a set of higher order piezoresistive coefficients, experimentally obtained for P-type diffused resistors aligned along the <110> direction on the (100) silicon surface. It was shown that the nonlinearity could be calculated by applying the stress values included in the large deflection effect to the third-order polynomial approximation. It was also shown that the optimum layout for the piezoresistors to minimize their nonlinearity could be decided for any a/h value by using the third-order approximation. The results shown in this paper will be applicable to designing high performance silicon pressure sensors.

ACKNOWLEDGEMENT

The authors gratefully acknowledge encouragement from Dr. H. Shiraki. They also thank S. Suwazono for device fabrication and M. Suda for his contribution in the device measurements.

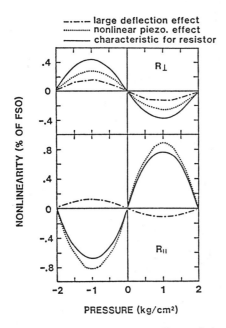

Fig. 8 Two different nonlinear effects (Calculation).

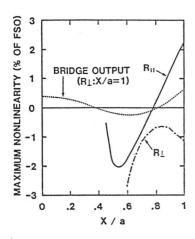

Fig. 9 Maximum nonlinearity as a function of x/a. $R_\perp$ and $R_{||}$ are laid out along the center line of a diaphragm.

REFERENCES

(1) H. Tanigawa, T. Ishihara, M. Hirata and K. Suzuki, 'MOS integrated silicon pressure sensor,' IEEE Trans. Electron Devices, vol. ED-32, No.7, pp.1191-1195, 1985.

(2) O. N. Tufte, P. W. Chapman and D. Long, 'Silicon diffused-element piezoresistive diaphragms,' J. Appl. Phys., vol.33, No.11, pp.3322-3327, 1962.

(3) D. R. Kerr and A. G. Milnes, 'Piezoresistance of diffused layers in cubic semiconductors,' J. Appl. Phys., vol.34, No.4, pp.727-731, 1963.

(4) J. Bretschi, 'A silicon integrated strain-gage transducer with high linearity,' IEEE Trans. Electron Devices, vol.ED-23, pp.59-61, 1976.

(5) L. B. Wilner, 'A diffused silicon pressure transducer with stress concentrated at transverse gages,' ISA Trans., vol.17, No.1, 1978.

(6) K. Yamada, M. Nishihara, S. Shimada, M. Tanabe, M. Shimazoe and Y. Matsuoka, 'Nonlinearity of the piezoresistance effect of p-type silicon diffused layers,' IEEE Trans. Electron Devices, vol.ED-29, No.1, 1982.

(7) S. Way, 'Uniformly loaded, clamped, rectangular plates with large deflection,' Proc. Fifth Int. Cong. Appl. Mech., pp.123-128, 1939.

THE VOLTAGE CONTROLLED DUTY-CYCLE OSCILLATOR;
BASIS FOR A NEW A-TO-D CONVERSION TECHNIQUE

Richard R. Spencer, Bruce M. Fleischer, Phillip W. Barth, and James B. Angell

Center for Integrated Systems
Stanford University, Stanford, California 94305

ABSTRACT

The Voltage Controlled Duty-Cycle Oscillator (VCDCO) is a simple circuit with a square-wave output whose duty-cycle depends only on the input voltage and temperature. The circuit is the basis for an A-to-D conversion technique developed for integrated silicon sensors. The accuracy and resolution of the technique are significantly better than previously published results. Because the output is independent of component values and matching, it is believed that this circuit will provide better long term stability than other methods. In addition, most of the circuitry required for a complete A-to-D conversion is external to the sensor itself, thus reducing the size of the sensor and allowing an array of sensors to make use of the same external circuitry.

INTRODUCTION

Many applications (*e.g.*, biomedical research, industrial controls) require small, stable, low power transducers [1]. In many hostile environments transducers suffer stability problems due to changing leakage currents resulting from moisture intrusion. Monolithic silicon sensors have the advantage that active circuitry can be added to combat this problem, without excessively complicating the process [2-3]. Previous efforts have concentrated on buffering the analog signal, or using conventional modulation [2]. The duty-cycle modulation implemented by the VCDCO is insensitive to leakage currents as well as changes in device parameters and provides an extremely linear function of the input voltage.

The VCDCO can be thought of as providing an output which has both pulse-width modulation (PWM) and pulse-period modulation (PPM). Although the input voltage could be recovered from either, that would require an accurate knowledge of the modulation constants. By using both types of modulation simultaneously, the input voltage can be determined independent of the modulation constants.

This implies that changes in these constants (*e.g.*, with aging) will not alter the basic resolution or accuracy of the value obtained for the input voltage.

A complete A-to-D converter comprises three distinct elements: the VCDCO, counters and sequence control logic, and logic for storage and computation. The VCDCO is small and does not require any precision components, tight matching, or trimming. The conversion is completed by external digital circuitry which measures t_1 and t_2 (see figure one) and calculates the input voltage based on either of the simple relationships given in equations (1) and (2).

CIRCUIT DESCRIPTION

The VCDCO circuit (see figure one) consists of an emitter-coupled multivibrator in which the two tail currents are set by an emitter-coupled pair (this circuit has been used before in monolithic waveform generators [4]). The input voltage sets the ratio of the tail currents, which in turn determines the duty cycle of the oscillator. The key idea here is that, except for the input offset of the transconductance stage (Q3 & Q4), all of the circuit variables (*e.g.*, R, C) cancel when any ratio involving t_1 and t_2 is taken (*e.g.*, t_1/t_2, $t_1/(t_1+t_2)$). This is evident from a simple analysis (shown in figure one) since t_1 and t_2 both contain the same multiplicative factor. The equations relating pulse-width (arbitrarily taken to be t_1) and pulse-period to input voltage are also shown in the figure. In practice, it is the times t_1 and t_2 (or t_1 & t_1+t_2) which are measured, and the input voltage is then determined from either of the equations shown below (derived from those shown in figure one).

$$v_{IN} = 2(kT/q)tanh^{-1}\{2t_1/(t_1+t_2) - 1\} \qquad (1)$$

$$v_{IN} = (kT/q)ln(t_1/t_2) \qquad (2)$$

The sensor temperature must be known in order to calculate v_{IN} from either equation (1) or (2). This does not impose any additional burden, however, since the

Reprinted from *Rec. of the 3rd Int. Conf. on Solid-State Sensors and Actuators*, 1985, pp. 49–52.

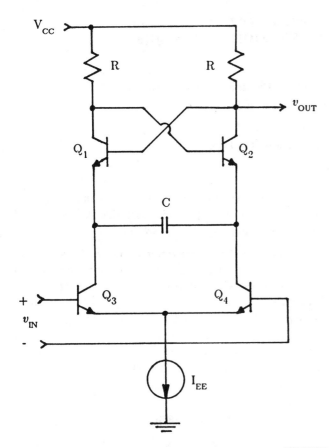

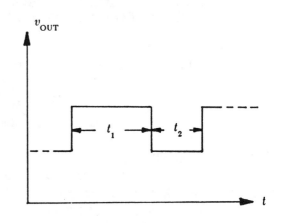

EQUATIONS

$$t_1 \approx 2(RCI_{EE})/I_{C4} \quad \text{and,} \quad t_2 \approx 2(RCI_{EE})/I_{C3}$$

$$\text{also,} \quad I_{C3}/I_{C4} = exp(qv_{IN}/kT)$$

Therefore;

$$\text{pulse-width} \equiv t_1 = 2RC\{1+exp(qv_{IN}/kT)\}$$

$$\text{pulse-period} \equiv t_1+t_2 = 8RCcosh^2(qv_{IN}/2kT)$$

FIGURE ONE

temperature must also be known in order to convert bridge voltage to pressure or stress (see equation (4)).

An alternate way of viewing the circuit is to think of the upper transistor pair as a differential comparator, with hysteresis, which monitors the voltage on the capacitor as it ramps. The times t_1 and t_2 are then used to compare the ramp rates of the capacitor voltage. This view of the circuit points out its similarity to a dual slope A-to-D converter. In contrast to a typical dual slope converter, the present method does not require a precise current source or a linear transconductance amplifier, but it does retain the advantage of signal integration and, therefore, reduced sensitivity to noise.

The VCDCO is ideally suited for sensors that use a piezoresistive bridge as the transducer element, such as pressure sensors [2-3], strain gauges, and tactile sensor arrays [6]. The bridge output provides both a differential input signal and the required DC bias for the VCDCO's input stage. One advantage of this method for an array of sensors is that the majority of the circuitry required for a complete A-to-D conversion can be shared by a number of different sensors.

CALIBRATION

Any piezoresistive bridge sensor has to be calibrated (if offset cancellation and temperature compensation are not included), and in each case the bridge offset voltage, which is the input to the VCDCO (v_{IN}), is related to the quantity being measured (*i.e.*, pressure or stress) by equation (3) [5].

$$v_{IN} = (\Delta R/R)V_B + v_{OFF} = GV_BP + v_{OFF} \qquad (3)$$

P is the parameter being measured, G is the gauge factor, or sensitivity, v_{OFF} is the bridge offset voltage, and V_B is the bridge bias. Typically G and v_{OFF} are functions of temperature and V_B is assumed constant. In practice, it is the temperature (T), and the times t_1 and t_2 which are measured. Using equations (2) and (3), P can be related to these quantities as shown in equation (4).

$$P = (kT/qGV_B)ln(t_1/t_2) - (v_{OFF}/GV_B) \qquad (4)$$

Equation (4) shows that P is a linear function of $ln(t_1/t_2)$. If G and v_{OFF} are linear functions of temperature, then four parameters are sufficient to determine P given measured values for T, t_1, and t_2. The simplest reasonable calibration procedure requires four measurements; two known inputs (P_1 & P_2) must be applied at each of two known temperatures. From this data the required four parameters can be derived. The offset voltage of the VCDCO's transconductance stage (Q3 & Q4 in figure one) is indistinguishable from the bridge offset and is automatically included during the calibration.

NONIDEAL EFFECTS

All of the equations presented thus far are based on a simple, first order, analysis of the VCDCO. When a more detailed analysis is performed there are additional terms present. If either equation (1) or (2) is used to estimate v_{IN} from measured values of t_1 and t_2 these additional terms will cause an error in the estimate. The largest error is caused by the non-zero transition times of the VCDCO output. These times are not proportional to t_1 and t_2, and so do not cancel when the ratio is taken. Assuming that the same constant (τ) is added to both times, the resulting error in the estimate of v_{IN}, using (1) or (2), can be calculated as a function of v_{IN} for different values of τ (see figure two).

The magnitude of the error caused by the transition times would be unacceptable for most applications. Fortunately, the calibration procedure used for the sensor will, at the same time, significantly reduce the magnitude of this error. From figure two it can be seen that this error is very nearly linear in v_{IN}, so, even if a simple linear calibration is used, the residual error (see figure three) is very small.

RESULTS

Results obtained from a prototype VCDCO, constructed using RCA transistor arrays and discrete passive components, validate the theory. Using the simple four point calibration to temperature, ten bit resolution and accuracy has been achieved with a 3mS conversion time and an LSB of $26\mu V$ in the presence of an input noise voltage of $690\mu V_{RMS}$ (measurement bandwidth = 100kHz). The basic v_{IN} versus $ln(t_1/t_2)$ relationship has been measured to be linear to within $\pm 5\ \mu V$ from -20 mV to +20 mV for the same breadboard.

The first monolithic VCDCO has been fabricated using an in-house, low power, bipolar, semi-custom chip and an external capacitor (figure four is a photograph of the circuit). Preliminary results demonstrate better than seven bits of resolution and accuracy, with an LSB of less than $130\mu V$, in a 2mS conversion time. The circuit uses a

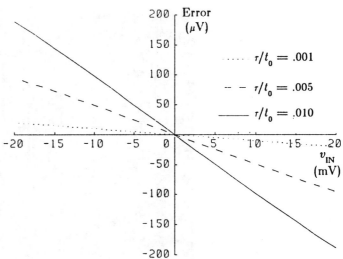

Fig. 2   Error voltage versus normalized, common-mode, additive error. $t_0 \equiv t_1(@v_{IN}=0)=4RC$

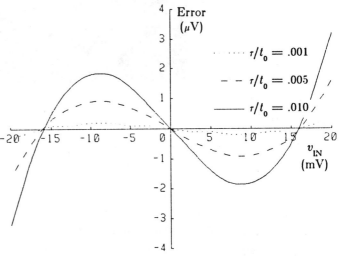

Fig. 3   Residual error voltage versus normalized, common-mode, additive error. $t_0 \equiv t_1(@v_{IN}=0)=4RC$

single 3 V supply and consumes 100 μW.

We expected slightly poorer performance from this monolithic VCDCO, than was obtained with the discrete version, due to its larger noise bandwidth [7], however, the performance observed is not presently limited by thermal noise induced jitter. At this time we cannot explain this data, but we believe that significantly better results will be obtained with a second generation VCDCO and a more sophisticated test setup.

315

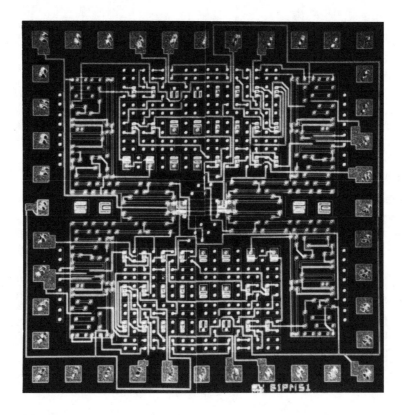

Fig. 4   Photograph of the semi-custom VCDCO. There are two VCDCOs on this die, which is $(2.6 \text{ mm})^2$.

ACKNOWLEDGEMENTS

This work is supported by the Biotechnology Resources Branch of the NIH Division of Research Resources under grant P41-RR-01086. We owe thanks to Nancy Latta, Jibreel Mustafa, Margaret Prisbe, and John Shott for assistance in layout, processing, and testing of the monolithic circuit.

REFERENCES

[1]   IEEE Transactions on Electron Devices, special issues on solid-state sensors; December 1979 and January 1982

[2]   John M. Borky, and Kensall D. Wise, "Integrated Signal Conditioning for Diaphragm Pressure Sensors", *ISSCC Digest*, 1979

[3]   Brent E. Burns, Phillip W. Barth, and James B. Angell, "Fabrication Technology for a Chronic in-vivo Pressure Sensor", *IEDM Digest*, 1984

[4]   Alan B. Grebene, "Monolithic Waveform Generation", *IEEE Spectrum*, April 1972

[5]   Samaun, Kensall D. Wise, James B. Angell "An IC Piezoresistive Pressure Sensor for Biomedical Instrumentation", *IEEE Transactions on Biomedical Eng.*, March 1973

[6]   Phillip W. Barth, Sharon L. Bernard, and James B. Angell, "Monolithic Silicon Fabrication Technology for Flexible Circuit and Sensor Arrays", *IEDM Digest*, 1984

[7]   Asad A. Abidi, and Robert G. Meyer, "Noise in Relaxation Oscillators", *IEEE Journal of Solid State Circuits*, Dec. 1983

Analysis, Design, and Performance of Micropower Circuits for a Capacitive Pressure Sensor IC

MICHAEL J. S. SMITH, LYN BOWMAN, STUDENT MEMBER, IEEE, AND
JAMES D. MEINDL, FELLOW, IEEE

Abstract —An integrated circuit has been designed, built, and tested as part of a capacitive pressure transducer. High-accuracy compact micropower circuits utilizing a standard bipolar IC process without any special components or trimming are used. This was achieved by examining the performance limits of two key circuits: a Schmitt trigger oscillator and a bandgap voltage reference. The sensor circuits consume 200 μW at 3.5 V, can resolve capacitance changes of 300 ppm, measure temperature to ±0.1°C over a limited temperature range, and presently occupy 4 mm² on a 2-mm×6-mm implanatable monolithic silicon pressure sensor. Further scaling of the sensor is discussed showing that a reduction of area by a factor of 4 is achievable.

I. Introduction

THERE is an urgent need in medical research for an implantable sensor that can measure pressure to 0.1 mmHg with a full scale (FS) of 300 mmHg (a resolution of 300-ppm FS). The sensor must be small (presently available sensors are 6 mm in diameter), stable and preferably self-calibrating, low power to allow for long battery implant lifetime, biocompatible for intravascular use, and inexpensive compared to commercially available sensors that cost approximately $1000. A companion article in the IEEE TRANSACTIONS ON BIOMEDICAL ENGINEERING [1] shows that a capacitive transducer best fulfills these needs and describes the sensor specifications, design requirements, measured results, and expected system performance, while this paper contains details of the micropower circuits used in the sensor IC.

The monolithic silicon sensor includes a capacitive transducer that is formed using complex micromachining techniques after circuit fabrication is complete [2]. To ensure a good overall yield we require very high circuit yield that dictates the use of a simple IC process. However, the desired system measurement accuracy would normally require the use of component trimming and/or thin-film resistors in the two key elements of the IC: a voltage reference and an oscillator. By examining the properties of these circuits and their impact on the system performance this paper describes how this dichotomy was resolved and the required specifications are achieved using a standard bipolar process.[1]

II. System Design

The measurement of the pressure sensing capacitance $C_{PRES}(P)$ (and thus pressure P) is based on a ratiometric scheme

$$\frac{C_{PRES}}{C_{REF}} = 1 + \frac{T(PRES) - T(ZERO)}{T(PSCALE) - T(ZERO)} \quad (1)$$

where C_{REF} is a fixed on-chip reference capacitor.[2] The period-coded outputs of an oscillator: $T(PRES)$, $T(REF)$, $T(PSCALE)$, and $T(ZERO)$ are time-multiplexed and passed to a custom FM transmitter (contained in a separate implanted electronics package), and decoded externally with a custom FM receiver/demodulator. Fig. 1 shows the IC implementation of such a scheme, which is designed to be closely compatible with an existing telemetry system [3]. During two cycles of the eight-cycle output the oscillator is connected to C_{PRES} which generates the signal $T(PRES)$. For the remaining six cycles I²L switches connect the oscillator to the fixed reference capacitor C_{REF}. These six cycles comprise three time slots containing three additional signals including $T(PSCALE)$ and $T(ZERO)$.[3] This ratiometric scheme reduces the effect of

Manuscript received April 9, 1986; revised July 30, 1986. This work was carried out at the Center for Integrated Electronics in Medicine at Stanford University and was supported by the National Institute of General Medical Sciences under PHS Research Grant 2P50 GM1940. M. J. S. Smith received support from IBM (United Kingdom) Laboratories Ltd.

M. J. S. Smith was with the IBM T. J. Watson Research Center, Yorktown Heights, NY 10598. He is now with the Department of Electrical Engineering, University of Hawaii at Manoa, Honolulu, HI 96822.

L. Bowman is with the Department of Electrical Engineering, Stanford University, Stanford, CA 94305.

J. D. Meindl was with the Department of Electrical Engineering, Stanford University, Stanford, CA 94305. He is now with Rensselaer Polytechnic Institute, Troy, NY 12180.

IEEE Log Number 8610950.

[1] The high electric fields created in the anodic bonding procedure used to fabricate the completed sensor presently preclude the use of a CMOS process.
[2] For the present design: $C_{REF} = 35$ pF; $C_{PRES} = 35$ pF at $P = 760$ mmHg (transducer sealed at atmospheric pressure); and $C_{PRES} = 70$ pF at $P = 1060$ mmHg (full scale).
[3] Two cycles are allocated for each signal to allow the currents in the oscillator to stabilize. The first corrupted cycle is not used for measurement.

Reprinted from *IEEE J. Solid-State Circuits*, vol. SC-21, no. 6, pp. 1045–1056, December 1986.

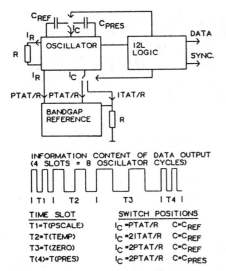

INFORMATION CONTENT OF DATA OUTPUT
(4 SLOTS = 8 OSCILLATOR CYCLES)

TIME SLOT	SWITCH POSITIONS	
T1=T(PSCALE)	I_C =PTAT/R	C=C_{REF}
T2=T(TEMP)	I_C =2ITAT/R	C=C_{REF}
T3=T(ZERO)	I_C =2PTAT/R	C=C_{REF}
T(4)=T(PRES)	I_C =2PTAT/R	C=C_{PRES}

Fig. 1. Block diagram of pressure transducer integrated circuit.

zero and sensitivity drift common to both capacitors, temperature dependence of the oscillator resistor, oscillator period offset owing to parasitic capacitance, and supply dependence of the oscillator.

A fourth signal T(TEMP) is generated by supplying the oscillator with a current derived from the bandgap reference voltage which is independent of absolute temperature (ITAT). The other oscillator signals are generated from a voltage which is proportional to absolute temperature (PTAT). The temperature T of the sensor is calculated in the external demodulator to allow temperature compensation of the pressure-sensitive and reference capacitors as follows:

$$\frac{T}{T_1} = \frac{I_{\text{PTAT}}}{I_{\text{ITAT}}} = \frac{T(\text{TEMP}) - T(\text{ZERO})}{T(\text{PSCALE}) - T(\text{ZERO})}. \quad (2)$$

A detailed description of the system design specifications can be found in [1]. In summary, the requirement to measure pressure to 0.1 mmHg with a typical sensor temperature dependence of 1 mmHg/°C demands that we measure temperature to ±0.1°C. Section III examines the limits to performance of the bandgap voltage reference, the key circuit component of the temperature measurement scheme. Knowing this scheme's inherent errors, analyzed in Appendix A, a trade-off can be made between the implanted chip circuit complexity and the amount of external signal processing necessary to achieve the desired measurement accuracy. The requirement to resolve 0.1 mmHg with a FS of 300 mmHg demands that we resolve capacitance changes of 300 ppm, placing stringent requirements on the oscillator performance which is examined in Appendix B.

III. ANALYSIS OF THE BANDGAP REFERENCE CIRCUIT

The performance of bipolar bandgap voltage reference circuits is well documented [4]–[15]. Their designs have used thin-film resistors and/or trimming and, as curvature correction techniques have been introduced, the reference

temperature coefficient (TC) has been steadily improved. Recently the desire to implement accurate bandgap references with standard CMOS processes has led to the examination of the effects of nonideal components [16]–[18]. We wish to use a standard bipolar process without using thin-film resistors, or any form of component trimming. We briefly review established theory and introduce new notation, which allows us to identify, clarify, and quantify the effects of component nonidealities. We examine each of the assumptions below, inherent in the derivation which follows, and then determine their effects on the reference TC.

1) Resistors are assumed insensitive to temperature. This is a good assumption for thin-film resistors normally used in bipolar references [6], [19], but a poor one for diffused and ion-implanted resistors that are often all that are available in a CMOS process [17], [18].

2) Temperature dependence of the bipolar device current gain is neglected.

3) We assume second-order effects can be neglected and that exact nulling of the first-order temperature coefficient can be achieved. It is the second-order effects that the so-called "curvature-corrected" references attempt to cancel [8], [12], [13], [17], [18], [20], [21].

4) We also assume that the output of the reference can be made exactly equal to that voltage which ensures a zero-reference TC. Manufacturing variation introduces errors that are normally removed using either laser or zener-diode component trimming [6], [9], [10], [15], [22].

A. Bandgap Reference Theory

For an nth-order corrected bandgap reference we wish the first n thermal derivatives of the reference voltage V_{REF} to vanish at temperature T_0. Then we must be able to express V_{REF} as[4]

$$V_{\text{REF}}(T) = V_{BE}(T) - \frac{V'_{BE}(T_0)}{1!}T$$
$$- \frac{V''_{BE}(T_0)}{2!}(T - T_0)^2 - \cdots \quad (3)$$

(prime denoting differentiation). Given[5]

$$V_{BE}(T) = V_T \ln(I_C/I_S) = V_g - V_T[(\gamma - \alpha)\ln T - \ln EG] \quad (4)$$

where $V_T = kT/q$, $I_S = E^{-1}T^\gamma \exp[-V_g/V_T]$, and $I_c = GT^\alpha$, E, G, γ, and α are constants assumed independent of temperature. Then noting $V'_T = k/q$

$$V'_{BE}(T_0) = \left[V'_g - V'_T[(1 + \ln T)(\gamma - \alpha) - \ln EG] \right]_{T_0}. \quad (5)$$

---

[4] Our construction of (3) illustrates how V_{BE} and a PTAT voltage are summed to provide first-order temperature correction and how second-order (and possibly higher) cancellation schemes correct about T_0.

[5] The notation follows Gray and Meyer [23]. We have assumed the emission coefficient n_E is unity; nonideality due to inverse Early effect, space-charge recombination, etc., can be important in very high precision references but has a small effect compared with the others treated here.

Direct substitution of (4) and (5) into (3) gives the bandgap voltage with first-order correction

$$V_{\text{REF}}(T_0) = \left[\left[V_g - T V_g'(T_0) \right] + V_T(\gamma - \alpha) \left[1 - \ln \frac{T}{T_0} \right] \right]_{T_0}. \tag{6}$$

B. Effect of the Temperature Coefficient of Resistance (TCR)

For reference circuits that generate the collector current in the bandgap cell by impressing a PTAT voltage across a resistor

$$I_C \propto \frac{R_0}{R} \frac{T}{T_0} \tag{7}$$

where (R/R_0) is the ratio of the value of the resistor R at temperature T relative to its value at T_0. This directly affects the reference TC because

$$^R V_{BE}' = V_{BE}' - V_T \frac{R'}{R} - V_T' \ln \frac{R}{R_0} \tag{8}$$

where $^R V_{BE}'$ denotes that the effect of the resistor temperature dependence is included. From (3) if we adjust the reference voltage such that

$$^R V_{\text{REF}}(T_0) = V_{BE}(T_0) - T_0 {}^R V_{BE}'(T_0)$$
$$= V_{\text{REF}}(T_0) + V_{T_0} T_0 \frac{R'(T_0)}{R_0} \tag{9}$$

the effect of the second term in (8) can be cancelled and $^R V_{\text{REF}}$ again has zero TC at T_0. Without this correction the reference TC[6] would be 138 ∓ 9.7 ppm/°C for a measured temperature coefficient of resistance (TCR) of 6725 ± 473 ppm/°C (1σ). The third term in (8) is zero at T_0 increasing in magnitude with $|T - T_0|$ and cannot be cancelled with first-order temperature correction. The effect of the resistor temperature coefficient on the second-order reference TC can be calculated as follows:

$$^R V_{BE}'' = V_{BE}'' - \frac{R'}{R} \left[2 V_T' - V_T \frac{R'}{R} \right] - V_T \frac{R''}{R}. \tag{10}$$

The nominal value of the bracket in the second term above is close to zero. However, the variation in TCR leads to an expected random variation in the second-order reference TC of approximately ± 0.054 ppm/°C^2. From measurements, the second-order TCR (R''/R) is less than ± 2.6 ppm/°C^2 which gives an additional random contribution to the second-order reference TC of smaller than ± 0.07 ppm/°C^2.

C. Effect of Finite Current Gain

Since we normally control the emitter current rather than the collector current in the bandgap circuit, the temperature coefficient of current gain also effects the reference TC. From an analysis identical to the above, the correction to the reference voltage to restore a zero TC at T_0 is $T_0 V_{T_0} \beta'/\beta(\beta + 1)$, which is less than 0.35 mV for[7] β greater than 100, and $(\beta'/\beta) = 4500$ ppm/°C at $T_0 = 300$ K. This correction voltage is much smaller than other errors in the nominal reference voltage and any second-order temperature dependence of β results in even smaller contributions to the reference TC.

D. Effect of Curvature

For a first-order corrected reference with ideal component behavior

$$V_{\text{REF}}'(T) = \left[V_g'(T) - V_g'(T_0) \right] - V_T'(\gamma - \alpha) \ln \frac{T}{T_0}$$

and

$$V_{\text{REF}}''(T) = V_g''(T) - \frac{V_T'}{T} (\gamma - \alpha). \tag{11}$$

Tsividis [20] points out that previous designs have ignored the second-order behavior or curvature of the bandgap voltage but proposes an equation for V_g which is linear above 300 K and which requires empirical determination of the parameter γ. Meijer [12] has conducted extensive measurements and determined that the most accurate approximation to the $I_C - V_{BE}$ characteristics can be made by assuming $V_g'' = 0$ and adjusting $V_g(0)$ and γ. An excellent discussion of this problem is given in [12, appendix B]. To obtain an upper bound on the second-order reference TC due to curvature we take $V_g'' = 0$, $\alpha = 1$, and a maximum value of $\gamma \simeq 4$ [8]. The systematic second-order TC due to curvature is then smaller than -0.7 ppm/°C^2 at $T_0 = 300$ K.

E. Effect of Manufacturing and Process Variations

Variations in $V_{\text{REF}}(T_0)$ can result from two sources:

1) variations in the first term in (3) due to variation in the constants E and G (we assume variations in γ, α, and V_g are small)[8]; and
2) variations in the coefficient of the second term in (3) which, depending on the actual circuit implementation, are caused by variations in device emitter sizes, resistor mismatch, etc.

These effects, from (3), give an error voltage V_E that is proportional to absolute temperature,[9] and a first-order reference TC of $V_E/(T_0 V_{\text{REF}})$. For the bipolar process used $V_{BE} = 547 \pm 7.3$ mV(1σ) at $I_E = 1$ μA, which gives a

---

[6] The first- and second-order reference TC's are defined as $V_{\text{REF}}'/V_{\text{REF}}$ and $V_{\text{REF}}''/V_{\text{REF}}$, respectively.

[7] Typical measured values for the bipolar process used (double-diffused n-p-n).

[8] Actually, because the $\ln T$ term in (4) is slowly varying with respect to the terms which are linear in T, the arguments in this section still hold, to a good approximation, for γ and α also.

[9] It can also be shown that the temperature at which the bandgap TC is zero is altered to T_1 where, $(T_1/T_0) = \exp[V_E/[(\gamma - \alpha) V_{T_0}]]$.

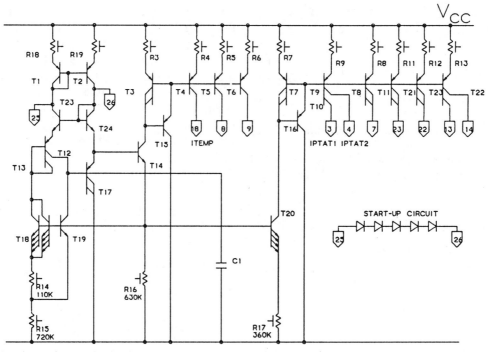

Fig. 2. Bandgap reference circuit schematic. $R14-R17$ are placed in a common tub connected to the base of $T18-T20$. All resistors are 250 kΩ unless shown.

random reference TC variation of ± 19 ppm/°C at $T_0 = 300$ K. The second source of error noted above is much harder to quantify and its magnitude can be deduced from the results on measured bandgap reference circuits.

IV. DESIGN OF THE BANDGAP REFERENCE CIRCUIT

The analysis in the previous section allows us to evaluate the effect of the bandgap reference circuit on system performance. The expected variation in the first-order bandgap reference TC is the rms sum of the individual random contributions listed above which amounts to an error of ± 21 ppm/°C (1σ) at $T_0 = 300$ K. Summing the systematic (algebraically) and random (rms) second-order reference TC components, the expected coefficient is -0.7 ± 0.09 ppm/°C². The measured range of the first-order reference TC is (-150)–50 ppm/°C. The difference between this figure and the predicted 99 percentile (3σ) range of ± 63 ppm/°C is due, firstly, to the fact that the spread in reference TC is not exactly centered on zero and, secondly, to the effects of device and resistor mismatch, etc., which could not be accounted for in the above analysis.

Taking the worst-case measured first-order reference TC of -150 ppm/°C and using a single-point calibration, it is shown in Appendix A that we can measure temperature to an accuracy of 0.1°C within a temperature range of $T_0 \pm 2.2$°C, which is acceptable only in a very limited range of applications.

Using a two-point calibration scheme and assuming the same first-order reference TC and a worst-case (3σ) second-order reference TC of -1 ppm/°C², the temperature measurement range may be increased to $T_0 \pm 18$°C with

the same accuracy. This is more than adequate for physiological use.

In summary, our analysis has shown that, for this application, the use of expensive trim procedures and special components in the bandgap reference is not only undesirable, but with proper circuit and system design, unwarranted and unnecessary.

A. Bandgap Circuit

The schematic of the bandgap reference is shown in Fig. 2, the design is chosen for compactness [6], [24]. Normally, to achieve better matching, an integral ratio would be used for $R14$ and $R15$ and for a given operating current a nonintegral emitter area ratio for $T18$ and $T19$ is then usually necessary. These devices can use unequal circular emitters which can also mitigate errors due to absolute V_{BE} variation by compensating for variations in the emitter masking step. Instead, multiple unit-value resistors with taps that allow centering of the reference TC by changing only the metal mask are used.[10] The micropower operation of the circuit of Fig. 2 together with restrictions on circuit area pose several problems discussed below.

B. Leakage Current

A large MOS capacitor (metal–emitter oxide–emitter–epi–substrate) is used to compensate the bandgap reference because the underlying junction capacitor (which has higher capacitance per unit area than the MOS capaci-

---

[10] The wafers can be metallized, the reference TC measured, the metal stripped, the mask changed, and then the wafers remetallized enabling the production spread to be centered using one lot.

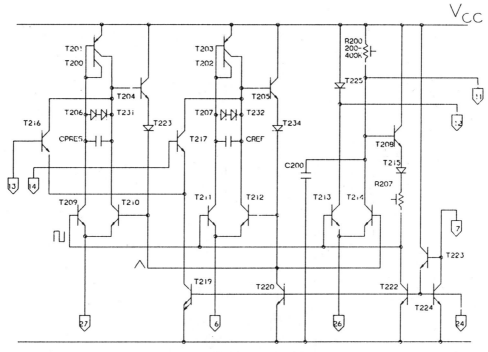

Fig. 3. Circuit schematic of oscillator core.

tor) cannot be used since the leakage current could imbalance the bandgap cell. However, some area is saved because the use of similarly sized collector tubs for $T18/19$ (so that collector-to-substrate leakage is equal) was found unnecessary. A Wilson configuration for the p-n-p cascode mirror ($T1/2$) which biases the cell from the bandgap voltage would reduce supply sensitivity and ease start-up [25], but with a worst-case end-of-life battery voltage of 3.5 V there is insufficient headroom available. A pinch resistor or epi-FET cannot be used for start-up, because the worst-case current flow in these devices is too high; instead a five-diode chain, carrying a small but predictable current, starts the bandgap mirror.

C. Ion-Implanted Resistors

High sheet ρ, 5-kΩ/sq, boron ion-implanted resistors were used to reduce the silicon area consumption. Their voltage characteristics can be approximated by the following equation:[11]

$$\frac{R}{R_{\text{NOMINAL}}} = 1 + 0.021(V_{\text{TUB}} - V_A) + 0.021(V_{\text{TUB}} - V_B)$$

$$(12)$$

where V_A and V_B are the terminal voltages and V_{TUB} is the epi tub voltage [26]. To combat resistor supply dependence, the most simple, reliable, and compact solution is to tie the bandgap resistor epi tub to the bandgap output and

thus minimize the variation of the bracketed terms in (12). Attempts to bias the tub from other circuit nodes are fraught with danger: the large capacitance makes the tub potential slow in rising at start-up, the tub to resistor junctions can become forward biased, and initiate latch-up.

Instability in large sheet ρ resistors is possible when certain kinds of overpassivation are used (most notably sputtered quartz). A metal shield plate covers critical resistors and pins the field-oxide surface potential at a known value. Contacts to this large aluminum shield plate are carefully placed to avoid silicon leaching. Dummy resistors are placed at each end of the tubs to improve matching. The techniques listed above result in across-tub matching of better than 0.1 percent.

V. ANALYSIS OF THE OSCILLATOR CIRCUIT

A compact low-power RC oscillator is required in this application. A Schmitt trigger oscillator is used because it has the advantage that one capacitor terminal is at ground making it easier to switch connections. The thermal and temporal behavior of the Schmitt trigger oscillator in the micropower regime has been analyzed in detail [27]. The oscillator period T can be expressed as[12]

$$T = T_0 F(G)$$
$$G = \frac{I_R R_L}{V_T}$$
$$T_0 = \frac{2 I_R R_L C}{I_C}.$$

$$(13)$$

---

[11]For a resistor whose epi tub is connected to the most positive terminal the equation predicts a fixed voltage dependence of 2.1 percent/V. Measurements show the actual dependence varies from 1.8 to 2.9 percent/V.

[12]The distinction between oscillator period and temperature, both denoted by T, is clear from the context.

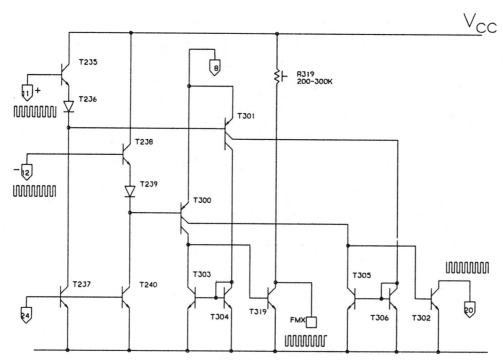

Fig. 4.   Circuit schematic of the data output stage.

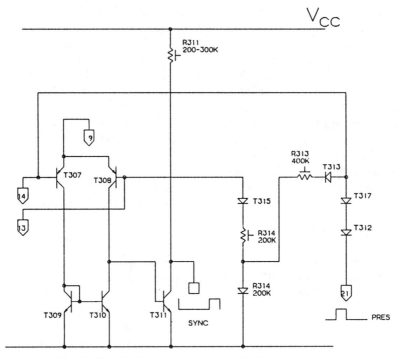

Fig. 5.   Circuit schematic of the synchronization output stage.

As explained in Section II the Schmitt trigger current I_R is derived from a PTAT voltage so that the current is PTAT/R. For the T(PRES), T(PSCALE), and T(ZERO) signals, the capacitor charging current I_C is also made PTAT/R. Since $F(G)$ is independent of temperature, the oscillator TC for these signals should be equal to that of the Schmitt trigger resistor R_L which measured results closely confirm.

VI.   Design of the Oscillator Circuit

To sense capacitance changes of 300-ppm FS, the cycle-to-cycle jitter of the oscillator must be below this level. In the present design, shown in Fig. 3, we take advantage of the theory presented in Appendix B by connecting a 6-pF MOS capacitor, C200, to the Schmitt trigger resistor node. This has the effect of reducing the oscillator jitter by

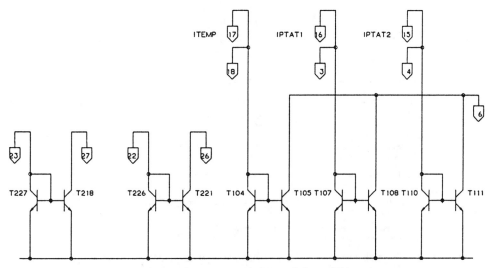

Fig. 6. Circuit schematic of multiplexer switches and bias mirrors.

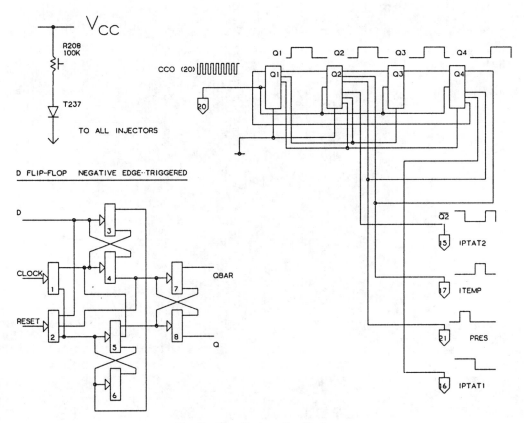

Fig. 7. Circuit schematic of logic counter.

reducing the noise bandwidth of the Schmitt trigger ($T213$–215). C200 also reduces the square-wave slew rate, thus cutting down capacitative feedthrough to the current bias chains. The C_{REF} section ($T209/10, T200/1$) and the C_{PRES} section ($T211/2, T202/3$) are switched by $T216/7$. Driven by I²L switches, $T206/7$ and $T231/2$ alternately clamp the emitter followers $T204$ or $T205$. $T225$ is smaller than a resistor and provides differential drive into the output stage; $T223/234$ also save area as level shifters.

VII. Design of the Output Stages, Multiplexer, and Logic

The output stage, shown in Fig. 4, drives a custom bipolar FM transmitter chip housed in a separate implanted electronics package, which also contains the lithium pacemaker battery. Buffers $T235/8$ are driven by the oscillator Schmitt trigger. Area is saved by using a configuration with split collector p-n-p's to drive $T319$ which is

IEEE JOURNAL OF SOLID-STATE CIRCUITS, VOL. SC-21, NO. 6, DECEMBER 1986

TABLE I
SUMMARY OF TEST RESULTS

Parameter	Value
Power consumption	$200 \mu W$ ($V_{cc} = 3.5V$)
Oscillator jitter	300 ppm (ave.)
Bandgap reference TC	$(-150) - 100$ ppm/°C $(20 - 60°C)$
Bandgap reference supply dependence	$114 - 626$ ppm/V ($V_{cc} = 3 - 4V$) 237 ppm/V (ave.)
Temperature measurement supply dependence	$0.05 - 0.2$ %/V ($18 - 45°C$, $V_{cc} = 3.4 - 4.0V$) 0.1 %/V (typ.)
Capacitance measurement supply dependence	$(-1.8) - (-2.5)$ %/V ($18 - 45°C$, $V_{cc} = 3.4 - 4.0V$) -2.0 %/V (typ.)
Capacitance measurement temperature dependence	$0.029 - 0.033$ %/°C $(18 - 45°C$, $V_{cc} = 3.4 - 4.0V$)
Capacitance measurement non-linearity variation	± 0.1 % ($C = 20 - 120pF$) $(25 - 45°C$, $V_{cc} = 3.4 - 4.0V$)

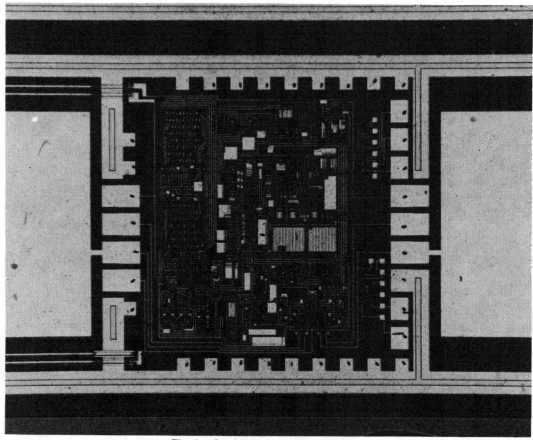

Fig. 8.   Capacitive pressure sensor IC die.

also clamped by the input stage of the custom FM transmitter and thus kept out of saturation.

The synchronization signal output stage, shown in Fig. 5, drives the differential switch in the oscillator through $T307/8$ and provides the sync. signal to the FM transmitter IC via $T311$. $T313/4/5$ and $R313/4$ maintain the correct bias points to prevent the differential pair switch in the oscillator from saturating. $T312/7$ reduce the signal swing, improving speed at the output of the I^2L gate that drives this circuit. The resistors in both of the output stages may be adjusted by metal mask options to accomodate different cable loadings.

The multiplexer, shown in Fig. 6, is driven by the I^2L counter, shown in Fig. 7. The standard mirrors, $T104 - T111$, use up little supply voltage and switch faster than Wilson or cascoded mirrors partly because diodes $T104/7/10$ clamp the swing at the I^2L outputs to approximately 0.5 V. The presence of a low-impedance shunt path at the bases of $T105/8/11$ also reduces current feedthrough from the logic. The Gray-code I^2L counter consists of four *D*-type flip-flops and can be shown to have the lowest possible power consumption per unit delay [26] (an asynchronous counter would occupy three-fourths the area but is not glitch-free). By arranging the switch timing carefully a configuration can be found which requires a maximum of three outputs (*Q* or NOT *Q*) from each *D*-type flip-flop, thus minimizing the cell footprint. The reset signals in the last two stages prevent unwanted states. $R208/T237$ provides injector bias in a simple but compact and effective way to maintain I^2L switching delay relatively independent of temperature [28].

VIII. RESULTS

A summary of test results is given in Table I. The absolute temperature measurement accuracy was found to be inseparable from the wafer hot-chuck accuracy; measurements made using the IC itself indicate temperature variations of $\pm 0.05°C$ (at 18°C) to $\pm 0.08°C$ (at 40°C). It is not clear at the present time how much of this error is due to the chip and how much to the hot-chuck controller.

The supply dependence of the capacitance measurement scheme is believed to be due to mismatch between the operation of the reference and pressure sections of the oscillator. It is, in fact, the largest source of error in the IC. However, to put this error in perspective, the Wilson–Greatbatch lithium pacemaker cells change in voltage by about 0.2 V over 90 percent of their lifetime. This results in a pressure error of only ± 0.2 percent over the life of the implant.

The temperature dependence of the capacitance measurement is accounted for by pressure calibration leaving a small supply-dependent residual error. The nonlinearity in the capacitance measurement is below 0.1 percent with the present minimum value of 35 pF for C_{PRES}. This is, strictly speaking, a random error superimposed on the systematic nonlinearity that is also accounted for in the pressure calibration [1].

The sensor IC die is shown in Fig. 8. The bottom plates of the capacitors can be seen along with the internal bonding pads and minimum circuit feature size is 10 μm. Currently, the transducer measures 2×6 mm², and the analysis presented in Appendix C shows that an areal reduction by a factor of four is straightforward.

IX. CONCLUSION

We have described a custom IC for use with a capacitive pressure sensor. By using a capacitive transducer and micropower circuit design the required specifications, a small, stable, accurate and inexpensive implantable pressure sensor for medical use can be met. The power consumption was reduced from 24 mW for the best commercially available implantable pressure transducer to 200 μW. Temperature measurement capability to $\pm 0.1°C$ over a limited temperature range allows accurate temperature compensation of the transducer.

APPENDIX A
ANALYSIS OF ERRORS IN THE RATIOMETRIC TEMPERATURE MEASUREMENT SCHEME

Without loss of generality,[13] let

$$I_{PTAT} \propto \frac{T}{T_1}, \quad I_{ITAT} \propto 1 + \xi$$

where

$$\xi = \xi'(T_1)(T - T_1) + \frac{1}{2!}\xi''(T_1)(T - T_1)^2 + \cdots \quad (A1)$$

and $\xi'(T_1)$ is the first-order and $\xi''(T_1)$ the second-order bandgap reference TC at T_1.[14] The external demodulator calculates temperature from the ratio S where

$$S(T) = \frac{I_{PTAT}}{I_{ITAT}} = \frac{T}{T_1(1 + \xi)}$$

$$= S(T_1) + S'(T_1)(T - T_1) + \frac{1}{2}S''(T_1)(T - T_1)^2 \quad (A2)$$

and

$$S'(T_1) = \left[\frac{1}{T_1} - \xi'(T_1)\right]$$

$$S''(T_1) = -2\xi'(T_1)S'(T_1) - \xi''(T_1). \quad (A3)$$

We now examine the effects of different calibration techniques and their effect on the temperature measurement accuracy. For a single-point temperature calibration we measure $S(T_1)$ and assume (incorrectly) that the temperature dependence of S is linear and that $S'(T_1) = S(T_1)/T_1 = 1/T_1$.[15] To first-order in $(T - T_1)$ the error in the temperature signal S_E is thus

$$S_E = \left\{ S(T_1) + \left[\frac{1}{T_1} - \xi'(T_1)\right](T - T_1) \right\}$$
$$- \left\{ S(T_1) + \frac{1}{T_1}(T_1 - T) \right\}. \quad (A4)$$

[13] The assumptions made are discussed below and in footnote 15.

[14] The temperature T_1 is determined by the temperature at which $I_{PTAT} = I_{ITAT}$ in the bandgap reference and is not necessarily equal to T_0 the temperature at which the reference TC is designed to be zero. Since these currents are derived from voltages $V_{PTAT} \propto V_T$ and $V_{ITAT} = V_{REF}$ (the bandgap reference voltage) we can estimate the expected variation of T_1: The reference is designed so that $I_{PTAT} = I_{ITAT} = 2$ μA and $V_{PTAT} = V_{ITAT} = V_{REF} = 1.259$ V at $T_0 = T_1 = 300$ K. If the error in the reference voltage V_{REF} is V_E (see Section III-E) then the expected variation in T_1 is $(V_E T_0)/V_{REF}$, since $V'_{PTAT} = V_{REF}/T_0$. From limited measurements made to date the variation is considerably less than this estimate. This is because of the strong correlation between V_E and V_{PTAT}, which were assumed independent, and, in fact, gives us a method to determine how much of the error in V_{REF} is due to absolute V_{BE} variation and how much to resistor and emitter area mismatch etc.

[15] We have assumed that we calibrate at T_1 and that $I_{PTAT} = I_{ITAT} = S(T_1) = 1$ at T_1. However, the analysis is also valid for $I_{PTAT} = I_{ITAT} = 1$ at T_2 and thus is valid for calibration at any temperature. The form of the analysis is almost identical to that shown, the key difference being the measurement of $S(T_2)$ and normalization by this quantity.

The error T_E in the calculated temperature is given by

$$T_E \simeq \frac{S_E}{S'(T_1)}. \tag{A5}$$

Since $S'(T_1) \simeq 1/T_1$ the temperature range $|T - T_1|$ over which we can measure temperature to a specified accuracy T_E is

$$|T - T_1| \simeq \frac{T_E}{|\xi'(T_1)T_1|}, \qquad \text{for single-point calibration.} \tag{A6}$$

For example, if $\xi' = -150$ ppm/°C we can measure temperature to an accuracy $T_E = 0.1$°C within a temperature range of ± 2.2°C.

With two-point calibration we still assume S varies linearly with temperature but we measure $S(T_1)$ and $S'(T_1)$. Then[16]

$$S_E = \frac{1}{2}S''(T_1)(T - T_1)^2 \tag{A7}$$

and

$$|T - T_1| \simeq \sqrt{\frac{T_E}{\left|\xi'(T_1) + \frac{1}{2}\xi''(T_1)T_1\right|}},$$

$$\text{for two-point calibration.} \tag{A8}$$

APPENDIX B
JITTER IN A MICROPOWER RELAXATION OSCILLATOR

This appendix extends the theoretical work of Abidi and Meyer [29] on relaxation oscillator jitter. If a capacitor C is alternately charged and discharged by a current I_C, consideration of noise correlation at the switching points gives the following expression for jitter J (in ppm) due to voltage noise E_n on the capacitor [29]:

$$J = \frac{\alpha \sqrt{6}\, E_n}{2 \Delta V_C}, \qquad 0.5 < \alpha < 1 \tag{B1}$$

where ΔV_C is the voltage swing across the capacitor. If the slew rate of the noise is much higher than the slope of the capacitor ramp rate the parameter α approaches 0.5.[17] This is because switching will always occur as the ramp crosses a positive noise peak. At higher currents with large values of timing capacitors the integration of the shot

noise in I_C by the capacitor is negligible [30]. In the micropower region of operation this is no longer true.

The integrated noise contribution of the shot noise $I_{nC}(t)$ of the capacitor charging current I_C is given by [30]

$$E_{nC} = \frac{1}{C} \int_0^{T_0/2} I_{nC}(t)\, dt = \frac{1}{C}\frac{\sqrt{S_{nC}T_0}}{2} \tag{B2}$$

where the spectral density of the shot noise is $S_{nC} = 2qI_C$ and is assumed to have a bandwidth much greater than $1/T_0$ where T_0 is the oscillator period

$$T_0 \simeq \frac{2\Delta V_C C}{I_C}. \tag{B3}$$

Considering now the Schmitt trigger noise contribution: for $I_R R_L \gg 2V_T$ the thermal noise in the Schmitt trigger resistor R_L is much less than the shot noise in I_R. We can estimate that the major contributions to shot noise come from the device that is turned on and the differential pair current source. Adding these correlated sources

$$E_{nR_L}^2 = 8qI_R R_L^2 \Delta f \tag{B4}$$

where Δf is the noise bandwidth at the resistor node which we estimate as

$$\Delta f = \frac{1}{4R_L C_{EQ}} \tag{B5}$$

and C_{EQ} is the equivalent node capacitance. Then

$$E_{nR_L}^2 \simeq \frac{2q\Delta V_C}{C_{EQ}} \qquad (\Delta V_C \simeq I_R R_L). \tag{B6}$$

We can refer this noise source directly to the oscillator capacitor node. From (B1)–(B6) the total rms jitter due to integrated shot noise in I_C and shot noise in I_R is thus

$$J = \alpha \sqrt{\frac{3q}{\Delta V_C}} \sqrt{\frac{1}{C} + \frac{1}{C_{EQ}}}. \tag{B7}$$

In the present design $C_{EQ} \simeq 12$ pF including parasitic device capacitance, the total predicted jitter is approximately 315 ppm (close to the measured value of 300 ppm) for $C = 50$ pF, $\Delta V_C = 0.5$ V, and assuming $\alpha = 1$. The rms contribution of the jitter given by the term involving C in the above equation is 138 ppm. This is the component normally ignored in higher power designs using a larger timing capacitor.

APPENDIX C
SCALING OF THE PRESENT TRANSDUCER AND CIRCUIT

Sander has examined in detail the fundamental limits to scaling a capacitive transducer [31]. We are more interested in reducing the lateral dimensions of the present sensor design and less interested in dealing with the difficulties of reducing the plate separation d_0 below the

---

[16] Calibration is at T_1 and $T_1 + \Delta T$, where the linear approximation error to $S(T_1 + \Delta T)$ is small compared to the temperature measurement error we wish to achieve (essentially we must determine an accurate value of $S'(T_1)$).

[17] There is no connection with α, the temperature exponent of the collector current in the bandgap cell. Recent work by Fleischer at Stanford has shown that for very high noise bandwidths α can be less than 0.5. However, for the circuit presented here $\alpha \simeq 1$.

present value of 1 μm, thus scaling at constant d_0 is more attractive. Scaling the sensor's remaining physical dimensions by a factor s gives

$$C \to \frac{C}{s^2} [d_0 \to d_0]. \tag{C1}$$

To maintain constant sensitivity we must keep $a^4/(h^3 d_0)$ constant [31], where h is the diaphragm thickness under the pressure sensitive capacitor and a^2 is the capacitor area. Then

$$h \to \frac{h}{s^{4/3}}. \tag{C2}$$

Turning now to the circuit design the following equations are of interest:

$$T_0 \propto \frac{C\Delta V_C}{I_C}$$

$$\Delta V_C \propto I_R R_L$$

$$J \propto \sqrt{\frac{1}{C\Delta V_C}}. \tag{C3}$$

These equations relate the oscillator period T_0, the capacitor voltage swing ΔV_C, and the oscillator jitter J to the capacitance C, the oscillator currents I_R and I_C, and the oscillator resistor R_L. The capacitor voltage swing $\Delta V_C \simeq 0.5$ V cannot easily be increased without increasing the battery voltage and/or altering the circuit design thus we should, if possible, scale with ΔV_C constant. Experience has shown that device currents should be kept above 10 nA; this figure is two orders of magnitude above the typical values of the leakage currents in the process used. The majority of the operating currents in the sensor IC are presently at the 1-μA level and we can therefore scale current downwards by up to two orders of magnitude.

To achieve a pressure measurement resolution of 300 ppm with the present oscillator period of 50 μs requires an external demodulator clock period of 50 μs $\times (300 \times 10^{-6})$ $= 15$ ns (60 MHz), which is close to the operating limit of commercially available logic circuits. In addition, the FM telemetry link is presently set at 80 MHz; we would thus like to maintain the oscillator period T_0 constant with scaling.

The above constraints determine the scaling laws shown in Table II. Notice we are free to choose how we scale I_R and R_L. In the Schmitt trigger oscillator the sections of the circuit carrying currents I_R and I_C are closely coupled. Ideally, therefore, I_R should scale with I_C

$$I_R \to \frac{I_R}{s^2}$$

$$R_L \to s^2 R_L. \tag{C4}$$

A practical limit to scaling is then the resistor size (we wish to avoid scaling the bandgap resistors which already occupy approximately 10–15 percent of the circuit area). If

TABLE II
SCALING LAWS: SCALING AT CONSTANT d_0, ΔV_c, AND T

Parameter	Scaling law
Plate separation	$d_0 \mapsto d_0$
Capacitor swing	$\Delta V_C \mapsto \Delta V_C$
Oscillator period	$T \mapsto T$
Diaphragm size	$a \mapsto a/s$
Capacitance	$C \mapsto C/s^2$
Diaphragm thickness	$h \mapsto h/s^{4/3}$
Capacitor current	$I_C \mapsto I_C/s^2$
Schmitt-trigger swing	$I_R R \mapsto I_R R$
Oscillator jitter	$J \mapsto sJ$

we scale only the oscillator circuit then we need only scale the oscillator resistor. At present $I_R = 2$ μA and $R_L = 300$ kΩ. It is reasonable to increase the value of R_L to 1.2 MΩ that is approximately twice the size of the largest resistor presently on the chip (in the bandgap reference). Increasing R_L by a factor of 4 corresponds to scaling by $s = 2$. To scale further requires us to break the scaling laws. Relaxing our constraint on d_0 and scaling d_0 by a factor σ, we have

$$d_0 \to \frac{d_0}{\sigma}$$

$$C \to \sigma C$$

$$I_C \to \sigma I_C$$

$$I_R \to \sigma I_R$$

$$R_L \to \frac{R_L}{\sigma}. \tag{C5}$$

Taking $\sigma = 2$ then allows us to scale by $s^2 = 8$ or $s = 2.8$; this would reduce the sensor width to below 1 mm.

ACKNOWLEDGMENT

Thanks are due to: M. Prisbe and J. Mustafa for helping the principal author with wafer fabrication; R. King for metal deposition; M. King and Z. Norris for optical and *E*-beam mask preparation; D. Roman and T. van Hooydonk for repairing key equipment; and the staff of the Center for Integrated Electronics in Medicine. IBM Corporation provided the facilities for the preparation of this article. G. Bronner offered helpful critical review.

REFERENCES

[1] M. J. S. Smith, L. Bowman, and J. D. Meindl, "Analysis, design, and performance of a capacitive pressure sensor IC," *IEEE Trans. Biomed. Eng.*, vol. BME-33, pp. 163–174, Feb. 1986.
[2] C. S. Sander, J. W. Knutti, and J. D. Meindl, "A monolithic capacitive pressure transducer with pulse-period output," *IEEE Trans. Electron. Devices*, vol. ED-17, pp 927–930, May 1980.
[3] S. J. Gschwend, J. W. Knutti, H. V. Allen, and J. D. Meindl, "A general-purpose implantable multichannel telemetry system for physiological research," *Biotelemetry Patient Monitg.*, vol. 6, pp. 107–117, 1979.
[4] R. J. Widlar, "New developments in IC voltage regulators," *IEEE J. Solid-State Circuits*, vol. SC-6, pp. 2–7, Feb. 1971.

327

[5] K. E. Kuijk, "A precision reference voltage source," *IEEE J. Solid-State Circuits*, vol. SC-8, pp. 222–226, June 1973.

[6] A. P. Brokaw, "A simple three-terminal bandgap reference," *IEEE J. Solid-State Circuits*, vol. SC-9, pp. 388–393, Dec. 1974.

[7] G. C. M. Meijer and J. B. Verhoeff, "An integrated bandgap reference," *IEEE J. Solid-State Circuits*, vol. SC-11, pp. 403–406, June 1976.

[8] G. C. M. Meijer and K. Vingerling, "Measurement of the temperature dependence of the IC(VBE) characteristics of integrated bipolar transistors, *IEEE J. Solid-State Circuits*, vol. SC-15, Apr. 1980.

[9] B. E. Amazeen, P. R. Holloway, and D. A. Mercer, "A complete single-supply microprocessor-compatible 8-bit DAC," *IEEE J. Solid-State Circuits*, vol. SC-15, pp. 1059–1070, Dec. 1980.

[10] C. R. Palmer and R. C. Dobkin, "A curvature corrected micropower voltage reference," in *ISSCC Dig. Tech. Papers*, 1981.

[11] R. J. van de Plassche and H. J. Schouwenaars, "A monolithic 14 bit A/D converter," *IEEE J. Solid-State Circuits*, vol. SC-17, pp. 1112–1117, Dec. 1982.

[12] G. C. M. Meijer, "Integrated circuits and components for bandgap references and temperature transducers," thesis, Delft Univ. Technol, Mar. 1982.

[13] G. C. M. Meijer, P. C. Schmale, and K. van Zalinge, "A new curvature-corrected bandgap reference," *IEEE J. Solid-State Circuits*, vol. SC-17, pp. 1139–1143, Dec. 1982.

[14] P. H. Saul and J. S. Urquhart, "Techniques and technology for high-speed D-A conversion," *IEEE J. Solid-State Circuits*, vol. SC-19, pp. 62–68, Feb. 1984.

[15] P. Menniti and S. Storti, "Low drop regulator with overvoltage protection and reset function for automotive environment," *IEEE J. Solid-State Circuits*, vol. SC-19, pp. 442–448, June 1984.

[16] E. A. Vittoz, "A low-voltage CMOS bandgap reference," *IEEE J. Solid-State Circuits*, vol. SC-14, pp. 573–577, June 1979.

[17] B-S. Song and P. R. Gray, "A precision curvature-compensated CMOS bandgap reference," *IEEE J. Solid-State Circuits*, vol. SC-18, pp. 634–643, Dec. 1983.

[18] J. Michejda and S. K. Kim, "A precision CMOS bandgap reference," *IEEE J. Solid-State Circuits*, vol. SC-19, pp. 1014–1021, Dec. 1984.

[19] C. A. Bittman, "Technology for the design of low-power circuits," *IEEE J. Solid-State Circuits*, vol. SC-5, Feb. 1970.

[20] Y. P. Tsividis, "Accurate analysis of temperature effects in IC-VBE characteristics with applications to bandgap reference sources," *IEEE J. Solid-State Circuits*, vol. SC-15, pp. 1076–1084, Dec. 1980.

[21] S. L. Lin and C. A. T. Salama, "A VBE(T) model with application to bandgap reference design," *IEEE J. Solid-State Circuits*, vol. SC-20, pp. 1283–1285, Dec. 1985.

[22] G. Erdi, "A precision trim technique for monolithic analog circuits," *IEEE J. Solid-State Circuits*, vol. SC-10, pp. 412–416, Dec. 1975.

[23] P. R. Gray and R. G. Meyer, *Analysis and Design of Analog Integrated Circuits*. New York: Wiley 1977.

[24] B. Gilbert, "A versatile monolithic voltage-to-frequency convertor," *IEEE J. Solid-State Circuits*, vol. SC-11, pp. 852–864, Dec. 1976.

[25] B. L. Hart, "Automatic start-up of complementary PTAT current generators," *Electron. Lett.*, vol. 18, pp. 776–777, Sept. 1982.

[26] M. J. S. Smith, "An integrated circuit for a biomedical capacitive pressure transducer," thesis, Stanford Univ., Stanford, CA, June 1985.

[27] M. J. S. Smith and J. D. Meindl, "Exact analysis of the Schmitt trigger oscillator," *IEEE J. Solid-State Circuits*, vol. SC-19, pp. 1043–1046, Dec. 1984.

[28] E. Bruun and O. Hansen, "Current regulators for I2L circuits to be operated from low-voltage power supplies," *IEEE J. Solid-State Circuits*, vol. SC-15, pp. 796–799, Oct. 1980.

[29] A. A. Abidi and R. G. Meyer, "Noise in relaxation oscillators," *IEEE J. Solid-State Circuits*, vol. SC-18, pp. 794–802, Dec. 1983.

[30] A. Abidi, "Effects of random and periodic excitations on relaxation oscillators," thesis, Univ. California, Berkeley, May 1981.

[31] C. S. Sander, "A bipolar-compatible monolithic capacitive pressure sensor," thesis, Stanford Univ., Stanford, CA, Dec. 1980.

AN MOS SWITCHED-CAPACITOR READOUT AMPLIFIER
FOR CAPACITIVE PRESSURE SENSORS

Y. E. Park and K. D. Wise

Department of Electrical and Computer Engineering
University of Michigan
Ann Arbor, Michigan 48109

ABSTRACT

This paper reports the development of an MOS switched-capacitor interface for use with capacitive pressure sensors. The overall circuit produces an output voltage proportional to capacitance change, is compatible with an on-chip ADC, and is ideally suited to array applications. The approach is also insensitive to ambient temperature variations and stray input capacitance. The theoretical limit on the minimum detectable signal charge (MDSC) is about 1 fC at room temperature and 10-15 fC over an uncompensated 0-100°C temperature range. Implemented in a LOCOS silicon-gate ED-NMOS process, the interface requires 0.7 mm² in 6 μm features. The power dissipation is 2 mw at 10 V. Wafer probe measurements have resolved values of MDSC of less than 25 fC.

INTRODUCTION

Silicon capacitive pressure sensors have been developed recently [1-3] as an alternative to silicon piezoresistive sensors. The capacitive devices have been shown to offer about ten times greater pressure sensitivity and ten times less temperature sensitivity when compared with their piezoresistive counterparts and promise to find wide application in areas such as transportation, health care, and robotics. Such devices use a thin silicon diaphragm as the moveable plate in a variable-gap capacitor which is completed by a metal electrode on an opposing glass plate. The principal problem in these capacitive transducers is the relatively low capacitance values per unit area (2 pF/mm²) and large resulting die sizes. A typical device of reasonable size offers a zero-pressure capacitance of only a few picofarads so that a resolution of one percent requires the reliable detection of capacitance shifts of the order of 50 fF or less. Thus, both the die size and performance of capacitive pressure sensors depend directly on the available readout circuitry.

While few monolithic circuits for such transducers have yet been reported, most approaches have been based on bipolar oscillator configurations in the past. Such schemes have generally exhibited rather high sensitivity to parasitics, required slow or awkward conversion processes back to analog voltage or digital BCD output formats, and been difficult to adapt to array applications. This paper reports a differential switched-capacitor interface for capacitive pressure sensors. The approach produces a voltage output directly, is compatible with an on-chip switch-capacitor ADC, and is ideally suited to array applications.

Figure 1 shows a cross-section of a typical silicon capacitive pressure sensor. The thin silicon diaphragm can be precisely controlled in thickness using a diffused boron etch-stop [4], while a thermocompression-bonded lead transfer, formed during the electrostatic bonding of the silicon to the glass plate, is used to connect the metal electrode on the glass to circuitry on the silicon. The reference cavity is typically 3-5 μm deep. Since the transducer itself is very simple here, there are few temperature drift mechanisms, and for a vacuum-sealed or relative device, the observed temperature coefficients are very low (0-50 ppm/°C) [1,4].

CIRCUIT OPERATION

The general circuit approach for interfacing with such a capacitive transducer is shown in Fig. 2. The transducer capacitance (C_X) is connected to a reference capacitor (C_R) as shown. These devices are driven by opposite clock phases, $\emptyset_1$ and $\emptyset_2$. If an array of transducers is used, $\emptyset_1$ can be decoded appropriately. When the clocks switch, a charge appears at the common capacitance node which is proportional to ($C_X - C_R$) and hence to applied pressure. This charge is integrated to produce an output voltage proportional to pressure. So long as the open loop

Reprinted from *Rec. of the IEEE Custom IC Conf.*, 1983, pp. 380–384.

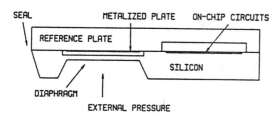

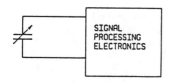

Fig. 1: Cross-section and Schematic Representation of a Silicon Capacitive Pressure Sensor.

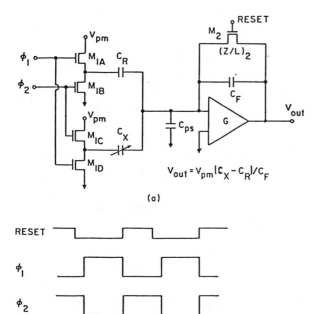

Fig. 2: General Circuit Approach to a Differential Switched-Capacitor Readout System for Capacitive Pressure Sensors.

gain of the amplifier is high, the output voltage is insensitive to stray input capacitance, amplifier offset, and temperature drift. The clock voltage swing (V_{pm}) is well defined and typically equal to the supply voltage. The structure is also relatively insensitive to process-related parameters.

The minimum detectable signal charge (MDSC) is a primary performance figure of merit in the implementation of this approach in MOS technology. Two main factors determine the MDSC limit: circuit noise and drift induced by ambient temperature variations. For a practical design such that

$$\omega_1 = (R_{on}C_R)^{-1}, \quad \omega_c \gg A_o\omega_o = g_{m1}/C_c \qquad (1)$$

and

$$\alpha^2 = (2C_R/C_F)^2 < 0.1 \qquad (2)$$

the total noise power expression at the output of the amplifier [5] can be simplified as

$$\overline{v_{on_T}^2} \sim kT(C_F^{-1} + C_C^{-1}) \qquad (3)$$

where R_{on} is the on-resistance of a single MOS switch, ω_c is the 1/f noise corner frequency of the op amp, and $A_o\omega_o$, g_{m1}, and C_C are the unity gain frequency, input stage transconductance, and frequency-compensation capacitance of the op amp, respectively. Equation (3) implies that for a fixed amount of silicon area (i.e., $C_F + C_C$ = a constant area), the minimum noise power occurs when the integration capacitance, C_F, is approximately equal to the compensation capacitance, C_C.

When the MOS reset device switches from the on-state to the off-state, the switching clock injects some charge from the gate of the reset device to the input sampling node through the gate-to-channel capacitance of the device. This injection charge results in an offset voltage at the output node which is often referred to as the "reset coupling noise voltage" or the "feedthrough error voltage". After compensation with a charge-canceling device [6], the equivalent reset noise charge (ENRC) is given by

$$ENRC = (h_1 - h_3)V_{Dsat}C_S \qquad (4)$$

where

$$h_1 = C_{GS}/(C_S + C_{GS}) \qquad (5)$$

and

$$h_3 = C_{GS(Sat)}/(C_S + C_{GS(Sat)}) \qquad (6)$$

330

Here, C_S is the value of the equivalent lumped capacitance at the input sampling node, $C_{GS} = (ZLC_{ox})/2$, $C_{GS(Sat)} = 4C_{GS}/3$, and $V_{D(Sat)}$ is the drain voltage where the MOS device enters saturation. It has been reported [7] that the combined use of a charge-canceling device and the balance switch technique can reduce the reset coupling noise two orders of magnitude as compared with no compensation. The correlated double-sampling technique [8] can also suppress this reset noise as well as the 1/f noise of the op amp.

Drift in the output voltage due to changes in ambient temperature can also be significant. For a vacuum-sealed reference cavity (no trapped gas expansion)[1], the temperature coefficients of both the reference and the pressure-variable transducers are low and well matched. Thus, the output drift can be expressed as

$$\frac{dV_{out}}{dT} = V_{pm} \frac{(C_X - C_R)}{C_F} (T_{C_R} - T_{C_F}) \qquad (7)$$

where V_{pm} is the pump voltage applied to the bottom plates of the transducers (see Fig. 2), T_{C_R} is the nominal temperature coefficient of C_R and C_X, T_{C_F} is the temperature coefficient of the gate-oxide capacitor C_F, and V_{out} is the output voltage. For example, for $T_{C_F} = 27$ ppm/°C [9] and T_{C_R} in the range 0-50 ppm/°C, the maximum expected equivalent drift charge over the 0-100°C temperature span is about 14 fC for a pressure equivalent to $(C_X - C_R) = 1$ pF, $C_F = 5$ pF, and $V_{pm} = 5$ V.

These results suggest that at high pressures and for widely-varying ambient temperatures, the lower limit for the MDSC is determined primarily by the mismatch in the temperature coefficients of the transducer (and its reference) and the gate-oxide integration capacitor. At lower pressures and in less temperature-variable environments, however, the total noise charge of the amplifier will set the MDSC.

EXPERIMENTAL RESULTS

This circuit approach was initially verified using commercial components with the results shown in Fig. 3. The output voltage is clearly proportional to the differential capacitance $(C_X - C_R)$ and varies inversely with the integration capacitance as expected. The scatter in the data for small values of differential capacitance was due to both the numerous parasitic elements in the discrete circuit implementation and to reset noise in the integrator.

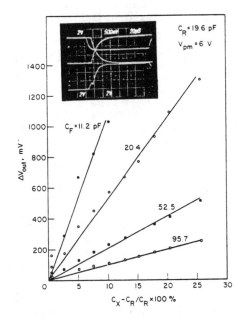

Fig. 3: Output voltage as a Function of Differential Capacitance for a Discrete Switched-Capacitor Interface Circuit.

Figure 4 shows a portion of a wafer containing capacitive pressure sensors. The diaphragm size is 2 mm X 2 mm. Both C_X and C_R are fabricated on the same chip and are identical except that only C_X has a selectively-thinned diaphragm. These devices can typically be matched to better than 0.5 percent. Concentric structures [2] in a single reference cavity could also be used to conserve die area and use the same gap.

Fig. 4: Backside view of an Array of Silicon Capacitive Pressure Transducers. C_X is shown above while C_R (with no diaphragm) is located below it.

The switched-capacitor interface circuit has been realized using a LOCOS silicon-gate ED-NMOS process. The circuitry employs three target device thresholds: -2.5 V, +0.2 V, and +0.8 V. Figure 5 shows the measured threshold voltages (total distribution ranges) for various implant doses over 18 wafers. Figure 6 shows an integrated realization of four complete readout amplifiers with test devices, while Fig. 7 shows the outputs of one such circuit. The open-loop amplifier gain is 1000 with a settling time of about 5 μsec into 30 pF and internal compensation. The circuit power dissipation is 2 mW at 10 V. A die area of 0.7 mm² is used for the interface circuit in 6 μm features. Wafer probe measurements have resolved MDSC values of less than 25 fC. Figure 8 shows a correlated double sampling chip developed to allow removal of reset noise from the sampled pressure signal. This readout scheme is being applied to a capacitive pressure sensor for biomedical applications and to a tactile imaging array for use in robotics.

Acknowledgements

The authors would like to thank J. Hile, J. Erskine, and F. Schauerte of the General Motors Research Laboratories for their assistance in this work and T. Mochizuki for many helpful discussions.

REFERENCES

1. Y. S. Lee and K. D. Wise, "A Batch-Fabricated Silicon Capacitive Pressure Transducer," IEEE Trans. Electron Devices, 29, pp. 42-47, January 1982.

2. W. H. Ko, M. H. Bao, and Y. D. Hong, "A High-Sensitivity Integrated-Circuit Capacitive Pressure Transducer," IEEE Trans. Electron Devices, 29, pp. 48-56, January 1982.

3. C. S. Sander, J. W. Knutti, and J. D. Meindl, "Monolithic Capacitance Pressure Transducer," IEEE Trans. Electron Devices, 27, p. 927, May 1980.

4. J. C.-M. Huang and K. D. Wise, "A Monolithic Pressure-pH Sensor for Esophageal Studies," 1982 IEDM Digest, pg.316.

5. B. Furrer and W. Guggenbühl, "Noise Analysis of Sampled-Data Circuits," IEEE Int. Symp. on Circuits and Syst., pp. 860-863, April 1981.

6. K. R. Stafford, P. R. Gray, and R. A. Blanchard, "A Complete Monolithic Sample/Hold Amplifier," IEEE Journal of Solid-State Circuits, 9, pp. 381-387, Dec. 1974.

7. L. A. Bienstman and H. J. DeMan, "An Eight-Channel 8-Bit Microprocessor-Compatible NMOS D/A Convertor with Programmable Scaling," IEEE Journal of Solid-State Circuits, 15, pp. 1051-1059, December 1980.

8. M. H. White, D. R. Lampe, F. C. Blaha, and I. A. Mack, "Characterization of Surface Channel CCD Image Arrays at Low Light Levels," IEEE Journal of Solid-State Circuits, 9, pp. 1-13, January 1974.

9. J. L. McCreary, "Matching Properties and Voltage and Temperature Dependence of MOS Capacitors," IEEE Journal of Solid-State Circuits, 16, pp. 608-616, Dec. 1981.

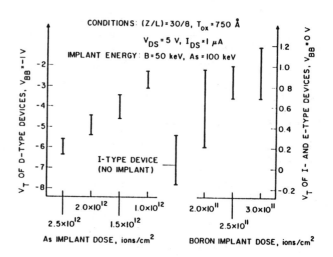

Fig. 5: Measured Threshold Voltages as a Function of Implant Dose for the ED NMOS Process.

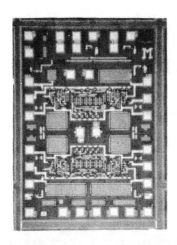

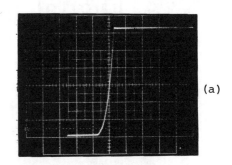

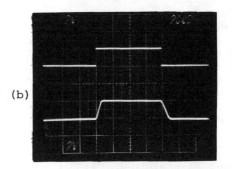

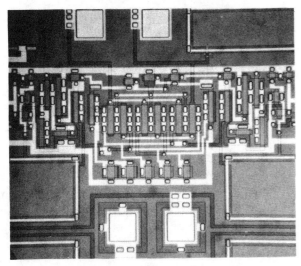

Fig. 6: Integrated Realization of Four Switched-Capacitor Readout Systems. Die Size is 2 x 2.7 mm.

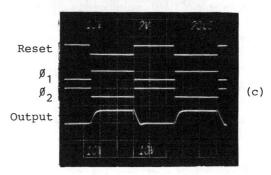

Reset

ϕ_1

ϕ_2

Output

(c)

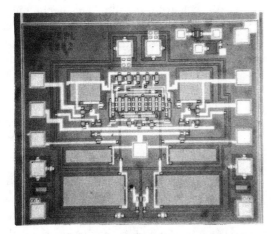

Fig. 8: A Correlated Double-Sampling Circuit for Reset-Noise Suppression. Die Size is 1.3 x 2.0 mm.

Fig. 7: Electrical Characteristics of the Charge-Detection System. (a) Open loop amplifier transfer characteristic (vertical: output voltage, 1 V/div; horiz.: input voltage, 15 mV/div). (b) Unity-gain pulse response (slew) (2 V/div and 20 μsec/div). (c) Charge-sensing operation (10 V/div except 2 V/div output, 20 μsec/div).

A Digital Readout Technique for Capacitive Sensor Applications

JOSEPH T. KUNG, STUDENT MEMBER, IEEE, HAE-SEUNG LEE, MEMBER, IEEE, AND
ROGER T. HOWE, MEMBER, IEEE

Abstract —The difference between two capacitors is measured digitally using a charge redistribution technique incorporating a comparator, MOS switches, a successive approximation register (SAR), and a digital-to-analog converter (DAC). The technique is insensitive to comparator offset and parasitic capacitance, and the effect of MOS switch charge injection is measured and canceled. Extensive measurements have been made from test chips fabricated in 3-μm CMOS technology. Detection of percent differences of less than 0.5 percent on 20–100-fF capacitors has been successfully demonstrated.

I. INTRODUCTION

THE measurement of capacitance difference is important for integrated sensors. Silicon structures sensitive to shear, acceleration, and pressure are examples of where a capacitive readout scheme is advantageous [1], [2]. In addition, changes in dielectric permittivity manifest themselves as a change in capacitance [3].

For integrated sensing structures, the readout capacitance can be on the order of tenths of a picofarad which complicates the capacitance detection circuitry. One detection method utilizes an oscillator which drives a capacitive bridge circuit. A change in capacitance relative to a reference capacitance produces an output voltage or shift in frequency which can be detected by an external circuit [4], [5]. For particular readout circuits, parasitic capacitances can cause an error in the measurement.

Recently, the advent of switched-capacitor techniques has led to new and innovative methods of capacitance detection [6], [7]. However, problems appear that are inherent to all switched-capacitor circuits. MOS switch charge injection, clock feedthrough, and circuit noise become major limiting factors in circuit performance.

Manuscript received June 3, 1987; revised February 2, 1988. This work was supported in part by the National Science Foundation under Contract ECS85-05145, by the Communication IC Consortium at M.I.T. (Analog Devices, AT&T Bell Labs., Digital Equipment Corporation, General Electric, Texas Instruments), and in part by IBM through an IBM Faculty Development Award held by R. T. Howe. MOSIS chip fabrication and other support was provided by DARPA under Grant N00014-80-C-0622.

J. T. Kung and H.-S. Lee are with the Microsystems Technology Laboratories, Department of Electrical Engineering and Computer Science, Massachusetts Institute of Technology, Cambridge, MA 02139.

R. T. Howe was with the Microsystems Technology Laboratories, Department of Electrical Engineering and Computer Science, Massachusetts Institute of Technology, Cambridge, MA 02139. He is now with the Berkeley Sensor & Actuator Center, Department of Electrical Engineering and Computer Sciences and the Electronics Research Laboratory, University of California, Berkeley, CA 94720.

IEEE Log Number 8821665.

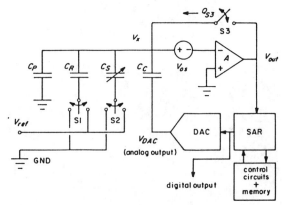

Fig. 1. Measurement system block diagram.

This paper describes a digital technique for measuring capacitance differences. It has its origins in charge redistribution A/D converters, and does not suffer from parasitic capacitance, op-amp offset, or charge injection problems.

II. THEORY

In 1979, an algorithm was developed that allowed calculation of ratio errors from a sequence of measurements based on charge redistribution [8]. This technique was implemented to study capacitor mismatch errors that cause linearity errors in charge redistribution A/D converters. MOS switch charge injection, however, was ignored due to the large size of the capacitors. Elimination of charge injection sources is crucial in obtaining higher resolution and smaller errors in these A/D converters. The self-calibration technique allowed higher resolution by eliminating errors caused by component mismatch and charge injection [9]. The technique can be equally applied to measure capacitance differences and random or controlled sources of charge injection. Since in sensor applications the sense capacitors may be much smaller than a picofarad, MOS charge injection causes a large error and must be canceled. The technique can measure errors due to capacitive mismatch, comparator offset, and charge injection and can compensate a system that has these errors. It is ideal for measuring capacitance differences (as in capacitive sensors) and reducing inherent circuit errors.

The basic circuit is shown in Fig. 1. It consists of the sense and reference capacitors C_S and C_R, respectively, the

Reprinted from *IEEE J. Solid-State Circuits*, vol. 23, no. 4, pp. 972–977, August 1988.

coupling capacitor C_C, five MOS switches, a voltage comparator, a digital-to-analog converter (DAC), a successive approximation register (SAR), and a memory register with associated logic capable of signal inversion. The nonidealities of the circuit appear as the offset of the comparator (V_{os}) and its finite gain (A), parasitic capacitance to ground (C_P), and switch charge injection (Q_{S3}). To better understand how these nonidealities are taken into account, an ideal system is first analyzed, then second-order effects are added later. In the assumption of an ideal circuit, $V_{os} = 0$, $C_P = 0$, $Q_{S3} = 0$, $A = \infty$, and DAC quantization error is negligible.

The measurement technique proceeds in two steps. In step 1, switch $S3$ is closed so that V_x is at ground. Switch $S1$ is set to V_{ref} and switch $S2$ is set to ground. The DAC output is also set to ground. The charge at the top node in this configuration is $Q_1 = -V_{ref}C_R$. The comparator is implemented so that when the feedback loop is closed with switch $S3$, V_x is forced to a virtual ground via the DAC. In step 2, $S3$ is opened, then $S1$ is set to ground and $S2$ is set to V_{ref}. The successive approximation search begins after this sequence and continues until the SAR reaches its quantization limit and stops. If the SAR, DAC, and voltage comparator are ideal, then the voltage from the DAC (V_{DAC}) precisely forces the top node voltage (V_x) to zero. The charge at the top node is thus $Q_2 = -V_{ref}C_S - V_{DAC}C_C$. By charge conservation, $Q_1 = Q_2$ and it follows that

$$V_{DAC} = \frac{V_{ref}(C_R - C_S)}{C_C}. \tag{1}$$

The output of the DAC produces a voltage proportional to the capacitance difference of C_S and C_R. Appropriate choices of V_{ref} and C_C can be made so that the maximum dynamic range of the DAC can be utilized. The parameter $\Delta C / C$ can be found by multiplying the numerator and denominator by C_R or C_S and rearranging so that

$$\frac{\Delta C}{C} = \left[\frac{V_{DAC}}{V_{ref}} \right]\left[\frac{C_C}{C} \right] \tag{2}$$

where ΔC is $C_R - C_S$ and C is a normalizing capacitance, typically either C_R or C_S. Notice that the result is a product of two ratios: a voltage ratio that can be measured easily and a capacitance ratio.

III. NONIDEALITIES

Several errors are introduced when the algorithm is implemented due to component nonidealities. Referring to Fig. 1, the comparator has an offset V_{os} and a finite gain A while the switch $S3$ injects a charge Q_{S3} when opened. A parasitic capacitance to ground C_P exists as well as a DAC quantization error of $\pm \frac{1}{2}$ LSB. It is found that the measurement algorithm can be implemented in either a closed-loop or open-loop topology. They differ only in that the closed-loop topology uses the feedback loop containing switch $S3$ and the open loop does not. The difference

between the two is that in closed loop, the voltage V_x is initially the offset of the comparator since it is connected like a voltage follower. Also, it is important to note that if the feedback path due to switch $S3$ is to cause $V_x = V_{os}$, the comparator must operate as a high-gain op amp and cannot be a regenerative latch comparator with only high or low digital outputs.

In open loop when the SAR/DAC feedback loop is initiated, the operation of the comparator can be limited to strictly digital output since it is never directly connected in negative feedback; rather the SAR/DAC generates the appropriate analog signal to the coupling capacitor as feedback. Usually, the comparator can be designed so that it can act as an op amp when the loop is closed [9]. In the open-loop topology, the top node is grounded in the first step so that the feedback loop through $S3$ is never established. Thus the comparator may always have digital outputs. Since the analog signal for measurement of small capacitors is usually small, a monolithic preamplifier can be used to buffer the voltage to an off-chip comparator in this configuration.

A. Quantization Error

The quantization error of the DAC contributes an error to the measurement. Assuming that V_{DAC} is in error by $\pm \frac{1}{2}$ LSB, the amount of error transferred to V_x can be easily shown to be

$$V_x = V_{os} \pm \frac{1}{2} \text{ LSB} \left[\frac{C_C}{C_P + C_R + C_S + C_C} \right] = V_{os} \pm \delta \left(\frac{1}{2} \text{ LSB} \right) \tag{3}$$

where δ is the capacitive divider ratio

$$\delta = \left[\frac{C_C}{C_P + C_R + C_S + C_C} \right] = \frac{C_C}{C_{total}}. \tag{4}$$

This is a simple capacitive divider. Any change in voltage ΔV at V_{DAC} results in a change in voltage $\delta \Delta V$ at V_x. Repeated use of the DAC's voltage output for subsequent measurements will accumulate this error in the worst case; however, averaging can reduce this problem.

B. Charge Injection

Charge injection can also be measured using this technique. Since the charge that $S3$ injects is independent of capacitor difference, it causes an error in the measurement. To correct for it, an additional step is added which will be denoted as the calibration cycle. The capacitive mismatch measurement will be denoted as the measurement cycle. The calibration and measurement cycles inherently eliminate the comparator offset in the closed-loop topology. In the open-loop topology, the offset is measured, then canceled.

The calibration procedure begins by measuring the charge injection. $S1$ is set to V_{ref} and both $S2$ and V_{DAC}

are set to ground. $S3$ is opened and the SAR/DAC is initiated. The calibration voltage at the output of the DAC can be shown to be

$$V_{DAC_{cal}} = \frac{-Q_{S3}}{C_C} \pm 2\Delta V \qquad (5)$$

in the closed-loop topology and

$$\dot{V}_{DAC_{cal}} = V_{os}\left[\frac{C_P + C_R + C_S + C_C}{C_C}\right] + \frac{-Q_{S3}}{C_C} \pm 2\Delta V \quad (6)$$

in the open-loop topology. $\pm \Delta V$ is the quantization error of the DAC. Charge injection must be measured since it upsets the charge conservation assumption made in the earlier ideal circuit analysis. If it is taken into account, then charge due to the capacitors can be accurately determined and hence so can the capacitance difference. Once the switch injection voltage is measured, it can be stored in a RAM. When the negative of $V_{DAC_{cal}}$ is applied to the coupling capacitor, a voltage at V_x is created that cancels out the error voltage generated by the switch injection charge. Alternatively, one can think of the DAC as creating a positive charge on the coupling capacitor that is just large enough to cancel the negative switch injection (assuming an NMOS switch). This is equivalent to analog voltage subtraction at V_x. Subtraction of the digital data is an alternate method of eliminating the charge injection error.

In the measurement cycle, $S3$ is closed, $S1$ is set to V_{ref}, $S2$ to ground, and V_{DAC} to the negative of the voltage measured during the calibration step. This switch sequence is exactly the same as in the ideal analysis except that V_{DAC} is at some voltage other than ground. $S3$ is then opened and the positions of $S1$ and $S2$ are reversed. The SAR/DAC is initialized, and the output of the DAC becomes

$$V_{DAC_{meas}} = \frac{V_{ref}(C_R - C_S)}{C_C} \pm 4\Delta V \qquad (7)$$

for both open-loop and closed-loop topologies. A disadvantage of the open-loop topology is that a large comparator offset may yield a calibration voltage larger than the DAC maximum voltage and calibration becomes impossible. The closed loop is preferred for this reason. For testing purposes, however, the open-loop topology is easier to implement.

C. Parasitic Capacitance

Parasitic capacitance imposes a constraint on the system. A large C_P reduces the divider ratio δ. If δ becomes too small, V_x becomes pinned by the large C_P and the DAC is not able to adjust V_x. The $\pm \Delta V$ error becomes limited by the capacitive divider ratio and not the quantization error because the minimum amount of control the DAC has is $\delta(\pm \Delta V)$ which may be smaller than the

comparator resolution. A smaller C_P increases δ but also increases the kT/C noise so that a trade-off is introduced.

Capacitance mismatch ratio for both topologies is

$$\frac{\Delta C}{C} = \left[\frac{V_{DAC_{meas}}}{V_{ref}}\right]\left[\frac{C_C}{C}\right] \pm \left[\frac{4\Delta V}{V_{ref}}\right]\left[\frac{C_C}{C}\right] \qquad (8)$$

where C is either C_R or C_S. It is the same as (2) except for the error term

$$\pm \left[\frac{4\Delta V}{V_{ref}}\right]\left[\frac{C_C}{C}\right]. \qquad (9)$$

This error term determines the minimum resolvable capacitance change for a single measurement. The effect of the parasitic capacitance appears as a constraint on the comparator resolution, as evidenced by (4).

D. Other Nonideal Effects

An MOS switch that is turned off can leak from the reverse-biased p-n junction at the body and source. Normal reverse saturation current from a p-n junction is typically 10 nA/cm² in MOS processes at room temperature. If the area of the MOS switch source region is 100 μm², the reverse leakage current is approximately 10 fA. If the switch is open for 100 μs, the charge transferred is approximately six electrons. This effect becomes significant only at low clock frequencies and at elevated temperatures. Leakage current introduces an extra charge source in the measurement and can be measured using this technique if the clock frequency is low enough.

Due to the proximity and similarity of the test capacitors used in this work, both the voltage and temperature coefficients of capacitance did not effect the measurement. Since each capacitor experiences similar conditions in the technique, the effects of temperature and voltage coefficient tend to track each other and appear as a common-mode error that cancels in the differential measurement.

The thermal noise generated by an MOS channel causes a variation in injected charge each time $S3$ is opened:

$$(V_n)^2_{rms} = \frac{kT}{C}. \qquad (10)$$

This noise is sampled on the capacitors when the switch is turned off. It can be shown that an MOS channel can be treated as a noiseless open circuit when turned off [10]. Increasing C can reduce this noise, but increases the parasitic capacitance if the sense and reference capacitors are already determined to be small. Complete cancellation is not possible with an individual measurement, but digital averaging can reduce its effect significantly since the noise is random.

IV. EXPERIMENTAL RESULTS

To test the theory of the charge redistribution technique on capacitance difference measurements, a test chip was designed and fabricated by MOSIS using a 3-μm p-well

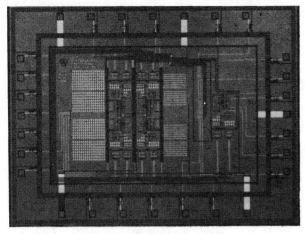

Fig. 2. Die photo of test chip.

TABLE I
TEST CAPACITOR SIZES[1]

C	Area (μm^2)	Capacitance Range (fF)
$C_{R_{large}}$	2132	75–110
$C_{S_{large}}$	2120	74–110
$C_{R_{small}}$	620	22–31
$C_{S_{small}}$	620	22–31
C_{C1}	1696	60–85
C_{C2}	848	30–42
C_{C3}	248	9–12
C_{P1}	31744	1100–1600
C_{P2}	108544	3800–5400

[1]Ranges are given since measurements on absolute capacitance values were never performed. Ranges are calculated from MOSIS vendors specifications.

CMOS technology. Its main purpose was to demonstrate difference measurements of metal/poly capacitors that are comparable in capacitance value to integrated sensing structures. An open-loop topology was implemented. An on-chip isolation amplifier was used to buffer the sensitive node to a comparator off-chip.

Three CMOS runs using MOSIS were made. Fig. 2 shows the die photograph of one of the chips. Five chips per run were obtained. As is shown in Table I, the test capacitors ranged from 20 to 100 fF. MOSIS vendors must meet requirements of metal one-to-poly/diffusion capacitances of 0.035–0.05 fF/μm^2 which indicates an oxide thickness range for the test capacitors of approximately 700–1000 Å. Measurements were also made for each set of sense and reference capacitors to test for any residual polarization [11]. The exact areas and estimated range of capacitance values for all capacitors are shown in Table I. Fig. 3 shows the positions of the circuits on the test chip. Circuits 1 and 3 have the same switch size as do circuits 2 and 4, while circuits 1 and 3 differ in parasitic capacitance as do circuits 2 and 4. Circuit 5 has no intentionally added parasitic capacitance.

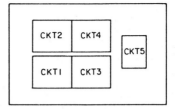

Fig. 3. Positions of circuits on test chip.

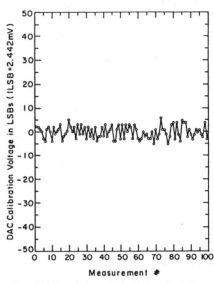

Fig. 4. Cancellation of charge injection and comparator offset.

Shown in Fig. 4 is a graph of calibration voltage output after cancellation. To measure the success of the calibration technique in eliminating charge injection, a calibration test was used. The calibration voltage was measured 100 times, then averaged. The negative of this value was then applied to the coupling capacitor and the calibration cycle was repeated. Cancellation of the charge injection should result in a DAC output voltage near 0 V. The data for all the chips show that on average the DAC calibration voltage is much less than 1 LSB, demonstrating the expected noise reduction from averaging.

The measurement data for one representative test chip are presented in Table II. A large amount of data were obtained since each of the 15 chips contained five separate circuits and each circuit had two sets of sense and reference capacitors. Since in sensor applications extensive averaging may not be possible due to speed considerations, the data represent a reasonable averaging of 16 times. In addition, residual polarization was observed to be negligible. Results compare favorably with previous work on MOS capacitors [11]. Standard deviations are for 16 measurements in time for one chip rather than between differing chips. The deviations of the measurements fall in the range of that expected from (5) and (7). We believe that the discrepancy in standard deviations between the negative and positive measurements is caused by slight asymmetrical noise coupling of the control signals to the test

TABLE II
MEASUREMENT DATA

Parameter	CKT1		CKT3		CKT5	
	C_{large}	C_{small}	C_{large}	C_{small}	C_{large}	C_{small}
Value (fF)	90	26	90	26	90	26
$V_{DAC+meas}$ (V)	0.415	3.66E-3	0.287	7.17E-3	0.258	4.43E-3
Std. dev. (LSB)	3.63	3.38	3.14	2.94	2.56	2.69
$\frac{\Delta C}{C}$ (%)	3.32	0.100	2.30	0.196	2.07	0.121
Std. dev. (%)	0.071	0.23	0.061	0.20	0.050	0.18
$V_{DAC-meas}$ (V)	-0.418	-2.44E-3	-0.287	-8.39E-3	-0.257	9.16E-4
Std. dev. (LSB)	2.28	1.63	3.00	2.33	2.17	1.89
$\frac{\Delta C}{C}$ (%)	-3.34	-6.67E-2	-2.30	-0.230	-2.05	2.51E-2
Std. dev. (%)	0.045	0.11	0.059	0.16	0.042	0.13

1. Test data averaged 16 times.

2. CKT1 has $C_P = 5$ pF, CKT3 has $C_P = 2$ pF, CKT5 has $C_P = 1$ pF

3. +meas denotes a positive measurement where C_R has V_{ref} across it in step 1.

4. -meas denotes a negative measurement where C_S has V_{ref} across it in step 1.

TABLE III
MEASUREMENT SYSTEM PARAMETERS

Switching speed	100 kHz
Successive Approximation Register	12-bit
Digitial-to-Analog Converter	12-bit
1 LSB	2.44 mV
Conversion speed	120 μsec
Positive analog supply	5 V
Negative analog supply	-5 V
Reference Voltage V_{ref}	+5.00000 V
Temperature	$\approx$ 25°C
Resolution	$\leq$ 0.05fF

chip. This is currently being corrected. Measurement system characteristics are shown in Table III.

The small capacitors differed by very little, on the order of 0.1–1 percent. This resolution limit was reached by averaging only 16 times. One of the larger capacitors was made approximately 3 percent larger than the other one. As shown in Table II, measurements resolved this difference. Since a large amount of theoretical and experimental work has been done on random MOS capacitor mismatches [12], [13], this paper did not attempt to seek correlations to previous work. From the data in Table II, standard deviations indicate that the resolution of a single measurement is close to 1000–1500 electrons, corresponding to nearly 0.05 fF in this study. Averaging can significantly increase the resolution but at a cost of increased time.

Two different switching sequences were used in this work to measure capacitance differences and residual polarization. Several alternate sequences will yield $V_x \propto \Delta C$. Table IV shows some of these sequences if the only avail-

TABLE IV
ALTERNATE SWITCHING SEQUENCES

Sequence number	V_{R1}	V_{S1}	V_{R2}	V_{S2}
1	0	V_{ref}	V_{ref}	0
2	0	0	V_{ref}	$-V_{ref}$
3	0	0	$-V_{ref}$	V_{ref}
4	0	$-V_{ref}$	$-V_{ref}$	0
5	V_{ref}	0	0	V_{ref}
6	V_{ref}	$-V_{ref}$	0	0
7	$-V_{ref}$	V_{ref}	0	0
8	$-V_{ref}$	0	0	$-V_{ref}$
9	V_{ref}	$-V_{ref}$	$-V_{ref}$	V_{ref}
10	$-V_{ref}$	V_{ref}	V_{ref}	$-V_{ref}$

able switching voltages are $\pm V_{ref}$ and ground. V_{R1} and V_{S1} are the voltages applied to the bottom plates of the reference and sense capacitors in step 1, respectively, while V_{R2} and V_{S2} are the voltages applied in step 2. Sequences 8–10 yield the same result as given by (7) while 9 and 10 yield (7) multiplied by two. The many possible switching sequences make it possible for the sensor designer to choose the appropriate sequence that best suits the particular application. For example, sequences 6 and 7 have 0 V across both capacitors in step 2 so that any change in either capacitor during the measurement cycle will not introduce an error. This may be important if both capacitors change during a sensing operation.

V. CONCLUSIONS

This paper demonstrates a technique that can measure capacitance differences with a resolution of 0.05 fF on capacitors in the 20–100-fF range in the presence of parasitic capacitances nearly 100 times larger. It is shown that nonideal effects such as charge injection, parasitic capacitance, and voltage and temperature coefficients are either negligible or can be calibrated. Junction leakage, threshold-voltage hysteresis, and capacitor hysteresis are shown to be negligible. Digital averaging can increase resolution but increases measurement time.

The charge redistribution technique measures capacitance differences and can be applied directly to sensor design. The technique is simple, requiring three capacitors, a voltage comparator, a successive approximation register, and a digital-to-analog converter. It provides extremely high resolution and an inherently digital readout. Its simplicity and compatibility with digital signal processing make it ideally suited for readout in sensor systems requiring a capacitance difference measurement.

ACKNOWLEDGMENT

The authors would like to thank the reviewers for their helpful comments.

REFERENCES

[1] M. A. Schmidt, R. T. Howe, S. D. Senturia, and J. H. Haritonidis, "A micromachined floating-element shear sensor," in *Tech. Dig., 4th Int. Conf. Solid-State Sensors and Actuators (Transducers '87)* (Tokyo, Japan), 1987, pp. 383–386.

[2] Y. S. Lee and K. D. Wise, "A batch-fabricated silicon capacitive pressure transducer with low temperature sensitivity," *IEEE Trans. Electron Devices*, vol. ED-29, pp. 42–48, Jan. 1982.

[3] N. F. Sheppard, D. R. Day, H. L. Lee, and S. D. Senturia, "Microdielectrometry," *Sensors and Actuators*, vol. 2, pp. 263–274, 1982.

[4] D. R. Harrison and J. Dimeff, "A diode-quad bridge circuit for use with capacitance transducers," *Rev. Sci. Instrum.*, vol. 44, pp. 1468–1472, Oct. 1973.

[5] C. S. Sander, J. W. Knutti, and J. D. Meindl, "A monolithic capacitive pressure transducer with pulse-period output," *IEEE Trans. Electron Devices*, vol. ED-27, pp. 927–930, May 1980.

[6] K. Watanabe and W. Chung, "A switched-capacitor interface for intelligent capacitive transducers," *IEEE Trans. Instrum. Meas.*, vol. IM-35, pp. 472–476, Dec. 1986.

[7] K. Watanabe and G. C. Temes, "A switched-capacitor digital capacitance bridge," *IEEE Trans. Instrum. Meas.*, vol. IM-33, pp. 247–251, Dec. 1984.

[8] J. L. McCreary and D. A. Sealer, "Precision capacitor ratio measurement technique for integrated circuit capacitor arrays," *IEEE Trans. Instrum. Meas.*, vol. IM-28, pp. 11–17, Mar. 1979.

[9] H. S. Lee, D. A. Hodges, and P. R. Gray, "A self-calibrating 15 bit CMOS A/D converter," *IEEE J. Solid-State Circuits*, vol. SC-19, pp. 813–819, Dec. 1984.

[10] R. Gregorian and G. C. Temes, *Analog MOS Integrated Circuits for Signal Processing.* New York: Wiley, 1986.

[11] H. Lee and D. A. Hodges, "A precision measurement technique for residual polarization in integrated circuit capacitors," *IEEE Electron Device Lett.*, vol. EDL-5, pp. 417–420, Oct. 1984.

[12] J. Shyu, G. C. Temes, and F. Krummenacher, "Random error effects in matched MOS capacitors and current sources," *IEEE J. Solid-State Circuits*, vol. SC-19, pp. 948–955, Dec. 1984.

[13] J. Shyu, G. C. Temes, and K. Yao, "Random errors in MOS capacitors," *IEEE J. Solid-State Circuits*, vol. SC-17, pp. 1070–1075, Dec. 1982.

Part 5
Selected Microsensors

ALTHOUGH comprehensive coverage of the many innovative concepts employed in microsensors is clearly impossible within the confines of this collection of reprints, it is useful to provide a sampling of some important and useful device ideas by considering specific microsensors. The Editors have selected the papers in this section for this purpose; there is no claim to completeness in the collection.

Pressure sensors are discussed in articles throughout this volume because they represent one of the most important and commercially successful segments of the microsensor field. The two papers selected here illustrate first, significant miniaturization (Chau and Wise), and second, the application of surface micromachining to the creation of small, densely spaced, planar pressure-sensor arrays (Guckel, Burns, and Rutigliano).

Because accelerometers have many important applications, and because packaging is relatively straightforward for these devices, microaccelerometer applications seem certain to expand in significance in the near future. This is one reason that microaccelerometers are given significant attention; another is because there are so many interesting ways to sense acceleration. Described in papers in this section are methods using an elastically supported proof mass whose deflection is read out either piezoresistively (Roylance and Angell; Benecke, Csepregi, Heuberger, Kühl, and Seidel), capacitively (Rudolf, Jornod, and Bencze), or piezoelectrically (Chen, Muller, Jolly, Halac, White, Andrews, Lim, and Motamedi). Resonant microstructures are likely to play an important role in sensing, as described by Howe's review paper on the subject. Specific applications to sensing acceleration by its effect on resonant structures are discussed in papers by Motamedi and by Satchell and Greenwood.

Magnetic microsensors are well established commercially. Sensors based on the Hall effect or magnetoresistance have played a major role in such diverse applications as computer-keyboard readout devices and position sensors for automotive-engine control. New device structures in which magnetic sensing capabilities are merged into transistor structures are described in a review paper by Baltes and Popović. As the integration of circuitry with the primary sensor increases, these merged device structures can be expected to assume greater importance.

The burgeoning field of robotics creates a strong demand for tactile sensors. Two particular approaches are presented here, a capacitive microsensor array (Chun and Wise), and a piezoresistively sensed array of force sensors (Petersen, Kowalski, Brown, Allen, and Knutti).

Perhaps the field that presents the greatest frustration to the microsensor designer is that of chemical sensors, where innovative ideas are many, but practical successes are relatively few. The promise of small size and low cost is attractive; however, the problems of packaging, of stability, and of selectivity are formidable. Nevertheless, the number of successful chemical microsensors is now beginning to grow. In previous sections, the use of chemically sensitive films for gas and vapor sensing have been illustrated, based both on resistivity changes and on mass changes sensed with acoustic waves. In this section, papers are presented that address several types of chemical microsensors based on the metal-oxide-semiconductor and related potentiometric structures: the palladium-gate hydrogen detector (Armgarth and Nylander), oxygen sensing with palladium/tin oxide (Kang, Xu, Lalevic, and Poteat), microelectronic ion sensors (Kelly and Owen), and light-addressable potentiometric biochemical sensors (Hafeman, Parce, and McConnell). Finally, the use of a microelectronic chip to measure dielectric and conductive properties of resins, plastics, adhesives, and fluids, and the correlation of these properties with chemical state of cure, is described in the paper by Sheppard, Day, Lee, and Senturia. This microdielectrometry technique has been used in a commercial chemical microsensor in which chemical changes are correlated with changes in electrical properties.

An Ultraminiature Solid-State Pressure Sensor for a Cardiovascular Catheter

HIN-LEUNG CHAU, MEMBER, IEEE, AND KENSALL D. WISE, FELLOW, IEEE

Abstract—An ultraminiature solid-state capacitive pressure sensor is described that can be mounted in a 0.5-mm OD catheter suitable for multipoint pressure measurements from within the coronary artery of the heart. The transducer consists of a silicon micro-diaphragm measuring $290 \times 550 \times 1.5$ μm surrounded by a 12-μm-thick silicon supporting rim, both defined by the boron etch-stop technique. The transducer process features a batch wafer-to-glass electrostatic seal followed by the silicon etch, which eliminates handling of individual small diaphragm structures until die separation and final packaging. A hybrid interface circuit chip provides a high-level output signal and allows the sensor to be compatible with use on a multisite catheter having only two leads.

I. INTRODUCTION

THERE are many applications in health care and in industrial process monitoring where it is desirable to measure pressure with an extremely small sensor so as not to disturb the system being monitored. For example, cardiovascular catheterization has become a major diagnostic tool in dealing with the cardiovascular system. In angioplasty (balloon pumping) [1] to treat occlusions in the coronary artery of the heart, there is presently no satisfactory means of judging the results on-line (as treatment is being administered). Existing catheter-tip pressure sensors are single point, not highly reliable, very expensive, and too large for use within the coronary artery.

Recent advances in silicon micromachining technology [2] have resulted in the development of a wide range of solid-state pressure sensors [3], [4]. In the past few years, the use of impurity-sensitive etch-stops [5], [6] and deposited diaphragm structures [7], [8] has allowed the high-yield fabrication of precise ultrathin diaphragms, and such techniques now make the significant miniaturization of solid-state pressure sensors feasible [9]. While ultraminiature pressure sensors using the piezoresistive approach have recently been reported [8], [10], capacitive pressure sensors are potentially more attractive for biomedical applications due to their much lower power requirements. However, improved sensor fabrication and packaging

Manuscript received January 11, 1988; revised July 18, 1988. Portions of this work were supported by the Semiconductor Research Corporation under Contract 86-07-085 and by a Rackham Fellowship from the University of Michigan.

H.-L. Chau was with the Center for Integrated Sensors and Circuits, University of Michigan, Ann Arbor, MI 48109-2122. He is now with the Foxboro Company, Foxboro, MA 02035.

K. D. Wise is with the Center for Integrated Sensors and Circuits, Solid-State Electronics Laboratory, Department of Electrical Engineering and Computer Science, University of Michigan, Ann Arbor, MI 48109-2122.

IEEE Log Number 882415.

TABLE I
SPECIFICATIONS FOR THE MULTISITE CATHETER SENSING SYSTEM

Parameter	Specification
Catheter Size	0.5 mm OD
Number of External Leads	2
Number of Sensing Sites	2
Number of Transducers/Site	2 (Pressure/Temperature)
Diaphragm Size	$290 \times 550 \times 1.5$ μm
Capacitor Plate Separation	2.0 μm
Pressure Sensitivity	850 ppm/mmHg @ $P_{applied} = 0$
Pressure Range	500 mmHg
Pressure Resolution	1 mmHg
Pressure Accuracy	2 mmHg
Signal Bandwidth	40 Hz
Signal Format	Supply Current Pulse Rate
Temperature Compensation	Performed Digitally Off-Chip using an On-Chip Temperature Transducer

techniques together with high-performance interface electronics are required to support scaling of the overall packaged device dimensions for use in an ultraminiature catheter.

This paper describes a capacitive pressure sensor that can be mounted in a 0.5-mm OD multisite cardiovascular catheter suitable for measuring blood pressure gradients inside the coronary artery of the heart. The specifications of this catheter sensing system are shown in Table I. The transducer contains a silicon micro-diaphragm measuring $290 \times 550 \times 1.5$ μm surrounded by a thicker supporting rim, both defined by the boron etch-stop technique. The present device is addressable, provides a high-level output signal, and is compatible with use on a multisite catheter having only two leads. It allows on-chip temperature measurement for purposes of compensation and for eventual flow measurements via thermodilution in cardiovascular applications.

II. PRESSURE TRANSDUCER FABRICATION

The construction of the multisite catheter sensing system is shown in Fig. 1. Two sensing sites separated by 5 cm are located along the 0.5-mm OD catheter and are connected in series by two wire leads. At each sensing site, an ultraminiature pressure transducer and a temperature transducer are provided. The detailed construction of the ultraminiature sensor at a particular site is shown in Fig. 2. The sensor consists of three chips: a glass substrate, a silicon transducer chip, and an interface circuit chip. The glass substrate contains grooves for lead attach-

Reprinted from *IEEE Trans. Electron Devices*, vol. 35, no. 12, pp. 2355–2362, December 1988.

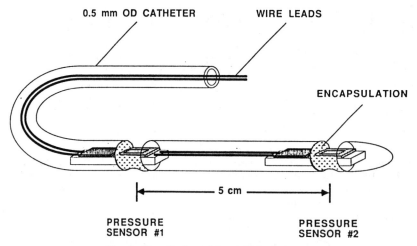

Fig. 1. Overall view of the multisite catheter.

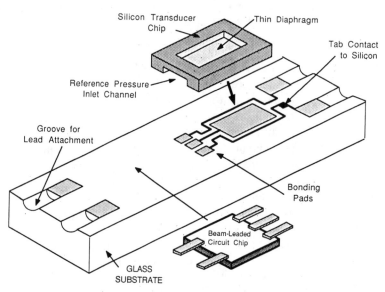

Fig. 2. Construction of a single ultraminiature sensing site.

ment and metallized areas for interconnects, lead contacts, and the active plate of the transducer capacitor. The transducer chip contains a 1.5-μm-thick silicon diaphragm surrounded by a 12-μm-thick rim. A pressure-variable capacitance is formed between the silicon diaphragm and metallization on the glass. The circuit chip contains all readout electronics.

A key part of the design and process deals with the formation of the transducer chip and the joining of this chip to the glass substrate. The process starts with a 250-μm-thick Corning 7740 borosilicate glass plate. Grooves 30-μm deep are first etched on the glass with concentrated hydrofluoric acid using a gold/chromium mask. Following complete metal removal, a second layer of gold/chromium is evaporated and patterned on the glass to form metallized areas for the pads, interconnects, and active capacitor plate. Many sensor sites are created in batch on a single sheet of glass. Processing of the silicon transducer chip is shown in Fig. 3. After an initial oxide growth of 0.5 μm, a 3-μm-deep silicon etch is performed to de-

fine the reference cavity and the eventual gap spacing for the capacitor. A complete oxide removal and regrowth to a thickness of 1.2 μm then follows. Next, a deep boron diffusion is performed at 1175°C to define the 12-μm-thick rim areas for the transducer. Oxide windows are now opened to allow a shallower boron diffusion to define the 1.5-μm-thick diaphragm. A dielectric is next deposited and patterned on the diaphragm to provide subsequent protection against shorts in the transducer. A combination of silicon dioxide and silicon nitride can be used for this purpose. Further study is required to understand and optimize the state of stress in the composite diaphragm. After complete dielectric layer removal over all areas other than the diaphragm, the silicon wafer is batch bonded using an electrostatic (anodic) seal to the glass plate. Alignment here is straightforward since the glass is transparent. Following bonding, the silicon-glass "wafer" is immersed in an anisotropic etchant for silicon (such as ethylenediamine/pyrocatechol/water (EDP)) and the silicon wafer is dissolved. Only the boron-doped portions of the wafer are

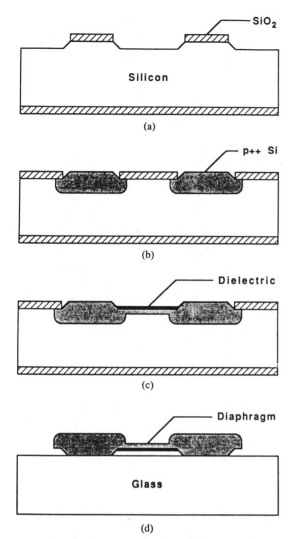

Fig. 3. Single-sided fabrication process for the silicon transducer chip. (a) KOH etch. (b) Long boron diffusion. (c) Short boron diffusion, dielectric layer deposition. (d) Oxide removal, electrostatic bonding to glass and EDP etch.

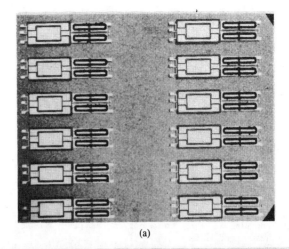

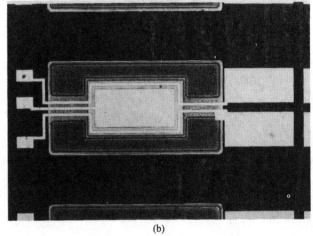

Fig. 4. Batch-fabricated ultraminiature capacitive pressure transducers. (a) Array of transducers on a glass substrate. (b) Single transducer taken through the glass.

retained. These areas do not etch in EDP nor do the interconnect metals, the glass or the dielectric layers. After the etch, the wafer is soaked in hot water followed by a dip in isopropyl alcohol. This process removes any residual EDP in the diaphragm cavity. As a result, a glass plate containing glass-silicon transducers is formed as shown in Fig. 4. Within each transducer, the silicon chip is electrically connected to the catheter ground line on the glass via an overlapping metal tab that is thermocompression bonded to the silicon during the electrostatic sealing process.

The transducer process utilizes single-sided processing of silicon wafers having normal thickness. It requires only four noncritical masking steps for the silicon and produces a very high yield. The rim and diaphragm thicknesses are set by the boron diffusion with a precision of better than 0.1 μm, while the lateral dimensions are controlled by lithography to a precision of better than 0.25 μm. The rim size is scalable but is typically 12 μm thick and 80 μm wide. This rim width is significantly smaller

than in conventional bulk-diaphragm pressure-sensing structures [3], [4], where the diaphragm is formed using an anisotropic etch from the back and the width of the rim is comparable to the wafer thickness (300 μm or more). Furthermore, in the present approach, the batch bonding of the wafer to the glass plate before wafer dissolution eliminates electrostatic diaphragm deflection problems during bonding as well as handling of the individual diaphragm structures until die separation and final packaging.

A typical measured pressure characteristic for the sensor is shown in Fig. 5. The reference cavity for this device was sealed at atmospheric pressure and 37°C. With a diaphragm size of 290 × 550 × 1.5 μm and a capacitor plate separation of 2.0 μm, the zero-pressure capacitance is 490 fF at 37°C. The pressure sensitivity at zero applied pressure is 0.41 fF/mmHg (850 ppm/mmHg), increasing to 1.39 fF/mmHg (2900 ppm/mmHg) at 500-mmHg applied pressure. This pressure sensitivity is less than its stress-free unsealed (gauge) value due to tensile stress in the boron-doped silicon diaphragm and the presence of the trapped gas itself. The sensitivity reduction is currently being modeled and quantified. The full-scale capacitance change (over a 500-mmHg pressure range) is

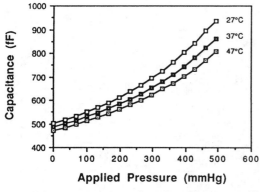

Fig. 5. Measured pressure response of an ultraminiature capacitive pressure transducer. The diaphragm size is 290 × 550 × 1.5 μm and the plate separation is 2.0 μm. The sensor reference cavity is sealed at ambient pressure and 37°C.

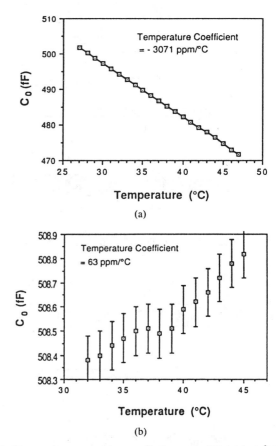

Fig. 6. Measured zero-pressure capacitance versus temperature for (a) an ambient-sealed transducer and (b) an unsealed device.

390 fF. By scaling the lateral and vertical diaphragm dimensions, the pressure sensitivity of the structure can be scaled over several orders of magnitude. The maximum nonlinearity in the pressure characteristics is 16 percent of full-scale output (FSO) and is principally due to the reciprocal relationship between the capacitor plate separation and the applied pressure. This large intrinsic nonlinearity can be corrected using external compensation as described below.

The temperature sensitivity of these capacitive pressure transducers is important and is strongly dependent on the technique used to seal the reference chamber. There are essentially three options available. The reference chamber (between the capacitor plates) can be sealed under controlled conditions similar to those expected in use (e.g., 760 mmHg and 37°C) using a glass or polymer to block the lead tunnels between the chamber and the external surface of the glass chip. If the metallization on the glass is dielectric coated, a metal eutectic seal might also be used. In either case, a fast-setting sealant of high viscosity is necessary to prevent the sealant from being drawn into the reference chamber. Glass dicing should be performed after sealing to avoid any possibility of particulates lodging in the chamber. Transducer sealing in a controlled ambient is the simplest from a process standpoint but incurs a substantial temperature sensitivity due to trapped gas expansion as noted in Fig. 5. Fig. 6(a) plots the measured zero-pressure capacitance C_0 for an ambient-sealed device. The temperature coefficient of C_0 is −3100 ppm/°C (−0.39 percent FSO/°C), while the temperature coefficient of pressure sensitivity is −3900 ppm/°C (−0.49 percent FSO/°C). Although these coefficients are large, the high pressure sensitivity of the capacitive transducer limits the equivalent pressure drift to about −5 mmHg/°C in the 500-mmHg pressure range.

As sealing alternatives, the reference chamber could also be vacuum sealed or could be referenced to the catheter lumen (where an open lumen is available). In either case, the gas expansion effects are removed and the temperature sensitivity of the transducer is largely eliminated. Fig. 6(b) shows the measured C_0 for an unsealed

(gauge) device as a function of temperature. The temperature coefficient of offset is about 60 ppm/°C (equivalent to 0.0075 percent FSO/°C or <0.1 mmHg/°C). For a vacuum-sealed device of the same area and plate spacing, however, a large pressure offset is incurred, requiring a thicker diaphragm and substantially reducing the pressure sensitivity of the device. This would degrade overall pressure resolution and accuracy. For an unsealed device, access to the reference chamber from the lumen must be provided by an access hole in the glass plate, complicating the overall process. For the measurements reported in this paper, we have used ambient-sealed devices, preferring to simplify the process and minimize the pressure offset while dealing with the resulting temperature coefficients. The very small temperature variations in the body (37°C ± 2°C or less) greatly relax the problems in this area. Temperature compensation is discussed in the next section.

III. Interface Electronics

The small size of the capacitance change associated with the sensor and the need to simplify packaging by minimizing external leads pose a challenge to the design of the circuit chip. An oscillatory-type circuit that requires only two external leads has been developed in which the supply current is pulse-period modulated by the applied pressure [11], [12]. The functional block diagram of this circuit is

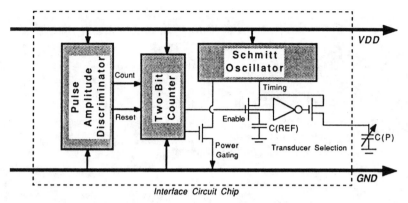

Fig. 7. Functional block diagram of the interface circuit chip.

shown in Fig. 7. A Schmitt trigger oscillator sets up an oscillation in the supply current. The transducer capacitance $C(P)$ or an on-chip reference capacitor $C(REF)$ serves as the timing capacitor. The output pressure signal is extracted by detecting the frequency of current variations over the power lines. Tradeoff between pressure signal bandwidth and resolution can be attained by altering the length of the sampling time. The reference capacitor, together with the temperature coefficient of the circuit, also serves as a transducer for on-chip temperature read-out.

Site and pressure/temperature transducer addressing is accomplished by superimposing clock pulses on the supply voltage. These pulses are detected by a pulse amplitude discriminator which in turn triggers a two-bit binary counter and allows one particular site and sensor on the catheter to be activated. The oscillator in the other site is disabled. Thus, multisite operation is possible. This approach is also compatible with expanded sensing on future multisite catheters. For example, the direct measurement of blood temperature through the thin diaphragm should allow estimation of blood flow via the thermodilution measurements. The use of thermally-based blood velocity/flow sensors [13] is also a possibility.

Fig. 8 shows a photomicrograph of a prototype circuit chip fabricated using a standard 10-μm enhancement/depletion (E/D) NMOS process. The circuit chip can be configured with beam leads to allow low-capacitance low-profile high-density interconnects to the transducer and the output leads. The present use of a hybrid arrangement has several important advantages: 1) the circuits can be processed using standard techniques and could be realized using a chip foundry; 2) the circuitry is not exposed to the high voltage needed for the electrostatic bonding process; and 3) working circuit chips can be selected for bonding to working sensors, thus improving yield. Fig. 9(a) shows the oscillator output voltage of the circuit chip. In response to clock pulses superimposed on the supply voltage, the circuit first gives an output oscillation that corresponds to the transducer capacitance and then switches to a slightly faster oscillation that corresponds to the reference capacitor. During the last two clock cycles, the

oscillator is deactivated to allow pressure readings to be taken from the other sensing site.

In the actual catheter sensing system, the pressure signal is extracted from the current in the supply lines. The supply current oscillation at a particular sensing site is shown in Fig. 9(b). It consists of sharp spikes (600-μA p-p over a 600-μA dc baseline) that are much higher in amplitude than the stand-by dc current (250 μA) that flows when the oscillator is inactive, thus allowing easy signal detection by external electronics during multisite operation. The oscillation period of the circuit with the reference capacitor is 1.54 μs at 37°C. The circuit sensitivity to capacitance change is 1.54 ns/fF (1000 ppm/fF) which, in connection with the ultraminiature pressure transducer, gives a sensitivity of 0.63 ns/mmHg (410 ppm/mmHg) for the overall pressure sensor. This sensitivity is sufficient to allow the desired pressure bandwidth (40 Hz) and resolution (1 mmHg) to be achieved. The temperature coefficient of offset and temperature coefficient of sensitivity of the circuit are −7800 and −2600 ppm/°C, respectively. No attempt was made to minimize these coefficients for the present interface circuit since, as noted above, the circuit is used as a temperature transducer for the purpose of compensating the pressure transducer. The supply voltage dependence of the circuit oscillation frequency is about 2 percent/V, which requires the supply voltage to be controlled at the eight-bit level. This is readily achieved. The present size of the prototype circuit chip (0.45 × 1.1 mm) is comparable with the transducer die size; however, by redesigning the circuit using a 3-μm CMOS technology, a significant improvement in die size can be achieved along with improved temperature sensitivity and a greatly reduced power consumption (Table II). A CMOS version of the circuitry is now being developed.

Fig. 10 shows a block diagram of the external readout electronics. A digital-to-analog converter driven by a microcomputer sets the supply voltage and the signaling levels for site and transducer addressing. The supply current variations are detected by a current-to-voltage converter and are subsequently converted into rectangular pulses by a comparator. The number of oscillations within a given

347

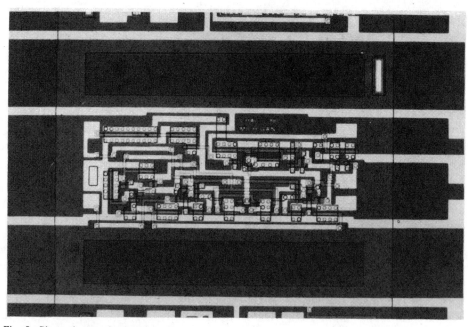

Fig. 8. Photomicrograph of the interface circuit chip. The chip size is 0.45 × 1.1 mm in 10-μm features.

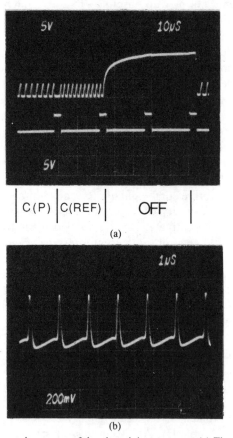

Fig. 9. Measured response of the ultraminiature sensor. (a) The oscillator output voltage in response to clock pulses superimposed on the supply voltage. In (b) the supply-current variation for zero applied pressure is shown at 200 μA/div.

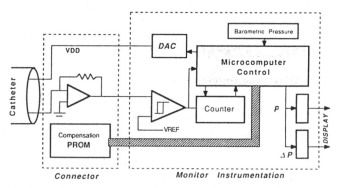

Fig. 10. Schematic diagram of the catheter with the external signal-processing system.

TABLE II
COMPARISON OF CIRCUIT PERFORMANCE BETWEEN THE PRESENT 10-μm
E/D NMOS DESIGN AND A NEXT-GENERATION 3-μm CMOS DESIGN
NOW IN DEVELOPMENT

Circuit Parameter	10μm NMOS	3μm CMOS
Die Size	0.45 x 1.1 mm	0.4 x 0.5 mm
Power Supply	5V	3V
Signaling Levels	8V Addressing	5V Addressing
	11V Reset	7V Reset
DC Supply Current	850 μA	<50μA
p-p Current Variation	600μA	50-100μA
Power Consumption	5 mW	<200 μW

sampling period is recorded by a counter. During measurement periods, the comparator output is also monitored by a separate circuit (not shown) so that in the event of any abnormal change in the supply current oscillations

the power supply to the catheter can be interrupted. This feature (together with the use of a grounded diaphragm, appropriate device encapsulation, and (for the next generation) micropower circuit operation) should ensure that the vital safety requirements associated with such instrumentation [14] can be satisfied.

Bonding of the circuit chips to the glass substrates can

be performed before or after the glass is sawed into individual dies for mounting in the catheter. The two required catheter leads are then attached and the ends of the glass support are inserted into the catheter, leaving only the silicon diaphragm exposed for pressure measurement. Thus, the only handling of small parts is when the circuit chips are bonded on the glass and as the final sensors are mounted on the catheter or other header.

The use of an external PROM provides an effective way of digitally compensating both the nonlinearity and the temperature coefficients of the catheter sensing system. The pressure P at a given site can be expressed as a polynomial function of the measured oscillation period S as

$$P = c_0 + c_1 S + c_2 S^2 + c_3 S^3.$$

The coefficient c_n's of the polynomial are temperature dependent and can be adequately modeled by a second-degree polynomial as

$$c_n = k_{0n} + k_{1n} T + k_{2n} T^2.$$

Thus, a total of twelve coefficients are used to describe each pressure sensor. These coefficients have been found adequate to compensate the sensor to within 1 mmHg over a pressure range of 500 mmHg and temperature range of 35 to 39°C. Since the on-chip temperature measurement is independent of pressure, only four coefficients are necessary to fit the temperature characteristics. Hence, with two sensing sites and two transducers per site, a total of 32 coefficients are required to compensate the catheter. These coefficients are readily stored in a 2-kbit PROM along with identification information for the external instrumentation system (Fig. 10).

The characterization data needed to generate the compensation coefficients can be generated automatically using a computer-controlled test station. In production, several catheters could be characterized in parallel to reduce testing cost. The test station should be capable of providing controlled pressure and temperature levels to within 1 mmHg and 0.25°C, respectively. At each test temperature, the catheter outputs are measured as a function of pressure, and the coefficients are generated from this data. In a commercial device, the compensation PROM would be built into the catheter connector and thus would be transparent to the user.

IV. Conclusion

A process has been described that is capable of fabricating ultraminiature capacitive pressure sensors that can

be mounted in a 0.5-mm OD catheter suitable for multipoint blood pressure measurements from within the coronary artery of the heart. The present device is addressable and is compatible with use on a multisite catheter having only two leads. The new design offers major improvements in size, performance, and potential cost as compared with previously reported catheter-tip pressure sensors and should make possible substantial advances in the clinical diagnosis and understanding of the cardiovascular system.

Acknowledgment

The authors would like to thank Dr. B. Pitt and Dr. W. O'Neill of the University of Michigan Medical School, Division of Cardiology, for their encouragement and advice throughout this project.

References

[1] A. R. Gruntzig, A. Senning, and W. E. Siegenthaler, "Nonoperative dilatation of coronary-artery stenosis—Percutaneous transluminal coronary angioplasty," *New England J. Med.*, vol. 301, pp. 61–68, July 1979.
[2] K. E. Petersen, "Silicon as a mechanical material," *IEEE Proc.*, vol. 70, pp. 420–457, May 1982.
[3] Y. S. Lee and K. D. Wise, "A batch-fabricated silicon capacitive pressure transducer with low temperature sensitivity," *IEEE Trans. Electron Devices*, vol. ED-29, pp. 42–48, Jan. 1982.
[4] S. C. Kim and K. D. Wise, "Temperature sensitivity in silicon piezoresistive pressure transducers," *IEEE Trans. Electron Devices*, vol. ED-30, pp. 802–810, July 1983.
[5] K. D. Wise and S. K. Clark, "Diaphragm formation and pressure sensitivity in batch-fabricated silicon pressure sensors," in *IEDM Tech. Dig.*, pp. 96–99, Dec. 1978.
[6] T. N. Jackson, M. A. Tischler, and K. D. Wise, "An electrochemical p-n junction etch-stop for the formation of silicon microstructures," *IEEE Electron Device Lett.*, vol. EDL-2, pp. 44–45, Feb. 1981.
[7] R. T. Howe, "Polycrystalline silicon microstructures," in W. H. Ko and D. G. Fleming, Eds., *Micromachining and Micropackaging of Transducers*. Amsterdam: Elsevier, 1985.
[8] H. Guckel, D. W. Burns, H. H. Busta, and J. F. Detry, "Laser-recrystallized piezoresistive micro-diaphragm sensor," in *Dig. Tech. Papers (Transducers '85) IEEE Int. Conf. Solid-State Sensors Actuators*, pp. 182–185, June 1985.
[9] H. -L. Chau and K. D. Wise, "Scaling limits in batch-fabricated silicon pressure sensors," *IEEE Trans. Electron Devices*, vol. ED-34, pp. 850–858, Apr. 1987.
[10] S. Sugiyama, T. Suzuki, K. Kawahata, K. Shimaoka, M. Takigawa, and I. Igarashi, "Micro-diaphragm pressure sensor," in *IEDM Tech. Dig.*, pp. 184–187, Dec. 1986.
[11] J. M. Borky and K. D. Wise, "Integrated signal conditioning for silicon pressure sensors," *IEEE Trans. Electron Devices*, vol. ED-26, pp. 1906–1910, Dec. 1979.
[12] J. C.-M. Huang, "A monolithic pH/pressure/temperature sensor for esophageal studies," Ph.D. dissertation, Univ. of Michigan, 1983.
[13] O. Tabata, H. Inagaki, and I. Igarashi, "Monolithic pressure-flow sensor," *IEEE Trans. Electron Devices*, vol. ED-34, pp. 2456–2462, Dec. 1987.
[14] *Essential Standards for Biomedical Equipment Safety and Performance*. Arlington, VA: Assoc. for the Advancement of Med. Instr., 1985.

DESIGN AND CONSTRUCTION TECHNIQUES FOR PLANAR POLYSILICON PRESSURE TRANSDUCERS WITH PIEZORESISTIVE READ-OUT

H. Guckel, D. W. Burns and C. R. Rutigliano
Wisconsin Center for Applied Microelectronics
Department of Electrical and Computer Engineering
1415 Johnson Drive
Madison, WI 53706

We have previously reported on experimental construction techniques for planar, vacuum sealed piezoresistive pressure transducers which are fabricated from polysilicon.[1] Fabrication methods have now been finalized and involve a polysilicon diaphragm which is reactive-sealed using a low strain LPCVD silicon nitride layer.[2] The nitride is also used for dielectric isolation of polysilicon strain sensors which are formed from a second polysilicon film. A stable, high yield processing sequence has been achieved.

In order to make this work useful a design method is needed which converts pressure transducer specifications to device structure. The design algorithm must predict plate stresses, strains and deflections. This is complicated by the non-zero, but reproduceable built-in strain field which has a first order effect on thin film plates. Virtual displacement techniques and a double trigonometric series are used to produce the necessary mechanical data for rectangular plates. The differential equation describing plate deflections w as a function applied pressure q and stresses due to the built-in strain field N_x, N_y, and N_{xy} is[3]

$$\frac{\partial^4 \omega}{\partial x^4} + 2\frac{\partial^4 \omega}{\partial x^2 \partial y^2} + \frac{\partial^4 \omega}{\partial y^4} = \frac{1}{D}(q + N_x\frac{\partial^2 \omega}{\partial x^2} + N_y\frac{\partial^2 \omega}{\partial y^2} + N_{xy}\frac{\partial^2 \omega}{\partial x \partial y})$$

If the deflection profile is known, the bending and twisting moments, and shear forces can be calculated. The boundary conditions are for the case of clamped edges. A solution of the form

$$\omega(x,y) = \sum_{m=1}^{\infty} \sum_{n=1}^{\infty} b_{mn}(1-(-1)^m \cos 2\frac{m\pi x}{a})(1-(-1)^n \cos 2\frac{n\pi y}{a})$$

where the coefficients b_{mn} are given by

$$b_{mn} = \frac{3q(1-\nu^2)a^4}{\pi^4 Eh^3(3m^4+2m^2n^2+3n^4)(1+\frac{9(1-\nu^2)(m^2+n^2)\varepsilon_o a^2}{\pi^2(3m^4+2m^2n^2+3n^4)h^2})}$$

satisfies the boundary conditions and converges to the deflected surface. The origin is located at the center of the plate. The formulation also predicts a critical built-in strain value, e_o, that will cause a plate of a given length to thickness ratio (a/h) to buckle. The algorithm produces plate dimensions to give the maximum strain sensitivity within the limits set by processing considerations. These limits include plate size due to buckling for compressive films, minimum and maximum thickness for the plate which is set by mechanical and patterning constraints, and maximum stresses before fracturing.

The built-in strain field modifies the deflections and strain sensitivity appreciably. The built-in strain is dependent on processing procedures. It can be determined and monitored during device fabrication by buckling of cofabricated doubly supported beams.[4]

The plate behavior is monitored by using polysilicon strain-sensitive resistors.[5] The mechanical and electrical response of these devices have been studied using cantilever beam techniques. The temperature behavior, gage factor and noise figure for the resitors have been measured. The ability to produce p- and n-doped structures with opposite gage factors is of particular interest.

Resistor data and mechanical data are combined to determine resistor placement. At least two design constraints occur: a fully active bridge configuration and the need to minimize the influence of optical alignment tolerances. The design algorithm accounts for both problems. Fully active bridges are realized using n- and p- resistors on the plate. The alignment problem is reduced considerably by extending the piezoresistors slightly beyond the plate edge and with turn-around points located at the zero-strain points on the plate. Calculation of strain sensitivities must include the variation of strain with position on the piezoresistive element as well as effects of geometrical changes during loading. The algorithm has been used to design absolute pressure transducers for 15 psi, 100 psi and 500 psi. The designs are based on built-in strains of -0.05% with 2 micron polysilicon thickness. Plate dimensions of 147, 100 and 73.8 micron, respectively, have been calculated and are currently being tested in fabricated transducers. Test results for the 1 atm gages show full-scale resistance changes of 1.2% for resistor values of 47-49kΩ.

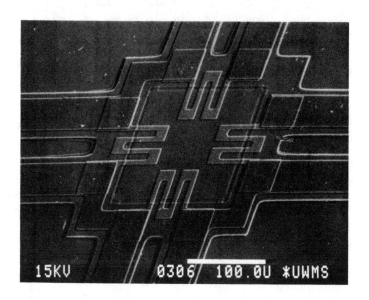

Fig. 1. An SEM of a bonded 100 psi transducer shows the serpentine piezoresistor configuration which minimizes placement error.

Reprinted from *Rec. of the IEEE Solid-State Sensors Workshop*, 1986.

References

[1] H. Guckel and D. W. Burns, "Planar Processed Polysilicon Sealed Cavities for Pressure Transducer Arrays," Technical Digest IEEE IEDM, 1984, p. 233.

[2] H. Guckel, D. K. Showers, D. W. Burns, C. R. Rutigliano and C. G. Nesler, "Deposition Techniques and Properties of Strain Compensated LPCVD Silicon Nitride Films", this conference.

[3] Timoshenko, Theory of Plates and Shells, Second Edition, McGraw-Hill Book Co., New York, 1959, Cha. 4.

[4] H. Guckel, T. Randazzo and D. W. Burns, "A Simple Technique for the Determination of Mechanical Strain in Thin Films with Applications to Polysilicon," J. Appl. Phys. 57(5), 1 March 1985, p. 1671-1675.

[5] E. Obermeier, "Polysilicon Layers Lead to a New Generation of Pressure Sensors," IEEE Transducers '85 Technical Digest, 1985 International Conference on Solid-State Sensors and Actuators, p. 430.

A Batch-Fabricated Silicon Accelerometer

LYNN MICHELLE ROYLANCE, MEMBER, IEEE, AND JAMES B. ANGELL, FELLOW, IEEE

Abstract—An extremely small batch-fabricatable accelerometer has been developed using silicon IC technology. The device, 3 mm long and weighing 0.02 g, is a simple cantilevered beam and mass structure sealed into a silicon and glass package. The fabrication of the accelerometer is described, and the theory behind its operation developed. Experimental results on sensitivity, frequency response, and linearity are presented and found to agree with theory. The accelerometer is capable of measuring accelerations from 0.001 to 50 *g* over a 100-Hz bandwidth, while readily implemented geometry changes allow these performance characteristics to be varied over a wide range to meet the needs of differing applications.

I. INTRODUCTION

INTEGRATED-CIRCUIT (IC) fabrication technology has permitted the development of an accelerometer weighing less than 0.02 g, in a 2 X 3 X 0.6-mm package. The accelerometer will detect accelerations down to 0.01 *g* over a 100-Hz bandwidth, with an upper acceleration limit of 50 *g*. These characteristics make the accelerometer ideal for applications requiring a very small and light transducer. Further, the versatile design allows the range to be varied readily over several orders of magnitude. The given limits meet the requirements of the biomedical applications such as measurement of heart wall motion for which the accelerometer was initially developed.

The goal of this work was to develop a transducer which meets the following specifications: 1) small size and mass; 2) sensitivity to accelerations as low as one-hundredth of the acceleration of gravity; 3) a bandwidth of 100 Hz; 4) an output stable over the limited range of temperatures encountered in biological environments; 5) an inert package; 6) an accuracy of around 1 percent; and 7) an output which is linear with acceleration. The device should be sensitive to only one component of acceleration.

II. ACCELEROMETER STRUCTURE

The accelerometer is a glass–silicon–glass sandwich; the details of this three-layer composition are shown in Fig. 1. The center layer is the heart of the device, a very thin silicon cantilevered beam surrounded by a 200-μm-thick rim. This rim provides a rigid support for one end of the beam, a region for contact pads, and mounting surfaces parallel to the plane of the beam. The beam widens at its free end into a rectangular paddle which supports a mass, either of some dense substance such as gold or of silicon. Fig. 2, a scanning-electron micrograph of the bottom side of the silicon element, clearly shows

Manuscript received May 22, 1979; revised July 30, 1979.

L. M. Roylance was with the Integrated Circuits Laboratory, Stanford University, Stanford, CA 94305. She is now with Hewlett-Packard Laboratories, Palo Alto, CA 94304.

J. B. Angell is with the Department of Electrical Engineering, Stanford University, Stanford, CA 94305.

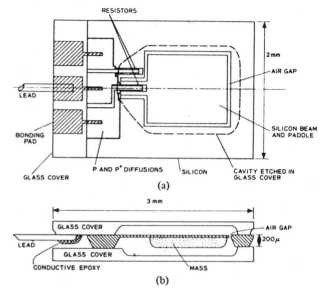

Fig. 1. Top and cross-section views of the accelerometer. (a) Top view. (b) Centerline cross section.

Fig. 2. SEM of backside of the accelerometer with a silicon mass after KOH etch.

the supporting rim, thin silicon beam, and integral silicon mass. A resistive half-bridge composed of two p-type resistors, one centered on the top surface of the beam and the other placed in an unstressed region of the rim, and three large p+ contact regions complete the silicon portion of the accelerometer. The beam resistor changes its value with acceleration due to the stress induced in the beam, while the second resistor is used for temperature compensation.

The top and bottom layers, both of glass, take the place of the TO-5 can or dual-in-line package used for standard IC's. A well etched into each glass cover allows the beam to deflect

Reprinted from *IEEE Trans. Electron Devices*, vol. ED-26, no. 12, pp. 1911–1917, December 1979.

freely up to a given distance—and hence acceleration—set by the depth of the wells. The glass covers are hermetically sealed to the thick silicon rim using anodic bonding [1], protecting the diffusions and, by creating a sealed cavity enclosing the fragile beam and mass, the cantilever. Three narrow fingers extending from metal pads on the top glass make contact to the resistors through p^+ diffusions. A cable leading to an amplifier and recorder can be attached to these pads where the top glass overhangs the silicon. With a nonconductive epoxy filling the region between the two glass covers and around the leads, the pad region is sealed and lead bond strength improved.

III. FABRICATION

Fabrication of the accelerometer is a batch process utilizing standard IC photolithographic and diffusion techniques in addition to the special etching techniques required to shape the silicon and glass. The silicon element and the top and bottom glass covers are fabricated separately in wafer form and then bonded together. The final steps are die separation and lead attachment.

The starting material is n-type (100) silicon, chosen because the preferred ⟨110⟩ direction for p-type piezoresistors coincides with the pattern orientation of anisotropic etchants such as KOH in silicon. Precise dimensional control can be obtained even with a large etch depth since the {111} planes are etched two orders of magnitude more slowly than {100} and {110} surfaces. The first step is to etch half a dozen widely spaced alignment holes completely through the wafer to obtain proper registration of patterns on the top and bottom surfaces. A 1.5-μm thermal oxide is grown and two photolithographies and diffusions done to form the 10 $\Omega/\square$ p^+ contacts and the 100 $\Omega/\square$ p resistors. The front oxide is stripped before drive-in to minimize surface steps, while the back oxide is preserved as the final etch mask. The remaining processing steps all concern the shaping of the beam, silicon mass (if present), scribe lines, and the window where the glass overhangs the silicon.

Using a thick densified layer of deposited silicon dioxide to protect the front, windows are opened in the backside oxide and the silicon is etched away around the beam and mass and in the region where the beam is to be thinned. The etch is stopped when the beam region is twice the desired final thickness. The sequence of operations is sketched in Fig. 3. A photolithography on the top surface of the partially etched wafer defines the air gap around the beam and the window opening. The final KOH etch, which etches the beam from the bottom and the air-gap regions from both top and bottom, is quenched the moment the silicon disappears from the large window openings. This visual endpoint gives very good control of the beam thickness provided the front and back surfaces of the wafer are parallel. Observed uniformity has been very good. The final step is stripping the remaining oxide.

The glass cover plates are also prepared in wafer form, from 200-μm-thick pieces of #7740 Pyrex glass, polished optically flat on one side. The type of glass is dictated by the silicon-glass bonding process which requires a glass which is slightly conductive at the bonding temperature and whose thermal expansion coefficient matches that of silicon. Unfortunately, this glass is not nearly as easy to etch as is silicon. Wells are etched in the top and bottom glass covers with a 30-percent

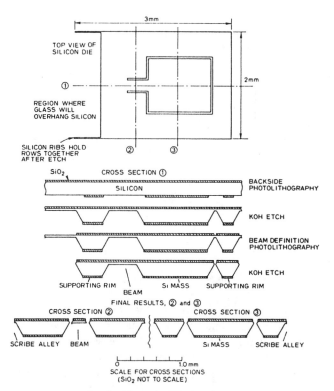

Fig. 3. Diagram of final etch steps.

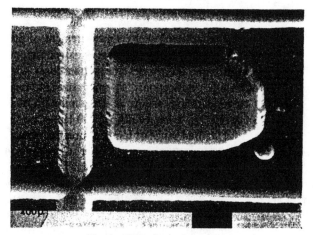

Fig. 4. SEM of top glass cover.

HNO_3, 70-percent HF mixture at 48°C and a chrome-gold etch mask. This procedure was found to give smooth, consistent results with minimal undercutting (Fig. 4). Once the masking layer is stripped, aluminum is deposited on the top glass and the metal bonding pads defined.

Final assembly of the accelerometer sandwich involves attaching gold masses if needed, and aligning and bonding the glass covers to the silicon. Only after completion of the bonding process are the individual accelerometers broken apart and handled one by one through the final phases—attaching a cable suited to the proposed application and applying insulation or some other protective coating. The heart of the assembly procedure is the anodic bonding technique which produces a hermetic and irreversible seal between silicon and glass. The glass and silicon are aligned, the temperature raised to about 400°C, and 600 V applied between the silicon and the glass. An advantage of this technique, in addition to its simplicity

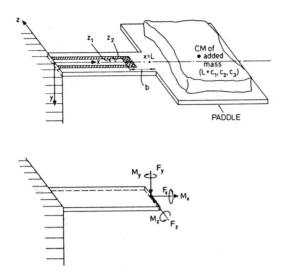

Fig. 5. Cantilevered beam and mass, with diagram of equivalent loads.

and lack of filler or "glue," is the visual inspection possible for a successful bond. Bonded areas appear dark gray, while unbonded regions are much lighter in color and show interference fringing. With the top and then bottom covers bonded, a dicing saw is used to separate the individual devices and leads are attached.

IV. ANALYSIS

Static Response

A detailed analysis [2] (see Appendix) of the accelerometer structure shown in Fig. 5 gives the fractional resistance change of the accelerometer due to an acceleration a_y

$$\frac{\Delta R}{R} = \left\{ \left[\frac{1}{2} (\Pi_{11} + \Pi_{12} + \Pi_{44}) - \left(s_{12} + \frac{1}{2} s_{44} \right) \right] \cdot \left[M \left(c_1 + \frac{L+b}{2} \right) \frac{c-h}{\mathcal{I}_z} \right] \right\} a_y = S_0 a_y \quad (1)$$

where L and h are the beam length and thickness, $\mathcal{I}_z$ the moment of inertia of the beam cross section about the z axis, c the distance from the centroid of the cross section to the bottom of the beam, M the mass loading the beam with its center of gravity at $(L + c_1, c_2, c_3)$, and Π_{ij} and s_{ij} the piezoresistive and elastic compliance coefficients referred to the cubic axes of silicon. The resistor extends from $x = 0$ to $x = L - b$. The accelerometer's sensitivity shows the desired linearity; further, the magnitude of the response can readily be controlled by varying the geometry.

Ideally (1) is the only term present for any acceleration a. The structure's effectiveness as a uniaxial accelerometer can be assessed by comparing the magnitudes of the responses to a_x and a_z with the desired sensitivity S_0. The only nonzero terms are due to the axial force F_x and the moment M_z produced by a_x

$$\frac{\Delta R/R \,|_{a_x}}{\Delta R/R \,|_{a_y}} \simeq \left\{ \frac{h}{12 \,(2c_1 + L + b)} - \frac{2c_2}{2c_1 + L + b} \right\} \frac{a_x}{a_y}. \quad (2)$$

The approximation assumes $h \ll w$, where w is the width of the beam. Clearly, an effort must be made to place the mass

symmetrically $(c_2 \ll 2c_1 + L + b)$ and to ensure that the beam thickness is small compared to its overall length if axial accelerations are to be ignored.

In deriving the foregoing results, a massless beam and perfect alignment centering the resistor and orienting it along a 110 direction were assumed. However, for the alignment tolerances and beam dimensions of interest the combined error involved is less than 1 percent worst case [3]. In addition, the added bending of a deflected beam due to an axial load can produce another undesired output term. However, calculations of the combined effects of such axial and lateral loads show that the error term is less than 0.4 percent for an axial load equal to 1 percent of the Euler load. For a thin beam and heavy mass (worst case), this loading is roughly equal to the safe operating range of the device.

In summary, the conditions necessary to minimize the undesirable responses to a_x and a_z are

$$h, c_2 \ll 2c_1 + L + b \qquad |z_r| \ll |h - c|$$
$$h \ll w \qquad\qquad \alpha_m \ll 1 \text{ rad} \quad (3)$$

where α_m and z_r are the angular and linear misalignments of the resistor. Similarly, the criterion for a large response to a_y can be expressed as

$$3M(2c_1 + L + b) \gg wh^2. \quad (4)$$

Since this condition is compatible with the constraints (3), the accelerometer design meets the criteria for a sensitive, uniaxial device whose response to a given acceleration can be both closely controlled and tailored to a particular task.

Dynamic Behavior

The primary purpose of this accelerometer is to measure time-varying accelerations, requiring an understanding of the variation in its behavior with the frequency of the excitation. For very low frequencies, the beam can follow the excitation without appreciable delay, so the discussion of the accelerometer's static response is directly applicable. At higher frequencies, however, the dynamic characteristics of the system produce phase delays and amplitude variations, depending on the natural frequencies of the device. A high-Q resonance (in an undamped accelerometer) corresponding to the first mode of lateral vibration of the beam dominates the behavior of the accelerometer. The other vibrational modes of the beam occur at much higher frequencies. Hence, the motion of the cantilevered beam and mass can be modeled as a simple two-pole spring and mass system.

The deflection of the beam is very nearly the deflection of a massless beam with a rigid distributed mass M at its end. For a static load F_y acting on the centroid of M, the deflection can be written as [4]

$$y = \begin{cases} \dfrac{F_y x^2}{2E\mathcal{I}_z} \left(L - \dfrac{x}{3} + c_1 \right), & 0 \leqslant x \leqslant L \\[2ex] \dfrac{F_y L^2}{2E\mathcal{I}_z} \left(\dfrac{2}{3} L + c_1 \right) + \left[\dfrac{F_y L^2}{2E\mathcal{I}_z} + \dfrac{F_y c_1 L}{E\mathcal{I}_z} \right] (x - L), & \\ & L < x \leqslant L + 2c_1 \end{cases} \quad (5)$$

where E is Young's Modulus. Note that the restoring force

(F_y) on the mass M is proportional to its displacement, giving simple harmonic motion. Applying the Rayleigh principle [5] to this deflection curve, a very good approximation to the resonant frequency ω_n is obtained (for M uniformly distributed)

$$\omega_n \simeq \sqrt{\frac{E \mathcal{I}_z}{ML^3} \frac{2 + 6f + f^2}{\frac{2}{3} + 4f + \frac{21}{2}f^2 + 14f^3 + 8f^4}} \tag{6}$$

where the second factor under the radical varies from three to one-tenth as $f = c_1/L$ varies between zero and two. Note that the key parameters affecting ω_n also play a large role in determining the sensitivity. This interrelationship limits the bandwidth which can be obtained for a given sensitivity. For typical accelerometer designs, ω_n is between 500 Hz and 5 kHz; the analysis of the preceding section is, therefore, valid over the 100-Hz bandwidth of interest.

Optimization

We now have the tools to explore the capabilities and limitations of this miniature accelerometer. Expressions for the dependence of the sensitivity, resonant frequency, and deflection of the accelerometer on its geometric and materials parameters have been developed which apply, in fact, to any accelerometer design consisting of a uniform cantilevered beam, a mass, and a stress-sensitive element to detect the stress in the beam. Therefore, much of the following discussion applies rather generally to small accelerometers using the cantilevered structure.

Three materials parameters, the magnitude of the piezoresistive coefficients in silicon, the density of the mass, and the fracture stress of silicon, set fundamental limits on the performance. In particular the relationship between the maximum stress in the beam and the applied acceleration places an upper limit on the sensitivity

$$S_0 \big|_{max} = \frac{\Pi_{eff}}{a_r} \sigma_{fracture} \left[\frac{2c_1 + L + b}{2(L + c_1)} \right]$$
$$\simeq \Pi_{eff} \frac{3.5 \times 10^9}{a_r} \simeq \frac{0.12}{a_r} \tag{7}$$

where a_r is the range in acceleration, $\sigma_{fracture}$ the fracture stress, Π_{eff} the effective piezoresistive coefficient, and where $2c_1 \gg L + b$.

The geometry of the accelerometer also has a profound effect on its performance. The most significant parameter is the beam thickness h, whose impact on sensitivity, resonant frequency, and beam deflection is plotted in Fig. 6 for typical values of h. The optimal choice for h is the smallest value consistent with the bandwidth and deflection constraints (set by the glass wells) in the design, since this choice gives both a large gain-bandwidth product and minimal transverse axis response. The minimum beam thickness, from fabrication and deflection considerations, is 5 μm. Although w and M also influence the performance, the beamwidth is set primarily by structural considerations, while the mass is a convenient way of setting the sensitivity. By selecting h and M a wide variety of accelerometers of differing ranges, sensitivities, and useful bandwidths can be built from the same basic structure. The

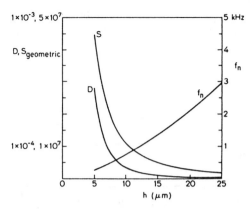

Fig. 6. Sensitivity S, resonant frequency f_n, and deflection D versus beam thickness h. Accelerometer dimensions: $w = 0.02$ cm, $c_1 = 0.06$ cm, $b = 0.01$ cm, $M = 4.5 \times 10^{-4}$ g (scale for S must be multiplied by Π_{eff} to obtain $\Delta R/R$).

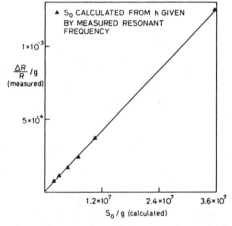

Fig. 7. Experimental versus theoretical geometric sensitivity. Slope = 3.5×10^{-11} cm^2/dyne = Π_{eff}.

current package size allows a dynamic range of 10^4 and ranges of 1 to 10^3 g.

V. EXPERIMENTAL RESULTS

The static responses of accelerometers of varying beam dimensions and masses were measured using the acceleration of gravity as a reference. By varying the orientation of the beam with respect to gravity, and comparing the outputs, the contributions of each of the three components of acceleration were determined. The observed outputs were linear in a_y and a_x, while no dependence on a_z was detected, as predicted by (1) and (2). The measured fractional resistance change due to a_y is graphed for several accelerometers in Fig. 7 against the prediction in (1), omitting the term in the piezoresistive coefficients. The points fall beautifully on a straight line through the origin with a slope of 3.5×10^{-11} cm^2/dyne, very close to the 4.5×10^{-11} calculated for Π_{eff}, using published values for the piezoresistive and elastic compliance coefficients.

Turning to the most significant transverse or off-axis responses, which are plotted in Fig. 8, two of the terms, e_1 and e_3, representing the effects of axial stress and of a beam of nonzero mass, respectively, are found to be extremely small even under worst case conditions. The third, and critical, component, e_2, arises from the moment M_z due to an acceleration

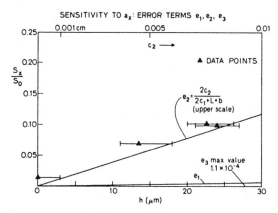

SENSITIVITY TO a_x: ERROR TERMS e_1, e_2, e_3

Fig. 8. Components of relative sensitivity to a_x; e_1, e_3 versus beam thickness h; e_2 versus c_2.

a_x. For a large mass mounted on one side of the paddle, this component is a substantial fraction of the y-axis sensitivity S_0, although it decreases to zero for a symmetrically mounted mass. Data from several accelerometers are plotted in Fig. 8 and show the trend predicted, although the rather large uncertainty in the magnitude of c_2 prevents assessment of the accuracy of the theory to better than about 20 percent.

To compare the two-pole model with the actual behavior of the accelerometer, the impulse responses of several accelerometers were determined experimentally.[1] Oscillograms of a typical impulse response can be found in Fig. 9. The damped sinusoidal behavior and the clean characteristics of the first few cycles indicate that the two-pole model is indeed a good choice. Values of the resonant frequency were found to agree very well with (6). Despite a factor of eighteen variation in sensitivity for the accelerometers tested, all had similar values for the damping factor ζ, Q, and the damping force F_d, corresponding to air damping of the beam plus any internal damping in the silicon. The exponential decay shown corresponds to $\zeta = 0.0046$, $Q = 109$, and $F_d = 0.065$ dy/dt (dynes).

Damping the beam resonance by adding a suitable fluid to the accelerometer cavity is a very attractive approach to minimize the impact of the resonance and increase the useful bandwidth. Four common laboratory fluids of roughly equal densities, acetone, methanol, deionized water, and isopropyl alcohol, were used to investigate the dependence of the accelerometer damping factor on fluid viscosity. The damping factor was found to vary linearly with fluid viscosity, as shown for one device in Fig. 10, while the viscosity needed to give 0.7 critical damping was found, by extrapolating the data points, to vary between 3 and 4 centipoise depending on the device. Some of the silicone oils have viscosities of this magnitude, and may well prove attractive to damp the accelerometer.

Damping the accelerometer will make it less susceptible to small fluctuations in ambient temperature, due to the increase in thermal mass, and may also help maintain the two resistors at the same temperature, both desirable effects. However, the

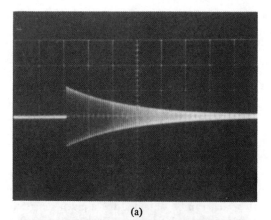

(a)

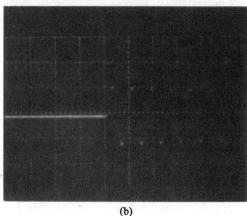

(b)

Fig. 9. Accelerometer impulse response. (a) 1 V/vert. div., 5 ms/horiz. div. (b) Same device, 1 V/vert. div., 0.5 ms/horiz. div.

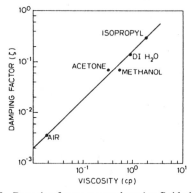

Fig. 10. Damping factor versus damping fluid viscosity.

beam and mass are totally immersed in the damping fluid, so that the buoyant force on the mass must be considered in determining the sensitivity of the damped accelerometer. In effect, the mass appearing in (1) is reduced by the mass of an equal volume of fluid. Although this change has little effect for a device with a gold mass, it can be important for a silicon mass device, particularly if the fluid density approaches the density of silicon.

Performance Limitations

Perceived accelerometer performance is affected by two inescapable properties of real systems, noise and temperature sensitivity. Thermal noise ultimately determines the useful

---

[1] The accelerometer was mounted on one face of a short piece of drill rod used as the weight for one pendulum of a dual-pendulum arrangement. A steel sphere weighting the other pendulum was released from a measured distance to swing into the opposite face of the rod; the resulting impact is an impulse on the time scale of the accelerometer.

range of the accelerometer. Measurements made on several accelerometers indicate that the resistors contribute only Johnson noise, so that fairly conventional amplifier noise analysis and design techniques are applicable to the accelerometer. Modeling the system as a Wheatstone bridge followed by a single amplifier stage, the minimum detectable acceleration a_{min} (for signal/noise = 1), assuming a low noise amplifier and a 100-Hz bandwidth, is

$$a_{min} = \frac{1}{1.5 \times 10^8 \sqrt{P_S} \, \Delta R/R} \tag{8}$$

where P_S is the power supplied to the bridge, $\Delta R/R$ is the factional resistance change per unit acceleration of the accelerometer, and $R = 7.5 \text{ k}\Omega$. For the most sensitive designs a_{min} is less than $0.001 \, g$.

The temperature dependence of the accelerometer output imposes another limitation on performance. Both the resistance differential between the sensing and temperature compensation resistors and the sensitivity $\Delta R/R$ are functions of temperature. The drift in the accelerometer output is due principally to the combination of the temperature coefficients of the diffused resistors and any mismatch between the resistors. However, since drift can be eliminated whenever it is not necessary to measure true dc acceleration, the thermal variation of accelerometer sensitivity is generally more significant. This variation is due to the temperature coefficient of the piezoresistive effect in silicon, and has been found to be between -0.2 and -0.3 percent per degree Celsius, in agreement with published values. Although this variation is not significant for the constant temperature environment of the body, for other applications both the sensitivity change and the drift can be compensated for by using more elaborate techniques on the temperature compensation resistor output.

The final concern in evaluating limitations on accelerometer performance is the linearity of the input–output characteristics. This linearity was verified experimentally by measuring outputs over the range ±1 g and peak responses to impulses of various magnitudes. Results from several accelerometers are plotted in Fig. 11 against the magnitude of the applied impulse, with the equivalent peak acceleration given for each device. The output characteristics are quite linear even though the data include points up to roughly one-fifth the acceleration at which the devices are expected to break. The nonlinearity is about ±2 percent of the maximum output, due to the bridge configuration and other sources such as the nonlinearity of the piezoresistive effect at high stress levels.

VI. Conclusion

The purpose of this section is twofold: to summarize the key characteristics of this accelerometer, and to evaluate what has been achieved in its fabrication. The characteristics of two miniature accelerometers, one fabricated with a gold mass and the other with silicon mass, are given in Table I. They are representative of the range of devices fabricated during this investigation. Immediately apparent are the very small size and mass of both accelerometers—a major goal. Further, most of the remaining device characteristics are more than adequate for the proposed applications, and compare favorably with

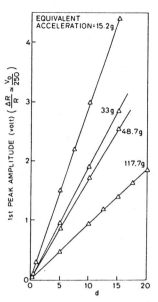

Fig. 11. Linearity of accelerometer response: impulse $\simeq 0.37 \, \omega_n d$.

TABLE I
CHARACTERISTICS OF THE MINIATURE ACCELEROMETERS

Property	Silicon Mass	Gold Mass
Size	2x3x0.6mm	
Mass (of accelerometer)	0.02 gm	0.02 gm
Range	± 200 g	± 40 g
Overrange	± 600 g	± 120 g
Sensitivity $\frac{\Delta R}{R}$/g	2×10^{-4}	1×10^{-3}
mV/g/V supply	0.05	0.25
Resonant Frequency	2330 Hz	1040 Hz
Transverse Sensitivity	10 %	2 %
Nonlinearity	$\pm$ 1% Full Scale	
Thermal Zero Shift	$\pm$ 1.4% FS/100°F	
Thermal Sensitivity Shift	$\pm$ 11%/100°F	
Resistance	7.5 kΩ	

similar (though much larger) commercial strain gauge accelerometers. The undesirable transverse sensitivity of the silicon mass device is due solely to the asymmetric loading of the mass, and hence the cost of the miniature accelerometer comes in sensitivity and in thermal sensitivity shift. Temperature compensation of the response should reduce the thermal sensitivity shift to close to ±1 percent/100°F in applications where thermal variations are important. In addition, the sensitivity of the present design can be doubled by going to a full bridge configuration with two active elements.

This investigation has demonstrated the feasibility of an extremely small batch-fabricated accelerometer. The performance and limitations of the miniature transducer have been thoroughly explored, and an understanding developed of the importance of symmetry in the design to minimize cross-axis responses. Accelerometers with sensitivities varying from $\Delta R/R = 5 \times 10^{-5}$ to 2×10^{-3} per g have been fabricated, allowing accelerations less than $0.001\ g$ to be detected. The miniature accelerometers compare very well with the small strain gauge accelerometers available commercially, while providing more than an order of magnitude reduction in volume and mass. The small size of this accelerometer, coupled with its performance and the low cost potential of batch fabrication, makes it extremely attractive for many applications.

APPENDIX

Analysis of the accelerometer response to an arbitrary input can be divided into two distinct areas; one concerns the electrical output due to a given stress distribution in the silicon, while the other concerns the stress distribution which results from an arbitrary load. For a cubic crystal such as silicon and an arbitrary (') coordinate system, the effect of stress on Ohm's Law can be written as

$$E'/\rho_0 = (1 + \Delta')J \qquad (A1)$$

where E' and J' are the electric field and current density, respectively, ρ_0 the unstressed resistivity, and[2]

$$\Delta' = \begin{bmatrix} \Delta\rho'_1/\rho_0 \\ \Delta\rho'_2/\rho_0 \\ \Delta\rho'_3/\rho_0 \\ \Delta\rho'_4/\rho_0 \\ \Delta\rho'_5/\rho_0 \\ \Delta\rho'_6/\rho_0 \end{bmatrix} = \begin{bmatrix} \Pi'_{11} & \Pi'_{12} & \cdots & \Pi'_{16} \\ \Pi'_{21} & \cdot & & \cdot \\ \Pi'_{31} & \cdot & & \cdot \\ \Pi'_{41} & \cdot & & \cdot \\ \Pi'_{51} & \cdot & & \cdot \\ \Pi'_{61} & \cdot & \cdots & \Pi_{66} \end{bmatrix} \begin{bmatrix} \sigma'_1 \\ \sigma'_2 \\ \sigma'_3 \\ \sigma'_4 \\ \sigma'_5 \\ \sigma'_6 \end{bmatrix} \qquad (A2)$$

Π'_{ii} and ρ'_i are the piezoresistive and resistivity coefficients, and σ'_i the stresses. For a resistor oriented along the 110 direction the resulting fractional resistance change is

$$\frac{\Delta R}{R} = \frac{1}{2}(\Pi_{11} + \Pi_{12} + \Pi_{44})\sigma'_x + \Pi_{12}\sigma'_y$$
$$+ \frac{1}{2}(\Pi_{11} + \Pi_{12} - \Pi_{44})\sigma'_z + \epsilon'_x - \epsilon'_y - \epsilon'_z \qquad (A3)$$

where the σ'_i are now the tensile stresses and ϵ'_i the strains along and perpendicular to the beam and resistor. The piezoresistive coefficients are referred to the cubic axes of silicon. (In deriving (1) the strains have been expressed in terms of the stresses and the elastic compliance coefficients s_{ij}.)

The stress in the resistor as a function of acceleration can be found by an analysis of the equivalent cantilever of Fig. 5 since the resistor does not extend into the region near the

transition between the beam and paddle where the actual distribution of shear and normal forces due to a load on the mass is significant [6]. Replacing these forces by their resultants, the load on the equivalent cantilever is

$$F_x = -Ma_x \qquad M_x = M(c_3 a_y - c_2 a_z)$$
$$F_y = -Ma_y \qquad M_y = M(c_1 a_z - c_3 a_x)$$
$$F_z = -Ma_z \qquad M_z = M(c_1 a_y - c_2 a_x) \qquad (A4)$$

where a_i, F_i, and M_i are the components of the acceleration and the resultant forces and moments. Applying the general theory of mechanics [7], [8] and the principle of superposition to the accelerometer's trapezoidal cross section and cubic anisotropy, a complete solution for the tensile stress distribution and an approximate solution for the shearing stresses is obtained [2]. The tensile stress σ'_x is the only nonzero stress term appearing in (A3).

$$\sigma_x = -a_x\left[\frac{M}{A} + \frac{Mc_2}{\mathcal{I}_z}y + \frac{Mc_3}{\mathcal{I}_y}z\right] + a_y\left[\frac{M(L + c_1 - x)}{\mathcal{I}_z}y\right]$$
$$+ a_z\left[\frac{M(L + c_1 - x)}{\mathcal{I}_y}z\right] \qquad (A5)$$

where A is the area of the beam cross section and $\mathcal{I}_i$ are the appropriate moments of inertia. The current flow in the resistor can be assumed to be confined to the surface of the beam, $y = c - h$, because of the doping profile of the diffusion. However, the stress in the beam must be averaged over the x and z excursions of the resistor to calculate the resulting fractional resistance change (1).

REFERENCES

[1] G. Wallis and D. I. Pomerantz, "Field-assisted glass-metal sealing," *J. Appl. Phys.*, vol. 40, no. 10, p. 3946, Oct. 1969.

[2] L. M. Roylance, "A miniature integrated circuit accelerometer for biomedical applications," Ph.D. dissertation, Department of Electrical Engineering, Stanford University, Stanford, CA, pp. 47–80, 1977.

[3] *Ibid.*, pp. 74–78.

[4] S. Timoshenko and D. H. Young, *Elements of Strength of Materials*, 5th ed. New York: Van Nostrand Reinhold, 1968, p. 197ff.

[5] W. T. Thomson, *Theory of Vibration with Applications.* Englewood Cliffs, NJ: Prentice-Hall, 1972, p. 200.

[6] Adhémar Jean Claude Barré de Saint-Venant, "Mémoire sur la torsion des prismes," in *Mémoires des Savants Étrangers*, vol. 14, 1885.

[7] S. Timoshenko and J. N. Goodier, *Theory of Elasticity*, 2nd. ed. New York: McGraw-Hill, 1951.

[8] S. G. Lekhnitskii, *Theory of Elasticity of an Anisotropic Elastic Body*, translated by P. Fern. San Francisco, CA: Holden-Day, 1963.

---

[2] If the cubic axes are used, all the Π_{ij} coefficients are zero except for the nine matrix elements in the upper left quadrant and the three along the lower diagonal.

A FREQUENCY-SELECTIVE, PIEZORESISTIVE SILICON VIBRATION SENSOR

W. Benecke, L. Csepregi, A. Heuberger, K. Kühl, H. Seidel

Fraunhofer-Institut für Mikrostrukturtechnik, D-1000 Berlin 33, Germany

ABSTRACT

A novel miniature silicon vibration sensor has been developed, consisting of an array of mechanical oscillators. This frequency-selective, piezoresistive sensor employs cantilevers with their natural frequencies placed adjacently. Design considerations with respect to frequency range and frequency resolution are presented together with experimental results. An outline of the fabrication process is given.

INTRODUCTION

The conventional arrangement of a vibration sensor consists of a single mass-spring system, where the deflection or the resulting stress of the spring is converted into an analogous electrical signal. This signal conversion can be achieved by means of piezoresistors [1], piezoelectric materials [2], or by a changing capacitance [3]. The usable bandwidth of such a system is restricted by its first resonant frequency, since at this point and in its vicinity, the sensitivity changes drastically, leading to a considerable nonlinearity of the sensor. Due to the inverse relationship between bandwidth and sensitivity, one usually has to accept a reduced sensitivity when a large bandwidth is needed [4]. This problem becomes increasingly important when the sensor is intended to operate in a higher frequency range.

A different approach for vibration measurement is to use a large number of mechanical oscillators with their natural frequencies placed adjacently. In contrast to the single mass-spring system, such a system operates at the natural frequencies of the individual oscillators, taking advantage of the high sensitivity and the low transverse axis response at these points. Another advantage of great importance is the possibility of obtaining a Fourier-transformed frequency spectrum directly, without requiring additional complex electronic circuitry.

The highly sophisticated silicon technology and anisotropic etching techniques make it possible to integrate a large number of miniature oscillators with the necessary accuracy and reproducibility. An example for such a device is illustrated by the scanning electron micrograph in Fig. 1. It consists of 50 miniature silicon cantilevers suspended over an etch cavity with outer dimensions of $1.8 \times 3.6 \text{ mm}^2$. Each cantilever has a polycrystalline silicon piezoresistor at its supported end where the largest mechanical strain occurs. This sensor was designed for a frequency range from 4 kHz to 14 kHz. Design considerations and an outline of the fabrication process for such a sensor are described in the following.

DESIGN CONSIDERATIONS

The basic structure chosen for the oscillators is a silicon cantilever with an additional gold mass at its free end. The conversion of the mechanical deformation of the cantilever into an electrical signal is obtained by means of a polycrystalline piezoresistor [5].

The starting point regarding the design is the definition of the frequency range and the frequency resolution required for the sensor. Based on these data, the necessary number of cantilevers

Fig. 1: Piezoresistive vibration sensor with silicon cantilevers varying in length between 600 μm and 1100 μm.

Reprinted from *Rec. of the 3rd Int. Conf. on Solid-State Sensors and Actuators*, 1985, pp. 105–108.

can be determined. The most important parameters to be considered for each cantilever are the damping, the resonant frequency and the sensitivity. The damping is very important because it determines the resonance full width, and thus the quality factor Q, of each individual oscillator. A high-resolution sensor requires low damping, which automatically leads to a high sensitivity. The most important contribution to the damping force is the viscosity of the surrounding medium, which in the present case is air. It is hardly possible to make an exact theoretical calculation of the frictional force acting on an oscillating cantilever; however, a good estimate can be given. The following equation for the full frequency width Γ at half-maximum power, which has been deduced from Stoke's theorem, has shown reasonable agreement with experimental data:

$$\Gamma = 3\pi\eta c \frac{\sqrt{A}}{m} \qquad (1)$$

where A is the surface area of the end mass m, η is the viscosity of the surrounding medium, and c is a constant in the order of unity. This factor depends on the geometric shape of the cantilever, and has to be determined experimentally. The first resonant frequency can be determind by an implicit formula, which has to be solved numerically [4]. This equation has the additional advantage that higher resonant frequencies can also be calculated.

The procedure used for the design of the vibration sensor starts by determining the thickness of the silicon cantilevers. Unless the desired frequency range is very high, this thickness is usually chosen as small as technologically feasible, in order to minimize the size of the entire device. A typical thickness was 4 μm. From the definition of the frequency range and the number of cantilevers, the required bandwidth of each individual cantilever is determined. This bandwidth is then adjusted by selecting the surface area A and the mass m according to Eq. (1). Next, the width of the cantilever is chosen as small as possible. The minimum width is determined by the space required for the piezoresistor. At this point, the only parameter remaining for adjustment of the resonant frequency is the length of the cantilever. It is chosen according to the formula given in reference [4].

DEVICE FABRICATION

The silicon etch rate of anisotropic etchants is very highly dependent on the crystal orientation and the boron doping concentration. These properties can be taken advantage of for precise shaping of structures in silicon, such as grooves, membranes, bridges, and cantilevers [6, 7].

An outline of the sequence of steps in the fabrication of the silicon vibration sensor is shown in Fig. 2. As starting material, 3", p-type silicon wafers with (100) orientation and a resistivity of 1-10 Ωcm were used. As a first step, a highly boron-doped epitaxial layer was grown. A typical boron concentration was 1×10^{20} atom/cm^3, which provides an excellent etch stop in anisotropic

etchant solutions [8]. The thickness of this layer determines the thickness of the cantilevers. In the next step, the wafers were thermally oxidized and a boron-doped polycrystalline layer, as well as a second SiO_2 layer, were deposited in a horizontal CVD reactor. The piezoresistors were then defined photolithographically and masked with another CVD-silicon dioxide layer. Next, the contact windows and the outline of the cantilevers were defined. After a chromium-gold metallization, the highly boron-doped epitaxial layer was removed in the unmasked areas by a $HF-HNO_3-CH_3COOH$ solution. The final etching of the cantilevers was done with an ethylenediamine-based anisotropic etchant. In order to ensure well-defined underetching of the cantilevers, the first photomask had to be aligned carefully with the (111) crystal planes at the beginning of the process.

Scanning electron micrographs of two vibration sensors with different geometries in regard to the end mass are shown in Figs. 3 and 4. In both cases, the sensor has 16 cantilevers, each of which has one piezoresistor and one reference resistor.

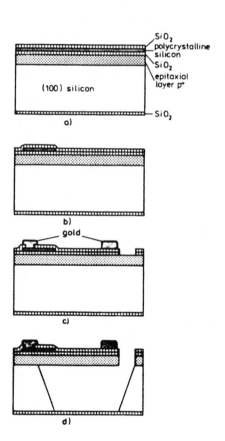

Fig. 2: Sequence of fabrication steps for a silicon vibration sensor: a) deposition of epitaxial layer and polycrystalline silicon, including passivation; b) definition of piezoresistor; c) opening of contact windows and etch cavity, metallization; d) anisotropic etching of cantilevers.

Fig. 3: Silicon vibration sensor with 16 cantilevers covering a frequency range from 4.2 kHz to 6.4 kHz.

Fig: 4: Silicon vibration sensor with 16 cantilevers, each having an extended end mass. The frequency range is from 0.87 kHz to 1.96 kHz.

EXPERIMENTAL RESULTS

In order to measure the performance of the devices, we used a vibration testing system operating in the range between 10 Hz and 15 kHz. The maximum obtainable acceleration was 10 g. The root-mean-square value of the acceleration was determined with a vibration analyser. The output signals of the individual cantilevers were amplified by a factor of 100 with a precision AC amplifier, and the output was obtained on a digital multimeter.

The output signals as a function of vibration frequency for the first eight cantilevers of the device shown in Fig. 4 are given in Fig. 5. For this measurement the acceleration was held constant at a root-mean-square value of 1 g. The frequency width Γ at half-maximum power was determined to be 10 - 12 Hz. The theoretical estimate resulting from Eq. (1) is 11.7 Hz. For this calculation, the constant c was set equal to one.

The sensor output signal as a function of acceleration measured at the resonant frequency of one cantilever is shown in Fig. 6. It is linear up to an acceleration of 4 g. For larger accelerations, a deviation from linearity is clearly visible. A possible reason for this effect is that the assumption of linear proportionality between frictional force and cantilever velocity, inherent in Eq. (1), is not true for large cantilever amplitudes. Instead, the damping force becomes proportional to a power of the velocity greater than one.

CONCLUSION

The cantilever with a polycrystalline piezoresistor has been found to be a suitable structure for a frequency-selective vibration sensor. For spe-

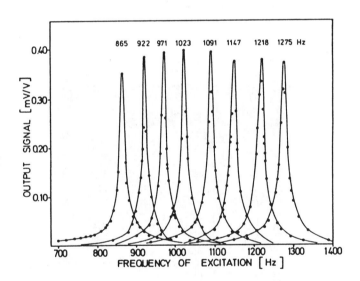

Fig. 5: Frequency response curve of the first eight cantilevers of the device shown in Fig. 4.

cial applications, however, other basic structures could also be chosen. Silicon bridges supported on both ends [7] are suitable for very high frequencies. At the lower end of the frequency spectrum, where cantilevers would tend to become extremely long, it is possible to fold them into a spiral. An example for this is shown in Fig. 7.

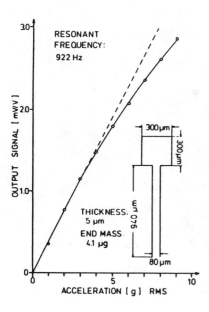

Fig. 6: Output voltage of a single cantilever as a function of acceleration, measured at its natural frequency of 922 Hz.

Fig. 7: Cantilever with a thickness of 5 μm and a total length of 4 mm folded into a spiral. Its resonant frequency is 100 Hz.

A high-resolution sensor with a large number of oscillators, such as the one shown in Fig. 1, requires a large number of output pins. Therefore, it would be very desirable to integrate electric circuitry directly on the chip. In this way, the required number of output pins could be drastically reduced.

The fields of applications for the sensors described in this paper include all areas where vibrations occur, for example, in connection with apparatus such as machine tools, automobiles, and aircraft, or in the field of acoustics. Among other applications, this type of sensor could prove to be useful for early detection of bearing damage in machines, for detection of knocking in automobile engines, and in connection with problems of speech recognition.

ACKNOWLEDGEMENT

We would like to thank R. Niessl and W. Kühl for their valuable assistance.

REFERENCES

(1) L. M. Roylance and J. B. Angell, A batch-fabricated silicon accelerometer, IEEE Trans. Electron Devices, ED-26 (1979) 1911-1917.

(2) P. Chen, et al., Integrated silicon microbeam PI-FET accelerometer, IEEE Trans. Electron Devices, ED-29 (1982) 27-33.

(3) F. Rudolf, A micromechanical capacitive accelerometer with a twopoint inertial-mass suspension, Sensors and Actuators, 4 (1983) 191-198.

(4) H. Seidel and L. Csepregi, Design optimization for cantilever-type accelerometers, Sensors and Actuators, 6 (1984) 81-92.

(5) E. Obermeier and H. Reichl, Polykristalline Siliziumschichten als Basismaterial für Sensoren, Sensoren-Technologie und Anwendung, NTG-Fachberichte, Vol. 79, VDE-Verlag, Berlin, 1982, p. 49.

(6) K. E. Petersen, Silicon as a mechanical material, Proc. IEEE, 70 (1982) 420-457.

(7) H. Seidel and L. Csepregi, Three-dimensional structuring of silicon for sensor applications, Sensors and Actuators, 4 (1984) 455-463.

(8) H. Seidel and L. Csepregi, Studies on the anisotropy and selectivity of etchants used for the fabrication of stress-free structures, Abstract 123, p. 194, The Electrochemical Society Extended Abstracts, Vol. 82-1, Montreal, Que., Canada, May 9-14, 1982.

Integrated Silicon Microbeam PI-FET Accelerometer

PAU-LING CHEN, RICHARD S. MULLER, SENIOR MEMBER, IEEE, RICHARD D. JOLLY, GREGORY L. HALAC,
STUDENT MEMBER, IEEE, RICHARD M. WHITE, FELLOW IEEE, ANGUS P. ANDREWS, MEMBER, IEEE,
TEONG C. LIM, AND M. E. MOTAMEDI, SENIOR MEMBER, IEEE

Abstract—Integrated accelerometers showing excellent linearity have been designed and fabricated using silicon planar technology, zinc-oxide sputtering, and anisotropic etching. Small cantilevered beam structures overcoated with piezoelectric ZnO films act as force transducers, and the electrical signal is directly coupled to the gate of a depletion-mode, p-channel MOS transistor. The accelerometers have a nearly flat response from very low frequencies until beam resonances become significant (above 40 kHz). The near-dc response results from completely isolating the piezoelectric film from electrical leakage paths. Measured performance has matched very well with theory. Theoretical analysis has been used to derive useful design tradeoffs.

Manuscript received May 7, 1981; revised August 26, 1981. This research was supported by the National Science Foundation under NSF Grant ENG-7822193 (Industry/University Cooperation Program) and partially under a Grant from Rockwell International Science Center, Thousand Oaks, CA.

P-L. Chen, R. S. Muller, G. L. Halac, and R. M. White are with the Department of Electrical Engineering and Computer Sciences, and the Electronics Research Laboratory, University of California, Berkeley, CA 94720.

R. D. Jolly was with the Department of Electrical Engineering and Computer Sciences, and the Electronics Research Laboratory, University of California, Berkeley. He is now with Hewlett-Packard Laboratories, Palo Alto, CA.

A. P. Andrews, T. C. Lim, and M. E. Motamedi are with Rockwell International Corporation, Thousand Oaks, CA 91360.

INTRODUCTION

EXCELLENT experimental performance has been observed on integrated accelerometer structures that were fabricated using capacitive PI-FET transducers [1] to detect strains in miniature cantilever beams. The beams are composite structures consisting of Si, SiO_2, ZnO, metal, and passivating oxide (Fig. 1). They are formed by anisotropically etching the silicon [2] from underneath the layered structures using an aqueous solution of ethylenediamine and pyrocatechol (EDP) in the manner described by Petersen [3], or else by a combination of backside and frontside etching of the wafer as described by Roylance and Angell [4]. The piezoelectric strain-sensing layers are directly coupled to depletion-type, p-channel MOS transistors. Dependent on the fabrication procedures employed, the total beam thicknesses are either below 5 μm (for top-surface etching), or else range to about 50 μm (for etching from both sides of the wafer).

Most of the previous work with thin-film ZnO transducers has been directed at SAW applications in which strain waves have characteristically very high frequencies (above 10 MHz). For an accelerometer, on the other hand, low frequency and

Reprinted from *IEEE Trans. Electron Devices*, vol. ED-29, no. 1, pp. 27–33, January 1982.

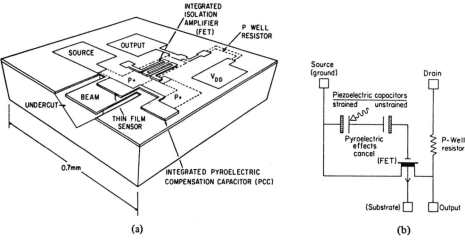

Fig. 1. (a) Integrated accelerometer formed by anisotropic etching and planar technology. The compensation ZnO capacitor is used to reduce temperature sensitivity by differentially removing the pyroelectric effect. (b) Equivalent circuit for the accelerometer shown in Fig. 1(a).

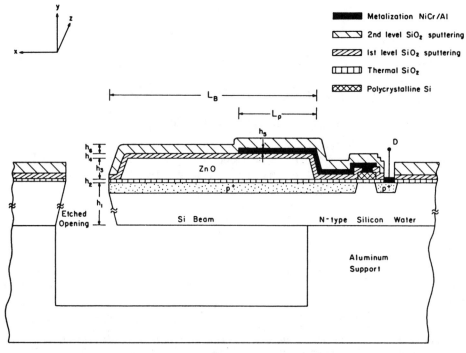

Fig. 2. Cross section of the beam portion of an accelerometer formed by two-sided etching.

even dc strain response is desirable. By encapsulating the ZnO film entirely within insulating layers that are thin relative to the Debye length within the piezoelectric ZnO, we have been able to obtain a near-dc response in the overall structure. The insulating layers consist of an underlying film of thermally grown SiO_2 and an overlaying film of sputtered SiO_2. This construction has led to a measured decay time for a stress-induced signal of approximately seven days even though the dielectric relaxation time of the sputtered ZnO is only about 1 ms [5].

THEORY

A cross section of the accelerometer is shown in Fig. 2. Its operating principles can be described as follows. When an inertial force strains the cantilever beam, the piezoelectric ZnO becomes polarized. The resultant surface charge on the ZnO is coupled capacitively to the gate of the p-channel, depletion-mode FET and an amplified output voltage is then obtained at the drain. The polarized charge density can be calculated from the equation

$$D_i = e_{ikl} S_{kl} + \epsilon_{ik}^S E_k \tag{1}$$

where D is the electrical displacement, E is the field, S is the strain, and e and ϵ are the piezoelectric and dielectric coefficients, respectively. The low-frequency piezoelectric response of the ZnO capacitor in the bending mode can be modeled by the circuit shown in Fig. 3 in which C_{d1} and C_{d2} represent the two SiO_2 capacitors which are sandwiched on either side of the ZnO film. If the Si beam thickness is not much greater than the composite ZnO layer thickness, the strain within the

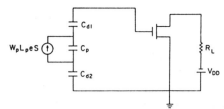

Fig. 3. Equivalent circuit of a simple (noncompensated) accelerometer.

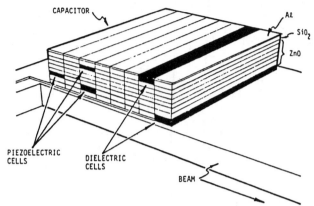

Fig. 4. Elemental contributors to the overall piezoelectric voltage.

piezoelectric film is not localized but is a function of x and y. Therefore, a distributed circuit model, as shown in Fig. 4, has to be used to analyze the structure. In Fig. 4, we divide the ZnO layer into infinitesimal cell stripes that are uniform across the beam (since the strain is uniform in that dimension). Straightforward mathematical analysis using (1) on the model of Fig. 4 and the equivalent circuit of Fig. 3 leads to the following expression for the induced piezoelectric voltage at the MOS gate:

$$V = \frac{W_P L_P e \overline{S}}{C_G + C_P + C_P C_G (C_{d1} + C_{d2})/C_{d1} C_{d2}} \qquad (2)$$

where W_P and L_P are the width and length of the ZnO capacitor layer, respectively, e is the piezoelectric constant, and C_P and C_G are the capacitances of the piezoelectric layer and of the FET gate, respectively. $\overline{S}$ is the strain averaged over the thickness and length (along the beam) of the ZnO piezoelectric capacitor.

The overall expression for the averaged strain is algebraically complicated, but quite readily derived. It shows that the principal design parameters specifying sensitivity and dynamic range are: the beam thickness, the ratio of the length of the piezoelectric capacitor to the beam length, the piezoelectric-layer thickness, the overall beam length, and the position and magnitude of any proof mass that might be used. Relevant design tradeoffs derived from the theoretical investigation are described in a separate section. We confine our discussion here to mentioning one important consideration in the design of a beam accelerometer, the role of the neutral axis in the beam. Since there is a null output if the neutral axis of the beam system is situated midway within the piezoelectric film, the designer must consider and avoid this eventuality as variations are made in the thicknesses and densities of overlaying films.

A simple case occurs when the Si layer is thick compared to the total thicknesses of all other layers forming the beam. In

this case, the strain at the clamped end is found to be [6]

$$\overline{S} = \frac{3 \rho a L_B^2}{E_Y h} \qquad (3)$$

where a is the acceleration of the beam end, ρ is the Si density, L_B the beam length, E_Y Young's modulus for Si, and h the thickness of the beam. The maximum g loading that the beam can sustain can be calculated by inserting the yield limit for Si ($S = 10^{-3}$) into (3). This maximum strain value, obtained by experiment, is roughly consistent with measurements of the fracture strength of Si [7] reported by Chen which showed a 50-percent failure probability for chemically polished Si wafers of 2.17, 2.78, and 4.96×10^8 N $\cdot$ m^{-2}, respectively, for twist, cylindrical bending, and biaxial stressing. The resonant frequency of the beam is given by the formula [8]

$$f \simeq 0.16 \sqrt{E_Y/\rho} \left[\frac{h}{L_B^2} \right]. \qquad (4)$$

All of the beams designed for the accelerometer studies have resonant frequencies above 40 kHz which is appreciably higher than the typical frequencies of interest.

DESIGN TRADEOFFS

A desktop computer has been programmed to evaluate design tradeoffs between the various microbeam parameters. An example of the useful design information that can be extracted from these programs, which are based on the distributed model of Fig. 4, is shown in Figs. 5 and 6. The figures refer to an accelerometer design that has been extensively studied theoretically, but not yet fabricated. The design is for an accelerometer that is specified to have a nominal scale factor of $\frac{1}{3}$ mV/g (at the gate of the FET), a 300g operating range, and a linear range of 100 mV. The peak strain sensitivity, constrained in the design to be less than 0.1 ppm/g, occurs at the beam root on the outside surface.

In the tradeoff curves, shown in Figs. 5 and 6, the design conditions satisfying the scale-factor constraint are shown by solid lines. The design limits imposed by the strain-sensitivity constraint are shown by dotted lines. A "design envelope" is defined by the intersections of these two types of lines, hence the region outside this envelope is labeled "forbidden zone" in the figures. The two figures display the same data, but in alternative formats which give different insights into the tradeoffs that can be made in designing the accelerometer.

In Fig. 5, the curves correspond to differing Si thicknesses. Each solid curve shows values of beam length and thickness of the ZnO film which yield the design scale factor. The curves terminate at the heavy solid curve where they intersect the corresponding strain-limit curve (dotted). This curve represents the boundary of the design envelope. Note that this boundary is essentially horizontal on the upper right corner, which indicates that the minimum thickness of the ZnO layer is about 1.2 μm, and this limit is reached at beam lengths of about 3 mm.

A similar limiting length is apparent in the curves of Fig. 6. The solid curves on this graph correspond to the simultaneous values of the Si and ZnO thicknesses which yield the design scale factor for fixed values of the beam length. As the beam lengths decrease from 3 to 1.1 mm, the curves move to the upper right, and eventually disappear after the 1-mm curve.

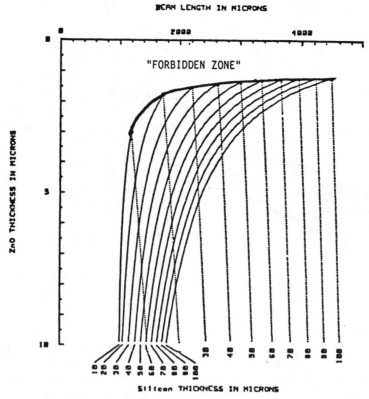

Fig. 5. Tradeoff curves illustrating the effect of variable ZnO thickness, Si-beam thickness, and beam length. All curves on the family have the same sensitivity.

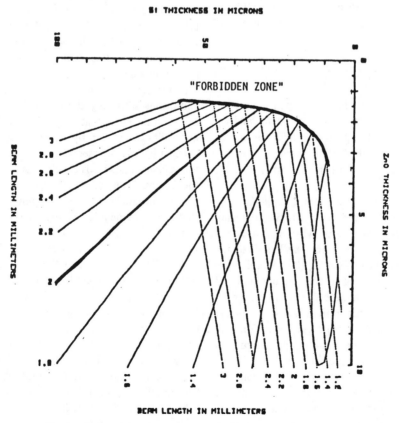

Fig. 6. Tradeoff curves for the same variables considered in Fig. 5.

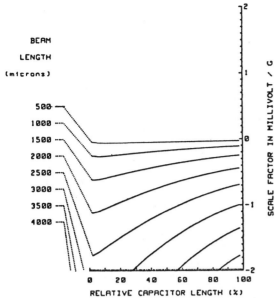

Fig. 7. Influence of the capacitor length on the accelerometer scale factor.

TABLE I
PROFILES OF BEAM LAYERS

Dimension µm	Material
(top) 1.5	Au-Ti
5.0	Sputtered SiO_2
0.4	Al-NiCr
0.5	Sputtered SiO_2
varied	ZnO
0.10	Thermal SiO_2
(bottom) varied	Si

This indicates that there is a lower limit on the beam length between 0.8 and 1 mm for the sensitivity required in this design.

The design parameters that are not varied in these tradeoffs are shown in Table I. These constant parameters include the thicknesses of other layers in the beam and the relative surface dimensions of the beam and piezoelectric capacitor.

The influence of the length of the piezoelectric capacitor on scale factor was evaluated independently. The design tradeoff curves for capacitor size are shown in Fig. 7. The thickness of the ZnO layer was assumed to be 2 µm and the Si layer was taken to be 6.25 µm. These curves show a decrease in scale factor as the capacitor length is made an increasing fraction of the beam length. This behavior results from the extra loading on the strain-induced signal by the parasitic capacitance in the relatively nonstrained region of the cantilever toward its free end. For a capacitor design length equal to 25 percent of the beam length, the scale factor is roughly 80 percent of the maximum value, and this design has been used generally.

FABRICATION

The fully integrated accelerometer structure was produced using a seven-mask IC process. Except for the ZnO sputtering and the anisotropic etching step, standard Si-gate, PMOS processing was employed. The substrate material was 10-Ω · cm n-type ⟨100⟩ Si. After following the conventional MOS IC fabrication steps through field-oxide growth, windows for the gate and active sensor region are opened and thin thermal SiO_2 is grown. Following a 35-keV threshold-adjusting B implant (dosage = 1.75×10^{12}), a polysilicon gate is deposited and patterned. The next step is a heavy implant of B through the thin oxide (BF_2 dosage = 1×10^{16} at 180 keV). This implant dopes the Si surface strongly p-type in order to stop the EDP etch and also to define the source and drain regions and to dope the polycrystalline Si gate.

A 2-µm film of piezoelectric ZnO is laid down by planar magnetron sputtering [9]. The ZnO is then patterned to have a slope of about 70° at its edge using a dilute combination of acetic and phosphoric acids in order to improve step coverage by subsequent depositions. An overlying sputtered SiO_2 layer prevents leakage of charge across the surface and helps to assure nearly dc response for the accelerometer. Metal connections consisting of Al and NiCr layers are evaporated and patterned, and a final thick SiO_2 layer is sputtered and overlain with a gold layer for the purpose of EDP etch resistance and surface passivation.

The process is IC compatible; neither the SiO_2 nor the ZnO sputtering have been observed to affect the FET parameters. Etching to delineate the Si beam [10] has been carried out in one of two ways. Direct etching in EDP results in a Si beam that is on the order of 1 to a few micrometers thick. A thicker beam can be made by using double-sided etching. For this latter process, the front surface is covered and etching begins from the back. This process continues for 2 h for a 50-µm-thick wafer and then the front pattern is uncovered so that etching proceeds from both sides until the beam is formed. The thickness of the beam is controlled by the width of the window opened to the EDP etch.

A modeling program has been found helpful to predict the shapes of EDP etch patterns in Si. The program makes use of an HP 9845 desktop computer and takes as input data the window dimensions opened to the EDP etch, the orientation of the Si wafer, and the temperature of the etch. The model was derived from experiments with different surface geometries and has been validated by stereographic SEM images. Its output is a plot of the predicted evolution of the etched surface geometry.

Fig. 8 shows a completed chip containing several accelerometers. On the chip are six sets of beam width/length ratios which have mask dimensions of: 70/235, 200/225, 275/510, 410/505, 500/570, and 950/1240 µm. The range in sizes was taken in order to achieve sensitivities to various *g* values and to determine processing constraints. The larger beams, processed by two-sided etching, have Si beam thicknesses of 50 µm and a ZnO layer thickness of 1.7 µm.

RESULTS

An Unholtz–Dickie Model 351 Recording Vibration Calibration system was used to investigate the performance of several of the beams. This system consists of a controlled shake table

(a)

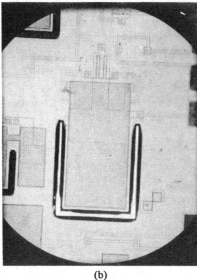

(b)

Fig. 8. (a) Chip photograph showing six versions of the accelerometer. Two structures are compensated with pyroelectric compensation capacitors, two are uncompensated, and two structures have no amplifiers on chip. (b) 950/1240 μm beam without temperature compensation.

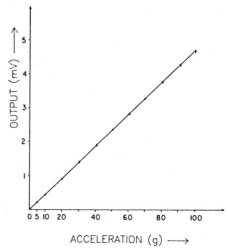

Fig. 9. Measured output (at the FET gate) versus acceleration in *g* for a 950/1240 μm beam.

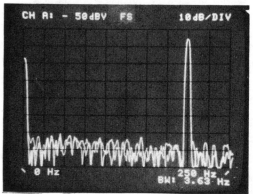

Fig. 10. Spectrum analyzer output pictures showing a 50*g* signal at 200 Hz.

which incorporates a carefully calibrated reference accelerometer. It can be programmed to test a specimen over a wide range of *g* values and frequencies. A series of measurements was carried out on the 950/1240 μm beam which had been designed to have a scale factor of 44 μV/*g*. The depletion-mode, p-channel MOSFET had a threshold voltage of +1.5 V and the load resistance was 1 kΩ. Fig. 9 is a plot of the measured output voltage (at the FET gate) at 200 Hz as the acceleration is increased from 5 to 100*g*. The indicated scale factor is 47 μV/*g*, which is very close to the theoretical design value. The response is also seen to be extremely linear—the measured non-linearity is lower than 0.8 percent. At accelerations less than 50*g*, the nonlinearity is below 0.2 percent. This extreme linearity is a very useful property of the accelerometer.

A Hewlett-Packard model 3582A spectrum analyzer was used to investigate signal and noise characteristics. The noise spectrum for the device appears to be white. Fig. 10 is an oscilloscope trace showing the spectrum of the output signal with an input acceleration of 50*g* applied at 200 Hz. The noise

at this *g* value is 60 dB below the signal or equivalent to 0.05*g* for a 3.63-Hz bandwidth.

The accelerometer was expected to have a nearly flat sensitivity from dc to frequencies that approach the beam resonance. To study the frequency response, the accelerometer was coupled to the spectrum analyzer through a series capacitor (in order to block the dc voltage at the MOS drain). The frequency response was then measured at 50 and 100 Hz, and then in 100-Hz intervals to 1 kHz at 5*g* input excitation. Fig. 11 is a superposition of the images from the spectrum analyzer obtained during this test. The coupling capacitor introduces a zero at roughly 100 Hz which is responsible for the low-frequency rolloff. The small apparent increase in sensitivity at higher frequencies is thought to be a result of mechanical resonance in a light-tight housing which enclosed the device. From these data, we can see that the equivalent noise at the input over the entire frequency range is 40 dB down from the 5*g* signal in the 14.5-Hz bandwidth passed by the spectrum analyzer. The beam used for these studies was designed to sense accelerations ranging to 1000*g*, and, therefore, not to have the highest sensitivity. The observed sensitivity could readily be improved by adding proof mass. The design, construction, and fabrication of beams that make use of added

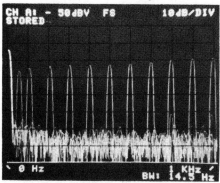

Fig. 11. Spectrum analyzer output showing the frequency response from 50 to 1000 Hz for a *5g* signal.

proof masses is presently under study and will be reported on at a later time.[1]

As described earlier, several of the accelerometer structures are partially temperature compensated by the addition of an unstrained ZnO capacitor to balance out the voltage produced by pyroelectricity in the ZnO. Earlier research had shown this effect to be the most influential cause of temperature sensitivity [11]. Preliminary studies on the beam accelerometers have shown that the compensating capacitor essentially cancels the pyroelectric effect, and the remaining temperature sensitivity is essentially that of the p-channel MOSFET. The percentage change in $I_D/^\circ$C has been measured as −0.16 percent.

An important result of this research is the demonstration that by isolating the ZnO and passivating the overall circuit with deposited SiO_2 layers, there can be an extremely slow decay in the piezoelectric response to a static load. By coupling the output of the piezoelectric film capacitor through a deple-

tion-mode MOS transistor, we have been able to achieve a near-dc response in the integrated accelerometer structure. Measurements of the response to an applied static load on the cantilever have indicated a time constant for the decay of the piezoelectrically induced voltage of seven days.

The results summarized in this section validate the design concepts for the integrated accelerometer, and point the way toward fully integrated systems in which on-chip signal processing can simplify the design of control systems.

REFERENCES

[1] S. H. Kwan, R. S. Muller, and R. M. White, "High frequency strain transducer for acoustic-wave signal processing," presented at the IEEE Device Research Conf., Cornell Univ., Ithaca, NY, June, 1977.

[2] R. D. Jolly and R. S. Muller, "Miniature cantilever beams fabricated by anisotropically etching of silicon," *J. Electrochem. Soc.*, vol. 127, p. 2750, Dec. 1980.

[3] K. E. Petersen, "Dynamic micromechanics on Si: Techniques and devices," *IEEE Trans. Electron Devices*, vol. ED-25, p. 2141, Oct. 1978.

[4] L. M. Roylance and J. B. Angell, "A batch-fabricated silicon accelerometer," *IEEE Trans. Electron Devices*, vol. ED-26, p. 1911, Dec. 1979.

[5] P. L. Chen, R. S. Muller, R. M. White, and R. Jolly, "Thin film ZnO-MOS transducer with virtually dc response," in *Proc. IEEE Ultrasonics Symp.* (Boston, MA, Nov. 5-7, 1980).

[6] *Machinery Handbook*, 19th Ed. New York: Industrial Press, 1973, p. 408.

[7] C. P. Chen, "Fracture strength of Si solar cells," Rep. DOE/JPL 1012-32, Jet Propulsion Laboratory, Calif. Inst. of Tech., Pasadena, Oct. 15, 1979.

[8] C. M. Harris, Ed. *Shock and Vibration Handbook*. New York: McGraw-Hill, 1976, pp. 7-31.

[9] T. Shiosaki, T. Yamamoto, A. Kawarata, R. S. Muller, and R. M. White, "Fabrication and characterization of ZnO piezoelectric films for sensor devices," in *1979 IEDM Tech. Dig.* (Washington, DC), pp. 151-154.

[10] R. D. Jolly and R. S. Muller, "Miniature cantilever beams fabricated by anisotropic etching of silicon," *J. Electrochem. Soc.*, vol. 127, pp. 2750-2754, Dec. 1980.

[11] S. H. Kwan, Ph.D. dissertation, Dep. Elec. Eng. Comput. Sci., Univ. of California, Berkeley, 1978.

[1] Measured results on an accelerometer with the geometry described in this paper except for a proof mass formed by unetched Si at the free end have a sensitivity of 1.5 mV/*g* and a flat response from 3 Hz to 3 kHz.

Acoustic Accelerometers

M. EDWARD MOTAMEDI

Abstract—The characteristics of a number of recently developed acoustic accelerometers are discussed. The accelerometers discussed employ resonant vibrating elements such as piezoelectric crystals and micromechanical cantilever beams containing either a piezoelectric capacitor, a piezoresistive circuit, or a surface acoustic wave (SAW) oscillator. Other accelerometers discussed employ a force rebalance technique using closed-loop servo electronics. Accelerometer sensitivities are limited by weight and frequency response constraints but can be as high as 300 mV/g for voltage-type accelerometers and 10 000 Hz/g for SAW accelerometers. The accelerometer measurand range is from 10 μg to 200 000 g with a dynamic range as high as 10^7 and scale-factor stability as high as 10^{-3} ppm in frequency. Accelerometers can be discrete, hybrid-integrated, or completely monolithic.

INTRODUCTION

THE NEED for less costly light-weight high-performance accelerometers is growing. Some of the demanding applications are in automated manufacture, automobiles and other vehicles, vibration and seismic monitoring, scientific measurement, and in military and space systems. In some of these applications, discrete or hybrid structures are acceptable. In others, where low weight, small size, and superior performance are essential, monolithic integration is desirable. It is possible to achieve self-calibrated accelerometers by including semicustom VLSI gate arrays with the accelerometer elements.

We will now discuss several accelerometer technologies, grouping devices under three headings: 1) bulk transducer piezoelectric accelerometers; 2) cantilever-beam accelerometers; and 3) magnetic force rebalance accelerometers. The cantilever-beam devices are of three types: piezoresistive, piezoelectric, and SAW oscillator devices.

BULK TRANSDUCER PIEZOELECTRIC ACCELEROMETER

This type of accelerometer is the oldest of the designs and it is based on piezoelectric effects [1]. Fig. 1 shows the two most commonly used structures. The piezoelectric element is sandwiched between a base material and an invariant seismic mass, which is forced by the piezoelectric crystal to follow the motion of the base. In many cases, crystalline quartz, either in its natural state or processed form, is used for piezoelectric elements. In Fig. 1, the piezoelectric crystals perform a dual function, acting as a precision spring to balance the applied force and gen-

Manuscript received August 29, 1986; revised November 4, 1986.

The author is with Rockwell International Corporation, Microelectronics Research and Development Center, Anaheim, CA.

IEEE Log Number 8612467.

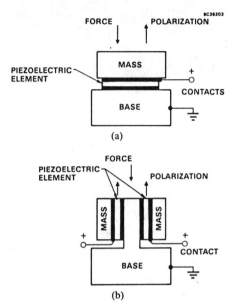

Fig. 1. Bulk transducer piezoelectric accelerometer. (a) Compression-type device. (b) Shear-type device.

erating an electrical signal proportional to deflection of the accelerometer. This kind of piezoelectric accelerometer is commonly used by many manufacturers.

Some potential applications of bulk piezoelectric accelerometers are scientific measurements, vibration monitoring, and transportation safety and control. Some commercial accelerometers of this type are Endevco Model 2215E, Sensotec JIF general-purpose accelerometer, and Dytran Model 3101. Some manufacturers have included signal-conditioning elements in the accelerometer either by a hybrid or semi-integrated technique; an example is the family of ICP accelerometers manufactured by PCB Piezotronics, Inc.

Piezoelectric accelerometers in discrete form can also offer excellent insulation from nonvibrating environments either by using an isolated base or shear-mode structure. Endevco isoshear accelerometers are made from this principle.

CANTILEVER-BEAM ACCELEROMETERS

This type of accelerometer uses a cantilever beam that is supported (or "clamped") at one end and has its other end free from all except gravitational forces. As noted earlier, there are three types of cantilever-beam accelerometers: piezoresistive, piezoelectric, and SAW oscillator devices.

In the first device, a cantilever beam fashioned typi-

Reprinted from *IEEE Trans. Ultrason. Ferroelec. Freq. Contr.*, vol. UFFC-34, no. 2, pp. 237–242, March 1987.

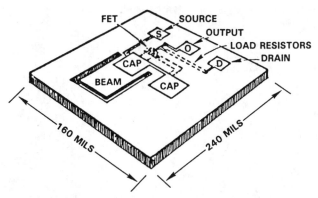

Fig. 2. Miniature accelerometer chip layout. Device has temperature compensation and input stage amplifier including on-chip load resistor.

Fig. 3. SiO$_2$ cantilever beams fabricated at Rockwell using top-surface etching. Pattern is designed for studying EDP anisotropic etching. Larger beam has dimension of 12 × 38 mils.

cally from single-crystal Si, as described later, deflects when the device is accelerated in a direction normal to the plane of the beam [2]. The deflection of the beam is inferred from change in the values of resistors diffused into the beam near its point of support. In the second type of device, a piezoelectric film deposited on the beam produces a voltage when the beam deflects as a result of acceleration [3], [4]. In both of these devices, IC's on the devices condition signals for output from the chip. In the third type, a SAW delay line or a resonator on a cantilever beam is connected to a feedback amplifier so that oscillation results [5]–[8]. The oscillation frequency is altered by the change of phase velocity caused by acceleration-induced strain.

The piezoelectric cantilever-beam devices provide larger outputs than the piezoresistive devices, but require deposition of a piezoelectric film. High-yield production techniques for fabricating these devices have, however, been developed during the past five years at Rockwell International. The SAW accelerometer is characterized by high resolution and high dynamic range. We shall now discuss just the piezoelectric and SAW cantilever devices.

Piezoelectric Cantilever-Beam Accelerometer

We will discuss a miniature monolithic accelerometer designed and fabricated to be compatible with standard VLSI technology at Rockwell International. Fig. 2 shows a schematic of the device. The accelerometer IC contains a small cantilever beam on which is a thin-film piezoelectric capacitor. The beam is produced by etching the Si "chip." Acceleration forces normal to the surface of the chip cause the beam to bend and increase strain in the piezoelectric capacitor. The piezoelectric effect converts this strain into an electrical charge that is proportional to the acceleration. The on-chip IC also includes 1) a temperature-compensating capacitor, which is electrically back-to-back and physically identical to the sensor capacitor, and 2) a p^+ load resistor to linearize the FET amplifier. The FET amplifier is a depletion-mode PMOS, which is completely passivated by Si glass to eliminate any possible surface charge leakage.

The important process steps of an accelerometer chip like the one shown in Fig. 2 are the formation of canti-

lever beam. The process involves micromachining, which is the three-dimensional sculpting of the silicon to form mechanical structure. In the following, we will discuss two major techniques for processing cantilever-beam structures.

In the first technique, the beam formation is only by a top-surface etching which is commonly a chemical process like a solution of ethylenediamine and pyrocatechol (EDP) to anisotropically etch the Si from underneath the layered structure [9]. The cantilever beam is a composite structure of metal, ZnO, SiO$_2$, and Si, with thicknesses ranging from 2 to 5 μm. When the beam thickness is required to be in the 2-μm range, the Si portion of the structure may be eliminated. Fig. 3 shows a picture of several SiO$_2$ cantilever beams with different beam dimensions and a thickness of 1 μm. The larger beam has a dimension of 12 × 38 mil. The beams are fabricated using the top-surface EDP process. The pattern is designed at Rockwell for studying anisotropic etching.

In the second technique of cantilever-beam formation, two etching steps, frontside and backside, are required. This process is called the two-sided etching technique. This method is commonly used when a beam thickness larger than 5 μm is required.

For high-g Si accelerometers, Rockwell has developed a two-sided etching process based on electrochemical control [10]. Fig. 4 shows cross sections of an Si cantilever beam designed for the two-sided etching process. For fabrication of such a structure, etch depth of practically whole wafer thickness is required, which makes process control of uniformity, sidewalls and etch-stop extremely difficult. To fabricate a cantilever beam such as shown in Fig. 4, a low-resistivity Si substrate is coated with two layers of epitaxial Si films. The top layer epi-

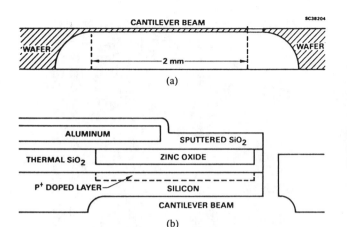

Fig. 4. Cross section of Si cantilever-beam accelerometer designed for two-sided etching. (a) Drawn to scale. (b) Vertical scale increased to show detail of composite structure.

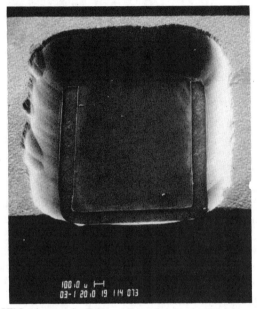

Fig. 5. SEM micrograph of Si cantilever beam processed by two-sided etching technique. Beam thickness is 25.4 μm and can survive 20 000 g acceleration level.

Fig. 6. Complete 3-in processed wafer of monolithic accelerometer with yield better than 60 percent (wafer is processed in VLSI lab, Rockwell).

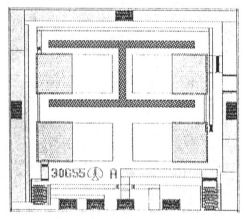

Fig. 7. Computer-generated plot of all frontside layers superimposed. Device includes temperature compensation, buffer amplifier, and cross-axis compensation. Chip size is 120 × 140 mils.

taxial Si is for the purpose of MOS circuitry and signal conditioning of the sensor. The buried layer epitaxial Si is very high resistivity for the purpose of electrochemical etching control. The backside etching is achieved by nitric-based solutions. Results of the etch profile demonstrate a perfect isotropic process. The detail of the process was reported earlier [10]. Fig. 5 shows SEM micrographs of a cantilever beam processed by the two-sided etching technique. The beam thickness is 25.4 μm, and it can survive a 20 000 g acceleration.

The piezoelectric cantilever-beam accelerometer has the attributes of small size and low weight, and it requires low power. It uses the low-cost photolithographic fabrication methods developed for the IC industry and is well-suited to high acceleration levels. Initial feasibility studies of the device at Rockwell concentrated on performance analysis and fabrication processing. The principal problem in earlier studies had been the incompatibility of

the processing methods for forming the beam for those forming the rest of the IC. We have now demonstrated the possibility of using the standard VLSI process and producing micromechanic structures like accelerometers with high yield and excellent performance.

A new VLSI-compatible micromechanic process was recently developed in our lab for two-sided cantilever-beam processing. In this technique, both frontside and backside etching is done by a dry controlled-etching process. Fig. 6 shows a complete 3-in processed wafer using two-sided dry plasma etching.

A new design was optimized to customer performance requirement for high-g applications by computer-aided design studies. This uses parametric models of the critical performance criteria as functions of physical dimensions of the beam and its composite elements. In this design, the beam dimension is 40 × 20 mils. Beam thickness is 5 mils, and the survival g-level requirement is 200 000 g. Fig. 7 shows a computer-generated plot of all frontside

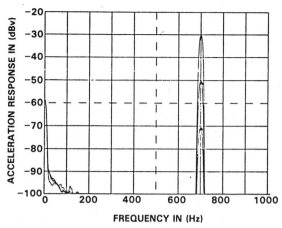

Fig. 8. Output response of typical monolithic accelerometer for three input levels; 1, 10, and 100 g. Results demonstrate excellent linearity over 40-dB dynamic range.

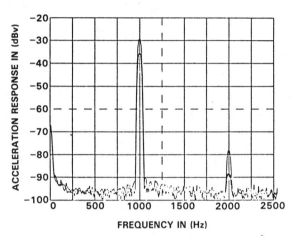

Fig. 9. Output response of typical monolithic accelerometer, g^2 error term is 50 dB below linear term.

design layers superimposed. Some new features of this design are 1) low-impedance output level ($<100\ \Omega$) for increasing signal transmission efficiency by using an on-chip buffer amplifier, and 2) cross-axis compensation by designing the sensor to be composed of two identical accelerometers with 180° relative rotation. The first feature is extremely important for high-g packaging, while the second feature is important to minimize the complexity of signal conditioning circuitry.

For testing the accelerometer device, an Unholtz-Dickie Model 351 dynamic shaker was used. This system consists of a controlled shake table which incorporates a carefully calibrated reference accelerometer. It can be programmed to test a device over side frequency ranges up to 10 kHz. The maximum g-level of this shaker machine is only 100 g. This level of acceleration is good enough for device characterization.

Two typical output responses of cantilever-beam piezoelectric accelerometers are shown in Figs. 8 and 9. Fig. 8 shows the output variation over a 40-dB dynamic range with inputs of 1-, 10-, and 100-g sinusoidal acceleration. Fig. 9 shows a g^2 error term about 50 dB below the linear term. The linearity is limited by the current PMOS design more than piezoelectric sensor response. The frequency of the lowest resonant mode of the typical cantilever beam was found to be at 8.4 kHz.

SAW Cantilever-Beam Accelerometer

Production SAW devices are now available for use in many high-performance and high-precision sensors. A cantilever-beam SAW accelerometer contains a SAW delay line or a resonator oscillator acting as a sensing element [7]–[9].

To construct the SAW accelerometer, a hybrid integration technique should be used. Since the sensor element is a SAW oscillator, a piezoelectric quartz substrate is the best choice considering its temperature stability and high-Q performance. When a high stability on the order of 10^{-3} ppm in frequency is required, a SAW resonator oscillator should be considered. In many applications when frequency stability is not critical, a delay line oscillator can be well-suited instead of a SAW resonator oscillator.

SAW accelerometer sensitivity of 10 μg has been achieved using a quartz cantilever-beam accelerometer with a proof mass included in the beam. A unique technology has been developed at Rockwell for forming a cantilever quartz beam shaped to a two-dimensional hammerhead structure.

For this process, quartz wafers which were polished on both sides were coated on both sides with chrome–gold films. The chrome–gold films were patterned into special geometries to form structures of hammerhead cantilever beams. The patterning is done using standard photolithography and a double-sided alignment technique. Chemical etching has been used to etch the quartz from the both sides with the gold acting as an etch mask. After successful etching, the beams are still connected together by small unetched bridge sections. These bridge sections can be lightly forced to separate the beams. The process is mostly reproducible, but there are some critical points. One critical point is the adherence of the gold film to the quartz to assure being a good mask for chemical etching. Another critical point is the quality of the surface finish of the polished quartz to minimize the anisotropic etching process.

When lower sensitivity on the order of 10^{-3} g is required, the cantilever beam of the SAW accelerometer can be a simple rectangular quartz substrate. In this case, the SAW structure, resonator, or delay line can be batch-processed on a 3-in quartz wafer and then diced to the required beam size. In the case of the SAW accelerometer with a beam shape designed to include a proof mass, it is sometimes cost-effective to produce individual substrates shaped to the beam structure in the manner explained earlier and process the SAW resonator on each substrate. This procedure is very expensive for production, but it is the price that one pays for high-sensitivity (10^{-5}–10^{-6} g) high-stability (10^{-3} ppm in frequency) accelerometers.

SAW technology offers in inherently digital acceleration sensor with simple two- to three-layer processing. The technology uses established planar photolithography for low-cost fabrication. The associated electronics contain only one active amplifier. This simplicity makes it an attractive candidate for digital sensor applications.

A new design of the SAW accelerometer is now under investigation in our laboratory. The major requirements

TABLE I
INDUSTRY SURVEY FOR LOW g AND HIGH-SENSITIVITY ACCELEROMETER[a]

COMPANY / PARAMETERS	ENDEVCO 1	ENDEVCO 2	SENSOTEC	SUNDSTRAND 1	DYTRAN	PCB 1	PCB 2
MODEL NO.	7705-1000	5210-100	JTF	QA-2000	3100A	308B15	302B03
INPUT RANGE (g)	250	10	5	25-150	50	50	16
OUTPUT SIGNAL	ANALOG VOLTAGE	ANALOG VOLTAGE	ANALOG VOLTAGE	ANALOG VOLTAGE	ANALOG VOLTAGE	ANALOG VOLTAGE	ANALOG VOLTAGE
SENSITIVITY mV/g	169	100	20	1.2	100	100	300
STABILITY (SHORT-TERM)	5×10^{-2}	NA	10^{-3}	NA	10^{-4}	NA	NA
TRANSVERSE SENSITIVITY (%)	3	3	5	2	5	5	5
g-SURVIVAL (g)	1000	2000	25	250	3000	10,000	5000
FREQ. RANGE (Hz)	1-2000	2-5000	10-200	40-300	1-3500	1-6000	1-7000
TEMP. RANGE (C°)	$-54 \sim 260$	$-160 \sim 125$	$-40 \sim 93$	$-55 \sim 95$	$-65 \sim 107$	$-73 \sim 120$	$-54 \sim 79$
DYNAMIC RANGE	NA	NA	10^5	10^6	NA	5×10^4	6.4×10^3
WEIGHT (gram)	120	140	28	80	57	55	39
VOLUME (cm^3)	12	20	5	16	2	12	14
POWER (mW)	NA	NA	25	600	480	76	76

[a]Data were obtained from manufacturers during the year 1986. NA indicates data not available.

of this design are 1) full-scale range is ± 10 g; 2) sensitivity of 10 μg; 3) scale factor of 5000 Hz/g; and 4) stability of 10^{-3} ppm in frequency. Several prototypes of this design have already been fabricated, assembled, and tested. Major problems are 1) facing device phase noise which makes it very difficult to achieve stability requirements; and 2) fabricating high-Q SAW resonators (Q in the range of 40 000).

To construct the sensor element of the cantilever beam accelerometer, a two-stage silicon-on-sapphire (SOS) hybrid amplifier has been assembled for use with the high-Q SAW resonator as a feedback element to provide a UHF SAW oscillator with low phase noise. The SAW resonator, fabricated with ST-cut quartz, is mounted in a cantilever beam configuration.

To optimize accelerometer phase-noise performance, a fundamental frequency oscillator or an oscillator of either one-half or one-fourth the fundamental frequency should be used. Phase-noise contributions in the oscillator circuit will arise not only from the resonator or SAW delay line but also from noise arising from the other circuit components. Of primary concern are the noise figures for the active devices in the oscillator, which may be either bipolar transistors or GaAs devices of the MESFET variety. Noise contributions are also made by passive circuit components. Careful selection must be made, for example, of the types of resistors and capacitors used in the amplifying stage of the oscillator circuit.

Force Rebalance Accelerometer

The force balance accelerometer is a simple instrument for measuring acceleration. In this kind of accelerometer, commonly, a flexure and proof mass assembly is supported in a plane perpendicular to the input axis. A central disk made of fused silica is associated as a part of the proof mass. A portion of the central disk is made conductive by a vapor-deposited metallic film to provide a differential capacitive pickoff circuit and to detect deflection

of the proof mass as the result of acceleration force. Through a high-current feedback loop, a magnetically induced rebalance force is generated to restore the proof mass to its null position. The rebalance current is proportional to acceleration input. Manufacturers have developed many different kinds of force rebalance accelerometers according to customer requirements and their cost-effective guidelines. As the requirements for sensitivity, dynamic range, size, transverse sensitivity, and input range become tied, the mechanical design becomes more complicated. The force rebalance accelerometer, considering the present available technology, is limited to be used only for medium-to-near high-sensitivity acceleration measurements (10^3–10^{-5} g). An input range as high as 500 g is feasible.

As an example, Sandstrand presently markets a force rebalance accelerometer (Model QA-2000 Q-Flex) that has 10 μg resolution when the input range is not higher than 25 g. Applications of the force rebalance accelerometer range from the aerospace industry to oceanography, and some guidance and inertial navigation systems.

Summary of Device Characteristics

We have discussed a number of different accelerometers and have commented on the advantages of monolithic accelerometers. Table I shows the results of a survey of commercially available bulk transducer piezoelectric accelerometers. The characteristics of this and other types of accelerometers discussed here are compared in the manner described by White [11] and are categorized in Table II.

Appendix
Glossary of Accelerometer Terms

Cross-Axis Coupling: the proportionality constant that relates a variation of accelerometer output resulting in a rotation error or true axes (in rad/g).

374

TABLE II
CLASSIFICATION SCHEME OF ACOUSTIC ACCELEROMETERS DISCUSSED

ACC. TYPE / SENSOR CLASS	PIEZOELECTRIC ACCELEROMETER (BULK TRANSDUCER)	CANTILEVER BEAM ACCELEROMETER (PIEZOELECTRIC)	CANTILEVER BEAM ACCELEROMETER (PIEZORESISTIVE)	CANTILEVER BEAM ACCELEROMETER (SAW OSCILLATOR)	FORCE-REBALANCE ACCELEROMETER
A. MEASURAND	ACCELERATION	ACCELERATION	ACCELERATION	ACCELERATION	ACCELERATION
B. TECHNOLOGICAL ASPECTS	PRECISION RUGGED SIMPLICITY SENSITIVITY LOW IMPEDANCE	RUGGED MINIATURE ON-CHIP INTEGRATION LOW COST HIGH TEMPERATURE	dc RESPONSE LOW IMPEDANCE LOW THERMAL TRANSIENTS HIGH-g HIGH FREQUENCY	HIGH SENSITIVITY HIGH STABILITY HIGH PRECISION RUGGED SEMIDIGITAL HI	PRECISION HI DYNAMIC RANGE LOW CROSS AXIS RESPONSE
C. DETECTION MEANS	MECHANICAL	MECHANICAL	MECHANICAL	MECHANICAL	MECHANICAL
D. CONVERSION PHENOMENA	ELASTO-ELECTRIC	ELASTO-ELECTRIC	ELASTO-ELECTRIC	ELASTO-ELECTRIC	ELASTO-MAGNETIC
E. SENSOR MATERIALS	INORGANIC INSULATOR	INORGANIC INSULATOR SEMICONDUCTOR	INORGANIC SEMICONDUCTOR	INORGANIC INSULATOR	INORGANIC INSULATOR
F. FIELD OF APPLICATION	ENVIRONMENT MANUFACTURING MILITARY	COMMECIAL MANUFACTURING MARINE MILITARY SPACE	ENVIRONMENT MANUFACTURING (TESTING) TRANSPORTATION	ENVIRONMENT MARINE MILITARY	MILITARY TRANSPORTATION
G. STRUCTURE	DISCRETE OR SEMI-IC	MONOLITHIC IC	DISCRETE OR MONOLITHIC	HYBRID SYSTEM	DISCRETE SYSTEM

Dynamic Range: the ratio of the maximum measurable acceleration to the minimum detectable acceleration.

g-Survival: the maximum acceleration that the device can handle with no noticeable change in performance characteristics, also referred to as "shock level," specified in *g*.

Input Range: the design range of acceleration, commonly referred to as the maximum value of the range. The minimum value can be determined knowing the dynamic range. Range is specified in *g*.

Linearity: maximum deviation from a least squares fit straight line through the data points over the measuring range, specified in percent.

Mechanical Resonance: frequency at which the internal structure of the accelerometer resonates, specified in kHz.

Power Consumption: power necessary for the accelerometer to function properly, specified in mW.

Resolution: A measure of the ability to delineate detail or distinguish between nearly equal values of a quantity. Also referred to as a "threshold"; lowest level of valid measurement. Resolution is generally specified in μg.

Scale Factor: the factor by which the number of scale divisions of the output signal should be multiplied to compute the value of measured acceleration.

Sensitivity: electrical output per unit input of acceleration, specified in mV/g for electrical only, for piezoresistive and piezoelectric accelerometers, or in frequency change (Δf) per g for oscillator accelerometers.

Stability: a fluctuation of accelerometer scale factor in ppm during the time of the measurement. If the measurement time is shorter than 10 s, it is called short-term stability.

Transverse Sensitivity: ratio of the output when acceleration is perpendicular to the sensitive axes to the output when acceleration is in the direction of a sensitive axis, specified in percent.

REFERENCES

[1] W. G. Cady, *Piezoelectricity*, vols. 1 and 2. New York: Dover, 1964.
[2] W. E. Rosvold and M. L. Stephens, Contract AFAL-TR-77-152, Wright-Patterson AFB, OH, 1977.
[3] P. L. Chen *et al.* "Integrated silicon microbeam PI-FET accelerometer," *IEEE Trans. Electron Devices*, vol. ED-29, pp. 27–33, 1982.
[4] M. E. Motamedi, A. P. Andrews, and E. Brower, "Accelerometer sensors using piezoelectric ZnO thin films," in *Proc. 1982 Sonics and Ultrasonics*, p. 303.
[5] P. Harteman and P. L. Meunier, "Surface acoustic wave accelerometers," in *Proc. 1981 IEEE Ultrasonics Symp.*, pp. 152–154.
[6] E. J. Stabled *et al.*, "Pressure and acceleration sensing of a SAW interferometer," in *Proc. 1981 IEEE Ultrasonic Symp.*, pp. 155–158.
[7] M. E. Motamedi, "Acoustic sensors," 1985 Rockwell International IR&D Brochure, Project 882.
[8] C. A. Erikson and D. Thoma, "SAW accelerometers: Integration of thick and thin film technologies," presented at the 1985 Ultrasonic, Ferroelectrics, and Frequency Control Symp.
[9] K. E. Petersen, "Dynamic micromechanics on Si: Techniques and devices," *IEEE Trans. Electron Devices*, vol. ED-25, p. 2141, 1978.
[10] M. E. Motamedi *et al.*, "Application of electrochemical etch-stop in processing silicon accelerometer," *Electrochem. Abstracts*, p. 188, May 1982.
[11] R. M. White, "A sensor classification scheme," *IEEE Trans.*, this issue.

SILICON MICROACCELEROMETER

Félix Rudolf, Alain Jornod, Philip Bencze
Centre Suisse d'Electronique et de Microtechnique S.A.
Maladière 71
2007 Neuchâtel, Switzerland

ABSTRACT

The concept, the fabrication techniques and experimental results of a new miniaturized high-sensitivity capacitive servoaccelerometer for space applications are presented. The accelerometer has 4 electrically switchable working ranges between ± .001 and ± 1 g. µg resolution has been demonstrated. The useful frequencies are 0 to 100 Hz and 0 to 3 kHz for the lowest and highest ranges respectively.

INTRODUCTION

The combination of force balancing concepts, well known from macroscopic high-precision accelerometers, with silicon micromechanical techniques has been used to realize highly-miniaturized, high performance accelerometers. In this paper an accelerometer with µg capabilities (1 µg = 9.81×10^{-6} m/s^2) for space applications is described. These devices called microaccelerometers, the term "micro" refering to their size as well as to the measurement range, will be used for instance to monitor the residual acceleration in microgravity experiments. The basic requirements are a measurement range from 1 µg to 1 g for frequencies from 0 to 10 Hz and 0 to 1 kHz respectively.

A conceptual study has shown that a capacitive approach such as described in [1] is most promising. Small-sized, light, low power consumption devices are very desirable features which have been obtained. The microaccelerometer is based on a silicon micromechanical chip, comprising a movable plate between fixed capacitor plates, and on electronics circuitry for plate position measurement and electrostatic force balancing. The high degree of miniaturization of the mechanical parts and the associated small inertial force make it possible to generate the necessary compensation forces with low voltages applied to the capacitors.

The electromechanical accelerometer chips are fabricated using a capacitive silicon sensor technology, based on silicon micromachining, anodic glass to silicon bonding and hermetic sealing of the capacitor cavities.

In its final form, the accelerometer chip and an electronics chip will be packaged in a miniature housing.

ACCELEROMETER CONCEPT

The servoaccelerometer concept is shown in Fig. 1. The mechanical part is formed by a plate suspended by spring elements designed in order to have only one degree of freedom for the movable plate. The corresponding rotation around a fixed axis has been found to be the optimum configuration. The measurement principle is well described by a linear model illustrated in Fig. 1, since rotation angles are very small. Plate position is measured with an AC-bridge of which one branch is formed by the capacitors located on either side of the movable plate. This ensures high resolution and symmetry of the electrostatic forces, which depend on the square of the applied voltage. In order to obtain a linear relation between the output voltage V_S and the acceleration, the linearizing scheme illustrated in Fig. 1 is used. V_O is a constant voltage which provides a reference potential. Different measurement ranges are selected by using electrodes of different sizes. The basic characteristics of the accelerometer system are given in Table I.

	Measurement range			
	I	II	III	IV
Measurement range [g]	± 1	± 0.1	± 0.01	± 0.001
Sensitivity [V/g]	1	10	100	1000
Frequency response [kHz]	0 to 3	0 to 1	0 to 1	0 to 0.1

Table I Accelerometer system characteristics

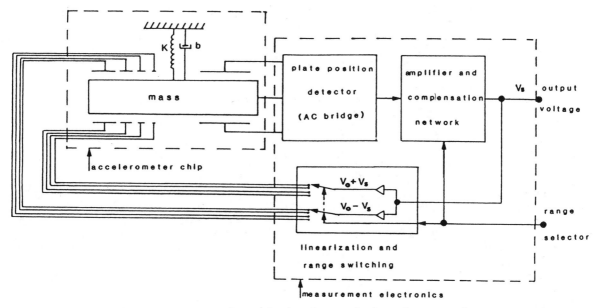

Fig. 1 Schematic representation of the force balancing accelerometer system.

The structure of the accelerometer chip is illustrated in Fig. 2. It is made of three plates bonded together. The bottom plate acts as a support for the lower set of fixed electrodes. The middle part comprises the movable plate, which is suspended to an outer frame by torsion bars. The top plate carries the upper electrodes. The upper fixed electrodes are connected to the metallization layer on the top side of the middle plate. Consequently all the electrodes can be bonded from the top side through dedicated openings in the top and in the middle plates.

The bottom and top plates are both manufactured by stacking a silicon wafer and a glass wafer of suitably matched thickness. This approach has been chosen in order to minimize the stress induced by the glass.

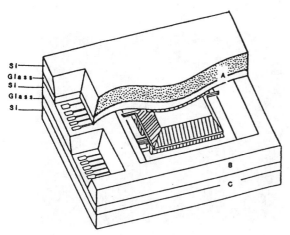

Fig. 2 Schematic drawing of the micro-accelerometer chip.

(A: top plate, B: middle plate, C: bottom plate)

Typical characteristics of the accelerometer chip are indicated in Table II. The open loop sensitivity is defined as

$$S_{\Delta C} = \frac{\Delta C}{\Delta a} = \frac{\Delta (C_1 - C_2)}{\Delta a}$$

C_1 and C_2 are the capacitance on either side of the movable plate.

Inertial mass :	4.6×10^{-6} kg
Torsion bar dimensions :	length: 200 µm
	width: 11 µm
	thickness: 9 µm
Distance between the electrodes at rest :	2 µm
Working capacitance on either side :	15.5 pF
Sensitivity $S_{\Delta C}$:	1000 pF/g
Measurement range :	± 0.007 g
Resonance frequency :	60 Hz
Chip dimensions :	7.5 mm x 4.5 mm

Table II   Characteristics of accelerometer chip (open loop design values)

FABRICATION OF THE ACCELEROMETER CHIP

Batch processing of the accelerometer includes three basic steps :

1. Silicon wafer processing of the middle plate.
2. Processing of bottom and top silicon/glass wafers.
3. Bonding, sealing and dicing of triple-stacked wafers.

The middle plate is fabricated with a 8 mask silicon micromachining process. Silicon is etched in a KOH solution.

All etching depths are time controlled. The thickness tolerance obtained with this method is ± 1.5 μm, which is sufficient, since the absolute value of the spring constant is not critical in a force balancing system.

The silicon-glass compound plates are fabricated by anodic bonding of glass and silicon wafers. The access holes to the contact pads are obtained by wet chemical etching of silicon and glass. The electrodes are deposited and patterned with a thin film process.

The three individual wafers are then bonded together by anodic bonding. For the final low-pressure sealing of the measurement cavity an indium-soldering technique is used.

All processing steps are performed on 3 inch wafers. 84 accelerometer chips (Fig. 3) are fabricated on a single wafer.

Fig. 3 Photograph of a finished chip.

EXPERIMENTAL RESULTS

Accelerometer chips with sensitivities ranging from 0.2 to 760 pF/g have been fabricated. The different sensitivities have been obtained by changing the thickness of the suspension bars. The chips were mounted on standard IC packages for characterization purposes.

Measurements are done with laboratory equipments, e.g. a high precision capacitance bridge, and with a dedicated electronics breadboard.

Acceleration stimulus was applied with a high precision tilt-test equipment with μ-rad resolution and with a commercially available vibrating table.

A typical static open loop response curve is shown in Fig. 4. The tilt angle is measured between the mechanical holder of the IC-housing and the horizontal plane. The curve shown in Fig. 4 includes offset due to the mechanical asymmetry of the movable plate, stray capacitances, the mechanical positioning error of the chip in the holder and the holder in its standard socket. A test procedure has been worked out, which allows to

measure the intrinsic offset and the positioning errors independently. This procedure has not been applied to the curve shown in Fig. 3 in order to illustrate the low overall offset.

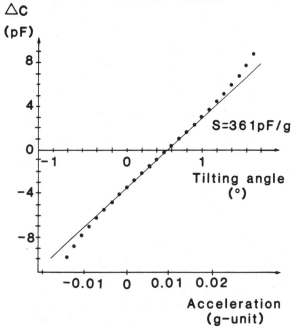

Fig. 4 Static open loop accelerometer response as measured by tilting the accelerometer. The equivalent acceleration is also shown.

Open loop frequency response is shown in Fig. 5. It has been measured using pseudo random excitation and by comparing the signals of the device under test with that of a reference piezoelectric accelerometer. The damping depends largely on the pressure surrounding the movable plate. In high vacuum the resonance has a Q factor up to 100. At atmospheric pressure, the devices are heavily damped. The pressure at which damping is critical depends on the sensitivity.

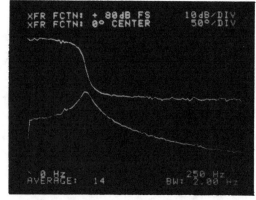

Fig. 5 Typical uncalibrated open loop frequency response. The transfer function of the amplitude (bottom) and the phase (top) are shown. The sensitivity of the device under test is 360 pF/g.

For high-sensitivity devices, critical damping is achieved at a pressure of about 10 Pa.

The accelerometer resolution is illustrated in Fig. 6. The output signal due to a dynamic acceleration stimulus of 1 μg at a frequency of 1 Hz is shown. The noise floor of about 0.3 μg/√Hz includes acceleration noise and the intrinsic noise of the device under test. It has been found that this noise level is essentially the same for devices with sensitivities ranging from 40 to 360 pF/g and for the reference seismic piezoelectric accelerometer.

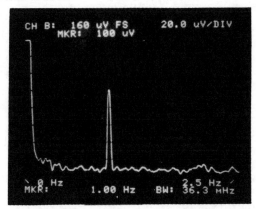

Fig. 6 Uncalibrated open loop response to a sin excitation with an amplitude of 1 μg at 1 Hz.

The temperature behaviour of accelerometers with an open loop measurement range of 0.5 g has been determined. The temperature coefficient of the sum of the capacitances on either side of the plate is 240 ppm/°C. The temperature coefficient of the open loop sensitivity is 300 ppm/°C. The total drift of the offset is 3 mg in the temperature range between 20°C and 70°C.

A static closed loop response curve is shown in Fig. 7. The device has an open loop sensitivity of 70 pF/g which is sufficient for μg-resolution. Thus with one single device, measurements between 1 μg and 1 g are possible, using electrical switching of the measurement electronics parameters.

DISCUSSION

Many of the characteristics of the silicon microaccelerometer such as the resolution, the dynamic range, the offset stability compare favorably with the performance of state of the art precision instruments commercially available. The introduction of micromechanical techniques allows to achieve high performance with a miniaturized and potentially low cost device.

There is still room for improvement of the accelerometer performance. Presently the extension of the measurement range into the sub-μg domain using multi-accelerometer assemblies and digital signal-conditionning (e.g. correlation techniques) is being studied.

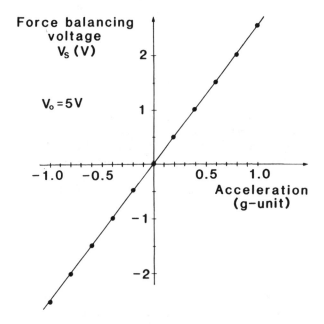

Fig. 7 Static closed loop response. A linearizing schema as indicated in Fig. 1 is used.

On the other hand, many applications have less stringent resolution requirements. In these cases the use of the same basic mechanical structure, but with much thicker suspension bars, and without force balancing loop might be a cost-effective solution. Measurement of accelerations up to 10^4 g is feasible.

Due to its wide range of acceleration and the high intrinsic precision, this capacitive microaccelerometer approach may suit many different applications ranging from high precision instruments to low cost consumer goods.

ACKNOWLEDGEMENTS

We would like to thank Philippe Roussel of ESTEC, Noordwijk, Holland for helpful suggestions and stimulating discussions. This work was performed under contract for the European Space Agency (ESA).

REFERENCES

[1] F. Rudolf, "A micromechanical capacitive accelerometer with a two-point inertial-mass suspension", Sensors and Actuators, 4 (1983), 191.

RESONANT MICROSENSORS

Roger T. Howe

Microsystems Technology Laboratories
Department of Electrical Engineering and Computer Science
Massachusetts Institute of Technology
Cambridge, Massachusetts 02139 USA

Abstract

Resonant sensors are attractive for precision measurements because of their high sensitivity to physical or chemical parameters and their frequency-shift output. This paper develops the theory of mechanical resonance and examines its implications for microfabricated resonant sensors. An equivalent circuit for a capacitively excited microbridge is derived and used to study the dynamic response of resonant microsensors.

Introduction

The frequency of a mechanical resonator is a highly sensitive probe for parameters that alter its potential or kinetic energy. A major class of measurement devices, termed *resonant sensors,* makes use of this phenomenon. Physical or chemical parameters can be sensed either by coupling loads to the resonator or by coating it with sensitive films.[1-3] Resonant sensors are attractive because of their high sensitivity and frequency-shift output. Over the past two decades, quartz mechanical resonators,[4,5] quartz bulk-wave resonators,[6,7] and surface-acoustic-wave oscillators[8,9] have been investigated extensively for precision sensing applications.

Recently, silicon microfabrication technology has been enhanced with a collection of chemical etching processes for micromachining mechanical structures.[10-12] Given this capability and the burgeoning interest in microfabricated sensors (microsensors), micromechanical resonant structures have been explored in several prototype sensors.[13-16] Resonant microsensors promise better reproducibility through well-controlled material properties and precise matching of micromachined structures. Furthermore, batch fabrication of resonant microsensors with on-chip sustaining amplifier and signal-processing circuitry should reduce manufacturing costs.

This paper develops the theory of sensors incorporating resonant mechanical structures, and examines its implications for designing microfabricated devices. Rayleigh's Method for finding resonant frequencies provides the analytical framework for understanding resonant microsensors for vapor concentration and pressure. Mechanical resonance can be excited and the resulting vibration detected in several ways. For the important case of electrostatic drive and detection, a linearized model is derived and then used to analyze the response of resonant sensors to time-varying parameters.

Mechanical Resonance

Figure 1 illustrates a tensioned bridge vibrating in its fundamental mode. The vertical deflection is

$$y(x,t) = Y_1(x)e^{j\omega_1 t}, \qquad (1)$$

where $Y_1(x)$ is the fundamental mode shape.

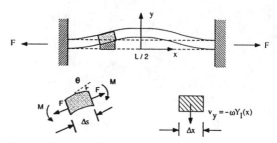

Fig. 1 Tensioned bridge executing harmonic motion

At the point of maximum deflection, the potential energy PE equals the total vibrational energy. The differential segments of the bridge are subjected to a bending moment and an axial load, both of which contribute to PE:

$$PE = \int_0^L \tfrac{1}{2} M\theta \, dx + \int_0^L F(\Delta s - \Delta x)dx$$

$$= \int_0^L \frac{EI}{2}\left[\frac{d^2 Y_1(x)}{dx^2}\right]^2 dx + \int_0^L \frac{F}{2}\left[\frac{dY_1(x)}{dx}\right]^2 dx, \quad (2)$$

where E is the Young's modulus and $I = Wt^3/12$ is the moment of inertia for the bridge of thickness t and width W. The maximum kinetic energy KE occurs a quarter-cycle later, when each segment passes through the centerline with a velocity $v_y(x) = -\omega_1 Y_1(x)$:

$$KE = \omega_1^2 KE' = \omega_1^2 \int_0^L \tfrac{1}{2}\rho Wt \, Y_1^2(x)dx, \quad (3)$$

where ρ is the density. By energy conservation, the first resonant frequency is given by $\omega_1 = (PE/KE')^{1/2}$. Rayleigh's Method for the tensioned bridge yields

$$\omega_1^2 = \frac{\displaystyle\int_0^L \frac{EI}{2}\left[\frac{d^2 Y_1(x)}{dx^2}\right]^2 dx + \int_0^L \frac{F}{2}\left[\frac{dY_1(x)}{dx}\right]^2 dx}{\displaystyle\int_0^L \frac{1}{2}\rho Wt \, Y_1^2(x)dx}. \quad (4)$$

Reprinted with permission from *Transducers '87, Rec. of the 4th Int. Conf. on Solid-State Sensors and Actuators*, 1987, pp. 843–848.

Other energy terms, such as shear potential energy, rotational kinetic energy, and self-induced stretching energy may be added to (4) if significant.[4]

Rayleigh's Method yields excellent approximations for the resonant frequency when reasonable assumptions are made for the mode shape. For example, a bridge may have varying width and material properties along its length. As long as the basic structure does not depart radically from that of a uniform bridge, the mode shape $Y_n(x)$ of the latter can be used in (4) with confidence that the estimated ω_n is reasonable.[17,18]

Rayleigh's Method is also important for resonant sensor design. In general, a parameter p can perturb the kinetic or the potential energy of the structure. The normalized sensitivity S is defined as the fractional change in the unperturbed resonant frequency ω_o due to an incremental change in the parameter, or $S = \omega_o^{-1}(d\omega_o / dp)$. Unless there is an offset or bias value of the parameter, the derivative is evaluated at $p = 0$. Substitution of Rayleigh's expression for resonant frequency yields

$$S = -\frac{1}{2}\frac{1}{KE_o'}\left[\frac{dKE'}{dp}\right]_{p=0} \frac{1}{2}\frac{1}{PE_o}\left[\frac{dPE}{dp}\right]_{p=0} \quad (5)$$

for parameters affecting either KE' or PE.

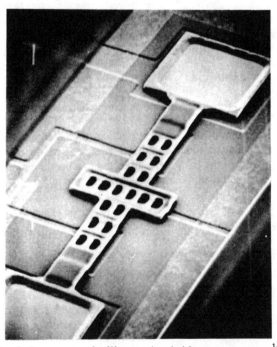

Fig. 2 Resonant polysilicon-microbridge vapor sensor[16]

The resonant-microbridge vapor sensor (Fig. 2) is sensitive to kinetic energy perturbations.[16] The 1.35 μm-thick polysilicon microbridge is driven electrostatically in its fundamental mode. A 150 nm-thick polymer film covers selected areas of the microbridge and absorbs chemically compatible organic vapors. The resulting mass increase perturbs the kinetic energy. According to (5), the sensitivity to vapor concentration c is

$$S = -\frac{1}{4KE_o'}\frac{d\rho_f}{dc}\int_0^L W_f h\, Y_1^2(x)\, dx, \quad (6)$$

where ρ_f is the density and W_f the width of the polymer film. This expression indicates that S can be increased by increasing the polymer thickness h relative to the polysilicon thickness, which contributes to KE_o'. In addition, the polymer film should be concentrated in the center of the bridge, where $Y_1(x)$ is maximum.

The axially loaded bridge (Fig. 1) is a resonant structure sensitive to potential energy perturbations. If the mode shape is not altered by the applied load F and there is no static bias load, then it follows from (5) that

$$S = \frac{1}{2}\frac{\frac{1}{2}\int_0^L\left[\frac{dY_1(x)}{dx}\right]^2 dx}{\frac{EI}{2}\int_0^L\left[\frac{d^2Y_1(x)}{dx^2}\right]^2 dx}. \quad (7)$$

The sensitivity of the axially loaded bridge is independent of applied load and proportional to the ratio of stretching to bending potential energies. This linear response to F has made axially loaded bridges popular in quartz resonant force and pressure sensors.[4,5]

Fig. 3 Silicon micromachined torsional resonator[15]

Figure 3 is an SEM of a silicon torsional resonator fabricated by anisotropic etching.[15] The bottoms of the pedestals supporting the resonator are embedded in a thin silicon diaphragm, formed by diffusing boron into the opposite side of the substrate. An unetched portion of the substrate, partially visible in Fig. 3, surrounds the pedestals and supports the thin diaphragm. This structure has been developed by Greenwood for use in an innovative resonant pressure sensor, shown in cross section in Fig. 4.[15] Electrodes on the glass top plate are used to electrostatically excite torsional modes in the structure, which is encapsulated in vacuum. The structure's resonant frequency is perturbed by external pressure, which stretches the silicon diaphragm and generates a tensile axial load on the torsional springs (Fig. 4).

Electromechanical Model

Micromechanical structures have been excited by means of thermal expansion[13], piezoelectric thin films[14], and capacitive coupling[15,16]. Detection of the resulting vibration has been accomplished using these effects, as well as by use of piezoresistors[13] and fiber optics.[14] In Nathanson's early work, deflection of a microcantilever modulated the current in an inversion layer.[19]

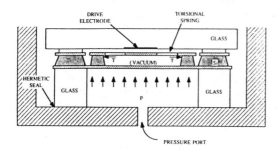

Fig. 4 Resonant pressure sensor (cross section)[15]

Of the various approaches, capacitive (electrostatic) excitation and detection are attractive for conducting microstructures. High electric fields are available at relatively low voltages, since micron or submicron electrode gaps are readily fabricated by surface micromachining.[11,12] Furthermore, electrostatic excitation does not require a composite resonant structure incorporating electrode layers or diffusions. For example, piezoelectric excitation requires the active film to be encapsulated in a sandwich of conducting and insulating films.[14]

This point has important implications for the control of residual stress. In constrained microstructures, such as bridges or diaphragms, residual stress generates a built-in axial load,[20] which shifts the resonant frequency ω_1 according to (4). If the residual stress relaxes over time, ω_1 will drift accordingly. For example, quartz resonant sensors have exhibited long-term calibration problems due to stress relaxation in the evaporated metal electrodes.[21] The advantage of capacitive drive and detection is that process optimization for controlling residual stress can focus on a single microstructural film.

The design of resonant structures with capacitive drive and detection involves complicated modeling of the electromechanical response.[16] It is therefore useful to derive a linearized equivalent circuit. The specific case of a microbridge excited in its fundamental mode (Fig. 5) and detected with a separate electrode is used for developing an equivalent two-port network.

The forward mechanical response $M_d(j\omega)$, defined as the average deflection over the drive electrode in response to the electrostatic force applied by the drive electrode, is the starting point for the derivation. For a uniform bridge, this transfer function is

$$M_d(j\omega) = \frac{\overline{Y}_d(j\omega)}{\overline{F}_d(j\omega)} = \frac{a_d^2 K^{-1}}{1 - (\omega/\omega_1)^2 + j(\omega/Q\,\omega_1)}, \quad (8)$$

where ω_1 is the first resonant frequency, Q is the quality factor, and $K = \omega_1^2 M$ is the effective spring constant (M is the bridge's mass.) The constant a_d enters twice: from the coupling between the distributed electrostatic force and the fundamental mode and from taking the average deflection over the drive electrode,[16]

$$a_d = 1.59\, l_d^{-1} [Y_1(L/2)]^{-1} \int_{drive} Y_1(x)\, dx, \quad (9)$$

where l_d is the length of the drive electrode (Fig. 5).

Rayleigh's Method can approximate ω_1 and K in (8). The Q of micromechanical structures is dominated by viscous damping at atmospheric pressure,[16,22] with typical values being less than 100. In vacuum, the structure exhibits its intrinsic quality factor Q_i, which is determined by internal losses. For crystalline silicon, Q_i ranges from about 10^4 for heavily doped etch-stop layers[15] to 10^5 for lightly doped material.[23] A high Q is desirable for isolating the mechanical resonance from dependence on the phase shift of the sustaining amplifier.[1]

The electrostatic force f_d on the microbridge can be found from the stored energy U_E in the drive capacitor:

$$f_d = \left[\frac{\partial U_E}{\partial \overline{y}_d} \right]_{v_d} \quad (10)$$

In phasor form, the component of f_d at the frequency of the drive voltage is

$$F_d(j\omega) = -\overline{C}_d d^{-1} V_P V_d(j\omega) + \overline{C}_d d^{-2} V_P^2 \overline{Y}_d(j\omega), \quad (11)$$

where it has been assumed that the deflection $\overline{y}_d$ is much less than the nominal gap width d. The transfer function $N_d(j\omega) = \overline{Y}_d(j\omega)/V_d(j\omega)$ can be found by substitution of (8) into (11).

Now that the deflection resulting from the drive voltage is known, all that is needed to find the drive admittance $Y_d(j\omega)$ is the current into the drive capacitor $C_d(t)$ with the sense port short-circuited ($V_s(j\omega) = 0$):

$$i_d(t) = \left[\frac{\partial q}{\partial v_d} \right]_{\overline{y}_d} \frac{\partial v_d}{\partial t} + \left[\frac{\partial q}{\partial \overline{y}_d} \right]_{v_d} \frac{\partial \overline{y}_d}{\partial t}. \quad (12)$$

By expanding the partial derivatives and assuming $\overline{y}_d \ll d$, the drive current can be expressed in phasor form:

$$I_d(j\omega) = j\omega \overline{C}_d V_d(j\omega) - j\omega \overline{C}_d V_P d^{-1} \overline{Y}_d(j\omega). \quad (13)$$

Finally, substitution of $N_d(j\omega)$ into (13) yields

$$Y_d(j\omega) = j\omega \overline{C}_d + \frac{j\omega g_d \overline{C}_d (1 - g_d)^{-1}}{1 - (\omega/\omega_1')^2 + j\omega/(Q'\omega_1')}. \quad (14)$$

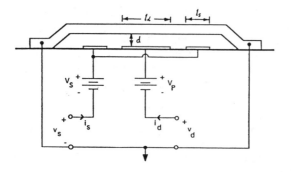

Fig. 5 Capacitively excited and sensed microbridge

The resonant frequency and quality factor have been perturbed by the *dc* bias voltage V_P on the drive electrode:

$$\omega_1' = \omega_1 (1-g_d)^{1/2}, \quad Q' = Q (1-g_d)^{1/2}, \quad (15)$$

where $g_d = a_d^2 \bar{C}_d V_P^2 K^{-1} d^{-2}$. The sign of g_d in (15) was incorrect in a previous analysis.[16]

The sense admittance $Y_s(j\omega)$ can be found using the same approach. The reverse mechanical response $M_s(j\omega)$ is identical to (8), except that the parameter a_d is replaced by

$$a_s = 1.59 \, l_s^{-1} [Y_1(L/2)]^{-1} \int_{sense} Y_1(x) \, dx, \quad (16)$$

where l_s is the length of half of the sense electrode, as shown in Fig. 5. For excitation of the fundamental mode, $a_s < a_d$ because sense electrode is located away from the center of the bridge where the maximum of $Y_1(x)$ occurs. The driving force f_s has the same form as (11), with the average sense capacitance $\bar{C}_s$ and the sense electrode bias voltage V_S replacing the $\bar{C}_d$ and V_P. The resulting expression for $Y_s(j\omega)$ differs from (14) only in that $\bar{C}_s$ replaces $\bar{C}_d$ and that the parameter $g_s = a_s^2 \bar{C}_s V_S^2 K^{-1} d^{-2}$ replaces g_d.

Evaluation of the forward transadmittance $Y_f(j\omega)$ (the ratio of short-circuit sense current $I_s(j\omega)$ to drive voltage $V_d(j\omega)$) is a straightforward extension of the above results. The average deflection over the sense electrodes $\bar{Y}_s(j\omega)$ is proportional to the sense current:

$$I_s(j\omega) = -j \omega \bar{C}_s V_S d^{-1} \bar{Y}_s(j\omega). \quad (17)$$

From (9) and (16), $\bar{Y}_s(j\omega)/\bar{Y}_d(j\omega)$ is equal to a_s/a_d. Substitution into (17) of $N_d(j\omega)$ then yields

$$Y_f(j\omega) = \frac{a_s V_S \bar{C}_s}{a_d V_P \bar{C}_d} Y_d'(j\omega) = \phi_d Y_d'(j\omega), \quad (18)$$

where $Y_d'(j\omega)$ is the second term in (14). A parallel analysis shows that the reverse transadmittance is

$$Y_r(j\omega) = \phi_s Y_s'(j\omega), \quad (19)$$

where $\phi_s = \phi_d^{-1}$ and $Y_s'(j\omega)$ is defined similarly to $Y_d'(j\omega)$.

Equivalent circuit elements can be extracted from (14), (18), and (19), leading to the linearized two-port model shown in Fig. 6. Current-controlled current generators couple the drive and sense ports by relating the motional currents i_1 and i_2 to generated currents in the other port.

Table I gives expressions for the equivalent circuit elements. By substitution of the definitions of g_d and g_s, the equivalent capacitors C_1 and C_2 are found to be proportional to the compliance K^{-1} of the bridge for small g_d and g_s. The equivalent inductors L_1 and L_2 are proportional to the bridge's mass M.

$C_1 = g_d \bar{C}_d (1-g_d)^{-1}$	$C_2 = g_s \bar{C}_s (1-g_s)^{-1}$
$L_1 = \omega_1^{-2} [g_d \bar{C}_d]^{-1}$	$L_2 = \omega_1^{-2} [g_s \bar{C}_s]^{-1}$
$R_1 = [g_d \omega_1 \bar{C}_d Q]^{-1}$	$R_2 = [g_s \omega_1 \bar{C}_s Q]^{-1}$

Table I. Equivalent circuit elements

The two-port circuit model in Fig. 6 provides insight into the electrostatic drive and detection of microstructures. For example, the sensitivity of ω_1' to drifts in the bias voltage V_P indicates that the stability of V_P is important for precision measurements. The model also clarifies the design constraints on the sustaining amplifier.

Dynamic Response

In the above sensitivity analysis, the perturbation to the potential or kinetic energy was considered to be time-independent. Since chemical and physical parameters vary with time, it is important to study the dynamic response of resonant sensors. In order to analyze a specific case, the vibrating microstructure is assumed to be capacitively driven and detected and modeled by the two-port equivalent circuit in Fig. 6. The sustaining amplifier is assumed to detect the *ac* sense current i_s by means of a shunt input, resulting in a simplification of the equivalent circuit. Figure 7 shows the circuit diagram for the electromechanical oscillator. A power-series approximation to the nonlinear transresistance $v_d(i_a)$ is used for purposes of analysis:

$$v_d = \alpha i_a + \beta i_a^2 + \gamma i_a^3, \quad (20)$$

where α, β, and γ are constants. This drive and detection scheme has the advantage that $V_d(j\omega_1')$ and $I_s(j\omega_1')$ are in phase, which simplifies the design of the sustaining amplifier.[16]

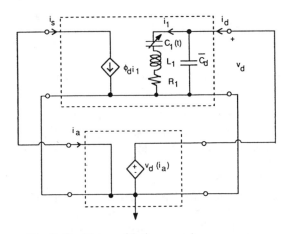

Fig. 7 Oscillator with time-varying parameter

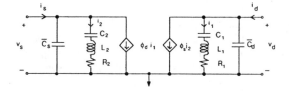

Fig. 6 Two-port equivalent circuit for Fig. 5

Sinusoidal perturbation of the compliance of the vibrating structure is represented in Fig. 7 by a time-varying equivalent capacitance:

$$C_1(t) = \bar{C}_1 + \Delta C_1 \cos(\omega_m t), \qquad (21)$$

where $\bar{C}_1$ is the average equivalent capacitance, ΔC_1 is the amplitude of the perturbation, and ω_m is the modulation frequency. From Table I, the resonant frequency is given by $\omega_1' = (L_1 \bar{C}_1)^{-1/2}$. The peak deviation $\Delta\omega_1'$ due to compliance perturbations is therefore

$$\frac{\Delta\omega_1'}{\omega_1'} = -\frac{1}{2}\frac{\Delta C_1}{\bar{C}_1}. \qquad (22)$$

The first step in analyzing the response is to write the differential equation for the current i_1, including the effect of the time-varying capacitor:

$$L_1 C_1(t)\frac{d^2 i_1}{dt^2} + [(R_1 + \alpha\phi_d)C_1 + L_1\frac{dC_1}{dt}$$

$$- 2\beta\phi_d^2 C_1 i_1 + 3\gamma\phi_d^3 C_1 i_1^2]\frac{di_1}{dt} + [(R + \alpha\phi_d)\frac{dC_1}{dt} + 1]i_1$$

$$+ [-\beta\phi_d^2 i_1^2 + \gamma\phi_d^3 i_1^3]\frac{dC_1}{dt} = 0. \qquad (23)$$

If the modulation is limited to a single set of sidebands, i_1 can be expressed generally by[24]

$$i_1(t) = I_1\cos(\omega_o t) \qquad (24)$$

$$+ \frac{A}{2}\cos[(\omega_o + \omega_m)t - \phi] + \frac{A'}{2}\cos[(\omega_o - \omega_m)t + \phi'],$$

where A, A', ϕ, and ϕ' are arbitrary amplitudes and phases. The steady-state amplitude I_1 and oscillation frequency ω_o are found by substituting the first term in (24) into (23) and requiring that the coefficient of the term at the carrier frequency vanish:[24]

$$I_1 = \left[\frac{4(R_1 + \alpha\phi_d)}{-3\gamma\phi_d^3}\right]^{1/2}, \quad \omega_o = (L_1\bar{C}_1)^{-1/2} = \omega_1'. \qquad (25)$$

From the amplitude expression, it follows that the linear constant α is negative with $|\alpha\phi_d| > R_1$ and that the cubic coefficient γ is positive. The latter represents the necessary rolloff in amplifier gain for large $|i_a|$.

Substitution of (24) into (23) and forcing the coefficients of the upper and lower sidebands to vanish results in $A = I_1(\Delta\omega_o/\omega_m)$, $A' = -A$, and $\phi = \phi'$, if $\omega_m, \Delta\omega_o \ll \omega_o$. Assuming further that the modulation index $\beta = \Delta\omega_o/\omega_m \ll 1$ (narrow-band modulation), then (23) can be reduced to

$$i_1(t) = I_1\cos[\omega_o t + \beta\sin(\omega_m t - \phi)], \qquad (26)$$

the standard form for a frequency-modulated (FM) carrier. A similar approach yields the same result for a time-varying inductor $L_1(t)$, which models a kinetic energy (mass) perturbation.

The expression for $i_1(t)$ in (26) is more general than suggested by the narrowband assumption. In many practical applications, the peak deviation $\Delta\omega_o$ is greater than the modulation frequency ω_m, resulting in $\beta > 1$ and wideband FM. By leaving the modulation index indeterminate

in the FM expression for $i_1(t)$ in (26), the differential equation (23) can be used to find an expression for β. Forcing the coefficients of the carrier terms to zero in the resulting equation, and keeping only terms of first order in ω_m/ω_o and $\Delta\omega_o/\omega_o$ (both of which are assumed small), the modulation index is found to be

$$\beta = \frac{\Delta\omega_o}{\omega_m}, \qquad (27)$$

which reproduces (23) without the narrowband assumption. Modulation of the equivalent inductor yields the same result for β.

In summary, resonant sensors represented by Fig. 7 respond to time-varying potential or kinetic energy perturbations by generating a frequency-modulated current $i_1(t)$ and drive voltage $v_d(t)$. Demodulation of these signals by differentiation using standard techniques will recover the perturbation without attenuation, in contrast to oscillators with perturbations in the equivalent resistance. In the latter case, the sidebands are identified with amplitude modulation, and are attenuated at high modulation frequencies.[24]

Conclusion

The theory developed in this paper supplies a framework for developing microfabricated resonant sensors. Careful consideration of the materials and structural constraints on resonant microsensors is essential to developing practical devices. Thin films are typically under residual stress,[11,20] which shifts the resonant frequency in constrained structures. Residual stress must be accounted for in sensor calibration, and may limit the ultimate performance specifications. The dimensional scale of resonant microstructures can also affect sensor design. For example, the low quality factors due to viscous damping at atmospheric pressure complicates sustaining amplifier design for resonant gas and vapor microsensors. Furthermore, the essentially two-dimensional planar process poses challenges for coupling mechanical loads to resonating microstructures. The innovative pressure sensor shown in Fig. 3 indicates that two-sided processing can be advantageous for this purpose.

A serious obstacle to progress is the lack of thorough studies on the process dependence of the mechanical properties of microstructural materials. In addition to the basic properties such as Young's modulus, Poisson's ratio, and residual stress, dynamic mechanical properties such as internal damping, stress relaxation, and fatigue are important for resonant microsensor design.

Acknowledgements

The author is grateful to J. C. Greenwood of STC Technology, Ltd., for providing the torsional resonator SEM. Discussions with M. A. Schmidt on resonant sensors have been valuable. L. Y. Pang and J. E. Goldsberry contributed to the modeling and dynamic response analysis. S. D. Senturia was helpful in suggesting the connection to his work on marginal oscillators. This work was supported by an IBM Faculty Development Award and by an NSF Presidential Young Investigator Award.

References

1. M. A. Schmidt and R. T. Howe, "Resonant Structures for Integrated Sensors," *Technical Digest,* IEEE Solid-State Sensor Workshop, Hilton Head, SC, June 2-5, 1986.

2. R. M. Langdon, "Resonator Sensors -- a Review," *J. Phys. E., Sci. Instrum.,* vol. 18, 1985, pp. 103-115.

3. T. Gast, "Sensors with Oscillating Elements," *J. Phys. E., Sci. Instrum.,* vol. 18, 1985, pp. 783-789.

4. W. C. Albert, "Vibrating Quartz Crystal Beam Accelerometer," *Proceedings,* 28th ISA International Instrumentation Symposium, 1982, pp. 33-44.

5. E. P. EerNisse, and J. M. Paros, "Practical Considerations for Miniature Quartz Resonator Force Transducers," *Proceedings,* 37th Annual Symposium on Frequency Control, 1983, pp. 255-260.

6. E. Karrer and R. Ward, "A Low-Range Quartz Resonator Pressure Transducer," *ISA Trans.,* vol. ·16, 1977, pp. 90-98.

7. J. Hlavay and G. G. Guilbault, "Application of the Piezoelectric Crystal Detector in Analytical Chemistry," *Analytical Chemistry,* vol. 49, 1977, 1890-1898.

8. H. Wohltjen, "Mechanism of Operation and Design Considerations for Surface Acoustic Wave Device Vapour Sensors," *Sensors and Actuators,* vol. 5, 1984, pp. 307-325.

9. S. J. Martin, K. S. Schweizer, A. J. Ricco, and T. E. Zipperian, "Gas Sensing with Surface Acoustic Wave Devices," *Technical Digest,* 3rd International Conference on Solid-State Sensors and Actuators, 1985, 71-73.

10. K. E. Petersen, "Silicon as a Mechanical Material," *Proc. IEEE,* vol. 70, 1982, pp. 420-457.

11. R. T. Howe, "Polycrystalline Silicon Microstructures," in *Micromachining and Micropackaging of Transducers,* C. D. Fung, P. W. Cheung, W. H. Ko, and D. G. Fleming, eds., Amsterdam, Elsevier, 1985, pp. 169-187.

12. H. Guckel, "Fabrication Techniques for Integrated Sensor Microstructures," *Technical Digest,* IEEE International Electron Devices Meeting, 1986, pp. 176-179.

13. T. S. J. Lammerink and W. Wlodarski, "Integrated Thermally Excited Resonant Diaphragm Pressure Sensor," *Technical Digest,* 3rd International Conference on Solid-State Sensors and Actuators, 1985, pp. 97-100.

14. J. G. Smits, H. A. C. Tilmans, and T. S. J. Lammerink, "Pressure Dependence of Resonant Diaphragm Pressure Sensors," *Proceedings,* 3rd International Conference on Solid-State Sensors and Actuators, 1985, pp. 93-96.

15. J. C. Greenwood, "Etched Silicon Vibrating Sensor;" *J. Phys. E., Sci. Inst.,* vol. 17, 1984, 650-652.

16. R. T. Howe and R. S. Muller, "Resonant-Microbridge Vapor Sensor," *IEEE Trans. on Electron Devices,* vol. ED-33, 1986, pp. 499-506.

17. S. Timoshenko and D. H. Young, *Vibration Problems in Engineering,* 3rd ed., D. Van Nostrand, New York, 1955.

18. C. M. Harris and C. E. Crede, *Shock and Vibration Handbook,* 2nd ed., McGraw-Hill, New York, 1976.

19. H. C. Nathanson, *et al,* "The Resonant Gate Transistor," *IEEE Trans. on Electron Devices,* vol. ED-14, 1967, pp. 117-133.

20. H. Guckel, T. Randazzo, and D. W. Burns, "A simple technique for the determination of mechanical strain in thin films with application to polysilicon," *J. Appl. Phys.,* vol. 57, 1985, pp. 1671-1675.

21. E. P. EerNisse, "Quartz resonator frequency shifts arising from electrode stress," *Proceedings,* 29th Annual Symp. on Frequency Control, 1976, pp. 1-4.

22. W. E. Newell, "Miniaturization of tuning forks," *Science,* vol. 161, 1968, pp. 1320-1326.

23. G. Kaminsky, "Micromachining of silicon mechanical structures," *J. Vac. Sci. Technol. B,* vol. 3, 1985, pp. 1015-1024.

24. M. S. Adler, S. D. Senturia, and C. R. Hewes, "Sensitivity of Marginal Oscillator Spectrometers," *Rev. Sci. Instrum.,* vol. 42, 1971, pp. 704-712.

Silicon microengineering for accelerometers

D W SATCHELL, BSc, CEng, FInstP, MIEE and **J C GREENWOOD**, BSc, MInstP
STC Technology Limited, Harlow, Essex

SYNOPSIS Silicon microengineering enables the excellent mechanical properties of silicon to be combined with electronic ones to produce accelerometers of good performance, small size and low cost. The design and fabrication of two types of analogue accelerometer, using this technique, are described. One employs implanted strain gauges to give a dc output, while the other has a strain-sensitive resonant structure which gives a varying frequency signal.

1. INTRODUCTION

All open-loop accelerometers consist of a seismic mass restrained by some kind of spring, and provided with a transducer to give an output related to the deflection of the spring. Closed loop or force-feedback accelerometers have the spring replaced by a servo loop, but an elastic suspension of good performance is still required. Both types require damping of the natural resonances, overload protection and cross-axis signal rejection. Only devices with analogue output will be described here; of these, the signal may take the form of amplitude variations of a dc current or voltage, or that of frequency modulation of an ac carrier.

A good accelerometer must, above all, show repeatability of both zero offset (bias) and sensitivity (span). This is because, while non-linearities can often be compensated by subsequent processing, zero shifts and hysteresis are difficult or impossible to compensate. Linearity is important, although a simple and well-defined power law response is often acceptable. Small size, light weight and low cost are, of course, of great practical significance.

Silicon microengineering provides the means to achieve all these aims. It exploits the excellent mechanical properties of silicon at the same time as the well-known electronic ones, and permits manufacture by batch processing methods to reduce labour cost. Very small and complex 3-dimensional structures, which could hardly be made by any other means, are possible. The seismic mass, spring suspension, electromechanical transducer, and sometimes the overload restraint can all be formed as integral parts of a single structure having dimensions scaled in micrometres.

2. PROPERTIES OF SILICON

Single-crystal silicon is very well known in electronic applications, but much less so as a mechanical engineering material. It has, however, some properties which make it an excellent choice for the mechanical parts of many types of transducer. The stress-strain relationship is linear up to very high strain levels, plastic deformation being essentially absent at moderate temperatures. Zero drift and hysteresis are therefore very small, and calibration tends to be unaffected by overload. The intrinsic strength is very high, $2 \times 10^9 \text{N/m}^2$, and the density is low, $2.3 \times 10^3 \text{ kg/m}^3$, therefore the strength weight ratio is high. Thermal conductivity is high, 157 W/m deg.C.

3. PRACTICAL ACCELEROMETERS

Figure 1 shows a dc analogue open-loop accelerometer (overall dimensions 15 mm x 12 mm) made by microengineering. Anisotropic etching, combined with more orthodox silicon processing, has been used to shape the structure from a conventional wafer of (100) silicon. A central block forms the seismic mass, restrained on two sides by thin cantilever beams supported by an outer frame. An ion-implanted piezoresistive strain-gauge bridge in the opposite sides of the beams provides an output when the mass is displaced normal to the wafer surface. Range and linearity are improved by the provision of slots close to the ends of the beams. These relieve the beams of longitudinal tension and permit them to function in a clamped-free mode. The basic element is bonded between two further silicon plates, which limit overload displacement and provide air damping. A cross-section of this composite structure is shown in Figure 2.

Figure 3 shows such an accelerometer mounted in a conventional DIL package, with associated buffer and temperature compensation circuits. Structures of this kind provide a low profile, low cost accelerometer for medium precision applications such as short-term guidance. More sophisticated packages, with greater rigidity and better reference surfaces, can be provided to exploit the basic performance in more demanding applications.

Reprinted by permission of the Council of the Institution of Mechanical Engineers from *Rec. of the Int. Conf. on the Mech. Technol. of Inertial Devices*, 1987, pp. 191–193. On behalf of the Institution of Mechanical Engineers (London).

Accelerometers of this type may typically have a performance as follows:

Range: 10 g

Overload: 200 g

Linearity:
0.5% error almost entirely second order. (Provision of a simple UAA correction chip reduces non-linearity by two orders of magnitude).

Temperature range:
-40°C to +120°C (uncompensated). Any 100 degrees within this range compensated to 2% by simple resistors: (much better with UAA compensation).

Zero stability and hysteresis:
Dependent on package, can be extremely good.

An accelerometer element giving a varying frequency output is shown in Figure 4. This device employs a resonating element in the form of a strain-sensitive double-ended tuning fork as both the restraining spring and the transducer. The fork is maintained in oscillation by thermal expansion effects, an embedded heater being driven at the resonant frequency (about 100 kHz) by an amplifier whose input is derived from a strain-gauge embedded in the structure. This device is substantially smaller than the one previously described, the overall size of the chip being 6 x 7 mm, and the length of the fork 2 mm. The seismic mass moves about a flexure hinge formed by thinning the silicon in the one etching process, the mechanical properties of silicon being almost ideal for this purpose. Strain in the fork generated by inertial forces changes the resonant frequency by about 2% at full scale, giving a sensitivity in the region of 100 Hz per g.

4. FABRICATION TECHNIQUE

The enabling technology behind all these devices is the anisotropic etching of single-crystal silicon. This is a wet etching process characterised by etch rates which differ widely along the different crystal axes. Knowledge of this behaviour and of the crystallography of the material, together with conventional photolithography and masking, permits the formation of these complex structures. When (100) silicon wafers are the starting material, as with the structures described here, etching produces deep pits with flat bottoms, rectangular planforms and planar sides inclined at 35.26 deg. to the wafer surface. These characteristics are determined by the crystal structure of the material, and are therefore independent of the precision of mask-making or processing. Undercutting is very small and is predictable. These features permit the fabrication of very small structures with high precision and repeatability. In addition, selected areas of the silicon may be rendered insoluble by doping or by the provision of biased p-n junctions. This facility adds greatly to the complexity which can be achieved.

In this paper only accelerometers have been discussed; however, silicon microengineering is a very flexible technology which can be applied to a wide range of other electromechanical devices.

ACKNOWLEDGEMENT

This work has been carried out with the support of the Procurement Executive, Ministry of Defence, sponsored by DCVD and RAE, and also STC Power & Microdevices Division (Hybrids Unit). The major contributions made by P.N. Egginton, C. Evestaff and R.C. Stern of STL, and K. Buchanan of STC, are gratefully acknowledged.

Fig 1   d.c. analogue open-loop accelerometer chip

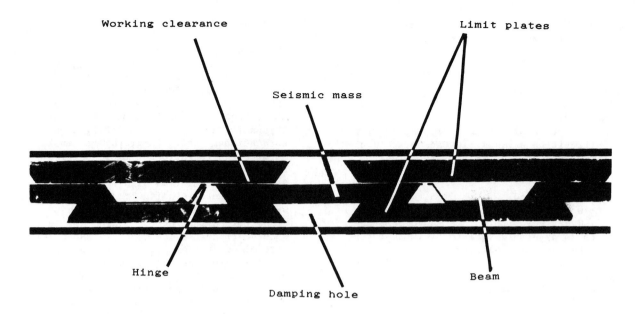

Fig 2   Cross-section of d.c. accelerometer chip

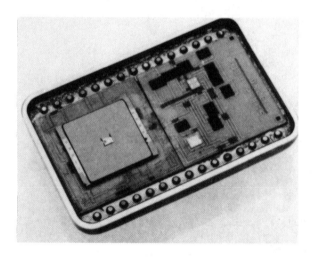

Fig 3   d.c. analogue accelerometer in DIL package

Fig 4   Resonant fork accelerometer chip

Integrated Semiconductor Magnetic Field Sensors

HENRY P. BALTES, MEMBER, IEEE, AND RADIVOJE S. POPOVIĆ, MEMBER, IEEE

A magnetic field sensor is an entrance transducer that converts a magnetic field into an electronic signal. Semiconductor magnetic field sensors exploit the galvanomagnetic effects due to the Lorentz force on charge carriers. Integrated semiconductor, notably silicon, magnetic field sensors, are manufactured using integrated circuit technologies. Integrated sensors are being increasingly developed for a variety of applications in view of the advantage offered by the integration of the magnetic field sensitive element together with support and signal processing circuitry on the same semiconductor chip. The ultimate goal is to develop a broad range of inexpensive batch-fabricated high-performance sensors interfaced with the rapidly proliferating microprocessor. This review aims at the recent progress in integrated silicon magnetic devices such as integrated Hall plates, magnetic field-effect transistors, vertical and lateral bipolar magnetotransistors, magnetodiodes, and current-domain magnetometers. The current development of integrated magnetic field sensors based on III–V semiconductors is described as well. Bulk Hall-effect devices are also reviewed and serve to define terms of performance reference. Magnetic device modeling and the incorporation of magnetic devices into an integrated circuit offering in situ amplification and compensation of offset and temperature effects are further topics of this paper. Silicon will continue to be aggressively exploited in a variety of magnetic (and other) sensor applications, complementary to its traditional role as integrated circuit material.

I. INTRODUCTION

Integrated solid-state sensors are currently in the limelight of scientific and engineering research [1]–[6]. Silicon has already revolutionized the way we think about electronics and is now in the process of altering conventional perceptions on sensors and transducers. This holds in particular in the case of magnetic field sensors: integrated silicon magnetic field sensors [1], [5], [6]–[9] can now be manufactured using standard integrated circuit (IC) processing technologies without invoking additional processing steps such as "micromachining" [4] or film deposition as in the case of most mechanical or chemical sensors.

Manuscript received July 12, 1985; revised May 16, 1986. The submission of this paper was encouraged after review of an advance proposal. This work was supported by the Natural Sciences and Engineering Research Council of Canada (NSERC).
H. P. Baltes is with the Department of Electrical Engineering, The University of Alberta, Edmonton, Alta., Canada T6G 2G7.
R. S. Popović is with Landis & Gyr Corporation, CH-6301 Zug, Switzerland.

MFS Applications

A magnetic field sensor (MFS) is an input transducer that is capable of converting the magnetic field H into a useful electronic signal. An MFS is also needed whenever a nonmagnetic signal is detected by means of an intermediary conversion into H ("tandem" transduction), e.g., the detection of a current through its magnetic field or the mechanical displacement of a magnet. Thus we can distinguish two main groups of MFS applications [8]: in direct applications, the MFS is part of a magnetometer [10]. Examples are earth magnetic field measurements, the reading of magnetic tapes and disks [11], the recognition of magnetic ink patterns of banknotes [12] and credit cards, or the control of magnetic apparatus. With respect to high-density magnetic recording, some of the recently devised integrated silicon MFSs are now able to compete with the traditional NiFe thin-film magnetoresistor devices [11]. In indirect applications, the magnetic field is used as an intermediary carrier for detecting nonmagnetic signals. Indirect applications include contactless switching [13], [14], linear and angular displacement detection (e.g., in automotive systems [1], [15]), potential-free current detection, and integrated wattmeters [14], [16].

The above applications require the detection of magnetic fields in the micro- and millitesla range, which can be achieved by integrated semiconductor sensors. Contactless switching (collectorless dc motor control, keyboards), displacement detection (proximity switch, crankshaft position sensor), and potential-free current detection seem to comprise the most important large-scale applications of MFS. It is for these large-scale applications that inexpensive, batch-fabricated integrated semiconductor sensors are highly desirable. It is unlikely that integrated silicon MFS will ever replace nuclear magnetic resonance (NMR) magnetometry (nanotesla) let alone the superconducting quantum interference devices (SQUID) required to resolve the picotesla signals occurring in biomagnetometry [17], [18].

MFS Specifications

Each MFS application comes with its specific sensor requirements, such as the required sensitivity and resolution:

Reprinted from *Proc. IEEE*, vol. 74, no. 8, pp. 1107–1132, August 1986.

switching and displacement detection applications involve permanent magnets with fields of about 5–100 mT, whereas 10 μT to 10 mT is the range of stray fields of magnetic domains in recording media. A conductor carrying a current of 1 A produces a magnetic field of the order of 100 μT at the surface of the conductor. The selection of the appropriate MFS technology depends on a number of specifications that may vary widely from one application to another. For instance, spatial resolution is crucial for high-density binary magnetic recording while linearity is not. A list of selection criteria for MFS design comprises the following items:

- availability of technology
- manufacturing cost
- environment to which sensor is exposed:
  - temperature
  - humidity and chemical stress
  - mechanical stress, vibrations
- sensor geometry (e.g., H parallel or perpendicular to chip surface)
- sensitivity, output signal level
- signal/noise, magnetic field resolution
- spatial resolution
- time resolution, frequency response
- linearity
- temperature coefficient of sensitivity
- offset, temperature dependence of offset
- power consumption, size, weight
- electrical input and output impedance
- stability, reliability, lifetime.

Most of the above concerns are shared by many other sensors as well as by analog integrated circuits. Once a specific MFS has been designed and manufactured, its performance with respect to the crucial specifications must be checked by appropriate measurements.

As an outstanding example, we mention a recent GaAs Hall device [19] fabricated by ion implantation and intended for magnetic flux meters operating between room temperature and 4 K. For this purpose, high-linearity, low-temperature coefficient, and small, temperature-independent offset are crucial requirements. Linearity and offset were checked by comparison with Hall devices calibrated by NMR magnetometry. The specifications were achieved by tight control of the device manufacturing technology and by exploiting the good temperature stability of GaAs. For other applications, it may be advantageous to adopt a less expensive alternative strategy, viz. silicon integrated circuit technology and compensation of temperature dependence, nonlinearity, and offset by appropriate integrated circuitry.

MFS Technologies

Most MFSs exploit the Lorentz force $F = qv \times B$ on electrons in a metal, a semiconductor, or an insulator in one way or another, where q denotes the electron charge, v the electron velocity, and B the magnetic induction. In view of the relation $B = \mu\mu_0 H$ with $\mu\mu_0$ denoting the magnetic permeability of the sensor material we can readily distinguish two major classes of MFS:

i) MFSs involving *high-permeability* (ferro- or ferrimag-

netic) materials, where $\mu \gg 1$ brings about a corresponding enhancement of sensitivity. Examples are MFS based on the magnetoresistance of NiFe thin films, the magnetostriction of the nickel cladding of an optical fiber, or the magneto-optic effects in garnets, and any MFS combined with a flux-concentrating device.

ii) MFSs using *low-permeability* (dia- or paramagnetic) materials, where $\mu \sim 1$ does not provide any appreciable "leverage." For example, all MFS based on galvanomagnetic effects in semiconductors (Si, GaAs, InSb) belong to this class.

Important high-resolution MFSs not fitting in the above groups are the NMR and SQUID magnetometers, which are beyond the scope of this paper [17].

Optoelectronic MFSs use light as an intermediary signal carrier. *Magnetooptic MFS* are based on the Faraday rotation of the polarization plane of linearly polarized light due to the Lorentz force on bound electrons. Useful MFSs can be realized by using optical-fiber coils providing a long light path and an accordingly large rotation per unit magnetic field. Magnetooptic current sensors for high-voltage transmission lines have been realized in this way [20], [21]. A much larger Faraday rotation angle per unit path length is obtained in optically transparent ferrimagnetic garnet materials [22]. The most promising optoelectronic MFSs are those using optical fibers with *magnetostrictive* jacketing material such as nickel or an optical fiber wound under tension onto a cylinder of magnetostrictive material. The strain transferred to the fiber from the magnetostrictive material results in a change of optical path length leading to a phase shift detected with an optical-fiber interferometer [23]. Minimum detectable fields of less than 1 nT have been reported [24].

Thin-metal film or wire MFSs are based on ferromagnetic materials. The low-magnetostriction alloy $Ni_{81}Fe_{19}$ is preferably used for thin-film MFSs. The most successful sensor effect is the *magnetoresistive* (MR) *switching* of anisotropic NiFe or NiCo films [7], [11], [25]–[27]. High spatial resolution applications require the reduction of the MR element down to micrometer dimensions, where Barkhausen noise may lead to a degradation of the MR switching characteristics. The problem can be overcome by using sandwich structures of two NiFe layers with proper orientation of easy axis separated by a silicon monoxide or conductive layer [11], [28]. This type of structure reduces demagnetizing fields and Barkhausen noise.

Semiconductor MFSs exploit the galvanomagnetic effects such as Hall voltage, carrier deflection, magnetoresistance, and magnetoconcentration, which are due to the action of the Lorentz force on the charge carriers (electrons and holes) [5], [7]–[10], [14], [29], [30]. In n-type semiconductor material, for instance, the magnitude of the sensor effects in question is controlled by the product of the electron mobility and the magnetic induction. Hence high electron mobility (given below in square meters per volt-second ($m^2/V \cdot s$) at room temperature) is crucial for achieving high sensitivity. Thus InSb (8 in bulk, 6.5 in film material) and InAs (3) seem to be superior to GaAs (0.5), let alone Si (0.14 in bulk material, 0.05 to 0.09 in the conducting channel of an n-type MOSFET). Generally, n-type semiconductor material is superior to p-type material because of the much lower hole mobility (0.05 for bulk Si, 0.04 for GaAs, 0.14 for

InSb) [7]. As pointed out by Zieren [8], mobility is not the only important semiconductor MFS parameter. The small bandgap of InSb (0.2 eV) and InAs (0.4 eV) is a drawback because of the intrinsic behavior prevailing at room temperature, which excludes other than magnetoresistor applications. In this respect, Si (1.12 eV) and GaAs (1.42 eV) are outstanding materials. Si can be used up to about 150°C and GaAs even up to about 250°C. Another important figure of merit for Hall devices is the Hall voltage in terms of the dissipated power, in terms of which Si ranks equal to InSb and InAs [8].

Si and (in the future) also GaAs offer the unique advantage of inexpensive batch fabrication by integrating one or several basic sensor elements together with appropriate support and signal processing circuitry in an advanced standard technology of established reliability, such as bipolar or CMOS technology. Indeed, a large number of integrated silicon MFSs have been realized recently which are designed following the design rules of a standard chip manufacturing process as offered by custom chip manufacturers and many university laboratories. Such integrated silicon MFSs are the main thrust of this review. MFS developments outside established mainstream IC technologies have to face the extra cost of developing specific manufacturing technologies and tools for mass production as well as appropriate test and reliability procedures. This development cost is usually beyond the financial possibilities of small and medium-size companies. This investment can be justified when the applicability of an integrated Si MFS can be clearly ruled out as, e.g., in the case of high operating temperature or very-high resolution requirements. Integrated GaAs MFSs seem to be the proper next choice if operating temperatures above 150°C are required. An example is the GaAs Hall IC chip developed by Siemens [5], [30]. A nonstandard MFS chip combining Si IC and MR thin-film technology ($Ni_{76} Co_{24}$) has been realized by Toyota for digital application [31].

Galvanomagnetic Effects in Semiconductors

The action of the Lorentz force manifests itself in the carrier-transport equations [8], [32]–[35]. We assume isotropic n-type material with zero temperature gradient. Let us denote the electron current density for $B = 0$ by $J_n(0)$. The diffusion approximation of the Boltzmann transport equation leads to [36], [37]

$$J_n(0) = \sigma_n E + q D_n \nabla n \qquad (1)$$

where $\sigma_n = q\mu_n n$ denotes the electronic conductivity for $B = 0$, E the electrical field, $q = 1.6 \times 10^{-19}$ A · s the magnitude of the electron charge, $D_n = \mu_n kT/Q$ the electron diffusion constant, n the electron density, and μ_n the electron drift mobility. In (1) the term $\sigma_n E$ describes the drift current and $q D_n \nabla n$ the diffusion current. For nonzero magnetic induction B, the electron current density $J_n(B)$ obeys the equation

$$J_n(B) = J_n(0) - \mu_n^*(J_n(B) \times B) \qquad (2)$$

where μ_n^* is the Hall mobility for electrons. The Hall mobility is proportional to the drift mobility μ_n, $\mu_n^* = r_n\mu_n$, with the scattering factor $r_n = \langle \tau_n^2 \rangle / \langle \tau_n \rangle^2$, where τ_n denotes the free time between collisions. The value of r_n is determined by the energy-band structure and the underlying scattering

processes. For n-type silicon, r_n is about 1.15 at room temperature for low donor concentrations [38], [39]. Equation (2) can be solved with respect to $J_n(B)$, viz.

$$J_n(B) = \Big[J_n(0) + \mu_n^*(B \times J_n(0)) + (\mu_n^*)^2 (B \cdot J_n(0)) B \Big] \Big[1 + (\mu_n^* B)^2 \Big]^{-1}. \qquad (3)$$

This equation comprises the *isothermal* galvanomagnetic effects for electrons. It accounts for the direct effects of temperature on carrier concentration, diffusion, and mobility, but does not include thermomagnetic or thermoelectric effects. An analogous equation holds for the hole current density. In general, the electron and hole current equations have to be solved together with the pertinent continuity equations for electrons and holes as well as Poisson's equation. In specific configurations characterized by the device geometry, doping, and boundary and operating conditions, the one or the other galvanomagnetic effect may prevail.

Equation (3) includes the action of the Lorentz force on both carrier drift (terms containing E) and diffusion (terms containing ∇n). Diffusion is important in the case of magnetoconcentration or space-charge effects occurring, e.g., in magnetodiodes (injection of both electrons and holes). If carrier concentration gradients can be neglected, as, e.g., in n-type slabs with ohmic contacts, (3) becomes

$$J_n(B) = \sigma_{nB} \Big[E + \mu_n^*(B \times E) + \mu_n^{*2}(B \cdot E) B \Big] \qquad (4)$$

where $\sigma_{nB} = \sigma_n [1 + (\mu_n^* B)^2]^{-1}$. If B is parallel to E, $B \times E = 0$ leads to $J_n(B) = \sigma_n E = J_n(0)$: no longitudinal galvanomagnetic effect is observed in isotropic semiconductors. For B perpendicular to E, $B \cdot E = 0$, we obtain

$$J_n(B) = \sigma_{nB} \Big[E + \mu_n^*(B \times E) \Big]. \qquad (5)$$

This equation describes the transverse galvanomagnetic effects in the case of negligible diffusion. In terms of $B = (0, 0, B)$, $E = (E_x, E_y, 0)$, and $J_n(B) = (J_{nx}, J_{ny}, 0)$, (5) reads

$$\left. \begin{array}{l} J_{nx} = \sigma_{nB}\big(E_x - \mu_n^* B E_y \big) \\ J_{ny} = \sigma_{nB}\big(E_y + \mu_n^* B E_x \big). \end{array} \right\} \qquad (6)$$

Two limiting cases are usually distinguished:

i) *Hall field:* It is assumed that the current density has only an x-component, i.e., $J_{ny} = 0$. This can be achieved approximately in a long, thin rod sample geometry with current electrodes at the small faces. Then the Hall field

$$E_y = -\mu_n^* B E_x = R_H J_{nx} B \qquad (7)$$

appears, where

$$R_H = -\mu_n^*/\sigma_n = -r_n/qn \qquad (8)$$

denotes the Hall coefficient. This results in a rotation of the equipotential lines by the Hall angle θ_H with

$$\tan \theta_H = E_y/E_x = -\mu_n^* B = \sigma_n R_H B. \qquad (9)$$

For a long Hall plate of thickness t carrying a current I, the Hall field produces the Hall voltage $V_H = R_H IB/t$. The corresponding Hall sensor has the sensitivity $V_H/IB = R_H/t = r_n/qnt$. Thus high sensitivity requires small carrier concentration n. This explains why semiconductors are more useful here than metals.

ii) *Carrier deflection and magnetoresistance:* Now zero Hall field, $E_y = 0$, is assumed. This condition can be realized approximately by a short sample of wide cross section with current electrodes at the large faces. The carrier deflection resulting from (6) is given by the ratio

$$-J_{ny}/J_{nx} = \mu_n^* B = \tan \theta_H. \tag{10}$$

The longer carrier drift paths lead to the geometrical magnetoresistance effect, viz.

$$(\rho_{nB} - \rho_n)/\rho_n = (\mu_n^* B)^2 \tag{11}$$

where $\rho_n = \sigma_n^{-1}$ is the resistivity for $B = 0$ and $\rho_{nB} = E_x/J_{nx} = \sigma_{nB}^{-1}$ the resistivity enhanced by the magnetic induction. This effect is very small for silicon, e.g., $\rho_{nB} \approx 1.02 \, \rho_n$ for B as dense as 1 T. Sensors based on this effect require InSb or InAs [12], [14], [40]. For bulk InSb, $\rho_{nB} \approx 1.02 \, \rho_n$ at $B = 18$ mT.

Equations (2)–(11) correspond to the leading terms of a weak field expansion and involve relative errors of $(\mu_n^* B)^2$ [8], [32]. They provide a good approximation $((\mu_n^* B)^2 \lesssim 0.1)$ for n-type Si below $B = 2$ T, n-type GaAs below $B = 600$ mT, and n-type InSb below $B = 40$ mT.

For p-type semiconductor material characterized by drift and Hall mobilities μ_p and $\mu_p^* = r_p \mu_p$, the Hall coefficient is $R_H = r_p/qp$, with p denoting the hole concentration. For low doping $r_p \approx 0.7$ is measured at room temperature [39]. The Hall angle is given by $\tan \theta_H = -\mu_p^* B$. The magnetoresistance effect is $\rho_{pB}/\rho_p = 1 + (\mu_p^* B)^2$, where $\rho_p = 1/q\mu_p p$. For mixed n- and p-type conduction, the Hall coefficient has the general form

$$R_H = -\left[r_n (\mu_n/\mu_p)^2 n - r_p p \right] \Big/ q \left[(\mu_n/\mu_p) n + p \right]^2. \tag{12}$$

Scope of Review

In the following section, bulk and integrated Hall devices are presented. These are by far the best explored MFSs and offer terms of performance reference for the integrated MFSs described in Sections III–VI. Magnetic field sensitive field-effect transistors (MAGFET) are the topic of Section III. This section includes Hall-type and the split-drain MAGFET devices. In Section IV, we present a variety of bipolar magnetotransistors (MT) of the vertical (VMT) and lateral (LMT) type. Next (in Section V), we discuss integrated magnetodiodes (MD), which can be realized in silicon-on-sapphire (SOS) or complementary metal–oxide–semiconductor (CMOS) technology. Horizontal and vertical current domain magnetometers (CDM) are presented in Section VI. They provide an electronic signal whose frequency is proportional to the magnetic induction B. In view of the limited performance of the MFS elements, their incorporation into appropriate integrated circuits is indispensable. This topic is addressed in Section VII. The full understanding of the MFS operating principles and the optimization of MFS device geometries can hardly be achieved without analytical, numerical, or equivalent circuit device modeling as outlined in Section VIII. Finally, in Section IX, we try to assess the existing integrated MFSs and to summarize present and future trends.

Review-type literature on integrated semiconductor MFSs seems to be scarce. Some of the older work is summarized in [7], [14], [29]. A recent condensed overview can be found in [9]. A multiauthor book on applications of the Hall effect has been published in 1980 [41]. Potential calculations in Hall plates are reviewed in [34]. Many references are found in the recent Ph.D. dissertations [8], [40] and bibliographies [42], [43]. A comparison of some integrated MFSs with other types of MFS can be found in the brief overview [17].

II. Bulk Hall-Effect Devices

The simplest, oldest, and best understood semiconductor magnetic field sensors are the Hall plates. They have been commercially available for more than 30 years, since the mid-1950s as discrete devices and since the early 1970s as monolithic silicon integrated circuits. Strong practical interest has motivated a lot of work towards better understanding and optimizing of these devices. This has led to an abundant literature on the subject, see, e.g., the books [44]–[46], review-type papers [47], [48], and references therein. Thus the Hall-type devices have reached a high degree of maturity. We include this section not only because of their practical importance, but also in order to define the standards by which other magnetic field sensors can be assessed.

We start with a brief review of the main properties of ideal Hall plates (i.e., geometrical symmetry, material homogeneity, absence of parasitic effects, etc.). Next we describe several approaches to integrate Hall devices into monolithic integrated circuits and some related new effects. Finally, we analyze various figures of merit of Hall-effect devices.

Ideal Hall Plates

The Hall plate is a magnetic field sensor exploiting the Hall effect practically in the same way as Hall has discovered it: it is a thin, usually rectangular plate of a relatively high resistivity semiconductor material, provided with four ohmic contacts (see Fig. 1). The *Hall voltage* V_H, ap-

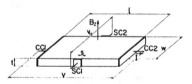

Fig. 1. Rectangular Hall plate. The Hall voltage $V_H \simeq R_H B_z I/t$ appears between the sensor contacts $SC1$ and $SC2$. R_H denotes the Hall coefficient, t the plate thickness, B_z the perpendicular component of the magnetic induction B, I the bias current supplied through the current contacts $CC1$ and $CC2$.

pearing between the sensor contacts $SC1$ and $SC2$, is approximately proportional to the product of the component of the magnetic induction B_z perpendicular to the plate plane and the sensor bias current I supplied via the current contacts $CC1$ and $CC2$ [46]

$$V_H = \frac{R_H}{t} G\left(\frac{\ell}{w}, \frac{s}{w}, \frac{y}{\ell}, \theta_H \right) B_z I. \tag{13}$$

Here R_H denotes the Hall coefficient, G the geometrical correction factor, t, ℓ, w, and s dimensions defined in Fig. 1, and θ_H the Hall angle as defined by (9). The Hall coefficient of a semiconductor with mixed conductivity (i.e., where both electrons and holes take part in the transport process) is given by (12). We recall that the value of the scattering parameters r_n and r_p are always close to unity, but they play an important role in the temperature behavior of the Hall plates.

The integrated Hall plates normally work under strong extrinsic conditions, e.g., $n \gg p$, so that (8) applies, and hence

$$|V_H| \simeq \frac{r_n}{qnt} GB_z I. \tag{14}$$

This equation explains why the Hall plates are usually thin and made of high-resistivity (low carrier concentration) material: the Hall voltage is inversely proportional to both carrier concentration n and plate thickness t.

The so-called geometrical correction factor G in (13) and (14) expresses the difference between the infinitely long Hall plate ($\ell/w \to \infty$) and the finite one, viz.,

$$G\left(\frac{\ell}{w}, \frac{s}{w}, \frac{y}{\ell}, \theta_H\right) \equiv V_H/V_{H\infty}. \tag{15}$$

It accounts for the short-circuiting effects caused by current and sensor contacts illustrated in Fig. 2. The short-circuiting

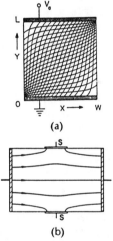

(a)

(b)

Fig. 2. Short-circuiting effects in Hall plate by current contacts (a) and sensor contacts (b). (a) Current density lines (ending at top and bottom current contacts) and equipotential lines (from left to right). Inclination of lines is due to magnetic induction perpendicular to plate [35], [53]. In the center, the equipotential lines are more inclined than close to current contacts, which are equipotential areas and shunt the Hall generated in their vicinity. (b) Short-circuiting of current lines ($B = 0$) by sensor contacts S and corresponding reduction of current density in plate center.

effects have been studied by many workers (see, e.g., [34], [49]–[51]). Under the assumptions that only one type of carrier is involved, there is no surface recombination, and the geometry of the plate is such that $\ell/w > 1.5$, $s/w < 0.18$, and $y = \ell/2$, the geometrical correction factor can be approximated by the following analytical function [51] G_L, viz.

$$G_L = \left[1 - \exp\left(-\frac{\pi}{2}\frac{\ell}{w}\frac{\theta_H}{\tan\theta_H}\right)\right] \cdot \left[1 - \frac{2}{\pi}\frac{s}{w}\frac{\theta_H}{\tan\theta_H}\right]. \tag{16}$$

It approaches unity if $\ell/w > 3$ and $s/w < 1/20$. For very small Hall plates and short samples ($\ell < w$) with point sensor contacts, ($s \ll \ell$), and $y = \ell/2$, the function G can be approximated by [52]

$$G_s \simeq 0.74 \frac{\ell}{w}. \tag{17}$$

Inspection of (14) and (16) shows the following: for efficient use of the available current, the Hall plate must be as long, as thin, and as low-doped as possible. On the other hand, the voltage drop across the plate

$$V = \frac{1}{q\mu_n n}\frac{\ell}{wt} I \tag{18}$$

may become unacceptably high. By inserting (18) into (14), we express the Hall voltage in terms of the voltage V applied between the supply contacts, viz.,

$$V_H = \mu_n^* \frac{w}{\ell} GB_z V. \tag{19}$$

This equation clearly shows the importance of high carrier mobility materials for efficient Hall plates.

The Hall voltage of the plate with constant bias voltage increases continuously with the aspect ratio w/ℓ. There is, however, a limiting value, which is obtained by inserting (17) into (19)

$$V_{H,0} \simeq 0.74\mu_n^* B_z V. \tag{20}$$

Combining (14) and (19), we obtain the Hall voltage in terms of the power $P = VI$ dissipated in the plate

$$V_H = r_n(\mu_n w/qnt\ell)^{1/2} GB_z P^{1/2}. \tag{21}$$

The product $G(w/\ell)^{1/2}$ shows a broad maximum of about 0.7 at $\ell/w \simeq 1.3$.

Apart from the simple, rectangular Hall plate analyzed up to now, many other shapes have been proposed (see Fig. 3). Using conformal mapping theory, Wick [49] demonstrated the invariance of Hall plate electrical efficiency with respect to geometry ("... there are no properties ... that cannot be obtained from a circle, square, or any other simple shape by proper size and position of the electrodes." [49]). Still, some of the shapes shown in Fig. 3 may have some *technological*

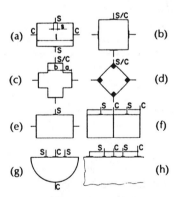

Fig. 3. Various shapes of Hall plates with current contacts C and sensor contacts S. Electronic performance is independent of geometry [49], but some shapes are easier to fabricate in IC technology than others.

advantages over the others. For example, due to limitations in photolithography, it is not easy to fabricate a small-size rectangular plate with (Fig. 3(a)) $s/\ell < 0.1$. The same geometrical factor G can, however, be achieved by a cross-shaped configuration (Fig. 3(c)) with $a/b > 0.5$ [51], which is much easier to realize.

The three-terminal device (Fig. 3(e)) allows the combination of the Hall and magnetoresistance effects [54]. (It should be noted that Wick's invariance theory [49] does not apply directly to correspondence between this device and other shown devices, since they have different numbers of contacts.) This device is particularly favorable when non-uniform magnetic fields are measured [40]. A combination of two three-terminal devices (Fig. 3(f)) [55] is useful in cases when both Hall contacts must be on one side of the plate. A similar result is achieved with semicircular (Fig. 3(g)) and half-plane (Fig. 3(h)) [56] devices; in the latter case *all* contacts are situated on one side of the plate. Devices which are invariant under rotation of 90° (such as those in Fig. 3(b), (c), (d)) allow the nonreciprocal property of the Hall effect to be utilized in circuit design.

Integrated Hall Devices

The integration of a new device into a monolithic circuit raises the crucial issue of compatibility of the device (in terms of material, fabrication technology, geometry, electric and thermal properties) with the rest of the circuit. Today only silicon, and to some extent, galium arsenide are routinely used to build monolithic integrated circuits. The question is, whether a Hall plate can be made compatible with silicon or galium arsenide integrated circuit technology.

In discussing the issue, we shall consider only monolithic integration approaches. We do not discuss hybrid solutions which require extensive post-processing of practically finished IC wafers, such as deposition of special thin films.

Material compatibility in this context means: do the materials usually present in integrated circuits show an efficient sensor effect? Efficiency of the Hall effect may be assessed by inspecting (14), (19), and (21). Equation (19) shows that neither silicon, nor GaAs, in view of their moderate carrier mobilities (Si: $\mu_n \leqslant 0.15$ m²/V · s, GaAs: $\mu_n \leqslant 0.85$ m²/V · s at 300 K), are the best materials for Hall plates operating at a given supply voltage. In this respect, much better materials would be InAs ($\mu_n \leqslant 3.3$ m²/V · s) and InSb ($\mu_n \leqslant 8$ m²/V · s). However, the advantage of high-mobility materials disappears if the plate has to operate at a given power level; see (21): the material figure of merit is now $(\mu_n/nt)^{1/2}$. Since high-mobility materials feature small bandgaps, there is no way to obtain, at room temperature, the low carrier densities of Si and GaAs. If the plate has a limited current supply, its efficiency does not depend on the carrier mobility at all; see (14). Therefore, large bandgap and technologically manageable materials have to be chosen. Silicon and GaAs are outstanding materials in that respect. Since in integrated circuits both low power dissipation and low current consumption are often very desirable, both Si and GaAs Hall plates are well suited for this environment.

Efficiency of the Hall effect is, of course, not the only selection criterion for optimal Hall plate material. Other important issues are noise, offset, and temperature behavior. However, none of these additional criteria will essentially influence the above conclusion, as we shall show below.

Technological compatability is the next important issue to be considered: is it possible to incorporate a Hall plate into an IC chip without too many (or any) additional processing steps? The answer is positive, as demonstrated by the following examples.

Fig. 4 shows a Hall plate realized as a part of the n-type epitaxial (collector) layer in a silicon integrated bipolar circuit process. This idea, proposed by Bosch [57], was basic

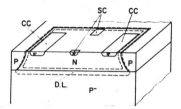

Fig. 4. Hall plate in bipolar IC technology with current contacts *CC* and sensors contacts *SC*, cf. Fig. 1. The n-type plate region is isolated from surrounding p-type material by depletion layer *DL*. Oxide and metal layers are not shown.

to a number of commercially successful products [47], [48]. The pertinent process sequence is obtained by reducing the sequence routinely used to make integrated bipolar transistors: by appropriate mask design, neither the buried layer nor the p (base) diffusion layer are made in the Hall plate area. Planar geometry of the plate is defined by p (isolation diffusion) and n⁺ (emitter diffusion) regions which are used as intermediate layers to provide ohmic contacts to the metal layer. The isolation between the Hall plate and the rest of the chip is achieved in the way common in monolithic ICs; namely, by reverse-biased p-n junctions surrounding the plate.

Attempts have also been made to combine an integrated Hall plate with active devices [58], [59]. For further details, we refer to the discussion of injection modulation in Section IV.

Usually, only n-type material is used as active region in integrated Hall devices. The reason is to achieve a maximal voltage-related efficiency; see (19) (in silicon, $\mu_n/\mu_p \approx 3$). Typical electron densities and thicknesses of the epitaxial layer are $n = 10^{15}$–10^{16} cm⁻³ and $t = 5$–10 μm, respectively. This yields nt products (see (14) and (21)) of 5×10^{11} cm⁻² to 10^{13} cm⁻². Incidentally, optimal values of the nt product for many Hall plate applications are in this range. Typical Hall plate dimensions are: $w \simeq 200$ μm, $\ell \simeq 200$–400 μm.

If the epitaxial layer resistivity is much higher than needed for the Hall plate, it may be reduced by local ion implantation. The implanted layers have much better uniformity than the original ones. For instance, phosphorus ion implantation with a dosage of 5×10^{12} cm⁻² has been used for this purpose [13].

Both epitaxial [60] and ion-implanted [30], [61] active layers have also been proposed for GaAs Hall plates. For example, silicon ions have been implanted with a dose of 9×10^{12} cm⁻² into semi-insulating GaAs substrate for this purpose [19].

All Hall plates described so far have the form of a plate merged in the chip surface. So they are sensitive to the magnetic field perpendicular to the chip plane. However, there are some applications where sensitivity to the magnetic field parallel to the device surface is preferred. Only recently, an integrated *vertical* Hall device serving such a purpose has been devised [56].

The vertical Hall device (see Fig. 5) is based on the following two ideas. First, the general plate shape is chosen in such a way that at least three out of the four contacts are available on the top chip surface, as shown in Fig. 3(g) and Fig. 3(h). Second, a deeply diffused p-type ring is used to

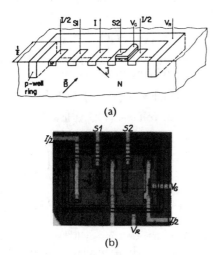

(a)

(b)

Fig. 5. (a) Vertical Hall device (VHD) in CMOS technology with geometry similar to Fig. 3(h). Active device region is n⁻-substrate surrounded by p-type ring. (b) Photomicrograph of VHD shown in Fig. 5(a).

define the device vertical geometry. The active device volume is part of the n-type substrate material. Apart from its bottom, it is isolated from its surroundings by the reverse-biased p-n junction ring substrate. The device was realized in standard bulk CMOS technology with p-ring (p-well) diffusion depth of 12 μm. The average active region (substrate) doping and the effective plate thickness were $n \simeq 10^{15}$ cm^{-3} and $t \simeq 12$ μm, respectively, making the nt product of 1.2×10^{12} cm^{-2}. It has been found [56] that the sensitivity is essentially not affected by the unusual geometry, in agreement with Wick's invariance considerations [49]. An example of the current density and voltage distributions in a vertical Hall device is shown in Fig. 6 [62]. A three-contact vertical Hall device has also been realized [172] and modeled [93].

We notice that in all these examples of integrated Hall plates application of the available fabrication steps yields only an approximation of the desired device geometry. Although this fact makes device design and optimization

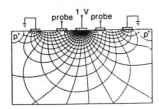

Fig. 6. Equipotential and current lines of VHD (cf. Fig. 5(a), (b)) obtained by numerical simulation [62], [93] for $I = 211$ μA, $V = 1$ V, $\mu_n^* B = 0.2$. Asymmetry of lines is due to Hall effect.

difficult, it does not seriously affect the functioning of the Hall device.

Electrical compatibility between Hall plate and IC chip is regarded here in connection with technological compatibility. More precisely, we look for possible parasitic electrical effects inherent to the p-n junction isolation utilized in devices described above. Two such effects have been reported in literature: effective plate thickness variation and leakage current.

The Hall plate thickness varies with voltage applied to the isolating junction due to the well known junction field effect (the p-n junction depletion-layer width depends on the applied reverse bias). The effective plate thickness is given by [63]

$$t = t_0 - t_{DL}(V_B) \qquad (22)$$

where t_0 denotes the so-called metallurgical plate thickness (defined by metallurgical p-n junction(s)), $t_{DL}(V_B)$ is the width of the part of the depletion layer(s) on the Hall plate side of the junction, and V_B denotes the bias voltage. This voltage dependence of the plate thickness causes a dependence of the Hall voltage on the applied junction bias. Similarly, the voltage drop (9) across the plate depends also on the junction bias. Moreover, due to this voltage drop, the effective junction bias is position-dependent and, consequently, so is the effective plate thickness. Our previous assumption of plate thickness uniformity does not hold and, strictly speaking, none of (14) or (16)–(21) is correct. However, in bulk Hall plates with nt product of more than 10^{12} cm^{-2}, this effect is rather weak, and we may still accept these equations as good approximations.

The variation of the effective plate thickness due to the junction field effect is always present as a parasitic effect and may be used as an efficient means of adjusting device sensitivity to the magnetic field [63]. An adjustment range as large as ± 25 percent has been achieved [56].

The influence of the junction leakage current on the integrated Hall plate performance has not been studied in detail up to this time. It was mentioned only that it contributed generation–recombination noise [47].

Figures of Merit

Sensitivity is the most important characteristic of any sensor. Absolute sensitivity of a Hall element is defined as

$$S_A \equiv \left| \frac{\partial V_H}{\partial B_z} \right|_{I=\text{const.}} \quad [\text{V/T}]. \qquad (23)$$

Since the Hall-effect device belongs to the group of modulating transducers [64], it is reasonable to define relative sensitivities as well, i.e., absolute sensitivity divided by the modulated physical quantity. In view of (14) and (19) we may define a supply-current-related sensitivity, viz.

$$S_{RI} \equiv \left| \frac{1}{I} \frac{\partial V_H}{\partial B_z} \right| = \frac{r_n G}{qnt} \quad [\text{V/A} \cdot \text{T} = \text{volt/ampere} \cdot \text{tesla}] \qquad (24)$$

and a supply-voltage-related sensitivity

$$S_{RV} \equiv \left| \frac{1}{V} \frac{\partial V_H}{\partial B_z} \right| = \mu_n^* \frac{w}{\ell} G \quad [\text{T}^{-1}]. \qquad (25)$$

We would like to stress that S_{RI} does not depend on the sensor material. With respect to the device shape, S_{RI} has a maximum value for *long* ($\ell \gg w$) devices, where $G \approx 1$. The corresponding $S_{RI,max}$ depends on the product nt and can become rather large for small nt. In integrated Hall plates, however, the minimum value of nt is limited by the junction field effect. A practical limit seems to be $nt \geq 5 \times 10^{11} cm^{-2}$, i.e.,

$$S_{RI,max} \leq 1250\, V/A \cdot T. \tag{26}$$

Typical experimental values range between $80\, V/A \cdot T$ [48] and $400\, V/A \cdot T$ [56]. When the doping profile is nonuniform and depends on the depth z in the device (see Fig. 1), as, e.g., in the case of ion-implanted Hall plates, then the role of nt is played by the dose

$$Q = \int_0^{z_j} n(z)\, dz \tag{27}$$

with z_j denoting the junction depth.

Voltage-related sensitivity, (25), depends on both material properties (such as mobility) and device design. Its optimal value is obtained in *short* Hall plates, the physical limit being (20)

$$S_{RV,max} \simeq 0.74\mu_n^*. \tag{28}$$

Taking $r_n \simeq 1$ and appropriate mobility values, this yields, at room temperature,

$$S_{RV,max} \simeq \begin{cases} 0.11\, T^{-1}, & \text{or 11 percent/T for Si} \\ 0.63\, T^{-1}, & \text{or 63 percent/T for GaAs.} \end{cases} \tag{29}$$

Typical values reported in the literature are $0.07\, T^{-1}$ for Si [48] and $0.2\, T^{-1}$ for GaAs [19], with T denoting the tesla unit.

So far we have considered only Hall plates operating in the low electric field region, where the mobility is constant and the carrier drift velocity v_d is proportional to the electric field $E = V/\ell$. However, at high electric fields, v_d reaches a saturation value and so does the Hall field. Nevertheless, all our conclusions about maximal achievable sensitivities essentially hold. An experimental silicon saturated velocity sensor has been reported [65] with a voltage-related sensitivity of $0.07\, T^{-1}$.

Recently [66], a method was devised to virtually eliminate the short-circuit effect caused by supply contacts (see Section III). In that case, operation at carrier saturation velocity does make sense. An overall voltage-related sensitivity as high as $0.14\, T^{-1}$ has been reported. This is above the theoretical limit of conventional silicon Hall plates (29), although the carrier mobility in the experimental device was less than half the maximal mobility.

In contrast to the behavior of silicon Hall devices discussed above, GaAs miniature Hall plates show a substantial increase in sensitivity [67] when working under high electric field. Unfortunately, this effect goes together with an even larger increase in noise.

Noise phenomena severely limit sensor performance. For most applications, the signal-to-noise ratio (SNR) rather than sensitivity alone is the dominant figure of merit of a sensor. Some results on rectangular Hall plates are presented in the following.

The voltage noise spectral density across the Hall contacts is given by [68]

$$S_V(f) = S_{V\alpha}(f) + S_{VT} \tag{30}$$

where $S_{V\alpha}$ stands for $1/f$ noise and S_{VT} for thermal (Nyquist) noise. According to the concept of bulk-generated $1/f$ noise [69], $S_{V\alpha}$ is given by a semi-empirical expression [70] which in our notation reads

$$S_{V\alpha} \simeq \alpha(V/\ell)^2 (2\pi nt f)^{-1} \log(w/s) \tag{31}$$

where α denotes the Hooge $1/f$ noise parameter [69], f the frequency, and log the natural logarithm. The thermal noise density is

$$S_{VT} \simeq 8kT(\pi\mu_n qnt)^{-1} \log(w/s) \tag{32}$$

with T now denoting absolute temperature. The SNR in a narrow bandwidth Δf around f is given by

$$\text{SNR}(f) \equiv V_H[S_V(f)\Delta f]^{-1/2} \tag{33}$$

where V_H is the Hall voltage, (19). At low frequencies, $1/f$ noise is dominant

$$\text{SNR}(f) \simeq \frac{\mu_n^*}{\sqrt{\alpha}} \left(\frac{2\pi nt\ell w}{\log(w/s)} \right)^{1/2} G\left(\frac{w}{\ell}\right)^{1/2} \left(\frac{f}{\Delta f}\right)^{1/2} B_z. \tag{34}$$

This relation holds for $f \ll f_c$, f_c being the corner frequency defined by $S_{V\alpha}(f_c) = S_{VT}$. Note that $f_c \propto V^2/T \propto I^2/T$, with T denoting absolute temperature.

The SNR at *low* frequencies will be large under the following conditions: a) The Hall plate is made of material with high mobility and low α-parameter. The Hooge α-parameter has long been assumed to be approximately a constant of the order of 10^{-3}. Recently, however, α-values as low as 10^{-7} have been reported [71]. The latest results [72] indicate even lower values, e.g., as low as 10^{-9}. This indicates that one can achieve higher SNR by selecting proper material, technology, and geometry for low α-values (with potential SNR improvement of up to 1000) rather than by selecting a high-mobility material (SNR improvement of about 100). b) The Hall plate contains a large number of carriers, $N = nt\ell w$, unless large dimensions imply also high α-values [72], [73]. c) The aspect ratio $\ell/w \simeq 1.3$ (see the comment following (21)).

Thermal noise dominates at *high* frequencies, $f \gg f_c$. The SNR increases with increase in device current and is limited only by the maximal acceptable heat dissipation P_{max}. Inserting (32) and the maximal value of V_H according to (21) into (33) we obtain

$$\text{SNR}_{max} \simeq 0.44\mu_n^* B_z (P_{max}/kT\Delta f \log(w/s))^{1/2}. \tag{35}$$

This equation clearly shows the advantage of using high-mobility materials for Hall devices operating at high frequencies.

Another way to express the noise-related quality of a magnetic field sensor is to define the equivalent input magnetic field noise spectral density

$$S_B \equiv \frac{d\langle B_N^2, eq\rangle}{df} = \frac{S_V(f)}{S_A^2} \tag{36}$$

where S_A denotes the absolute sensitivity (23).

The detection limit for the magnetic induction in a frequency range between f_1 and f_2 is given by

$$B_{min} \equiv \left[\int_{f_1}^{f_2} S_B\, df \right]^{1/2} \tag{37}$$

or by (34) and (35), with SNR = 1, $\Delta f = f_2 - f_1 \ll f_1$ and $f = f_1 + \Delta f/2$. The detection limit of Hall elements at $f \gg f_c$

and $\Delta f = 1$ Hz with $\mu_n^* = 6$ m^2/V $\cdot$ s, $\ell/w = 2$, $w/s = 4$, and $P_{max} = 0.5$ W has been assessed theoretically to be $B_{min} \simeq 4 \times 10^{-11}$ T [68]. The measured S_B of a silicon vertical Hall device operated at $I = 0.5$ mA and $P \simeq 1$ mW at 100 Hz is 3×10^{-13} T^2/Hz (1/f noise) and at 100 kHz is 10^{-15} T^2/Hz (thermal noise) [74].

Offset voltage is a static or very slowly varying output voltage of a Hall element in the absence of a magnetic field. It causes a serious problem in static and low-frequency magnetic field measurements, since the offset voltage cannot be distinguished from the magnetic-field-proportional useful signal. The offset is usually characterized by an equivalent magnetic induction $B_{o,eq}$ corresponding to the offset voltage V_o, viz.

$$B_{o,eq} \equiv V_o/S_A \tag{38}$$

where S_A denotes the absolute sensitivity (23).

The major causes of offset in integrated Hall elements are imperfections in the fabrication process (such as misalignment and fluctuations in material characteristics) and piezoresistive effects. Both effects can be represented using a simple bridge circuit model of the Hall plate [75] (see Fig. 7). The offset voltage caused by the shown asymmetry of

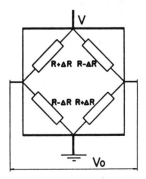

Fig. 7. Bridge circuit model of Hall plate. Resistance variations ΔR produced by alignment tolerance or piezoresistive effect lead to bridge asymmetry resulting in offset voltage V_o.

the bridge is

$$V_o = \frac{\Delta R}{R} V. \tag{39}$$

Using (38), (39), and (25) we obtain the equivalent offset field, viz.

$$B_{o,eq} = \frac{\Delta R}{R} \frac{1}{\mu_n^*} \frac{\ell}{wG}. \tag{40}$$

This equation demonstrates once more the importance of high mobility. For many applications, a usual requirement is $B_{o,eq} < 10$ mT. For a silicon Hall plate, with $\mu_n^* \simeq 0.15$ m^2/V $\cdot$ s and $w/\ell \simeq 1$, (40) leads to the requirement $\Delta R/R < 10^{-3}$. Such a precision is not a trivial task even for the highly developed current IC processing.

In [76], offset in silicon integrated horizontal Hall plates due to the fabrication tolerances has been studied. It is found that errors in geometry, caused by etching randomness and rotation alignment errors, are the major sources of offset voltage. For a plate with an area of 0.2 mm^2 it was found that a rather flat optimum in aspect ratio exists, namely, $\ell/w_{opt} \simeq 0.8 \cdots 1.7$, where an average etching

tolerance of 0.1 μm produces an average equivalent magnetic induction offset of 6 mT.

The offset voltage is also seriously affected by mechanical stress in the sensor chip introduced during packaging. Using the square-type bridge circuit model of the Hall plate (Fig. 7), Kanda and Migitaka [75], [77] found the maximum offset voltage to be

$$V_o = \frac{(\pi_\ell - \pi_t) XV}{2(3 - 2w/\ell)} \tag{41}$$

where π_ℓ and π_t are the longitudinal and transverse piezoresistance coefficients parallel to the bridge sides (i.e., corresponding to the direction in the plate plane which makes an angle of $\pi/4$ with the direction of the device current flow) and X is the stress applied. Using (38) and (25), we find the maximum of the equivalent offset field to be

$$B_{o,eq} = \frac{(\pi_\ell - \pi_t) X}{\mu_n^* 2(3 - 2w/\ell) wG/\ell}. \tag{42}$$

This equation shows effects of geometry and material parameters. An optimal geometry, yielding a minimal $B_{o,eq}$ turns out to show an aspect ratio $\ell/w \simeq 1.5$. As regards material, the best choice for a silicon Hall device is to place the plate in the (110) crystal plane so that current flows in the $\langle 100 \rangle$ direction. For such a plate, theory and experiment agree on an equivalent offset field of $B_{o,eq} \simeq 8.4$ mT at a strain of 10^{-4} [78]. Conversely, the Hall device with a $\langle 110 \rangle$ current direction in the (100) plane yields a highly sensitive strain gauge [78].

Offset can be further reduced by compensation [47], [48], trimming, or calibration, using for example, an offset-controlling gate [79]. The last two techniques are expensive, since they require an individual treatment of every sensor. Yet, they do not give perfect results, since the offset depends also on temperature and temperature gradient and varies with time. Recently, a substantial reduction of the equivalent offset field and its drift has been reported for horizontal integrated Hall plates [82]. This was achieved by a combination of the compensation method [47], [48] and addition of an implanted p-type surface layer, which makes the Hall region similar to a pinch resistor. Equivalent offset fields as low as 1 mT have been demonstrated in this way.

It is interesting to note that slowly varying offset in some devices can be related to 1/f noise [80], [73] about which many experimental facts have been established down to frequencies as low as 10^{-6} Hz [81]. Part of the offset phenomena might be explained in terms of very-low-frequency 1/f noise.

Linearity error is defined as the ratio

$$LE = (V_H - V_H^0)/V_H^0 \tag{43}$$

where V_H is the measured Hall voltage and V_H^0 is usually assumed to be the best linear fit to the measured values. If the device is used as a magnetic field sensor, i.e., working at a constant current bias, V_H^0 is defined as

$$V_H^0 = \frac{V_{HL}}{B_L} B, \quad I = \text{const} \tag{44}$$

where V_{HL} and B_L are the values shown in Fig. 8. If the Hall device works as a four-quadrant multiplier around the point $B_z = 0$, $I = 0$, the appropriate definition of the "linearized" Hall voltage output reads

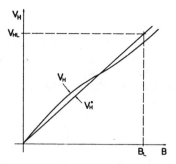

Fig. 8. Nonlinearity of Hall MFS: hypothetical measured curve V_H and best linear fit V_H^0.

$$V_H^0 = \frac{V_{HL}}{B_L I_L} BI \tag{45}$$

where V_{HL}, B_L, and I_L have an analogous meaning in a three-dimensional plot. On the other hand, $V_H(B, I)$ can be expanded into a McLaurin series and approximated by the first nonzero term, viz.

$$V_H^0 \approx \left[\frac{\partial^2 V_H(B, I)}{\partial B \, \partial I} \right]_{B=0, I=0}. \tag{46}$$

The second derivative in (46) is equal to the relative sensitivity S_{RI} (24) at $B = 0$ and $I = 0$. This means that a linear sensor working as a current–field multiplier must have a relative sensitivity S_{RI} that is independent of the magnetic field and the current. An analogous line of thought applies to sensors working as voltage–field multipliers and leads to the requirement of magnetic-field- and voltage-independent relative sensitivity S_{RV}.

Most of the published work on nonlinearity in Hall elements [44]–[46] deals with magnetic field measurements, i.e., $I = $ const as in (43) and (44). This type of nonlinearity is mostly due to the short-circuit effects described by (14) and (16). Nonlinearity can be minimized by loading the sensor output with a proper valued resistor. One of the lowest linearity error values, LE $= \pm 3 \times 10^{-4}$ in the induction range of less than 1 T and at room temperature, has been achieved by an implanted GaAs cross-shaped Hall device [19].

The linearity error of a multiplier (43) with (45), may be also caused by the junction field effect or by thermal effects. It seems that none of these has been studied in detail.

Temperature coefficients of the sensitivity are defined by

$$TC \equiv \frac{1}{S} \frac{\partial S}{\partial T} \tag{47}$$

where S denotes the absolute sensitivity (23) or one of the relative sensitivities (24) and (25) and T the absolute temperature. In the following discussion of the temperature coefficients, we assume that the carrier concentration is constant, i.e., that the material of the sensor active region stays in the temperature saturation range where $n = N_D$ is temperature independent, N_D being the donor concentration. We also assume a very small Hall angle and we neglect the junction field effect.

If the Hall plate is supplied by constant current, we obtain from (25) and (47) the pertinent temperature coefficient

$$TC_I = \frac{1}{r_n} \frac{\partial r_n}{\partial T}. \tag{48}$$

Hence the temperature coefficient of the sensitivity S_{RI} is equal to that of the Hall scattering factor r_n. Their values are as follows: for silicon with donor concentrations of $1.75 \times 10^{14} \, \text{cm}^{-3}$ and $2.1 \times 10^{15} \, \text{cm}^{-3}$, $|(1/r_n)(\partial r_n/\partial T)| \simeq 10^{-3} \, \text{K}^{-1}$ in the temperature range from 200 to 400 K [39]; for GaAs with electron concentrations in the range between $3 \times 10^{15} \, \text{cm}^{-3}$ and $1.6 \times 10^{17} \, \text{cm}^{-3}$, $|(1/r_n)(\partial r_n/\partial T)| < 0.3 \times 10^{-3} \, \text{K}^{-1}$ in the temperature range from 200 to 320 K [83].

If the Hall plate is supplied by a constant voltage, (25) and (47) yield

$$TC_V = \frac{1}{\mu_n^*} \frac{\partial \mu_n^*}{\partial T}. \tag{49}$$

Since the integrated Hall plates are usually made of low-doped material, the temperature dependence of the Hall mobility is dominated by a strong temperature influence on the drift mobility, because we have $\mu_n \propto T^{-2.4}$ in the phonon-scattering region [36]. The corresponding temperature coefficient around room temperature is $(1/\mu_n)(\partial \mu_n/\partial T) \simeq -8 \times 10^{-3} \, \text{K}^{-1}$.

The experimental temperature coefficients correspond roughly to those quoted above: $TC_I \simeq +0.8 \times 10^{-3} \, \text{K}^{-1}$ and $TC_V \simeq -4.5 \times 10^{-3} \, \text{K}^{-1}$ is found for silicon integrated Hall plates in the temperature range ($-20°C$, $+120°C$) [47], [48]; and $TC_I \simeq +0.3 \times 10^{-3} \, \text{K}^{-1}$ for a GaAs Hall plate [19].

The thermal variations in Hall voltage may be compensated by a proper variation in biasing conditions, i.e., by increasing supply voltage or decreasing supply current with increase in temperature [44]–[46]. An alternative approach was proposed recently [84] in which the junction field effect is used to compensate the variations in sensitivity.

III. MOSFET MAGNETIC FIELD SENSORS

The MOS (metal–oxide–semiconductor) structure has been recognized as an attractive candidate for integrated MFS as early as 1966 [85]. The surface inversion layer or channel can be readily used as an extremely thin Hall plate; such a Hall element is technologically and electrically compatible with the MOSFET (MOS field-effect transistor). The acronym MAGFET is sometimes used to refer to the family of MOSFET magnetic sensors. These devices exploit the Hall effect or the deflection of carriers in the inversion layer.

MOS Hall Plates

The basic device structure is shown in Fig. 9. A MOS Hall plate is a MOSFET with the usual source (S) and drain (D) regions and the channel (Ch), on top of which the gate (G) is located, but has, in addition, two Hall probe regions (H), which are made simultaneously with the S and D diffusion. The channel length and width are denoted by L and W, respectively. The position of the Hall probes is described by their distance y from the source.

Let us first consider an n-channel MOS Hall plate working in the *linear* region of the pertinent MOSFET characteristics. This means $V_G - V_T \gg V_D$ with V_G, V_T, and V_D denoting the gate, threshold, and drain–source voltage, respectively. In this case, the channel charge density, $Q_{ch} = $

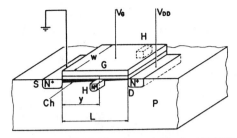

Fig. 9. Hall-plate MAGFET. The channel Ch of a MOSFET is used as extremely thin Hall plate. Two Hall-voltage probes H are added to the usual MOSFET structure.

$C_{ox}(V_G - V_T)$, is approximately constant across the channel; C_{ox} denotes the oxide capacitance per area in the gate region. Under these conditions, MOS and bulk Hall plates are equivalent. With respect to the sensor performance, Q_{ch} has the same meaning as the area charge density qnt in bulk Hall plates (14). In bulk Hall plates we can control this charge density by the junction field effect and in MOS Hall plates by the MOS field effect. By analogy with (24), the *current-related sensitivity* of the "linear" MOS Hall plate is given by

$$S_{RI} = r_n G / Q_{ch} = r_n G / C_{ox}(V_G - V_T) \qquad (50)$$

where G denotes the geometrical factor with the same meaning and value as (15) for bulk Hall plates. By proper choice of device parameters (C_{ox} and V_T) and operating conditions (V_G), very small Q_{ch} and, accordingly, very large S_{RI} can be achieved. This provides the "extremely thin Hall plate" referred to earlier.

Linear operating conditions, however, limit the drain voltage and current to rather small values, leading to unpractically small Hall voltage signals. Therefore, the MOS Hall plate has to operate at higher voltage, i.e., in the *triode* region (slightly below pinchoff) or even in saturation. This leads to the following additional effect: operating at $V_D \lesssim V_{sat}$, where $V_{sat} \simeq V_G - V_T$ denotes the drain saturation voltage, the channel is much thinner and has much higher resistivity in the vicinity of the drain region than close to the source region. Therefore, shunting effects due to the drain region are much less severe than those close to the source region. Moreover, the Hall voltage increases for larger y/L because of the same effect. This is described in the relationship [86], [87]

$$V_H = G W \mu_n^* B I_D / \beta L \left[(V_G - V_T)^2 - 2 y I_D / \beta L\right]^{1/2} \qquad (51)$$

with drain current

$$I_D = \beta \left[(V_G - V_T)V_D - V_D^2/2\right] \qquad (52)$$

and $\beta = \mu_n C_{ox} W/L$. Hence, the optimal position y_{opt} of the Hall probe contacts is not at $y = L/2$ as in bulk and linear CMOS plates, but somewhat closer to the drain region: $L/2 < y_{opt} < L$. Moreover, y_{opt}/L increases with V_D until saturation, $V_D = V_{sat}$, is reached. On the other hand, offset due to alignment tolerances increases as y approaches L. Therefore, the practical optimum Hall contact position is usually chosen as $0.7 \leq y_{opt}/L \leq 0.8$ [86]–[90]. Finally, optimal MOS Hall plates are somewhat shorter than the corresponding bulk Hall plates, e.g., $(W/L)_{opt} = 1.2$ [87].

In analogy with (25), the *voltage-related sensitivity* of MOS Hall plates is $S_{RV} = \mu_n^* G W/L$. Thus the use of the thin

channel instead of the bulk plate offers no advantage. On the contrary, S_{RV} is smaller because $\mu_{channel}^* \leq \mu_{bulk}^*/2$. S_{RV} decreases further if $V_D > V_{sat}$, because part of the applied voltage, $V_D - V_{sat}$, idly drops over the drain depletion region, while the Hall voltage stays at the value reached for $V_D = V_{sat}$ [85]. Also, $1/f$ noise increases due to surface recombination. For circuit convenience, MOS Hall plates are usually biased such that $V_G = V_D$. Under this bias, enhancement-type devices [85] work in "shallow" saturation and depletion-type devices [87] work in the triode region, but close to saturation.

The *pre-1980 MAGFET* investigations are reviewed in [8], such as the theoretical treatments [86], [88]. The n-channel MOS Hall plate measurements [89] demonstrate the invariance of the Hall voltage to gate geometry as predicted by [49]. A typical early n-channel Hall plate [90] with $L = W$ and $y/L = 0.8$ achieved 0.035/T or 10^3 V/A · T with 20-mT offset equivalent field and 0.2 mT/K temperature coefficient.

The recent *MOS Hall plate by Hirata and Suzuki* [87] is part of a silicon chip for a noncontact keyboard switch. The n-channel depletion-type Hall device is integrated together with an amplifier on one chip by silicon-gate NMOS technology. Offset voltage due to piezoresistance is reduced by annealing (after die bonding). Offset due to Hall probe misalignment is kept small by choosing a large device size. A number of different channel lengths $L = 100$ and 600 μm, aspect ratios $W/L = 0.5$, 1, and 2, and probe positions $y/L = 0.5, \cdots, 0.9$ were investigated. *Optimal device parameters* and operating conditions are the following: $L = 600$ μm, $W/L = 1.2$, $y/L = 0.7$, gate voltage $V_G = 5$ V, drain voltage $V_D = 5$ V, resulting in drain current $I_D = 0.5$ mA. This device shows 640 V/A · T or 0.064/T sensitivity and -0.4 percent/K temperature coefficient. The average offset equivalent field is 14 mT. Linearity is good below $B = 0.1$ T. No data on noise are available.

Elimination of short-circuit effects irrespective of the aspect ratio W/L is achieved by a novel MAGFET design devised by Popović [66]. To this end, the conventional (highly conducting) drain and source regions are replaced by so-called *distributed current sources* which inject the current directly into a short Hall plate in a uniform manner. In view of the extremely high output resistance of the current sources, the Hall plate behaves as if it were infinitely long even when $W/L \gg 1$. Nevertheless, the actual short-channel length L allows high electric fields to be generated with conventional supply voltages.

A Hall MAGFET with $W/L = 10$ and $L = 10$ μm was made [66] in standard double-polysilicon CMOS technology. The distributed current sources are realized by one double-gate NMOS transistor (replacing the conventional source region) and by an array of p-channel current sources (replacing the drain region). Sensitivities up to $S_A = 0.7$ V/T, $S_{RI} = 4 \times 10^3$ V/A · T, and $S_{RV} = 0.14$/T with 5-V supply voltage are obtained in this way. The *physical limit* of the absolute sensitivity is estimated as $S_{A,max} \approx W v_{d,sat}$, where $v_{d,sat}$ denotes the saturation drift velocity. With $v_{d,sat} \approx 10^5$ m/s and $W = 100$ μm this leads to 10 V/T. The above mentioned physical limit does not depend on the Hall mobility. Thus it does not depend strongly on the type of material, since all relevant materials have a carrier saturation velocity of the same order of magnitude.

Split-Drain MOS Devices

The split-drain (SD) MAGFET is a MOSFET with two or three adjacent drain regions ($D1$, $D2$ in Fig. 10) which share the drain current. A magnetic field, perpendicular to the chip surface, causes deflection of the current lines in the channel region, as illustrated by the numerical modeling results shown in Fig. 11.

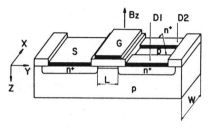

Fig. 10. Split-drain MAGFET. Magnetic induction perpendicular to chip surface produces current imbalance in drains $D1$, $D2$.

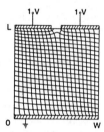

Fig. 11. Numerical modeling [35], [93] of split-drain MAGFET operation. Current lines (ending at the bottom source and top dual-drain contacts) are deflected towards the left drain by the Lorentz force, resulting in increased current in the left drain at the expense of the right drain.

This current deflection is one of the basic galvanomagnetic effects (see (10)). It eventually leads to an asymmetry in the drain currents, which is a measure of the magnetic field strength. The relative sensitivity of the split-drain devices at small magnetic induction is defined in analogy with (25), viz.

$$S_R = \left| I_D^{-1} \, \partial(I_{D1} - I_{D2})/\partial B \right|_{B=0} \quad [\text{1/T} = \text{1/tesla}] \quad (53)$$

i.e., the derivative of the relative current imbalance with respect to the magnetic induction, taken at zero induction. Because of the small Hall mobility in the channel region, S_R is rather limited. Nevertheless, high absolute sensitivities can be obtained with devices operating in the saturation region and whose drains are loaded by high resistances. In this case the useful signal is the differential voltage between the two drains.

Complementary SD MAGFET pairs [91] can be readily made in CMOS technology. A complementary pair consists of a dual-drain n-channel and a corresponding p-channel device. The SD MAGFET pair is incorporated into an integrated circuit configuration reminiscent of the well-known CMOS differential amplifier (see Section VIII). The action of the SD MAGFET is enhanced and stabilized in this way. Sensitivities as large as $S_A = 1.2$ V/T and $S_{RI} = 1.2 \times 10^4$ V/A · T are accomplished at 10-V supply voltage and 100-μA current [91].

Triple-drain MAGFET devices feature a split of the drain region into three separate subregions D_1, D_2, and D_3. Already Fry and Hoey [88] recognized the superior relative sensitivity of the triple-drain structure over the dual-drain structure. Recent numerical modeling results [92], [93] predict a 50-percent increase of the current imbalance between the two outer drains ($L = W = 100 \ \mu$m) in comparison with the corresponding dual-drain device. This result is supported by experiments [94].

IV. MAGNETOTRANSISTOR

A magnetotransistor (MT), also referred to as "magistor" or "magnistor," is a *bipolar* transistor whose design and operating conditions are optimized with respect to magnetic field sensitivity of the collector current I_C. Vertical (VMT) or lateral (LMT) device structures can serve this purpose. The useful signal is either $I_C = I_C(B)$ or the corresponding voltage drop $V_C = V_C(B) = R_C \cdot I_C(B)$ across the collector load resistor (see Fig. 12(a)). The absolute and

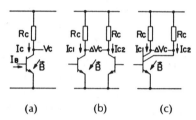

Fig. 12. Simple magnetotransistor circuits. Magnetic induction causes variation of collector currents and, hence, collector voltages,

relative MT small induction sensitivities S_A and S_R are usually defined as the pertinent derivatives at zero induction, viz.

$$S_A \equiv |\partial V_C/\partial B|_{B=0} = R_C |\partial I_C/\partial B|_{B=0} \quad [\text{V/T}] \quad (54)$$

and

$$S_R \equiv \left| V_C^{-1}(\partial V_C/\partial B) \right|_{B=0} \quad [\text{T}^{-1} = \text{1/tesla}]. \quad (55)$$

For sufficiently large collector output resistance $r_C \gg R_C$, definition (55) can be replaced by

$$S_R \equiv \left| I_C^{-1}(\partial I_C/\partial B) \right|_{B=0} \quad [\text{T}^{-1}] \quad (56)$$

also known as transduction efficiency. While (54) and (55) characterize the magnetic-field-sensitive circuit, (56) describes the MT element. The above definitions are similar to (23) to (25) for the Hall plate. It is convenient to use two integrated MTs (Fig. 12(b)) or a double-collector MT (Fig. 12(c)) as a differential pair. Definitions (54) and (56) apply if I_C is replaced by $I_{C1} + I_{C2}$ and $\partial I_C/\partial B$ by $\partial(I_{C1} - I_{C2})/\partial B$, where I_{C1} and I_{C2} denote the individual collector currents.

Important milestones in MT development were the magnetic field modulation of current gain [95], [96] and the double-collector VMT [97]–[99], with $S_R \approx 0.03/$T [99] (up to 0.05/T for a four-collector version [99]). Early lateral structures [58], [100] merge a planar Hall plate with a lateral two-collector transistor [100] or two vertical transistors [58], yielding $S_R = 0.5/$T [58]. The so-called parallel-stripe LMT [101] is sensitive to the magnetic field parallel to the chip surface. $S_R \approx 0.08/$T has been reported for an n-p-n version of this device [102]. In a recent optimized design [103], S_R is proportional to the current supplied and is as high as 4/T with 0.5-mA supply current. Another recent CMOS LMT reaches 1.5/T [104]. The pre-1982 MT work is reviewed in [8].

In contrast to the bulk Hall and MAGFET devices presented in Sections II and III, the MT sensitivities reported hitherto cover a surprisingly wide range, viz. between 10^{-2}/T and 4/T. This large sensitivity spread indicates that the understanding of MT operation is a rather involved problem. Indeed, galvanomagnetic effects may combine with effects inherent in bipolar transistor action in many ways, resulting in a variety of operating principles. Three combined effects basic to MT operation have been identified hitherto:

i) The *MT deflection effect* refers to the action of the Lorentz force on *minority* carriers in the base region, base-collector depletion layer, and low-doped collector region (where they turn into majority carriers) of the MT.

ii) The *MT Hall effect* refers to any action of the Hall electric field generated by the Lorentz force acting on *majority* carriers in the base region.

iii) The *MT magnetoconcentration effect* results from a competition of the Lorentz force on *both types* of carriers with their tendency to maintain zero space charge by adjusting their concentration.

In general, these three effects coexist, but the one or the other may prevail under specific design and operating conditions and eventually provide useful sensor action, as described in the following.

MT Deflection Effect

This effect prevails in the split-collector-contact vertical n-p-n transistor devised by Zieren and Duyndam [99] (see Fig. 13). Electrons injected from the heavily doped emitter move generally downwards through the base and low-doped collector regions until they reach the two buried collector contacts. With zero magnetic field, the collector current I_C is split into $I_{C1} = I_{C2} = I_C/2$, provided that the structure is perfectly symmetrical. With a field B parallel to the chip surface and perpendicular to the main current flow (i.e., perpendicular to the drawing plane in Fig. 13), the current beam is deflected under the Hall angle (9), (10) and a small current imbalance $\Delta I_C = I_{C1} - I_{C2}$ results. This simple model readily leads to the sensitivity estimate

$$S_R \approx 2\mu_n^* L/W_E \qquad (57)$$

where L denotes the emitter–collector distance and W_E the emitter width (see Fig. 13). Measured sensitivities range from 0.03 to 0.05/T and are about 2.5 times smaller than predicted by (57). This discrepancy may be due to the

current beam spread and the deflection efficiency reduction discussed in the next paragraph. A four-collector version of this MT is capable of measuring the magnitude and direction of the vector $B = (B_x, B_y, 0)$ in the chip plane (vector sensor) [99], [105]. A recent three-dimensional (3-D) vector sensor uses vertical and lateral components of the collector current [173].

S_R cannot be arbitrarily increased by increasing L/W_E, because the electron spread would invalidate the model underlying (57). Moreover, the carrier deflection becomes less effective in long ($L \geq W_E$) configurations, where it is eventually canceled by the Hall field developing across the electron beam. Deflection prevails only in short ($L \leq W_E$) configurations (see Section I), where the Hall electric field is shunted by the contacts. The latter argument is valid only for the majority carriers in strongly extrinsic samples such as the collector region of the MT shown in Fig. 13, or in depletion layers [98].

Minority-carrier deflection, occurring, for example, in the MT base region, requires a separate discussion. Here, the deflection efficiency is not linked to sample geometry. The majority carriers cancel the Hall field that the minority-carrier flow would produce, in the same way as they cancel any other space-charge effect. Thus sensitivity can be increased by using minority-carrier deflection in a long-*base* region. Values of 0.07 [100], 0.31 [107], 1.2 [108], and 1.5/T [104], [106] have been reported for such MT.

The last mentioned example is the CMOS-compatible LMT devised by Popović and Widmer [104], [106]. Its structure, shown in Fig. 14, is similar to a conventional lateral

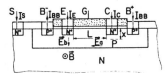

Fig. 14. Cross section of lateral magnetotransistor (LMT) in CMOS technology [104], [106]. Magnetic induction B (perpendicular to drawing plane) modulates the current ratio I_C/I_S. Modulation of I_C is used as sensor signal.

bipolar transistor with accelerating field in the base region ("drift-aided" LMT). The device is located in the p-well, which serves as base region. Two base contacts, B^+ and B^-, help to apply the accelerating voltage. The two n^+ regions, at the lateral base-length distance L apart from each other, serve as emitter (E) and primary collector (C). The substrate S works as secondary collector.

Due to the accelerating field E_a, the electrons are injected into the base region mostly by the emitter's right-hand side wall, they drift mostly along the base length, and get collected by the collector C, producing the collector current I_C. Some of them, however, diffuse downwards, get collected by the secondary collector S and produce the substrate current I_S. A magnetic field will produce a change in the ratio I_C/I_S. When the field is directed into the figure plane, the Lorentz force deflects the paths of the electrons towards the substrate junction and therefore less electrons contribute to the collector current. The reverse-biased base–substrate junction acts as a high recombining surface similar to that used in magnetodiodes (see Section V).

Building on the theory outlined in Section I, estimates of

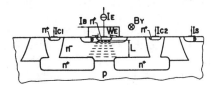

Fig. 13. Cross section of n-p-n vertical magnetotransistor (VMT) made in bipolar IC technology [8], [99]. Magnetic induction B_y causes asymmetry in collector currents, I_{C1}, I_{C2}.

the sensitivity under low injection condition, $p \gg n$, can be obtained for two operating limits. For small E_a, diffusion prevails and $S_R \approx \mu_n^* L / x$, where L denotes the emitter–collector distance and x is a geometrical fitting parameter, $x_{jn} < x < x_{jp}$, see Fig. 14. For strong E_a, $S_R \approx (\mu_p^* + \mu_n^*) L / x$. The terms containing μ_n^* describe the pure deflection, while the term with μ_p^* accounts for some additional, "indirect" deflection of the minority carriers by the Hall electric field generated by the majority carriers. The sensitivity estimates are supported by experiments [104], [106] with L/x between 10 μm/6 μm and 60 μm/6 μm. This confirms the role of minority-carrier deflection and its independence of sample geometry!

A deflection-type LMT sensitive to a magnetic field perpendicular to the chip surface is shown in Fig. 15. This is

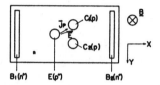

Fig. 15. Drift-aided LMT [100]. Magnetic induction B leads to rotation of electric field E with respect to x-axis and rotation of current density J_b with respect to E.

the drift-aided lateral double-collector p-n-p transistor devised by Davies and Wells [100]. Emitter and both collectors are embedded in an n-type Hall plate, which forms the base region. The base may be part of the n-type epitaxial layer, so that the device is compatible with standard bipolar IC technology. In spite of their structural dissimilarity, the devices shown in Figs. 14 and 15 share the operating principle of minority-carrier deflection, and their sensitivity is described by similar equations. Surprisingly, however, the difference in doping type (n-p-n versus p-n-p) does matter. Because of $\mu_n^* \geq 3\mu_p^*$ in silicon, direct deflection of minority carriers (electrons) makes the major contribution to the sensor action in the n-p-n device, whereas indirect deflection of minority carriers (holes) by the Hall field generated by the majority carriers (electrons) is dominant in the p-n-p device. This is only one example of the Hall field action, which is discussed in more detail below.

MT Hall Effect

Let us consider the base region of a drift-aided LMT such as shown in Fig. 15. For simplicity, we assume operation in the forward active regime and strong accelerating field, so that minority-carrier diffusion can be neglected. In the presence of a magnetic field, a Hall electric field $E_H = R_H J_b \times B$ develops (see Section I) with J_b denoting the total current density in the base region. Under *low*-level injection conditions, the Hall coefficient R_H is determined by the majority-carrier concentration, viz. $R_H = -r_n/qn_b$ for p-n-p transistors and $R_H = r_p/qp_b$ for n-p-n transistors, where n_b and p_b denote the pertinent carrier concentrations in the base region. Thus the base region behaves as an extrinsic Hall plate.

Under *high*-injection conditions, however, both carrier concentrations n_b and p_b must be taken into account [110]. Inserting $p_b \approx n_b$ into (12) and assuming $r_n \approx r_p \approx 1$, we

obtain

$$R_H \approx (qp_b)^{-1}(1 - \mu_n^*/\mu_p^*)(1 + \mu_n^*/\mu_p^*)^{-1}. \qquad (58)$$

Since p_b is proportional to the injected current, this leads to $R_H \propto I_C^{-1}$, i.e., the Hall field eventually decays with increasing collector current. Moreover, R_H of the base region of an n-p-n transistor changes its sign when going from low ($R_H > 0$) to high ($R_H < 0$ since $\mu_n^* > \mu_p^*$) injection. (A similar effect occurs in p-type semiconductor samples when intrinsic condition is established by heating.)

Depending on the position in the base region and the relative directions of the minority-carrier current and the Hall field, the action of the Hall field on the minority carriers can result in three different effects, viz.

i) contribution to the minority-carrier deflection (already mentioned above as "indirect" deflection),
ii) modulation of the base transport factor,
iii) injection modulation.

The *contribution to deflection* (i) has been studied in detail [100], [104], [106], [111], [112]. When B is perpendicular to the p-n-p device plane, the action of the Hall field E_H will result in a rotation of the total electric field E in the base region by the Hall angle (9) for electrons. The injected hole current, however, in addition rotates by the angle $\theta_H = \arctan(\mu_p^* B)$ with respect to E. Thus the resulting deflection angle of the minority (hole) current-density vector with respect to the x-axis (see Fig. 15) is given by the angle

$$\theta_p = \arctan(\mu_n^* B) + \arctan(\mu_p^* B). \qquad (59)$$

This deflection is appreciable provided that i) the Hall field is not prevented by shunting effects or high-injection conditions and ii) majority- and minority-current densities have essentially parallel directions for $B = 0$.

Modulation of the base transport factor (ii) plays a role in the parallel-stripe LMT [101], [102] shown in Fig. 16, where the latter of the above conditions is not fulfilled. In the

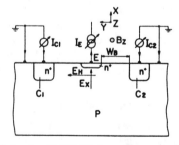

Fig. 16. Parallel-stripe LMT [101]. Prevailing operating principle depends on design and operating conditions.

active part of the base region (under the emitter and between the two collectors $C1$, $C2$) the majority carriers drift generally along the x-axis, whereas most of the minority carriers contributing to the collector currents diffuse laterally along the y-axis. Thus the Hall field in the base region does not contribute appreciably to the minority-carrier deflection, but rather modulates the transport factor [113]: In the base region close to $C1$, electron diffusion is hindered by $E_H \| y$, while diffusion is enhanced close to $C2$. This increases the current gain for $C2$ at the expense of $C1$ and results in a current imbalance $I_{C1} - I_{C2}$. Modulation of the base-transport factor occurs in most MT structures in one way or another [106], [114] and is also basic to the early MT work [95], [96]. A common feature of the two effects (i) and (ii) is the involvement of minority carriers all over the base region.

The term *injection modulation* introduced by Vinal and Masnari [102] refers to the action of the Hall field on the process of injecting minority carriers into the base region. This effect has been studied by many investigators [58], [102], [103], [110], [115]–[117] and was first exploited in the sensor devised by Takamiya and Fujikawa [58], see Fig. 17.

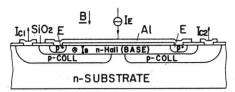

Fig. 17. Cross section of merged combination of Hall plate and two bipolar transistors [58]. Injection modulation is the prevailing operating principle [8].

This is a merged combination of a horizontal Hall plate and a pair of vertical bipolar transistors. Two p$^+$-emitters are diffused into the n-type Hall plate instead of the usual Hall contacts. Two separate n-collectors embrace the Hall plate, isolating it almost completely from the n-substrate. The Hall plate serves as the common base of the two vertical p-n-p transistors. Since the transistor pair is a part of a differential amplifier, the device is called "differential amplification magnetic sensor" (DAMS). In operation, the Hall plate biasing current I_B flows along the base region; its direction is perpendicular to the drawing in Fig. 17. With B perpendicular to the chip surface, a Hall voltage appears across the base region and is directly converted into a difference in injection level between the two emitters, which are kept at the same potential. This finally results in an imbalance between the two collector currents. A double-diffused version (D^2DAMS) of this device has been reported recently [59]. Of course, the device shown in Fig. 17 can be viewed as a Hall plate combined with bipolar transistors. It is reported here for the purpose of clearly demonstrating the MT injection modulation principle which states: The Hall voltage, generated by the magnetic field acting on the *majority* carriers moving in the base region, modulates the minority-carrier injection efficiently at low injection level. At high injection level, the sensitivity decreases as was discussed above (58).

The role of injection modulation is less obvious [118], [119] in other MT structures such as the parallel-stripe LMT shown in Fig. 16. In order to enhance the effect in question, Popović and Baltes [103] modified the parallel-stripe LMT as follows. The emitter was purposely designed to yield low efficiency, allowing a larger base (majority-carrier) current and thus a larger Hall field. Moreover, the base region was surrounded by a deep diffusion ring in order to confine the base current in a thin slab (parallel to the drawing plane in Fig. 16). Between any points A and B located close to the emitter sidewalls (see Fig. 18), the Hall voltage is [103]

$$V_{A-B}^H \approx R_H B I_E / t \qquad (60)$$

where R_H denotes the Hall coefficient at low injection level, I_E the emitter current, and t the effective base stripe width in B-direction. Note that $I_E \approx I_B$ because of the low

Fig. 18. Hall voltage modulating carrier injection between points A and B is obtained by integration of Hall field $E_H = R_H J_B \times B$ along line AB [103].

emitter efficiency. Hall voltage in the emitter region is negligible because of high doping density and small aspect ratio (see Section II). The Hall voltage (13) produces an asymmetrical emitter–base bias, which leads to asymmetrical injection of minority carriers. Thus the resulting collector currents are

$$I_{C1,C2} \approx I_S \exp\left[q\left(V \pm V_{A-B}^H/2\right)kT\right] \qquad (61)$$

where I_S denotes the collector saturation current, V the equivalent emitter–base biasing voltage at $B = 0$, and T the absolute temperature. The corresponding sensitivity reads

$$S_R = qV_{A-B}^H / BkT = r_n I_E / n_b t k T. \qquad (62)$$

With $n_b \approx 10^{15}$ cm^{-3}, $t \approx 37$ μm, and $I_E = 0.5$ mA, one obtains $V_{A-B}^H/B \approx 100$ mV/T and $S_R = 4/T$ at room temperature, in agreement with the experimental results [103]. This high sensitivity is obtained only if the collector current is kept small ($I_C \lesssim 1$ μA in [103]). Otherwise, the sensitivity decreases, $S_R \propto I_C^{-1}$.

MT Magnetoconcentration Effect

The *magnetoconcentration effect* (iii) becomes relevant when indirect deflection and injection modulation are less effective, i.e., under high injection condition. In Fig. 19 we reconsider the parallel-stripe LMT [101] and emphasize the base-spreading resistance R between the base contacts and

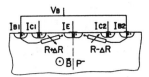

Fig. 19. Magnetoconcentration effect as operating principle of parallel-stripe LMT [101]. Magnetic induction B causes asymmetrical carrier injection leading to asymmetrical modulation $\pm \Delta R$ of base-spreading resistances R resulting in collector current imbalance.

the emitter junction. These resistances are modulated by the excess carrier junction. Assume that some of the previous MT effects incidentally produce a small imbalance, ΔR, between the two base-spreading resistances. This leads to an according imbalance in the base currents I_{B1}, I_{B2}. This results in an asymmetrical bias of the emitter–base junction, asymmetrical minority-carrier injection, and finally a further increase of the imbalance ΔR, which thus can become rather large! This effect amplifies the previous MT effects and compensates their reduced efficiency under high-injection conditions (see also Sections V and VI).

Figures of Merit

The *relative sensitivity* (56) of MT devices can be as high as 1.5/T [104] or 4/T [103]. This seems far superior to typical silicon Hall plates. However, the devices [103], [104] require large bias current not accounted for in definition (56). Their sensitivity looks less impressive and approaches that of silicon Hall plates if relative sensitivity is redefined [8] as

$$S_R^T \equiv \left| I_{tot}^{-1} \left(\partial I_C / \partial B \right) \right|_{B=0}. \tag{63}$$

I_{tot} denotes the sum of all currents needed to bias the MT, e.g., $I_{tot} = I_E + I_{BB}$ in the MT shown in Fig. 14. As opposed to the well-investigated Hall plates, MT devices still seem to have a large potential for further optimization in view of the variety of effects involved.

Noise affecting the collector current is $1/f$ noise and shot noise. The mean square shot noise current in a frequency band Δf is given by

$$\langle i_{CS}^2 \rangle = 2q I_C \Delta f. \tag{64}$$

The equivalent input magnetic field noise spectral density

$$S_B = d\langle B_{n,eq}^2 \rangle / df = S_R^{-2} I_C^{-2} d\langle i_{CN}^2 \rangle / df \tag{65}$$

has been analyzed [104] for the MT shown in Fig. 14. In (65), $d\langle i_{CN}^2 \rangle / df$ denotes the collector-current-noise spectral density, and S_R the relative sensitivity (56). Considering only shot noise and using $S_R = \mu_n^* L/x$, the estimate

$$S_B \leq 2qx^2 / \mu_n^* L^2 I_C \tag{66}$$

is obtained. Device geometry and biasing conditions can be optimized with respect to minimum S_B. For devices with $L = 300\ \mu m$ and $I_{tot} \approx 600\ \mu A$, $S_B^{1/2} < 0.4 \times 10^{-6}\ T/Hz^{1/2}$ was found for frequencies as low as 10 Hz.

To the best of our knowledge, very little has been published on MT $1/f$ noise. It may be speculated that some MT devices are superior to Hall plates in this respect in view of the small value of the Hooge α-parameter observed in bipolar transistors [72]. Recent experiments [120] show that the MT [104], [106] exhibits up to 100 times smaller S_B below 10 Hz than the vertical Hall device [56].

Offset in MT is due to the same basic effects as offset in Hall plates (see Section II). It can be expressed in terms of an equivalent magnetic induction $B_{0,eq}$ corresponding to the underlying collector current offset. In general, higher values of $B_{0,eq}$ are reported for MTs than for Hall plates. For example, the MT shown in Fig. 13 is prone to misalignment between the buried layer and the emitter masks, resulting in $B_{0,eq}$ as high as 1 T [121]. However, the sensitivity variation method [122], [123] provides a unique way to reduce offset in these devices by more than 95 percent. The offset current is eliminated electronically from two magnetic field measurements made with different absolute MT sensitivities. Variation of the sensitivity is obtained through variation of the collector–base voltage.

Linearity strongly depends on the underlying operating principle. MT devices based on carrier deflection (Figs. 13–15) are expected to be as linear as Hall plates. Indeed, linearity errors as small as 5×10^{-3} have been observed for

fields up to 1 T [99]. MTs based on injection modulation are inherently nonlinear at higher magnetic fields, cf. (60) and (61). For example, $B < 30$ mT must be observed in order to keep the linearity error below 6×10^{-2}.

Temperature coefficients may vary somewhat with the operating principle used, but are expected to be close to those of Hall plates, since $\mu_{n,p}^*$ or $r_{n,p}$ are always involved. This expectation is supported by experiments. For example, a value of -6.2×10^{-3}/K has been measured [99], which is close to that of TC_V of the Hall plate (49).

Frequency response of MT devices is limited by capacitance charging times and base transit times as in ordinary bipolar transistors. For the MT shown in Fig. 16, cutoff frequencies of the sensitivity range between 1 and 10 MHz [124]. Drift-aided MT might be somewhat better, but will probably not work faster than a Hall plate.

V. MAGNETODIODES

The magnetoconcentration (MC) effect modulates the resistance in the base region of bipolar lateral magneto-transistors under high-injection conditions, as outlined in Section IV. Magnetoconcentration plays a much more dominant role in magnetodiodes (MD). The MD operating principle is the superposition of the MC effect and double-injection phenomena. The electrons and holes injected by the end contacts of a semiconductor slab are deflected by magnetic field towards the same surface of the semiconductor, where the recombination process is responsible for the carrier-density modulation. The resulting transverse carrier gradient directly influences the diode current [125], [126].

Since the pioneering work of Stafeev [127] and Karakushan [128] about 25 years ago, a considerable amount of work has been devoted to the MD family of magnetic sensors. A book covering much of the older work on these devices has been published in 1975 [114]. An excellent introduction to magnetodiode sensors is found in an overview by Cristoloveanu [129]. MD sensors have been commercially available since about 15 years; see, e.g., [130]. Their sensitivity drops for magnetic field frequencies above 10 kHz.

The general magnetodiode structure is shown in Fig. 20.

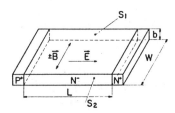

Fig. 20. Basic magnetodiode structure. Carriers injected by heavily doped p^+ and n^+ regions are deflected by Lorentz force against low-recombining (S_1) or high-recombining (S_2) surface.

Carriers are injected from the n^+ and p^+ regions into the low-doped semiconductor slab of length L, width W, and thickness b, where they drift under the action of the electric field E. The semiconductor slab has a low- and high-recombining surface with recombination rates s_1, s_2. Depending on the direction of the magnetic induction, B or $-B$, perpendicular to E, both electrons and holes are deflected by virtue of the Lorentz force, towards the surface S_1 or the surface S_2 of the slab. This leads to a carrier-concentration gradient perpendicular to E and B and finally to a magnetic field modulation of the diode current–voltage characteristics. Crucial requirements for the magnetodiode operation are, e.g., the difference between the recombination rates s_1 and s_2, the slab geometry (thickness b should be of the order of the ambipolar diffusion length), and high injection current in order to achieve the quasi-intrinsic condition.

Ge, Si, as well as GaAs have been used for the practical realizations of a first generation of *discrete* magnetodiodes. The difference in surface recombination rate was achieved through a difference in mechanical surface roughness (e.g., by grinding one surface and polishing and etching the other). A generation of *integrated* magnetodiodes was developed within the SOS (silicon-on-sapphire) integrated circuit technology (see, e.g., [131]–[133] and references therein). These devices elegantly exploit an inherent weakness of the SOS technology, i.e., the high recombination rate of the Si/Al_2O_3 interface, and use the Si/SiO_2 interface as the low recombination surface. A remaining disadvantage is the poor reproducibility of the Si/Al_2O_3 interface. SOS technology is not a standard technology available everywhere.

It is desirable to exploit the powerful magnetodiode effect in a sensor that can be manufactured by a standard bulk IC process such as CMOS. This can be realized by a structure reminiscent of a bipolar transistor, but whose operation is essentially that of a magnetodiode. A reverse-biased p-n (collector) junction plays a role similar to that of the high-recombining Si/Al_2O_3 interface of the SOS MD, i.e., recombination is replaced by collection [134], [135]. Sensors of this type have been manufactured using a standard CMOS process [135]. Its sensitivity is up to 25 V/T at 10-mA current consumption. This compares well with the previous SOS MD.

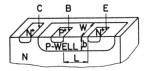

Fig. 21. Integrated magnetodiode made by standard p-well CMOS process [135] with reverse-biased p-well–substrate junction acting as high-recombining surface.

Fig. 21 shows the basic device geometry in the case of a p-well process. The p-well is used as the base active region, the substrate serves as collector region, and n^+ and p^+ source and drain diffusions are used to create the emitter, base, and collector contact regions. A Si/SiO_2 interface (not shown in Fig. 21) is located on top of the p-well between the B and E contacts and provides the low-recombining surface. The corresponding n-well devices have been fabricated as well and show higher sensitivity than p-well devices, in particular with a long base (126 μm). The device is operated under high injection condition. The EB junction is forward-biased and the BC junction is reverse-biased. In contrast to usual bipolar transistor operation, the useful signal is obtained between the E and B terminals, and the active (i.e., magnetic-field-sensitive) region is the lightly doped base region between the chip surface, E and B contacts, and the collector junction.

VI. CARRIER-DOMAIN MAGNETOMETERS

A carrier domain in semiconductor material is a region of high nonequilibrium carrier density. The domain actually consists of an electron–hole plasma, since the concentrations of both types of carriers are approximately equal as a consequence of the charge-neutrality condition. Localization of the domain is enforced by a suitable potential distribution or by some positive feedback mechanism. The concept of carrier domain [136] has been demonstrated in a bipolar transistor with elongated emitter. Spatial positive feedback can be used for domain enhancement in a p-n-p-n structure [137]. Domain formation has also been realized in an n-p-n structure working in the breakdown region [138].

Carrier-domain magnetometers (CDM) are semiconductor magnetic field sensors exploiting the action of the Lorentz force on the charge carriers moving in the domain. Under this force, the entire carrier domain eventually travels through the semiconductor, or a domain migration caused by some other effect is modulated. In each case, detection of the domain migration provides information on the magnetic field. Vertical four-layer [137], [139] CDMs as well as circular four-layer [140]–[146] and circular three-layer [147] CDMs have been investigated.

The *vertical four-layer CDM* [139] shown in Fig. 22 has a structure similar to that of the MT shown in Fig. 13. Both

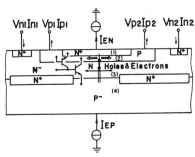

Fig. 22. Cross section of vertical four-layer carrier-domain magnetometer (CDM) [139], equivalent to merged combination of n-p-n and p-n-p transistors, as indicated by circuit symbols.

devices are vertical bipolar transistors with split buried collector contacts and can be fabricated using bipolar IC technology, but differ in the mode of operation. In MT operation, the junction between p-substrate and n-epitaxial layer is always reverse-biased and serves as an isolation layer only. In CDM operation, this junction is forward-biased for the purpose of domain formation.

The n-p-n-p structure can be viewed as a merged combination of n-p-n and p-n-p transistors that share a common base–collector junction (2/3 in Fig. 22). In operation, this junction is always reverse-biased. Both transistors are biased by emitter current sources and operate in the forward active region. Electrons are injected into the base of the n-p-n transistor (layer 2) and collected in the p-n-p base (layer 3), which simultaneously serves as the n-p-n collector. Similarly, holes are injected from the p-n-p emitter (layer 4) and collected in the n-p-n base (layer 2). Due to the lateral voltage drops in both bases, the carrier injection occurs in two opposite small spots of the base–emitter junctions, i.e., a carrier domain forms. The domain consists of electrons and holes moving in opposite directions and thus can be viewed as a current filament.

A magnetic field perpendicular to the plane of Fig. 22 produces a displacement of the domain. If, for example, the domain moves to the right, both right-hand base currents, I_{p2} and I_{n2} increase, while both left-hand base currents, I_{p1} and I_{n1}, decrease. This current modulation indicates the domain displacement and hence the presence of the magnetic field. On the other hand, the asymmetry produced by the domain displacement eventually brings about a restoring force that prevents further displacement of the domain. The sensitivity of this device [139] is 3 mA/T with 10-mA current drive. In terms of the relative sensitivity (63) this amounts to $S_R^T = 0.3/T$, which is among the highest sensitivity figures reported for silicon MFS hitherto. The equivalent noise magnetic field is 5 μT (no frequency range is given). The high temperature coefficient (about 3 percent/K) is a disadvantage of this device.

The *circular four-layer CDM* proposed by Gilbert [140] and realized by Manley and Bloodworth [141] is shown in Fig. 23. The purpose of the circular structure is to avoid the

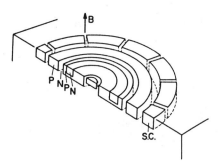

Fig. 23. Circular four-layer CDM [140], [141] with four layers labeled *P, N, P, N*. Current pulses appear at segmented collectors *S.C.* with frequency proportional to magnetic induction *B*.

restoring forces that would limit the migration of the domain. Thus the domain travels around the circumference of the structure. The frequency of this rotation is proportional to the magnetic field. This generation of an ac output signal is a unique feature of the circular CDM. The device basically consists of a ring-shaped lateral four-layer n-p-n-p structure. Carrier-domain formation and localization as well as the action of the magnetic field rely on the same principles as in the vertical four-layer structure described above. The four layers in the circular structure alternate in the radial rather than in the vertical direction. The axis along the domain points through the center of the circle. The actual domain position is detected by monitoring currents at the segmented outer collectors. The vertical n-p-n transistor is essentially parasitic.

A magnetic induction *B* perpendicular to the device surface produces a Lorentz force acting on both types of carriers in tangential direction. This results in a domain rotation, whose estimated [142] frequency is

$$f = d\mu_p^* B / 2\pi t_p R. \tag{67}$$

Here *d* denotes the effective width of the n-p-n transistor base region, t_p the p-n-p transistor base region charging time, *R* the device radius, and $\mu_p^* B$ the Hall angle of the holes. Frequency (indicating the magnitude of *B*) and direction (orientation of *B*) of the rotation are obtained from monitoring the current pulses appearing in the segmented outer collector whenever the domain passes by.

Experiments [141]–[145] show that the linear dependence (67) between the output frequency *f* and the magnetic induction *B* holds within 1-percent accuracy provided that $B \geq 0.2$ T. The domain rotation stops if *B* drops below a threshold value, which is between 0.1 and 0.4 T. It is believed that the domain may stick to a preferred location as a consequence of geometrical imperfections due to alignment tolerances. Sensitivities are between 10 and 100 kHz/T at about 10 mA supply current. Beside the high-threshold field, another drawback of this device is its large temperature coefficient: the rotation frequency is proportional to $T^{-4.4}$, i.e., $f^{-1}(\partial f/\partial T) \simeq -0.015$. This is about twice the temperature coefficient TC_V of the silicon Hall plate (49). Temperature compensation by linear variation of the CMD bias with temperature [146] allows to keep the rotation frequency within ± 2 percent of the central value over $-18°C < T < 102°C$ with 57 kHz/T sensitivity.

A *circular three-layer CDM* as devised by Popović and Baltes [147] is shown in Fig. 24. The sensor has the structure of a circular lateral bipolar n-p-n transistor surrounded by four voltage probing contacts $S_1, \cdots, S_4$. The diameter of the

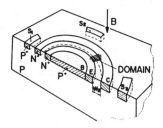

Fig. 24. Circular three-layer CDM [147], with carrier domain in the base region of circular, lateral n-p-n transistor operating in breakdown regime. Frequency of domain rotation is modulated by magnetic induction *B* and detected at contacts $S1$ to $S4$.

collector junction is 500 μm and the width of the base region 8 μm. The device operates in the collector–emitter breakdown regime with short-circuited emitter and base contacts. The carrier domain appears for sufficiently large supply current (about 30 mA), which is supplied by an external current source. The internal feedback involved in the transistor breakdown mechanism confines the current domain to a narrow sector of the base region. In contrast to the above four-layer CDM, no threshold magnetic induction is required, since the domain rotates spontaneously with $f = 280$ kHz at $B = 0$, the direction of the rotation being incidental. The domain rotation is detected from voltage pulses appearing at the contacts $S_1, \cdots, S_4$ whenever the domain passes by.

The angular frequency of the carrier domain rotation is modulated by a magnetic field perpendicular to the device planar surface. The rotation frequency is enhanced or lowered, depending on the relative directions of *B* and the domain rotation. The sensitivity is about 250 kHz/T around zero magnetic induction. The rotation direction is reversed when the frequency drops down to about 150 kHz. The overall plot of rotation velocity versus magnetic induction shows a hysteresis loop.

VII. INCORPORATION OF MFS INTO INTEGRATED CIRCUITS

Input transducer (sensor), signal processor, and output transducer (actuator) are three principal components of an information processing system. The central part of the system, the signal processor, is definitely integrated. This brings about a strong quest for integrating the input transducer on the signal processing chip or a suitable part of the signal processing circuitry on the sensor chip. Such integrated combinations of sensor elements and processing circuitry are often referred to as "smart sensors" [1]–[3], [6], [148].

Distinct advantages of smart sensors are *low cost* (use of well-established batch IC fabrication), *high performance* (*in situ* compensation of sensor nonlinearity, temperature coefficient, and offset, as well as prevention of disturbances), and *high system reliability*. This statement, of course, holds as well for magnetic field sensors and is the reason why this review aims at *integrated silicon* MFS. The important issue of technological compatibility of the various MFSs with standard IC processing is addressed in each of the preceding sections. Now we consider the MFS as an electronic device in interaction with its direct electronic circuit environment such as bias circuitry and first amplifier stage. More remote circuit functions such as A/D conversion, data storage, or output conditioning, are beyond the scope of this review.

Biasing Circuitry

Semiconductor MFSs are modulating transducers [64] and thus require a source of electric energy. Most MFSs are biased either by a constant voltage or by a constant current source. Some MFSs, e.g., Hall plates, can function with each of these biasing methods, but the choice of bias may affect the sensor performance; see (48) and (49) in Section II. Other MFSs function only if biased in a specific way. For example, the vertical CDM shown in Fig. 22 needs two current sources (emitters) and at least one voltage source (base contacts). Another example is the MOS Hall plate [66] supplied by distributed current sources: here the outstanding sensitivity achieved is mainly due to the special biasing technique; see Section III.

The choice of the biasing method may also affect the first sensor signal amplification stage, as in the case of the Hall plate with current source (CS) bias shown in Fig. 25. The temperature-dependent voltage drop over the Hall plate or, worse, the application of an *ac current*, result in large excursions of the common-mode voltage at the sensor output. This requires a complicated structure of the instrumentation amplifier A for the differential sensor signal [149]. The problem can be circumvented by the biasing method [16] shown in Fig. 26. Here, the common-mode voltage results only from the input offset voltage of the operational amplifier OA and thus a rather simple amplifier A is sufficient. Moreover, this biasing scheme allows to double the current through the Hall sensor in the base of an ac sensor bias current.

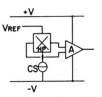

Fig. 25. Conventional current-biasing and signal-amplification circuitry for Hall plate produces high common mode at amplifier input.

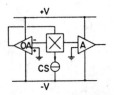

Fig. 26. Preferable biasing and signal amplification circuitry for Hall plate [16]: left sensor contact is virtually at ground so that full Hall voltage appears at right Hall contact against ground.

Amplifier

Even under modest signal processing requirements the MFS chip should definitely include some amplification boosting the small sensor signal above the noise level of the subsequent circuitry. Here, "noise" is meant to comprise both fluctuations and interfering signals. Conventional and "merged" amplifiers may serve this purpose.

Conventional amplifiers are used whenever the sensor is self-contained, i.e., when appropriate bias is sufficient for its operation and a signal suitable for direct amplification is delivered. An example is the Hall plate, where the Hall voltage appears directly between the Hall contacts and a differential amplifier is sufficient; see Figs. 25 and 26. Suitable amplifier designs are available [150].

Merged amplifiers can be designed when the sensor element lends itself to being part of an amplifier. Examples are the split-drain MAGFET when working in saturation and bipolar magnetotransistors. In view of their extremely high output resistance, high-load resistors (see Fig. 12) may be used in order to obtain large output voltages directly at the MT drains or collectors. Overall absolute sensitivities as high as 470 V/T have been achieved in this way [101]. The DAMS [58] is one of the early merged MFS–amplifier combinations; see Fig. 17.

An example of a fully integrated MFS–amplifier circuit [91] is shown in Fig. 27. The circuit structure is reminiscent of the well-known CMOS differential amplifier with dynamic load. Each of the usual two complementary MOS transistor pairs is replaced by a single split-drain MAGFET (see Fig. 10). The MAGFET drains are cross-coupled. A magnetic field normal to the chip surface affects the currents in both split-drain transistors, leading to an increase of the current in one drain at the expense of the current in the other drain. The drain with increasing current of the one transistor is connected to the drain with decreasing current of the complementary transistor, and *vice versa*. Thanks to this cross coupling in connection with the usual dynamic load technique, small imbalances in the drain currents (typically a few percent per tesla) lead to large output voltage variations. Four different functions are simulta-

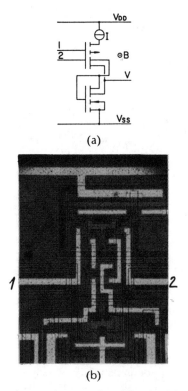

(a)

(b)

Fig. 27. Merged magnetic sensor/amplifier circuit in CMOS technology [91] with two complementary split-drain MAGFET devices replacing the usual two pairs of MOS transistors in a CMOS differential amplifier. Sensitivity is 1.2 V/T at 10 V and 100 μA supply.

neously provided by the split-drain transistor pair: magnetic field sensing, working point control, differential-to-single-ended conversion, and gain. An absolute sensitivity of about 1.2 V/T at 10 V supply voltage and 0.1 mA bias is obtained in this way. Considering the entire circuit as a sensor and applying the definitions (24) and (25), we obtain the sensitivities $S_{RI} = 1.2 \times 10^4$ V/A $\cdot$ T and $S_{RV} = 0.12$/T. These figures are larger than those for any stand-alone bulk or MOS Hall plate. It has to be borne in mind that the CMOS sensor shown in Fig. 5 internally amplifies the signal of its sensing element together with the respective noise. Therefore, the SNR of this sensor is not necessarily superior to that of the stand-alone sensing element.

The merged-amplifier strategy is straightforward in an IC environment. The advantages over the conventional-amplifier approach are less total supply current and lower equivalent input noise for a given supply current. On the other hand, it is easier to optimize a sensing element and an amplifier separately than simultaneously.

As in all electronic systems designed for low-power operation, noise most severely limits the performance of integrated MFS. Even when relatively large SNR may be obtained for the sensing element [68], amplifier noise may still be a limiting factor; often the rather small sensor signal is comparable to the equivalent input noise voltage of the amplifier. As illustrated in [74], even a low-noise amplifier [150] can deteriorate the noise behavior of an integrated Hall sensor or may become the major current consumer on the sensor chip.

Correction Circuits

Efficient *in situ* compensation of various sensor imperfections can be provided by additional devices or circuits available on the MFS chip.

Offset compensation of integrated Hall plates can be achieved by connecting two or four identical square-shaped devices in an orthogonal coupling arrangement [82]. Offset due to both piezoresistivity and misalignment is reduced by a factor of ten in this way. An IC implementation of the sensitivity-variation approach to offset reduction [122], [123] for bipolar MT has been reported recently [151].

Temperature compensation for MFS is usually approached by measurement of the chip temperature and corresponding variation of the biasing conditions. The Hall device lends itself to temperature detection through the thermal variation of the resistance between its current contacts. At approximately constant bias current, this change of resistance produces a proportional variation of the voltage drop over the device, which may be used to control the device current [44], the effective device thickness [84], or to correct offset [152]. Temperature-dependent biasing conditions are also used for the rotating carrier domain magnetometer [146].

A recent novel approach to temperature compensation [153] is based on continuous measurement of the CDM sensitivity (Fig. 22) and corresponding correction of the amplifier gain. The sensitivity is measured using a known magnetic induction generated by an integrated flat coil driven by a current source.

Magnetically Controlled Oscillators

The integration of split-drain MAGFET with voltage-controlled [174] and current-controlled [175] oscillators has been achieved recently. The output voltage of the resulting magnetic-field-sensitive integrated circuit is frequency-modulated by the input magnetic signal.

VIII. MODELING

Device modeling means to produce a mathematical representation (model) of a device. A device model allows one to *analyze* the device operation, provided that the model adequately incorporates the pertinent physical effects, possibly in a very complex manner [37]. A model for the purpose of device *simulation*, such as a curve-fitting model describing some device characteristics, is less demanding and may address the underlying physics in a heuristic or qualitative way. Our main interest here is in modeling for device analysis. *Semiconductor* device modeling under the assumption of zero magnetic induction is a highly developed field of research [37], [154]–[156]. Much of the current work in this area involves the numerical solution of the nonlinear system of partial differential equations (PDE) comprising Poisson's equation, the continuity equations, and the current relations [37]. Modeling of integrated semiconductor magnetic field sensors (MFS) requires the solution of a modified PDE system that accounts for the influence of the Lorentz force on carriers under nonzero magnetic induction [8], [35], [53], [62], [92], [93], [157]–[161].

Why Modeling?

As we have seen in Sections II–VI, sensitivity estimates in terms of simple expressions based on qualitative concepts of MFS operation are viable only in specific limiting cases of magnetic field orientation, device geometry, material parameters, and operating conditions. In general, more realistic situations, device modeling based on the solution of the underlying carrier equations of motion is indispensable for understanding and optimizing the MFS operation.

The development of a new semiconductor device involves several design and fabrication cycles until a specification goal is reached. Device modeling can substantially reduce the number of costly trial-and-error steps in this development [37]. Device modeling is, however, not only driven by development cost reduction, but also by scientific curiosity. Modeling provides unique insight into the functioning of devices by means of the distributions of various physical quantities in the interior of the device, which are not readily accessible to experiments (see, e.g., Figs. 2(a), 6, and 11).

Semiconductor MFS Modeling

Much of the older work in this area is devoted to strongly extrinsic Hall plates, where the voltage distribution is described by a potential satisfying the Laplace equation in two dimensions [49]. Solutions can be found using conformal mapping techniques, the finite-difference approximation, or the boundary-element method [34]. Recent Hall plate modeling addresses the influence of finite contacts [162] and the response to nonuniform magnetic fields [163].

Since the fundamental paper by Pfleiderer [125], analytical modeling of the magnetoconcentration effect and the discrete and SOS magnetodiodes has been pursued, notably by the Grenoble Group [126], [164]. Numerical modeling results are now available for the magnetotransistor [8] shown in Fig. 13, the vertical Hall device [62], [93], [160] shown in Figs. 5 and 6, and the split-drain MAGFET [35], [92], [93], [159]–[161], see Figs. 10 and 11.

Beside the analytical or numerical solution of the underlying carrier equations of motion, a third type of approach is to simulate an equivalent network or circuit. For instance, the channel region of a MOS device can be represented by a net of identical circuit cells. This equivalent circuit can be simulated using existing computer programs such as SPICE. Various MAGFET sensors have been modeled in this way [165], [166].

Basic Equations

Under the usual assumptions [37], but admitting a not too large magnetic induction B, the carrier motion in a semiconductor MFS is approximately described by the following nonlinear PDE system [32]–[35], [53], [125]:

$$\Delta\Psi = (q/\epsilon)(p - n + N) \tag{68}$$

$$\nabla \cdot J_n - q\partial n/\partial t = qR(\Psi, n, p) \tag{69}$$

$$\nabla \cdot J_p + q\partial p/\partial t = -qR(\Psi, n, p) \tag{70}$$

$$J_n = J_n^0 - \mu_n^* J_n \times B \tag{71}$$

$$J_p = J_p^0 + \mu_p^* J_p \times B \tag{72}$$

where

$$J_n^0 = J_n(B = 0) = \sigma_n E + qD_n\nabla n \tag{73}$$

$$J_p^0 = J_p(B = 0) = \sigma_p E - qD_p\nabla p. \tag{74}$$

Equations (71) and (73) are identical with (1) and (2) of Section I. Notation is as in Section I, (1) and (12). Moreover, Ψ denotes the electrical potential, viz. $E = -\text{grad } \Psi$, ϵ the (uniform) electrical permittivity, N the ionized net impurity concentration, and $R = R(\Psi, n, p)$ the net recombination rate.

We notice that Poisson's equation (68) and the continuity equations (69) and (70) have the same form as for $B = 0$. The Lorentz force manifests itself in the current relations (71) and (72). Obviously the usual current relations are recovered in the limit $B = 0$. We recall that (71) and (72) result from a weak magnetic field expansion [32], [33] and are meaningful for $\mu_n^* B$ and $\mu_p^* B \lesssim 0.3$. By substituting the current relations (71) and (72) into the continuity equations (69) and (70), we can obtain a system of three PDEs with dependent variables Ψ, n, and p. For numerical treatment, these PDEs must be implemented with values of physical models for ϵ, N, R, μ_n^*, μ_p^*, σ_n, σ_p, D_n, and D_p. Some of these parameters may depend on n, p, Ψ, and moreover, on impurity concentrations (doping profiles) and temperature.

The PDE system (68)–(72) simplifies in special cases, e.g., B perpendicular to J_n, J_p, and E, negligible concentration gradients, or negligible concentration of one kind of carriers, as discussed in Section I. We stress that the full system including both types of carriers must be considered in, e.g., intrinsic material, magnetotransistors under high injection condition, or magnetodiodes with double injection of carriers, where magnetoconcentration effects prevail.

Solutions of (68) to (72) are subject to appropriate *boundary and interface conditions* based on physical models and operating conditions. Examples are constant electrical potential at ohmic contacts or vanishing normal components of the current densities at semiconductor–insulator boundaries (unless surface recombination has to be taken into account). Such "natural" boundary conditions are usually not sufficient to make the finite device domain under consideration self-contained. "Artificial" boundary conditions justified by physical or mathematical plausibility (and sometimes disputed) have to be imposed in addition.

Examples

Numerical MFS modeling is more involved than $B = 0$ semiconductor device modeling. In view of the term $J_n \times B$ in (71), at least two-dimensional modeling is required. Three-dimensional MFS modeling has not been achieved up to this time, but would be highly desirable in many cases. Moreover, the vector B invalidates some of the symmetries facilitating the numerical solution of the PDE system in the case $B = 0$. Finally, the proper choice of the boundary conditions, notably the artificial ones, is more crucial than in the case $B = 0$. For instance, the Lorentz force may drive carriers against a boundary.

Finite-difference [35], [53], [157]–[159] and, more recently, finite-element [160] schemes have been used for the discretization of the PDE system (68) to (72) in two dimensions, with **B** perpendicular to the current densities. The complex technicalities [35], [53] of the numerical solution procedure are beyond the scope of this review. The results provide electric potential and current distribution of the type shown in Figs. 2(a) and 11.

Magnetoconcentration effects are demonstrated in the modeling results for intrinsic or nearly intrinsic Hall plates and p-i-n diodes [35], [53], [158]. For example, Fig. 28 shows the equipotential and current lines of an intrinsic silicon Hall plate, sized 4 μm by 4 μm, under $B = 1$ T at temperature $T = 500$ K with 1-V applied voltage V_a. In contrast to

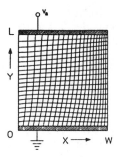

Fig. 28. Numerical modeling of intrinsic Hall plate [35], [53], [158] showing magnetoconcentration: current lines (ending on top and bottom contacts) and equipotential lines.

the extrinsic plate (Fig. 2(a)), current crowding close to the right-hand side open boundary is featured by Fig. 28. This corresponds to the carrier-concentration gradient across the sample shown in Fig. 29.

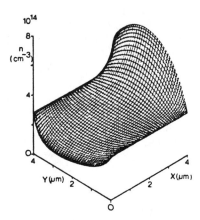

Fig. 29. Numerical modeling of intrinsic Hall plate [35], [53], [158] showing magnetoconcentration: electron concentration.

Split-drain MAGFET devices operating in the linear region were modeled using the same numerical procedure [35], [92], [93]. Current density and carrier concentration are integrated over the channel depth in order to obtain a two-dimensional problem. A typical result are the equipotential and current lines shown in Fig. 11. The current imbalance between the two drains produced by the (perpendicular) magnetic induction is demonstrated. The corresponding surface charge-density distribution is shown in

Fig. 30. The modeling results allow to predict the sensitivity of split-drain MAGFET devices and to choose an optimal device geometry [93], [159]. Combining the sensitivity results with current MOSFET noise models, the minimum detectable induction can be predicted as well [161].

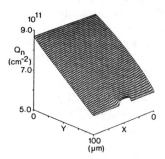

Fig. 30. Numerical modeling of split-drain MAGFET [35], [92], [93]: surface charge distribution (cf. Fig. 11 for equipotential and current lines).

The *vertical Hall device* [56] was modeled in two dimensions using a Green's function approach [62], [93] and the finite-element discretization [160] of the underlying PDE system (ALBERTINA program). Typical equipotential and current lines are shown in Fig. 6. Although the sensitivity calculated compares well with the experiment, the essential role of the p-well ring (Fig. 5) cannot be elucidated by two-dimensional modeling, but would require a three-dimensional treatment. This is beyond the present state of numerical MFS modeling.

Nonuniform magnetic induction, such as $B = (0, 0, B_z(x, y))$, leads to complex equipotential and current line patterns. This has recently been modeled for Hall plates [176] and split-drain MAGFETs [177] under locally inverted induction produced by, e.g., magnetic bubbles.

IX. Conclusions

A variety of integrated semiconductor MFSs and related circuitry and modeling has been reviewed in this paper. The magnetic field sensitivity of all these devices is based on the action of the Lorentz force on the carriers and exploits the Hall effect (*Hall field* or *carrier deflection*) in one way or another. *Magnetoconcentration* of carriers plays a role under intrinsic conditions, high-injection conditions, or double injection, and surface or interface recombination. The combination of Hall field, carrier deflection, or magnetoconcentration with the various basic (**B** = 0) device operating principles brings about the large number of different semiconductor MFSs considered.

There are no integrated Si or GaAs MFSs based on *magnetoresistance*, because the effect is too small in these materials. Integrated InSb magnetoresistive sensors, however, have been realized by Kataoka [12], [14]. They are successfully applied in displacement transducers (potentiometers), inclinometers, and vending machines (magnetic pattern verification of banknotes printed with magnetic ink). The relative resistivity increase is up to 15/T. Thus a sensitivity of 1.5×10^4 V/A · T is achieved with 1-kΩ magnetoresistive InSb elements.

A few integrated silicon MFSs seem not to fit into the device categories introduced in Sections II–VI, viz., the unijunction transistor (UJT) [167], [168], the junction FET

[169], the magnetic avalanche transistor [170], and the polar MOS sensor [171]. The UJT biased for operation close to the critical point (i.e., zero differential resistance) is sensitive to alternating magnetic fields [167].

In Sections II–VI we have analyzed the operation of selected device structures in simple, limiting cases, which allow to focus on one physical effect at a time. The investigation of mixed-mode operation and correspondingly optimized new device structures remains a challenging task, for which numerical MFS modeling provides an indispensable tool.

Which out of the many integrated silicon MFSs is the best? The answer to this question, as usual, depends on what is required by a specific application. Considering only MFSs compatible with standard IC technology (MOS, bipolar), we compile the following list for quick orientation.

The general requirement of high sensitivity favors MT devices (Figs. 14–17) using combined effects in the base region and the vertical CDM (Fig. 22). If high sensitivity at very low current is required, the split-drain MAGFET circuit (Fig. 27) is a good choice. Low noise at low frequency ($1/f$ region) is a strength of the MT shown in Fig. 14, whereas low noise at high frequencies is offered by bulk Hall plates such as shown in Figs. 4 and 5(a). Small offset is provided by horizontal integrated Hall plates (Fig. 4) in the combination [82]. Hall plates are the best choice for high linearity and high operating frequency. A small temperature coefficient is achieved with Hall plates supplied by a current source (Fig. 26).

By and large, integrated Hall plates seem to constitute the most mature family of integrated semiconductor MFSs. On the other hand, the CMOS-compatible LMT group seems to offer the best research value with respect to both scientific investigation and application orientation. The complex combinations of galvanomagnetic effects and bipolar action in the MT need further elucidation and offer a hardly exploited potential for device optimization. We give the lateral MT preference in view of the growing importance of CMOS technology, notably the current research efforts in analog CMOS integrated circuitry.

Acknowledgment

The authors wish to thank many colleagues and collaborators who supported our research in the MFS field and the accomplishment of this review: W. Allegretto (Edmonton, Canada), J.-L. Berchier (Zug, Switzerland), S. Cristoloveanu (Grenoble, France), P. Haswell (Edmonton), S. Kataoka (Nara, Japan), A. M. J. Huiser (Edmonton), A. Krause (Zug), S. Middelhoek (Delft, The Netherlands), A. Nathan (Edmonton), J. Petr (Zug), M. Rudan (Bologna, Italy), H. G. Schmidt-Weinmar (Edmonton), P. Schwendimann (Bern, Switzerland), A. W. Vinal (Raleigh, NC), E. A. Vittoz (Neuchâtel, Switzerland), R. Widmer (Zürich, Switzerland), and T. Zajc (Zug). The manuscript was prepared by June Swanson.

References

[1] W. G. Wolber and K. D. Wise, "Sensor development in the microcomputer age," *IEEE Trans. Electron Devices*, vol. ED-26, pp. 1864–1874, 1979.

[2] S. Middelhoek and D. J. W. Noorlag, "Signal conversion in solid-state transducers," *Sensors and Actuators*, vol. 2, pp. 211–228, 1982.

[3] W. H. Ko and C. D. Fung, "VLSI and intelligent transducers," *Sensors and Actuators*, vol. 2, pp. 239–250, 1982.

[4] K. E. Petersen, "Silicon as a mechanical material," *Proc. IEEE*, vol. 70, pp. 420–457, 1982.

[5] W. Heywang, *Halbleiter-Elektronik, Volume 17: Sensorik*. Berlin, Germany: Springer-Verlag, 1984.

[6] R. Allan, "Sensors in silicon," *High Technol.*, vol. 4, no. 9, pp. 43–50, 1984.

[7] J. H. Fluitman, "A survey of solid state magnetic field sensors," in *Summer Course 1982: Solid-State Sensors and Transducers, Volume II*, W. Sansen and J. Van der Spiegel, Eds. Leuven, Belgium: Katholieke Universiteit, 1982, pp. IX-1–23.

[8] V. Zieren, "Integrated silicon multicollector magnetotransistors," Ph.D. dissertation, Delft University of Technology, Delft, The Netherlands, 1983.

[9] H. P. Baltes, "Magnetic sensors," in *Trends in Physics 1984*, (Proc. 6th Gen. Conf. Europ. Phys. Soc., Prague, Czechoslovakia, Aug. 27–31, 1984, J. Janta and J. Pantoflíček, Eds., Europ. Phys. Soc.), 1985, pp. 606–611.

[10] S. Foner, "Review of magnetometry," *IEEE Trans. Magn.*, vol. MAG-17, pp. 3358–3363, 1981.

[11] A. W. Vinal, "Considerations for applying solid state sensors to high density magnetic disc recording," *IEEE Trans. Magn.*, vol. MAG-20, pp. 681–686, 1984.

[12] S. Kataoka, "New magnetoresistive sensors," *Sensors and Actuators*, to be published.

[13] Y. Kanda, M. Migitaka, H. Yamamoto, H. Morozumi, T. Okabe, and S. Okazaki, "Silicon Hall-effect power IC's for brushless motors," *IEEE Electron Devices*, vol. ED-29, pp. 151–154, 1982.

[14] S. Kataoka, "Recent developments of magnetoresistive devices and applications," *Circulars Electrotech. Lab.*, no. 182, (Tokyo, Japan, Electrotech. Lab.), 1974.

[15] W. G. Wolber, "Automotive engine control sensors '80," in *Sensors for Automotive Systems* (SP-458). Warrendale, PA: Soc. Automotive Eng., 1980, pp. 63–77.

[16] K. Matsui, S. Tanaka, and T. Kobayashi, "GaAs Hall generator application to a current and watt meter," in *Proc. 1st Sensor Symp. 1981*, S. Kataoka, Ed. (Tokyo, Japan, Inst. Elec. Eng. of Japan, 1982), pp. 37–40.

[17] J. E. Lenz, "Magnetic sensors," *Scientific Honeyweller*, vol. 6, no. 1, pp. 16–25, 1985.

[18] G. L. Romani, S. J. Williamson, and L. Kaufman, "Biomagnetic instrumentation," *Rev. Sci. Instrum.*, vol. 53, pp. 1815–1850, 1982.

[19] T. Hara, M. Mihara, N. Toyoda, and M. Zama, "Highly linear GaAs Hall devices fabricated by ion implantation," *IEEE Trans. Electron Devices*, vol. ED-29, pp. 78–82, 1982.

[20] S. C. Rashleigh and R. Ulrich, "Magneto-optic current sensing with birefringent fibers," *Appl. Phys. Lett.*, vol. 34, pp. 768–770, 1979.

[21] A. Papp and H. Harms, "Magneto-optical current transformer. 1: Principles," *Appl. Opt.*, vol. 19, pp. 3729–3834, 1980.

[22] J.-P. Castéra and G. Hepner, "Device for modulating optical radiation by a variable magnetic field," U.S. Patent 4 236 782, 1980.

[23] R. E. Jones, J. P. Willson, G. D. Pitt, R. H. Pratt, K. W. H. Foulds, and D. N. Batchelder, "Detection techniques for measurement of DC magnetic fields using optical fibre sensors," in *Optical Fibre Sensors* (IEE Conf. Publ. No. 221), D. E. N. Davies, Ed. London, England: Inst. Elec. Eng., 1983, pp. 33–37.

[24] A. Dandridge, A. B. Tveten, G. H. Sigel, Jr., E. G. West, and T. G. Giallorenzi, "Optical fibre magnetic field sensors," *Electron. Lett.*, vol. 11, pp. 408–409, 1980.

[25] T. R. McGuire and R. I. Potter, "Anisotropic resistance in ferromagnetic 3d alloys," *IEEE Trans. Magn.*, vol. MAG-11, pp. 1018–1038, 1975.

[26] D. A. Thompson, L. T. Romankiw, and A. F. Maydas, "Thin film magnetoresistors in memory, storage, and related applications," *IEEE Trans. Magn.*, vol. MAG-11, pp. 1039–1050, 1975.

[27] *International Magnetics* (*INTERMAG*) *Conference Proceed-*

ings, B. E. MacNeal, R. E. Fontana, Jr., and J. C. Smits, Eds. (*IEEE Trans. Magn.*, Session BE—Magnetic Sensors, vol. MAG-20, pp. 954–974, 1984).

[28] J.-L. Berchier, K. Solt, and T. Zajc, "Magnetoresistive switching of small permalloy sandwich structures," *J. Appl. Phys.*, vol. 55, pp. 487–492, 1984.

[29] H. Weiss, "Utility and futility of semiconductor effects," *Festkörperprobleme*, vol. 14, pp. 39–66, 1974.

[30] E. Pettenpaul, J. Huber, H. Weidlich, W. Flossman, and U. v. Borcke, "GaAs Hall devices produced by local ion implantation," *Solid-State Electron.*, vol. 24, pp. 781–786, 1981.

[31] T. Usuki, S. Sugiyama, M. Takeuchi, T. Takeuchi, and I. Igarashi, "Integrated magnetic sensor," in *Proc. 2nd Sensor Symp., 1982*, S. Kataoka, Ed. (Tokyo, Japan, Inst. Elec. Eng. of Japan), pp. 215–217, 1982.

[32] O. Madelung, *Introduction to Solid State Theory*. Berlin, Germany: Springer-Verlag, 1978, ch. 4.

[33] K. Seeger, *Semiconductor Physics*. Berlin, Germany: Springer-Verlag, 1982, chs. 4, 7, 8.

[34] G. DeMey, "Potential calculations in Hall plates," *Adv. Electron. Electron Phys.*, vol. 61, pp. 1–62, 1983.

[35] A. Nathan, "Numerical analysis of MOS magnetic field sensors," M.Sc. thesis, Univ. of Alberta, Edmonton, Alta., Canada, 1984.

[36] S. M. Sze, *Physics of Semiconductor Devices*, 2nd ed. New York, NY: Wiley, 1981, ch. 1.

[37] S. Selberherr, *Analysis and Simulation of Semiconductor Devices*. Vienna, Austria: Springer-Verlag, 1984.

[38] P. Norton, T. Braggins, and H. Levinstein, "Impurity and lattice scattering parameters as determined from Hall and mobility analysis in n-type silicon," *Phys. Rev. B*, vol. 8, pp. 5632–5653, 1973.

[39] Landolt-Börnstein, *Numerical Data and Functional Relationships in Science and Technology* (vol. III/17a, *Semiconductors*). Berlin, Germany: Springer-Verlag, 1982, pp. 380–381, and references therein.

[40] Y. Sugiyama, "Fundamental research on Hall effects in inhomogeneous magnetic fields," *Res. Electrotech. Lab.*, no. 838 (Tokyo, Japan, Electrot. Lab.), 1983.

[41] C. L. Chien and C. R. Westgate, *The Hall Effect and Its Applications*. New York, NY: Plenum, 1980.

[42] T. Lammerink, *Ontwikkeling van een Geïntegreerde Eenzijdige Hall Transducer. Deel II: Literatuuronderzoek.* Twente, The Netherlands: Technische Hogeschool, 1982 (92 references).

[43] S. Kordić, "Review of silicon magnetic field sensor," unpublished (89 references).

[44] F. Kuhrt and H. J. Lippmann, *Hallgeneratoren*. Berlin, Germany: Springer-Verlag, 1968.

[45] H. Weiss, *Physik und Anwendung Galvanomagnetischer Bauelemente*. Braunschweig, Germany: F. Vieweg & Sohn, 1969.

——, *Structure and Applications of Galvanomagnetic Devices*. Oxford, England: Pergamon, 1969.

[46] H. H. Wieder, *Hall Generators and Magnetoresistors*. London, England: Pion Ltd., 1971.

[47] J. T. Maupin and M. L. Geske, "The Hall effect in silicon circuits," in C. L. Chien and C. R. Westgate, *The Hall Effect and Its Applications*. New York, NY: Plenum, 1980, pp. 421–445.

[48] G. S. Randhawa, "Monolithic integrated Hall devices in silicon circuits," *Microelectron. J.*, vol. 12, pp. 24–29, 1981.

[49] R. F. Wick, "Solution of the field problem of the germanium gyrator," *J. Appl. Phys.*, vol. 25, pp. 741–756, 1954.

[50] H. J. Lippmann und F. Kuhrt, "Der Geometrieeinfluss auf den Hall-Effekt bei rechteckigen Halbleiterplatten," *Z. Naturforschung*, vol. 13a, pp. 474–483, 1958.

[51] J. Hoesler und H. J. Lippmann, "Hallgeneratoren mit kleinem Linearisierungsfehler," *Solid-State Electron.*, vol. 11, pp. 173–182, 1968.

[52] F. Kuhrt and H. J. Lippmann, *Hallgeneratoren*. Berlin, Germany: Springer-Verlag, 1968, p. 76.

[53] L. Andor, H. P. Baltes, A. Nathan, H. G. Schmidt-Weinmar, "Numerical modeling of magnetic-field sensitive semiconductor devices," *IEEE Trans. Electron Devices*, vol. ED-32, pp. 1224–1230, 1985.

[54] S. Kataoka, H. Yamada, Y. Sugiyama, and H. Fujisada, "New galvanomagnetic device with directional sensitivity," *Proc. IEEE*, vol. 59, p. 1349, 1971.

[55] J. H. J. Fluitman, "Hall-effect device with both voltage leads on one side of the conductor," *J. Phys. E: Sci. Instrum.*, vol. 13, pp. 783–785, 1980.

[56] R. S. Popović, "The vertical Hall-effect device," *IEEE Electron Device Lett.*, vol. EDL-5, pp. 357–358, 1984.

[57] G. Bosch, "A Hall device in an integrated circuit," *Solid-State Electron.*, vol. 11, pp. 712–714, 1968.

[58] S. Takamiya and K. Fujikawa, "Differential amplification magnetic sensors," *IEEE Trans. Electron Devices*, vol. ED-19, pp. 1085–1090, 1972.

[59] R.-M. Huang, F.-S. Yeh, and R.-S. Huang, "Double-diffusion differential-amplification magnetic sensor," *IEEE Trans. Electron Devices*, vol. ED-31, pp. 1001–1004, 1984.

[60] A. Thanailakis and E. Cohen, "Epitaxial gallium arsenide as Hall element," *Solid-State Electron.*, vol. 12, pp. 997–1000, 1969.

[61] T. Inada, T. Ohkubo, M. Kitahara, Y. Kanda, and T. Hara, "GaAs Hall effect devices fabricated by ion implantation technique," *Electron Lett.*, vol. 14, pp. 503–505, 1978.

[62] A. M. J. Huiser and H. P. Baltes, "Numerical modeling of vertical Hall-effect devices," *IEEE Electron Device Lett.*, vol. EDL-5, pp. 482–484, 1984.

[63] T. Janicki and A. Kobus, "Hall generator with variable active-layer thickness," *Electron. Lett.*, vol. 3, pp. 373–374, 1967.

[64] S. Middelhoek and D. J. W. Noorlag, "Silicon micro-transducers," *J. Phys. E: Sci. Instrum.*, vol. 14, pp. 1343–1352, 1981.

[65] Y. Takahana, K. Miyanchi, K. Tsuruta, and K. Tsuboi, "A saturation velocity magnetic sensor," in *ISSCC Dig. Tech. Papers* (Philadelphia, PA), vol. 24, pp. 42–43, 1981.

[66] R. S. Popović, "A MOS Hall device free from short-circuit effect," *Sensors and Actuators*, vol. 5, pp. 253–262, 1984.

[67] Y. Sugiyama and S. Kataoka, "S/N study of micro-Hall sensors made of single crystal InSb and GaAs," in *Transducers '85* (1985 Int. Conf. on Solid-State Sensors and Actuators, Dig. of Tech. Papers, W. H. Ko and K. D. Wise, Eds.), pp. 308–311, 1985, IEEE catalog no. 85 CH 2127-9.

[68] T. G. M. Kleinpenning, "Design of an ac micro-gauss sensor," *Sensors and Actuators*, vol. 4, pp. 3–9, 1983.

[69] F. N. Hooge, "1/f noise is no surface effect," *Phys. Lett.*, vol. 29A, pp. 139–140, 1969.

[70] T. G. M. Kleinpenning and L. K. J. Vandamme, "Comment on 'Transverse 1/f noise in InSb thin films and the SNR of related Hall elements'," *J. Appl. Phys.*, vol. 50, p. 5547, 1979.

[71] L. K. J. Vandamme, "Is the 1/f noise parameter α constant?" in *Noise in Physical Systems and 1/f Noise*, M. Savelli, G. Lecoy, and J.-P. Nougier, Eds. Amsterdam, The Netherlands: Elsevier, 1983, pp. 183–192.

[72] A. van der Ziel, P. H. Handel, X. Zhu, and K. H. Duh, "A theory of the Hooge parameters of solid-state devices," *IEEE Trans. Electron Devices*, vol. ED-32, pp. 667–671, 1985.

[73] L. K. J. Vandamme, personal communication.

[74] R. S. Popović, "A CMOS Hall effect integrated circuit," in *Proc. 12th Yugoslav Conf. on Microelectronics, MIEL '84* (Nis, Yugoslavia, 1984), vol. I, pp. 299–307.

[75] Y. Kanda and M. Migitaka, "Effect of mechanical stress on the offset voltage of Hall devices in Si IC," *Phys. Status Solidi (a)*, vol. 35, pp. K115–K118, 1976.

[76] G. Björklund, "Improved design of Hall plates for integrated circuitry," *IEEE Trans. Electron. Devices*, vol. ED-25, pp. 541–544, 1978.

[77] Y. Kanda and M. Migitaka, "Design consideration for Hall devices in Si IC," *Phys. Status Solidi (a)*, vol. 38, pp. K41–K44, 1976.

[78] Y. Kanda and A. Yasukawa, "Hall-effect devices as strain and pressure sensors," *Sensors and Actuators*, vol. 2, pp. 283–296, 1982.

[79] R. J. Braun, H. D. Chai, and W. S. Ebert, "FET Hall transducer with control gates," *IBM Tech. Discl. Bull.*, vol. 17, pp. 1895–1896, 1974.

[80] E. H. Vittoz, personal communication.

[81] F. N. Hooge, T. G. M. Kleinpenning, and L. K. J. Vandamme,

"Experimental studies on 1/f noise," *Rep. Prog. Phys.*, vol. 44, pp. 479–532, 1981.

[82] (No author given), "Improved Hall devices find new uses, orthogonal coupling yields sensitive products with reduced voltage offsets and low drift," *Electronics Week*, pp. 59–61, Apr. 29, 1985.

[83] Ref. [39, p. 534] and references herein.

[84] J. L. Berchier and R. S. Popović, "Sensitivity of the vertical Hall-effect device," presented at the 14th European Solid-State Device Research Conf. ESSDERC '84, Lille, France, 1984, also in *Europhysics Conf. Abstracts*, vol. 8F, pp. 265–266, 1984.

[85] R. C. Gallagher and W. S. Corak, "A metal-oxide-semiconductor (MOS) Hall element," *Solid-State Electron.*, vol. 9, pp. 571–580, 1966.

[86] G. R. Mohan Rao and W. N. Carr, "Magnetic sensitivity of a MAGFET of uniform channel current density," *Solid-State Electron.*, vol. 14, pp. 995–1001, 1971.

[87] M. Hirata and S. Suzuki, "Integrated magnetic sensor," in *Proc. 1st Sensor Symp.*, S. Kataoka, Ed. (Tokyo, Japan, Inst. Elec. Eng. of Japan, 1982), pp. 37–40.

[88] P. W. Fry and S. J. Hoey, "A silicon MOS magnetic field transducer of high sensitivity," *IEEE Trans. Electron Devices*, vol. ED-16, pp. 35–39, 1969.

[89] R. S. Hemmert, "Invariance of the Hall effect MOSFET to gate geometry," *Solid-State Electron.*, vol. 17, pp. 1039–1043, 1974.

[90] A. Yagi and S. Sato, "Magnetic and electrical properties of n-channel MOS Hall-effect device," *Jap. J. Appl. Phys.*, vol. 15, pp. 655–661, 1976.

[91] R. S. Popović and H. P. Baltes, "A CMOS magnetic field sensor," *IEEE J. Solid-State Circuits*, vol. SC-18, pp. 426–428, 1983.

[92] A. Nathan, A. M. J. Huiser, H. P. Baltes, and H. G. Schmidt-Weinmar, "A triple-drain MOSFET magnetic field sensor," *Can. J. Phys.*, vol. 63, pp. 695–698, 1985.

[93] A. Nathan, A. M. J. Huiser, and H. P. Baltes, "Two-dimensional numerical modeling of magnetic field sensors in CMOS technology," *IEEE Trans. Electron Devices*, vol. ED-32, pp. 1212–1219, 1985.

[94] A. Nathan, W. Allegretto, H. P. Baltes, P. Haswell, and A. M. J. Huiser, "Experimental versus numerically modeled Hall sensors in CMOS technology," in *1985 Can. Conf. on VLSI, Tech. Dig.*, C. A. T. Salama, Ed. (Toronto, Ont. Canada, Nov. 4–5, 1985), pp. 134–137.

[95] C. Bradner Brown, "High frequency operation of transistors," *Electronics*, vol. 23, pp. 81–83, 1950.

[96] I. Melngailis and R. H. Rediker, "The madistor—A magnetically controlled semiconductor plasma device," *Proc. IRE*, vol. 50, pp. 2428–2435, 1962.

[97] E. C. Hudson, Jr., "Semiconductive magnetic transducer," U.S. Patent 3 389 230, June 18, 1968.

[98] J. B. Flynn, "Silicon depletion layer magnetometer," *J. Appl. Phys.*, vol. 41, pp. 2750–2751, 1970.

[99] V. Zieren and B. P. M. Duyndam, "Magnetic-field-sensitive multicollector n-p-n transistors," *IEEE Trans. Electron Devices*, vol. ED-19, pp. 83–90, 1982.

[100] L. W. Davies and M. S. Wells, "Magneto transistor incorporated in a bipolar IC," in *Proc. ICMCST* (Sidney, Australia, 1970), pp. 34–35.

[101] I. M. Mitnikova, T. V. Persiyanov, G. I. Rekalova, and G. Shtyubner, "Investigation of the characteristics of silicon lateral magnetotransistors with two measuring collectors," *Sov. Phys.—Semicond.*, vol. 12, pp. 26–28, 1978.

[102] A. W. Vinal and N. A. Masnari, "Magnetic transistor behaviour explained by modulation of emitter injection, not carrier deflection," *IEEE Electron Device Lett.*, vol. EDL-3, pp. 203–205, 1982.

[103] R. S. Popović, and H. P. Baltes, "Dual-collector magnetotransistor optimized with respect to injection modulation," *Sensors and Actuators*, vol. 4, pp. 155–163, 1983.

[104] R. S. Popović and R. Widmer, "Sensitivity and noise of a lateral bipolar magnetotransistor in CMOS technology," in *IEDM Tech. Dig.*, pp. 568–571, Dec. 1984.

[105] V. Zieren and S. Middelhoek, "Magnetic-field vector sensor based on a two-collector transistor structure," *Sensors and Actuators*, vol. 2, pp. 251–261, 1982.

[106] R. S. Popović and R. Widmer, "Magneto-transistor in CMOS technology," *IEEE Trans. Electron Devices*, vol. ED-33, 1986, to be published.

[107] L. Halbo and J. Haraldsen, "The magnetic field sensitive transistor: A new sensor for crankshaft angle position," *Trans. ASE*, vol. 89, p. 701, 1981.

[108] R. S. Popović and H. P. Baltes, "Enhancement of sensitivity of lateral magnetotransistors," *Helvetica Phys. Acta*, vol. 55, pp. 599–603, 1982.

[109] L. W. Davies and M. S. Wells, "Magneto-transistor incorporated in an integrated circuit," *Proc. IREE Australia*, pp. 235–238, June 1971.

[110] R. S. Popović and H. P. Baltes, "An investigation of the sensitivity of lateral magnetotransistors," *IEEE Electron Device Lett.*, vol. EDL-4, pp. 51–53, 1983.

[111] I. M. Vikulin, N. A. Kanishcheva, and M. A. Glauberman, "Influence of the Hall emf on the sensitivity of a two-collector magnetotransistor," *Sov. Phys.—Semicond.*, vol. 11, p. 340, 1977.

[112] G. I. Rekalova, D. M. Kozlov, and T. V. Persiyanov, "Magnetic induction transducers based on silicon planar transistors," *IEEE Trans. Magn.*, vol. MAG-17, pp. 3373–3375, 1981.

[113] A. G. Andreou and C. R. Westgate, "The magnetotransistor effect," *Electron. Lett.*, vol. 20, pp. 699–701, 1984.

[114] V. I. Stafeev and E. I. Karakushan, *Magnetodiodes* (in Russian). Moscow, USSR: Science Press, 1975, ch. 10, pp. 173–187.

[115] A. W. Vinal and N. A. Masnari, "Bipolar magnetic sensors," in *IEDM Tech. Dig.*, pp. 308–311, Dec. 1982.

[116] _____, "Operating principles of bipolar transistor magnetic sensors," *IEEE Trans. Electron Devices*, vol. ED-31, pp. 1486–1494, 1984.

[117] Ch. S. Roumenin and P. T. Kostov, "Optimized emitter-injection modulation magnetotransistor," *Sensors and Actuators*, vol. 6, pp. 19–33, 1984.

[118] V. Zieren, S. Kordić, and S. Middelhoek, "Comment on 'Magnetic transistor behavior explained by modulation of emitter injection, not carrier deflection'," *IEEE Electron Device Lett.*, vol. EDL-3, pp. 394–395, 1982.

[119] A. W. Vinal and N. A. Masnari, "Response to 'Comment on 'Magnetic transistor behavior explained by modulation of emitter injection, not carrier deflection','" *IEEE Electron Device Lett.*, vol. EDL-3, pp. 396–397, 1982.

[120] R. Widmer and R. S. Popović, "Optimization of signal-to-noise ratio of lateral magnetotransistors," presented at the Swiss Physical Society Spring Meeting, Bern, Switzerland, Apr. 5–6, 1984.

[121] V. Zieren, "Geometrical analysis of the offset in buried collector vertical magnetotransistor," *Sensors and Actuators*, vol. 5, pp. 199–206, 1984.

[122] S. Kordić, V. Zieren, and S. Middelhoek, "A novel method for reducing the offset of magnetic field sensors," *Sensors and Actuators*, vol. 4, pp. 55–61, 1983.

[123] Y.-Z. Xing, S. Kordić, and S. Middelhoek, "A new approach to offset reduction in sensors: The sensitivity variation method," *J. Phys. E: Sci. Instrum.*, vol. 17, pp. 657–663, 1984.

[124] A. S. Andreou and C. R. Westgate, "AC characterization and modeling of lateral bipolar magnetotransistors," in *IEDM Tech. Dig.*, pp. 564–567, Dec. 1984.

[125] H. Pfleiderer, "Magnetodiode model," *Solid-State Electron.*, vol. 15, pp. 335–353, 1972.

[126] S. Cristoloveanu, "Magnetic field and surface influences on double injection phenomena in semiconductors," *Phys. Status, Solidi (a)*, vol. 64, pp. 683–695 and vol. 65, pp. 281–292, 1981.

[127] V. I. Stafeev, "Modulation of diffusion length as a new principle of operation of semiconductor devices," *Sov. Phys.–Solid State*, vol. 1, p. 763, 1959.

[128] E. I. Karakushan and V. I. Stafeev, "Magnetodiodes," *Sov. Phys.–Solid State*, vol. 3, p. 493, 1961.

[129] S. Cristoloveanu, "L'effet magnetodiode et son application aux capteurs magnetiques de haute sensibilité," *L'Onde Electrique*, vol. 59, pp. 68–74, 1979.

[130] (No author given), "Die Magnetodiode. Ein magnetfeld-

empfindliches Bauelement für die Steuer- und Regeltechnik," *Elektronik*, vol. 22, pp. 95–98, Nov. 5, 1982.

[131] O. S. Lutes, P. S. Nussbaum, and O. S. Aadland, "Sensitivity limits in SOS magnetodiodes," *IEEE Trans. Electron Devices*, vol. ED-27, pp. 2156–2157, 1980.

[132] A. Mohaghegh, S. Cristoloveanu, and J. De Pontcharra, "Double-injection phenomena under magnetic field in SOS films: A new generation of magnetosensitive microdevices," *IEEE Trans. Electron Devices*, vol. ED-28, pp. 237–242, 1981.

[133] A. Chovet, S. Cristoloveanu, A. Mohaghegh, and A. Dandache, "Noise limitations of magnetodiodes," *Sensors and Actuators*, vol. 4, pp. 147–153, 1983.

[134] K. Fujikawa and S. Takamiya, "Magnetic-to-electric conversion semi-conductor device," U.S. Patent 3 911 468, Oct. 7, 1975.

[135] R. S. Popović, H. P. Baltes, and F. Rudolf, "An integrated silicon magnetic field sensor using the magnetodiode principle," *IEEE Trans. Electron Devices*, vol. ED-31, pp. 286–291, 1984.

[136] B. Gilbert, "New planar distributed devices based on a domain principle," in *IEEE ISSCC Tech. Dig.*, p. 166, 1971.

[137] G. Persky and D. J. Bartelink, "Controlled current filaments in PNIPN structures with application to magnetic field detection," *Bell Syst. Tech. J.*, vol. 53, pp. 467–502, 1974.

[138] R. S. Popović and H. P. Baltes, "New oscillation effect in a semi-conductor device," in *Europhysics Conf. Abstracts*, vol. 7b, p. 223, 1983.

[139] J. I. Goicolea, R. S. Muller, and J. E. Smith, "Highly sensitive silicon carrier domain magnetometer," *Sensors and Actuators*, vol. 5, pp. 147–167, 1984.

[140] B. Gilbert, "Novel magnetic field sensors using carrier domain rotation: proposed device design," *Electron. Lett.*, vol. 12, pp. 608–610, 1976.

[141] M. H. Manley and G. G. Bloodworth, "Novel magnetic field sensor using carrier domain rotation: Operation and practical performance," *Electron. Lett.*, vol. 12, pp. 610–611, 1976.

[142] ——, "The carrier-domain magnetometer: A novel silicon magnetic field sensor," *Solid-State and Electron Devices*, vol. 2, pp. 176–184, 1978.

[143] ——, "The design and operation of a second-generation carrier-domain magnetometer device," *Radio Electron. Eng.*, vol. 53, pp. 125–132, 1983.

[144] S. Kirby, "The characteristics of the carrier domain magnetometer," *Sensors and Actuators*, vol. 4, pp. 25–32, 1983.

[145] D. R. S. Lucas and A. Brunnschweiller, "Recent studies of the carrier domain magnetometer," *Sensors and Actuators*, vol. 4, pp. 33–43, 1983.

[146] S. Kirby, "Temperature compensation technique for the carrier domain magnetometer," *Sensors and Actuators*, vol. 3, pp. 373–384, 1983.

[147] R. S. Popović and H. P. Baltes, "A new carrier-domain magnetometer," *Sensors and Actuators*, vol. 4, pp. 229–236, 1983.

[148] S. Middelhoek and A. C. Hoogerwerf, "Smart sensors: When and where?", *Sensors and Actuators*, vol. 8, pp. 39–48, 1985.

[149] J. M. Huijsing, "Integrated circuits for accurate linear analogue electric signal processing," Ph.D. dissertation, Delft University of Technology, Delft, The Netherlands, 1981.

[150] E. A. Vittoz, "MOS transistors operated in lateral bipolar mode and their application in CMOS circuits," *IEEE J. Solid-State Circuits*, vol. SC-18, pp. 273–279, 1983.

[151] S. Kordić and P. C. M. Van der Jagt, "Theory and practice of electronic implementation of the sensitivity-variation offset-reduction method," *Sensors and Actuators*, vol. 8, pp. 197–217, 1985.

[152] B. Wilson and B. E. Jones, "Feed-forward temperature compensation for Hall effect devices," *J. Phys. E: Sci. Instrum.*, vol. 15, pp. 364–366, 1982.

[153] J. Goicolea, R. S. Muller, and J. E. Smith, "An integrable silicon carrier-domain magnetometer with temperature compensation," in *Transducers '85* (1985 Int. Conf. on Solid State Sensors and Actuators, Digest Tech. Papers, W. H. Ko and K. D. Wise, Eds.), 1985, pp. 300–303, catalog no. 85CH2127-9.

[154] W. L. Engl. H. K. Dirks, and B. Meinerzhagen, "Device modeling," *Proc. IEEE*, vol. 71, pp. 10–33, 1983.

[155] M. Kurata, *Numerical Analysis for Semiconductor Devices*. Lexington, MA: Heath, 1982.

[156] M. S. Mock, *Analysis of Mathematical Models of Semiconductor Devices*. Dublin, Ireland: Boole, 1983.

[157] H. P. Baltes, L. Andor, A. Nathan, and H. G. Schmidt-Weinmar, "Two-dimensional numerical analysis of a silicon magnetic field sensor," *IEEE Trans. Electron Devices*, vol. ED-31, pp. 996–999, 1984.

[158] H. G. Schmidt-Weinmar, L. Andor, H. P. Baltes, and A. Nathan, "Numerical modeling of silicon magnetic field sensors: Magnetoconcentration effects in split-metal-contact devices," *IEEE Trans. Magn.*, vol. MAG-20, pp. 975–977, 1984.

[159] A. Nathan, L. Andor, H. P. Baltes, and H. G. Schmidt-Weinmar, "Modeling of a dual-drain MOS magnetic field sensor," *IEEE J. Solid-State Circuits*, vol. SC-20, pp. 819–821, 1985.

[160] W. Allegretto, Y. S. Mun, A. Nathan, and H. P. Baltes, "Optimization of semiconductor magnetic field sensors using finite element analysis," in *Proc. NASECODE IV Conf.* Dublin, Ireland: Boole, 1985, pp. 129–133.

[161] H. P. Baltes, A. Nathan, D. R. Briglio, and Lj. Ristić, "Optimization of split-drain MAGFET geometries with respect to 1/f noise," in *Noise in Physical Systems and 1/f Noise*, A. D'Amico and P. Mazzetti, Eds. Amsterdam, The Netherlands: North Holland, 1986, pp. 497–500.

[162] W. Versnel, "Analysis of symmetrical Hall plates with finite contacts," *J. Appl. Phys.*, vol. 52, pp. 4659–4666, 1981.

[163] J. H. J. Fluitman, "On the calculation of the response of (planar) Hall-effect devices to inhomogeneous magnetic fields," *Sensors and Actuators*, vol. 2, pp. 155–170, 1981/1982.

[164] S. Cristoloveanu, A. Chovet, and S. Lakeou, "Carrier concentration under high magnetic fields," *J. Phys. C: Solid State Phys.*, vol. 16, pp. 927–938, 1983.

[165] C. Arnold, "Computer simulation of conductivity and Hall effect in inhomogeneous inversion layers," *Surface Sci.*, vol. 113, pp. 239–243, 1982.

[166] R. S. Popović, "Numerical analysis of MOS magnetic field sensors," *Solid-State Electron.*, vol. 28, pp. 711–716, 1985.

[167] J. Brini and G. Kamarinos, "The unijunction transistor used as a high sensitivity magnetic sensor," *Sensors and Actuators*, vol. 2, pp. 149–155, 1981/1982.

[168] S. L. Agrawal and R. Swami, "Theoretical analysis of a magneto-unijunction transistor," *J. Phys. D: Appl. Phys.*, vol. 14, pp. 283–291, 1981.

[169] Y. Ruan, "Theoretical analysis of the Hall effect of the junction field effect transistor" (in Chinese), *Chinese J. Semicond.*, vol. 1, pp. 107–120, 1980.

[170] A. W. Vinal, "A magnetic sensor utilizing an avalanching semiconductor device," *IBM J. Res. Develop.*, vol. 25, pp. 196–201, 1981.

[171] N. D. Smirnov, C. S. Roumenin, and I. G. Stoev, "Polar MOS —A magneto sensitive element," *Sensors and Actuators*, vol. 2, pp. 187–193, 1981/1982.

[172] C. S. Roumenin and P. T. Kostov, "Tripole Hall sensor," *C. R. Acad. Bulgare des Sci.*, vol. 38, pp. 1145–1148, 1985.

[173] S. Kordić, "Integrated 3-D magnetic sensor based on an n-p-n transistor," *IEEE Electron Device Lett.*, vol. EDL-7, pp. 196–198, 1986.

[174] A. R. Cooper and J. E. Brignell, "Electronic processing of transducer signals: Hall effect as an example," *Sensors and Actuators*, vol. 7, pp. 189–198, 1985.

[175] I. A. McKay, A. Nathan, I. M. Filanovsky, and H. P. Baltes, "A magnetically controlled oscillator in CMOS technology," in *Proc. 1986 IEEE Custom Integrated Circuits Conf.* (Rochester, NY, May 12–14, 1986), pp. 30–33.

[176] A. Nathan, W. Allegretto, H. P. Baltes, and Y. Sugiyama, "Numerical modeling of Hall devices for nonuniform magnetic induction," in *Proc. 6th Sensor Symp., 1986*, S. Kataoka Ed. (Tokyo, Japan, Inst. Elec. Eng. of Japan), to be published.

[177] H. P. Baltes, W. Allegretto, A. Nathan, and Y. Sugiyama, "Carrier transport in semiconductor magnetic sensors under locally inverted fields," in *Tech. Dig. 1986 Solid State Sensor Workshop*, D. Eddy, Ed. (Hilton-Head Island, SC, June 2–5, 1986), paper no. 3, catalog no. CH-2258-2/86.

A CAPACITIVE SILICON TACTILE IMAGING ARRAY

K. J. Chun and K. D. Wise

Solid-State Electronics Laboratory
Department of Electrical and Computer Engineering
University of Michigan
Ann Arbor, Michigan 48109

ABSTRACT

This paper describes an 8x8-element tactile imager based on a silicon capacitive transducer. The imager consists of an X-Y organized array of capacitive force sensors on 2mm centers fabricated using integrated-circuit process technology and a simple electronic interface. The capacitive transducer array uses a diaphragm formed using a boron etch-stop, and metal plates isolated from silicon substrate. The array is read out using a switched-capacitor charge integrator giving a resolution of more than eight bits. The readout scheme permits off-chip electronics to be used so that the fabrication sequence requires only five noncritical masks. Each capacitive cell has a no-load capacitance of 1.6 pF, a low-force sensitivity of 15mV/gm, and a maximum operating force of about 45 gms per element. The operating force can be scaled over several orders of magnitude without changing the process or lateral array dimensions. The array is addressed as a memory, with an access time of less than 20μsec for 8 pixels.

INTRODUCTION

A wide number of tactile array sensors are now being developed to provide industrial robots with information on the distribution and amount of contact force (pressure) between the workpiece and the robot gripper. While vision sensors have existed for several years and are becoming increasingly sophisticated, tactile imagers are only now emerging for use in applications ranging from assembly tasks to surface texture measurement. This paper reports a silicon capacitive tactile imager which offers high resolution, high stability, a simple electronic interface, and can be easily scaled over a wide operating force range.

A wide variety of tactile sensors have been reported recently, based on both digital and analog readout schemes [1-3]. The earliest approaches to such sensors were based on the use of carbon-impregnated foam-rubber pads, which change their resistance when compressed by applied force. These sensors were relatively simple but exhibited serious stability problems, both over temperature and over time. The majority of more recent development efforts are based on the use of piezoelectric polymeric films. These structures are still relatively simple; however they lack DC response, are difficult to scale to different force ranges, and are not noted for high stability. In most such approaches, the stability of the polymeric film or pad sets the operating performance limits of the overall sensor, and sensor development requires a simultaneous optimization over both the electrical and the mechanical characteristics of the pad material. This typically forces compromises in the performance of the pad and is generally more difficult than optimization of the electrical or mechanical characteristics separately. Pad instabilities have been a serious problem in development of high-performance tactile imagers.

Semiconductor-based sensors are now widely used for the measurement of pressure in a variety of applications They offer advantages over alternative approaches in their low cost, high sensitivity, small size, wide dynamic range and high stability. Both piezoresistive and capacitive structures are being used. Piezoresistive devices offer higher linearity and somewhat simpler packaging than capacitive devices; however, capacitive pressure sensors are about an order of magnitude more pressure sensitive for a given device size and more than an order of magnitude less sensitive to temperature. Both of these sensor types are naturally fabricated in array form, and thus both are candidates for use in tactile imagers.

The requirements on tactile imagers are still evolving and are certainly application dependent so that a variety of array structures are likely to find eventual use. Typical performance requirements for tactile sensors intended for the high-performance end of the applications spectrum include at least 6 bits of resolution (grey scale) per pixel, an element spacing of 1-2mm to mimic the two-point resolution and pattern discrimination of the human finger-tip, high stability, and ruggedness in a difficult operating environment. A wide operating temperature range and fast response time are required.

ARRAY STRUCTURE AND FABRICATION

Figure 1 shows the cross-section and the layout of the capacitive tactile imaging array which is the subject of this paper. The basic cell is formed between a selectively-etched, boron-doped thin silicon diaphragm, which moves in response to applied force, and a metalized pattern on an opposing glass substrate to which the silicon substrate is electrostatically bonded. Silicon dioxide is used to isolate the transducers plates on the silicon from the substrate and allow them to function as isolated row lines. In the layout, row conductors are run vertically across the silicon wafer in slots which are simple extensions of the capacitive gap recess. Metal column lines run horizontally on the glass under recesses in the silicon, expanding to form capacitor plates over the cell areas. Thus, a simple X-Y

* This work was supported by the Air Force Office of Scientific Research under Contract No. F49620-82-C-0089.

Reprinted from *Rec. of the 3rd Int. Conf. on Solid-State Sensors and Actuators*, 1985, pp. 22–25.

capacitive keyboard is formed which has precisely controlled dimensions and a force sensitivity set by the thickness of the silicon plate.

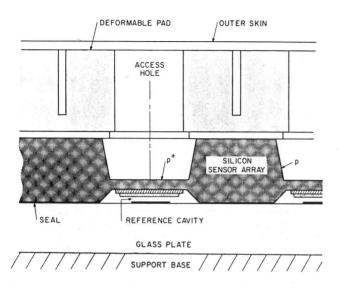

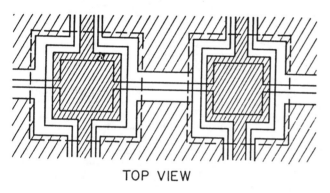

TOP VIEW

Figure 1.   Cross-Section of the Silicon-Based Tactile Imager (Top) and the Layout of Two Cells (Bottom).

A perforated cover plate is overlayed on the array for protection, and this plate is in turn covered by a compliant, replaceable pad and outer skin. This pad can be slit to decrease blooming of applied local force if the pad compliance is high. The access holes coupling force to the cell are filled with a substance (such as silastic) which acts as a force transmitter. Effort has been made here to minimize the importance of the pad on determining the performance of the overall array, since such pads are known to be the performance-limiting factor in most reported designs. The dominant structure in determining overall force sensitivity is the silicon diaphragm, whose properties can be well controlled, easily scaled for various applications, and are known to be stable over time and free of hysteresis. While pad designs vary widely for applications which range from enveloping grasp to surface texture measurement, the important feature of this basic cell structure is that the pad is used only for force transmission and plays a relatively minor mechanical role [4].

Since the overall array structure must be physically large, it is essential that the fabrication process be kept simple and compatible with achieving high yield. In order to fabricate the array, a silicon wafer is selectively etched on the front side to form a recessed area approximately $6\,\mu m$ deep. The wafer is then subjected to a deep boron diffusion followed by the deposition of dielectrics ($S_iO_2/S_{i_3}N_4$) for isolation between substrate and row lines. Polysilicon is deposited and patterned to form the row lines, and the wafer is subsequently etched selectively from the back to form thin ($10\,\mu m$) diaphragms behind the front recesses. The dielectrics are next stripped from the non-recessed portions of the wafer to expose silicon for electronic bonding, and contact areas are opened to the polysilicon. Finally, contacts are made to the polysilicon at the edge of the array. The second plate is formed by a metalization pattern on the glass to which the silicon is electrostatically bonded. The thickness of the diaphragm is controlled by the depth of the boron diffusion and the electrostatic bond is hermetic, irreversible, and free from deformation. It is particularly important to note that the dielectric under the polysilicon rows is immediately covered by the poly and is not exposed to any etching, so that the possibility of pinholes over the large-area row lines is minimized. The overall process has been designed to ensure high yields and requires a total of five noncritical masks.

READOUT SYSTEM DESIGN

Figure 2 shows a block diagram of the tactile transducer matrix and its interface circuitry along with signal conditioning electronics. The tactile imager lays out naturally in an X-Y organization and appears much like a memory, with row lines on the silicon and orthogonal column lines on the glass. A counter generates row and column addresses which are fed to a row decoder and multiplexer, respectively. These addresses are available to the microprocessor along with the pressure data from the A-D converter. The row addresses select one of eight rows, and following the read operation the eight outputs from the row cells are simultaneously available from the column detection amplifiers using MOS switched-capacitor interface circuitry [5]. The readout signals are processed in parallel up to the multiplexer, and in serial from A-D converter. The data during one frame time is read by the NSC 800 microprocessor, compensated for any nonuniformities in offset or sensitivity and then stored in memory used for subsequent real-time data processing by a host machine.

Since the readout amplifiers must be able to detect small variations in capacitance (a few femtofarads), high performance readout circuitry is needed. The basic organization of the readout scheme is shown in Fig. 3 for one column. The output voltage after integration is [4]

$$V_o = V_P \frac{(C_X - C_R)}{C_F} = V_P \frac{\Delta C}{C_F}$$

where C_X and C_R are the transducer and reference capacitors, and C_F is the integration capacitor. The suppression of parasitic capacitance C_{PS} is very important since it allows electronics to be eliminated from the array itself, simplifying the fabrication process. Note that C_{PS} is approximately equal to $(N-1)C_X$, where N is the number of cells per column, e.g., C_{PS} is the capacitance of all non-selected cells. Since the readout is dependent on ΔC but not on the no-load capacitance C_o, scaling of array dimensions is somewhat simplified. Also, the output voltage is not dependent on the open-loop gain of the integrator or its temperature coefficient, which enables the overall readout system to be quite

416

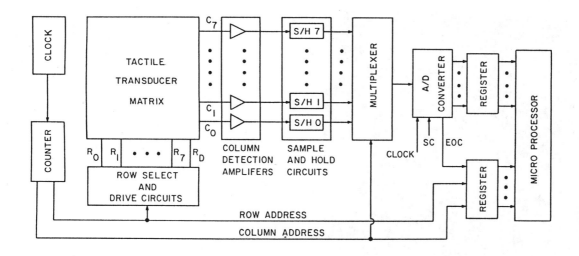

Figure 2.   Block Diagram of the Readout System for the Capacitive Tactile Imager.

temperature insensitive. Any effects due to integrator offset voltage can be eliminated using correlated double sampling [5]. A four-chip hybrid of the present 8x8 array might be used for larger arrays with no significant reduction in array performance.

TACTILE ARRAY PERFORMANCE

Several 8x8-element tactile imaging arrays have been fabricated, implementing readout electronics with commercial integrated circuits as shown in Fig. 4. The clock and address generating circuitry on the left side of the board is physically separated from the data processing circuitry. A wire-wrap board is used to reduce parasitic capacitance and to simplify the connections between the array and electronics. This implementation is being used to measure the performance of the fabricated arrays for testing purposes, and a new design is being developed for mounting on robot grippers. In the new design, only a tactile array and readout amplifiers are mounted on the support base inside the gripper. The diaphragm size is 1.2mm x 1.2mm x 10μm, and the array active area is 1.6 x 1.6 cm. Figure 5 shows

the output voltage from a typical cell as a function of applied force. Localized force at the center of the diaphragm has higher sensitivity and lower dynamic range than uniform force applied to both the diaphragm and support rim. Silicone rubber (400μm thick) was used to couple force to the array. The force sensitivity is 15mV/gm at low load and 100mV/gm at maximum load, with a maximum operating force of 45g/element for this design. The array is capable of a resolution and reproducibility exceeding 8 bits, and the operating force range can be varied over several orders of magnitude by controlling the diaphragm thickness.

The boron etch-stop provides a plate thickness which is highly uniform across the array so that the primary variations in sensitivity from cell to cell arise due to variations in the cavity depth. Measurements using a surface profilometer have shown a separation of $4\mu m \pm 0.2\mu m$ on a typical wafer, leading to a zero-pressure capacitance of $C_o=1.6\pm0.08pF$. These variations can be measured when the array is not contacting an object and subsequent data can be corrected during readout. Since rubber transmits force in every direction, coupling between adjacent cells is an important parameter in the pad design. Figure 6 shows the ratio of the measured crosstalk to primary output in the array when load is applied on an adjacent cell with no load on the measured cell. The force range is normalized to the maximum operating force. The crosstalk is only a few percent of the primary output, and while it represents one type of noise source for vertical force detection, it can also provide useful information regarding shear forces and slip.

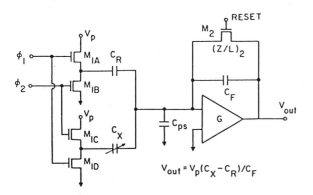

$$V_{out} = V_p(C_X - C_R)/C_F$$

Figure 3.   A Switched-Capacitor Column Readout Amplifier.

Figure 4. A Wire-Wrap Implementation of the Readout System.

REFERENCES

[1] M. H. Raibert, "An All Digital VLSI Tactile Array Sensor," Proc. International Conference on Robotics, Atlanta, Ga. pp. 314-319, March 1984.

[2] W. D. Hillis, "A High-Resolution Imaging Touch Sensor," The International Journal of Robotics Research, vol.1, No.2, pp. 33-44, Summer 1982.

[3] R. A. Boie, "Capacitive Impedance Readout Tactile Image Sensor," Proc. International Conference on Robotics, Atlanta, Ga. pp. 370-378, March 1984.

[4] K. Chun, and K. D. Wise, "A High-Performance Silicon Tactile Imager Based on a Capacitive Cell," Proc. SME Conf. on Robotics Research, Bethlehem, Pa. MS84-494, August 1984.

[5] Y. Park, and K. D. Wise, "An MOS Switched-Capacitor Readout Amplifier for Capacitive Pressure Transducer," Proc. Custom IC Conference, pp. 380-384, May 1983.

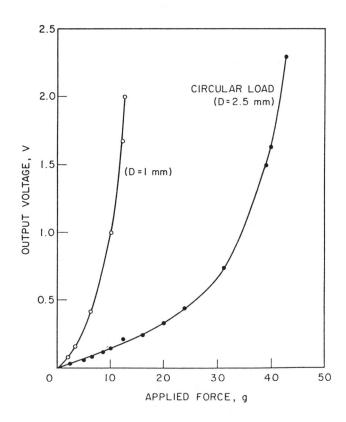

Figure 5. The Output Characteristic of a Typical Tactile Cell for Two Different Force Contact Areas.

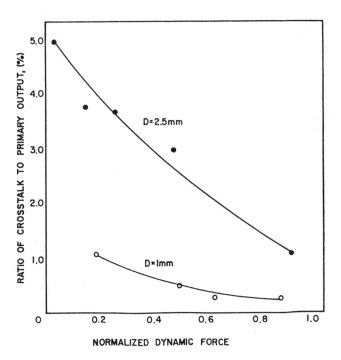

Figure 6. The Ratio of Nearest Neighbor Crosstalk to Primary Output for Two Different Force Contact Areas.

A FORCE SENSING CHIP DESIGNED FOR
ROBOTIC AND MANUFACTURING AUTOMATION APPLICATIONS

Kurt Petersen, Carl Kowalski, Joseph Brown,
Henry Allen, and Jim Knutti

Transensory Devices, Fremont, CA

ABSTRACT

A novel, silicon-based force sensor has been developed which can be used either as single element sensors or can be readily configured into dense tactile sensor arrays. The individual devices are composed of two parts: a micro-machined silicon chip bonded to a micro-machined glass substrate. The silicon chip contains transduction and addressing circuitry on one side and is machined with a force concentrating bump on the backside. The glass substrate contains bonding pad metallization and recessed areas into which the silicon chip can be deflected. A special bonding pad arrangement provides each device with an X-Y wiring capability. These devices and the tactile sensing arrays which they make possible open new dimensions in the field of force sensing.

INTRODUCTION

As computers continue to expand into robotics, automated manufacturing systems, and instrumentation, the need for computer-compatible and addressable sensors has become acute. In many potential applications, practical implementations of force sensor arrays are limited because adequate sensor systems with the required reliability, stability, addressability, and performance are simply not available. In this presentation, a unique, solid-state force sensor structure will be described in which a micro-machined silicon transducer chip is bonded to a matching micro-machined glass substrate. As a force is applied to the device, the silicon element deflects into a cavity etched into the attached glass substrate. Circuitry integrated in the silicon transduces the force into an electrical signal which can then be transmitted off the chip when it is addressed through the control line.

Most other silicon-based tactile sensor arrays reported to date have been designed with the idea that EACH chip should contain a number of sensor locations, perhaps with the outputs of the entire 2-dimensional array multiplexed through common on-chip circuitry. This approach is certainly viable. In practice, however, the problems of fabricating such large arrays with reasonable yield and close sensitivity matching between all the sensing locations may force the costs of such integrated arrays to become prohibitively high.

In contrast to most other silicon-based force transducer arrays, the device described here overcomes these yield and cost issues since the individual elements were designed as independent devices which could be assembled and wired into virtually any array size simply by mounting the chips side by side on any bare substrate surface. This fabrication technique makes it possible to pre-test, characterize, and pre-sort the chips (if necessary) prior to final array construction.

SENSOR FABRICATION

Fabrication of the sensor proceeds in four major parts: 1) definition of circuitry on the wafer using conventional, commercial IC processes; 2) micro-machining of the glass substrate and patterning of the interconnect metallization lines; 3) aligning and anodic bonding of the two wafers into a composite wafer; and, finally, 4) micro-machining of the wafer to define the force-concentrating mesas. The process was designed to use batch-fabrication as much as possible, to be insensitive to process variables, to optimize yield, and to maximize long-term stability and reliability of the sensor.

During the fabrication of the on-board circuitry, many of the same etching and diffusion steps which are used to define the various circuit layers are also used simultaneously to produce the surfaces necessary for anodic bonding, the regions designed for over-force protection, and the piezo-resistors employed in force transduction. After the wafers have completed the standard IC processing steps, only one additional oxide etch step is required on the silicon wafer prior to anodic bonding to the matching glass substrate.

Micro-machining of the glass substrate consists of etching the 100 mm diameter Pyrex plates with HF-based solutions, using deposited Cr/Au thin film masking layers to protect unetched regions. Specially polished glass wafers have been developed to ensure the etched surfaces are smooth and free from

Reprinted from *Rec. of the 3rd Int. Conf. on Solid-State Sensors and Actuators*, 1985, pp. 30–32.

pits and scratches. In addition, special care has been taken during the preparation of the metal masking films to eliminate undesired undercutting and to minimize pinholes. As expected for ideal isotropic etching, a 50 micron deep etch results in a 50 micron undercut beneath the composite metal masking layer. After etching, a corrosion-resistant metallization is deposited and patterned to provide gold-based wire bonding pads, metal feed-thru cross-over lines, and a ground-shield pattern to protect the circuitry during anodic bonding.

The two wafers are aligned and clamped together on a special fixture, then anodically bonded at a temperature near 350 C with a 1200 volt negative bias applied to the glass plate. Since both glass and silicon wafers are patterned, alignment is easily accomplished by viewing through the transparent glass substrate. As first reported by Sander and Knutti (1), the metal ground-shield on the glass protects the circuitry while the voltage is applied.

A mechanical machining process is employed to selectively thin the silicon wafer from the backside. During the thinning operation, most of the silicon is removed except a 520 micron thick pedestal region centered over a 115 micron thick silicon diaphragm. The central pedestal, measuring 600 x 600 microns, serves as a force-concentrating "bump", transferring externally applied forces into deflections of the diaphragm. These deflections are detected by the silicon piezo-resistors and the on-board electronics. The electronic circuitry itself is located directly beneath the thick pedestal to isolate it from the stress sensitive regions of the structure.

SENSOR PERFORMANCE

Figure 2 shows typical measurements on a tactile force-sensing chip over the range of 0 to 2 lbs. The data indicates a linearity within the 1% F.S. accuracy design goal. Repeated testing and cycling has shown that other errors due to drift and hysteresis are also well within the accuracy specifications.

A tradeoff during design is that of maximizing sensitivity while not sacrificing overforce protection. The thinner the diaphragm is, the higher the sensitivity - but at the expense of overforce protection. Designed to provide a nominal 50 mv/lb scale factor, the overforce protection has been measured to be in excess of 10 lbs. This enhanced overforce endurance is the result of positive stops micro-machined into the glass substrate.

Although better than elastomeric approaches, the sensor itself exhibits an offset temperature coefficient of 0.08 per cent of full scale output per degree C. For most applications this is not a problem. If the temperature of the sensors in an array can be stabilized, the offset variations can be eliminated either electronically or in software.

SENSOR ARRAY IMPLEMENTATION

An important feature of the tactile sensor is that the device was designed to be operated either as a single force-sensing element or assembled into dense, touch-sensing arrays. In the array format (shown in Figure 4), the individual tested die are butted together and the adjacent output pads are wiered directly together. Relatively simple, computer-controlled X-Y matrix addressing can now be employed to extract the information. Note that the sensor chips can be fully tested, characterized, and sorted prior to fabrication of the array, thereby greatly improving reliability and yield. Electronic X-Y addressing modules have been designed and built for use with array configurations up to 10 x 10. Data is taken from each element of the array, multiplied by a calibration factor stored in ROM for each element, and formatted for serial output to a host computer through an R5-232 port. Arrays of 3 x 3 tactile elements are currently being shipped for evaluation.

CONCLUSION

This paper has described a new design for a flexibly configurable force sensing cell and has shown how it could be used to form arrays or matrices of force sensing sites in robotic or automated manufacturing applications. A merging of ideas is now necessary to bring tactile sensing to the same status that vision systems have achieved in robotics. The widespread commercial availability of these advanced silicon tactile sensors will help to increase the use of touch sensing in intelligent and adaptive automation applications.

1) C.S. Sander, J.W. Knutti, and J.D. Meindl, "A monolithic capacitive pressure sensor with pulse-period output," IEEE Transactions on Electron Devices, vol. ED-27, p. 927, 1980.

Figure 1   Scanning Electron Micrograph of
             Tactile Force Sensor. The black
             bar is 400 microns long.

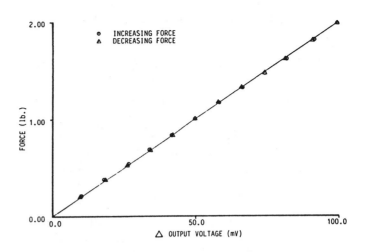

Figure 2   Typical output response
             characteristics of force sensor.

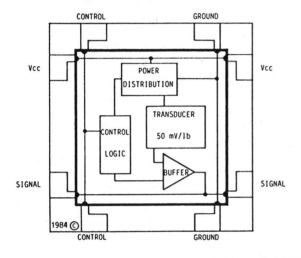

Figure 3   Schematic representation of force
             sensor showing circuit functional
             elements as well as the layout
             and distribution of the 8 contact
             pads.

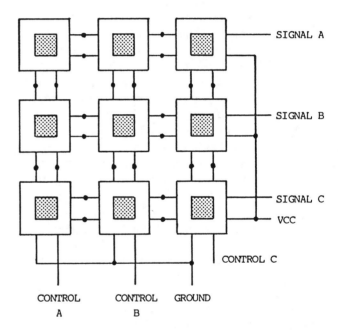

Figure 4   Typical interconnection
             scheme for a 3x3 force
             sensor array. Activation
             of Control A, for example,
             will cause signal lines A,
             B, and C to transmit data
             from the left-most column
             of the sensor array.

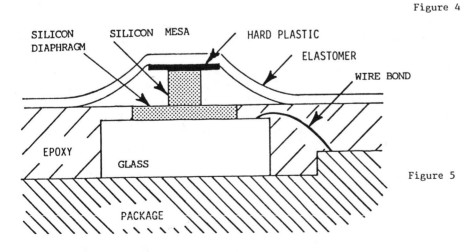

Figure 5   Typical packaging
             scheme for a single
             element tactile
             sensor chip.

Section 5.5: Chemical Sensors

A stable hydrogen-sensitive Pd gate metal-oxide semiconductor capacitor

M. Armgarth and C. Nylander

Laboratory of Applied Physics, Linköping Institute of Technology, Department of Physics and Measurement Technology, S-581 83 Linköping, Sweden

(Recieved 2 March 1981; accepted for publication 6 April 1981)

A palladium gate metal-oxide semiconductor device has serious hydrogen-induced drift problems. We have shown that this drift can be eliminated through the introduction of a thin alumina layer between the metal and the silicon dioxide. This makes it possible to use Pd metal oxide semiconductor devices as stable and accurate sensors for hydrogen.

PACS numbers: 85.30.Tv, 73.40.Qv, 06.70.Dn

The Pd gate metal-oxide semiconductor (MOS)-transistor first presented in this journal by Lundström *et al.*[1] shows a sensitivity to hydrogen and other hydrogen-containing gases. The gas sensitivity is due to the following processes. Hydrogen molecules adsorb on the outer surface of the palladium gate and dissociate. Hydrogen atoms then diffuse rapidly through the metal film and adsorb at the metal-oxide interface, changing the metal work function.[1,2] This is measured as a shift of the drain current through the transistor or as a shift of the flat-band voltage of a MOS capacitor.

The PdMOS is an excellent leak detector, but has not been very useful as a monitor of hydrogen concentration because of serious drift problems.[3] When the device is exposed to hydrogen, a fast shift of the flat-band voltage is followed by a slow one as shown in Fig. 1. The kinetics of this drift are difficult to characterize in a quantitative manner, since the time constants involved in the process range from seconds to several hours.

We show that this drift can be eliminated by a thin layer of alumina between the palladium gate and the silicon dioxide.

Sample preparation, except deposition of the alumina layer, followed a standard procedure. The samples were made of *p*-type silicon wafers with 100-nm thermally grown dry oxide. A part of the wafers were coated with 10 nm of aluminum, thermally evaporated. The alminum was oxidized at 500 °C in oxygen, resulting in a transparent film of alumina upon the silicon dioxide. Palladium dots were then evaporated through a metal mask onto the alumina parts of the wafers as well as on the untreated parts as shown in Fig. 2.

Typical hydrogen response for the samples are shown in Fig. 1. The devices were held at 150 °C and were exposed to 100 ppm of hydrogen in oxygen. MOS-structures on the alumina parts of the wafers showed a fast shift and no drift whatsoever. MOS structures on the reference parts of the wafers showed serious drift.

The alumina samples gave stable and reproducible shifts of the flat-band voltage. One of these samples was exposed to 100 ppm of hydrogen in oxygen for 18 h without showing any drift. Only small fluctuation were observed, probably due to instabilities in our gas mixing system.

It is seen in Fig. 1 that the hydrogen-induced shift is less for the alumina samples than for the normal PdMOS samples. There are two possible explanations for this. First of all, the number of adsorption sites for hydrogen may not be the same for the two different metal-insulator interfaces. The change in work function for a given hydrogen pressure would therefore differ between the two samples. We believe, however, that the main difference is due to a quick onset of the drift at the Pd-SiO$_2$ interface.

The experimental results give us strong reasons to believe that the drift phenomena in hydrogen sensitive PdMOS capacitors is due to some process in the silicon dioxide. It has earlier been shown[3] that ion drift in the silicon dioxide is affected during hydrogen exposure. Recently we have observed that contamination of the silicon dioxide with sodium hydroxide increases the hydrogen-induced drift markedly.

A more thorough description of the experiments together with some recent experiments and a theoretical model for hydrogen-induced drift in PdMOS structures will be presented elsewhere.

The most important result is, however, that through introduction of a thin layer of alumina between the palladium gate and the silicon dioxide, PdMOS devices become

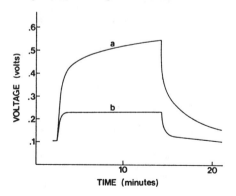

FIG. 1. Flat-band voltage shift when the ambient is changed from 10 to 100 ppm of hydrogen in oxygen. The device was held at a temperature of 150 °C. (a) Normal PdMOS capacitor. (b) Alumina PdMOS capacitor.

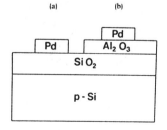

FIG. 2. Type of samples prepared for the experiments described in text. (a) A normal PdMOS. (b) A PdMOS with an intermediate layer of alumina.

stable and accurate sensors for hydrogen.

We wish to thank A. Spetz for helping us to prepare the samples and Prof. Lundström and Dr. Svensson for helpful comments on the manuscript. This work has been sponsored by the Swedish Board for Technological Development.

[1] I. Lundström, S. Shivaraman, C. Svensson, and L. Lundqvist, Appl. Phys. Lett. **26**, 55 (1975).

[2] K. I. Lundström, M.S. Shivaraman, and C.M. Svensson, J.Appl. Phys. **46** 3876 (1975)

[3] C. M. Svensson, Proceedings of the International Topical Conference, Raleigh, North Carolina, 1980 (unpublished).

A STUDY OF DETECTION MECHANISM IN THE Pd-SnO$_x$ MIS Oxygen Sensor

W. P. Kang, J. F. Xu[*], B. Lalevic
Department of Electrical and Computer Engineering, Rutgers University
Piscataway, NJ 08854, U.S.A.
T. L. Poteat
AT&T, Bell Laboratories, Murray Hill, NJ 07978, U.S.A.

ABSTRACT

Triangular voltage sweep method combined with high frequency C-V technique was used to study the detection mechanism and the influence of O$_2$ adsorption on ionic charge in the adsorptive SnO$_x$ layer of Pd-SnO$_x$ MIS device. The results have shown a decrease in area under the current peak in J$_p$-V plot and a shift of current peak toward positive gate bias voltages with increasing O$_2$ partial pressure or temperature. These results indicated that the adsorbed O$_2$ ions are "associated" with positive ions in the SnO$_x$ oxide layer thus reducing the total ionic charge and producing a shift in the flat-band voltage of the MIS capacitor. The two methods applied produced similar quantitative results concerning the number of adsorbed oxygen ions and subsequent shift in the ionic charge centroid.

INTRODUCTION

We have previously reported[1-3] on the device characteristics of the Pd-SnO$_x$-Si$_3$N$_4$-SiO$_2$-Si-Al O$_2$ gaseous sensors with the highly resistive SnO$_x$ layer as oxygen adsorptive element. The device is based on the shift in flat-band voltage ΔV_{fb} of the MIS capacitor upon O$_2$ adsorption in the SnO$_x$ layer.

The following static and transient parameters of the device were determined: a) the shift ΔV_{fb} as a function of the O$_2$ partial pressure at a given temperature or as a function of temperature at different O$_2$ pressure; b) adsorption and desorption time constants and $\Delta V_{fb}/\Delta t$ as function of pressure, temperature and applied field. Unique properties of the MIS devices as compared to other types of O$_2$ detector are: it operates at room temperature and above; it is more sensitive at 300°K where it can detect <0.1 torr of O$_2$ than other device at high temperature and due to its electroadsorptive properties it can be kept in room ambient without adsorpting oxygen.

The device detection mechanism is based on the change in ionic charge of the adsorptive SnO$_x$ layer in the MIS structure. The adsorptive tin oxide is non-stoichometric, in that it is deficient in oxygen atoms. Charge neutrality is maintained by the presence of some Sn II ions (Sn^{2+}) in place of Sn IV ions (Sn^{4+}). These Sn II ions act as electron donors, and the material is predominantly an n-type semiconductor. During the O$_2$ adsorption process, chemisorbed oxygen molecules capture electrons from the conduction band and form O$_2^-$, O$^-$ and O^{2-} ions. Further changes in the concentration of ions charge in the SnO$_x$ layer can occur through the interaction of diffusing O$_2$ with the impurity ions in SnO$_x$, which leads to a reduction in the effective ion charges in the SnO$_x$ layer. This change in ionic charge in SnO$_x$ film is reflected in the depletion region of the silicon and thus produces a shift in the flat-band voltage of the MIS capacitor. This effect is essentially the same as that of impurity ions e.g. sodium in SiO$_2$ have on conventional MOS devices.

It is therefore of importance to determine the change in ionic density in SnO$_x$ layer upon oxygen adsorption, in order to define detection mechanism. We have measured these charges as function of several parameters by using the Triangular Voltage Sweep (TVS) method and by the C-V technique as it applies to the MIS capacitors.

EXPERIMENTAL METHOD AND RESULTS

Experiments were performed on three types of device structure: 1) Pd-SiO$_2$-Si-Al; 2) Pd-SnO$_x$-SiO$_2$-Si-Al; and Pd-SnO$_x$-Si$_3$N$_4$-Si-Al. In order to make comparison with the previously reported TVS results on SiO$_2$, devices were fabricated on wafer divided into two segments; one segment contained Pd-SiO$_2$-Si-Al capacitors while the other segment was of the structure Pd-SnO$_x$-SiO$_2$-Si-Al. Detailed description of the devices including all relevant dimensions was previously reported. Sample under testing was placed inside a vacuum chamber, equipped with a heated stage. Oxygen is introduced through a leak-rate valve.

TVS METHOD

In the T.V.S. method, the MOS structure is held at a constant elevated temperature while a linear voltage ramp is applied to the gate and the resulting gate current measured against

_____

[*] A Visiting Scientist from East China Normal

Reprinted with permission from *Transducers '87, Rec. of the 4th Int. Conf. on Solid-State Sensors and Actuators*, 1987, pp. 610–613.
Copyright © 1987 by the Institute of Electrical Engineers of Japan.

gate bias as the mobile ions drift from one electrode to the other. The triangular ramp voltage on its reverse sweep drives the ions back to the other electrode and the measured gate current changes direction. Experimentally, the correct mobile ionic charge density can be obtained when a) the system is in the state of quasi-equilibrium, i.e. when the voltage sweep rate at a given temperature is properly selected and b) the observed polarization current J_p is attributable only to the ionic charge sweep. The total current during the voltage sweep consists of three components: 1) steady state electronic conduction current J_e; 2) displacement current J_c due to the charging of MOS capacitor and 3) polarization current J_p due to the motion of mobile ions. The three current components are shown in the Si-SiO$_2$-Si MOS Capacitor.[4]

The polarization current J_p can be expressed in a simplified form as:

$$J_p = -R_s (Q/L) d\bar{X}/dV \qquad (1)$$

where R_s is the constant voltage sweep rate, Q is the total space charge, $\bar{X}$ is the centroid of the space charge and L is the thickness of the insulating layer. Polarization current has a maximum value of J_{pmax} at the voltage V_m for which, $d^2\bar{X}/dV^2 = 0$. Thus, in a symmetrical structure J_p will peak at the zero applied voltage. Based on the above considerations, the following information can be obtained:

a) Total ionic charge from the area under the polarization current peak,

b) Static charge profile as a function of the applied potential.

c) A shift in V_m indicates a change in the position of the space charge centroid and thus the redistribution of the ionic space charge in the insulator. Information on the charge distribution can be extracted by fitting a theoretical model to the observed curves.

Total ionic charge distributed in the oxide film can be calculated from the area under the current peak. Integrating eq. (1) one obtains

$$Q_T = 1/R_s \int_{V_1}^{V_2} J_p \, dV \qquad (2)$$

where V_1 is the gate bias voltage when mobile ions leave from oxide-Si interface and V_2 is the gate bias when ions arrived at Pd-SnO$_x$ interface.

RESULTS

TVS measurements were performed in the temperature range from T = 178°C to 306°C, with a constant sweeping rate R_s at

R_s = 4.68 mV/sec to 10 mV/sec for the Pd-SnO$_x$-SiO$_2$-Si-Al structure. In this temperature region, R_s is chosen in the range where the normalized area, $1/R_s [\int J_p dV]$ under the J_{pmax} is independent of the sweep rate, and also where the voltage V_m corresponding to J_{pmax} is rate independent. Thus the conditions for a quasi-equilibrium were satisfied during the measurements.

Preliminary TVS measurements were performed on the Pd-SiO$_2$-Si-Al structure at 240°C and with voltage sweep R_s = 12.5 mV/sec. The resulting I-V plot is shown in the insert of Fig. 1. The measurements were performed in vacuum and at the oxygen partial pressure of 10 torr. It can be seen from Fig. 1 that current peaks and the area under the peak are the same in both directions of voltage sweep for the device in vacuum and at 10 torr O$_2$ pressure. This indicates that the same density of mobile charges is transferred in both directions across the oxide. The density of mobile charges in SiO$_2$ layer calculated from the peak in J_p-V plot is approximately 1.2×10^{10} ions/cm^2 in vacuum and the change in ionic density and V_m upon exposure to 10 torr of O$_2$ is insignificantly small corresponding approximately to 3×10^7 ions/cm^2 and 0.12 mV respectively. This is in agreement with the previous C-V measurement, where the shift in flat-band voltage of Pd-SiO$_2$-Si-Al capacitor was below detectable range.

Typical experimental TVS J-V plots for Pd-SnO$_x$-SiO$_2$-Si-Al capacitor in vacuum and upon O$_2$ adsorption and T = 226°C and R_s = 4.68 mV/sec are shown in Fig. 1. The measurements were performed as a function of O$_2$ partial pressure in the range of 4 to 10 torr.

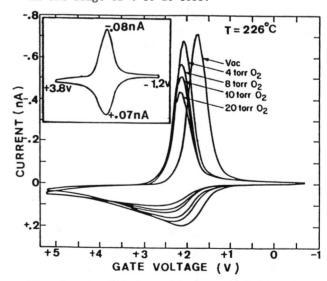

Fig. 1. Typical Triangular Voltage Sweep (TVS) J-V plots for several O$_2$ partial pressure.

These plots show well defined ionic polarization peaks and the same area under the peaks in both sweeping directions. They exhibit, however, several characterisitcs: a) a shift in

V_m from V=0 as it was the case for Pd-SiO$_2$-Si-Al capacitor to V_m = 1.75 V under the vacuum condition. Moreover, the area under the current peak corresponds to the impurity ion concentration in SnO$_x$ of 2×10^{13} ions/cm^2 i.e. Three order of magnitude higher than ionic density in SiO$_2$. Therefore most of the mobile ionic charge is concentrated in SnO$_x$ layer, consequently changing the coordinate of charge centroid and producing the observed shift in V_m. b) Asymmetrical features appear in both voltage sweep directions. The mobile ionic current peaks in the J_p-V plots are sharp during the sweep from positive to negative polarity of gate electrode. The current peaks in the opposite direction are very broad. c) A decrease in J_p peak and the area under the peak with increasing oxygen pressure from 4 to 20 torr. d) A shift in voltage V_m at which J_{pmax} occurs toward more positive voltage with increasing O$_2$ pressure.

DISCUSSION

A decrease in J_{pmax} and the area under the peak on the adsorption of O$_2$ and further observed decrease with increasing O$_2$ pressure indicate a decrease in the mobile ionic charge density. The observed decrease can be calculated from the relation

$$\Delta N = Q/q = (R_s q)^{-1}[\int J_{pvac}dV - \int J_{pO_2} dV] \quad (3)$$

where J_{pvac} is the area under the current peak in vacuum and J_{pO_2} is the area under the peak for a given oxygen pressure. Assuming that the adsorbed oxygen ion bounds one impurity ion in SnO$_x$, the number of adsorbed oxygen ion as a function of O$_2$ partial pressure is given in Fig. 2 at two different temperatures. It is clearly seen in Fig. 2 that there is a

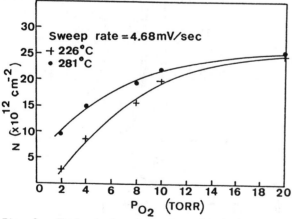

Fig. 2. Number of oxygen ion adsorbed N varies with O$_2$ partial pressure P$_{O_2}$ at several temperatures obtained from TVS method.

saturation trend in the number of adsorbed oxygen ions with increasing O$_2$ pressure. This is in agreement with the results obtained by the C-V measurements of O$_2$ adsorption where

a saturation trend in ΔV_{fb} with increasing O$_2$ pressure is observed at the same pressure level.

The dependence of the space charge centroid coordinate $\bar{X}$ on applied bias was calculated from Fig. 1. This dependence is presented in Fig. 3 for the Pd-SnO$_x$-SiO$_2$-Si-Al device in a vacuum and at 10 torr of O$_2$ pressure. The curves shown in Fig. 3 are quite similar to the $\bar{X}$ vs V and d$\bar{X}$ vs V curves for SiO$_2$ previously reported except that the center of symmetry i.e. V_m has moved from V = 0 to V = 1.7 and 2.17 for device in vacuum and in 10 torr of O$_2$ respectively. This would indicate that the sweep of ionic charges is limited to the interfaces between Pd-SnO$_x$ and SnO$_x$-SiO$_2$.

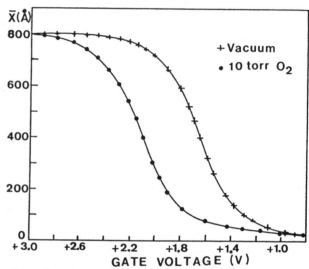

Fig. 3. Space charge centroid coordinate X vs applied gate bias V.

The above analysis indicated that the ionized impurities in SnO$_x$ are interacting with the adsorbed O$_2$ ions. The observed decrease in mobile ionic charge density indicate that positive mobile ions are "associated" with adsorbed O$_2$ ions. The adsorbed O$_2$ ions are distributed near the SnO$_x$ surface as a negative space charge which may hold on positive mobile ions. Hence, a more positive gate bias is required to sweep ions away from the Pd-SnO$_x$ interface, and less negative gate bias for pulling these mobile ions up from the SnO$_x$-SiO$_2$ interface. The position of polarization current peaks for the gate bias in both directions as a function of temperature are given in Fig. 4. There are two sets of curves as shown Fig. 4. Curves A and B indicate the change in position of mobile ions peaks with gate bias as a function of the temperature and in absence of oxygen while curves C and D indicate the change in position of positive mobile ions peaks with gate bias as the temperature varies in .4 torr oxygen. Curves A and C represent a motion of ions away from the Pd electrode while curves B and D represent a motion of positive ions away from the SnO$_x$-SiO$_2$ interface.

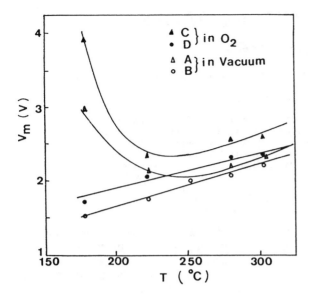

Fig. 4. Polarization current peaks at V_m vs temperature T.

It can be seen in Fig. 4 that curves B and D appear consistently approximately linear in the investigated temperature region, however, curves A and C are approximately linear only at high temperature. The departure from the linear region indicates that at lower temperature much stronger electric fields are required to move the ions away from the Pd electrode. The ions appears to be more strongly bound near the metal-SnO_x than at the SnO_x-SiO_2 interface. Further shift towards metal-SnO_x interface occurs after oxygen adsorption.

The ionic charge density obtained by the T.V.S. method can be used to calculate the shift in flat-band voltage ΔV_{fb} and compare it with the observed shift in the C-V plots. Experimtal J_p~V plot measured at T = 175°C and R_s = 4.63 mV/sec was used for this purpose. The density of oxygen ionic charge adsorbed upon exposure to 4 torr O_2 partial pressure calculated from equation is 1.338×10^{-6} col/cm^2. V_{fb} is given by:

$$V_{fb} = (C_i d)^{-1} \rho_m [\int_0^{X_1} x\,dx] \qquad (4)$$

where ρ_m is the density of ionic charges and C_i is the capacitance per unit area in the accumulation region, and assuming that X_1 is approximately equal the SnO_x thickness of 800 A.

Substituting the data for ρ_m obtained from T.V.S. measurement and the value of the device capacitance ($C_i = 11.5 \times 10^{-9}$ f/cm^2) to equation 4, the calculated ΔV_{fb} equals to 0.486 V. This is in good numerical agreement with the results obtained from C-V measurement, where the flat-band voltage shift at 4 torr O_2 pressure equals to 0.435 V.

CONCLUSION

Oxygen detection mechanism of Pd-SnO_x MIS device has been investigated by T.V.S. method. The results from J_p~V plot indicated that adsorbed oxygen ions interact with existing impurity mobile ions in SnO_x, causing a decrease in space charge density in the oxide layer of device, and bounding mobile ions near to the Pd-SnO_x interface. These results are in good numerical agreement with previously reported C-V measurements.

REFERENCES

1. J.F. Xu, W.P. Kang, B. Lalevic and T.L. Poteat, "Steady State and Transient Behavior of Pd-SnO_x MIS Oxygen Sensors," Proceedings of 2nd International Meeting on Chemical Sensor, Bordeaux, France, July 1986.

2. J.F. Xu, W.P. Kang, B. Lalevic and T.L. Poteat, "Steady State and Transient Behavior of Pd-SnO_x MIS Oxygen Sensors," Proceedings of Solid State Sensors Workshop, Hilton-Head Island, SC, U.S.A., June 1986.

3. W.P. Kang, J.F. Xu, B. Lalevic and T.L. Poteat, "Pd-SnO_x MIS Capacitor as a New Type of O_2 Gaseous Sensor," to be published in May 1986 issue of IEEE Electron Device Letters.

4. N.J. Chou, "Application of Triangular Voltage Sweep Method to Mobile Charge Studies in MOS Structure," J. Electrochem. Soc., 601-609, 1971.

Microelectronic ion sensors: A critical survey

R.G. Kelly, Ph.D., A.M.I.E.E., and Prof. A.E. Owen, F.R.S.E.

Indexing terms: *Semiconductor devices and materials, Measurement and measuring*

Abstract: Microelectronic ion sensors based on monolithic silicon integrated circuit (IC) and hybrid circuit technologies have been the subject of considerable research and development over the past 15 years. This paper reviews the conceptual background and history of both kinds of device, comparing their operation with those of conventional ion-selective electrodes and coated-wire electrodes. Attention is focused on the interfacial processes involved in the ion-sensing mechanism of microelectronic devices, with particular reference to the significance of models based on either ideally blocking or nonblocking mechanisms. This is a matter which has important theoretical and practical consequences for both silicon IC-based microelectronic sensors such as the ion-sensitive field-effect transistor (ISFET) and hybrid circuit based devices. Practical problems associated with effective encapsulation and the provision of a suitable reference electrode are also considered.

1 Introduction

The direct measurement of the hydrogen ion activity (pH) of aqueous solutions, by means of the glass membrane electrode, has been a valuable technique in analytical chemistry and process monitoring for many years [1]. Since the mid-1960s there has been a growing interest in extending electrode measurements to a wider range of ions. Extensive research by chemists has led to the development of many new types of electroactive materials for use as membranes in ion-selective electrodes (ISE) [2, 3]. Examples include: nonporous, sparingly soluble salts (e.g. silver halides sensitive to Cl^-, Br^- etc. and single crystal lanthanum fluoride sensitive to F^-); electroneutral macrocyclic ion carriers (e.g. valinomycin dissolved in plasticiser/PVC); enzymes (sensitive to penicillin, for example). The glass electrode itself has also been progressively refined, and new glass compositions with sensitivity to ions other than H^+ have been devised [3]. Materials like iridium oxide and zirconia have also been studied as alternative types of pH sensor to the glass electrode [3].

Alongside this research into new electroactive *materials*, which is aimed at improving the chemical properties of the sensor, especially its selectivity to a specific ion, there has been interest in improving the physical and mechanical properties of the electrode *structures*. The conventional ISE (Fig. 1) employs a membrane of the electroactive material interposed between the unknown test solution and an internal reference solution of known composition. An internal reference electrode makes electrical connection to the reference solution, and the measurement system is completed by a second reference electrode system contacting the external solution under test. Substitution of the internal solution and reference electrode by a direct solid-state (e.g. metallic) contact to the inner surface of the membrane (Fig. 2a) leads to a more easily manufactured, robust device of smaller size (requiring smaller volumes of analyte) and capable of withstanding exposure to elevated temperature and pressure. Solid-state contacts to many types of electroactive material have been investigated, e.g. silver wire to a pellet or cast membrane of silver halide [4], evaporated metal films to glass membranes [5, 6] and the so called 'coated-wire electrodes' (CWE) [7], in which a polymer based sensing material is dip coated on to a metal wire, a particularly simple method of fabrication (Fig. 2b).

ISEs generally have a very high output resistance

Paper 4160I (E3), received 6th August 1985

The authors are with the Department of Electrical Engineering, University of Edinburgh, The King's Buildings, Edinburgh EH9 3JL, United Kingdom

(typically 10^7 to 10^{10} Ω), and there are obvious advantages in situating the input stage of the voltmeter as closely as possible to the electrode itself. Arrangements using conventional circuit methods [8] and thick-film hybrid microcircuits [9, 10] have been described.

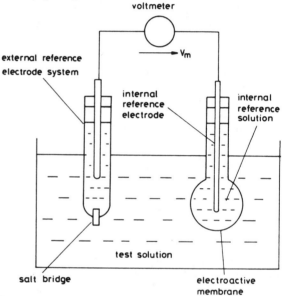

Fig. 1 *Conventional ion-selective electrode system*

In parallel with these developments in ISE techniques, deriving principally from the world of chemistry, there has been a growing awareness in the electronics field that sensor technology has lagged greatly in comparison with the dramatic progress of VLSIC-based information-processing methods [11]. The sensor now represents the weak link in many measurement systems. The idea of fabricating integrated sensor/signal conditioning systems using the methods of silicon-wafer fabrication is especially attractive, and silicon based sensors for pressure, acceleration and many other variables have been devised [12]. In many cases, however, the special requirements of the sensor device impose modifications to the standard silicon process which detract greatly from the cost advantages of using this highly refined technology. The appropriateness of attempting the total integration of the sensor on to silicon has therefore been questioned [10, 13, 14]. An alternative approach is to use hybrid-circuit (thin film or thick film) methods to fabricate the sensor itself, signal conditioning being carried out by standard or custom-designed integrated circuits bonded on to the hybrid substrate [14, 15]. This approach is especially relevant to chemical

Reprinted with permission from *IEE Proc.*, vol. 132, part I, no. 5, pp. 227–236, October 1985.

sensors which incorporate electroactive materials foreign to silicon technology and for which there are especially severe problems of encapsulating the device against a very unfriendly environment of conducting, corrosive, aqueous solutions.

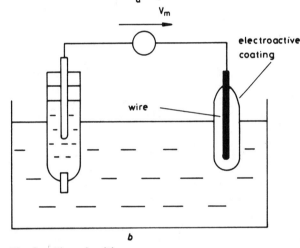

Fig. 2 *Electrode solid-state*

a Contact membrane type
b Coated wire configuration

Historically, however, the development of microelectronic ion sensors has been heavily focused on semiconductor techniques, especially the ion-sensitive field effect transistor (ISFET) [16]. Research into hybrid approaches to ion sensing has been relatively limited, although thick-film technology has been used for gas sensors [14]. ISFET development has been almost exclusively oriented to the particular requirements of biomedical instrumentation, including *in vivo* patient monitoring, and steps toward commercial production for this market have been announced [17]. Nevertheless, the performance of current devices is marred by long-term instability (output drift), and reservations about ISFETs have been expressed recently [3, 18]. Their application in other fields, such as pollution monitoring or process control, has received less attention.

Theoretical and practical issues relevant to the ISFET and alternative types of device will be discussed in the following Sections.

2 Conventional membrane ISEs

In the arrangement of Fig. 1, the essential sensing process takes place at the interface between the outer surface of the membrane and the analyte solution. Transfer of charge between the solution and the surface layers of the membrane (e.g. by an ion-exchange reaction) gives rise to an interfacial potential. If the charge transfer process is sufficiently selective to the measured ion, with respect to other ions in the test solution, we have the basis of a useful ISE. In order to make contact to the inner surface of the membrane it is usual to make use of the same charge-transfer process. The interfacial potentials are functions of the activities (which are related to the concentrations) of the participating ion in the analyte and reference solutions. The former is the quantity to be measured while the latter is maintained constant. Assuming that the participating ion reaches a state of thermodynamic equilibrium across each membrane/solution interface, it can be shown [2] that the total membrane potential, V_m, is a function of the activities, a_1, a_2, of the ion in the analyte and reference solutions

$$V_m = \frac{RT}{nF} \ln \frac{a_1}{a_2}$$

where n denotes the number of unit charges on the ion, and R, T and F have their usual meanings. If a_2 is constant (the reference solution), then the graph of V_m against $\log_{10} a_1$ is linear with a slope of 58 mV per decade change in activity at room temperature, for a singly charged ion. This is the well known 'Nernst equation' which is characteristic of processes in equilibrium and is closely obeyed for most conventional ISEs. In order to measure V_m, using an electronic instrument, it is necessary to define the potentials of the metal wires at the meter inputs with respect to the inner and outer reference solutions. The processes which take place at the reference electrodes must therefore involve a reaction between electrons in the metal and ionic species in the solution; i.e. a redox reaction. The reference solution then fulfils the additional function of maintaining a constant activity of the ion participating in the reference-electrode reaction, in order to ensure a constant reference-electrode potential. In practice, it is often impracticable to place the external reference electrode directly into the analyte solution because it would not be possible to ensure the constant activity of the reference ion. It is common practice, therefore, to contain the external reference electrode within its own reference solution which is connected to the analyte through a 'salt bridge'. The salt bridge introduces an additional 'liquid-junction' potential into the measurement, but, by proper design, it is possible to minimise it [2, 19]. (Nevertheless, the constancy of the liquid-junction potential is an important source of uncertainty in precise measurements.) For the case of a glass membrane pH electrode, the entire electrochemical cell of Fig. 1 may be represented by:

Hg | HgCl | KCl   ‖ analyte ‖ glass membrane ‖
      (sat.)   liquid
                junction

                            internal
                            reference | AgCl | Ag
                            solution
                            H⁺, Cl⁻

in which the inner and outer reference electrodes are assumed to be of the silver/silver chloride and calomel types, respectively. The concentrations of H⁺ and Cl⁻ ions in the internal reference solution are both constant. It is usually assumed that the charge-transfer reactions at the membrane surfaces, and at the reference electrodes, must reach a state of thermodynamic equilibrium for the measured potential to be stable. Nevertheless, the cell of Fig. 1 does include nonequilibrium, but steady-state, potentials at the liquid junction and, possibly, within the bulk membrane inside the phase boundaries.

3 Coated-wire electrodes and related devices

We shall now suppose that the internal reference electrode and reference solution of Fig. 1 are replaced by a direct metallic contact to the inner surface of the membrane, either by coating the membrane material on to a wire, or by depositing a metal contact on to a previously formed membrane, e.g. of glass (Fig. 2). The cell may now be represented by:

$$Hg \,|\, HgCl \,|\, KCl \quad \| \text{ analyte } \| \text{ membrane } | \text{ metal}$$
$$\text{(sat.)} \qquad \text{liquid}$$
$$\text{junction}$$

In the case of glass, and many other types of membrane which are known to be purely ionic conductors, it is difficult to identify a mechanism for ion/electron interaction at the metal contact that is analogous to the processes taking place within the internal reference system of Fig. 1. There is a widely held view that inadequate definition of the contact process implies that the 'standard potential' (E_0) of the electrode will be irreproducible, even for devices fabricated under similar conditions, and will also be liable to long-term drift. Cattrall and Hamilton [7] have recently discussed this issue with particular reference to polymer membrane CWEs. They note that some sort of electrode process must take place at the metal contact (and be stable, at least in the short term) in order to explain the observed performance of CWEs. They go on to suggest various possible mechanisms and to review several methods which have been proposed to improve the thermodynamic definition of the metal reference contact.

It is sometimes difficult to compare the experimental data on long-term stability reported by different laboratories because inconsistent methods of specification are used, e.g. mV/hour over an unspecified period or total excursion of output voltage over a stated time. Furthermore, the measurement conditions are not always clearly stated, so that it is not easy to judge how closely they correspond to practical measurement situations, with regard to temperature variation for example. It is widely accepted, however, that CWEs do suffer from drift and poor reproducibility [7], although they might well be analytically useful in applications where appropriate recalibration is possible. As Cattrall and Hamilton have pointed out, though, the observed drifts might originate from sources other than the metal/membrane contact. The absorption of water by the membrane or leaching out of the electroactive component are examples.

Work on metal-contacted pH glass membranes has been carried out in our own laboratory [7]. The range of E_0 values for 'similar' devices was as much as 60 mV. Individual devices, held at constant temperature, showed a random drift, but the total excursion over five weeks was no more than 11 mV in some cases. Temperature variations or exposure to small polarising currents caused large drifts, however, from which recovery was slow and incomplete. Various methods for improving the contact to glass membranes have been proposed, e.g. by using metal/metal oxide or metal/metal halide combinations [20, 21].

In cases where the sensor membrane itself can support electronic as well as ionic conductivity, it is easier to identify thermodynamically acceptable mechanisms for the contact process. Single crystal membranes of silver halides,

contacted by silver metal, have been discussed by Buck and Shepard [4], for example. This situation is quite different from that of the ionically conducting materials like glass.

4 Hybrid microelectronic ion sensors

The concept of the CWE may be taken a stage further if the sensor material is deposited directly on to a thick- or thin-film hybrid-circuit substrate, the contact metal having previously been deposited by conventional methods for conductor deposition [13]. Signal conditioning electronics may be implemented on the same substrate, also by well established methods (Fig. 3).

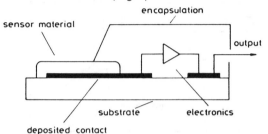

Fig. 3 *Cross-section of hybrid microelectronic ion sensor*

Microelectronic methods would be expected to give better reproducibility of the film dimensions and composition, as compared with dipcoating etc., and the entire device is amenable to mass production at low cost. *In situ* signal conditioning will simplify the external instrumentation and give greater convenience to the user. The hybrid-circuit processes involve less sophisticated technology than silicon integrated-circuit processes; capital costs are lower and economic production is possible for medium volume markets. For many sensor applications (*in vivo* measurements being an exception) extreme miniaturisation is not essential, and the hybrid sensor will be small compared with conventional ISEs, even though the device feature sizes are larger than those of silicon circuits. Furthermore, the hybrid fabrication processes are less susceptible to contamination where nonstandard materials are introduced as the sensor membranes. From the practical and economic points of view, therefore, the hybrid approach is an attractive alternative to silicon for many sensor applications.

From the theoretical standpoint, a hybrid sensor which is based upon the principle of the direct measurement of cell potential is electrochemically identical to the coated-wire types of device discussed in Section 3. It is subject to the same questions regarding the definition of the potential determining process at the internal contact and hence the long term stability of E_0.

The use of hybrid-circuit methods to fabricate a fluoride ion sensor has been reported by Fjeldly, Nagy and Stark [10]. Reversible solid-state contacts were achieved by using a multilayer arrangement, e.g.

$$Ag \,|\, AgF \,|\, LaF \,|\, F^- \text{ in solution}$$

Fluoride ions are in equilibrium at the AgF/LaF_3 interface, whereas electrons are in equilibrium across the Ag/AgF interface. (The thermodynamics of metal/salt interfaces had previously been discussed by Buck and Shepard [4] with reference to conventional ISEs based on pellets or membranes of salts with direct metal connection.)

The 'hybrid' sensors due to Fjeldly *et al.* [10] incorporated membranes prepared using conventional (non-microelectronic) methods and connected by a short lead to a thick-film hybrid amplifier circuit. Other research groups [22, 23], including our own, have addressed themselves to the deposition of pH sensitive glasses by thick-film methods. This raises greater problems with regard to the chemical and thermal compatability between the substrate, conductor and sensor phases at the high temperatures used in the firing process. (Similar problems have been overcome in developing 'standard' thick-film circuit processes.) Adverse effects on the sensor properties of the glass, due to physical or chemical changes in the glass resulting from the printing and firing processes, are also possible. (Loss of sodium from the glass is one example.) Visually satisfactory films have been produced, but we have experienced some difficulty in obtaining the stable and reproducible Nernstian pH responses characteristic of bulk glass membranes; research into this aspect is continuing.

5 Extended-gate FET sensor

Notwithstanding the above mentioned arguments in favour of the hybrid approach, it is possible to take the process of miniaturisation a stage further by depositing the electroactive material directly on to the gate metal of the input FET of the signal conditioning amplifier. Conceptually, this is merely a matter of reducing the length of the metal connection in Figs. 2 or 3, and this introduces no change whatsoever to the electrochemical structure of the cell. A device of this kind, in which the gate metal of the FET is extended in order to separate the sensor and transistor areas, and hence simplify device ecapsulation, has been reported [24] (Fig. 4).

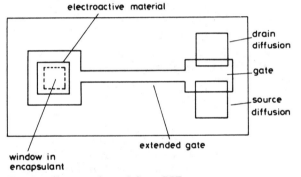

Fig. 4 *Plan view of extended gate FET sensor*

6 ISFET devices

The ISFET was the first reported and most extensively researched type of 'chemically sensitive semiconductor device' (CSSD) and predated the hybrid and extended-gate devices described in Sections 4 and 5. The first ISFETs, described by Bergveld [25, 26], were directly derived from basic MOSFET structures by simply eliminating the gate metal and exposing the silicon-dioxide gate insulator to the electrolyte solution. The silicon dioxide was expected to be pH sensitive by analogy with silicate glasses (Fig. 5). The devices showed a measureable sensitivity to both Na^+ and H^+, although the magnitude of the change in the 'equivalent gate voltage' (calculated from the slope of the drain current against pH graph) was less than the Nernstian value characteristic of a process in thermodynamic equilibrium.

Bergveld's original results were obtained by measuring the device's drain current without using any form of external reference electrode. It was claimed to be an advantage of ISFETs that they could be used without a reference electrode, and a theoretical model to support this conten-

tion was proposed by Bergveld, De Rooij and Zemel [27]. The model was based on the well known dependence of the threshold voltage, V_t, of an MOS transistor on the density of charged surface states located at the silicon/

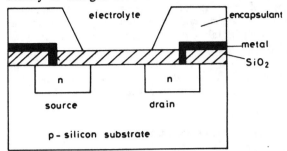

Fig. 5 *Cross-section of early ISFET design using SiO₂ gate insulator as the sensor material*

oxide interface. The sensing effect was seen as a change in V_t due to a modification of the surface state density, N_{ss}, as a consequence of species diffusing from the electrolyte through the oxide to the silicon surface.

Janata *et al.* [28, 29] and Kelly [30], however, argued that a reference electrode would be necessary, on both thermodynamic and electrical circuit grounds, for stable operation. To use an ISFET without a reference electrode was equivalent to measuring a single halfcell potential, which is thermodynamically impossible, In any case, the floating solution would act as a gate and would modulate the channel conductivity (in addition to any V_t effect) as a result of electrostatic pick up. It was suggested [28] that Bergveld's original results might have been due to a failure of the device's encapsulation, allowing the potential of the silicon substrate to be referenced to that of the solution by an unintentional electrical leakage path. (The problem of encapsulation is a recurrent one in this field and will be returned to below.) It has also been shown that thermally grown SiO_2 often develops microcracks, which can allow a direct electrolytic contact to the underlying silicon substrate [31]. The surface-state mechanism, as the origin of ISFET ion sensitivity, could not explain the observed response times to solution activity changes, which were too fast to be consistent with diffusion through the SiO_2 [32]. (This process might contribute to long-term drift, though.) Furthermore, an early refinement of Bergveld's original ISFET was the introduction of a silicon nitride layer as an ion barrier between the SiO_2 and the silicon [28]. The surface-state mechanism may thus be discounted. The reference-electrode controversy continued for several years, but it is now generally accepted that a reference is required for proper ISFET operation [33].

Subsequent development of ISFETs divided into two main themes. Some workers, notably Janata's group at the University of Utah, adopted a variety of 'conventional' ion-sensitive materials, especially those based on polymers, for use with a standard ISFET substrate. ISFETs sensitive to a range of ions were thus developed [34–36], but most of them employed substantially similar device structures, the differences between them being mainly due to the chemical properties of the overlying films of sensor material. Other workers focused their attention on inorganic sensor materials which can be deposited by processes more directly compatible with conventional silicon fabrication. It was soon discovered that, perhaps surprisingly, silicon nitride functioned as a pH sensor[37]. Subsequent work included the deposition, e.g. by chemical vapour deposition, of alumina and alumino-silicates [38]. ISFETs with ZrO_2 and Ta_2O_5 gates have also been described [39, 40]. It must be borne in mind that most of these materials have not previously been employed in conventional ISEs. Their chemical properties as sensor materials are therefore not well established, e.g. selectivity, susceptibility to 'poisoning', extent of dissolution over an extended period etc. The use of these materials has necessitated the introduction of new theoretical models to explain the sensing process itself, notably the 'site-binding model' for interfacial potentials, derived from colloid chemistry [41]. The development of a novel device *structure*, the ISFET, has thus promoted the introduction of new sensor *materials* and given a new impetus to research into sensor *mechanisms*.

7 ISFET operating mechanism

Fig. 6 depicts a 'floating gate ISFET'. It is electrochemically identical to the extended-gate FET described above,

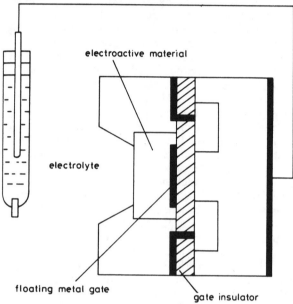

electroactive material

electrolyte

floating metal gate

gate insulator

Fig. 6 *Floating gate ISFET*

except that the gate metallisation has been shortened to the minimum possible, and we may regard the device as a conventional FET with a film of electroactive material deposited directly on to the metal gate. Conditions below the gate metal in Fig. 6 are identical to those in a conventional MOSFET; the potential of the gate with respect to the substrate determines the electric field strength in the oxide and the silicon surface potential, and hence the carrier concentrations in the channel and the channel conductivity (Fig. 7). The gate potential is the sum total of the various electronic and electrochemical processes that take place in the external path between the silicon substrate and the gate metal; i.e. metal/silicon contact potential, reference-electrode halfcell potential, liquid-junction potential, potential drop in the bulk solution, solution/membrane interfacial potential, bulk-membrane potential and membrane/metal interfacial potential. The arrangement is electrochemically identical to that of a coated-wire electrode working into a MOSFET amplifier. Hence, if it is assumed that charge transfer (Faradaic) processes take place at the membrane interfaces, the device is subject to precisely the same considerations with regard to long-term stability as were discussed in Sections 3 and 4 in connection with the CWE and hybrid types of device.

Quite early in the history of ISFET development, however, Janata and Moss [29] raised the radically different possibility that an ISFET could operate in a stable

manner even if it incorporated a polarisable or 'blocking' interface. An ideal blocking interface is one across which no charge transfer can take place; its electrical nature is purely capacitive, therefore. Janata and Moss questioned

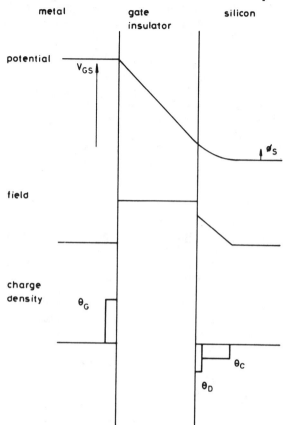

Fig. 7 *Variation of electrical potential, field and charge in a MOSFET or floating gate ISFET (inversion mode conditions)*

V_{GS} = gate-substrate voltage
ϕ_S = silicon surface potential
θ_G = gate charge
θ_D = depletion charge
θ_C = mobile channel charge

the commonly held view that an interface without a definitive redox couple is thermodynamically undefined and hence unstable. They pointed out that, *in the absence of any charge transfer*, the ideally polarised interface is in a state of true equilibrium at any potential; i.e. it has one more degree of freedom than the usual Nernstian interfaces. They further observed that a conventional FET device incorporates a blocked interface (the gate insulator) and is nevertheless stable.

An exhaustive theoretical discussion of polarisable electrodes was given by Lauks [42, 43]. Assuming that a polarisable membrane/metal interface can be implemented in practice, Lauks' arguments apply equally to the floating gate ISFET (Fig. 6) or to the CWE or hybrid device. In the case of the floating gate ISFET, however, it is further argued that as the gate metal is electrically isolated (having a blocked contact on one side and an ideal gate insulator on the other), it may therefore be dispensed with to give the usual ISFET structure of Fig. 8. This structure may be represented electrochemically by the cell:

metal A	Pt H$_2$, 1 atm.	reference solution $a_{H^+}=1$	analyte solution A$^+$			
		ISM A$^+$	insulator	semi-conductor	metal A

in which the external reference system is assumed to be a standard hydrogen electrode and liquid junction; ISM denotes the ion-sensitive membrane phase. Using this cell, Lauks derives the total potential between the metal A and a point just inside the ISM, adjacent to the insulator interface. He points out that this is 'not a true potential in the

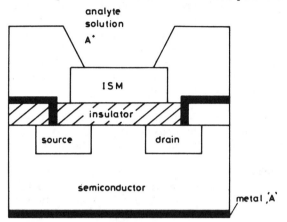

Fig. 8   *Typical ISFET structure*

The ion-sensitive material (ISM) is deposited directly on to the gate insulator, the gate metal being eliminated. The ISM-insulator interface is assumed to be blocked.

traditional sense since the ISM in contact with the solution cannot be considered as an electrode' [43]. As a consequence, the derived expression for cell potential contains terms relating to the variation in chemical and electrical potentials between the surface and bulk regions of the membrane. Hence, the cell potential cannot be derived as a function of the solution ion activity by purely thermodynamic arguments. Lauks goes on to introduce various assumptions about the charge distribution in the cell, for different classes of ion-sensitive membrane, and hence predicts Nernstian behaviour in some cases and non-Nernstian responses in others. It has been pointed out [16], however, that for one of these cases (a polymer gate film doped with neutral charge carriers), the predictions from Lauks' model are at variance with published experimental results.

Like Janata and Moss [29], Lauks [43] assumes that the basic ion-sensing mechanism at the membrane/solution interface involves a nonblocking (charge transfer) process. The polarisable interface is taken to be at the membrane/insulator boundary or at the boundary between a hydrated surface region of the membrane and an insulating bulk region. In contrast, it has also been suggested [44] that *totally* blocked systems of the type:

$$\text{electrolyte} \,|\, \text{insulator} \,|\, \text{semiconductor}$$

$$\text{blocked} \quad \text{blocked}$$

might be used as chemical sensors. Siu and Cobbold [41] compared the limiting cases of ideally blocked and ideally unblocked electrolyte/insulator interfaces and went on to consider the intermediate case of a blocked interface in which there is adsorption of charged species at the insulator surface. (Lauks [42] specifically excluded such effects from his model). The surface adsorption process is ion selective, in contrast to an ideally blocked interface which would respond to the total ionic strength of the solution. With reference to the particular case of silicon dioxide as the insulator, Siu and Cobbold [41] adopted the 'site binding model' from colloid chemistry to describe the surface effects. This model assumes that surface SiOH sites can exhibit amphoteric behaviour in respect of the poten-

tial determining ion (i.e. H^+) and that binding of counter ions to charged surface sites can also occur:

$$SiOH + H^+ \rightleftharpoons SiOH_2^+$$

$$SiOH \rightleftharpoons SiO^- + H^+$$

$$SiOH_2^+ + Cl^- \rightleftharpoons SiOH_2Cl$$

$$SiO^- + Na^+ \rightleftharpoons SiONa$$

The site binding model also predicts non-Nernstian responses, and it has been used to explain the measured performances of ISFETs using films of the inorganic oxides, SiO_2, ZrO_2 and Ta_2O_5 as the sensing material [39]. The site binding model assumes that the bulk membrane is impermeable to ions and charge free. It is a pure dielectric through which the potential varies linearly (Fig. 9). Questions about the variation of electrochemical potential of species within the membrane, discussed previously

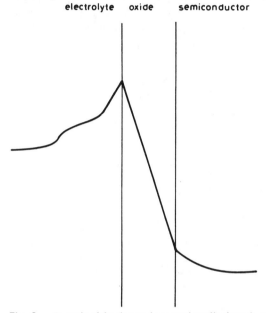

Fig. 9   *Example of the electrical potential profile through an electrolyte-oxide–oxide-semiconductor system as predicted using the site binding model to describe reaction at the electrolyte-oxide interface*

After Siu and Cobbold [41]

in connection with Lauks' model, do not therefore arise in this case.

The theoretical models discussed above show that the silicon surface potential can be related to solution ion activity for ISFETs incorporating blocked interfaces. We must next consider whether such devices are likely to be sufficiently stable for use as practical sensors. Clearly, the first requirement is that there must be no unintended Faradaic process which could allow charge transfer across the 'blocked' interface; this point was acknowledged by Lauks [43].

Secondly, in cases where the bulk membrane is permeable to ions (i.e. the majority of conventional ion-selective materials), the 'nonthermodynamic' nature of Lauks' model leads to an equation for cell potential which includes terms related to the chemical potentials of ions in the bulk membrane. Now, the classical theory of the glass membrane electrode, for example, shows that a diffusion potential exists within the bulk membrane [45]. It can be shown, however, that the total value of this diffusion potential is determined by the concentrations of the diffusing species just inside the membrane surfaces. These boundary conditions are, in turn, determined by the ionic concentrations in the adjacent aqueous phases, through the rapid and reversible phase-boundary ion-exchange processes. This leads to the important conclusion that, although the diffusion process is not in a state of thermodynamic equilibrium, the diffusion *potential* will be constant once the phase-boundary processes have reached equilibrium [46]. If the reference solution is replaced by a polarisable contact, however, the boundary condition is no longer well defined and the question of potential drift arises. (It is accepted, however, that the diffusion constants in materials like glass, at room temperature, are sufficiently small that the rate of drift might be slow enough to permit practical use of the device as a sensor.)

Thirdly, in cases where the electrolyte/membrane interface is blocked, we must consider the electrical-circuit implications of a measurement in which the voltage source, i.e. the blocked interface, has an infinite output resistance and the measuring instrument, i.e. the insulator-silicon FET structure, has infinite input resistance; that is to say, both interfaces are essentially capacitive. Janata [47] has discussed this problem using the circuit model of Fig. 10.

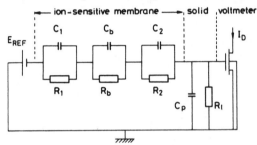

Fig. 10 *Equivalent circuit model for an ion sensor with solid-state internal contact*

After Janata [47]

The parallel combinations of resistance and capacitance, $R_1 C_1$ and $R_2 C_2$, represent the membrane/solution and membrane/solid interfaces, respectively. The resistances tend to infinity for the case of an ideally polarisable interface. $R_b C_b$ represents the bulk region of the membrane. Parasitic capacitance C_p and leakage resistance R_l associated with the voltmeter input are also represented. Janata [47] points to the significance of the exchange-current densities at the interfaces and notes that a 'good sensor cannot be made out of a bad membrane'; i.e. one in which the exchange-current density at the solution interface is low, i.e. $R_1 \to \infty$. The practical difficulties of making measurements using a blocked solution interface are also underlined, in connection with attempts to measure interfacial charge using an ISFET with a gold gate in contact with sodium fluoride solution which is known to be a good approximation to a blocked interface [47].

On the other hand, as Janata [47] also points out, if the exchange current at interface 1 is sufficiently large to represent a 'good' membrane, we may consider two extreme cases for interface 2. If it is nonblocking, then the operation of the sensor depends on charge exchange processes and its stability will depend on the attainment of equilibrium conditions across the interface or, at least, a steady-state condition which is sufficiently slowly changing for practical purposes. The leakage resistance in Fig. 10 is then subject only to the practical requirement that it must be large compared with $(R_1 + R_2 + R_b)$. If R_2 is large, however, i.e. interface 2 is blocking, the requirements for stable operation become more stringent; i.e. the parasitic capacitance must be small and constant and the leakage resistance must be infinitely high. It follows that the practical problems will increase with the length of the connecting lead between the membrane and the voltmeter input.

The ISFET is therefore claimed to have practical superiority over the coated-wire or hybrid sensors, because, in its case, the lead length reduces to zero [47]. Like Lauks [43], however, Janata [47] accepts that there is no difference, *in principle*, between the various types of device. In the case of a hybrid device the 'connecting lead' comprises a few millimetres of metal film, and, in the opinion of the present authors, it seems unlikely that any additional parasitic effects associated with it would be sufficient to render the device significantly less practical than the corresponding ISFET structure. Difficulties might arise, however, with a CWE arrangement which uses an appreciable length of cable to connect into a conventional voltmeter.

The present authors doubt that devices incorporating a blocking interface will be capable of stable performance over a prolonged period under practical conditions. The published models for device operation require stringent assumptions about the complete absence of Faradaic processes across the blocked interface and about the elimination of parasitic electrical leakage effects. In addition, if there is a significant thickness of ion-permeable membrane between the solution interface and the polarisable interface, questions arise about diffusion potentials and the effective control of boundary concentrations for the diffusing species at the polarisable interface. These problems are avoided if it is assumed that the solution/ISM interface is also blocked (e.g. in the site binding model), but this implies the existence of an electrostatic field in the bulk membrane (analogous to the oxide field in a conventional MOSFET), and the problem of threshold drift, due to the motion of any small residual ionic charge, must then be considered. We also note that it is misleading to argue, as Janata and Moss [29] have, that the insulator of a conventional metal-gate MOSFET is a blocked contact that is theoretically equivalent to the blocked interface in an ISFET. In the metal-gate device of Fig. 11a, the gate-substrate potential V_{GS}, is determined by the EMF due to

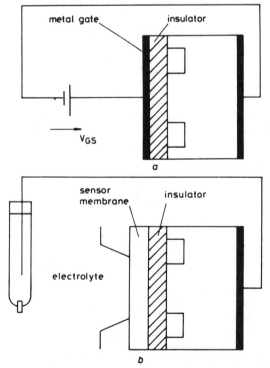

Fig. 11 *Electrochemical relationship between metal gate FET and ISFET structures*

a MOSFET   b ISFET

the battery. A detailed consideration of the internal operation of the battery allows V_{GS} to be calculated in terms of its anode and cathode halfcell potentials. Each of these halfcells is a true electrode at which a redox process (ion-electron interaction) takes place. Hence, the theoretical derivation of the whole cell potential is based on the overall electrochemical equilibrium of *electrons* in the silicon and *electrons* in the gate metal. In the case of the ISFET of Fig. 11b, however, the model due to Lauks [43], discussed above, derives the potential between the silicon and a point inside the membrane on the basis of the equilibrium between *electrons* in the silicon and *ions* in the membrane. As observed by Lauks, it is not a true electrode. We note here that a silver halide membrane (case (ii) in Lauks' [43] paper) represents a special case. As pointed out by Buck and Hackleman [44], there is internal equilibrium between ions and electrons in a silver halide. Hence, in this case, the system *is* analogous to the conventional MOSFET and gate bias battery.

We believe that the best approach to designing stable sensors will be by incorporating nonblocking processes at both the electrolyte and the solid interfaces to the electroactive material. The problems involved were considered in Section 3 on coated-wire electrodes, and research is continuing into the use of more complex, multilayer, contact systems to overcome them.

8 ISFET performance

It is difficult to make precise comparisons, with regard to long-term stability, between devices produced by different laboratories, owing to the lack of a standard method of quantifying drift which was noted previously. It is widely accepted, however, that ISFETs do suffer from instability problems [3, 10, 48–50]. Bousse and Bergveld [50] have identified three elements in the response of ISFETs to ion activity changes, namely fast and slow components, both of which are related to solution ion activity, and a third, long-term drift, which is not related to solution activity. It is not yet clear whether this drift is a consequence of fundamental theoretical issues, as discussed in Section 7, or whether (as suggested by Lauks [43]) it merely represents inadequacies in the current state of the technology of device fabrication, similar to the threshold-voltage drift problems which were experienced in the early days of MOSFET development and subsequently overcome. We note that drift could result from slow second-order processes at the electrolyte interface itself (e.g. leaching of electroactive material, water penetration, poisoning etc.), or as a consequence of degradation of the device encapsulation leading to the development of electrical leakage conductances.

An ISFET-based sensor system for intravascular measurements is being developed commercially by Cordis Europa N.V. The system incorporates a programmable read-only memory (PROM) to store the parameters of the individual ISFET device, as determined by *in vitro* calibration before use [17].

9 Practical issues

The commercial development of ISFET sensors has been slow, in spite of extensive research. Work on these devices has been heavily oriented towards biomedical applications which have particular requirements that cannot always be met by conventional electrodes and for which the ISFET shows great promise. In the context of industrial applications, however, a new sensor must be able to compete with existing ISEs, in terms of both price, performance and reliability, and must be amenable to production by commercially acceptable methods with good reproducibility and reasonable yield.

A particularly intractable problem with all types of chemical sensor is the encapsulation of the device. The requirements are more demanding than those of packaging conventional electronic devices, because the environment is especially severe (total immersion in corrosive and conducting liquids) and because the sensor cannot be completely 'potted' as it must incorporate a window where the electroactive material is exposed to the analyte solution. Furthermore, extremely high insulation levels must be maintained at the input stage of the amplifier. Against this background, the progress which has been made in various research laboratories to bring these devices to their present level of performance is undoubtedly impressive. However, most of the methods proposed have employed coatings of various polymeric materials, applied manually. Little has been done, to the authors' knowledge, to evaluate their effectiveness under conditions of industrial use or to assess their feasibility as production methods.

The problem of encapsulation applies to all types of sensor, but will be most severe with ISFETs owing to their inherently small size and the high cost of nonstandard process variations. A particular difficulty is the requirement to make contact to metal bonding pads on the same face of the chip as the exposed window, and in close proximity to it. Methods for implementing conducting paths through the silicon wafer, so that contacts can be made on the reverse side, have been described, but they involve highly nonstandard processing techniques [51, 52]. The ceramic substrate of a hybrid device will be more adaptable in this respect.

It has already been shown that ISFETs, like any other type of ion sensor, require an external reference electrode for proper operation. For those applications in which the concentration of a reference ion is known to be constant in the analyte solution, a differential measurement can be made using a pair of solid-state sensors, one being selective to the measurand ion, the other to the reference ion [10]. (A third, metal, electrode is also required as a 'solution ground' to reduce common-mode signals when both ion sensors are high-impedance devices.) In many applications, however, it is impracticable to control the activity of the reference ion in the analyte solution itself, and so a separate reference electrode system is used, electrically connected to the analyte by a liquid junction, e.g. a ceramic-plug salt bridge (see Fig. 1). At the present time there is no established method for implementing a reference system of this kind in a solid-state form compatible with the microelectronic ion sensor, and this will detract from the advantages of using the new sensors in many cases. The Cordis ISFET uses a miniaturised form of a liquid filled Ag/AgCl reference electrode [17]. A reference system built directly on to an ISFET chip has also been described [53]. It uses a small quantity of buffered agarose gel encapsulated in epoxy resin with a 200 μm length of glass capillary serving as the liquid junction. An attempt to implement a reference system more compatible with semiconductor fabrication methods has been described recently [54]. It employs an ISFET in which the gate is covered by an insulating polymer which is claimed to be ion blocking and essentially free of surface sites. More research will be required into ways of further reducing the site density achieved in practice and, if successful, this would represent a significant step towards the practical application of solid-state ion sensors. We note, however, that conventional reference systems have been refined over many years to bring them to their present state of development, and yet they are still regarded as one of the principal sources of uncertainty, especially in precise measurements.

10 Conclusions

There can be no doubt as to the potential advantages of applying microelectronic methods to the fabrication of solid-state ion sensors with integral signal conditioning electronics. Substantial progress has been made in the ISFET field, although it has been largely focused on biomedical applications. The hybrid approach may offer practical advantages in other applications, especially where the market size does not warrant the high cost of variations to a standard semiconductor process.

The development of theoretical models for ISFET operation was slow until about 1980, and, in the authors' opinion, there remain serious fundamental questions about the long-term stability of devices which incorporate blocked interfaces. Equally, much remains to be learned about how to implement stable unblocked contacts in solid-state form. Poor reproducibility between nominally similar devices, and long-term drift of operating parameters, have been widely experienced in practice, but it has been claimed that their effects can be minimised by appropriate calibration procedures.

ISFET development has stimulated complementary research into the ion-sensitive properties of materials hitherto unused for this purpose, but further evaluation will be required here also.

It remains to be seen whether existing methods of device encapsulation will be adequate under practical conditions of use and amenable to commercial production; further development will probably be required.

The full potential of the new sensors will not be realised until a compatible reference-electrode system is implemented in solid-state form.

References

1 EISENMAN, G. (Ed.): 'Glass electrodes for hydrogen and other cations' (Dekker, 1967)
2 KORYTA, J.: 'Ion-selective electrodes' (Cambridge University Press, 1975)
3 ARNOLD, M.A., and MEYERHOFF, M.E.: 'Ion-selective electrodes', *Anal. Chem.*, 1984, **56**, pp. 20R–48R
4 BUCK, R.P., and SHEPARD, V.R.: 'Reversible metal/salt interfaces and the relation of second kind and 'all-solid-state' membrane electrodes', *ibid.*, 1974, **46**, pp. 2097–2103
5 GUIGNARD, J-P., and FRIEDMAN, S.M.: 'Construction of ion-selective glass electrodes by vacuum deposition of metals', *J. Appl. Physiol.*, 1970, **29**, pp. 254–257
6 KELLY, R.G.: 'Microelectronic approaches to transducers for chemical activity measurement'. Ph.D. thesis, University of Edinburgh, 1979
7 CATTRALL, R.W., and HAMILTON, I.C.: 'Coated-wire ion-selective electrodes', *Ion-Selective Electrode Rev.*, 1984, **6**, pp. 125–172
8 LANGMAIER, J., STULIK, K., and KALVODA, R.: 'Some potentiometric sensors with low output impedance', *Anal. Chim. Acta*, 1983, **148**, pp. 19–25
9 MELLOR, T.J., HASKARD, M., and MULCAHY, D.E.: 'A teflon-graphite ion-selective probe with integrated electronics', *Anal. Lett.*, 1982, **15**, pp. 1549–1555
10 FJELDLY, T.A., NAGY, K., and STARK, B.: 'Solid-state differential potentiometric sensors', *Sens. & Actuators*, 1983, **3**, pp. 111–118
11 MIDDELHOEK, S., NOORLAG, D.J.W., and STEENVOORDEN, G.K.: 'Silicon and hybrid microelectronic sensors', *Electrocompon. Sci. & Tech.*, 1983, **10**, pp. 217–229
12 Special issue on 'Solid-state sensors, actuators and interface electronics', *IEEE Trans.*, 1979, **ED-26**, (12), pp. 1861–1984
13 KELLY, R.G., JORDAN, J.R., and OWEN, A.E.: 'Microelectronic ion-selective electrodes', *Proc. Analyt. Div. Chem. Soc.*, 1977, **14**, pp. 338–340
14 VELASCO, G., SCHNELL, J.Ph., and CROSET, M.: 'Thin solid-

15 MORTEN, B., PRUDENZIATI, M., and TARONI, A.: 'Thick-film technology and sensors', *ibid.*, 1983, **4**, pp. 237–245
16 SIBBALD, A.: 'Chemical-sensitive field-effect transistors', *IEE Proc. I, Solid-State & Electron Dev.*, 1983, **130**, (5), pp. 233–244
17 BERGVELD, P.: 'Biomedical applications of ISFETs'. IEE Colloquium Digest 1985/54, pp. 2/1-2/4
18 CZABAN, J.D.: 'Electrochemical sensors in clinical chemistry: yesterday, today, tomorrow', *Anal. Chem.*, 1985, **57**, pp. 345A–356A
19 MATTOCK, G., and BAND, D.M.: 'Interpretation of pH and cation measurements', *in* EISENMAN, G. (Ed.): 'Glass electrodes for hydrogen and other cations' (Dekker, 1967), pp. 9–49
20 US Patent 1 260 065.
21 US Patent 4 312 734.
22 AFROMOWITZ, M.A., and YEE, S.S.: 'Fabrication of pH-sensitive implantable electrode by thick film hybrid technology', *J. Bioeng.*, 1977, **1**, pp. 55–60
23 LEPPAVUORI, S.I., and ROMPPAINEN, P.S.: 'The use of hybrid microelectronics in the construction of ion-selective electrodes', *Electrocompon. Sci. & Tech.*, 1983, **10**, pp. 129–133
24 VAN DER SPIEGEL, J., LAUKS, I., CHAN, P., and BABIC, D.: 'The extended gate chemically sensitive field effect transistor as multi-species microprobe', *Sens. & Actuators*, 1983, **4**, pp. 291–298
25 BERGVELD, P.: 'Development of an ion-sensitive solid-state device for neurophysiological measurements', *IEEE Trans.*, 1970, **BME-17**, pp. 70–71
26 BERGVELD, P.: 'Development, operation and application of the ion-sensitive field-effect transistor as a tool for electrophysiology', *ibid.*, 1972, **BME-19**, pp. 342–351
27 BERGVELD, P., DEROOIJ, N.F., and ZEMEL, J.N.: 'Physical mechansims for chemically sensitive semiconductor devices', *Nature*, 1978, **273**, pp. 438–443
28 MOSS, S.D., JANATA, J., and JOHNSON, C.C.: 'Potassium ion-sensitive field-effect transistor', *Anal. Chem.*, 1975, **47**, pp. 2238–2243
29 JANATA, J., and MOSS, S.D.: 'Chemically sensitive field-effect transistors', *Biomed. Eng.*, 1976, **11**, pp. 241–245
30 KELLY, R.G.: 'Microelectronic approaches to solid-state ion-selective electrodes', *Electrochim. Acta*, 1977, **22**, pp. 1–8
31 COHEN, R.M., HUBER, R.J., JANATA, J., URE, R.W., and MOSS, S.D.: 'A study of insulator materials used in ISFET gates', *Thin Solid Films*, 1978, **53**, pp. 169–173
32 VLASOV, Y.G., BRATOV, A.V., and LETAVIN, V.P.: 'Investigation of the ion selectivity mechanism of hydrogen ion-sensitive field-effect transistors', *Analyt. Chem. Symp. Ser.*, 1981, **8**, (Ion-Selective Electrodes, 3), pp. 387–396
33 BERGVELD, P., and DEROOIJ, N.F.: 'The history of chemically sensitive semiconductor devices', *Sens. & Actuators*, 1981, **1**, pp. 5–15
34 McBRIDE, P.T., JANATA, J., COMTE, P.A., MOSS, S.D., and JOHNSON, C.C.: 'Ion-selective field-effect transistors with polymeric membranes', *Anal. Chim. Acta*, 1978, **101**, pp. 239–245
35 SHIRAMIZU, B., JANATA, J., and MOSS, S.D.: 'Ion-selective field-effect transistors with heterogeneous membranes', *ibid.*, 1979, **108**, pp. 161–167
36 OESCH, U., CARAS, S., and JANATA, J.: 'Field-effect transistors sensitive to sodium and ammonium ions', *Anal. Chem.*, 1981, **53**, pp. 1983–1986
37 MATSUO, T., and WISE, K.D.: 'An integrated field-effect electrode for biopotential recording', *IEEE Trans.*, 1974, **BME-21**, pp. 485–487
38 ABE, H., ESASHI, M., and MATSUO, T.: 'ISFETs using inorganic gate thin films', *ibid.*, 1979, **ED-26**, pp. 1939–1944
39 AKIYAMA, T., UJIHIRA, Y., OKABE, Y., SUGANO, T., and NIKI, E.: 'Ion-sensitive field-effect transistors with inorganic gate oxide for pH sensing', *ibid.*, 1982, **ED-29**, pp. 1936–1941
40 SOBCZYNSKA, D., and TORBICZ, W.: 'ZrO₂ gate pH-sensitive field effect transistors', *Sens. & Actuators*, 1984, **6**, pp. 93–105
41 SIU, W.M., and COBBOLD, R.S.C.: 'Basic properties of the electrolyte-SiO₂-Si system: physical and theoretical aspects', *IEEE Trans.*, 1979, **ED-26**, pp. 1805–1815
42 LAUKS, I.R.: 'pH measurements using polarisable electrodes', *ibid.*, 1979, **ED-26**, pp. 1952–1959
43 LAUKS, I.: 'Polarisable electrodes', *Sens. & Actuators*, 1981, **1**, pp. 261–288
44 BUCK, R.P., and HACKLEMAN, D.E.: 'Field-effect potentiometric sensors', *Anal. Chem.*, 1977, **49**, pp. 2315–2321
45 EISENMAN, G.: 'The origin of the glass-electrode potential', *in* EISENMAN, G. (Ed.): 'Glass electrodes for hydrogen and other cations' (Dekker, 1967), pp. 133–173
46 CONTI, C., and EISENMAN, G.: 'The non-steady-state membrane potential of ion exchangers with fixed sites', *Biophys. J.*, 1965, **5**, pp. 247–256
47 JANATA, J.: 'Electrochemistry of chemically sensitive field effect transistors', *Sens. & Actuators*, 1983, **4**, pp. 255–265

48 VAN DER SCHOOT, B.H., BERGVELD, P., BOS, M., and BOUSSE, L.J.: 'The ISFET in analytical chemistry', *ibid.*, 1983, **4**, pp. 267–272

49 LAUKS, I., YUEN, M.F., and DIETZ, T.: 'Electrically freestanding IrO$_x$ thin film electrodes for high temperature, corrosive environment pH sensing', *ibid.*, 1983, **4**, pp. 375–379

50 BOUSSE, L., and BERGVELD, P.: 'The role of buried OH sites in the response mechanism of inorganic-gate pH-sensitive ISFETs', *ibid.*, 1984, **6**, pp. 65–78

51 US Patent 4 232 326

52 WEN, C-C., CHEN, T.C., and ZEMEL, J.N.: 'Gate-controlled diodes for ionic concentration measurements', *IEEE Trans.*, 1979, **ED-26**, pp. 1945–1951

53 COMTE, P.A., and JANATA, J.: 'A field-effect transistor as a solid-state reference electrode', *Anal. Chim. Acta*, 1978, **101**, pp. 247–252

54 MATSUO, T., and NAKAJIMA, H.: 'Characteristics of reference electrodes using a polymer gate ISFET', *Sens. & Actuators*, 1984, **5**, pp. 293–305

Light-Addressable Potentiometric Sensor for Biochemical Systems

Dean G. Hafeman, J. Wallace Parce, Harden M. McConnell

Numerous biochemical reactions can be measured potentiometrically through changes in pH, redox potential, or transmembrane potential. An alternating photocurrent through an electrolyte-insulator-semiconductor interface provides a highly sensitive means to measure such potential changes. A spatially selectable photoresponse permits the determination of a multiplicity of chemical events with a single semiconductor device.

POTENTIOMETRIC MEASUREMENTS are commonly used in biophysical and biochemical studies. Examples include pH measurements with glass electrodes and redox measurements with metal electrodes. Such potentiometric measurements usually involve high input impedance measuring devices so as not to disturb the chemical equilibrium. High-impedance measurements are also used in determinations of transmembrane potentials. There is much interest in the miniaturization of devices for these measurements. A frequently studied device is the chemically sensitive field-effect transistor (CHEMFET), in which the gate region of a field-effect transistor is made sensitive to chemical events through their effect on the gate potential. The CHEMFET was first described in 1970 (*1*) and has been the subject of several reviews (*2–4*). Here we describe a simple alternative semiconductive structure, the light-addressable potentiometric sensor (LAPS). Brief descriptions of some aspects of this work have been presented elsewhere (*5–7*).

Both LAPS and CHEMFET are insulated semiconductor devices that respond to surface potentials at an electrolyte-solid interface through the effect of such potentials on electric fields within the semiconductor. Surface potentials can be established by chemically selective surfaces that acquire electrical charge in response to changes in chemical properties of an electrolyte. In addition, the electric fields within the semiconductor can be modulated by transmembrane potentials when the membranes are appropriately positioned relative to the insulated semiconductor surface. Attractive features of the LAPS include potentiometric stability and the ability to address different

Molecular Devices Corporation, 3180 Porter Drive, Palo Alto, CA 94304.

regions of the semiconductor with light rather than with fixed wires or other current paths. As discussed below, potentiometric stability implies sensitivity in biochemical determinations such as enzyme-linked immunoassays. The ability to optically address different spatial regions of a sensor surface allows multiple potentiometric measurements to be made with a single semiconductor device.

Figure 1a shows the insulated surface of a thin flat plate of *n*- or *p*-type silicon in contact with an electrolyte. The insulator separating the electrolyte from silicon is a layer of silicon oxynitride approximately 1000 Å thick. The direct current through this insulating layer is negligible, less than a picoampere per square centimeter under the experimental conditions described below. In Fig. 1a, a potential Ψ is shown applied from the silicon plate to a Ag/AgCl controlling electrode. This controlling electrode also serves as a reference electrode in that it fixes the potential from the variable potential source to solution. The sign and magnitude of Ψ can be adjusted so as to deplete the semiconductor of majority charge carriers at the insulator interface. In this state the semiconductor produces a transient photocurrent in response to transient illumination either from above or below the silicon plate. An intensity-modulated light source, such as one of the light-emitting diodes (LEDs) A through D in Fig. 1a, gives rise to an alternating photocurrent through the indicated circuit. The amplitude of the alternating current (I) is measured with a low-impedance ac ammeter. Many configurations of semiconductor, light source, controlling electrode, and reference electrode (if needed) are possible (Fig. 1b). The value of I depends on the applied bias potential Ψ (Fig. 2). This photocurrent varies from a minimum value of near 0 under forward bias conditions to a maximum value limited by

the minimum attainable depletion layer capacitance under reverse bias conditions (*8*). Electrical circuits can be devised that sweep the bias potential with time while simultaneously multiplexing a series of LEDs that illuminate different sites on the silicon surface. We have measured the surface potential at nine different sites once every second using this approach. This time includes the time required to compute the potential Ψ_i, which is the potential where the slope ($dI/d\Psi$) is maximum, and present the values of the Ψ_i on a personal computer. In Fig. 1a different surface structures or chemistries, or both, at different positions are indicated schematically. Such structures include those that are sensitive to pH, redox potential, or transmembrane potential.

Previous work has shown that silicon oxynitride is pH-sensitive over a large pH range. Such measurements have been performed with the oxynitride on the gate region of an FET (*9*) or alternatively by monitoring the voltage dependence of the capacitance of the semiconductor-insulator interface (*10–12*). This pH sensitivity is due

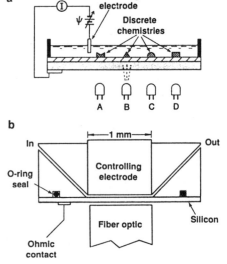

Fig. 1. Light-addressable semiconductor sensors. (**a**) A silicon plate with a surface insulator of oxynitride (diagonal lines) in contact with an electrolyte is photoresponsive to the light emitting diodes A, B, C, and D. The resultant alternating photocurrent I in the external circuit depends on the applied bias potential Ψ. Different chemistries located on different regions of the insulating surface produce variations in the local surface potential that can be determined by selective illumination with one or another of the light-emitting diodes. (**b**) For high-sensitivity measurements of enzyme activity it is advantageous to localize the enzyme molecules in a small volume so that they are present at a high effective concentration. Volumes as small as a nanoliter can be achieved with a configuration such as the one illustrated. The controlling electrode acts as a piston. It can be raised to allow fluid to be passed over the sensor surface and then lowered to create the small reaction volume.

Reprinted with permission from *Science*, vol. 240, pp. 1182–1185, May 1988.

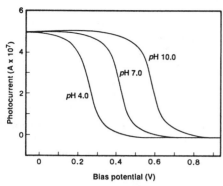

Fig. 2. Alternating photocurrent (I) as a function of bias potential (Ψ) for different values of pH. The photocurrent for a circuit such as that illustrated in Fig. 1 depends on the bias potential. Positive values of Ψ indicate that the controlling electrode is biased positively with respect to the silicon. The data illustrated are for n-type silicon, where the lower voltages (to the left) correspond to the depletion condition for the semiconductor at the insulator interface. For p-type silicon the shape of the curve is reversed, left to right. The surface in contact with the electrolyte is silicon oxynitride, which is pH sensitive.

to the proton binding capacity of Si–O and Si–NH$_2$ groups on the silicon oxynitride surface (*13*). Figure 3 shows that the pH response over a range from 2 to 12 of the LAPS device at an oxynitride site is Nernstian.

Redox potential measurements can be made by depositing pads of metallic gold 5000 Å thick over the insulating silicon oxynitride coating on silicon plates. When the electrolyte solution contains a redox pair such as ferricyanide-ferrocyanide, the potential of the gold is determined by the ratio of the concentrations of these two species, in accordance with the Nernst equation. Thus intensity-modulated illumination of a region of the semiconductor beneath the gold pad produces an alternating photocurrent similar to that observed with the pH sensing device. In this case, however, Ψ_i responds to changes in redox potential of the electrolyte.

When a membrane is interposed between the controlling-reference electrode and the silicon, a transmembrane potential will add in series with Ψ and thus affect the photoresponse. Transmembrane potentials are also often Nernstian and can be conveniently considered along with pH and redox. Figure 3 shows a Nernstian potential response arising from potassium ions when a valinomycin-containing membrane on top of an oxynitride-coated silicon plate is placed in contact with solutions containing various concentrations of potassium chloride.

Figure 4 illustrates the response that is observed when two chemically distinct regions are illuminated simultaneously with a single intensity-modulated light source. One region is a silicon oxynitride surface and the other is a gold pad deposited on top of oxynitride. The intensity-modulated light beam is expanded in diameter so that the illuminated area is approximately twice as large as the gold spot. Upon sweeping the value of Ψ, two inflections in the I versus Ψ curve are seen. The potential at the inflection point seen near -0.1 V is responsive to pH. The potential at the inflection point seen near -0.9 V is responsive to redox potential. When the solution contains a redox buffer, such as a mixture of ferri- and ferrocyanide, the potential from the gold to solution is fixed, and changes in the photoresponse from the oxynitride region can be used to measure changes in the pH of the solution. The signal from under the gold-covered area acts as a reference. When the solution is buffered with respect to pH and a redox chemical reaction takes place, the potential from the gold to solution will change. This change in solution redox potential can be measured by using the signal from the oxynitride region as a reference. This internal reference mechanism allows for pH and redox measurements to be carried out simultaneously at different illuminated sites. In this case the controlling electrode can be simplified (for example, a piece of wire in contact with solution) since it need not act as a reference.

An area of application of the LAPS device is in the high-sensitivity measurements of enzyme activity as used, for example, in enzyme-linked immunoassays. Appropriate enzymes are those that produce pH changes or changes in redox potential. For simplicity we equate sensitivity with the number of enzymes that can be detected with quantitative accuracy, as this is often the criterion of practical interest. Two features of the LAPS device contribute to high sensitivity. First, the device has high potentiometric stability. A drift in surface potential of less than 0.1 μV per second can be achieved, which corresponds to 1.7 micro pH units per second. Second, when the enzymes to be detected are trapped or immobilized near the semiconductor surface, the buffering effect of the electrolyte can be minimized by reducing the volume of the solution in contact with the insulated semiconductor surface. Figure 1b illustrates a cell designed to provide a small fluid volume in contact with the insulator surface. The sensitivity of a small volume system of this sort to detect an enzyme that produces a pH change following substrate turnover can be calculated as follows. Let the enzyme turnover number to produce or consume protons be n (with units of moles of protons per second per mole of enzyme), and let e be the number of moles of enzyme in the sample chamber. The rate of pH change of solution is

$$d(pH)/dt = en/(Bv + bs) \qquad (1)$$

where B is the volumetric buffer capacity of the electrolyte, v is the volume of the solution, b is surface buffer capacity of the chamber surfaces, and s is the surface area of the chamber in contact with the electrolyte. The volumetric buffer capacity of the solution is

$$B = 2.303\,[a - (a^2/c)] \qquad (2)$$

where c is the molar concentration of the buffer and a is the molar concentration of the acidic species of the buffer pair. An equivalent expression for b is obtained in which the concentrations of buffer species are given in terms of surface concentration or in units of moles per square centimeter. According to Eq. 1, the sensitivity to detect a given number of enzymes from a pH change can be increased by decreasing the volume or volumetric buffer capacity. Decreasing the volumetric buffer capacity by reducing the buffer concentration below 1 mM or adjusting the initial solution pH to be far from the buffer pK is generally not practical, as substantial pH drifts will occur due to equilibration of atmospheric CO$_2$ with solution. For a fixed buffer concentration in the electrolyte, the rate of pH change can be increased by reducing the electrolyte volume v until the surface buffer capacity of the chamber becomes dominant, as indicated in Eq. 1. The data given in Fig. 5 illustrate the determination of surface buffer capacity of a sample cell. In this experiment, the sample volume is estimated to be of the order of 1 nl, and the surface buffer is about

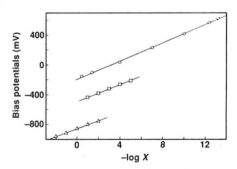

Fig. 3. Nernstian responses to electrolyte composition. This graph shows plots of potential (Ψ_i) at maximum photocurrent slope (maximum $dI/d\Psi$) for three different chemically sensitive surfaces on the semiconductor. For $X = [H^+]$, the proton concentration (data points are given by circles) shows that the photoresponse is Nernstian over a pH range greater than from pH 2 to 12. The sensing surface is silicon oxynitride. For $X =$ [ferricyanide/ferrocyanide], the data points are given by triangles, and the surface is gold over silicon oxynitride. For $X = [K^+]$, data points are given by squares, and the sensing surface is a membrane over the oxynitride. The membrane is 66% polyvinylchloride, 33% dioctyladipate, and 1% valinomycin [see (*14*)].

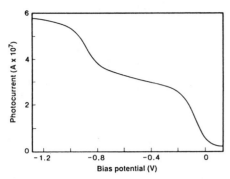

Fig. 4. Biphasic response for a binary sensing surface. In this experiment, a single light-emitting diode illuminates an area under a surface that is approximately 50% silicon oxynitride in contact with the electrolyte and 50% gold on silicon oxynitride in contact with the electrolyte. The solution is buffered at pH 7.4 and contains 5 mM ferricyanide and 5 mM ferrocyanide. The biphasic response originates from the different surface potentials for the two cases. The inflection on the left is associated with the gold surface (redox potential) and that on the right is associated with the silicon oxynitride surface (pH).

1 pmol, which limits the sensitivity according to Eq. 1; about 10,000 molecules of an active enzyme such as urease, $n = 5,870$ (15), should give a rate of pH change significantly above the background drift stated above.

Enzyme-linked redox chemistry can be used for a number of biochemical and immunochemical assays (16–18). The theoretical sensitivity for a redox measurement is limited by the metal-electrolyte interfacial capacitance, which is of the order of 20 μF cm^{-2}. This capacitance must be charged by exchange of electrons to or from the redox species. The requirement for charging this interfacial capacitance in order to generate a change in surface potential determines the ultimate theoretical sensitivity of this sensor for the measurement of changes in both redox potential and pH.

The experiments described in this report have all been performed with an illumination area of about 1 mm^2. For the silicon used, the minority-carrier diffusion length is on the order of 1 mm, which means that the measured photocurrent is obtained from an area of approximately 1 mm^2 even if the surface chemistry of interest is confined to a smaller area. High spatial resolution can in principle be achieved in several ways. The silicon can be masked, for example, with a thick insulating layer, so as to allow only a very small area of the silicon oxynitride surface to come in contact with the aqueous medium. Silicon can be doped with gold to reduce the minority carrier lifetime and thus reduce diffusion lengths. The frequency of modulation of the illumination source can be increased. Phase-sensitive detection of the photocurrent will also in-

crease spatial resolution.

For the purpose of discussing signal-to-noise ratio in the LAPS device, the signal is defined as the slope of the midpoint region of the photocurrent versus potential curve (Fig. 2), and noise is the time-dependent variation of photocurrent amplitude at that potential. The signal, or slope of the photocurrent curve, is a function of the absolute amplitude of the photocurrent and the width of the curve given by the first derivative of photocurrent versus potential. Typical full-widths at half-height for first derivative photoresponse curves are approximately 0.1 V at low photocurrents. The photocurrent increases linearly with increasing illumination intensity up to the point where the alternating photopotential generated across the insulator approaches the width of the photoresponse curve. At this point, a further increase in illumination intensity results in an increase in width of the photoresponse curve and thus tends to negate the increase in signal due to an increase in photocurrent. Therefore the optimum photocurrent I is approximated by the equation

$$I \sim EC/t \qquad (3)$$

where E is the width of the photoresponse curve (0.1 V), C is the capacitance of the oxynitride insulator (0.05 μF cm^{-2}), and t is the illumination time per modulation cycle (0.05 msec for 10-kHz square-wave modulation). This optimum photocurrent is ~1 μA mm^{-2}. In general, to optimize signal-to-noise ratio the input impedance of the photocurrent amplifier is matched to the sensor impedance. In the case of the LAPS, however, this strategy results in cross talk between various chemistry sites. To minimize this cross talk, most of the photocurrent must flow into the input amplifier. This is accomplished by making the input impedance of the amplifier small with respect to the sensor impedance. In this configuration, the major source of noise is the equivalent input voltage noise of the first operational amplifier (current to voltage converter). Because the sensor acts as a shunting capacitor from this input to ground, the noise is essentially proportional to frequency in the frequency range of interest (about 10 kHz). Varying the illumination modulation frequency in this range does not alter significantly the signal-to-noise ratio, as the signal also is proportional to frequency (Eq. 3).

Two features, the planar surface of the device and the ease with which multiplicity can be achieved by addressing discrete sensing sites with light, make the device described in this report an ideal candidate for use as a signal transducer in biosensors. In biosensors, the most difficult component to control is the biological or biochemical com-

ponent. Enzymatic reactions are temperature-sensitive, proteins tend to denature and become inactive as a function of time and their environmental history, and even at the level of manufacturing, different preparations of the same biological material often result in different levels of biological activity. For these reasons a robust biosensor should have a variety of "on-board" biochemical calibrators. The light-addressable aspect of the LAPS makes sensing a multiplicity of on-board calibrators a relatively straightforward task. To date we have measured the kinetics of reactions run simultaneously at as many as 23 sites on a single sensor with only two electrical leads from the controlling

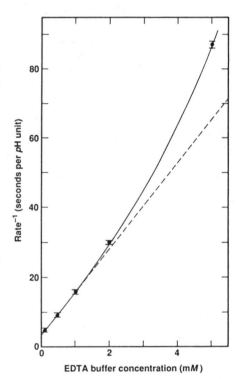

Fig. 5. Determination of sensor-surface buffer capacity. The rate of pH change as a function of substrate buffer concentration was determined using urease as the pH changing enzyme. Urease was nonspecifically adsorbed to the surface of the silicon. The solution contained 16 mM urea, 150 mM NaCl, and various concentrations of EDTA (pH 5.7) as indicated. The sensing chamber was a cylinder with a diameter of 3 mm and a height of approximately 0.2 μm. The height of the chamber could be temporarily increased to allow for introduction of fresh substrate and buffer. Two separate measurements were performed at each buffer concentration, except for 0.1 mM EDTA, where four measurements were taken. The data are corrected for an apparent first-order loss of urease activity with successive buffer replacements in the flow cell. The error bars indicate the range of the data points. The solid line connects the data points. The dashed line is a linear fit to the first three data points. Deviation of the points from the linear fit at 2 mM and more strongly at 5 mM EDTA is probably due to inhibition of urease due to accumulation of ammonia. For urease the inhibition constant for ammonia is 10 mM (19).

electronics to the sensor. Furthermore, by partially covering the surface with metal, separate pH and redox measurements can be made at each site to control for potentiometric stability of the system if one or the other of these solution parameters is made stable at each site by an appropriate buffer. Another feature of the LAPS that lends convenience to the fabrication of biosensors is its planar surface. The flat polished sensing surface has two significant attributes. First, it is easy to create fluid seals to maintain the aqueous solution only in contact with the insulator surface. For high-sensitivity assays, microscopic fluid leaks between the aqueous compartment and the electrical contact on the back side of the semiconductor can result in substantial potentiometric drifts. For some sensor configurations, a resistive path of 100 gigaohms through the fluid leak into a small volume can result in substantial drifts. Second, the flat surface makes it possible to generate very small, defined aqueous volumes. These volumes can be as small as a nanoliter. The advantage of using small volumes for performing high-sensitivity assays is discussed above. We have de-

tected one attomole (600,000) of enzyme molecules adsorbed to filter paper 100 μm thick. The time required to make this determination was approximately 2 minutes.

The principal advantage of biosensors, such as the one described here, is the ease with which miniaturization can be achieved. This miniaturization in turn facilitates multiplicity and high sensitivity. Specific applications of this methodology to enzyme-linked immunochemical assays for therapeutic drugs, hormones, and bacterial pathogens will be given elsewhere.

REFERENCES AND NOTES

1. P. Bergveld, *IEEE Trans. Biomed. Eng.* **BME-19**, 70 (1970).
2. J. Janata and R. J. Huber, in *Ion-Selective Electrodes in Analytic Chemistry*, H. Freiser, Ed. (Plenum, New York, 1980), pp. 107–174.
3. J. N. Zemel, *Anal. Chem.* **47**, 255A (1975).
4. G. F. Blackburn, in *Biosensors Fundamentals and Applications*, A. P. F. Turner, I. Karube, G. S. Wilson, Eds. (Oxford Univ. Press, Oxford, 1987), pp. 481–530.
5. H. M. McConnell, J. W. Parce, D. G. Hafeman, *Electr. Soc. Abstr.* **87**, 2272 (1987).
6. _____, *Proc. Electrochem. Soc.* **87**, 292 (1987).
7. D. G. Hafeman, J. W. Parce, H. M. McConnell, *Proceedings of the 2nd International Meeting on Chem-*
ical Sensors, J. L. Aucouturier *et al.*, Eds. (Bordeaux, 1986), p. 69.
8. For a general reference on semiconductor photo-responses, see S. N. Sze, *Physics of Semiconductor Devices* (Wiley, New York, 1981), pp. 362–430.
9. L. Bousse, P. de Rooij, P. Bergveld, *IEEE Trans. Electr. Dev.* **ED-30**, 1263 (1983).
10. Y. G. Vlasov, A. J. Bratov, V. P. Letavin, in *Ion-Selective Electrodes 3*, vol. 8 of the *Analytical Chemistry Symposium Series*, E. Pungor, Ed. (Elsevier, Amsterdam, 1981), pp. 387–397.
11. F. Chauvet, A. Amari, A. Martinez, *Sensors Actuators* **6**, 255 (1984).
12. M. T. Pham and W. Hoffmann, *ibid.* **5**, 217 (1984).
13. L. Bousse and J. D. Meindl, *ACS Symp. Ser.* **323**, 79 (1986).
14. U. Oesch, D. Ammann, W. Simon, *Clin. Chem.* **32**, 1448 (1986).
15. R. L. Blakeley and B. J. Zerner, *J. Mol. Catal.* **23**, 263 (1984).
16. W. U. de Alwis, B. S. Hill, B. I. Meiklejohn, G. S. Wilson, *Anal. Chem.* **59**, 2688 (1987).
17. D. S. Wright, H. B. Halsall, W. R. Heineman, *ibid.* **58**, 2995 (1986).
18. F. A. Armstrong, H. A. O. Hill, N. J. Walton, *Q. Rev. Biophys.* **18**, 261 (1986).
19. L. Goldstein, M. Levy, L. Shemer, *Biotechnol. Bioeng.* **25**, 1485 (1963).
20. We are greatly indebted to G. Pontis, J. Kercso, and L. Bousse for helpful discussions and technical assistance. Supported in part by the Army Research Office and Defense Advanced Research Projects Agency.

8 February 1988; accepted 20 April 1988

MICRODIELECTROMETRY

NORMAN F. SHEPPARD, DAVID R. DAY, HUAN L. LEE and STEPHEN D. SENTURIA

Department of Electrical Engineering and Computer Science, and Center for Materials Science and Engineering, Massachusetts Institute of Technology, Cambridge, MA 02139 (U.S.A.)

Abstract

The low-frequency dielectric properties of resins provide a useful tool for characterizing both the curing process and the fully cured material. We have used silicon microfabrication technology to develop a miniaturized dielectric sensor that combines a planar interdigitated electrode structure with a pair of matched field-effect transistors to achieve sensitivity at frequencies as low as 1 Hz. The microdielectrometer "chip" can be implanted in a specimen, or a small sample of material (a few milligrams) can be placed on the active area of the device. When combined with an off-chip electronic feed-back system, the device can be used to measure the complex dielectric constant of the sample material either as it cures, or after cure. The device is capable of operating in excess of 200 °C, making it useful for a wide variety of curing and post-cure studies. Device calibration is based on a two-dimensional computer model which has been experimentally confirmed by experiments, and the use of the data to follow changes in the dominant low-frequency dielectric relaxation during the cure of a model epoxy resin system is presented.

1. Introduction

This paper presents a new microelectronic technique for the measurement of low-frequency dielectric properties of materials [1]. Measurements of dielectric properties of polymers and other materials are used in a variety of applications [2]. This paper is directed particularly toward measuring the dielectric changes that occur in various polymer resin systems during the cure process.

The conventional method for measuring dielectric properties is to use a parallel-plate capacitor filled with or embedded in the sample. This method has the ideal geometry, but is susceptible to expansion or contraction that might occur during heating or curing.

The technique that we call "microdielectrometry" differs from the conventional measurement in several ways. First, it uses as sense electrodes a pair of very small planar interdigitated electrodes fabricated as part of a silicon integrated circuit (the "microdielectrometer" chip). This electrode geometry, while much less efficient than the parallel-plate geometry in terms of electric field coupling between the electrodes, can be manufactured with high precision using microelectronic techniques, and therefore can yield an electrode pattern with known and reproducible calibration. For example, the electrode geometry is not affected by contraction or expansion of the material being monitored. Second, the technique incorporates high impedance amplifiers in the form of depletion-mode MOSFETs built into the microdielectrometer chip, achieving a sensitivity improvement that more than compensates for the relatively inefficient electrode structure, and permits operation of the device down to 1 Hz, an advantage for studying slow relaxations in materials. Finally, with the use of a specially designed electronic feed-back circuit, such potential sources of problems as FET temperature and pressure dependences can be cancelled out, permitting the microdielectrometer chip to be used over a wide temperature range (up to 250 °C), even when implanted in a bulk specimen of curing material.

Section 2 of this paper contains a description of the microdielectrometer device and accompanying measurement system. Section 3 illustrates how raw data obtained during the cure of diglycidyl ether of bisphenol-A (DGEBA) and m-phenylene diamine (MPDA), a two-component epoxy-amine system, can be converted into the conventional real and imaginary parts of the dielectric constant by the use of calibration curves developed from two-dimensional computer simulations of the device. Finally, Section 4 demonstrates how the resulting dielectric constant data can be interpreted to obtain a dynamical relaxation time which is shown to be strongly correlated with the behavior of the viscosity during cure.

2. The microdielectrometer chip and measurement system

The microdielectrometer chip contains a planar interdigitated electrode and two depletion-mode metal–oxide–semiconductor field-effect transistors (MOSFETs). A top view of the most critical portion of the device is shown in Fig. 1. The outer electrode is called the driven gate, and is connected to a normal bonding pad. The inner electrode is called the floating gate, and extends over the channel region of one of the FETs as shown. This particular combination of a driven gate and a floating gate controlling a MOSFET is called a floating-gate charge-flow transistor (CFT) [3]. The same device has been referred to elsewhere as a surface-impedance measurement device (SIM) [4].

The silicon dioxide layer between the floating gate and the silicon substrate as well as between the two gates effectively isolates the floating gate, with the result that the primary electrical connection between the driven and

Reprinted with permission from *Sensors and Actuators*, vol. 2, pp. 263–274, 1982. Copyright © 1982 by Elsevier Sequoia.

reference FET and feed-back circuit serves to provide both thermal and pressure compensation of the measurement.

The microdielectrometer chip as presently designed is 75 mils square, and has a total of eight contacts. The device is mounted on to a standard TO-8 transistor header, and wire bonded. Other more compact mounting and packaging methods could also be used. The device can be used either by placing a small sample of material over the electrodes or by implanting the entire device into a bulk specimen. In either case, the device can readily be placed in an oven for isothermal or ramped temperature studies.

A block diagram of the instrumentation system is shown in Fig. 3. The device and interface circuit are connected to a computer-controlled function generator, and the magnitude and phase of the floating-gate voltage are measured with a gain-phase meter, also linked to the computer. The transfer function between the driven gate and the floating gate depends only on the real and imaginary parts of the dielectric constant (ϵ' and ϵ''), and not explicitly on the the frequency of the sinusoid. Therefore, there is a unique mapping between measured gain and phase, on the one hand, and ϵ' and ϵ'' on the other hand. At present, the graphics display on the HP-85 computer used to run the system can be set up for real-time display of either the gain-

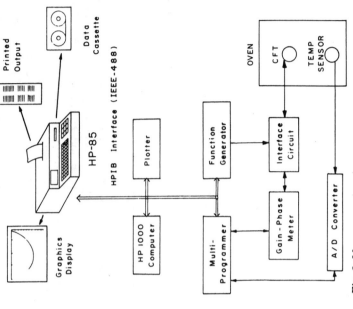

Fig. 3. Measurement system.

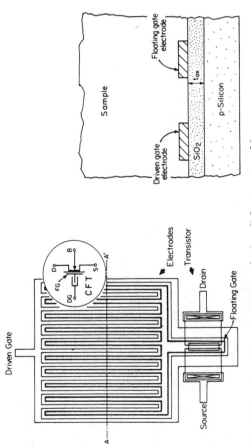

Fig. 1. Top view of sensor portion of microdielectrometer chip.

Fig. 2. Cross-section of electrode portion of microdielectrometer chip.

floating gates is through the sample material under study, which is placed over the electrodes as in the schematic cross-section in Fig. 2. Because of the distributed nature of the device–sample combination, there is no simple lumped circuit equivalent for the device. Approximate models based on distributed RC transmission lines have been discussed elsewhere [5, 6], and a more accurate model based on a numerical two-dimensional solution of Laplace's equation has been developed [7]. The results of this numerical modelling are described in Section 3 below.

Operation of the device is as follows: a sinusoidal voltage applied to the driven gate causes time-varying current (both conduction and displacement current) to flow through the sample toward the floating gate. This current charges the capacitance between the floating gate and the substrate, and the resulting time-varying charge on the floating gate modulates the conductance of the FET channel. Thus, the primary measurement consists of determining the magnitude and phase of the charge on the floating gate produced by a sinusoidal waveform applied to the driven gate. Clearly, this will depend on the dielectric properties of the sample.

In order that the measurements do not depend on the electrical characteristics of the FETs (which are subject to manufacture-related variations as well as temperature and pressure dependences), two identical FETs are fabricated on the microdielectrometer chip. The second reference FET is connected in a specially designed feed-back interface circuit which permits inference of the floating gate voltage by applying to the gate of the reference FET exactly that voltage required to make the two FET source currents identical. The details of this circuit are described elsewhere [4]. Use of the phase characteristic or of the dielectric loss factor, ϵ''. Other components of the system include a link to a larger computer (HP-1000), and access to line printers, cassette storage, and a graphics plotter. All data are stored in files on cassette, permitting very flexible data analysis procedures with no transcription or re-formatting required.

3. Dielectric data extraction

Experimental results have been obtained for a stoichiometric mix of DGEBA and MPDA cured at 60 and 100 °C. Microdielectrometer chips with oxide thicknesses of both 540 nm and 1000 nm were used to monitor the cure. The data extraction will be illustrated with results for the thicker oxide.

Data are collected simultaneously at seven frequencies as a function of cure time. The gain-phase data are converted into ϵ' and ϵ'' in real time using a calibration obtained through a two-dimensional numerical solution to Laplace's equation with a complex amplitude for the potential. Details of the calibration calculation are available elsewhere [7].

Figure 4 shows a typical set of calculated calibration curves obtained for a device with a 1000 nm oxide. Several features can be noted. For a perfect dielectric ($\epsilon'' = 0$), the gain-phase points lie on the gain axis (zero phase shift). A check of the calculated calibration was made by measuring the high-frequency transfer function of the device in air ($\epsilon' = 1.0$) and was found to be within 1 dB of the calculated value. In addition, various materials with dielectric constants ranging from 2 to 7 were checked and were found to be in reasonable agreement with the calculation (± 1 dB). Examination of the

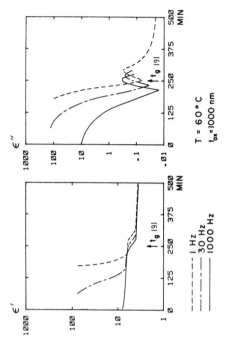

Fig. 5. Time dependence of ϵ' and ϵ'' for DGEBA/MPDA cured at 60 °C.

calibration curves shows that near the origin (corresponding to early in cure), the various curves crowd together. This means that the measurement is most prone to error when the data are near the origin, and that small gain offsets or spurious sources of either magnitude or phase errors might produce problems early in cure. Indeed, we have had problems with the calibration for gains above −3 dB, due to a spurious conduction path in the present chip design which creates magnitude and phase errors when the resin is highly conductive, and due to a small gain offset (of order 1 dB) between the floating gate CFT and the reference FET. The details of the origins of these errors, their implications for the interpretation of measurements early in cure, and ways to remove the errors in future designs are discussed elsewhere [8]. For the present discussion, we have elected to ignore all data closer to the origin than −3 dB, and have compensated for the gain offset by a small correction for each device which makes the final high-frequency dielectric constant agree with the result of a parallel-place capacitor measurement on fully cured material. In no case is the gain correction more than 1.2 dB.

The time dependences of ϵ' and ϵ'' at three different frequencies for a DGEBA/MPDA sample cured at 60 and 100 °C are plotted in Figs. 5 and 6 respectively. Times to gelation, t_g, for this resin system are also shown for reference 9. Comparison between Figs. 5 and 6 shows that the behavior of the dielectric constant is the same, but occurs over a longer time scale as the temperature is decreased, due to the thermal activation of the curing reaction.

The results in Figs. 5 and 6 show that two relaxations are occurring: a large one prior to gelation, and a smaller one after gelation. The frequency dependence demonstrates that the 1 Hz measurement is more sensitive to these relaxations than higher frequencies, which is consistent with a model of dipole orientation. Increased frequency or extent of cure makes it more

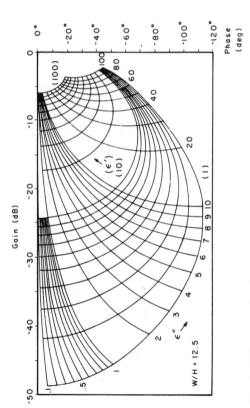

Fig. 4. Calibration plot showing contours of constant permittivity (ϵ') and loss factor (ϵ'').

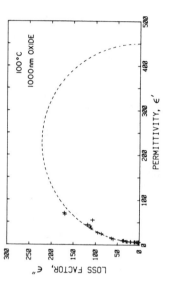

Fig. 6. Time dependence of ϵ' and ϵ'' for DGEBA/MPDA cured at 100 °C.

- - - - - 1 Hz
—·—·— 30 Hz
————— 1000 Hz

$T = 100$ °C
$t_{ox} = 1000$ nm

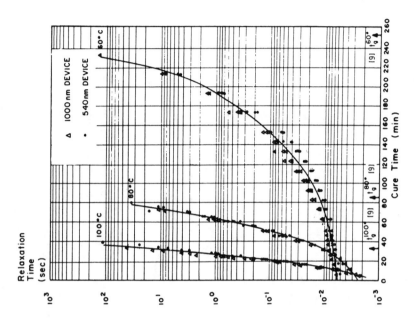

Fig. 7. Cole–Cole plot for data of Fig. 6 before gelation (dashed line represents Debye model fit).

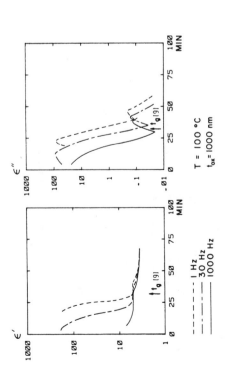

Fig. 8. Relaxation time *versus* cure time for all experiments.

difficult for the dipoles to respond to the alternating electric field. This leads us to attribute the first relaxation to the increasing viscosity of the liquid phase as the epoxy polymerizes and approaches gelation, and the second to further crosslinking in the gel phase. The principal conclusion is that reliable low-frequency dielectric data can be obtained with this device, and with the improved low-frequency sensitivity, direct comparison of dielectric and mechanical properties, which are typically measured at low frequencies, becomes possible.

4. Interpretation with a relaxation time model

The dielectric data of Figs. 5 and 6 show a great deal of structure. In particular, prior to the classical gelation time, there is a strong dielectric relaxation evident. A good way to examine relaxation time models is to plot ϵ'' *versus* ϵ' (the so-called Cole–Cole plot [10]), using frequency as the plotting parameter. If a Debye model for dipole orientation holds and if only the relaxation time changes with time, then the data lie on a semicircle. The position of a point on the semicircle is determined by the product of the angular frequency and the relaxation time, so knowledge of the frequency permits determination of the relaxation time.

Figure 7 shows a Cole–Cole plot for the data of Fig. 6 prior to gelation. The points correspond to the values of ϵ' and ϵ'', and the dashed curve represents the case for an ideal Debye model having a limiting low-frequency dielectric constant of 450 and a limiting high-frequency dielectric constant of 6. Agreement with the Debye model is seen to be very good. It should be emphasized that although it appears that only a small portion of the semi-circle is covered by the data, the fit to the ideal Debye model is justified by the fact that the approach of the data to the ϵ' axis is fully perpendicular [10].

445

For each temperature, two devices were used with different oxide thicknesses. The relaxation time was determined at each point in cure prior to gelation, and the results are plotted in Fig. 8 for each of the six devices tested (two oxide thicknesses at each of three temperatures). For any one device, the agreement between the relaxation times determined from different frequencies at the same elapsed cure time was within 20%. The agreement between the relaxation times determined from the devices with different oxide thicknesses at the same cure temperature was within 50%. Since the relaxation time changes by many orders of magnitude during cure, this degree of agreement is already very good. Nevertheless, it suggests that further refinements of the calibration may be required.

The time dependence of the relaxation time resembles the time dependence of the viscosity of thermosetting systems. The slope of a plot of log viscosity *versus* time breaks to a much steeper slope as cure proceeds, and is attributed to chain entanglement among the branched polymers formed early in cure [11]. Another important similarity to viscosity is that the relaxation time at the start of cure increases as the temperature decreases. In fact, we have found an excellent correspondence between the time dependence of the relaxation time, as determined in our experiments, and the viscosity data reported by Kamal [12], by simply scaling the relaxation time axis appropriately (see Fig. 9). (A similar scaling of dielectric relaxation time and viscosity has been reported by Denney in a very different chemical system [13].) The viscosity data and relaxation time data for various temperatures form a smooth progression. As confirmation of the correspondence between the two, we have plotted in Fig. 10 an Arrhenius plot for the time to reach 400 poise or 8 seconds relaxation time. The points fall on one straight line characterized by an activation energy of 11.5 kcal/mole, indi-

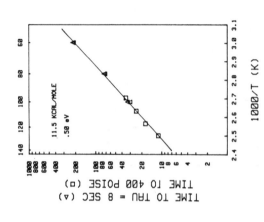

Fig. 10. Arrhenius plot of time in minutes to reach either a fixed relaxation time or a fixed viscosity.

cating that the same process is responsible for the behavior of both the viscosity and the dielectric relaxation time.

5. Discussion

The results of this paper have been presented in fairly condensed form, and have been organized to emphasize the measurement technique, and the kinds of data that can be obtained with it. Clearly, there is more work to be done on device design, device packaging and device calibration, particularly near the gain-phase origin. Nevertheless, we have found that the microdielectrometry technique is quite capable of measuring dielectric properties (ϵ' and ϵ'') of curing resins down to frequencies of 1 Hz. Furthermore, it is possible to extract from low-frequency dielectric data a quantity (the low-frequency dielectric relaxation time) which appears to carry the same information as bulk viscosity. This relaxation time can be measured entirely electrically, and the data-processing needed to convert gain-phase data to relaxation time is modest enough to be performable on a real-time basis. Thus, one can anticipate being able to monitor the viscosity of a curing system, *in situ*, and in real time. Given the role played by the viscosity in the determination and control of cure cycles for composites, the microdielectrometer should prove an advantageous addition to the family of cure monitoring methods.

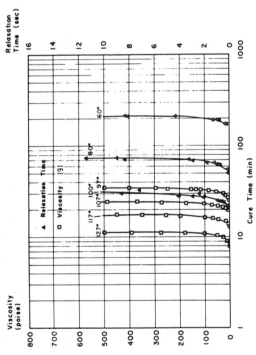

Fig. 9. Comparison between time dependence of relaxation time and viscosity.

Acknowledgements

This work was supported in part by the Office of Naval Research. Particular thanks is due Dr. Leighton H. Peebles, Jr., of ONR, who originally suggested that our work on MOS sensing devices might be usefully applied to cure monitoring problems. Device fabrication was carried out in the M.I.T. Microelectronic Laboratory, a Central Facility of the Center for Materials Science and Engineering which is sponsored in part by the National Science Foundation under contract DMR-78-24185. Some of the instrumentation used in the microdielectrometry system was purchased under NSF contract ENG-7717219. Resin samples used in these experiments were obtained from Dr. N. Schneider of the Army Materials and Mechanics Research Center.

References

1 N. F. Sheppard, S. L. Garverick, D. R. Day and S. D. Senturia, Microdielectrometry: a new method for in situ cure monitoring, *Proc. 26th SAMPE Symposium, Los Angeles, U.S.A., April, 1981*, pp. 65 - 76.

2 P. Hedvig, *Dielectric Spectroscopy of Polymers*, Wiley, New York, 1977.

3 S. D. Senturia, *Charge-Flow Transistors*, U.S. Patent, 4 316 140, 16 Feb. 1982.

4 S. L. Garverick and S. D. Senturia, An MOS device for AC measurement of surface impedance with application to moisture monitoring, *IEEE Trans. Electron Devices, ED-29* (1982) 90.

5 S. D. Senturia, N. F. Sheppard, S. Y. Poh and H. R. Appelman, The feasibility of electrical monitoring of resin cure with the charge-flow transistor, *Polymer Eng. Sci., 21* (1981) 113.

6 R. S. Jachowicz and S. D. Senturia, A thin-film capacitance humidity sensor, *Sensors and Actuators, 2* (1981/82) 171.

7 H. L. Lee, *S.M. Thesis*, Massachusetts Institute of Technology, 1982.

8 N. F. Sheppard, *S.M. Thesis*, Massachusetts Institute of Technology, 1981.

9 S. Sourour and M. R. Kamal, Differential scanning calorimetry of epoxy cure: isothermal cure kinetics, *Thermochimica Acta, 14* (1976) 41.

10 K. S. Cole and R. H. Cole, Dispersion and absorption in dielectrics, *J. Chem. Phys., 9* (1941) 341.

11 F. G. Musatti and C. W. Macosko, Rheology of network forming systems, *Polymer Eng. Sci., 13* (1973) 236.

12 M. R. Kamal, S. Sourour and M. Ryan, Integrated thermo-rheological analysis of the cure of thermosets, *Technical Papers, 31st Annual Technical Conference, Society of Plastics Engineers, Montreal, Canada, May 1973*, p. 187.

13 D. J. Denney, Viscosities of some undercooled liquid alkyl halides, *J. Chem. Phys., 30* (1959) 159 and D. J. Denney, Dielectric properties of some liquid alkyl halides, *J. Chem. Phys., 27* (1957) 259.

Part 6
Selected Bibliography

THE bibliography in this section cites, in addition to the papers comprising this volume, references that are selected to broaden perspectives within the subject areas of the collection. Many cited papers would appropriately be reprinted if space were available to accommodate them. A far more comprehensive bibliography on microsensors (containing roughly ten times as many listings as the bibliography printed here) has been compiled and is available on computer diskette as described in the Preface.

The selections in the following bibliography are divided using the subject areas that were introduced in the reprint collection. Within each section, entries are listed alphabetically by first-named author.

Bibliography

Review Papers

Barth, P. W. and J. B. Angell, "Miniature silicon sensors: Present status and future prospects," *Proc. 1st Sensor Symp.,* pp. 1–10, Tsukuba Science City, Japan, June 1981.

Giachino, J. M., "Smart sensors," *Sensors and Actuators,* vol. 10 (3) & (4), pp. 239–248, Nov./Dec. 1986.

Greenwood, J. C., "Silicon in mechanical sensors," *J. Phys. E: Scientific Instruments,* vol. 21 (12), pp. 1114–1128, December 1988.

Igarashi, I., "New technology of sensors for automotive applications," *Sensors and Actuators,* vol. 10 (3) & (4), pp. 181–194, Nov./Dec. 1986.

Igarashi, I. and T. Takeuchi, "Automotive sensors," *Proc. 2nd Sensor Symp.,* pp. 149–156, Tsukuba Science City, Japan, June 1982.

Kataoka, S., "Research and development project of future electron devices and intelligent sensors," *Proc. 2nd Sensor Symp.,* pp. 1–6, Tsukuba Science City, Japan, June 1982.

Ko, W. H., "Solid-state physical transducers for biomedical research," *IEEE Trans. Biomed. Eng.,* vol. BME-33 (2), pp. 153–162, February 1986.

Kobayashi, T., "Solid-state sensors and their applications in consumer electronics and home appliances in Japan," *Sensors and Actuators,* vol. 9 (3), pp. 235–248, May 1986.

Middelhoek, S. and A. C. Hoogerwerf, "Classifying solid-state sensors: The 'sensor effect cube'," *Sensors and Actuators,* vol. 10 (1) & (2), pp. 1–8, Sept./Oct. 1986.

Middelhoek, S. and A. C. Hoogerwerf, "Smart sensors: When and where?," *Tech. Dig., Transducers '85, Third Int. Conf. on Solid-State Sensors and Actuators,* pp. 2–7, Philadelphia, PA, June 1985.

Muller, R. S., "Strategies for sensor research," *Tech. Dig., Transducers '87, 4th Int. Conf. on Solid-State Sensors and Actuators,* pp. 107–111, Tokyo, Japan, June 1987.

White, R. M., "A sensor classificiation scheme," *IEEE Trans. Ultrason., Ferroelec., Freq. Contr.,* vol. UFFC-34, pp. 124–126, March 1987.

Wise, K. D., "Integrated sensors: Key to future VLSI systems," *Proc. 6th Sensor Symp.,* pp. 1–12, Tsukuba Science City, Japan, May 1986.

Yamasaki, H., "Approaches to intelligent sensors," *Proc. 4th Sensor Symp.,* pp. 69–76, Tsukuba Science City, Japan, June 1984.

Microsensor Fabrication Technology: Materials

Binder, J., W. Henning, E. Obermeier, H. Schaber, and D. Cutter, "Laser-recrystallized polysilicon resistors for sensing and integrated circuits applications," *Sensors and Actuators,* vol. 4 (4), pp. 527–536, December 1983.

Bomchil, G., R. Herino, and K. Barla, "Formation and oxidation of porous silicon for silicon on insulator technologies," in *Energy Beam-Solid Interactions and Transient Thermal Processing.* Les Ulis, France: Les Editions de Physique, 1985, pp. 463–474.

Ebata, T., M. Sekimoto, T. Ono, K. Suzuki, J. Matsui, and S. Nakayama, "Transparent x-ray lithography masks," *Japan. J. Appl. Phys.,* vol. 21 (5), pp. 762–767, 1982.

Erskine, J. C., "Polycrystalline silicon-on-metal strain gauge transducers," *IEEE Trans. Electron Devices,* vol. ED-30 (7), pp. 796–801, July 1983.

French, P. J. and A. G. R. Evans, "Polycrystalline silicon strain sensors," *Sensors and Actuators,* vol. 8 (3), pp. 219–226, November 1985.

Guckel, H., D. W. Burns, H. A. C. Tilmans, D. W. DeRoo, and C. R. Rutigliano, "Mechanical properties of fine grained polysilicon—The repeatability issue," *Tech. Dig., IEEE Solid-State Sensor and Actuator Workshop,* pp. 96–99, Hilton Head, SC, June 1988.

Guckel, H., D. W. Burns, H. A. C. Tilmans, C. C. G. Visser, D. W. DeRoo, T. R. Christenson, P. J. Klomberg, J. J. Sniegowski, and D. H. Jones, "Processing conditions for polysilicon films with tensile strain for large aspect ratio microstructures," *Tech. Dig., IEEE Solid-State Sensor and Actuator Workshop,* pp. 51–54, Hilton Head, SC, June 1988.

Guckel, H., D. W. Burns, C. R. Rutigliano, D. K. Showers, and J. Uglow, "Fine grained polysilicon and its application to planar pressure transducers," *Tech. Dig., Transducers '87, 4th Int. Conf. on Solid-State Sensors and Actuators,* pp. 277–282, Tokyo, Japan, June 1987.

Guckel, H., T. Randazzo, and D. W. Burns, "A simple technique for the determination of mechanical strain in thin films with applications to polysilicon," *J. Appl. Phys.,* vol. 57 (5), pp. 1671–1675, 1985.

Lober, T. A., J. Huang, M. A. Schmidt, and S. D. Senturia, "Characterization of the mechanisms producing bending moments in polysilicon micro-cantilever beams by interferometric deflection measurements," *Tech. Dig., IEEE Solid-State Sensor and Actuator Workshop,* pp. 92–95, Hilton Head, SC, June 1988.

Maseeh, F., M. A. Schmidt, M. G. Allen, and S. D. Senturia, "Calibrated measurements of elastic limit, modulus, and the residual stress of thin films using micromachined suspended membranes," *Tech. Dig., IEEE Solid-State Sensor and Actuator Workshop,* pp. 84–87, Hilton Head, SC, June 1988.

Mequio, C., R. H. Coursant, and J. M. Tellier, "Characterization of piezoplastics," *Sensors and Actuators,* vol. 14 (1), pp. 1–8, May 1988.

Mercer, H. D. and R. L. White, "Photolithographic fabrication and physiological performance of microelectrode arrays for neural stimulation," *IEEE Trans. Biomed. Eng.,* vol. BME-25 (6), pp. 494–500, November 1978.

Muller, R. S., "Heat and strain-sensitive thin-film transducers," *Sensors and Actuators,* vol. 4, pp. 173–182, December 1983.

Nagaune, F., K. Tezuka, Y. Kakuhara, K. Kamimura, Y. Onuma, and T. Homma, "Piezoresistive elements using

451

polycrystalline germanium films prepared by photo assisted chemical vapor deposition," *Proc. 6th Sensor Symp.,* pp. 13–16, Tsukuba Science City, Japan, May 1986.

Obermeier, E., P. Kopystynski, and R. Nießl, "Characteristics of polysilicon layers and their application in sensors," *Tech. Dig., IEEE Solid-State Sensors Workshop,* Hilton Head, SC, June 1986.

Onuma, Y., K. Kamimura, and Y. Homma, "Piezoresistive elements of polycrystalline semiconductor thin films," *Sensors and Actuators,* vol. 13 (1), pp. 71–78, January 1988.

Petersen, K. E., "Silicon as a mechanical material," *Proc. IEEE,* vol. 70, pp. 420–457, May 1982.

Schubert, D., W. Jenschke, T. Uhlig, and F. M. Schmidt, "Piezoresistive properties of polycrystalline and crystalline silicon films," *Sensors and Actuators,* vol. 11 (2), pp. 145–156, March 1987.

Sekimoto, M., H. Yoshihara, and T. Ohkubo, "Silicon nitride single-layer x-ray mask," *J. Vac. Sci. Technol.,* vol. 21 (4), pp. 1017–1021, 1982.

Senturia, S. D., "Microfabricated structures for the measurement of mechanical properties and adhesion of thin films," *Tech. Dig., Transducers '87, Fourth Int. Conf. on Solid-State Sensors and Actuators,* pp. 11–16, Tokyo, Japan, June 1987.

Smith, R. L., S.-F. Chuang, and S. D. Collins, "Theoretical model of the formation morphologies of porous silicon," *J. Electron. Mat.,* vol. 16 (6), pp. 533–541, November 1988.

Tabata, O., K. Kawahata, S. Sugiyama, H. Inagaki, and I. Igarashi, "Internal stress and Young's modulus measurements of thin films using micromachining technology," *Tech. Dig. 7th Sensor Symp.,* pp. 173–176, Tokyo, Japan, May 1988.

Tai, Y. C. and R. S. Muller, "Fracture strain of LPCVD polysilicon," *Tech. Dig., IEEE Solid-State Sensor and Actuator Workshop,* pp. 88–91, Hilton Head, SC, June 1988.

Tai, Y., C. Mastrangelo, and R. Muller, "Thermal conductivity of heavily doped LPCVD polysilicon," *Tech. Dig., 1987 Int. Electron Devices Meeting,* pp. 278–281, Washington, DC, December 1987.

Microsensor Fabrication Technology:
Crystalline Semiconductor Micromachining

Bassous, E., "Fabrication of novel three-dimensional microstructures by the anisotropic etching of (100) and (110) silicon," *IEEE Trans. Electron Devices,* vol. ED-25, (10), pp. 1178–1185, October 1978.

Bean, K. E., "Anisotropic etching of silicon," *IEEE Trans. Electron Devices,* vol. ED-25 (10), pp. 1185–1193, October 1978.

BeMent, S. L., K. D. Wise, D. J. Anderson, K. Najafi, and K. L. Drake, "Solid-state electrodes for multichannel multiplexed intracortical neuronal recording," *IEEE Trans. Biomed. Eng.,* vol. BME-33 (2), pp. 230–241, 1986.

Chang, S.-C. and D. B. Hicks, "Mesa structure formation using potassium hydroxide and ethylene diamine based etchants," *Tech. Dig., IEEE Solid-State Sensor and Actuator Workshop,* pp. 102–103, Hilton Head, SC, June 1988.

Clark, L. D., Jr. and D. J. Edell, "KOH:H(2)O etching of (110) Si, (111) Si, SiO(2), and Ta: An experimental study," *Proc. IEEE Micro Robots and Teleoperators Workshop,* Hyannis, MA, November 1987.

Edell, D. J., "A peripheral nerve information transducer for amputees: Long-term multichannel recordings from rabbit peripheral nerves," *IEEE Trans. Biomed. Eng.,* vol. BME-33 (2), pp. 163–174, February 1986.

Esashi, M., H. Komatsu, T. Matsuo, M. Takahashi, T. Takishima, K. Imabayashi, and H. Ozawa, "Fabrication of catheter-tip and sidewall miniature pressure sensors," *IEEE Trans. Electron Devices,* vol. ED-29 (1), pp. 57–63, January 1982.

Guckel, H., J. Uglow, M. Lin, D. Denton, J. Tobin, K. Euch, and M. Juda, "Plasma polymerization of methyl methacrylate: A photoresist for 3D applications," *Tech. Dig., IEEE Solid-State Sensor and Actuator Workshop,* pp. 9–12, Hilton Head, SC, June 1988.

Hirata, M., K. Suzuki, and H. Tanigawa, "Silicon diaphragm pressure sensors fabricated by anodic oxidation etch-stop," *Sensors and Actuators,* vol. 13 (1), pp. 63–70, January 1988.

Hok, B., C. Ovren, and E. Gustafsson, "Batch fabrication of micromechanical elements in GaAs—Al(x)Ga(1 − x)As," *Sensors and Actuators,* vol. 4 (3), pp. 341–348, November 1983.

Ikeda, K., H. Kuwayama, T. Kobayashi, T. Watanabe, T. Nishikawa, and T. Yoshida, "Three dimensional micromachining of silicon resonant strain gage," *Tech. Dig. 7th Sensor Symp.,* pp. 193–196, Tokyo, Japan, May 1988.

Johnson, R. G. and R. E. Higashi, "A highly sensitive silicon chip microtransducer for air flow and differential pressure sensing applications," *Sensors and Actuators,* vol. 11 (1), pp. 63–72, January 1987.

Kaminsky, G., "Micromachining of silicon mechanical structures," *J. Vac. Sci. Technol.,* vol. B3 (4), pp. 1015–1024, 1985.

Kendall, D. L. and G. R. de Guel, "Orientations of the third kind: The coming of age of (110) silicon," *Workshop on Micromachining and Micropackaging of Transducers,* pp. 107–130, Cleveland, OH, November 1984.

Kim, S. J., M. Kim, and W. J. Heetderks, "Laser-induced fabrication of a transsubstrate microelectrode array and its neurophysiological performance," *IEEE Trans. Biomed. Eng.,* vol. BME-32 (7), pp. 497–502, July 1985.

Kloeck, B. and N. de Rooij, "A novel four electrode electrochemical etch-stop method for silicon membrane formation," *Tech. Dig., Transducers '87, Fourth Int. Conf. on Solid-State Sensors and Actuators,* pp. 116–119, Tokyo, Japan, June 1987.

Kuhn, L., E. Bassous, and R. Lane, "Silicon charge electrode array for ink jet printing," *IEEE Trans. Electron Devices,* vol. ED-25 (10), pp. 1257–1260, October 1978.

Mehregany, M. and S. D. Senturia, "Anisotropic etching of

silicon in hydrazine,'' *Sensors and Actuators,* vol. 13 (4), pp. 375–390, April 1988.

Najafi, K., K. D. Wise, and T. Mochizuki, ''A high-yield IC-compatible multichannel recording array,'' *IEEE Trans. Electron Devices,* vol. ED-32 (7), pp. 1206–1211, July 1985.

Nakamura, M., K. Murakami, H. Nojiri, and T. Tominaga, ''Novel electrochemical micro-machining and its application for semiconductor acceleration sensor IC,'' *Tech. Dig., Transducers '87, Fourth Int. Conf. on Solid-State Sensors and Actuators,* pp. 112–115, Tokyo, Japan, June 1987.

Palik, E. D., V. M. Bermudez, and O. J. Glembocki, ''Ellipsometric study of bias-dependent etching and the etch-stop mechanism for silicon in aqueous KOH,'' *Workshop on Micromachining and Micropackaging of Transducers,* pp. 135–151, Cleveland, OH, November 1984.

Petersen, K., P. Barth, J. Poydock, J. Mallon Jr., and J. Bryzek, ''Silicon fusion bonding for pressure sensors,'' *Tech. Dig., IEEE Solid-State Sensor and Actuator Workshop,* pp. 144–147, Hilton Head, SC, June 1988.

Petersen, K. E., ''Dynamic micromechanics on silicon: Techniques and devices,'' *IEEE Trans. Electron Devices,* vol. ED-25 (10), pp. 1241–1250, October 1978.

Poteat, T. L., ''Submicron accuracies in anisotropic etched silicon piece parts—A case study,'' *Workshop on Micromachining and Micropackaging of Transducers,* pp. 151–158, Cleveland, OH, November 1984.

Reisman, A., M. Berkenblit, S. A. Chan, F. B. Kaufman, and D. C. Green, ''The controlled etching of silicon in catalyzed ethylene-diamine-pyrocatechol-water solutions,'' *J. Electrochem. Soc.,* vol. 126 (8), pp. 1406–1414, August 1979.

Sarro, P. M., H. Yashiro, A. W. van Herwaarden, and S. Middelhoek, ''An infrared sensing array based on integrated silicon thermopiles,'' *Tech. Dig., Transducers '87, Fourth Int. Conf. on Solid-State Sensors and Actuators,* pp. 227–230, Tokyo, Japan, June 1987.

Seidel, H. and L. Csepregi, ''Advanced methods for the micromachining of sensors,'' *Tech. Dig. 7th Sensor Symp.,* pp. 1–6, Tokyo, Japan, May 1988.

Seidel, H., ''The mechanism of anisotropic silicon etching and its relevance for micromachining,'' *Tech. Dig., Transducers '87, Fourth Int. Conf. on Solid-State Sensors and Actuators,* pp. 120–125, Tokyo, Japan, June 1987.

Esashi, M., S. Shoji, and T. Matsuo, ''A new three-dimensional lithographic technique and its applications to the fabrication of micro probe sensors,'' *Tech. Dig., Transducers '87, Fourth Int. Conf. on Solid-State Sensors and Actuators,* pp. 91–94, Tokyo, Japan, June 1987.

Stoev, I., S. Simov, E. Simova, N. Bonnet, and G. Balossier, ''Scanning electron microscopy of the geometry of V-groove structures,'' *Sensors and Actuators,* vol. 12 (1), pp. 1–8, July 1987.

Takahashi, K. and T. Matsuo, ''Integration of multi-micro-electrode and interface circuits by silicon planar and three-dimensional fabrication technology,'' *Sensors and Actuators,* vol. 5 (1), pp. 89–100, January 1984.

White, R. M. and S. W. Wenzel, ''Inexpensive and accurate two-sided semiconductor wafer alignment,'' *Sensors and Actuators,* vol. 13 (4), pp. 391–396, April 1988.

Wu, X. P. and W. H. Ko, ''A study on compensating corner undercutting in anisotropic etching of (100) silicon,'' *Tech. Dig., Transducers '87, Fourth Int. Conf. on Solid-State Sensors and Actuators,* pp. 126–129, Tokyo, Japan, June 1987.

Wu, X., Q. Wu, and W. H. Ko, ''A study on deep etching of silicon using EPW,'' *Tech. Dig., Transducers '85, Third Int. Conf. on Solid-State Sensors and Actuators,* pp. 291–294, Philadelphia, PA, June 1985.

Microsensor Fabrication Technology:
Surface (Thin-Film) Micromachining

Ehrfeld, W., R. Goetz, D. Muenchmeyer, W. Schleb, and D. Schmidt, ''LIGA process: Sensor construction techniques via X-ray lithography,'' *Tech. Dig., IEEE Solid-State Sensor and Actuator Workshop,* pp. 1–4, Hilton Head, SC, June 1988.

Fan, L. S., Y. C. Tai, and R. S. Muller, ''Pin joints, gears, springs, cranks, and other novel micromechanical structures,'' *Tech. Dig., Transducers '87, Fourth Int. Conf. on Solid-State Sensors and Actuators,* pp. 849–852, Tokyo, Japan, June 1987.

Guckel, H., D. W. Burns, C. R. Rutigliano, D. K. Showers, and J. Uglow, ''Fine grained polysilicon and its application to planar pressure transducers,'' *Tech. Dig., Transducers '87, Fourth Int. Conf. on Solid-State Sensors and Actuators,* pp. 277–282, Tokyo, Japan, June 1987.

Guckel, H. and D. Burns, ''Fabrication techniques for integrated sensor microstructures,'' *Tech. Dig., 1986 Int. Electron Devices Meeting,* pp. 176–179, Los Angeles, CA, December 1986.

Guckel, H., D. K. Showers, D. W. Burns, C. K. Nesler, and C. R. Rutigliano, ''Deposition techniques and properties of strain compensated LPCVD silicon nitride films,'' *Tech. Dig., IEEE Solid-State Sensor and Actuator Workshop,* Hilton Head, SC, June 1986.

Guckel, H. and D. W. Burns, ''A technology for integrated transducers,'' *Tech. Dig., Transducers '85, Third Int. Conf. on Solid-State Sensors and Actuators,* pp. 90–92, Philadelphia, PA, June 1985.

Guckel, H., D. W. Burns, H. H. Busta, and D. F. Detry, ''Laser-recrystallized piezoresistive micro-diaphragm sensors,'' *Tech. Dig., Transducers '85, Third Int. Conf. on Solid-State Sensors and Actuators,* pp. 182–185, Philadelphia, PA, June 1985.

Guckel, H. and D. Burns, ''Planar processed polysilicon sealed cavities for pressure transducer arrays,'' *Tech. Dig., 1984 Int. Electron Devices Meeting,* pp. 223–225, San Francisco, CA, December 1984.

Howe, R. T., ''Surface micromachining for microsensors and microactuators,'' *J. Vac. Sci. Technol.,* vol. B-6 (6), pp. 1809–1813, Nov./Dec. 1988.

Howe, R. T., ''Polycrystalline silicon microstructures,'' *Workshop on Micromachining and Micropackaging of Transducers,* pp. 169–188, Cleveland, OH, November 1984.

Howe, R. T. and R. S. Muller, "Polycrystalline silicon micromechanical beams," *Extended Abstracts, Electrochemical Society Spring Meeting,* vol. 82-1, pp. 184–185, Montreal, Quebec, Canada, May 1982.

Lober, T. A. and R. T. Howe, "Surface micromachining processes for electrostatic microactuator fabrication," *Tech. Dig., IEEE Solid-State Sensor and Actuator Workshop,* pp. 59–62, Hilton Head, SC, June 1988.

Muller, R. S., "Technologies and materials for microsensors and actuators," *Tech. Dig. 7th Sensor Symp.,* pp. 7–12, Tokyo, Japan, May 1988.

Nathanson, H. C., W. E. Newell, R. A. Wickstrom, and J. R. Davic, "The resonant gate transistor," *IEEE Trans. Electron Devices,* vol. ED-14 (3), pp. 117–133, 1967.

Schmidt, M. A., R. T. Howe, and S. D. Senturia, "Surface micromachining of polyimide/metal composites for a shear-stress sensor," *Proc., IEEE Micro Robots and Teleoperators Workshop,* Hyannis, MA, November 1987.

Sugiyama, S., T. Suzuki, K. Kawahata, K. Shimaoka, M. Takigawa, and I. Igarashi, "Micro-diaphragm pressure sensor," *Tech. Dig., IEEE Int. Electron Devices Meeting,* pp. 176–179, Los Angeles, CA, December 1986.

Microsensor Fabrication Technology:
Integration with Electronic Fabrication
Processes

Barth, P. W. and J. B. Angell, "Fabrication technology for a chronic in-vivo pressure sensor," *Tech. Dig., IEEE Int. Electron Devices Meeting,* pp. 210–212, Los Angeles, CA, December 1984.

Borky, J. M. and K. D. Wise, "Integrated signal conditioning for silicon pressure sensors," *IEEE Trans. Electron Devices,* vol. ED-26 (12), pp. 1906–1910, December 1979.

Bousse, L., J. Shott, and J. D. Meindl, "A CMOS process for the fabrication of ISFETs and associated circuits," *Tech. Dig., Transducers '87, Fourth Int. Conf. on Solid-State Sensors and Actuators,* pp. 99–102, Tokyo, Japan, June 1987.

Burns, B. E. and G. Stemme, "A CMOS integrated silicon gas-flow sensor with pulse-modulated output," *Sensors and Actuators,* vol. 14 (3), pp. 293–304, July 1988.

Howe, R. T. and R. S. Muller, "Resonant-microbridge vapor sensor," *IEEE Trans. Electron Devices,* vol. ED-33 (4), pp. 499–506, April 1986.

Kataoka, S., "An attempt towards an artificial retina: Three-dimensional IC technology for an intelligent image sensor," *Tech. Dig., Transducers '85, Third Int. Conf. on Solid-State Sensors and Actuators,* pp. 440–442, Philadelphia, PA, June 1985.

Najafi, K., K. D. Wise, and T. Mochizuki, "A high-yield IC-compatible multichannel recording array," *IEEE Trans. Electron Devices,* vol. ED-32 (7), pp. 1206–1211, July 1985.

Nakamura, M., K. Murakami, H. Nojiri, and T. Tominaga, "High-resolution silicon pressure imager with CMOS processing circuits," *Tech. Dig., Transducers '87, Fourth Int. Conf. on Solid-State Sensors and Actuators,* pp. 112–115, Tokyo, Japan, June 1987.

Parameswaran, M., H. P. Baltes, and A. M. Robinson, "A polysilicon microbridge fabrication using standard CMOS technology," *Tech. Dig., IEEE Solid-State Sensor and Actuator Workshop,* pp. 148-151, Hilton Head, SC, June 1988.

Polla, D. L., R. S. Muller, and R. M. White, "Integrated multisensor chip," *IEEE Electron Device Lett.,* vol. EDL-7 (4), pp. 254–256, April 1986.

Smith, R. L. and D. C. Scott, "An integrated electrochemical sensor," *IEEE Trans. Biomed. Eng.,* vol. BME-33 (2), pp. 83–90, February 1986.

Sugiyama, S., K. Kawahata, M. Abe, H. Funabashi, and I. Igarashi, "High-resolution silicon pressure imager with CMOS processing circuits," *Tech. Dig., Transducers '87, Fourth Int. Conf. on Solid-State Sensors and Actuators,* pp. 444–447, Tokyo, Japan, June 1987.

Suzuki, K., T. Ishihara, M. Hirata, and H. Tanigawa, "Nonlinear analysis on CMOS integrated silicon pressure sensor," *Tech. Dig., IEEE Int. Electron Devices Meeting,* pp. 137–140, Washington, DC, December 1985.

Takahashi, K. and T. Matsuo, "Integration of multi-micro-electrode and interface circuits by silicon planar and three-dimensional fabrication technology," *Sensors and Actuators,* vol. 5 (1), pp. 89–99, 1984.

Tanigawa, H., T. Ishibara, M. Hirata, and K. Suzuki, "MOS integrated silicon pressure sensor," *IEEE Trans. Electron Devices,* vol. ED-32 (7), pp. 1191–1195, July 1985.

Microsensor Fabrication Technology:
Microsensor Packaging

Bowman, L. and J. D. Meindl, "The packaging of implantable integrated sensors," *IEEE Trans. Biomed. Eng.,* vol. BME-33 (2), pp. 248–255, February 1986.

Bowman, L., J. M. Schmitt, and J. D. Meindl, "Electrical contacts to implantable integrated sensors by CO(2) laser-drilled vias through glass," *Workshop on Micromachining and Micropackaging of Transducers,* pp. 79–84, Cleveland, OH, November 1984.

Ko, W. H. and T. Spear, "Packaging of implantable electronics: Past, present, and future developments." In *Implantable Sensors for Closed-Loop Prosthetic Systems,* pp. 259–304. Mount Kisco, NY: Futura, 1985.

Ko, W. H., J. T. Suminto, and G. J. Yeh, "Bonding techniques for microsensors," *Workshop on Micromachining and Micropackaging of Transducers,* pp. 41–62, Cleveland, OH, November 1984.

Naruse, Y., L. Bowman, and J. D. Meindl, "Corrosion protection for implantable integrated sensors by CO(2) laser processing of glass and silicon," *Workshop on Micromachining and Micropackaging of Transducers,* pp. 63–78, Cleveland, OH, November 1984.

Petersen, K. E., P. Barth, J. Poydock, J. Brown, J. Mallon, Jr., and J. Bryzek, "Silicon fusion bonding for microsensors," *Tech. Dig., IEEE Solid-State Sensor and Actuator Workshop,* pp. 144–147, Hilton Head Island, SC, June 1988.

Senturia, S. D. and R. L. Smith, "Microsensor packaging and system partitioning," *Sensors and Actuators,* vol. 15 (3), pp. 221–234, November 1988.

Senturia, S. D. and D. R. Day, "Packaging considerations for microdielectrometer and related chemical sensors," *Workshop on Micromachining and Micropackaging of Transducers,* pp. 29–40, Cleveland, OH, November 1984.

Smith, R. L. and S. D. Collins, "Micromachined packages for chemical microsensors," *IEEE Trans. Electron Devices,* vol. ED-35 (6), pp. 787–792, June 1987.

Spear, T., A. Leung, and W. H. Ko, "Packaging of an intracranial pressure telemetering unit for chronic implantation," *Workshop on Micromachining and Micropackaging of Transducers,* pp. 85–106, Cleveland, OH, November 1984.

Microsensor Transduction Principles:
Electronic Devices as Microsensors

Goicolea, J. I. and R. S. Muller, "A silicon magneto-coupler using a carrier-domain magnetometer," *Tech. Dig., IEEE Int. Electron Devices Meeting,* pp. 276–279, Washington, DC, December 1985.

Meijer, G. C. M., A. J. M. Boomkamp, and R. J. Duguesnoy, "An accurate biomedical temperature transducer with on-chip microcomputer interfacing," *IEEE J. Solid-State Circuits,* vol. 23 (6), pp. 1405–1410, December 1988.

Senturia, S. D., "The role of the MOS structure in integrated sensors," *Sensors and Actuators,* vol. 4 (4), pp. 507–526, December 1983.

Timko, M. P., "A two-terminal integrated circuit temperature transducer," *IEEE J. Solid-State Circuits,* vol. SC-11 (6), pp. 784–788, December 1976.

Zieren, V. and P. M. Duyndam, "Magnetic-field-sensitive multicollector npn transistors," *IEEE Trans. Electron Devices,* vol. ED-29 (1), pp. 83–89, January 1982.

Microsensor Transduction Principles:
Microstructures for Thermal Isolation

Bouwstra, S., P. Kemna, and R. Legtenberg, "Thermally excited resonating membrane mass flow sensors," *Eurosensors '87 Conf. Proc.,* pp. 109–113, Cambridge, England, September 1987.

Choi, I. H. and K. D. Wise, "A silicon-thermopile-based infrared sensing array for use in automated manufacturing," *IEEE Trans. Electron Devices,* vol. ED-33 (1), pp. 72–79, January 1986.

Crary, S. B., "Thermal management of integrated microsensors," *Sensors and Actuators,* vol. 12 (4), pp. 303–312, Nov./Dec. 1987.

Demarne, V., A. Grisel, and R. Sanjines, "Comparison of the thermomechanical behaviour and power consumption between different integrated thin film gas sensor structures," *Tech. Dig., Transducers '87, Fourth Int. Conf. on Solid-State Sensors and Actuators,* pp. 605–609, Tokyo, Japan, June 1987.

Herwaarden, A. W. van and P. M. Sarro, "Floating-membrane thermal vacuum sensor," *Sensors and Actuators,* vol. 14 (3), pp. 259–268, July 1988.

Huff, M. A., S. D. Senturia, and R. T. Howe, "A thermally isolated microstructure suitable for gas sensing applications," *Tech. Dig., IEEE Solid-State Sensor and Actuator Workshop,* pp. 47–50, Hilton Head, SC, June 1988.

Huijsing, J. H., J. P. Schuddemat, and W. Verhoef, "Monolithic integrated direction-sensitive flow sensor," *IEEE Trans. Electron Devices,* vol. ED-29 (1), pp. 133–136, January 1982.

Johnson, R. G. and R. E. Higashi, "A highly sensitive silicon chip microtransducer for air flow and differential pressure sensing applications," *Sensors and Actuators,* vol. 11 (1), pp. 63–72, January 1987.

Mastrangelo, C. H. and R. S. Muller, "A constant-temperature gas flowmeter with a silicon micromachined package," *Tech. Dig., IEEE Solid-State Sensor and Actuator Workshop,* pp. 43–46, Hilton Head, SC, June 1988.

Petersen, K., J. Brown, and W. Renken, "High-precision high-performance mass flow sensor with integrated laminar flow microchannels," *Tech. Dig., Transducers '85, Third Int. Conf. on Solid-State Sensors and Actuators,* pp. 361–363, Philadelphia, PA, June 1985.

Sarro, P. M., H. Yashiro, A. W. van Herwaarden, and S. Middelhoek, "An integrated thermal infrared sensing array," *Sensors and Actuators,* vol. 14 (2), pp. 191–202, June 1988.

Stemme, G., "A CMOS integrated silicon gas-flow sensor with pulse-modulated output," *Sensors and Actuators,* vol. 14 (3), pp. 293–304, July 1988.

Stemme, G. N., "A monolithic gas flow sensor with polyimide as thermal insulator," *IEEE Trans. Electron Devices,* vol. ED-33 (10), pp. 1470–1474, October 1986.

Tai, Y.-C. and R. S. Muller, "Lightly doped polysilicon bridge as an anemometer," *Tech. Dig., Transducers '87, Fourth Int. Conf. on Solid-State Sensors and Actuators,* pp. 360–363, Tokyo, Japan, June 1987.

Zemel, J. N., J. R. Frederick, P. Hesketh, and B. Gebhart, "Pyroelectric anemometry," *Proc. 4th Sensor Symp.,* pp. 1–8, Tsukuba Science City, Japan, June 1984.

Microsensor Transduction Principles:
Sensitive Films

Evans, N. J., G. G. Roberts, and M. C. Petty, "Effects of hydrogen gas on palladium/lb film/silicon MIS devices," *Eurosensors '87 Conf. Proc.,* pp. 131–132, Cambridge, England, September 1987.

Fu, C. W., D. A. Batzel, S. E. Rickert, W. H. Ko, C. D. Fung, and M. E. Kenney, "Langmuir–Blodgett films of unsymmetrical axially substituted phthalocyanine as a gas sensor," *Tech. Dig., Transducers '87, Fourth Int. Conf. on Solid-State Sensors and Actuators,* pp. 583–584, Tokyo, Japan, June 1987.

Grate, J. W., A. W. Snow, D. S. Ballantine, Jr., H. Wohltjen, M. H. Abraham, R. A. McGill, and P. Sasson, "The use of GC partition coefficients and solubility properties to understand and predict SAW vapor sensor behavior," *Tech. Dig., Transducers '87, Fourth Int. Conf. on Solid-State Sensors and Actuators,* pp. 579–582, Tokyo, Japan, June 1987.

Heiland, G. and D. Kohl, "Problems and possibilities of oxidic and organic semiconductor gas sensors," *Tech. Dig., Transducers '85, Third Int. Conf. on Solid-State Sensors and Actuators,* pp. 262–263, Philadelphia, PA, June 1985.

Katsube, T., I. Lauks, and J. N. Zemel, "pH-sensitive sputtered iridium oxide films," *Sensors and Actuators,* vol. 2 (4), pp. 399–410, September 1982.

Komiyama, H., A. Hayashi, T. Yasuda, A. Yanase, K. Tanaka, T. Maejima, and S. Nagai, "Gas sensing by ultrafine metal particle dispersions," *Tech. Dig., Transducers '87, Fourth Int. Conf. on Solid–State Sensors and Actuators,* pp. 595–598, Tokyo, Japan, June 1987.

Nakamoto, S., N. Ito, T. Kuriyama, and J. Kimura, "A lift-off method for patterning enzyme-immobilized membranes in multibiosensors," *Sensors and Actuators,* vol. 13 (2), pp. 165–172, February 1988.

Polla, D. L. and R. S. Muller, "Zinc-oxide thin films for integrated-sensors applications," *Tech. Dig., IEEE Solid-State Sensors Workshop,* Hilton Head Island, SC, June 1986.

Roberts, G. G., "Transducer and other applications of Langmuir–Blodgett films," *Sensors and Actuators,* vol. 4 (2), pp. 131–146, October 1983.

Siegel, M. W., L. Wong, R. J. Lauf, and B. S. Hoffheins, "Dual gradient thick-film metal oxide gas sensors," *Tech. Dig., Transducers '87, Fourth Int. Conf. on Solid–State Sensors and Actuators,* pp. 599–604, Tokyo, Japan, June 1987.

Takada, T. and K. Komatsu, "O(3) gas sensor of thin film semiconductor In(2)O(3)," *Tech. Dig., Transducers '87, Fourth Int. Conf. on Solid-State Sensors and Actuators,* pp. 693–696, Tokyo, Japan, June 1987.

Tsurumi, S., K. Mogi, and J. Noda, "Humidity sensors of Pd–ZnO diodes," *Tech. Dig., Transducers '87, Fourth Int. Conf. on Solid-State Sensors and Actuators,* pp. 661–664, Tokyo, Japan, June 1987.

Windischmann, H. and P. Mark, "A model for the operation of a thin-film SnO(x) conductance-modulation carbon monoxide sensor," *J. Electrochem. Soc.,* vol. 126, pp. 627–633, 1979.

Xin, Y. Q., M. Hirata, and R. Yosomiya, "Quaternized 4-vinylpyridine-styrene copolymer-perchlorate complexes as a humidity sensor materials," *Tech. Dig., Transducers '87, Fourth Int. Conf. on Solid-State Sensors and Actuators,* pp. 669–672, Tokyo, Japan, June 1987.

Microsensor Transduction Principles:
Ultrasonic Microsensors

Hanna, S. M., "Magnetic field sensors based on SAW propagation in magnetic films," *IEEE Trans. Ultrason., Ferroelec., Freq. Contr.,* vol. UFFC-34 (2), pp. 191–194, March 1987.

Martin, S. J., A. J. Ricco, and R. C. Hughes, "Acoustic wave devices for sensing in liquids," *Tech. Dig., Transducers '87, Fourth Int. Conf. on Solid-State Sensors and Actuators,* pp. 478–481, Tokyo, Japan, June 1987.

Martin, S. J., A. J. Ricco, D. S. Ginley, and T. E. Zipperian,

"Isothermal measurements and thermal desorption of organic vapors using SAW devices," *IEEE Trans. Ultrason., Ferroelec., Freq. Contr.,* vol. UFFC-34 (2), pp. 143–148, March 1987.

Motamedi, M. E., "Acoustic accelerometers," *IEEE Trans. Ultrason., Ferroelec., Freq. Contr.,* vol. UFFC-34 (2), pp. 237–242, March 1987.

Nakazawa, M., A. Ballato, and T. Lukaszek, "An ultralinear stress-compensated temperature sensor," *IEEE Trans. Ultrason., Ferroelec., Freq. Contr.,* vol. UFFC-34 (2), pp. 270–277, March 1987.

Shimotahira, H., S. Inagaki, K. Miyagi, and Y. Kawano, "SAW force sensor utilizing ZnO thin film," *Proc. 6th Sensor Symp.,* pp. 73–78, Tsukuba Science City, Japan, May 1986.

Venema, A., E. Nieuwkoop, M. J. Vellekoop, M. S. Nieuwenhuizen, and A. W. Barendsz, "Design aspects of SAW gas sensors," *Sensors and Actuators,* vol. 10 (1) & (2), pp. 47–64, Sept./Oct. 1986.

Vetelino, J. F., R. K. Lade, and R. S. Falconer, "Hydrogen sulfide surface acoustic wave gas detector," *IEEE Trans. Ultrason., Ferroelec., Freq. Contr.,* vol. UFFC-34 (2), pp. 149–156, March 1987.

Wenzel, S. W. and R. M. White, "A multisensor employing an ultrasonic Lamb-wave oscillator," *IEEE Trans. Electron Devices,* vol. 35 (6), pp. 735–743, June 1988.

Wohltjen, H., "Surface acoustic wave microsensors," *Tech. Dig., Transducers '87, Fourth Int. Conf. on Solid-State Sensors and Actuators,* pp. 471–477, Tokyo, Japan, June 1987.

Wohltjen, H., A. W. Snow, R. W. Barger, and D. S. Ballantine, "Trace chemical vapor detection using SAW delay line oscillators wave devices," *IEEE Trans. Ultrason., Ferroelec. Freq. Contr.,* vol. UFFC-34 (2), pp. 172–178, March 1987.

Wohltjen, H., "Mechanism of operation and design considerations for surface acoustic wave device vapour sensors," *Sensors and Actuators,* vol. 5 (4), pp. 307–326, July 1984.

Zellers, E. T., R. M. White, and S. W. Wenzel, "Computer modelling of polymer-coated ZnO/Si surface-acoustic-wave and Lamb-wave chemical sensors," *Sensors and Actuators,* vol. 14 (1), pp. 35–46, May 1988.

Microsensor Circuit Interfaces

Atkinson, J. K., "An SDLC based implementation of an instrumentation field bus," *Eurosensors '87 Conf. Proc.,* pp. 29–31, Cambridge, England, September 1987.

Brignell, J. E., "Digital compensation of sensors," *J. Phys. E: Scientific Instruments,* vol. 20 (9), pp. 1097–1102, September 1987.

Kolling, A., P. Bergveld, and E. Seevinck, "CMOS oscillator as universal heart for a sensor chip," *Eurosensors '87 Conf. Proc.,* pp. 182–183, Cambridge, England, September 1987.

Lian, W. J. and S. Middelhoek, "Flip-flop sensors: A new class of silicon sensors," *Sensors and Actuators,* vol. 9 (3), pp. 259–268, May 1986.

Middelhoek, S., P. J. French, J. H. Huijsing, and W. J. Lian, "Sensors with digital or frequency output," *Sensors and Actuators,* vol. 15 (2), pp. 119–134, October 1988.

Najafi, N., K. W. Clayton, W. Baer, K. Najafi, and K. D. Wise, "An architecture and interface for VLSI sensors," *Tech. Dig., IEEE Solid-State Sensor and Actuator Workshop,* pp. 76–78, Hilton Head Island, SC, June 1988.

Sansen, W. and M. Steyaert, "VLSI technology and sensor interface circuits." In *Implantable Sensors for Closed-Loop Prosthetic Systems,* pp. 21–31. Mount Kisco, NY: Futura, 1985.

Microsensor Circuit Interfaces:
Piezoresistive Interface

Bryzek, J., R. Mayer, and P. Barth, "New generation of disposable blood pressure sensors brings on-chip digital laser trimming," *Tech. Dig., IEEE Solid-State Sensor and Actuator Workshop,* pp. 121–122, Hilton Head, SC June 1988.

Bryzek, J., "Modeling performance of piezoresistive pressure sensors," *Tech. Dig., Third Int. Conf. on Solid-State Sensors and Actuators,* pp. 168–173, Philadelphia, PA, June 1985.

Chau, H. L. and K. D. Wise, "Scaling limits in batch-fabricated silicon pressure sensors," *IEEE Trans. Electron Devices,* vol. ED-34 (4), pp. 850–858, April 1987.

Chau, H. L. and K. D. Wise, "Noise due to Brownian motion in ultrasensitive solid-state pressure sensors," *IEEE Trans. Electron Devices,* vol. ED-34, p. 859, April 1987.

Hirata, M., T. Ishihara, K. Suzuki, and H. Tanigawa, "An integrated silicon pressure sensor with NMOS operational amplifier," *Proc. 4th Sensor Symp.,* pp. 237–244, Tsukuba Science City, Japan, June 1984.

Spencer, R. R., B. M. Fleischer, P. W. Barth, and J. B. Angell, "A theoretical study of transducer noise in piezoresistive and capacitive silicon pressure sensors," *IEEE Trans. Electron Devices,* vol. 35 (8), pp. 1289–1298, August 1988.

Spencer, R. R., B. M. Fleischer, P. W. Barth, and J. B. Angell, "The voltage-controlled duty-cycle oscillator: Basis for a new A-to-D conversion technique," *Tech. Dig., Third Int. Conf. on Solid-State Sensors and Actuators,* pp. 49–52, Philadelphia, PA, June 1985.

Suzuki, K., T. Ishihara, M. Hirata, and H. Tanigawa, "Nonlinear analysis on CMOS integrated silicon pressure sensor," *Tech. Dig., IEEE Int. Electron Devices Meeting,* pp. 137–140, Washington, DC, December 1985.

Microsensor Circuit Interfaces:
Capacitive Interface

Ko, W. H., B.-X. Shao, C. D. Fung, W.-J. Shen, and G.-J. Yeh, "Capacitive pressure transducers with integrated circuits," *Sensors and Actuators,* vol. 4 (3), pp. 403–412, November 1983.

Krummenacher, F., "A high-resolution capacitance-to-frequency converter," *IEEE J. Solid-State Circuits,* vol. SC-20(3), pp. 666–670, June 1985.

Kung, J. T., R. T. Howe, and H.-S. Lee, "A digital readout technique for capacitive sensor applications," *IEEE J. Solid-State Circuits,* vol. 23 (4), pp. 972–977, August 1988.

Park, Y. E. and K. D. Wise, "An MOS switched-capacitor readout amplifier for capacitive pressure sensors," *Proc., IEEE Custom IC Conf.,* pp. 380–384, Rochester, NY, May 1983.

Sander, C. S., J. W. Knutti, and J. D. Meindl, "A monolithic capacitive pressure sensor with pulse-period output," *IEEE Trans. Electron Devices,* vol. ED-27 (5), pp. 927–930, May 1980.

Smith, M. J. S., L. Bowman, and J. D. Meindl, "Analysis, design, and performance of a capacitive pressure sensor IC," *IEEE Trans. Biomed. Eng.,* vol. BME-33 (2), pp. 163–174, February 1986.

Wise, K. D., "Circuit technique for integrated solid-state sensors," *Proc., IEEE Custom IC Conf.,* pp. 436–440, Rochester, NY, May 1983.

Yeh, G. J., I. Dendo, and W. H. Ko, "Switched capacitor interface circuit for capacitive transducers," *Tech. Dig., Transducers '85, Third Int. Conf. on Solid-State Sensors and Actuators,* pp. 60–64, Philadelphia, PA, June 1985.

Selected Microsensor Applications:
Pressure Sensors

Andres, M. V., K. W. H. Foulds, and M. J. Tudor, "Sensitivity of a frequency-out silicon pressure sensor," *Eurosensors '87 Conf. Proc.,* pp. 18–19, Cambridge, England, September 1987.

Bryzek, J., "Approaching performance limits in silicon piezoresistive pressure sensors," *Sensors and Actuators,* vol. 4 (4), pp. 669–678, December 1983.

Burns, B., P. Barth, and J. Angell, "Fabrication technology for a chronic in-vivo pressure sensor," *Tech. Dig., 1984 Int. Electron Devices Meeting,* pp. 210–212, San Francisco, CA, December 1984.

Chau, H. L. and K. D. Wise, "An ultraminiature solid-state pressure sensor for a cardiovascular catheter," *Tech. Dig., Transducers '87, Fourth Int. Conf. on Solid-State Sensors and Actuators,* pp. 344–347, Tokyo, Japan, June 1987.

Clark, S. K. and K. D. Wise, "Pressure sensitivity in anisotropically etched thin-diaphragm pressure sensors," *IEEE Trans. Electron Devices,* vol. ED-26 (12), pp. 1887–1896, December 1979.

Greenwood, J. C., "Etched silicon vibrating sensor," *J. Phys. E: Scientific Instruments,* vol. 17, pp. 650–652, 1984.

Ikeda, K., H. Kuwayama, T. Kobayashi, T. Watanabe, T. Nishikawa, and T. Yoshida, "Silicon pressure sensor with resonant strain gages built into diaphragm," *Tech. Dig. 7th Sensor Symp.,* pp. 55–58, Tokyo, Japan, May 1988.

Ishihara, T., K. Suzuki, S. Suwazono, and M. Hirata, "Silicon diaphragm pressure sensor with CMOS integrated peripheral circuits," *Proc. 6th Sensor Symp.,* pp. 17–22, Tsukuba Science City, Japan, May 1986.

Kawamura, Y., K. Sato, T. Terasawa, and S. Tanaka, "Si cantilever-oscillator as a vacuum sensor," *Tech. Dig., Transducers '87, Fourth Int. Conf. on Solid-State Sensors and Actuators,* pp. 283–286, Tokyo, Japan, June 1987.

Ko, W. H., "Solid-state capacitive pressure transducers," *Sensors and Actuators,* vol. 10 (3) & (4), pp. 303–320, Nov./Dec. 1986.

Ko, W. H., M.-H. Pao, and Y.-D. Hong, "A high-sensitivity integrated-circuit capacitive pressure transducer," *IEEE Trans. Electron Devices,* vol. ED-29 (1), pp. 48–56, January 1982.

Ko, W. H., J. Hynecek, and S. F. Boettcher, "Development of a miniature pressure transducer for biomedical applications," *IEEE Trans. Electron Devices,* vol. ED-26 (12), pp. 1896–1905, December 1979.

Neumeister, J., G. Schuster, and W. von Munch, "A silicon pressure sensor using MOS ring oscillators," *Sensors and Actuators,* vol. 7 (3), pp. 167–176, July 1985.

Royer, M., J. O. Holmen, M. A. Wurm, O. S. Aadland, and M. Glenn, "ZnO on Si integrated acoustic sensors," *Sensors and Actuators,* vol. 4 (3), pp. 357–362, November 1983.

Smits, J. G., H. A. C. Tilmans, T. S. J. Lammerink, H. Guckel, D. W. Burns, and C. R. Rutigliano, "Design and construction techniques for planar polysilicon pressure transducers with piezoresistive readout," *Tech. Dig., IEEE Solid-State Sensor and Actuator Workshop,* pp. 93–96, Hilton Head Island, SC, June 1986.

Sugiyama, S., M. Takigawa, and I. Igarashi, "Operation temperature limit of silicon pressure sensor with diffused piezoresistors," *Proc. 4th Sensor Symp.,* pp. 231–236, Tsukuba Science City, Japan, June 1984.

Suzuki, K., T. Ishihara, M. Hirata, and H. Tanigawa, "Nonlinear analyses on CMOS integrated silicon pressure sensor," *Tech. Dig., 1985 Int. Electron Devices Meeting,* pp. 137–139, Washington, DC, December 1985.

Tabata, O., H. Inagaki, and I. Igarashi, "Monolithic pressure-flow sensor with a thermal isolation structure," *Tech. Dig., Transducers '87, Fourth Int. Conf. on Solid-State Sensors and Actuators,* pp. 340–343, Tokyo, Japan, June 1987.

Tilmans, H. A. C., K. Hoen, H. Mulder, J. van Vuuren, and G. Boom, "Resonant diaphragm pressure measurement system with ZnO on Si excitation," *Sensors and Actuators,* vol. 4 (4), pp. 565–572, December 1983.

Warkentin, D. J., J. H. Haritonidis, M. Mehregany, and S. D. Senturia, "A micromachined microphone with optical interference readout," *Tech. Dig., Transducers '87, Fourth Int. Conf. on Solid-State Sensors and Actuators,* pp. 291–294, Tokyo, Japan, June 1987.

Selected Microsensor Applications:
Acceleration Sensors

Aske, V. H., "An integrated silicon accelerometer," *Scientific Honeyweller,* vol. 8 (1), pp. 53–58, Fall 1987.

Benecke, W., L. Csepregi, A. Heuberger, K. Kuhl, and H. Seidel, "A frequency-selective piezoresistive silicon vibration sensor," *Tech. Dig., Transducers '85, Third Int. Conf. on Solid-State Sensors and Actuators,* pp. 105–108, Philadelphia, PA, June 1985.

Chen, P. L., R. S. Muller, and A. P. Andrews, "Integrated silicon PI-FET accelerometer with proof mass," *Sensors and Actuators,* vol. 5 (2), pp. 119–126, February 1984.

Chen, P.-L., R. S. Muller, R. D. Jolly, G. L. Halac, R. M. White, A. P. Andress, T. C. Lim, and M. E. Motamedi, "Integrated silicon microbeam PI-FET accelerometer," *IEEE Trans. Electron Devices,* vol. ED-29 (1), pp. 27–33, January 1982.

Greenwood, J. C., "Resonant silicon sensors at STL," *Eurosensors '87 Conf. Proc.,* pp. 14–15, Cambridge, England, September 1987.

Hok, B. and K. Gustafsson, "Vibration analysis of micromechanical elements," *Sensors and Actuators,* vol. 8 (3), pp. 235–244, November 1985.

Howe, R. T., "Resonant microsensors," *Tech. Dig., Transducers '87, Fourth Int. Conf. on Solid-State Sensors and Actuators,* pp. 843–848, Tokyo, Japan, June 1987.

Motamedi, M. E., "Acoustic accelerometers," *IEEE Trans. Ultrason., Ferroelec., Freq. Contr.,* vol. UFFC-34 (2), pp. 237–242, March 1987.

Petersen, K. E., A. Shartel, and N. F. Raley, "Micromechanical accelerometer integrated with MOS detection circuitry," *IEEE Trans. Electron Devices,* vol. ED-29 (1), pp. 23–27, January 1982.

Roylance, L. M. and J. B. Angell, "A batch-fabricated silicon accelerometer," *IEEE Trans. Electron Devices,* vol. ED-26 (12), pp. 1911–1917, December 1979.

Rudolf, F., A. Jornod, and P. Bencze, "Silicon microaccelerometer," *Tech. Dig., Transducers '87, Fourth Int. Conf. on Solid-State Sensors and Actuators,* pp. 395–398, Tokyo, Japan, June 1987.

Rudolf, F., "A micromechanical capacitive accelerometer with a two-point inertial-mass suspension," *Sensors and Actuators,* vol. 4 (2), pp. 191–198, October 1983.

Sandmaier, H., K. Kuhl, and E. Obermeier, "A silicon based micromechanical accelerometer with cross acceleration sensitivity compensation," *Tech. Dig., Transducers '87, Fourth Int. Conf. on Solid-State Sensors and Actuators,* pp. 399–402, Tokyo, Japan, June 1987.

Satchell, D. W. and J. C. Greenwood, "Silicon microengineering for accelerometers," *Tech. Dig., Int. Conf. on the Mechanical Technology of Inertial Devices,* pp. 191–193, Newcastle, England, April 1987.

Seidel, H. and L. Csepregi, "Design optimization for cantilever-type accelerometers," *Sensors and Actuators,* vol. 6 (2), pp. 81–92, October 1984.

Terry, S., "A miniature silicon accelerometer with built-in damping," *Tech. Dig., IEEE Solid-State Sensor and Actuator Workshop,* pp. 114–116, Hilton Head, SC, June 1988.

Selected Microsensor Applications:
Magnetic Sensors

Baltes, H. P. and R. S. Popovic, "Integrated semiconductor magnetic field sensors," *Proc. IEEE,* vol. 74 (8), pp. 1107–1132, August 1986.

Burghartz, J. and W. von Munch, "Optimization of lateral

magnetotransistors with integrated signal amplification," *Sensors and Actuators,* vol. 11 (1), pp. 91–98, January 1987.

Goicolea, J. I., R. S. Muller, and J. E. Smith, "Highly sensitive silicon carrier-domain magnetometer," *Sensors and Actuators,* vol. 5 (2), pp. 146–168, February 1984.

Guvenc, M. G., "Finite element analysis of bipolar magneto-sensor structures with curved junction boundaries," *Tech. Dig., Transducers '87, Fourth Int. Conf. on Solid-State Sensors and Actuators,* pp. 515–518, Tokyo, Japan, June 1987.

Ishida, M., H. Fujiwara, T. Nakamura, M. Ashiki, Y. Yasuda, A. Yoshida, T. Ohsakama, and Y. Kawase, "Silicon magnetic vector sensors for integration," *Proc. 4th Sensor Symp.,* pp. 79–84, Tsukuba Science City, Japan, June 1984.

Kordic, S., "Sensitivity of the silicon high-resolution 3-dimensional magnetic-field vector sensor," *Tech. Dig., IEEE Int. Electron Devices Meeting,* pp. 188–191, Los Angeles, CA, December 1986.

Maenaka, K., T. Ohgusu, M. Ishida, and T. Nakamura, "Realization of omni-directional integrated magnetic sensor," *Tech. Dig. 7th Sensor Symp.,* pp. 43–46, Tokyo, Japan, May 1988.

Maenaka, K., T. Ohgusu, M. Ishida, and T. Nakamura, "Integrated magnetic sensors detecting x, y and z components of the magnetic field," *Tech. Dig., Transducers '87, Fourth Int. Conf. on Solid-State Sensors and Actuators,* pp. 523–526, Tokyo, Japan, June 1987.

Misra, D., T. R. Viswanathan, and E. L. Heasell, "A novel high gain MOS magnetic field sensor," *Sensors and Actuators,* vol. 9 (3), pp. 213–222, May 1986.

Nathan, A., W. Allegretto, W. B. Joerg, and H. P. Baltes, "Numerical modeling of bipolar action in magnetotransistors," *Tech. Dig., Transducers '87, Fourth Int. Conf. on Solid-State Sensors and Actuators,* pp. 519–522, Tokyo, Japan, June 1987.

Nojima, H., S. Kataoka, S. Tsuchimoto, M. Nagata, R. Kita, H. Shintaku, E. Ohno, and N. Hashizume, "Improvement in sensitivity of novel magnetic sensor using Y-Ba-Cu-O ceramic superconductor film," *Tech. Dig., IEEE Int. Electron Devices Meeting,* pp. 892–893, San Francisco, CA, December 1988.

Popovic, R. S., H. P. Baltes, and F. Rudolph, "An integrated silicon magnetic field sensor using the magnetodiode principle," *IEEE Trans. Electron Devices,* vol. ED-31 (3), pp. 286–291, March 1984.

Sugiyama, Y. and S. Kataoka, "S/N study of micro-Hall sensors made of single crystal InSb and GaAs," *Sensors and Actuators,* vol. 8 (1), pp. 29–38, September 1985.

Zieren, V. and B. P. M. Duyndam, "Magnetic-field-sensitive multicollector n-p-n transistors," *IEEE Trans. Electron Devices,* vol. ED-29 (1), pp. 83–90, January 1982.

Selected Microsensor Applications: Tactile Sensors

Allen, H. V., J. W. Knutti, M. L. Dunbar, A. J. Crabill, and J. A. Valdovinos, "A sub-miniature load cell configurable for multiplexed tactile arrays," *Tech. Dig., Transducers '87, Fourth Int. Conf. on Solid-State Sensors and Actuators,* pp. 448–450, Tokyo, Japan, June 1987.

Barth, P. W., M. J. Zdeblick, Z. Kuc, and P. A. Beck, "Flexible tactile sensing arrays for robotics: Architectural robustness and yield considerations," *Tech. Dig., IEEE Solid-State Sensor and Actuator Workshop,* Hilton Head, SC, June 1986.

Barth, P. W., S. L. Bernard, and J. B. Angell, "Flexible circuit and sensor array fabricated by monolithic silicon technology," *IEEE Trans. Electron Devices,* vol. ED-32 (7), pp. 1202–1206, July 1985.

Chun, K. J. and K. D. Wise, "A capacitive silicon tactile imaging array," *Tech. Dig., Transducers '85, Third Int. Conf. on Solid-State Sensors and Actuators,* pp. 22–25, Philadelphia, PA, June 1985.

Chun, K. and K. D. Wise, "A high-performance silicon tactile imager based on a capacitive cell," *IEEE Trans. Electron Devices,* vol. ED-32 (7), pp. 1196–1201, July 1985.

Esashi, M. and H. Hebiguchi, "Flexible silicon tactile imager," *Tech. Dig. 7th Sensor Symp.,* pp. 197–200, Tokyo, Japan, May 1988.

Fan, L. S., R. White, and R. Muller, "Mutual capacitive normal- and shear-sensitive tactile sensor," *Tech. Dig., 1984 Int. Electron Devices Meeting,* pp. 220–222, San Francisco, CA, December 1984.

King, A. A. and R. M. White, "Tactile sensing array based on forming and detecting an optical image," *Sensors and Actuators,* vol. 8 (1), pp. 49–64, September 1985.

Petersen, K., C. Kowalski, J. Brown, H. Allen, and J. Knutti, "A force sensing chip designed for robotic and manufacturing automation applications," *Tech. Dig., Transducers '85, Third Int. Conf. on Solid-State Sensors and Actuators,* pp. 30–32, Philadelphia, PA, June 1985.

Polla, D., W. Chang, R. Muller, and R. White, "Integrated zinc oxide-on-silicon tactile sensor array," *Tech. Dig., 1985 Int. Electron Devices Meeting,* pp. 133–136, Washington, DC, December 1985.

Raibert, M. H. and J. E. Tanner, "Design and implementation of a VLSI tactile sensing computer," *Int. J. Robotics Res.,* vol. 1, pp. 3-18, 1982.

Sugiyama, S., K. Kawahata, M. Abe, H. Funabashi, and I. Igarashi, "High-resolution silicon pressure imager with CMOS processing circuits," *Tech. Dig., Transducers '87, Fourth Int. Conf. on Solid-State Sensors and Actuators,* pp. 444–447, Tokyo, Japan, June 1987.

Suzuki, K., K. Najafi, and K. D. Wise, "A 1024-element high-performance silicon tactile imager," *Tech. Dig., IEEE Int. Electron Devices Meeting,* pp. 674–677, San Francisco, CA, December 1988.

Tanie, K., "Advances in tactile sensors for robots," *Proc. 6th Sensor Symp.,* pp. 63–68, Tsukuba Science City, Japan, May 1986.

Selected Microsensor Applications: Chemical Sensors

Armgarth, M., U. Ackelid, and I. Lundstrom, "Enhanced selectivity of catalytic gate metal-oxide-semiconductor field effect transistors by temperature scan,'' *Tech. Dig., Transducers '87, Fourth Int. Conf. on Solid-State Sensors and Actuators,* pp. 640–643, Tokyo, Japan, June 1987.

Armgarth, M. and C. Nylander, ''A stable hydrogen-sensitive Pd gate metal-oxide semiconductor capacitor,'' *Appl. Phys. Lett.,* vol. 39 (1), pp. 91–92, 1981.

Bergveld, P., ''ISFETs for physiological measurements.'' In *Implantable Sensors for Closed-Loop Prosthetic Systems,* pp. 89–104. Mount Kisco, NY: Futura, 1985.

Bergveld, P., ''Development, operation, and application of the ion-sensitive field-effect transistor as a tool for electrophysiology,'' *IEEE Trans. Biomed. Eng.,* vol. BME-19 (5), pp. 342–351, 1972.

Brown, R., R. J. Huber, D. Petelenz, and J. Janata, ''An integrated multiple-sensor chemical transducer,'' *Tech. Dig., Transducers '85, Third Int. Conf. on Solid-State Sensors and Actuators,* pp. 125–127, Philadelphia, PA, June 1985.

Chang, S.-C. and D. B. Hicks, ''Tin oxide microsensors,'' in *Fundamentals and Applications of Chemical Sensors,* American Chemical Society Symposium Series, vol. 309, pp. 58–70. Washington, DC: American Chemical Society, 1986.

Chang, S.-C. and D. B. Hicks, ''Tin oxide microsensors on thin silicon membranes,'' *Tech. Dig., IEEE Solid-State Sensors Workshop,* Hilton Head, SC, June 1986.

Choi, S.-Y., K. Takahashi, M. Esashi, and T. Matsuo, ''Stability and sensitivity of MISFET hydrogen sensors,'' *Tech. Dig., Transducers '85, Third Int. Conf. on Solid-State Sensors and Actuators,* pp. 232–234, Philadelphia, PA, June 1985.

Fare, T., A. Spetz, M. Armgarth, and I. Lundstrom, ''Quasi-static and high frequency C(V)-response of thin platinum metal-oxide-silicon structures to ammonia,'' *Sensors and Actuators,* vol. 14 (4), pp. 369–386, August 1988.

Fung, C. D., P. W. Cheung, and W. H. Ko, ''A generalized theory of an electrolyte-insulator-semiconductor field-effect transistor,'' *IEEE Trans. Electron Devices,* vol. ED-33 (1), pp. 8–18, January 1986.

Heiland, G. and D. Kohl, ''Problems and possibilities of oxidic and organic semiconductor gas sensors,'' *Sensors and Actuators,* vol. 8 (3), pp. 227–234, November 1985.

Howe, R. T. and R. S. Muller, ''Resonant-microbridge vapor sensor,'' *IEEE Trans. Electron Devices,* vol. ED-33 (4), pp. 499–506, April 1986.

Janata, J. and R. H. Huber, ''Chemically sensitive field effect transistors,'' in *Ion-Selective Electrodes in Analytical Chemistry, II,* p. 107, 1985.

Kang, W. P., J. F. Xu, B. Lalevic, and T. L. Poteat, ''Sensing behavior of Pd-SnO (x) MIS structure used for oxygen detection,'' *Sensors and Actuators,* vol. 12 (4), pp. 349–366, Nov./Dec. 1987.

Kang, W. P., J. F. Xu, B. Lalevic, and T. L. Poteat, ''A study of detection mechanism in the Pd-SnO (x) MIS oxygen sensor,'' *Tech. Dig., Transducers '87, Fourth Int. Conf. on Solid-State Sensors and Actuators,* pp. 610–617, Tokyo, Japan, June 1987.

Kondo, H., H. Takahashi, K. Saji, T. Takeuchi, and I. Igarashi, ''Thin film limiting current-type oxygen sensor,'' *Proc. 6th Sensor Symp.,* pp. 251–256, Tsukuba Science City, Japan, May 1986.

Krey, D., K. Dobos, and G. Zimmer, ''An integrated CO-sensitive MOS transistor,'' *Sensors and Actuators,* vol. 3 (2), pp. 169–178, March 1983.

Lundstrom, I., M. Armgarth, A. Spetz, and F. Winquist, ''Gas sensors based on catalytic metal-gate field-effect devices,'' *Sensors and Actuators,* vol. 10 (3) & (4), pp. 399–422, Nov./Dec. 1986.

Matsuo, T., ''Present and future chemical sensors,'' *Proc. 3rd Sensor Symp.,* pp. 13-16, Tsukuba Science City, Japan, June 1983.

Miyahara, Y., K. Tsukada, and H. Miyagi, ''Characteristics of FET type oxygen sensor,'' *Proc. 6th Sensor Symp.,* pp. 261–268, Tsukuba Science City, Japan, May 1986.

Mokwa W., K. Dobos, and G. Zimmer, ''Palladium-gate MOS devices for arsine detection,'' *Sensors and Actuators,* vol. 12 (4), pp. 333–340, Nov./Dec. 1987.

Morrison, S. R., ''Mechanism of semiconductor gas sensor operation,'' *Sensors and Actuators,* vol. 11 (3), pp. 283–288, April 1987.

Murakami, N., K. Takahata, and T. Seiyama, ''Selective detection of CO by SnO(2) gas sensor using periodic temperature change,'' *Tech. Dig., Transducers '87, Fourth Int. Conf. on Solid-State Sensors and Actuators,* pp. 618–621, Tokyo, Japan, June 1987.

Polla, D. L., R. M. White, and R. S. Muller, ''Integrated chemical-reaction sensor,'' *Tech. Dig., Transducers '85, Third Int. Conf. on Solid-State Sensors and Actuators,* pp. 33–36, Philadelphia, PA, June 1985.

Poteat, T. L. and B. Lalevic, ''Transition metal-gate MOS gaseous detectors,'' *IEEE Trans. Electron Devices,* vol. ED-29 (1), pp. 123–129, January 1982.

Prohaska, O., W. Chu, M. Patil, F. Kohl, P. Goiser, F. Olcaytug, G. Urban, A. Jachimowicz, J. LaManna, K. Pirker, and R. Vollmer, ''Multiple chamber-type probe for biomedical application,'' *Tech. Dig., Transducers '87, Fourth Int. Conf. on Solid-State Sensors and Actuators,* pp. 812–815, Tokyo, Japan, June 1987.

Prohaska, O. J., F. Olcaytug, P. Pfundner, and H. Dragaun, ''Thin-film multiple electrode probes: Possibilities and limitations,'' *IEEE Trans. Biomed. Eng.,* vol. BME-33 (2), pp. 223–229, February, 1986.

Schubert, P. J. and J. H. Nevin, ''A polyimide-based capacitive humidity sensors,'' *IEEE Trans. Electron Devices,* vol. ED-32 (7), pp. 1220–1224, July 1985.

Schwager, F. J., L. J. Bousse, L. Bowman, and J. D. Meindl, ''Chemical multisensors with selective encapsulation of ion-selective membranes,'' *Tech. Dig., IEEE Solid-State Sensors Workshop,* Hilton Head, SC, June 1986.

Senturia, S. D., S. L. Garverick, and K. Togashi, ''Monolithic integrated circuit implementations of the charge-flow transistor oscillator moisture sensor,'' *Sensors and Actuators,* vol. 2 (1), pp. 59–72, August 1981.

Shinohara, H., T. Chiba, and M. Aizawa, ''Enzyme microsensor for glucose with an electrochemically synthesized enzyme-polyaniline film," *Sensors and Actuators,* vol. 13 (1), pp. 79–86, January 1988.

Shoji, S., M. Esashi, and T. Matsuo, "Prototype miniature blood gas analyser fabricated on a silicon wafer," *Sensors and Actuators,* vol. 14 (2), pp. 101–108, June 1988.

Smith, R. L. and D. C. Scott, "An integrated sensor for electrochemical measurements," *IEEE Trans. Biomed. Eng.,* vol. BME-33 (2), pp. 83–90, February 1986.

Terry, S. C., J. H. Jerman, and J. B. Angell, "A gas chromatographic air analyzer fabricated on a silicon wafer," *IEEE Trans. Electron Devices,* vol. ED-26 (12), pp. 1880–1886, December 1979.

Toko, K., K. Hayashi, K. Yamafuji, and S. Iiyama, "Sensing of sweet and umami substances with synthetic lipid membranes," *Tech. Dig. 7th Sensor Symp.,* pp. 127–130, Tokyo, Japan, May 1988.

Wen, C.-C., T. C. Chen, and J. N. Zemel, "Gate-controlled diodes for ionic concentration measurement," *IEEE Trans. Electron Devices,* vol. ED-26 (12), pp. 1945–1951, December 1979.

Wilson, A. and R. A. Collins, "Electrical characteristics of planar phthalocyanine thin film gas sensors," *Sensors and Actuators,* vol. 12 (4), pp. 389–404, Nov./Dec. 1987.

Zaromb, S. and J. R. Stetter, "Theoretical basis for identification and measurement of air contaminants using an array of sensors having partly overlapping selectivities," *Sensors and Actuators,* vol. 6 (4), pp. 225–244, December 1984.

Author Index

Subject Index

465

467

Editors' Biographies

Richard S. Muller (S'57–M'58–M'62–SM'70–F'88) received the degree of Mechanical Engineer from Stevens Institute of Technology, Hoboken, NJ, and the M.S./E.E. and Ph.D. degrees from the California Institute of Technology.

He joined the Department of Electrical Engineering and Computer Sciences at the University of California, Berkeley, in 1962, where he is now Professor as well as Co-Director (with Richard M. White) of the Berkeley Sensor & Actuator Center, a National Science Foundation/industry/university cooperative research center. He is the author, together with T. I. Kamins, of *Device Electronics for Integrated Circuits,* second edition (Wiley, 1986).

Dr. Muller is a Fellow of the IEEE and has been awarded NATO and Fulbright Research Fellowships at the Technical University, Munich, Germany. He is Chairman of the Sensors Advisory Board and is a member of the Advisory Committee for the IEEE Electron Devices Society. He is past chair of the Committee on Sensors and Actuators for the IEEE International Electron Devices Meeting, and Chairman of the Steering Committee for the biennial TRANSDUCER Conference. He will serve as the General Chairman for TRANSDUCERS '91, an IEEE-sponsored conference, to be held in June, 1991 in San Francisco.

Roger T. Howe (S'79–M'84) was born in Sacramento, CA, on April 2, 1957. He received the B.S. degree in physics from Harvey Mudd College, Claremont, CA, in 1979, and the M.S. and Ph.D. degrees in electrical engineering from the University of California, Berkeley, in 1981 and 1984, respectively.

During the 1984–1985 academic year, he was on the faculty of Carnegie-Mellon University, Pittsburgh, PA. From 1985 to 1987 he was an Assistant Professor of Electrical Engineering at the Massachusetts Institute of Technology, Cambridge. In 1987 he joined the Department of Electrical Engineering and Computer Sciences at the University of California, Berkeley, where he is an Associate Professor of Electrical Engineering and the Associate Director of the Berkeley Sensor & Actuator Center. His research interests include microsensors and microactuators, micromachining processes, and integrated-circuit design.

Dr. Howe is a member of the Materials Research Society and The Electrochemical Society.

Stephen D. Senturia (M'77) received the B.A. degree in physics *summa cum laude* from Harvard University, Cambridge, MA, in 1961, and the Ph.D. degree in physics in 1966 from the Massachusetts Institute of Technology, Cambridge.

He is presently Professor of Electrical Engineering at MIT, having joined the department immediately after completing his education. In 1982, he founded Micromet Instruments, Inc., and is currently serving on its Board of Directors. He is author or coauthor of more than 130 scientific papers, and is coauthor of *Electronic Circuits and Applications*, an introductory electronics text published by Wiley in 1975. His research interests have included semiconductor physics and the application of the methods of physical measurement to practical problems. The charge-flow transistor for use with thin-film sensor materials and the microdielectrometer for monitoring chemical reactions in resins, adhesives, plastics, and rubber, are among his recent developments. His current principal research activity is the use of microfabricated structures both for microsensor and microactuator applications, and for materials research, particularly on the role of polymers in microelectronics.

Dr. Senturia is the Solid-State Sensors Associate Editor of the *IEEE Transactions on Electron Devices.* He is co-winner of the IR-100 Award for his work on automated reclamation of urban solid waste, and of the 1988 Arthur K. Doolittle Prize of the Division of Polymeric Materials Science and Engineering of the American Chemical Society. He is a member of Phi Beta Kappa, Sigma Xi, the American Chemical Society, the Adhesion Society, and SAMPE.

Rosemary L. Smith (S'77–S'78–M'79) received the B.S. degree in biomedical electronics engineering from the University of Rhode Island, Kingston, in 1977, and the M.S. and Ph.D. degrees in bioengineering from the University of Utah, Salt Lake City, in 1979 and 1982, respectively.

From 1982 to 1984, she was an Assistant Professor in the Department of Electrical Engineering at Drexel University, Philadelphia. In October 1984, she joined the Chemical Sensor Department at the Centre Suisse d'Electronique et Microtechnique in Neuchatel, Switzerland, as a Visiting Scientist. From 1986 to 1988, she held the position of Sinclair Visiting Assistant Professor in the Department of Electrical Engineering at the Massachusetts Institute of Technology. She joined the faculty of the University of California, Davis, in September of 1988. Her research interests encompass a number of disciplines, including electrochemistry, biology, and solid-state materials science, as well as microelectronics and device physics. Many of her current research projects are related to the creation of new microsensor-based systems. She is also continuing a fundamental study of porous silicon, its nitridation to silicon nitride, its fractal growth, and electronic properties.

Richard M. White (M'63–F'72) was born in Denver, CO. He received the A.B., A.M., and Ph.D. degrees in engineering sciences and applied physics from Harvard University in 1951, 1952, and 1956, respectively.

From 1956 to 1962 he participated in microwave-component research at the General Electric Microwave Laboratory in Palo Alto, CA. In 1962, he joined the Department of Electrical Engineering and Computer Sciences at the University of California, Berkeley. He has been primarily concerned with teaching and research in solid-state electronics, with particular emphasis on ultrasonics and sensors. His publications and patented inventions concern sensors, ultrasonic phenomena and devices, thermoelastic effects, and other topics such as stroboscopic scanning-electron microscopy and microwave electronics. He coauthored the text and reference book, *Solar Cells: From Basics to Advanced Systems,* McGraw–Hill, 1984, and he has served as an Editor of the *IEEE ElectroTechnology Review*, a publication highlighting new developments in electrical engineering and computer science. He is Co-Director of the Berkeley Sensor & Actuator Center, an NSF/industry/university cooperative research center.

Prof. White was a Co-Guest Editor of the March, 1987 special issue on acoustic sensors, *IEEE Transactions on Ultrasonics, Ferroelectrics, and Frequency Control (UFFC)*, for which he received the IEEE UFFC Society's award for Outstanding Contributions to the UFFC Transactions. In October 1988, Dr. White received that Society's Achievement Award for his contributions to the field of ultrasonics in photoacoustics, surface acoustic wave devices, and sensors. He was a Guggenheim Fellow in 1968–1969 and received the IEEE Cledo Brunetti Award in 1986. He is a member of the American Physical Society, Sigma Xi, and the American Association for the Advancement of Science.